# Normal Forms, Bifurcations and Problems in Differential Equatio

# NATO Science Series

*A Series presenting the results of scientific meetings supported under the NATO Science Programme.*

The Series is published by IOS Press, Amsterdam, and Kluwer Academic Publishers in conjunction with the NATO Scientific Affairs Division

*Sub-Series*

| | |
|---|---|
| **I. Life and Behavioural Sciences** | IOS Press |
| **II. Mathematics, Physics and Chemistry** | Kluwer Academic Publishers |
| **III. Computer and Systems Science** | IOS Press |
| **IV. Earth and Environmental Sciences** | Kluwer Academic Publishers |
| **V. Science and Technology Policy** | IOS Press |

The NATO Science Series continues the series of books published formerly as the NATO ASI Series.

The NATO Science Programme offers support for collaboration in civil science between scientists of countries of the Euro-Atlantic Partnership Council. The types of scientific meeting generally supported are "Advanced Study Institutes" and "Advanced Research Workshops", although other types of meeting are supported from time to time. The NATO Science Series collects together the results of these meetings. The meetings are co-organized bij scientists from NATO countries and scientists from NATO's Partner countries – countries of the CIS and Central and Eastern Europe.

**Advanced Study Institutes** are high-level tutorial courses offering in-depth study of latest advances in a field.
**Advanced Research Workshops** are expert meetings aimed at critical assessment of a field, and identification of directions for future action.

As a consequence of the restructuring of the NATO Science Programme in 1999, the NATO Science Series has been re-organised and there are currently Five Sub-series as noted above. Please consult the following web sites for information on previous volumes published in the Series, as well as details of earlier Sub-series.

http://www.nato.int/science
http://www.wkap.nl
http://www.iospress.nl
http://www.wtv-books.de/nato-pco.htm

**Series II: Mathematics, Physics and Chemistry – Vol. 137**

# Normal Forms, Bifurcations and Finiteness Problems in Differential Equations

edited by


## Yulij Ilyashenko

Department of Mathematics,
Cornell University, Ithaca, NY, US.A.
and Department of Mathematics,
Independent and Moscow State Universities and Steklov Institute,
Moscow, Russia

and

## Christiane Rousseau

Département de Mathématiques et de Statistique,
Université de Montréal, Montréal, Québec, Canada

Technical Editor:

## Gert Sabidussi

Département de Mathématiques et de Statistique,
Université de Montréal, Montréal, Québec, Canada



**Kluwer Academic Publishers**

Dordrecht / Boston / London

Published in cooperation with NATO Scientific Affairs Division

Proceedings of the NATO Advanced Study Institute on
Normal Forms, Bifurcations and Finiteness Problems in Differential Equations
Montreal, Canada
8–19 July 2002

A C.I.P. Catalogue record for this book is available from the Library of Congress.

ISBN 1-4020-1928-9 (HB)
ISBN 1-4020-1929-7 (PB)

---

Published by Kluwer Academic Publishers,
P.O. Box 17, 3300 AA Dordrecht, The Netherlands.

Sold and distributed in North, Central and South America
by Kluwer Academic Publishers,
101 Philip Drive, Norwell, MA 02061, U.S.A.

In all other countries, sold and distributed
by Kluwer Academic Publishers,
P.O. Box 322, 3300 AH Dordrecht, The Netherlands.

*Printed on acid-free paper*

---



Printed in the Netherlands.

To the Memory of

Andrei Bolibrukh and Vladimir Lazutkin

# Table of Contents

# Preface

The qualitative theory of differential equations, created by Poincaré at the end of the 19th century had major developments as a branch of the theory of dynamical systems during the 20th century. It is one of the very active areas of the new millenium and is an inter-disciplinary field, both by its applications in science (dynamical systems provide models in physics, in particular celestial mechanics, biology, etc.) and by its interaction with other mathematical disciplines: algebra, algebraic geometry, complex analysis.

The book presents in an elementary way a number of recent significant developments in the theory of differential equations, some widely scattered in the literature and some new, which have not previously appeared in book form. The common denominator is the theory of planar vector fields (real or complex); most recent significant developments in this field are covered in the book. Indeed, to solve problems in planar vector fields many sophisticated tools and theories have been built and are still being developed, whose field of application goes far beyond the initial subject. A second common feature is the study of bifurcations of dynamical systems. The book also presents several problems that are likely to be studied in the coming years.

The following subjects are covered: new tools for local and global properties of systems and families of systems, nonlocal bifurcations, finiteness properties of Pfaffian functions and of differential equations, geometric interpretation of the Stokes phenomena, analytic theory of ordinary differential equations and complex foliations, applications to Hilbert's 16th problem.

(1) *The study of differential equations, or of families of differential equations in general position.* This study is particularly important in the context of mathematical modelling where the "observable" situations are the generic ones. When one studies systems depending on parameters the study of bifurcations of small codimension is a key to the understanding of the different types of behavior.

(2) *Finiteness problems in differential equations.* Many of these questions were motivated by Hilbert's 16th problem (which asks for the maximum number and relative position of limit cycles of a polynomial vector field in the plane), the subproblems of Hilbert's problem or the related problems such as the Riemann-Hilbert problem, Hilbert-Arnold problem, infinitesimal Hilbert problem, and so on.

The book is intended for graduate students, postdocs and researchers in dynamical systems. It requires a minimum background of a one-year course on nonlinear differential equations.

Let us now describe in more detail the different contributions which we can group under large themes.

*Abelian integrals and complex differential equations.* In the context of Hilbert's 16th problem and polynomial vector fields it is natural to consider bifurcations leading to the creation of limit cycles by perturbation of ovals of Hamiltonian or integrable vector fields. While the phenomenon is reasonably well understood for regular ovals and singular ovals with one hyperbolic saddle point, two new difficulties occur when we approach more degenerate singular cycles. First, the bounds for the number of zeroes of Abelian integrals explode when critical values of the Hamiltonian coallesce. Second, the number of limit cycles is no longer bounded by the number of zeroes of the Abelian integrals.

The paper by Novikov and Yakovenko is a first step to address the first problem. Abelian integrals are studied through the integrable linear Pfaffian systems of which they are solutions. A solution proposed is to consider these systems in the multidimensional setting where the independent variable and the parameters play the same role. As a first step in this direction the paper presents the theory of integrable linear Pfaffian systems, and explores the possibility of getting rid of exploding residues by suitable meromorphic gauge transformations, thus opening the hope to get bounds for the number of zeroes of Abelian integrals even in the neighborhood of critical values.

The paper by Roussarie and Caubergh discusses the second problem and shows that the number of bifurcating limit cycles can indeed exceed the number of zeroes of the Abelian integrals. In some cases, however, a finite codimension of the Abelian integral can lead to a finite upper bound for the local cyclicity.

The paper of Gabrielov on the complexity of computations with Pfaffian and Noetherian functions presents an essential tool for solving these problems. It describes global finiteness results for the number of isolated solutions of a system of Pfaffian equations and the local finiteness properties of Noetherian functions.

A section of the paper of Dumortier and de Maesschalck deals with the study of Abelian integrals for Liénard equations in relation to the unfolding of singularities of codimension 3 and 4.

*Normal forms, Stokes pheonomena, problems of analytic classification and their geometric meaning.* Four papers discuss situations where we encounter divergent changes of coordinates: all of them are related to the theme of normal forms. While the paper of Écalle illustrates the importance of good normal forms for proving finitess results, in particular the Dulac conjecture, the paper of Gelfreich exhibits the geometric meaning of the divergence of normalizing series. Interaction between normal forms and bifurcation theory has usually been in the direction of normal forms as tools for studying bifurcation problems. The papers of Glutsyuk and Rousseau show the existence of a reverse direction by putting forward the role of bifurcation theory to understand the divergence of the normalizing series.

The paper of Gelfreich discusses the geometric interpretation of the Stokes phenomenon (divergence of the normalizing series) as the splitting of complex separatrices in the case of generic bifurcations of strongly resonant periodic orbits of analytic Hamiltonian systems with two degrees of freedom: a quantitative measure of the splitting can be derived from an analytic modulus.

The paper of Glutsyuk discusses the geometry of the unfolding (in the Poincaré direction) of a linear system of differential equations with an irregular singular point. In the particular case of a system of Poincaré rank 1 Glutsyuk shows that commutators of appropriate noninteger powers of the monodromy operators tend to Stokes operators.

The paper of Rousseau discusses a modulus of analytic classification for a generic unfolding of a generic parabolic fixed point where the phenomenon of "parametric resurgence" is observed: indeed the divergence of the normalization yields the nonlinearizability of the bifurcating singular points for special types of resonance of the singular points.

The paper of Écalle covers several themes of the book. It presents the simplification of the proof of non-accumulation of limit cycles for a polynomial vector field which comes from the notion of well-behaved convolution average. It also makes the distinction between the continuous prenormalization and the discontinuous normalization of local objects. It sheds new light on the KAM theorem of survival of invariant tori as supermultiple small denominators do not exist, allowing a simplification of KAM theory. The last part of the paper opens new horizons by outlining a structure unifying three theories which seem far apart: The Lie algebra ARI, the analysis of "parametric resurgence" in physics, and the arithmetic of the Multiple Zeta Values.

*Singular perturbations and multiple time scale systems.* Singular perturbed 2-dimensional systems are discussed at large with an emphasis on geometric methods. The theory is enlightened by a deep discussion of examples, such as the regular and forced van der Pol system.

The paper of Dumortier and de Maesschalck presents new developments allowing to treat more degenerate systems than the van der Pol equation which can produce multiple limit cycles. An essential tool is the refinement of the method of the blow-up of the family.

The paper by Guckenheimer focuses on slow-fast systems mainly with one fast and two slow variables. It presents a detailed analysis of a *smooth* periodically forced van der Pol system, with the emphasis on the study of the canard solutions.

The paper of Ilyashenko discusses the slow-fast extensions of classical local and nonlocal bifurcations, by means of generalized true slow surfaces.

*Systems having complicated behavior.* The paper of Kaloshin deals with the global organization of the space of $C^r$ diffeomorphisms on a compact manifold according to the number of their periodic points. This space has a very rich structure. Indeed, Artin-Mazur diffeomorphims (having at most an exponential number of periodic points of period $n$) are dense, while outside an open set of hyperbolic diffeomorphisms supergrowth is $C^1$-dense and $C^r$-generic.

The paper of Ilyashenko discusses the geometry of attractors and new nonlocal bifurcations exhibited by higher dimensional vector fields. The examples presented illustrate the following heuristic principle: *at a first glance properties of random dynamical systems seem much richer than those of classical (deterministic) dynamical systems. Yet, any effect observed for a random dynamical system may be found as well for a classical dynamical system in a phase space of higher dimension.*

*Algebro-geometric methods and bifurcations methods in polynomial vector fields.* Hilbert's 16th problem is at the heart of polynomial vector fields and is a recurrent theme of this book. The papers of Françoise and Schlomiuk discuss the power of algebro-geometric methods to study bifurcations and finiteness problems for polynomial vector fields.

The paper of Françoise contains a nice historical part on the Liénard equations and its generalizations and aplications. It also contains a quantitative version of Bautin's approach to the center-focus problem with an estimate of the radius of convergence of the first return map in the Liénard equation.

The paper of Schlomiuk introduces algebro-geometric tools to build a framework for some global studies on planar polynomial vector fields. They make use of the theory of invariants of Sibirskii. Applications to bifurcation diagrams in quadratic vector fields are also described.

Two closely related parts of the paper of Ilyashenko discuss polynomial and analytic dynamical systems. In particular, different topics related to Hilbert's 16th problem are presented in the paper.

The last section of the paper of Rousseau discusses the progress of the program to prove the finiteness part of Hilbert's 16th problem for quadratic vector fields by explicitely stating what is done and what remains to be done.

The reader will also find some items which are not contained in the main themes above. For instance, the paper of Ilyashenko states a large number of problems related to the different themes of the book. A significant number of the results appearing in his paper were obtained by undergraduate students! The paper of Dumortier and de Maesschalck contains the classifications of singularities of $C^\infty$ vector fields in $\mathbb{R}^3$ up to codimension 4 and an explicit proof that the stratification thus obtained is non-algebraic.

In the period of preparation of the conference, Vladimir Lazutkin agreed to be a lecturer at the seminar. He was one of the leaders of the mathematical school of St. Petersburg. But he passed away just before his 60th birthday.

During the ASI, although he was already ill, Andrei Bolibrukh gave a wonderful lecture course on the Riemann-Hilbert problem and its generalizations for various classes of complex linear differential equations. His illness worsened after the summer school and he could not write his lecture notes. Andrei Bolibrukh struggled with his illness like a soldier and kept high spirits all the time. He was one of the key figures in the Moscow mathematical life and at the same time an admirable human being. He passed away on November 11, 2003.

This volume is dedicated to the memory of Andrei Bolibrukh and Vladimir Lazutkin.

We are grateful to all participants of the ASI for their stimulating interest. It is also a great pleasure to thank Gert Sabidussi who did a fantastic job in the preparation of this book and Arik Loinaz who dealt with the LaTeX files.

Yulij Ilyashenko and Christiane Rousseau

1
2
3
4
5
6
7
8
9
10
11
12
13
14
15
16
17
18
19
20
21
22
23
24
25
26
27
28
29
30
31
32
33
34
35
36
37
38
39
40
41
42
43
44
45
46
47
48
49
50
51
52
53
54
55
56

## Key to group picture

1
2 Popa
3
4 Fishkin
5 Montealegre
6
7 Alvarez
8 Rabinovich
9 Zupanovic
10 Valeev
11 Zell
12 Curto
13 J.-L. David
14 Damgov
15 Marin
16 Elizarov
17 Sanz
18 Iliev
19 Gavrilov
20 de Maeschalck
21 Schultz
22
23 Neirynck
24 Del Blanco
25 Gorodetski
26 Pavlov
27
28 Villarini
29 Carletti
30 Pereira
31 Paul
32 Caubergh
33 Cozma
34 Mozo-Fernandez
35 Hur
36 Grau
37 Kleptsyn
38 Karpenkov
39 Golenichtcheva
40 Llibre
41 Teyssier
42 Christopher
43 G. David
44 Sabidussi
45 Gelfreich
46 Rousseau
47 Françoise
48 Écalle
49 Bolibrukh
50 Yakovenko
51 Ilyashenko
52 Gabrielov
53 Glutsyuk
54 Roussarie
55 Dumortier
56 Schlomiuk

## Participants

Mohammed AL-REFAI
Department of Mathematics and Statistics
Jordan Univ. of Science & Technology
P.O. Box 3030
22110 Irbid
Jordan
m_alrefai@yahoo.com

Maria J. ALVAREZ
Departament de Matemàtiques
Edifici B
Universitat Autònoma de Barcelona
08193 Bellaterra (Barcelona)
Spain
jalvarez@math.uab.es

Andrei BOLIBRUKH
Steklov Mathematical Institute (MIAN)
Gubkina St. 8
117966 Moscow
Russia

Pietro-Luciano BUONO
Centre de recherches mathématiques
Université de Montréal
C.P. 6128, succ. Centre-ville
Montréal, QC H3C 3J7
Canada
buono@crm.umontreal.ca

Timoteo CARLETTI
Dipartimento di Matematica "U. Dini"
Università degli Studi di Firenze
Viale Morgagni, 67/A
50134 Firenze
Italy
carletti@math.unifi.it

Gabriel CHÊNEVERT
Dép. de mathématiques et de statistique
Université de Montréal
C.P. 6128, succ. Centre-ville
Montréal, QC H3C 3J7
Canada
chenever@dms.umonteral.ca

Colin CHRISTOPHER
Department of Mathematics & Statistics
University of Plymouth
PL4 8AA Plymouth
UK
c.christopher@plymouth.ac.uk

Caroline COUTU
Dép. de mathématiques et de statistique
Université de Montréal
C.P. 6128, succ. Centre-ville
Montréal, QC H3C 3J7
Canada
coutu@dms.umonteral.ca

Dimitru COZMA
Department of Physics & Mathematics
Tiraspol State University
5 Gh. Iablocichin Str.
2069 Kishinev
Moldova
cozma@usm.md

Josep CURTO
Departament de Matemàtiques
Edifici B
Universitat Autònoma de Barcelona
08193 Bellaterra (Barcelona)
Spain
jcurto@math.uab.es

Vladimir DAMGOV
Bulgarian Academy of Science
Space Research Institute
6 Moscowska St.
P.O. Box 799
1000 Sofia
Bulgaria
vdamgov@bas.bg

Jesús M. DEL BLANCO
Dpto de Algebra, Geometrica y Topologia
Facultad de Ciencias
Universidad de Valladolid
Prado de la Magdalena, s/n
47005 Valladolid
Spain
jblanco@agt.uva.es

Stanislav DUBROVSKIY
Department of Mathematics
567 Lake Hall
Northeastern University
Boston, MA 02115
USA
dubr@neu.edu

Pavel ELIZAROV
Centre of Ergonomics Research
Tver' State University
Hrustal'naja Str. 2–4
170021 Tver'
Russia
pavele@cals.ru
pavele@tvcom.ru

Alexey FISHKIN
Faculty of Mechanics and Mathematics
Moscow State University
Vorobjevy Gory
119899 Moscow
Russia
fishkin@mccme.ru

Lubomir GAVRILOV
UFR MIG
Université de Toulouse 3
118, route de Narbonne
31062 Toulouse Cédex 04
France
gavrilov@picard.ups-tise.fr

Tatiana GOLENICHTCHEVA-KOUTOUZOVA
Faculty of Mechanics and Mathematics
Moscow State University
Vorobjevy Gory
119899 Moscow
Russia
tania@mccme.ru

Anton GORODETSKI
Department of Mathematics
Independent Moscow University
Bolshoy Vlas'evski per. 11
121002 Moscow
Russia
asgor@mccme.ru

Andrzej GRANAS
Dép. de mathématiques et de statistique
Université de Montréal
C.P. 6128, succ. Centre-ville
Montréal, QC H3C 3J7
Canada
granas@dms.umonteral.ca

Maite GRAU
Departament de Matemàtica
Escola Universitàrià Politècnica
Universitat de Lleida
Avda. Jaume II, 69
25006 Lleida
Spain
mtgrau@matematica.udl.es

Radu HAIDUC
Department of Mathematics
101 Malott Hall
Cornell University
Ithaca, NY 14853
USA
haiduc@math.cornell.edu

Seok HUR
GDSD — Imagerie Géom. & Biomédicale
Université P. M. Curie — Paris VI
175, rue du Chevaleret
75013 Paris
France
emptystone@yahoo.com

Alexey IGNATYEV
Inst. of Applied Mathematics & Mechanics
Donetsk State University
Roza Luxemburg Str. 74
83114 Donetsk 114
Ukraine
ignat@iamm.ac.donetsk.ua

Iliya ILIEV
Inst. of Mathematics and Informatics
Bulgarian Acad. of Sci. G. Bonchev
Bl. 8
1113 Sofia
Bulgaria
iliya@math.bas.bg

Oleg KARPENKOV
Department of Mathematics
Independent Moscow University
Bolshoy Vlas'evski per. 11
121002 Moscow
Russia
karpenk@mccme.ru

Victor KELPTSYN
Department of Mathematics
Independent Moscow University
Bolshoy Vlas'evski per. 11
121002 Moscow
Russia
kleptsyn@mccme.ru

Jean-Phillipe LESSARD
Dép. de mathématiques et de statistique
Université de Montréal
C.P. 6128, succ. Centre-ville
Montréal, QC H3C 3J7
Canada
lessardj@dms.umonteral.ca

Jaume LLIBRE
Departament de Matemàtiques
Edifici B
Universitat Autònoma de Barcelona
08193 Bellaterra (Barcelona)
Spain
jllibre@math.uab.es

Pavao MARDESIC
Lab. de Topologie — UMR 5584 CNRS
Université de Bourgogne
9, ave. Alain Savary
B.P. 47870
21078 Dijon Cédex
France
mardesic@u-bourgogne.fr

David MARIN
Lab. de Topologie — UMR 5584 CNRS
Université de Bourgogne
9, ave. Alain Savary
B.P. 47870
21078 Dijon Cédex
France
marin@topolog.u-bourgogne.fr
david@math.uab.es

Sigrid MARKERT
CERN
CH-1211 Genève 23
Switzerland
smarkert@cern.ch

Vadim MOLDAVSKY
Department of Mathematics
101 Malott Hall
Cornell University
Ithaca, NY 14853
USA
vsm7@cornell.edu

Claire MOUDA
Department of Mathematics
Weizmann Institute of Science
P.O. Box 26
76100 Rehovot
Israel
clairem@wisdom.weizmann.ac.il

Jorge MOZO-FERNÁNDEZ
Dpto Matemática Aplicada Fundamental
ETS de Arquitectura
Universidad de Valladolid
Avda. Salamanca, s/n
47014 Valladolid
Spain
jmozo@maf.uva.es

Koen NEIRYNCK
Limburgs Universitair Centrum
Universitaire Campus
3590 Diepenbeek
Belgium
koen.neirynck@luc.ac.be

Dmitry NOVIKOV
Department of Mathematics
1395 Mathematical Sci. Bldg.
Purdue University
West Lafayette, IN 47907
USA
dmitry@math.purdue.edu

Emmanuel PAUL
Laboratoire E. Picard, UMR 5580
UFR MIG
Université Paul Sabatier
118 route de Narbonne
31062 Toulouse Cédex
France
paul@picard.ups-tlse.fr

Dmitry PAVLOV
Faculty of Mechanics and Mathematics
Moscow State University
Vorobjevy Gory
119899 Moscow
Russia
dvpavlov@mail.ru

Mihail POPA
Inst. Mathematics and Computer Science
Academy of Sciences of Moldova
Academiei Str. 5
2028 Chisinau
Moldova
popam@math.md

Boris RABINOVICH
Department of Mathematics
Weizmann Institute of Science
P.O. Box 26
76100 Rehovot
Israel
borisra@wisdom.weizmann.ac.il

Olivier ROUSSEAU
Dép. de mathématiques et de statistique
Université de Montréal
C.P. 6128, succ. Centre-ville
Montréal, QC H3C 3J7
Canada
rousseao@dms.umontreal.ca

Fernando SANZ
Dpto de Algebra, Geometrica y Topologia
Facultad de Ciencias
Universidad de Valladolid
Prado de la Magdalena, s/n
47005 Valladolid
Spain
fsanz@agt.uva.es

Peter F. SCHULTZ
Dept. Mathematics and Computer Science
Clarkson University
Potsdam, NY 13699-5815
USA
pschultz@clarkson.edu

Loic TEYSSIER
UFR — Mathématiques
Univ. des Sciences et Techniques de Lille
Cité Scientifique
59655 Villeneuve-d'Ascq Cédex
France
teyssier@agat.univ-lille1.fr

Valera VALEEV
Department of Math. & Comp. Physics
Institute of Physics
St. Petersburg University
Uliyanovskaya 1/1, Petrodvorets
198504 St. Petersburg
Russia
valeev@mph.phys.spbu.ru

Bruno VALLET
Département de Mathématiques
Bâtiment 425
Univeristé Paris-Sud
91405 Orsay Cédex
France
bruno.vallet@math.u-psud.fr

Massimo VILLARINI
Dpto di Matematica Pura e Applicata
Università di Modena
via Campi 213/B
41100 Modena
Italy
villarini@unimo.it

Samuel ZAIDMAN
Dép. de mathématiques et de statistique
Université de Montréal
C.P. 6128, succ. Centre-ville
Montréal, QC H3C 3J7
Canada
zaidman@dms.umontreal.ca

Thierry ZELL
Department of Mathematics
1395 Mathematical Sci. Bldg.
Purdue University
West Lafayette, IN 47907
USA
tzell@math.purdue.edu

Yulin ZHAO
Centre de recherches mathématiques
Université de Montréal
C.P. 6128, succ. Centre-ville
Montréal, QC H3C 3J7
Canada
zhao@crm.umontreal.ca

Haiping ZHU
School of Mathematics
Georgia Institute of Technology
Atlanta, GA 30332
USA
zhu@math.gatech.edu

Vesna ZUPANOVIC
Department of Applied Mathematics
Faculty of Electrical Engineering
& Computing
University of Zagreb
Unska 3
10000 Zagreb
Croatia
vesna.zupanovic@fer.hr

## Contributors

Magdelena CAUBERGH
Limburgs Universitair Centrum
Universitaire Campus
3590 Diepenbeek
Belgium
magdalena.caubergh@luc.ac.be

Peter DE MAESSCHALCK
Limburgs Universitair Centrum
Universitaire Campus
3590 Diepenbeek
Belgium
peter.demaesschalck@luc.ac.be

Freddy DUMORTIER
Limburgs Universitair Centrum
Universitaire Campus
3590 Diepenbeek
Belgium
freddy.dumortier@luc.ac.be

Jean ÉCALLE
Laboratoire Analyse Harmonique
Département de Mathématiques
Université de Paris-Sud XI
15, rue Georges Clémenceau
91405 Orsay Cédex
France
jean.ecalle@math.u-psud.fr

Jean-Pierre FRANÇOISE
GDSD — Imagerie Géom. & Biomédicale
Université P. M. Curie — Paris VI
175, rue du Chevaleret
75013 Paris
France
jpf@ccr.jussieu.fr

Andrei GABRIELOV
Dept. of Mathematics and Geophysics
Purdue University
West Lafayette, IN 47907-1395
USA
agabriel@math.purdue.edu

Vassili GELFREICH
The Mathematics Institute
University of Warwick
Coventry CV4 7AL
UK

Alexey GLUTSYUK
U.M.P.A.
École Normale Supérieure de Lyon
46, Allée d'Italie
69364 Lyon Cédex 07
France
alexey.glutsyuk@umpa.ens-lyon.fr

John GUCKENHEIMER
Department of Mathematics
Center of Applied Mathematics
Cornell University
Ithaca, NY 14853
USA
gucken@cam.cornell.edu

Yulij ILYASHENKO
Department of Mathematics
Cornell University
Thaca, NY 14853
USA
yulij@math.cornell.edu

Steklov Mathematical Institute (MIAN)
Gubkina St. 8
117966 Moscow
Russia
yulijs@mccme.ru

Vadim KALOSHIN
Department of Mathematics 253–37
California Inst. of Technology
Pasadena, CA 91125
USA
kaloshin@caltech.edu

Robert ROUSSARIE
Lab. de Topologie — UMR 5584 CNRS
Université de Bourgogne
9, ave. Alain Savary
B.P. 47870
21078 Dijon Cédex
France
roussari@u-bourgogne.fr

Christiane ROUSSEAU
Dép. de mathématiques et de statistique
Université de Montréal
C.P. 6128, succ. Centre-ville
Montréal, QC H3C 3J7
Canada
rousseac@dms.umontreal.ca

Dana SCHLOMIUK
Dép. de mathématiques et de statistique
Université de Montréal
C.P. 6128, succ. Centre-ville
Montréal, QC H3C 3J7
Canada
dasch@dms.umontreal.ca

Nicolai VOROBJOV
Dept. of Computer Science
University of Bath
Bath BA2 7AY
United Kingdom
nnv@maths.bath.ac.uk

Sergei YAKOVENKO
Dept. Mathematics & Theor. Mechanics
Weizmann Institute of Science
P.O. Box 26
76100 Rehovot
Israel
yakov@wisdom.weizmann.ac.il

# Relations between Abelian integrals and limit cycles

Magdalena CAUBERGH

*Limburgs Universitair Centrum*
*Universitaire Campus*
*3590 Diepenbeek*
*Belgium*

Robert ROUSSARIE

*Laboratoire de Topologie, UMR 5584 CNRS*
*Université de Bourgogne*
*9 ave Alain Savary*
*21 078 Dijon Cédex*
*France*

**Abstract**

Limit cycles bifurcating in an unfolding from a *regular Hamiltonian cycle,* are in general directly controlled by the zeroes of the associated Abelian integral. Our purpose here is to investigate to what extent the Abelian integral allows one to study the limit cycles which bifurcate from a *singular Hamiltonian cycle.* We focus on the study of the 2-saddle cycles unfoldings. We show that the number of bifurcating limit cycles can exceed the number of zeroes of the related Abelian integral, even in generic unfoldings. However, in the case where one connection remains unbroken in the unfolding, we show how a finite codimension of the Abelian integral leads to a finite upper bound on the local cyclicity.

## 1 The problem

We consider $\mathcal{C}^\infty$ unfoldings of planar vector fields $(X_\lambda)_\lambda$, where $\lambda = (\bar{\lambda}, \varepsilon)$, $\bar{\lambda} \in (\mathbb{R}^k, 0)$, $\varepsilon \in [0, \varepsilon_0[$. We suppose that $X_{(\bar{\lambda},0)}$ is a Hamiltonian vector field $X_H$. So, there is a $\mathcal{C}^\infty$ function $H$ such that

$$X_{(\bar{\lambda},0)} = X_H = -\frac{\partial H}{\partial y}\frac{\partial}{\partial x} + \frac{\partial H}{\partial x}\frac{\partial}{\partial y}. \tag{1.1}$$

Hence we can write

$$X_\lambda = X_H + \varepsilon Y_{\bar{\lambda}} + o(\varepsilon), \quad \varepsilon \to 0. \tag{1.2}$$

We shall also consider the associated unfolding $(\omega_\lambda)_\lambda$ of dual 1-forms which can be written as

$$\omega_\lambda = dH + \varepsilon \nu_{\bar{\lambda}} + o(\varepsilon), \quad \varepsilon \to 0.$$

We suppose that $X_H$ is a Hamiltonian vector field of *center type.* This means that there exists an annulus $A$ of periodic orbits contained in $H^{-1}(]0, h_1[)$. More precisely, we assume



*Y. Ilyashenko and C. Rousseau (eds.),*
*Normal Forms, Bifurcations and Finiteness Problems in Differential Equations,* 1–32.

that for each $h \in ]0, h_1[$, one has a periodic orbit $\gamma_h$ of $X_H$ (also called Hamiltonian cycle), that belongs to the level curve $\{H = h\}$. The annulus $A$ is the union $A = \bigcup\{\gamma_h \mid h \in ]0, h_1[\}$. We assume that one of the connected components of the boundary of $A$ is a singular cycle, which we shall call $\gamma_0$, contained in the level curve $\{H = 0\}$. In Fig. 1 we have pictured such annuli. A singular cycle is present in the boundary in cases (b) and (c). The name "center" comes from the fact that for quadratic vector fields, a center is also present in the boundary, as it is pictured in the figure.

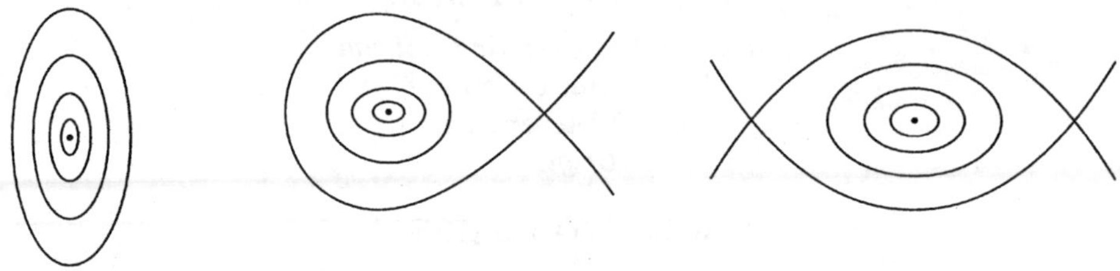

(a) (b) (c)

Figure 1: Closure of period annuli

We want to study the limit cycles in $(X_\lambda)_\lambda$ which can bifurcate from $X_H$. Recall that a limit cycle is an isolated periodic orbit. Of special interest is the maximum number of limit cycles $\gamma$ that can arise after small perturbations of the smooth vector field $X_H$ in any small neighborhood of a given $\gamma_h$, for $h \in [0, h_1[$, i.e. which bifurcate from a regular cycle or from the polycycle at the boundary of the annulus of regular cycles. This number is called the *cyclicity* of $(X_\lambda)_\lambda$ along $\gamma_h$ for $\lambda = 0$, and it is denoted by $\mathrm{Cycl}(X_\lambda, (\gamma_h, 0))$. For a vector field unfolding $(X_\lambda)_\lambda$ defined along some compact invariant set $\Gamma$ of the vector field $X_0$, a precise definition of the cyclicity $\mathrm{Cycl}(X_\lambda, (\Gamma, 0))$ is given by:

**1.1 Definition** Let $(X_\lambda)_\lambda$ be a family defined along some compact invariant set $\Gamma$ of the vector field $X_0$. The *cyclicity* of the unfolding $(X_\lambda, (\Gamma, 0))$ is defined as the number

$$\mathrm{Cycl}(X_\lambda, (\Gamma, 0)) = \limsup_{\lambda \to 0,\ \delta \to 0} N(\lambda, \delta),$$

where $N(\lambda, \delta)$ is the number of limit cycles of $X_\lambda$, at a Hausdorff distance of $\Gamma$ less than $\delta$.

A major question is to give sufficient conditions under which the cyclicity $\mathrm{Cycl}(X_\lambda, (\gamma_h, 0))$ can be finite and, in this case, to find an explicit estimation of this cyclicity.

When $\tilde{h} \in ]0, h_1[$, i.e. when $\gamma_{\tilde{h}}$ is a regular Hamiltonian cycle, the Poincaré map $P_\lambda(h)$ is a smooth function of $(h, \lambda)$ near $(\tilde{h}, 0)$. One can expand the displacement function $\delta_\lambda(h) = P_\lambda(h) - h$ in terms of the small parameter $\varepsilon$. The first term in this expansion is the integral unfolding:

$$I_{\bar{\lambda}}(h) = \int_{\gamma_h} \nu_{\bar{\lambda}}$$

which we call the *Abelian integral* associated to the vector field unfolding $(X_\lambda)_\lambda$. It is wellknown that the knowledge of the Abelian integral is in general sufficient to study the limit cycles bifurcating from the Hamiltonian cycle $\gamma_{\bar{h}}$. In Section 2 below we recall briefly this wellknown relation between the Abelian integral and the limit cycles in this *regular case.*

The rest of the paper deals with the bifurcation along the singular cycle $\gamma_0$, the so-called *singular case.* We shall limit ourselves to the simple cases of a hyperbolic singular cycle with just one hyperbolic saddle, called a *saddle-loop*, and with two hyperbolic saddles, called a 2-*saddle cycle.* As we also explain below, the Abelian integral extends as a continuous function to the value 0 corresponding to the singular cycle. The central question we want to investigate in this note is the following:

*To what extent does the Abelian integral $I_{\bar{\lambda}}(h)$ allow one to study the limit cycles which bifurcate from the singular cycle?*

We shall see that in the regular case, results on the zero-set of $I_{\bar{\lambda}}$ can be transferred to results on the set of limit cycles of $X_\lambda$. However, in the singular case, this transfer is not that obvious.

First, the Abelian integral is no longer differentiable at the limit point 0. Nevertheless, its type of non-differentiability reduces to the presence of linear logarithmic terms (Section 3). We make this precise, using the notion of formal series asymptotic to a function unfolding. It is possible to define a notion of codimension for the integral $I_0$ which is a bound for the number of zeros of $I_{\bar{\lambda}}$ bifurcating from 0.

Secondly, the return map is not defined at the singular points. We can consider its continuous extension, but in general, this extension has not enough differentiability to permit using the traditional techniques, derived from the implicit function theorem, as in the regular case. However, in the case of a saddle loop (also called homoclinic loop), the problem of transfer can be dealt with under certain genericity conditions on the Abelian integral (Section 5). We could say that the limit cycles are shadowed by the zeroes of the corresponding Abelian integral. In particular, the cyclicity of $\mathrm{Cycl}(X_\lambda, (\gamma_0, 0))$ is bounded by the codimension of the Abelian integral $I_0$.

In general, the transfer will not work even for the more simple and generic unfoldings of the 2-saddle cycle , as soon as the unfolding breaks the two connections, and one will find limit cycles not being shadowed by any zero of the Abelian integral (Section 4). In fact, in this general case, there is no reasonable expansion for the return map to be compared with the Abelian integral expansion. On the other hand, if the unfolding breaks just one connection, it is possible to expand in a natural way the return map and to compare it with the Abelian integral expansion (Section 6). Even in this case, the transfer does not work completely, because in the asymptotic development of the displacement map governing the limit cycles, one encounters not only terms tending to $h^n$ and $h^n \log h$ for $\varepsilon \to 0$, but also terms tending to $h^n \log^m h$ $(0 \leq m \leq n)$. These terms cannot be encountered in the related Abelian integral, but they play a significant role. However, in this case it is again possible to obtain a bound on the cyclicity it terms of the codimension of the Abelian integral.

This note is based on a forthcoming paper by Dumortier and Roussarie [DR] in which the reader will find more details and technicalities. In particular it is proved in [DR], that generic 4-parameter and 5-parameter unfoldings can produce three zeroes of the Abelian integral and four limit cycles. This proof is very long and technical and is not reproduced here. One will also find in this reference explicit polynomial examples of such generic unfoldings.

## 2 The regular case

In the regular case, we assume that $X_H$ contains an annulus $A$ of periodic orbits. We can take a section $\Sigma$ transverse to the Hamiltonian cycle $\gamma_h$, for all $h \in \Sigma_0 = \Sigma \cap A$. Here, $\Sigma$ is parametrized by $h$, the Hamiltonian value. The associated family $(P_\lambda)_\lambda$ of first-return mappings (or Poincaré mappings) is $\mathcal{C}^\infty$:

$$P : \Sigma_0 \times U \to \Sigma : h \mapsto P_\lambda(h),$$

where $U = U_0 \times [0, \varepsilon_0]$. It is clear that limit cycles of $X_\lambda$ correspond to isolated zeroes of the displacement mapping $\delta_\lambda \equiv P_\lambda - Id$. Since the vector field $X_H$ is of center type, the displacement mapping is divisible by $\varepsilon$, and so we can write for some smooth function $\bar{\delta}_{(\bar{\lambda},\varepsilon)}(h)$

$$\delta_{(\bar{\lambda},\varepsilon)}(h) = \varepsilon \bar{\delta}_{(\bar{\lambda},\varepsilon)}(h). \tag{2.1}$$

More precisely, we have the following well-known formula:

$$\delta_{(\bar{\lambda},\varepsilon)}(h) = -\varepsilon I_{\bar{\lambda}}(h) + o(\varepsilon), \quad \varepsilon \to 0, \tag{2.2}$$

where

$$I_{\bar{\lambda}}(h) = \int_{\gamma_h} \nu_{\bar{\lambda}}.$$

In this note, the smooth function $I_{\bar{\lambda}}$ will be called the *Abelian integral* of the unfolding $(X_\lambda)_\lambda$, even if $H, \nu_{\bar{\lambda}}$ are not polynomials. Observe that, up to sign, the Abelian integral corresponds to the first Melnikov function (i.e. the first order term in $\varepsilon$ of $\delta$):

$$M_1(h, \bar{\lambda}) = -I_{\bar{\lambda}}(h).$$

For $\varepsilon = 0$, $X_{(\bar{\lambda},\varepsilon)}$ has no limit cycles; as a consequence, limit cycles of $X_\lambda$ correspond to isolated zeroes of $\bar{\delta}_\lambda$. From (2.1) and (2.2) we have

$$\bar{\delta}_\lambda(h) = -I_\lambda(h) + O(\varepsilon), \quad \varepsilon \to 0.$$

Therefore, the equation $\bar{\delta}_\lambda(h) = 0$ is a $\mathcal{C}^\infty$ perturbation in $\varepsilon$ of the equation $I_{\bar{\lambda}}(h) = 0$, and we have that

$$\mathrm{Cycl}(X_\lambda, (\gamma_{h_0}, 0)) = \limsup_{\lambda \to 0,\ h \to h_0} \{\text{number of isolated zeroes } h \text{ of } \bar{\delta}_\lambda\}.$$

Now, we define the notion of cyclicity of the Abelian integral and the notion of a generic $k$-parameter unfolding:

**2.1 Definition** The *cyclicity* of an isolated root $h_0$ of $I_0$ in the unfolding $I_{\bar{\lambda}}$ at $\bar{\lambda} = \bar{\lambda}_0$ is defined as the number

$$\mathrm{Cycl}(I_\lambda, (h_0, \bar{\lambda}_0)) = \limsup_{\bar{\lambda} \to \bar{\lambda}_0,\ h \to h_0} \{\text{number of isolated zeroes } h \text{ of } I_{\bar{\lambda}}\}.$$

**2.2 Definition** Let: $I_{\bar{\lambda}}(h) = \sum_{i=0}^{k+1} \alpha_i(\bar{\lambda})(h - h_0)^i + O((h - h_0)^{k+2})$ be the finite Taylor expansion at $h_0 \in \Sigma_0$.

(1) One says that $I_0$ has *finite codimension* at $h_0$ if $h_0$ is a root of order $k+1$ of $I_0$, i.e.,

$$\alpha_0(0) = \alpha_1(0) = \cdots = \alpha_k(0) = 0, \quad \alpha_{k+1}(0) \neq 0.$$

The number $k$ is called the *codimension* of $I_0$ at $h_0$ and is written $\mathrm{cod}_{h_0}(I_0)$.

(2) The $k$-parameter unfolding $(I_{\bar{\lambda}}, 0)$ is called *generic* at $h_0$ if and only if the mapping

$$(\mathbb{R}^k, 0) \to (\mathbb{R}^k, 0) : \bar{\lambda} \mapsto (\alpha_0(\bar{\lambda}), \ldots, \alpha_{k-1}(\bar{\lambda}))$$

is a germ of a diffeomorphism at $(\mathbb{R}^k, 0)$.

**2.3 Remark** We will see in Section 3 that this definition of a generic unfolding differs from the one in the singular case. The reason is that in the latter case 0, corresponding to the polycycle, is a marked point. In the regular case, 0 is not fixed, and it does not play an important role; therefore, we can eliminate one parameter as we will see in next theorem.

The condition that all zeroes of $I_0$ have finite order will be called the *non-degeneracy hypothesis.* The following theorem says that, under the non-degeneracy hypothesis, all results about the limit cycles of $X_\lambda$ can be deduced from the knowledge of the zeroes of the Abelian integral. This theorem is a direct consequence of the implicit function theorem, Rolle's theorem and the preparation theorem.

**2.4 Theorem** (1) *If $h_0$ is a simple zero of $I_0$ (i.e. $\mathrm{cod}_{h_0}(I_0) = 0$), then $\bar{\delta}_\lambda$ has exactly one simple zero $h$ close to $h_0$ for all $\bar{\lambda}$ close enough to $(0,0)$. This means the birth of a hyperbolic limit cycle of $X_\lambda$ from $\gamma_{h_0}$. More precisely:*

$$\mathrm{Cycl}(I_{\bar{\lambda}}, h_0) = \mathrm{Cycl}(X_\lambda, \gamma_{h_0}) = 1.$$

(2) *If $\mathrm{cod}_{h_0}(I_0) = k$, then there are at most $k$ limit cycles that bifurcate from $\gamma_{h_0}$. More precisely:*

$$\mathrm{Cycl}(I_{\bar{\lambda}}, h_0) \le \mathrm{Cycl}(X_\lambda, \gamma_{h_0}) \le k.$$

(3) *If $(I_{\bar{\lambda}}, 0)$ is a generic $k$-parameter unfolding, then the germ $(I_{\bar{\lambda}}, 0)$ is $C^\infty$-contact equivalent to the versal one:*

$$P_\beta(x) = \beta_0 + \beta_1 x + \cdots + \beta_{k-1}x^{k-1} \pm x^{k+1}.$$

*We will call this unfolding the catastrophe model. Moreover,*

$$\mathrm{Cycl}(I_{\bar{\lambda}}, h_0) = \mathrm{Cycl}(X_\lambda, \gamma_{h_0}) = k.$$

By *$C^\infty$-contact equivalence* we mean that there exists a $C^\infty$ diffeomorphism $\bar{\lambda} \mapsto \beta(\bar{\lambda})$, $\beta(0) = 0$, such that $I_{\bar{\lambda}}$ has the same number of zeroes (counted with multiplicity) as $P_{\beta(\bar{\lambda})}$. Moreover, for any $\varepsilon > 0$ small enough, the diagram of bifurcation of $X_\lambda$ is $C^\infty$ conjugated to the diagram of $I_{\bar{\lambda}}$: there exists a $C^\infty$ mapping $\varphi_\varepsilon(\bar{\lambda})$ sending the diagram of $I_{\bar{\lambda}}$ on the diagram of $\bar{\delta}_{(\bar{\lambda},\varepsilon)}$. The latter is the bifurcation diagram for the limit cycles of $X_\lambda$. Note that a limit cycle $\gamma_h$ of $X_\lambda$ is said to be of order $k$, if it corresponds to a root $h$ of $\bar{\delta}_\lambda$ of codimension $k$.

Let us consider now the singular case. We suppose that $H(x,y)$ is a smooth function defined on some neighborhood $U$ of a polycycle $\Gamma$ that is bounded and hyperbolic. This means that $\Gamma$ is a compact connected curve, made by $k$ saddle-type Morse points of $H$, say $s_1, \ldots, s_k$; thus the eigenvalues of the Hessian of $H$ in $s_i$ are 1 and $-1$ $(i = 1, \ldots, k)$. We also suppose that $\Gamma$ is contained in the zero level curve,

$$\Gamma \subset \{H = 0\}.$$

Moreover, for $h > 0$ small enough, one has a regular cycle $\gamma_h \subset \{H = h\} \cap U$, and $\gamma_h$ tends toward $\Gamma$ in the Hausdorff topology when $h \to 0$. We assume that we are given a smooth family of 1-forms $\nu_{\bar{\lambda}}$ defined on $U$ with parameter $\bar{\lambda} \in (\mathbb{R}^p, 0)$:

$$\omega_{(\bar{\lambda},\varepsilon)} = dH + \varepsilon \nu_{\bar{\lambda}} + o(\varepsilon), \quad \varepsilon \to 0.$$

As above, let us associate the unfolding of integral functions (called its Abelian integral) defined by:

$$I(h, \bar{\lambda}) = I_{\bar{\lambda}}(h) = \int_{\gamma_h} \nu_{\bar{\lambda}}$$

for $(h, \bar{\lambda}) \in [0, h_0[ \times U$. Clearly, the function $I_{\bar{\lambda}}$ is smooth for $h > 0$. As we have said in the introduction, we want to investigate in what extension $I_{\bar{\lambda}}$ can be used to study the limit cycles bifurcating from $\Gamma$. This note intends to be an introduction to this problem which is essentially open. The problem comes from the fact that the continuous extension of $I_{\bar{\lambda}}$ at $h = 0$ is not differentiable. Nevertheless, its type of non-differentiability reduces to the presence of linear logarithmic terms as we prove in the Section 3. We make this precise, using the notion of formal series asymptotic to a function. The subsequent sections are devoted to the asymptotic theory of the displacement map which control the limit cycles and its relations with the Abelian integral.

## 3 Abelian integrals near hyperbolic polycycles

In this section we describe some general results on Abelian integrals near hyperbolic polycycles, stressing the notion of *Chebychev asymptotic scale of functions* that we shall simply call *asymptotic scale of functions*. In this note, we will always work with the differential operator $\nabla = h\frac{\partial}{\partial h}$.

**3.1 Definition** An *asymptotic scale of functions* is a sequence of functions $\mathcal{F} = \{f_i\}_{i \in \mathbb{N}}$, $f_i$ smooth on $]0, h_0[$, for some $h_0 > 0$, continuous at 0, such that

(1) $f_0 \equiv 1$ and no $f_i$ has zeroes near 0 in $]0, h_0[$. Moreover,

$$\frac{f_{i+1}}{f_i}(h) = o(1), \quad h \downarrow 0,$$

and $\nabla(f_{i+1}/f_i)$ has constant sign near 0 in $]0, h_0[$.

(2) One can construct the sequences $\mathcal{F}_0, \mathcal{F}_1, \ldots, \mathcal{F}_j, \ldots$ inductively: $\mathcal{F}_0 \equiv \mathcal{F}$, and if for $j \in \mathbb{N}$ the sequence $\mathcal{F}_j = \{f_i^j\}_{i \in \mathbb{N}}$ is defined on some interval $[0, h_j[$, then the sequence $\mathcal{F}_{j+1} = \{f_i^{j+1}\}_{i \in \mathbb{N}}$ is defined by

$$f_i^{j+1} = \frac{\nabla f_{i+1}^j}{\nabla f_1^j}, \quad \forall i \in \mathbb{N}$$

on some interval $[0, h_{j+1}[$. (This supposes that $\nabla f_1^j(h) \neq 0$ for $h \in [0, h_{j+1}[$.)

(3) The sequences $\mathcal{F}_j$, $j \in \mathbb{N}$, have the same property as $\mathcal{F}_0$: $f_0^j \equiv 1$, $f_{i+1}^j/f_i^j(h) = o(1)$ and $\nabla(f_{i+1}^j/f_i^j)$ has constant sign near 0 in $]0, h_j[$.

**3.2 Example** The sequences $\mathcal{T} = \{h^i\}_{i\in\mathbb{N}}$, $\mathcal{L} = \{1, h\log h, h, \dots, h^i, h^{i+1}\log h, \dots\}$ are asymptotic scales of functions.

**3.3 Definition** Let $\mathcal{F} = \{f_i\}_{i\in\mathbb{N}}$ be an asymptotic scale.

(1) We say that the series $\hat{f} = \sum_{i=0}^{\infty}\alpha_i f_i$ is *asymptotic to the germ* of $f$ at $0 \in \mathbb{R}^+$ (and that $f$ *expands in the asymptotic scale* $\mathcal{F}$) if for any $N \in \mathbb{N}$ :

$$\Phi_{N+1}(h) = f(h) - \sum_{i=0}^{N}\alpha_i f_i(h) = O(f_{N+1}(h)), \quad h \to 0 \tag{3.1}$$

has the *remainder-property of order* $N$, shortly denoted by property $(R)_N$, i.e.

$$\frac{\Phi_{N+1}}{f_N}(h) \to 0, \quad \text{when} \quad h \to 0,$$

and if one defines inductively the functions $\Phi_{N+1}^{j+1}$, $j = 0, \dots, N-1$, by

$$\Phi_{N+1}^{j+1} = \frac{\nabla\Phi_{N+1}^j}{\nabla f_1^j}, \quad \text{with} \quad \Phi_{N+1}^0 = \Phi_{N+1},$$

one has that

$$\frac{\Phi_{N+1}^j}{f_{N-j}^j} \to 0, \quad \text{when} \quad h \to 0;$$

in particular $\Phi_{N+1}^N(h) = o(1)$, when $h \to 0$.

(2) If the function $f$ also depends on a parameter $\bar{\lambda}$, we will suppose that the coefficients $\alpha_i$ are functions of $\bar{\lambda}$ and that the estimation order $O(f_{N+1}(h))$ in (3.1) is uniform in $\bar{\lambda}$. We shall say that the unfolding $f_{\bar{\lambda}}$ *expands in the asymptotic scale* $\mathcal{F}$.

We will now prove that any Abelian integral $I_{\bar{\lambda}}$ expands in the asymptotic scale of functions $\mathcal{L}$.

**3.4 Proposition** *The family $I_{\bar{\lambda}}$ has a Dulac series, linear in* $\log h$. *This means that there exists a formal series*

$$\hat{I}_{\bar{\lambda}}(h) = \sum_{i=0}^{\infty}(a_i(\bar{\lambda})h^i + b_i(\bar{\lambda})h^{i+1}\log h) \tag{3.2}$$

*where the $a_i, b_i$ are smooth coefficients in $\bar{\lambda}$ such that $\hat{I}_{\bar{\lambda}}$ is asymptotic to $I_{\bar{\lambda}}$ in the scale $\mathcal{L}$, in the sense of Definition* 3.3.

**Proof** Take two transverse sections on each separatrix $\Gamma$ connecting $p_i$ and $p_{i+1}$, near $p_i$ and $p_{i+1}$ respectively. These sections divide the polycycle $\gamma_h$ into regular curves $\rho_h^i$ and non-regular curves $\tau_h^i$ $(i = 1, \dots, n)$. Therefore, the Abelian integral can be written as the sum of two terms:

$$I_{\bar{\lambda}}(h) = \int_{\gamma_h}\nu_{\bar{\lambda}} = \sum_{i=1}^{n}\int_{\rho_h^i}\nu_{\bar{\lambda}} + \sum_{i=1}^{n}\int_{\tau_h^i}\nu_{\bar{\lambda}}.$$

Clearly, the integrals $\int_{\rho_h^i} \nu_{\bar\lambda}$ are $C^\infty$ in $h$, $\forall i = 1, \ldots, n$. Hence, it suffices to prove that $\forall i = 1, \ldots, n$: the integral $\int_{\tau_h^i} \nu_{\bar\lambda}$ can be written as (3.2). Fix $i \in \{1, \ldots, k\}$; we now denote $\tau_h^i$ by $\tau_h$ and denote the corresponding saddle by $p$. Since the saddle is a Morse point of $H$, we can take normal coordinates such that $H(x, y) = xy$, $\forall |x|, |y| \leq 2$. We can suppose that the chosen sections are the segments $\{x = 1\}$ and $\{y = 1\}$, respectively. We can write

$$\nu_{\bar\lambda} = f_{\bar\lambda}(x, y)\, dx + g_{\bar\lambda}(x, y)\, dy$$

for $C^\infty$ functions $f_{\bar\lambda}, g_{\bar\lambda}$ defined on a neighborhood of $|x| \leq 1$, $|y| \leq 1$. By Taylor's theorem, we can write

$$f_{\bar\lambda}(x, y) = \sum_{\substack{0 \leq j \leq n \\ 0 \leq i \leq n-1}} a_{ij}(\bar\lambda) x^i y^j + x^n \sum_{j=0}^{n} h_j(x, \bar\lambda) y^j \tag{3.3}$$

$$+ y^{n+1} \sum_{i=0}^{n-1} g_i(y, \bar\lambda) x^i + (xy)^n y G_{\bar\lambda}(x, y), \tag{3.4}$$

for smooth functions $h_j, g_i, G_{\bar\lambda}$. It is easy to see that integrals of the second and third term in the right-hand-side of (3.3) are smooth in $h$; indeed, $\forall j = 0, \ldots, n$:

$$\begin{aligned} \int_{\tau_h} x^n y^j h_j(x)\, dx &= \int_h^1 x^n (hx^{-1})^j h_j(x)\, dx \\ &= h^j \int_h^1 x^{n-j} h_j(x)\, dx \end{aligned}$$

and $\forall i = 0, \ldots, n-1$:

$$\begin{aligned} \int_{\tau_h} x^i y^{n+1} g_i(x)\, dx &= \int_1^h y^{n+1-i} h^i g_i(y)(hy^{-2})\, dy \\ &= -h^{i+1} \int_h^1 y^{n-i-1} g_i(y)\, dy. \end{aligned}$$

The integral of the first term in the right-hand-side of (3.3) gives a sum of terms in $h^j$ and $h^j \log h$; indeed, $\forall i = 0, \ldots, n-1$, $\forall j = 0, \ldots, n$:

$$\begin{aligned} \int_{\tau_h} x^i y^j\, dx &= h^j \int_h^1 x^{i-j}\, dx \\ &= \begin{cases} -h^j \log h & \text{if } j = i+1 \\ \dfrac{h^j - h^{i+1}}{i - j + 1} & \text{if } j \neq i+1 \end{cases}. \end{aligned}$$

The integral of the fourth term has only a contribution in the remainder:

$$\begin{aligned} \int_{\tau_h} (xy)^n y G_{\bar\lambda}(x, y)\, dx &= h^n \int_{\tau_h} y G_{\bar\lambda}(x, y)\, dx \\ &= h^n \int_0^{-\log h} e^{-t} G_{\bar\lambda}(he^t, e^{-t}) he^t\, dt \\ &= O(h^{n+1} \log h), \quad h \downarrow 0. \end{aligned}$$

By interchanging the roles of $x$ and $y$, one obtains analogous expressions for $\int_{\nu_h} g_{\bar{\lambda}}(x,y)\,dy$. □

**3.5 Definition** Let $\mathcal{F}=\{f_i\}_{i\in\mathbb{N}}$ be an asymptotic scale. Suppose that $f$ is asymptotic to the series

$$\hat{f} = \sum_{i=0}^{\infty} \alpha_i f_i.$$

Then, we say that $f$ has *finite codimension* if not all $\alpha_i$ are zero. If $\alpha_1 = \cdots = \alpha_{l-1} = 0$, $\alpha_l \neq 0$, then we say that $f$ is of *codimension* $l$ in the asymptotic scale $\mathcal{F}$ (this will be denoted by $\operatorname{codim}_{\mathcal{F}} f = l$).

**3.6 Remark** The notion of codimension depends on the asymptotic scale that one considers. It may happen that $f_0$ expands in some scale, but that $f_{\bar{\lambda}}$ needs a larger scale; then we shall define the codimension in this larger scale. For instance, we shall say that $f_{\bar{\lambda}}$, defined by

$$f_{\bar{\lambda}}(h) = \bar{\lambda}_1 + \bar{\lambda}_2 h \log h + h$$

is of codimension 2 (in the scale $\mathcal{L}$), although the function $f_0(h) = h$ is of codimension 1 as a differentiable function (i.e. in the scale $\mathcal{T}$).

The cyclicity of an unfolding $f_{\bar{\lambda}}$ that expands in an asymptotic scale is related to its codimension:

**3.7 Proposition** *Let $\mathcal{F}$ be an asymptotic scale of functions and $f_{\bar{\lambda}}$ an unfolding that expands in this scale with a finite codimension. Then*

$$\operatorname{Cycl}(f_{\bar{\lambda}}, (0,0)) \leq \operatorname{codim}_{\mathcal{F}} f_{\bar{\lambda}}.$$

*In particular, if the Abelian integral $I_{\bar{\lambda}}$ has finite codimension in $\mathcal{L}$, then*

$$\operatorname{Cycl}(I_{\bar{\lambda}}, (0,0)) \leq \operatorname{codim}_{\mathcal{L}} I_{\bar{\lambda}}.$$

**Proof** Suppose that $\operatorname{codim}_{\mathcal{F}} f_{\bar{\lambda}} = q$, we can write

$$f_{\bar{\lambda}}(h) = \sum_{i=0}^{q-1} \alpha_i(\bar{\lambda}) f_i + \alpha_q(\bar{\lambda}) f_q (1 + o(1)), \quad h \downarrow 0,$$

where $\alpha_q(0) \neq 0$. After a procedure of derivation-division we obtain smooth functions $f^0, f^1, \ldots, f^q$, defined by induction as follows:

$$\begin{aligned} f^0 &= f_{\bar{\lambda}}, \\ f^{j+1} &= \frac{\nabla f^j}{\nabla f_1^j}, \quad j = 0, 1, \ldots, q-1, \quad \text{locally for} \quad (h, \bar{\lambda}) \sim (0,0). \end{aligned}$$

In this way, we have

$$f^j(h) = \sum_{i\geq 0} \alpha_{i+j} f_i^j = \alpha_j + o(1), \quad h \downarrow 0;$$

in particular,

$$f^q(h) = \alpha_q + o(1), \quad h \downarrow 0.$$

As a consequence, $f^q_{\bar{\lambda}}(h) \neq 0$ in a neighbourhood of $(0,0)$. Applying inductively Rolle's theorem, it then follows that $f_{\bar{\lambda}}$ has at most $q$ zeroes in a neighborhood of $0 \in ]0, h_0]$ and for $\bar{\lambda}$ close to 0. □

Another interesting property for any asymptotic sequence is the Chebychev property (see [M]):

**3.8 Definition** Let $\mathcal{F} = \{f_i\}_{i\in\mathbb{N}}$ be a sequence of functions defined on some interval $[0, h_0]$ (for some $h_0 > 0$), such that $f_i$ is smooth on $]0, h_0]$ for all $i \in \mathbb{N}$. Then we say that $\mathcal{F}$ *has the Chebychev property at* 0 if and only if for any $n \in \mathbb{N}$ and any $a = (a_0, a_1, \ldots, a_{n-1}) \in \mathbb{R}^n$, the function

$$f_a \equiv \sum_{i=0}^{n-1} a_i f_i + f_n$$

has at most $n$ zeroes on $[0, h_n]$ for some $0 < h_n < h_0$.

In fact, any generic unfolding expanding in any asymptotic scale of functions has a similar bifurcation diagram at $0 \in \mathbb{R}^+$. Let us make this notion of genericity more precise.

**3.9 Definition** Let $\mathcal{F} = \{f_i\}_{i\in\mathbb{N}}$ be an asymptotic scale of functions. Let $f_{\bar{\lambda}}$ be an unfolding at $0 \in \mathbb{R}^+$, $\bar{\lambda} = (\bar{\lambda}_0, \ldots, \bar{\lambda}_{q-1}) \in (\mathbb{R}^q, 0)$, expanding in $\mathcal{F}$, say

$$f_{\bar{\lambda}}(h) = \sum_{i=0}^{q} \alpha_i(\bar{\lambda}) f_i(h) + o(f_q).$$

We say that $f_{\bar{\lambda}}$ is a *generic* unfolding if

(1) $\mathrm{codim}_{\mathcal{F}}\, f_{\bar{\lambda}} = q$ (i.e. $\alpha_q(0) \neq 0$);
(2) the mapping

$$(\bar{\lambda}_0, \ldots, \bar{\lambda}_{q-1}) \mapsto (\alpha_0(\bar{\lambda}), \ldots, \alpha_{q-1}(\bar{\lambda}))$$

is a local diffeomorphism at 0.

Observe again that this definition of generic unfolding differs from the one in the regular case: if $\alpha_0 = \cdots = \alpha_{q-1} = 0$, $\alpha_q \neq 0$, then the codimension is $q$. Now, one has the following result [M]:

**3.10 Theorem** *Let $f_{\bar{\lambda}}$ be a generic unfolding of codimension $q$ in an asymptotic scale of functions with the Chebychev property. Then, for $h_q > 0$ small enough, the diagram of bifurcation of the zeroes of $f_{\bar{\lambda}}$ is topologically equivalent on $[0, h_q[$ to the diagram of bifurcation of the universal polynomial of degree $q$:*

$$P^{\pm}_{\beta}(h) = \beta_0 + \cdots + \beta_{q-1} h^{q-1} \pm h^q$$

*(+ or − depending on the sign of $f_q(h)$ for $h > 0$ near 0).*

Here topologically equivalent means that there exists a local homeomorphism $\beta : (\mathbb{R}^q, 0) \to (\mathbb{R}^q, 0)$ given by $\bar\lambda \mapsto (\beta_0(\bar\lambda), \dots, \beta_{q-1}(\bar\lambda))$ such that $f_{\bar\lambda}$ and $P^{\pm}_{\beta(\bar\lambda)}$ have the same zeroes on $[0, h_q[$.

**3.11 Example** Let us consider $I_{\bar\lambda}(h) = \bar\lambda_0 + \bar\lambda_1 h \log h + h + \cdots$ with $\bar\lambda = (\bar\lambda_0, \bar\lambda_1)$ a codimension 2, generic unfolding of Abelian integrals. The bifurcation diagram is topologically equivalent to the one of $P_\beta(h) = \beta_0 + \beta_1 h + h^2$. The diagram of $I_{\bar\lambda}(h)$ includes the $\bar\lambda_1$-axis (a zero of $I_{\bar\lambda}$ crosses the origin) and a line $D$ of double zeroes. The difference with the diagram of $P_\beta$ is that $D$ has a flat contact with the $\bar\lambda_1$-axis, while the corresponding line of $P_\beta$ is just an arc of a parabola.

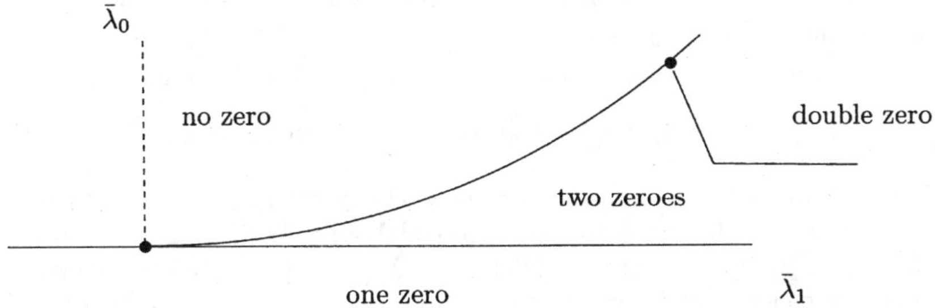


Figure 2: Bifurcation diagram for a generic unfolding of codimension 2

# 4 Generic unfoldings of a 2-saddle cycle

In this section, we will consider the simpler generic unfoldings with two parameters of the 2-saddle cycle. As the two connections are broken, we shall see that not all limit cycles can be traced by the zeroes of the corresponding Abelian integral. In fact, under specific generic conditions given by Mourtada, the cyclicity of the Hamiltonian 2-saddle cycle is at least 2. However, these generic conditions imply that the codimension of the Abelian integral is equal to 1, and then there is exactly one zero of it that bifurcates from $\{h = 0\}$. The bifurcation diagram for generic 2-parameter unfoldings of 2-saddle cycles is described for instance in [DRS]. Here we will only recall the results.

## 4.1 Limit cycles not shadowed by zeroes of the Abelian integral

We consider a hyperbolic 2-saddle cycle $\Gamma$. Let $s_1, s_2$ be the singular points of $\Gamma$. We call $\Gamma_1$ (resp. $\Gamma_2$) the local stable separatrix of $\Gamma$ at $s_1$ (resp. $s_2$). Let us consider an unfolding $X_\lambda$ with $\lambda \in \mathbb{R}^p$ near $0 \in \mathbb{R}^p$ $(p \geq 2)$, that is defined near the 2-saddle cycle $\Gamma$ of $X_0$.

**4.1 Definition** Let $s$ be a saddle point of $X_\lambda$. Then the *ratio of hyperbolicity* $r(\lambda)$ of the saddle point $s$ for $X_\lambda$ is defined as the ratio of the absolute value of the negative eigenvalue divided by the positive eigenvalue at $s$.

Let $r_1(\lambda)$ (resp. $r_2(\lambda)$) denote the ratio of hyperbolicity of the saddle $s_1$ (resp. $s_2$) for $-X_\lambda$ (resp. $X_\lambda$).

**4.2 Definition** Let us choose a transverse section $\Sigma_0$ to the separatrix $\Gamma_0$ of $\Gamma$ of $X_0$, connecting the saddle $p_1$ with $p_2$. Then, for $\lambda$ near 0, the local stable separatrix of $p_2$ cuts $\Sigma_0$ transversally at a point $S(\lambda)$, and the local unstable separatrix of $p_1$ cuts $\Sigma_0$ at $U(\lambda)$. The *breaking parameter* $b$ of $\Gamma_0$ is then defined as the function $b(\lambda) = S(\lambda) - U(\lambda)$.

Take sections $\Sigma_i$ transversal to $\Gamma_i$, $i = 1, 2$. Let $\Delta_1$ (resp. $\Delta_2$) denote the transition mapping from $\Sigma_2$ to $\Sigma_1$ for $-X_\lambda$ (resp. $X_\lambda$) near $s_1$ (resp. $s_2$). Denote the breaking parameter at $\Sigma_i$ by $b_i$, $i = 1, 2$.

Let us now recall the definition of *genericity of Mourtada*: a 2-parameter unfolding $X_\lambda$ is called *generic* if

(1) the mapping $\lambda \to (b_1(\lambda), b_2(\lambda))$ is a local diffeomorphism at $0 \in \mathbb{R}^2$ (in particular, one can suppose that $\lambda = b = (b_1, b_2)$);

(2) the following generic conditions on the values $r_i(0)$, $i = 1, 2$, are satisfied:

$$r_1(0) \neq 1, \qquad r_2(0) \neq 1, \qquad r_2(0)/r_1(0) \neq 1.$$

It has been proved by Mourtada that the cyclicity of any generic 2-parameter unfolding of a 2-saddle cycle is bounded by the universal bound $M(2) = 2$. Let us now return to the unfolding of a Hamiltonian 2-saddle cycle. We can translate the generic conditions of Mourtada into generic conditions for the Hamiltonian unfolding: a perturbation $X_\lambda$ of $X_H$ is said to be a *generic 2-parameter unfolding* if

(1) $\lambda = (\bar{\lambda}, \varepsilon) \in \mathbb{R}^3$ with $\bar{\lambda} \in (\mathbb{R}^2, 0)$ and $\varepsilon \in [0, \varepsilon_0[$;

(2) the breaking parameters have the form $b_i(\varepsilon, \bar{\lambda}) = \varepsilon\beta_i$, $i = 1, 2$, and $\bar{\lambda} = \beta = (\beta_1, \beta_2)$;

(3) the hyperbolicity ratios have the form $r_i(\beta, \varepsilon) = 1 + \varepsilon\alpha_i(\beta, \varepsilon)$, $i = 1, 2$, and the following conditions are supposed on $\alpha_i(0)$, $i = 1, 2$:

$$\alpha_1(0) \neq 0, \qquad \alpha_2(0) \neq 0, \qquad \alpha_2(0) - \alpha_1(0) \neq 0.$$

(This means that the generic conditions of Mourtada are satisfied on the ratios $r_i(\beta, \varepsilon)$, for $\varepsilon > 0$).

The first return mapping is given by

$$P = \Delta_2 \circ \Delta_1^{-1}.$$

The associated displacement mapping $\delta = P - Id$ can be written as

$$\begin{aligned} \delta &= (\Delta_2 - \Delta_1) \circ \Delta_1^{-1} \\ &= \Delta \circ \Delta_1^{-1}, \quad \text{where} \quad \Delta = \Delta_2 - \Delta_1. \end{aligned}$$

Therefore, we consider the difference function $\Delta$ rather than the displacement mapping $\delta$.

Since for $\varepsilon = 0$, the vector field $X_H$ is of center type, we can write

$$\Delta = \varepsilon\bar{\Delta}, \qquad \delta = \varepsilon\bar{\delta}, \qquad \Delta_1(x, \bar{\lambda}, \varepsilon) = x + O(\varepsilon), \qquad \varepsilon \to 0;$$

therefore, the limit of $\bar{\Delta}$ is again the Abelian integral when $\varepsilon \to 0$ :

$$\bar{\Delta}_{|\varepsilon=0} = \bar{\delta}_{|\varepsilon=0} = I_{\bar{\lambda}}.$$

As a consequence, the Abelian integral has the following expansion:

$$I_{\bar{\lambda}}(x) = \beta_2 - \beta_1 + (\alpha_2(\beta,0) - \alpha_1(\beta,0))x \log x + O(x), \quad x \to 0.$$

Now, the fact that the reduced breaking parameters $\beta_1$ and $\beta_2$ are assumed to be independent parameters together with the generic condition $\alpha_1(0) - \alpha_2(0) \neq 0$, imply that the Abelian integral $I_{\bar{\lambda}}$ is a generic unfolding of codimension 1. By Theorem 3.10, we know that there is exactly one zero of the Abelian integral that bifurcates from $\{x = 0\}$.

On the other hand, for any fixed small $\varepsilon > 0$, the unfolding $X_{(\beta,\varepsilon)}$ is generic in the sense of Mourtada, and therefore it can have a cyclicity equal to 2. In other words, one can find a sequence $(\beta^n_\varepsilon)_{n\in\mathbb{N}}$ converging to $0 \in \mathbb{R}^2$ such that each $X_{(\beta^n_\varepsilon,\varepsilon)}$ has two limit cycles which converge to $\Gamma$ when $n \to \infty$. Given a sequence $(\varepsilon_m)_m$ with $\varepsilon_m \downarrow 0$ $(m \to \infty)$, we can extract by the diagonal method a subsequence $(\beta^{n(m)}_{\varepsilon_m})_{m\in\mathbb{N}}$ converging to $0 \in \mathbb{R}^2$, such that the corresponding two limit cycles converge to $\Gamma$ when $m \to \infty$. Then the cyclicity of the Hamiltonian unfolding $\mathrm{X}_{(\beta,\varepsilon)}$ is at least 2. As a conclusion, when one breaks more than one connection, the Abelian integral $I_\beta$ is a very bad approximation of the displacement mapping $\bar{\delta}_{(\beta,\varepsilon)}$, which enters in the expression of the Poincaré mapping

$$P_{(\beta,\varepsilon)}(x) = x + \varepsilon\bar{\delta}_{(\beta,\varepsilon)}(x).$$

The Abelian integral has only one zero, while the displacement mapping has at least two zeroes that correspond to limit cycles bifurcating from $\Gamma$. *Hence, not all limit cycles can be traced by the zeroes of the corresponding Abelian integral, when the unfolding breaks more than one connection of the* 2*-saddle cycle.*

### 4.2 Bifurcation diagram and remarks

The study of the bifurcation diagram of such a generic unfolding was carried out in [DRS], and we just recall the results. Let us suppose that

$$\alpha_1(0),\ \alpha_2(0) > 0, \qquad \alpha_2(0) - \alpha_1(0) > 0. \tag{4.1}$$

The bifurcation diagram of the limit cycles is given by the one for the zeroes of the function $\bar{\Delta}_{(\bar{\lambda},\varepsilon)}$. As we remarked in Section 3, for $\varepsilon = 0$, the function $\bar{\Delta}_{(\beta,0)}$ is equal to the Abelian integral $I_\beta$. The diagram of bifurcation for $I$ reduces to a curve $Z$ tangent to $\{\beta_1 = \beta_2\}$ at the origin. One finds one zero for $\beta$ below this line and none for $\beta$ above.

For $\varepsilon > 0$, the curve $Z$ has a continuation consisting of an arc $R$ starting at the origin in the first quadrant $\{\beta_1 > 0, \beta_2 > 0\}$ and of a second arc $L$ starting at the origin in the quadrant $\{\beta_1 < 0, \beta_2 < 0\}$. Along $L$ one has a connection at the point $s_2$ (left connection) and along $R$ a connection at $s_1$ (right connection). Then, the bifurcation diagram is completed by an arc $D$ of double (semi-stable) limit cycles. This arc $D$ starts at the origin in the first quadrant, above the arc $R$. Moreover, it can be proved that these different arcs have a vertical tangent at the origin. One finds one limit cycle below the curve $L \cup R$, no limit cycle above the curve $L \cup D$, and two limit cycles inside the tongue between $D$ and $R$. Of course one has to draw these different figures for different values of $\varepsilon$ as one unique figure in the 3-parameter space $(\beta_1, \beta_2, \varepsilon)$. The curves are then organized in three surfaces called again $L, R, D$. This is shown in Fig. 3 (under the conditions (4.1)).

An important point to observe is that the bifurcation diagram is no longer *trivial in the $\varepsilon$-direction* as is the case for the generic unfoldings of a regular Hamiltonian cycle, a regular center or a saddle loop, where the first two cases follow from catastrophe theory and the third one has been treated by P. Mardešić, as will be recalled in Section 5.

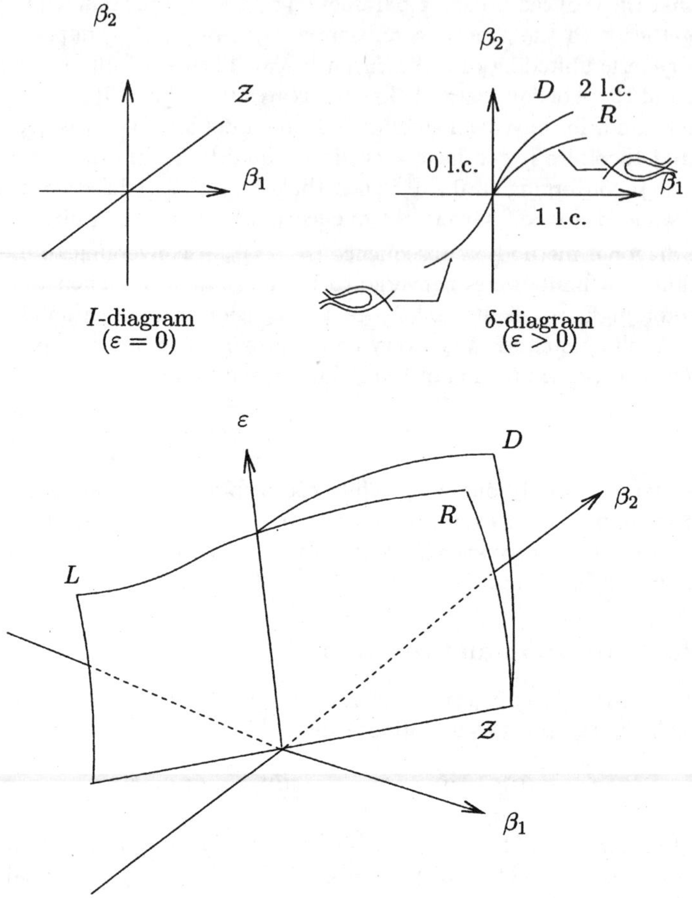


Figure 3: Bifurcation diagram for the 2-saddle cycle generic unfolding

This lack of triviality corresponds to the fact that the bifurcation diagram of the Abelian integral (found at $\varepsilon = 0$) is not preserved as a bifurcation diagram of the limit cycles: half of the curve $Z$ is split into a tongue, containing more limit cycles than the number of zeroes of the Abelian integral.

This can be seen as the *existence of limit cycles escaping from the validity domain of the Abelian integral.* For any $x_0 > 0$ , there exists an interval in $\varepsilon$, let us say $[0, \varepsilon(x_0)[$, and an interval $[x_0 - \eta(x_0), x_0 + \eta(x_0)] \subset R^+_*$ where all the limit cycles are related to zeroes of the Abelian integral (as a consequence of the implicit function theorem). One can give an estimate of $\varepsilon(x_0)$, using the properties of the Mourtada-type functions. For a fixed value of the parameter $\beta_2 > 0$ one finds:

$$\varepsilon(x_0) \sim \frac{1}{\beta_2} x_0 \mid \log x_0 \mid^{-1} .$$

This condition defines the *domain of validity of the Abelian integral or I-domain*, as a tongue in the space $(x_0, \varepsilon)$ between the graph of the function $\varepsilon(x_0)$ and the $x_0$-axis.

Consider now the *outer limit cycle* $\gamma_{\text{out}}$, i.e. the one corresponding to the smallest root $x_{\text{out}}$ of the equation for the limit cycles, when the parameter $\beta$ is chosen inside the tongue between $R$ and $D$. A direct calculation shows that $x_{\text{out}} \sim \varepsilon\beta_2$. Then, for a fixed $\beta_2 > 0$, one sees that the zero $x_{\text{out}}$ escapes from the $I$-domain: it corresponds to an arc $(\varepsilon, x_{\text{out}}(\varepsilon))$ which approaches to $(0,0)$ outside the $I$-domain, when $\varepsilon \to 0$.

**4.3 Remark** More generally, one can consider a $k$-parameter unfolding of a $k$-saddle cycle in a generic way. The associated Abelian integral remains of codimension 1, although there are more than $k$ limit cycles that bifurcate from the saddle [Mo]. As a conclusion we can say that almost none of the limit cycles can be traced by the zeroes of the Abelian integral in this general case.

# 5 Saddle loop

## 5.1 Expansion of the displacement function

We first recall results about unfoldings of a saddle loop (also called a saddle connection), which is a singular cycle containing precisely one hyperbolic singular point. In this case, the Abelian integral $I_{\bar{\lambda}}$ is a good approximation for the map $\bar{\delta}_\lambda$, which determines the limit cycles. In particular, the codimension of $I_{\bar{\lambda}}$ is an upper bound for the number of limit cycles.

We consider a saddle connection $\Gamma$ at a hyperbolic saddle point $p$. Let $X_\lambda$ be a smooth unfolding of the Hamiltonian vector field $X_H$, $\omega_\lambda$ being its dual form unfolding, and let $P(h, \lambda)$ be the return map on $\sigma$, defined for $(h, \lambda) \in [0, h_0[ \times U$. We suppose that $p$ is a saddle point of $X_\lambda$ for any $\lambda \in U$. Let $\sigma$ (resp. $\tau$) be transverse sections to the local stable (resp. unstable) separatrix of $p$, contained in $\Gamma$. Suppose that $\Gamma \subset \{H = 0\}$ and that $\{H \geq 0\}$ corresponds to the side where the return map $P_\lambda : \sigma \to \sigma$ is defined. Denote the Dulac mapping from $\sigma$ in $\tau$ by $D$, and the regular transition from $\tau$ in $\sigma$ by $R$. Then $P = R \circ D$.

We call $\delta(h, \lambda) = P(h, \lambda) - h$, the associated displacement function. We assume that $\lambda = (\bar{\lambda}, \varepsilon)$ and that $X_{(\bar{\lambda}, 0)} \equiv X_H$. This implies that

$$\delta(h, \lambda) = \varepsilon\bar{\delta}_\lambda(h) = \varepsilon\bar{\delta}(h, \bar{\lambda}, \varepsilon) \tag{5.1}$$

for some smooth function $\bar{\delta}$ that we call the reduced displacement function. Using the Melnikov formula, we have:

$$\bar{\delta}(h, \bar{\lambda}, 0) = I(h, \bar{\lambda}) = \int_{\gamma_h} \bar{\omega}_{\bar{\lambda}}, \tag{5.2}$$

where

$$\omega_\lambda = dH - \varepsilon\bar{\omega}_{\bar{\lambda}} + o(\varepsilon). \tag{5.3}$$

An asymptotic expansion for $\bar{\delta}$ at any order $N$ in $h$ was derived in [R1]. We briefly recall this result. Let $r(\lambda)$ be the ratio of hyperbolicity of the saddle point $p$ of $X_\lambda$; we can write $r(\lambda) = 1 + \varepsilon\alpha(\bar{\lambda}, \varepsilon)$. We introduce the *compensator*

$$\omega_\beta(h) = \begin{cases} \dfrac{h^\beta - 1}{\beta} & \text{for } \beta \neq 0 \\ \log h & \text{for } \beta = 0 \end{cases}.$$

Here $\omega = \omega_{\varepsilon\alpha}$. We have now the following asymptotic expansion for $\bar{\delta}(h, \bar{\lambda}, \varepsilon)$:

**5.1 Theorem** *There exists a sequence of germs of smooth coefficients $\alpha_i, \beta_i, i \in \mathbb{N}$, at $\bar{\lambda} = 0, \varepsilon = 0$, such that, for any $N \in \mathbb{N}$*

$$\begin{aligned} \bar{\delta}(h, \bar{\lambda}, \varepsilon) = \beta_0 &+ \alpha_1[h\omega + \varepsilon\eta_1] + \beta_1[h + \varepsilon\mu_1] + \cdots \\ &+ \beta_N[h^N + \varepsilon\mu_N] + \alpha_{N+1}[h^{N+1}\omega + \varepsilon\eta_{N+1}] + \Phi_{N+1}(h, \bar{\lambda}, \varepsilon) \end{aligned} \tag{5.4}$$

*where $\Phi_{N+1}$ is of class $C^N$ and $N$-flat at $h = 0$, $\forall\lambda$, i.e.,*

$$\Phi_N(0, \bar{\lambda}, \varepsilon) = \cdots = \frac{\partial^N \Phi_N}{\partial h^N}(0, \bar{\lambda}, \varepsilon) = 0 \quad \textit{for all} \quad (\bar{\lambda}, \varepsilon).$$

*The functions $\eta_i, \mu_i$ are polynomials in $h$ and $h\omega$ with valuation strictly greater than the leading term of the bracket, and their coefficients are polynomials in the functions $\alpha_i, \beta_i$. (The valuation is the weakest degree of the monomials in a polynomial.)*

**5.2 Remarks** (1) The function $\varepsilon\beta_0$ is the $h$-coordinate on $\sigma$ of the first intersection of the unstable separatrix, and it is the breaking parameter of the connection $\Gamma$.

(2) One has $\alpha_1 = (1 + O(\varepsilon))\alpha$, and the other parameters $\alpha_i$ are related to the normal form of the unfolding $X_\lambda$ at the saddle point $p$.

The expansion in the theorem is made in a sequence $\mathcal{W}$ of functions (the brackets) of the variable $h$ and depending on the parameter $(\bar{\lambda}, \varepsilon)$:

$$\mathcal{W} = \{1, [h\omega + \cdots], [h + \cdots], \ldots, [h^n + \cdots], [h^{n+1}\omega + \cdots], \ldots\}. \tag{5.5}$$

This sequence $\mathcal{W}$ coincides for $\varepsilon = 0$ with the logarithmic asymptotic scale

$$\mathcal{L} = \{1, h\log h, h, \ldots\};$$

the expansion (5.4) in the theorem coincides for $\varepsilon = 0$ with the expansion of the Abelian integral:

$$\bar{\delta}(h, \bar{\lambda}, 0) = I(h, \bar{\lambda}) = \sum_{i=0}^{N}(\bar{\beta}_i(\bar{\lambda})h^i + \bar{\alpha}_{i+1}(\bar{\lambda})h^{i+1}\log h) + o(h^N), \tag{5.6}$$

where $\bar{\beta}_i(\bar{\lambda}) = \beta_i(\bar{\lambda}, 0)$ and $\bar{\alpha}_{i+1}(\bar{\lambda}) = \alpha_{i+1}(\bar{\lambda}, 0)$. This leads to the introduction of the following definitions.

**5.3 Definition** Let $\{f_i\}_{i\in\mathbb{N}}$ be an asymptotic scale and let $\{F_i\}_{i\in\mathbb{N}}$ be a sequence of functions with parameter $\lambda$, defined for $(h,\lambda) \in [0,h_0[\times V_0$, where $\lambda = (\bar{\lambda},\varepsilon) \in V_0 = \bar{V}_0 \times [0,\varepsilon_0[$, such that

$$f_i(h) \equiv F_i(h,\bar{\lambda},0), \quad \forall (h,\bar{\lambda}) \in [0,h_0[\times \bar{V}_0.$$

Then we say that $\{F_i\}_{i\in\mathbb{N}}$ is an *asymptotic scale deformation* or a *deformation of the asymptotic scale* $\{f_i\}_{i\in\mathbb{N}}$ if it fulfills the following conditions:

(1) $\forall i \in \mathbb{N}$ : $F_i$ is continuous on $[0,h_0[\times V_0$, the functions $h \to F_i(h,\lambda)$ are smooth for $h \neq 0$ and $\forall \lambda \in V_0$, and $F_0(h,\lambda) \equiv 1$;

(2) $\forall i \in \mathbb{N}$: $F_i(h,\lambda) \neq 0$ for $h > 0$ small enough and

$$\frac{F_{i+1}}{F_i}(h,\bar{\lambda},\varepsilon) \to 0 \quad \text{when} \quad (h,\varepsilon) \to (0,0), \quad \text{uniformly on } \bar{\lambda};$$

moreover, $\nabla(\frac{F_{i+1}}{F_i})(h,\lambda)$ has constant sign for $h > 0$ small enough.

**5.4 Definition** (*Simple deformation of asymptotic scale* or *simple asymptotic scale deformation*) Let $\{f_i\}_{i\in\mathbb{N}}$ be an asymptotic scale and let $\{F_i\}_{i\in\mathbb{N}}$ be a sequence of functions with parameter $\lambda$, defined for $(h,\lambda) \in [0,h_0[\times V_0$, where $\lambda = (\bar{\lambda},\varepsilon) \in V_0 = \bar{V}_0 \times [0,\varepsilon_0[$, such that

$$f_i(h) \equiv F_i(h,\bar{\lambda},0), \quad \forall (h,\bar{\lambda}) \in [0,h_0[\times \bar{V}_0.$$

Then we say that $\{F_i\}_{i\in\mathbb{N}}$ is a simple deformation of the asymptotic scale $\{f_i\}_{i\in\mathbb{N}}$ if it fulfills the following conditions:

(1) $\{F_i\}_{i\in\mathbb{N}}$ is an asymptotic scale deformation of $\{f_i\}_{i\in\mathbb{N}}$;

(2) one can construct inductively a sequence $\{\mathcal{D}_j\}_{j\in\mathbb{N}}$ such that $\forall j \geq 0$,

$$\mathcal{D}_j = \{F_i^j\}_{i\in N},$$

where $\forall i \in \mathbb{N}$ : $F_i^j$ is defined on $[0,h^j[\times V_j$, $V_j = \bar{V}_j \times [0,\varepsilon_j[$ (for some sequences:

$$\cdots \leq h_{j+1} \leq h_j \leq \cdots \leq h_0, \qquad \cdots \subset \bar{V}_{j+1} \subset \bar{V}_j \subset \cdots, \qquad \cdots \leq \varepsilon_{j+1} \leq \varepsilon_j \leq \cdots \leq \varepsilon_0).$$

These sequences $\mathcal{D}_j$, $j \in \mathbb{N}$, are defined by induction:

$$F_i^0 = F_i, \quad \forall i \in \mathbb{N}, \tag{5.7}$$

$$F_i^{j+1} = \frac{\nabla F_{i+1}^j}{\nabla F_1^j}, \quad \forall i \in \mathbb{N} \tag{5.8}$$

for $h \in [0,h_{j+1}[$ and $i \geq 0$. (One chooses the $h_{j+1}$ such that $\nabla F_1^j(h,\lambda) \neq 0$ on $[0,h_{j+1}[\times V_{j+1}$.) Each sequence $\mathcal{D}_j$, $j \geq 0$, is an asymptotic scale deformation of $\mathcal{F}_j = \{f_i^j\}_{i\in\mathbb{N}}$ on $[0,h_j[\times V_j$.

**5.5 Remark** A simple asymptotic scale deformation is an asymptotic scale deformation which generates infinitely many new asymptotic scale deformations by the inductive application of the algorithm of derivation-division given by the formula (5.8) (derivation by $\nabla$ followed by the division by the first function of the sequence).

The sequence $\mathcal{W}$ is a simple deformation of the asymptotic scale $\mathcal{L} = \{1, h\log h, h, \dots\}$ associated to the Abelian integral. This fact was proved in [R1] (without introducing this terminology). Let us notice that Theorem 5.1 produces an expansion of the reduced displacement function $\bar{\delta}(h, \bar{\lambda}, \varepsilon)$ in this simple asymptotic scale deformation at any order, in the following sense:

**5.6 Definition** Let $F(h, \bar{\lambda}, \varepsilon)$ be a function defined for $(h, \bar{\lambda}, \varepsilon) \in [0, h_0[\times \bar{V}_0 \times [0, \varepsilon_0[$, and let $\mathcal{D} = \{F_i\}_{i\in\mathbb{N}}$ be a deformation of the asymptotic scale $\{f_i\}_{i\in\mathbb{N}}$; $\mathcal{D}$ is also supposed to be defined on $[0, h_0[\times \bar{V}_0 \times [0, \varepsilon_0[$. We say that $F$ has an *expansion* in $\mathcal{D}$ at *order* $N$ if

(1) there exist smooth functions $\alpha_i(\bar{\lambda}, \varepsilon)$, $0 \le i \le N$, defined on $\bar{V}_0 \times [0, \varepsilon_0[$, and a remainder $\Phi_{N+1}(h, \bar{\lambda}, \varepsilon)$ such that:

$$F(h, \bar{\lambda}, \varepsilon) = \sum_{i=0}^{N} \alpha_i(\bar{\lambda}, \varepsilon) F_i(h, \bar{\lambda}, \varepsilon) + \Phi_{N+1}(h, \bar{\lambda}, \varepsilon) \tag{5.9}$$

for all $(h, \bar{\lambda}, \varepsilon) \in [0, h_0[\times \bar{V}_0 \times [0, \varepsilon_0[$.

(2) Moreover, $\Phi_{N+1}(h, \bar{\lambda}, \varepsilon)$ has the *remainder-property of order* $N$ (for the given simple asymptotic scale deformation $\mathcal{D}$) denoted by property $(R)_N$ for short, i.e.

$$\frac{\Phi_{N+1}}{F_N}(h, \bar{\lambda}, \varepsilon) \to 0, \quad \text{when} \quad (h, \varepsilon) \to (0,0), \quad \text{uniformly in } \bar{\lambda},$$

and if one defines inductively the functions $\Phi^{j+1}_{N+1}$, $j = 0, \dots, N-1$, by

$$\Phi^{j+1}_{N+1} = \frac{\nabla \Phi^j_{N+1}}{\nabla F^j_1}, \quad \text{with} \quad \Phi^0_{N+1} = \Phi_{N+1},$$

one has that

$$\frac{\Phi^j_{N+1}}{F^j_{N-j}} \to 0, \quad \text{when} \quad (h, \varepsilon) \to (0,0), \quad \text{uniformly in } \bar{\lambda};$$

in particular $\Phi^N_{N+1}(h, \bar{\lambda}, \varepsilon) = o(1)$, when $(h, \varepsilon) \to (0,0)$, uniformly in $\bar{\lambda}$.

**5.7 Remark** If $\mathcal{D}$ is a simple asymptotic scale deformation, the property $(R)_N$ simply expresses that the remainder has a similar asymptotic behavior as the function $F_{N+1}$. For expansions in the simple asymptotic scale deformation $\mathcal{W}$ appearing in Theorem 5.1, this property was obtained in [R1] by considering an expansion (5.4) at order $2N+1$. This allows us to write the remainder $\Phi_N$ in (5.9) as the sum of the terms in the principal part of (5.4) from order $N+1$ to order $2N+1$ plus the remainder $\Psi_{2N+1}$ of (5.4), which is a function of class $C^{2N+1}$ and flat at $x = 0$ at the order $2N+1$. Clearly, the remainder $\Phi_N$ has the property $(R)_N$.

## 5.2 Finite cyclicity

Let us suppose that $F$ has an expansion at any order in a given simple asymptotic deformation $\mathcal{D} = \{F_i\}$ of the scale $\{f_i\}$. Recall that the codimension of $f(h, \bar{\lambda}) = F(h, \bar{\lambda}, 0)$ at $h = 0$,

$\bar{\lambda} = 0$, is defined as the order $q$ of the first non-zero coefficient $\bar{\alpha}_j(0)$ in the series:

$$\widehat{f}(h, \bar{\lambda}) = \sum_{i=0}^{\infty} \bar{\alpha}_i(\bar{\lambda}) f_i(h), \tag{5.10}$$

where $\bar{\alpha}_i(\bar{\lambda}) = \alpha_i(\bar{\lambda}, 0)$. As such, $\text{cod}(f) = q$ if and only if $\bar{\alpha}_0(0) = \cdots = \bar{\alpha}_{q-1}(0) = 0$ and $\bar{\alpha}_q(0) \neq 0$.

It is easy to extend the proof given in Proposition 3.7 for $f(h, \bar{\lambda})$ to the function $F(h, \bar{\lambda}, \varepsilon)$ itself. Such a proof is given in [R1] when $F$ expands in the sequence $\mathcal{W}$, but the same proof works for any simple asymptotic scale deformation. So, we have the following result:

**5.8 Theorem** *Let $F(h, \bar{\lambda}, \varepsilon)$ be a function unfolding as above, with an expansion at any order $N$ in a simple deformation $\{F_i\}_{i \in \mathbb{N}}$ of the asymptotic scale $\{f_i\}_{i \in \mathbb{N}}$. Let us suppose that the codimension of $f(h, \bar{\lambda}) = F(h, \bar{\lambda}, 0)$ is finite at $h = 0$, $\bar{\lambda} = 0$. Then*

$$\text{Cycl}(F, (0,0)) \leq \text{cod}(f_{0\cdot}) \tag{5.11}$$

As a corollary, one can deduce the following result that is proved in [R1] for unfoldings of saddle loops:

**5.9 Theorem** *Let $X_\lambda$ be a perturbation of a Hamiltonian vector field along a saddle loop $\Gamma$ and let $I_{\bar{\lambda}}$ be the corresponding Abelian integral unfolding. If* $\text{cod}(I_{\bar{\lambda}}) < \infty$*, then*

$$\text{Cycl}(X_\lambda, \Gamma) \leq \text{cod}(I_0). \tag{5.12}$$

### 5.3 Bifurcation diagram

For the bifurcation diagram (when $\mathcal{F} = \mathcal{L}$), we just mention the following result of Mardešić [M], which is very hard to prove (contrary to the regular case and the case of $I_{\bar{\lambda}}$ itself). The problem comes from the fact that $F_i$ depends on the parameter $\alpha_1$ through $\omega$ $(\alpha_1 = \alpha)$.

**5.10 Theorem** *Let $X_\lambda$ be a perturbation of a Hamiltonian vector field, along a saddle loop, with $\lambda = (\bar{\lambda}, \varepsilon)$. If the corresponding Abelian integral $I_{\bar{\lambda}}$ is a generic unfolding with $k$ parameters (in the sense of Definition 3.9), then the bifurcation set* $\text{Diagram}(\delta_\lambda)$ *is topologically equivalent to* $\text{Diagram}(I_{\bar{\lambda}}) \times [0, \varepsilon_0]$ *for $\varepsilon_0$ small enough (i.e., the bifurcation sets are homeomorphic by a homeomorphism of the form $(\bar{\lambda}, \varepsilon) \to (G(\bar{\lambda}, \varepsilon), \varepsilon)$).*

It is an open question if the proof of Mardesic is valid for any $\mathcal{F}$ with the Chebychev property and any simple deformation $\mathcal{D}$.

## 6 Unfoldings of the 2-saddle cycle breaking just one connection

### 6.1 The 2-saddle cycle

In the previous section, we have seen that for a saddle loop, the limit cycles of a perturbation $X_\lambda$ of a Hamiltonian vector field are completely related to the zeroes of the corresponding Abelian integral. It is legitimate to define the codimension of the saddle loop unfolding $X_\lambda$ to be the codimension of the associated Abelian integral unfolding, and to say that the unfolding $X_\lambda$ is generic if and only if the unfolding of the Abelian integral is generic. We shall see in this section, that this will no longer be the same as soon as the number of saddles is two or greater, even if the unfolding breaks only one connection. The notions of codimension as well as genericity will not be the same for a vector field unfolding and its related Abelian integral unfolding.

In this section we want to study the limit cycles that can be created by a 2-saddle cycle $\Gamma$ in a system (1.2) such that the phase portrait of the Hamiltonian vector field $X_H$ (1.1), which we unfold, is given in Fig. 4. The parameter $\lambda = (\bar{\lambda}, \varepsilon)$ is in a neighborhood of $(0,0) \in \mathbb{R}^k \times \mathbb{R}^+$. Our study merely deals with a neighborhood of the 2-saddle cycle $\Gamma$. We shall limit ourselves to perturbations that leave one of the connections unbroken. But besides this restriction, we shall work as generally as possible. Near the saddles $s_1$ and $s_2$ we can use normalizing coordinates (see [R1]), denoted respectively by $(x,y)$ and $(z,w)$. In Fig. 4, we represent some transverse sections $C_1, C_2, C_3, C_4$, corresponding respectively to $\{y=1\}$, $\{x=1\}$, $\{w=1\}$ and $\{z=1\}$ in the normalizing coordinates; $\{w=0\}$ corresponds to points on the local stable separatrix of $s_2$, and $\{x=0\}$ corresponds to points on the unbroken connection.

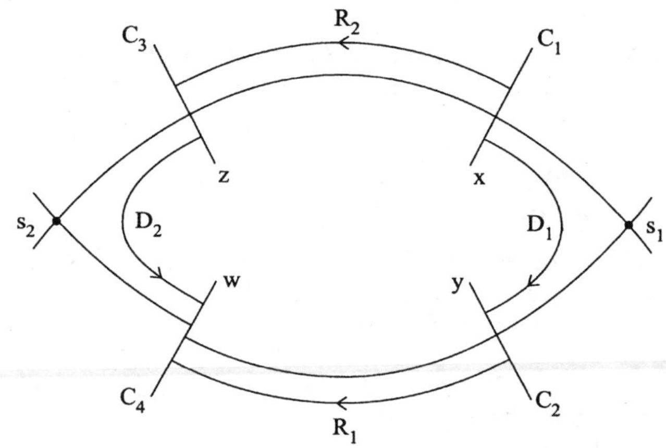


Figure 4: Transverse sections and related transition maps

## 6.2 A first asymptotic expansion for the difference function $\Delta$

For this 2-saddle problem it is convenient to replace the displacement map on a transverse section by the difference function $\Delta$ between two sections. More concretely, we consider the transition $\Delta_1$ for $X_\lambda$ from $C_1$ to $C_4$, and the transition $\Delta_2$ for $-X_\lambda$ from $C_1$ to $C_4$. The difference function $\Delta$ is defined by:

$$\Delta(x,\lambda) = \Delta_2(x,\lambda) - \Delta_1(x,\lambda), \quad \text{for} \quad x \in C_1,\ x > 0 \text{ and } \lambda = (\bar{\lambda}, \varepsilon). \tag{6.1}$$

Now we have to consider two compensators $\omega_1$ and $\omega_2$ associated to $X_\lambda$ and $-X_\lambda$, respectively at the saddle points $s_1$ and $s_2$:

$$\omega_1(x,\bar{\lambda},\varepsilon) = \frac{x^{\varepsilon\alpha_1}-1}{\varepsilon\alpha_1} \quad \text{and} \quad \omega_2(z,\bar{\lambda},\varepsilon) = \frac{z^{\varepsilon\alpha_2}-1}{\varepsilon\alpha_2}, \tag{6.2}$$

where the ratios of hyperbolicity of the vector fields $X_\lambda$ and $-X_\lambda$ at the points $s_1$ and $s_2$ are equal to $1+\varepsilon\alpha_1(\bar{\lambda},\varepsilon)$ and $1+\varepsilon\alpha_2(\bar{\lambda},\varepsilon)$, respectively.

Now, using [R1] again, one obtains for $\Delta_1$ and $\Delta_2$ asymptotic expansions at any order, in terms of $\omega_1(x,\bar{\lambda},\varepsilon)$ and $\omega_2(x,\bar{\lambda},\varepsilon)$, respectively, similar to the expansions for the map $\bar{\delta}$ in Section 5 (Theorem 5.1).

For $\varepsilon = 0$ the vector field is Hamiltonian, and the Hamiltonian function $H$ is equal to $xy$ and $zw$, respectively, in the normalizing coordinates near the saddle points $s_1$ and $s_2$. It follows that $\Delta_1 - \mathrm{Id}, \Delta_2 - \mathrm{Id}$ and $\Delta$ are divisible by $\varepsilon$. We write:

$$\Delta_i(x,\bar{\lambda},\varepsilon) = x + \varepsilon\bar{\Delta}_i(x,\bar{\lambda},\varepsilon) \quad \text{and} \quad \Delta = \varepsilon\bar{\Delta} = \varepsilon(\bar{\Delta}_2 - \bar{\Delta}_1), \tag{6.3}$$

where $\bar{\Delta}_i, \bar{\Delta}$ have the same type of expansion as $\Delta$. But now there is no reason to preserve the special grouping of terms in brackets appearing in the formula (5.4) for $\bar{\delta}$, and having the leading terms linear in the compensator. The asymptotic expansions that we can write for $\bar{\Delta}$ just have the property that their principal parts are polynomials in $x, x\omega_1, x\omega_2$ at any order $N$:

**6.1 Theorem** *Let $\bar{\Delta}$ be the reduced difference function, associated to an unfolding of a 2-saddle cycle leaving one connection unbroken as above. There exists a sequence of germs at $\bar{\lambda} = 0$, $\varepsilon = 0$, of smooth functions $\alpha_{ijk}(\bar{\lambda},\varepsilon)$, $i,j,k \in \mathbb{N}$, such that for any $N \in \mathbb{N}$, one has the following expansion at order $N$:*

$$\bar{\Delta}(x,\bar{\lambda},\varepsilon) = \sum_{i+j+k=0}^{N} \alpha_{ijk}(\bar{\lambda},\varepsilon)x^i(x\omega_1)^j(x\omega_2)^k + \Psi_N(x,\bar{\lambda},\varepsilon). \tag{6.4}$$

*The remainder $\Psi_N$ is of class $C^N$ and flat at order $N$ in $x=0$, for all $(\bar{\lambda},\varepsilon)$.*

**6.2 Remark** In fact, each monomial in the principal part of the expansion (6.4) contains at most one of the compensators, but this is without importance for the study that we want to make below.

Now the important *difference between the saddle loop case* which we have considered in the previous paragraph *and the 2-saddle case* is the following: *we have now two different compensators converging toward the same function* $\log x$, *when* $\varepsilon \to 0$. The limit of $\bar{\Delta}$ is again the Abelian integral when $\varepsilon \to 0$ : $\bar{\Delta}(x,\bar{\lambda},0) = I_{\bar{\lambda}}(x)$. Taking $\varepsilon = 0$ in $\bar{\Delta}$, every monomial $x^i(x\omega_1)^j(x\omega_2)^k$, corresponding to a given value of $i$ and $j+k=\ell$, converges towards the same function $x^{i+l}(\log x)^\ell$. This *degeneracy phenomenon* prevents us to use expansion (6.4) to prove directly a result similar to Theorem 5.8. Hence, we have to rearrange this expansion in order to avoid this degeneracy phenomenon; for this purpose, we will introduce new compensators in the next section. First, in the following subsection, we will state the result of finite cyclicity in case the unfolding has finite codimension. Afterwards in the next section, we continue with the preparation and the proof of this theorem.

### 6.3 Finite cyclicity result

We adopt the following lexicographic order on $\mathbb{N} \times \mathbb{N}$ : $(i',j') \succ (i,j)$ if and only if: $i' > i$ or $i' = i$ and $j' < j$. This order corresponds to the order of flatness at 0 of the sequence $f_{ij} = x^i \log^j x$.

As remarked before, we have that $\bar{\Delta}(x, \bar{\lambda}, 0) \equiv I(x, \bar{\lambda})$. In the previous subsection, we have already seen that $\bar{\Delta}$ expands in a deformation of an *enlarged* asymptotic scale $\mathcal{L}^e = \{x^i \log^j x : 0 \le j \le \inf\{1, i\},\ i \in \mathbb{N}\}$ with $0 \le j \le \inf\{1, i\}$. However, the terms $x^i \log^j x$, $j \ge 2$, must disappear in the expansion of $I$. The consequence will be that now the Abelian integral cannot completely control the zeroes of $\bar{\Delta}$, and hence neither the limit cycles of $X_\lambda$ bifurcating from $\Gamma$.

**6.3 Definition** The integral $I_{\bar{\lambda}}$ is said to be of *finite codimension* if one of the coefficients in the formal expansion of $I_0$ is non-zero. Moreover, in this case,

(1) the *codimension of* $I_{\bar{\lambda}}$, $cod_{\mathcal{L}}(I_{\bar{\lambda}})$, is defined to be the smallest order of such a non-zero coefficient in the asymptotic scale $\mathcal{L}$ for $\bar{\lambda} = 0$ and $\varepsilon = 0$;

(2) the *codimension of* $X_\lambda$, $\operatorname{cod} X_\lambda$, is defined as the order of the same coefficient in the asymptotic scale of functions $\mathcal{L}^e = \{x^i \log^j x : 0 \le j \le \inf\{1, i\},\ i \in \mathbb{N}\}$. This means that we define $\operatorname{cod} X_\lambda$ to be equal to $\operatorname{cod}_{\mathcal{L}^e}(I_{\bar{\lambda}})$.

**6.4 Proposition** *Suppose that the Abelian integral $I_{\bar{\lambda}}$ is of finite codimension, $\operatorname{cod} I_{\bar{\lambda}} = q < \infty$. Then the relation between these two numbers is given by*

$$\operatorname{cod} X_\lambda = \begin{cases} 2p + p(p-1)/2, & \text{if } q = 2p \\ 2p + 1 + p(p+1)/2, & \text{if } q = 2p+1 \end{cases}. \tag{6.5}$$

**6.5 Theorem** *Let $X_\lambda$ be a $C^\infty$ unfolding of a 2-saddle cycle, breaking just the lower connection. If $\operatorname{cod} X_\lambda$ is finite, then*

$$\operatorname{Cycl}(X_\lambda, \Gamma) \le \operatorname{cod} X_\lambda. \tag{6.6}$$

To prove the above theorem we need to produce some simple deformation of asymptotic scale in order to apply Theorem 5.8. This will be done in the next section in a number of steps.

## 7 Technicalities and proof of the finite cyclicity result

### 7.1 Introduction of new compensators

To remove the degeneracy caused by the two different compensators, we will introduce new compensators, leading to a new expansion of $\bar{\Delta}$. Let us consider the regular transitions $R_1, R_2$ and the transitions $D_1, D_2$ near the saddle points $s_1, s_2$ respectively (these last transitions are sometimes called Dulac transitions; Fig. 4). Now, one has:

$$\Delta_1 = R_1 \circ D_1, \qquad \Delta_2 = D_2 \circ R_2 \tag{7.1}$$

and

$$\Delta = \Delta_2 - \Delta_1 = D_2 \circ R_2 - R_1 \circ D_1. \tag{7.2}$$

First, we will remove the degeneracy phenomenon caused in the first three terms of the expansion of $\bar{\Delta}$ by introducing a first new compensator $\omega_{2-1}$. Next, we shall eliminate the compensator $\omega_2$ introducing a new compensator $\omega_{21}$. This compensator, fortunately enough, will disappear in the principal parts of functions in which we shall finally write an expansion for $\bar{\Delta}$.

To begin with, we shall write some short expansions for the different maps. In the normalizing coordinates $(z, w)$ we can apply a (parameter depending) similarity inducing:

$$\frac{dR_2}{dx}(0, \bar{\lambda}, \varepsilon) = 1, \quad \text{i.e.} \quad R_2(x) = x(1 + O(x)), \quad x \to 0. \tag{7.3}$$

On the other hand, we write:

$$R_1(y) = \varepsilon u_0 + u_1 y(1 + O(y)), \quad y \to 0, \tag{7.4}$$

with $u_1 = 1 + \varepsilon u$. We omit again the $\lambda$-dependence. From now on, all the coefficients we introduce, such as $u, u_0, u_1, \alpha_0, \alpha_1, \ldots$ are supposed to be smooth functions of the parameter $(\bar{\lambda}, \varepsilon)$. The Dulac maps have the following expansions:

$$D_1(x) = x + \varepsilon\alpha_1 x\omega_1(x, \varepsilon\alpha_1) + O(x^2\omega_1^2), \tag{7.5}$$

$$D_2(z) = z + \varepsilon\alpha_2 z\omega_2(z, \varepsilon\alpha_2) + O(z^2\omega_2^2). \tag{7.6}$$

We start with the expansion of $D_2 \circ R_2$. From (7.6), it follows that

$$D_2 \circ R_2(x) = R_2(x) + \varepsilon\alpha_2 R_2(x)\omega_2(R_2(x), \varepsilon\alpha_2) + O(R_2^2(\omega_2 \circ R_2)^2). \tag{7.7}$$

Here

$$\begin{aligned} \omega_2 \circ R_2(x) &= \frac{R_2(x)^{\varepsilon\alpha_2} - 1}{\varepsilon\alpha_2} = \frac{x^{\varepsilon\alpha_2}(1 + O(x))^{\varepsilon\alpha_2} - 1}{\varepsilon\alpha_2} \\ &= (1 + O(x))^{\varepsilon\alpha_2}\omega_2(x, \varepsilon\alpha_2) + \frac{(1 + O(x))^{\varepsilon\alpha_2} - 1}{\varepsilon\alpha_2}. \end{aligned} \tag{7.8}$$

Hence

$$\omega_2 \circ R_2(x) = \Psi_1(x)\omega_2(x, \varepsilon\alpha_2) + \Psi_2(x), \tag{7.9}$$

where $\Psi_1$ and $\Psi_2$ are smooth functions with

$$\Psi_1(x) = 1 + O(x) \quad \text{and} \quad \Psi_2(x) = O(x). \tag{7.10}$$

Again, we do not mention the dependence of $\omega_2, \Psi_1, \Psi_2$ on $(\bar{\lambda}, \varepsilon)$. From (7.7), (7.10), (7.3) and (7.9) we get

$$D_2 \circ R_2(x) = x + \varepsilon\alpha_2 x\omega_2(x, \varepsilon\alpha_2) + O(x^2\omega_2^2). \tag{7.11}$$

On the other hand, from (7.4) and (7.5) we obtain directly the expansion of $\Delta_1$:

$$R_1 \circ D_1(x) = \varepsilon u_0 + \varepsilon\alpha_1 u_1 x\omega_1(x, \varepsilon\alpha_1) + u_1 x + O(x^2\omega_1^2). \tag{7.12}$$

Combining (7.11) and (7.12):

$$\bar{\Delta} = -u_0 - \alpha_1(1 + \varepsilon u)x\omega_1(x, \varepsilon\alpha_1) + \alpha_2 x\omega_2(x, \varepsilon\alpha_2) - ux + O(x^2|\omega|^2), \tag{7.13}$$

where $|\omega| = \sup\{|\omega_1|, |\omega_2|\}$. From Theorem 6.1, we know that the remainder $O(x^2|\omega|^2)$ can be expressed at any order $N$ as $Q_N(x, x\omega_1, x\omega_2) + \Phi_N(x)$, where $Q_N$ is a polynomial of valuation $\geq 2$ in $x, x\omega_1$ and $x\omega_2$ and $\Phi_N$ is $C^N$ and is $N$-flat at $x = 0$. Therefore, putting $\beta = -u_0$ in (7.13), we obtain

$$\bar{\Delta} = \beta - \alpha_1(1 + \varepsilon u)x\omega_1(x, \varepsilon\alpha_1) + \alpha_2 x\omega_2(x, \varepsilon\alpha_2) - ux + Q_N + \Phi_N. \tag{7.14}$$

From this equality, we see that the degeneracy phenomenon in the first terms corresponds to the fact that $\omega_1 \mid_{\varepsilon=0} \equiv \omega_2 \mid_{\varepsilon=0} \equiv \log x$. We remove this degeneracy by introducing a *new compensator:*

$$\omega_{2-1}(x, \bar{\lambda}, \varepsilon) = \omega(x, \varepsilon(\alpha_2 - \alpha_1)). \tag{7.15}$$

Of course by this definition, we have that $\omega_{2-1}(x, \bar{\lambda}, \varepsilon) = \log x$ if $\varepsilon(\alpha_2 - \alpha_1) = 0$. From the definition of $\omega$, we observe that

$$\varepsilon\alpha_i x\omega_i = x^{1+\varepsilon\alpha_i} - x, \quad i = 1, 2. \tag{7.16}$$

Then we have:

$$\begin{aligned} -\varepsilon\alpha_1(1+\varepsilon u)x\omega_1 + \varepsilon\alpha_2 x\omega_2 &= (-\varepsilon\alpha_1 x\omega_1 + \varepsilon\alpha_2 x\omega_2) - \varepsilon u(\varepsilon\alpha_1 x\omega_1) \\ &= x^{1+\varepsilon\alpha_1}(x^{\varepsilon(\alpha_2-\alpha_1)} - 1) + \varepsilon ux - \varepsilon ux^{1+\varepsilon\alpha_1} \\ &= \varepsilon(\alpha_2 - \alpha_1)x^{1+\varepsilon\alpha_1}\omega_{2-1} + \varepsilon ux - \varepsilon ux^{1+\varepsilon\alpha_1} \\ &= \varepsilon((\alpha_2 - \alpha_1)x^{1+\varepsilon\alpha_1}\omega_{2-1} + ux - ux^{1+\varepsilon\alpha_1}) \end{aligned} \tag{7.17}$$

Hence, the expansion of $\bar{\Delta}$ in (7.14) can be written as

$$\bar{\Delta} = \beta + (\alpha_2 - \alpha_1)x^{1+\varepsilon\alpha_1}\omega_{2-1} - ux^{1+\varepsilon\alpha_1} + Q_N + \Phi_N. \tag{7.18}$$

Let us note that the three first functions in the expansion (7.18) converge to $1, x\log x$ and $x$, respectively, so that we *no longer have a degeneracy in the first terms.* We are only left with the treatment of the function $Q_N$, where the degeneracy comes from the presence of both $\omega_1$ and $\omega_2$. The idea now is to eliminate one of these two compensators, for instance $\omega_2$. For this, we introduce a *second new compensator*:

$$\omega_{21}(x, \bar{\lambda}, \varepsilon) = \frac{\omega_2 - \omega_1}{\varepsilon(\alpha_2 - \alpha_1)} \quad \text{if} \quad \varepsilon(\alpha_2 - \alpha_1) \neq 0. \tag{7.19}$$

**7.1 Remark** In the next subsection we shall study the properties of $\omega_{21}$ and describe its continuous extension along the set $\{\varepsilon(\alpha_2 - \alpha_1) = 0\}$.

From (7.19), it follows that $\omega_2 = \omega_1 + \varepsilon(\alpha_2 - \alpha_1)\omega_{21}$; substituting this relation in $Q_N$, we obtain

$$Q_N(x, x\omega_1, x\omega_2) = Q_N(x, x\omega_1, x\omega_1) + \varepsilon(\alpha_2 - \alpha_1)x\omega_{21}F_N(x, x\omega_1, x\omega_{21}), \tag{7.20}$$

where $F_N$ is a polynomial in $x, x\omega_1, x\omega_{21}$ of valuation 1 ($F_N(0,0,0) = 0$). Define

$$R_N(x_1, x_2) = Q_N(x_1, x_2, x_2);$$

then $R_N$ is a polynomial of valuation $\geq 2$. We can now rewrite the expansion given in (7.18) as follows:

$$\bar{\Delta} = \beta + (\alpha_2 - \alpha_1)[x^{1+\varepsilon\alpha_1}\omega_{2-1} + \varepsilon x \omega_{21} F_N] - u x^{1+\varepsilon\alpha_1} + R_N(x, x\omega_1) + \Phi_N(x) \qquad (7.21)$$

Finally, we have removed the degeneracy phenomenon up to any given order. Summarizing, we have proved the following theorem:

**7.2 Theorem** *Let $X_{(\bar{\lambda},\varepsilon)}$ be any unfolding of Hamiltonian perturbation type, along a 2-saddle cycle $\Gamma$. We suppose that the unfolding leaves one connection unbroken. Then there exists a sequence of functions $F_{ij}(x,\bar{\lambda},\varepsilon)$, $0 \leq j \leq i$, which are polynomials in $x, x\omega_1, x\omega_{2-1}, x\omega_{21}$ with coefficients smooth in the parameter $(\bar{\lambda},\varepsilon)$, $F_{ij}(x,\bar{\lambda},0) = f_{ij}(x) = x^i \log^j x$, such that for any $N$ one has the following expansion:*

$$\bar{\Delta} = \sum_{0 \leq j \leq i \leq N} \alpha_{ij}(\bar{\lambda},\varepsilon) F_{ij}(x,\bar{\lambda},\varepsilon) + \Psi_N(x,\bar{\lambda},\varepsilon). \qquad (7.22)$$

*The coefficients $\alpha_{ij}$ are smooth functions of the parameter and the remainder $\Psi_N$ is $C^N$ and flat of order $N$ at $x = 0$, for all $(\bar{\lambda},\varepsilon)$.*

In the next subsection, we shall deduce from $F_{ij}$, a new simple deformation of the asymptotic scale $f_{ij}(x) = x^i \log^j x$, $-2 \leq j \leq i$ and $i \geq 0$. This will allow us to apply Theorem 5.8 to obtain the finite cyclicity result (Theorem 6.5 ) for the unfolding $X_\lambda$.

## 7.2 Properties of the different compensators

We recall here the definitions of the different compensators and give some useful properties :

$$\omega_\alpha : \mathbb{R} \to \mathbb{R} : x \mapsto \omega_\alpha(x) = \begin{cases} \dfrac{x^\alpha - 1}{\alpha}, & \text{if } \alpha \neq 0 \\ \log x, & \text{if } \alpha = 0 \end{cases}.$$

Then we have

$$\omega_i = \omega_{\varepsilon\alpha_i},\ i = 1,2, \quad \text{and} \quad \omega_{2-1} = \omega_{\varepsilon(\alpha_2-\alpha_1)}.$$

Recall also that $\forall x > 0$:

$$\omega_{21}(x) = \frac{\omega_2(x) - \omega_1(x)}{\varepsilon(\alpha_2 - \alpha_1)}, \quad \text{if} \quad \varepsilon(\alpha_2 - \alpha_1) \neq 0.$$

Now we want to extend $\omega_{21}$ continuously to $\varepsilon(\alpha_2 - \alpha_1) = 0$. Therefore, we introduce two analytic functions:

$$\Phi : \mathbb{R} \to \mathbb{R} : u \mapsto \Phi(u) = \begin{cases} \dfrac{e^u - 1}{u}, & \text{if } u \neq 0 \\ 1, & \text{if } u = 0 \end{cases}$$

and

$$\Psi : \mathbb{R}^2 \to \mathbb{R} : (u,v) \mapsto \Psi(u,v) = \begin{cases} \dfrac{\Phi(u) - \Phi(v)}{u - v}, & \text{if } u \neq v \\ \dfrac{d\Phi}{du}(u), & \text{if } u = v \end{cases}. \qquad (7.23)$$

These functions allow the following nice expressions for the different compensators. Firstly, we have

$$\omega_\alpha(x) = \log x \cdot \Phi(\alpha \log x). \tag{7.24}$$

Secondly, we have for $\varepsilon(\alpha_1 - \alpha_2) \neq 0$,

$$\begin{aligned} \omega_{21}(x) &= \log x \cdot \frac{\Phi(\varepsilon\alpha_2 \log x) - \Phi(\varepsilon\alpha_1 \log x)}{\varepsilon(\alpha_2 - \alpha_1)} \\ &= \log^2 x \cdot \Psi(\varepsilon\alpha_1 \log x, \varepsilon\alpha_2 \log x) \end{aligned} \tag{7.25}$$

Expression (7.25) can now be used to extend $\omega_{21}$ continuously at $\varepsilon(\alpha_2 - \alpha_1) = 0$ by the definition of $\Psi$ in (7.23).

**7.3 Proposition** $\lim_{(x,\alpha)\to(0,0)} \omega_\alpha(x)^{-1} = 0.$

**Proof** We show that $\omega_\alpha(x)$ becomes arbitrarily large when $x$ and $\alpha$ are chosen small enough. Of course, we can suppose that $0 < x < 1$. If $\alpha < 0$, then $\alpha \log x \geq 0$. Hence $\Phi(\alpha \log x) \geq 1$, and therefore

$$|\omega_\alpha(x)| = |\log x| \cdot \Phi(\alpha \log x) \geq |\log x|.$$

If $\alpha > 0$, then $\alpha \log x < 0$, then

$$|\omega_\alpha(x)| = \frac{1 - x^\alpha}{\alpha} = \frac{1 - e^{\alpha \log x}}{\alpha}.$$

Both the numerator and denominator in this expression may go to 0, when $(x, \alpha) \to 0$. Therefore, we split up into two cases: $\alpha \log x < -M$ and $\alpha \log x \geq -M$, for a fixed small $M > 0$. Consider first the case that $\alpha \log x < -M$. Then, $e^{\alpha \log x} \leq e^{-M}$. Hence, $|\omega_\alpha(x)| \geq (1 - e^{-M})/\alpha$. Now we consider the case that $\alpha \log x \geq -M$. From the mean value theorem it follows that

$$\frac{1 - e^{\alpha \log x}}{\alpha |\log x|} = e^c \quad \text{for a certain} \quad c \in ]\alpha \log x, 0[.$$

As a consequence,

$$|\omega_\alpha(x)| = e^c |\log x| \geq e^{-M} |\log x|.$$

□

## 7.3 Derivation properties

In the derivation-division algorithm that is used in the proofs of next subsection, the algebra $O$ of functions defined below, will be very useful.

**7.4 Definition** Let $f$ be a germ of a function defined on $(\mathbb{R}^+ \setminus \{0\}, 0) \times (\mathbb{R}^{p+1}, 0)$.

(1) We say that $f \in \mathcal{F}$ if there exists $\tau > 0$ such that $f(x, \lambda) = O(x^\tau)$, $x \to 0$, and if it is obtained as the quotient of linear combinations of generalized monomials

$$x^{l+\varepsilon\beta} \omega_1^m \omega_2^p \omega_{2-1}^q \omega_{21}^r \log^s x \quad \text{for some} \quad \beta \in C^\infty(\lambda),\ \ell, m, p, r, s \in \mathbb{Z},$$

with coefficients in $C^\infty(\lambda)$, and such that the denominator is of the form $1 + o(1)$, $x \to 0$.

(2) We say that $f$ belongs to the set $O$ if $\forall N \in \mathbb{N}$, there exist functions $\phi_N \in \mathcal{F}$, $\Phi_N$ differentiable of class $\mathcal{C}^N$ and $N$-flat in $x$ at $x=0$, i.e. $\forall\lambda$:

$$\frac{\partial^j}{\partial x^j}\Phi_N(0,\lambda)=0,\quad 0\le j\le N,$$

such that

$$f(x,\lambda)=\phi_N(x,\lambda)+\Phi_N(x\lambda).$$

(3) We write $f=g+O$ if $f-g\in O$.

For instance, the function

$$F_{11}(x,\lambda)=x^{1+\varepsilon\alpha_1}\omega_{2-1}(x\,\lambda)+\varepsilon x\omega_{21}(x,\lambda)F_N(x,x\omega_1,x\omega_{21}),$$

appearing in expression (7.21), can be written as

$$F_{11}(x,\lambda)=x^{1+\varepsilon\alpha_1}\omega_{2-1}(x,\lambda)(1+O). \tag{7.26}$$

In fact, we can write

$$F_{11}(x,\lambda)=x^{1+\varepsilon\alpha_1}\omega_{2-1}(1+\varepsilon x^{-\varepsilon\alpha_1}\omega_{2-1}^{-1}\omega_{21}F_N(x,x\omega_1,x\omega_{21})).$$

As $F_N$ has valuation $\ge 1$ in $(x,x\omega_1,x\omega_{21})$, one has that

$$x^{-\varepsilon\alpha_1}\omega_{2-1}^{-1}\omega_{21}F_N(x,x\omega_1,x\omega_{21}))=O(x^\tau),\quad x\downarrow 0,$$

for any $\tau$, $0<\tau<1$.

Recall that we work with the differential operator $\nabla=x\frac{\partial}{\partial x}$. Denote the space of germs of smooth functions at $\lambda=0\in\mathbb{R}^{p+1}$ by $\mathcal{C}^\infty(\lambda)$. It is easy to verify that

$$\nabla\omega_\alpha(x)=x^\alpha$$

and, for $f(x)=x^{\ell+\varepsilon\beta}\omega_{\varepsilon\alpha}(x)^m$ where $\alpha,\beta\in\mathcal{C}^\infty(\lambda)$, $m,\ell\in\mathbb{Z}$, $\ell\ne 0$, that

$$\nabla f=(\ell+\varepsilon\beta)\cdot f\cdot\left(1+\varepsilon\frac{\alpha m}{\ell+\varepsilon\beta}+\frac{m}{\ell+\varepsilon\beta}\omega_{\varepsilon\alpha}^{-1}\right). \tag{7.27}$$

From formula (7.27) and Proposition 7.3, it is clear that the algebra $\mathcal{F}$ is closed under derivation by the operator $\nabla$. The same is true for the set $O$:

**7.5 Proposition** *The set $O$ is an algebra closed under derivation by $\nabla$. This means that $O+O\subset O$, $O\cdot O\subset O$ and $\nabla O\subset O$.*

**Proof** As $\mathcal{F}$ is an algebra, the set $O$ is also an algebra as a consequence of the following observation: if $\phi\in\mathcal{F}$ and $\Phi$ is differentiable of class $\mathcal{C}^N$ and $N$-flat in $x$ at $x=0$, for $\forall\lambda$, then $\phi\Phi$ is differentiable of class $\mathcal{C}^{N-1}$ and $(N-1)$-flat in $x$ at $x=0$, for all $\lambda$. Now the property for $O$ under the derivation for the operator $\nabla$ follows from the fact that a similar property is satisfied by $\mathcal{F}$. □

To avoid having to write positive coefficient functions of the parameter, we introduce the following definition:

**7.6 Definition** Suppose that $f$ and $g$ are germs at $(0,0)$ of functions of the variables $x>0$ and $\lambda$. Then we will write $f \approx g$ if and only if there exists a continuous function $c(\lambda)$ with $c(0)>0$ such that $f(x,\lambda)=c(\lambda)g(x,\lambda)$.

The proofs in the next section will be based on the following rules of derivation:

**7.7 Lemma** *Let us consider $\ell, m \in \mathbb{Z}$, $\ell \neq 0$ and $\alpha, \beta \in C^\infty(\lambda)$. Suppose $G$ is a rational function with coefficients in $C^\infty(\lambda)$ such that $G(0)_{|\varepsilon=0}=0$. One has:*

(1) *If $F=x^{\ell+\varepsilon\alpha}(1+O)$, then $\nabla F=(\ell+\varepsilon\alpha)x^{\ell+\varepsilon\alpha}(1+O)$, i.e. $\nabla F \approx F(1+O)$.*

(2) *If $F=\omega_{\varepsilon\alpha}^{\ell}(1+G(\omega_{\varepsilon\alpha}^{-1}))(1+O)$, then*

$$\nabla F=\ell\omega_{\varepsilon\alpha}^{\ell-1}x^{\varepsilon\alpha}(1+\tilde{G}(\omega_{\varepsilon\alpha}^{-1}))(1+O),$$

*where $\tilde{G}$ is rational function such that $\tilde{G}_{|\varepsilon=0}(0)=0$.*

(3) *If $F=x^{\ell+\varepsilon\beta}\omega_{\varepsilon\alpha}^{m}(1+G(\omega_{\varepsilon\alpha}^{-1}))(1+O)$, then*

$$\nabla F=(\ell+\varepsilon\beta)x^{l+\varepsilon\beta}\omega_{\varepsilon\alpha}^{m}(1+\tilde{G}(\omega_{\varepsilon\alpha}^{-1}))(1+O),$$

*where $\tilde{G}$ is rational function such that $\tilde{G}_{|\varepsilon=0}(0)=0$. This means that*

$$\nabla F \approx F(1+\bar{G}(\omega_{\varepsilon\alpha}^{-1}))(1+O).$$

**Proof** To begin, let us notice that if $G$ is like in the enunciate, one has that $1+G+O=(1+G)(1+O)$. This remark is very useful in the following computations.

(1) This fact follows from the observation that $\nabla x^{\ell+\varepsilon\beta}=(\ell+\varepsilon\beta)x^{\ell+\varepsilon\beta}$ and since $\nabla O \subset O$.

(2) $\nabla F=\nabla[\omega_{\varepsilon\alpha}^{\ell}(1+G)](1+O)+O\omega_{\varepsilon\alpha}^{\ell}(1+G)$. Let us consider the first term:

$$\nabla\left[\omega_{\varepsilon\alpha}^{\ell}(1+G(\omega_{\varepsilon\alpha}^{-1}))\right]=\ell\omega_{\varepsilon\alpha}^{\ell-1}x^{\varepsilon\alpha}\left[1+G-\frac{1}{\ell}\frac{dG}{du}(\omega_{\varepsilon\alpha}^{-1})\omega_{\varepsilon\alpha}^{-1}\right]=\ell\omega_{\varepsilon\alpha}^{\ell-1}x^{\varepsilon\alpha}[1+\tilde{G}] \qquad (7.28)$$

where $\tilde{G}(u)=G(u)-\frac{1}{\ell}\frac{dG}{du}(u)u$. Let us consider now the second term:

$$O\omega_{\varepsilon\alpha}^{\ell}(1+G)=\ell\omega_{\varepsilon\alpha}^{\ell-1}x^{\varepsilon\alpha}(1+\tilde{G})O$$

as $\frac{1}{\ell}\omega_{\varepsilon\alpha}x^{-\varepsilon\alpha}\frac{1+G}{1+\tilde{G}}O \subset O$. Summing the two terms, we have the desired formula.

(3) We have:

$$\nabla(x^{\ell+\varepsilon\beta}\omega_{\varepsilon\alpha}^{m})=(\ell+\varepsilon\beta)\cdot x^{\ell+\varepsilon\beta}\omega_{\varepsilon\alpha}^{m}\cdot\left(1+\varepsilon\frac{\alpha m}{\ell+\varepsilon\beta}+\frac{m}{\ell+\varepsilon\beta}\omega_{\varepsilon\alpha}^{-1}\right),$$

$$\nabla(1+G(\omega_{\varepsilon\alpha}^{-1}))=-\frac{\partial G}{\partial u}(\omega_{\varepsilon\alpha}^{-1})\omega_{\varepsilon\alpha}^{-2}x^{\varepsilon\alpha}=-\frac{\partial G}{\partial u}(\omega_{\varepsilon\alpha}^{-1})(\omega_{\varepsilon\alpha}^{-2}+\varepsilon\alpha\omega_{\varepsilon\alpha}^{-1}). \qquad (7.29)$$

Using these two formulas in the computation of $\nabla F$ we obtain the desired formula in a similar way as in (2).

□

## 7.4 A simple asymptotic scale deformation

Probably the sequence $\{F_{ij}\}$ is not a simple asymptotic scale deformation. In fact, if one tries to prove it, at some point one has to control linear combinations of $x^{\varepsilon\alpha_1}$ and $x^{\varepsilon(\alpha_2-\alpha_1)}$ (coming from the derivations of $\omega_1$ and $\omega_{2-1}$, respectively). But, fortunately enough, after three applications of the algorithm of derivation-division to $\bar{\Delta}$ (with the operator $\nabla$ used instead of $\frac{\partial}{\partial x}$), one can make the compensator $\omega_{2-1}$ disappear from the formulas. In fact it remains hidden in terms $O$.

**7.8 Proposition** *After three steps in the division-derivation algorithm (each step consisting of the division by a function, smooth and positive for $x>0$, followed by the derivation by $\nabla$, starting with the function $\bar{\Delta}$), one obtains a new function, $\bar{\Delta}^3$, with expansions at any order in an asymptotic scale deformation given by*

$$\{x^i\omega_1^j(1+G_{ij}(\omega_1^{-1}))(1+O),\ i\geq 2,\ 0\leq j\leq i\}, \tag{7.30}$$

*where the functions $G_{ij}(u)$ are rational functions of $u$ and $G_{ij|\varepsilon=0}(0)=0$.*

**Proof** Denote the scale defined in the expansion of $\bar{\Delta}$ given by formula (7.21) by $\{F_{ij}\}$. It suffices to define the three successive steps on the successive asymptotic scale deformations that one obtains, beginning with the scale $\{F_{ij}\}$. The elements of each scale will be defined up to the relation $\approx$.

*First step:* One divides by $x^{1+\varepsilon\alpha_1}$ the sequence $\{F_{ij}\}$, given as

$$\{1,\ x^{1+\varepsilon\alpha_1}\omega_{2-1}(1+O),\ x^{1+\varepsilon\alpha_1},\ x^i\omega_1^j,\ i\geq 2,\ 0\leq j\leq i\} \tag{7.31}$$

followed by a derivation by $\nabla$. One obtains the sequence

$$\{x^{-1-\varepsilon\alpha_1},\ x^{\varepsilon(\alpha_2-\alpha_1)}(1+O),\ 0,\ x^{i-1-\varepsilon\alpha_1}\omega_1^j(1+G_{ij}^1(\omega_1^{-1}),\ x^i\omega_1^j\},$$

again for $i\geq 2$, $0\leq j\leq i$, and rational functions $G_{ij}^1$ such that $G^1_{ij|\varepsilon=0}(0)=0$.

*Second step:* One divides by the second function $x^{\varepsilon(\alpha_2-\alpha_1)}(1+O)$ in the above scale, followed by a derivation by $\nabla$. One obtains:

$$\{x^{-1-\varepsilon\alpha_2}(1+O),\ 0,\ 0,\ x^{i-1-\varepsilon\alpha_2}\omega_1^j(1+G_{ij}^2(\omega_1^{-1}))(1+O)\} \tag{7.32}$$

for $i\geq 2$, $0\leq j\leq i$, and rational functions $G_{ij}^2$ such that $G^2_{ij|\varepsilon=0}(0)=0$.

*Third step:* One divides by the first function $x^{-1-\varepsilon\alpha_2}(1+O)$ in the above scale, followed by a derivation by $\nabla$. One obtains:

$$\{0,\ 0,\ 0,\ x^i\omega_1^j(1+G_{ij}(\omega_1^{-1}))(1+O),\ i\geq 2,\ 0\leq j\leq i\} \tag{7.33}$$

for rational functions $G_{ij}$ such that $G_{ij|\varepsilon=0}(0)=0$. We have maintained the 0 in the successive scales to give more transparency to the operations. If we now apply the successive steps to the function $\bar{\Delta}$ itself, expanded at any order, we obtain expansions of the resulting function $\bar{\Delta}^3$, also at any order, in the scale obtained in the third step. □

We now show that we can deduce a simple asymptotic scale deformation from the scale obtained in the previous proposition:

**7.9 Proposition** *Consider a sequence* $\{x^i\omega_1^j(1+G_{ij}(\omega_1^{-1}))(1+O),\ i\geq 2,\ 0\leq j\leq i\}$, *where the* $G_{ij}(u)$ *are rational functions of* $u$ *with* $G_{ij|\varepsilon=0}(0)=0$. *After division by the first function of the scale,*

$$x^2\omega_1^2(1+G_{22}(\omega_1^{-1}))(1+O),$$

*one obtains a sequence*

$$\mathcal{W}^e=\{x^i\omega_1^j(1+\tilde{G}_{ij}(\omega_1^{-1}))(1+O),\ i\geq 0,\ -2\leq j\leq i\},$$

*(where the* $\tilde{G}_{ij}(u)$ *are rational such that* $\tilde{G}_{ij}(0)_{|\varepsilon=0}=0$*), which is a simple deformation of the asymptotic scale* $\mathcal{L}^e=\{x^i\log^j x,\ i\geq 0,\ -2\leq j\leq i\}$.

**Proof** First observe that for $\varepsilon\to 0$, the sequence $\mathcal{W}^e$ reduces to the asymptotic sequence $\mathcal{L}^e$.

The first condition in Definition 5.3 is trivially verified. The quotient of two consecutive functions in the sequence $\mathcal{W}^e$ takes the form $\omega_1^{-1}(1+G(\omega_1^{-1}))(1+O)$ or $x\omega_1^{\ell}(1+G(\omega_1^{-1})(1+O)$, where $G$ is a rational function, $G(0)=0$, and $\ell\in\mathbb{N}$, $\ell\geq 3$. These ratios go to zero when $(x,\varepsilon)\to 0$, uniformly in $\bar{\lambda}$. The derivation by $\nabla$ of these ratios gives functions equivalent (in the $\approx$ sense) to $-\omega_1^{-2}x^{\varepsilon\alpha_1}(1+\tilde{G})(1+O)$ and $x\omega_1^{\ell}(1+\tilde{G})(1+O)$, respectively, with a constant sign for $x>0$ small enough. This gives the second condition.

We now proceed to prove that this asymptotic scale deformation is simple, the second condition in Definition 5.4. To simplify the notation, we will just write the principal term for each function in the different sequences. The principal term $P$ of any of these functions is in fact multiplied by a term $M=(1+G(\omega_1^{-1}))(1+O)$. The operation of division and derivation transforms this term into a similar one. Precisely, from Lemma 7.7 (3), one has that $\nabla(PM)\approx(\nabla P)\tilde{M}$, where $\tilde{M}$ is similar to $M$. Moreover, we indicate the functions just up to the relation $\approx$. We begin with the sequence $\mathcal{W}^e=\Delta_0$:

$$\Delta_0=\{1,\omega_1^{-1},\omega_1^{-2},x\omega_1,x,x\omega_1^{-1},x\omega_1^{-2},x^2\omega_1^2,\dots\}. \tag{7.34}$$

Application of the operator $\nabla$ gives

$$\{0,\omega_1^{-2}x^{\varepsilon\alpha_1},\omega_1^{-3}x^{\varepsilon\alpha_1},x\omega_1,x,x\omega_1^{-1},\dots\}. \tag{7.35}$$

We now divide by the first function to obtain

$$\Delta_1=\{1,\omega_1^{-1},x^{1-\varepsilon\alpha_1}\omega_1^3,x^{1-\varepsilon\alpha_1}\omega_1^2,\dots\}. \tag{7.36}$$

Next (after derivation by $\nabla$ and division), we will find $\Delta_2=\{1,x^{1-2\varepsilon\alpha_1}\omega^5,\dots\}$, $\Delta_3=\{1,\omega_1^{-1},\omega_1^{-2},\omega_1^{-3},\dots\}$, and so on. More generally, each sequence $\Delta_i$ begins with 1, each following function contains a finite number of terms (perhaps none) with principal part $\omega^{-\ell}$, $\ell\in\mathbb{N}\setminus\{0\}$, followed by terms with principal parts $x^{\ell+\varepsilon k_i\alpha_1}\omega^{\ell+p_i}$, $\ell\in\mathbb{N}\setminus\{0\}$, and $k_i\in\mathbb{Z}$, $p_i\in\mathbb{N}$, depending just on the sequence. Therefore, just like the sequence $\Delta_0$, each sequence $\Delta_i$ satisfies the conditions of Definition 5.4, provided one restricts $x$ to some interval $[0,x_i[$. □

### 7.5 Proof of the finite cyclicity result

**Proof of Theorem 6.5** First let us suppose that $q = \operatorname{cod} X_\lambda \geq 3$. Then, by Theorem 7.2, we can write

$$\bar{\Delta} = \sum_{i=0}^{q} a_i F_i + \Psi,$$

where $\Psi$ has the remainder property of order $q$, with $a_q(0) \neq 0$. The derivation-division algorithm applied three times as in Proposition 7.8 gives a new function $\bar{\Delta}^3$. This last function $\bar{\Delta}^3$ expands in the simple deformation of asymptotic scale $\mathcal{W}^e = \{F_i^3 : i \in \mathbb{N}\}$:

$$\bar{\Delta}^3 = \sum_{i=0}^{q-3} a_i^3 F_i^3 + \Psi^3,$$

where $a_{q-3}^3(0) \neq 0$ (since $a_q(0) \neq 0$), and $\Psi$ has the remainder property of order $q-3$, since $\Psi$ has the remainder property $q$. This last property is equivalent to the fact that one can write $a_{q-3}^3(\lambda)F_{q-3}^3 + \Psi^3 = a_{q-3}^3(\lambda)F_{q-3}^3(1+O)$. Hence $\bar{\Delta}^3$ has codimension equal to $q-3$ in the scale $\mathcal{W}^e$. Then, as follows from Theorem 5.8, the number of zeroes of $\bar{\Delta}^3$ is less than $q-3$. By Rolle's theorem, the number of zeroes of $\bar{\Delta}$ itself is bounded by $q$, since this function is obtained after three steps in the derivation-division algorithm.

Now let us suppose that $q = \operatorname{cod} X_\lambda \leq 2$. From formula (7.21), one can write the expansion of $\bar{\Delta}$ at order 2:

$$\bar{\Delta} = \alpha_0 + \alpha_{11} x^{1+\varepsilon\alpha_1}\omega_{2-1}(1+O) + \alpha_{10} x^{1+\varepsilon\alpha_1}(1+O), \tag{7.37}$$

where one of the three coefficients $\alpha_0(\bar{\lambda},\varepsilon), \alpha_{11}(\bar{\lambda},\varepsilon), \alpha_{10}(\bar{\lambda},\varepsilon)$ is not zero at $(0,0)$.

If $\alpha_0(0,0) \neq 0$, the cyclicity is trivially equal to 0. Let us suppose that $\alpha_0(0,0) = 0$ and $\alpha_{11}(0,0) \neq 0$. An application of the operator $\nabla$ to the sequence $\{F_{ij}\}$ gives the sequence

$$\{0, x^{1+\varepsilon\alpha_1}\omega_{2-1}(1+O), \dots\}, \tag{7.38}$$

and the division by the first term gives a sequence $1, \omega_{2-1}^{-1}(1+O), \dots$, and the expansion

$$\bar{\Delta}^1 = \alpha_{11} + O(\omega_{2-1}^{-1}) = \tilde{\alpha}_{11}(1+o(1)).$$

As the coefficient $\tilde{\alpha}_{11}(0,0) = \alpha_{11}(0,0) \neq 0$, it follows that $\bar{\Delta}^1$ near $\{x = 0\}$ is locally different from zero. Then the cyclicity is less than 1. Finally, let us suppose that $\alpha_0(0,0) = \alpha_{11}(0,0) = 0$ and that $\alpha_{10}(0,0) \neq 0$. Two applications of the algorithm of derivation-division gives a function $\bar{\Delta}^2 = \tilde{\alpha}_{10}(1+O)$ with a coefficient $\tilde{\alpha}_{10}(0,0) \neq 0$. Hence, the cyclicity is less than 2. □

The striking point to notice in this result is that the number of limit cycles bifurcating from $\Gamma$ is no longer bounded in general by the codimension of the Abelian integral, as it was the case for unfoldings of saddle loops. This phenomenon appears for $\operatorname{cod} I_{\bar{\lambda}} \geq 3$, because we see that $\operatorname{cod}(I_{\bar{\lambda}}) = \operatorname{cod} X_\lambda$ if $\operatorname{cod} I_{\bar{\lambda}} \leq 2$, but $\operatorname{cod}(X_\lambda) = 4$ if $\operatorname{cod} I_{\bar{\lambda}} = 3$.

This is illustrated in [DR] by giving a generic unfolding $X_\lambda$ with four parameters creating four limit cycles but only three zeroes of the corresponding Abelian integral. This can be explained in terms of the $\varepsilon$-expansion of the function $\Delta$. The first order term in $\varepsilon$ is given by the Abelian integral, which expands in the monomials $x^i$ and $x^i \log x$. At the order $\varepsilon^2$, there appear new monomials $x^i \log^j x$ with $2 \leq j \leq i$. In fact , such an expansion in $\varepsilon$ will introduce divergent quantities in $x = 0$ at order $\geq 3$ in $\varepsilon$: this shows up clearly if, in formula (7.22) we try to introduce the limit $\log x$ of the compensators instead of the genuine compensators. The introduction of the compensators, which unfold the function $\log x$, is a way to avoid the divergence of the $\varepsilon$-expansion at $x = 0$.

## References

[DR] F. Dumortier, R. Roussarie, Abelian integrals and limit cycles, Laboratoire de Topologie de l'Université de Bourgogne, prépublication n° 288, July 2002.

[DRS] F. Dumortier, R. Roussarie, and J. Sotomayor, Generic 3-parameter families of vector fields, unfoldings of saddle, focus and elliptic singularities with nilpotent linear part, in: *Bifurcation of Planar Vector Fields: Nilpotent Singularities and Abelian Integrals* (F. Dumortier et al., eds.), Lecture Notes in Math. **1480**, Springer-Verlag, Berlin-Heidelberg-New York, 1991, 1–164.

[H] G. H. Hardy, *Orders of Infinity*, Hafner Publishing Co., 1971.

[JR] P. Joyal, C. Rousseau, Saddle quantities and applications, *J. Differential Equations* **78** (2) (1989), 374–399.

[M] P. Mardešić, *Chebychev Systems and the Versal Unfolding of the Cusps of Order n*, Travaux en Cours **57**, Hermann, Paris, 1998.

[Mo] A. Mourtada, Cyclicité finie des polycycles hyperboliques de champs de vecteurs du plan : mise sous forme normale, in: *Bifurcations of Planar Vector Fields* (J. P. Françoise and R. Roussarie, eds.), Lecture Notes in Math. **1455**, Springer-Verlag, Berlin-Heidelberg-New York, 1990, 272–314.

[R1] R. Roussarie, On the number of limit cycles which appear by perturbation of separatrix loop of planar vector fields, *Bol. Soc. Brasil. Mat.* **17** (2) (1986), 67–101.

[R2] R. Roussarie, *Bifurcations of Planar Vector Fields and Hilbert's Sixteenth Problem*, Progr. in Math. **164**, Birkhäuser, Boston, 1998.

# Topics on singularities and bifurcations of vector fields

Freddy DUMORTIER
Peter DE MAESSCHALCK
*Department WNI*
*Limburgs Universitair Centrum*
*Universitaire Campus*
*B-3590 Diepenbeek*
*Belgium*

**Abstract**

These notes are essentially meant to be a continuation of the lecture notes [D]. Results are presented that have mostly been obtained after 1993. In a first section we describe the classification of singularities of smooth vector fields in real 3-space up to codimension 4. Besides giving the description, attention goes to the different techniques that have been used. A second section deals with the study of the unfoldings of planar singularities and how this relates to the study of polynomial Liénard equations and of Abelian integrals. The last section deals with the study of singular perturbations for 2-dimensional vector fields, essentially from a geometric point of view. Throughout the whole text considerable emphasis is put on the one hand on blow up of singularities, and on the other hand on rescaling of families and its generalization: blow up of families.

## 1 Introduction

These lectures are a continuation of the lectures that have been delivered in Montréal in 1993. The contents of the related course notes [D] are essentially supposed to be known. In the current notes, some items will be recalled, either rapidly or in more generality. The results that we present have mostly been obtained after 1993.

In Section 2 we describe the classification of singularities of $C^\infty$ vector fields in $\mathbf{R}^3$ up to codimension 4. Besides giving this description, attention goes to the different techniques that have been used. All the necessary information concerning the analogous study of 2-dimensional vector fields can be found in [D].

Section 3 deals with the study of the unfoldings of planar singularities and how this relates to the study of polynomial Liénard equations and of Abelian integrals. Without aiming at being complete we try to give an overview of interesting results obtained since 1993. This part can be seen as a continuation and a completion of the presentation made in [D].

Section 4 deals with singular perturbations for 2-dimensional vector fields, essentially from a geometric point of view. In 1993 we already described how to make a geometric treatment of the canard phenomenom in van der Pol's equation. We will now present some generalizations that have been obtained since then.



*Y. Ilyashenko and C. Rousseau (eds.),*
*Normal Forms, Bifurcations and Finiteness Problems in Differential Equations,* 33–86.

Throughout the whole text a lot of emphasis is put on one hand on blow up of singularities and on the other hand on rescaling of families and its natural generalization: blow up of families. Especially the last technique is quite interesting and by the end of the course notes we hope that the reader will have become acquainted with it.

## 2 Classification of singularities in $\mathbf{R}^3$ up to codimension 4

Although the classification which we present depends only on finite differentiability we will nevertheless work with $C^\infty$ vector fields. The vector fields can of course be analytic, but at the chosen level of the presentation this would not make a difference. For simplicity we will always keep the singularity at $0 \in \mathbf{R}^3$.

Let $X$ be a $C^\infty$ vector field on $\mathbf{R}^3$,

$$\dot{x} = X(x), \quad \text{with} \quad X(0) = 0. \tag{2.1}$$

The singularity of a vector field is characterized by the germ of $X$ at 0.

*Main question:* How does the phase portrait of $X$ look like in sufficiently small neighbourhoods of the origin?

In trying to classify the phase portraits, we use the following equivalence relation:

**2.1 Definition** Two singularities $(X, 0)$ and $(Y, 0)$ are (locally) $C^0$-equivalent (we write $(X, 0) \sim_{C^0} (Y, 0)$) if there exist neighbourhoods $V_1$ and $V_2$ of 0 and a homeomorphism $h\colon (V_1, 0) \to (V_2, 0)$ sending orbits of $X|_{V_1}$ to orbits of $Y|_{V_2}$ in a sense-preserving way, i.e.

$$\begin{aligned}&\forall x \in V_1 \text{ and } \forall t \in \mathbf{R} \text{ such that } X_{[0,t]}(x) \subset V_1,\\ &\text{there exists a } t' \in \mathbf{R} \text{ with } tt' > 0 \text{ and } h(X_{[0,t]}(x)) = Y_{[0,t']}(h(x)).\end{aligned}$$

Note that this definition is coarser than $C^0$-conjugacy or $C^r$-equivalence, $r \geq 1$, where in the former we require the equivalence to respect the flow, and in the latter we require the equivalence to be of class $C^r$.

A systematic study of the singularities of vector fields starts by looking at the linear approximation $DX(0) = j_1 X(0)$. Let $L = DX(0)$, then the vector field

$$\dot{x} = Lx \tag{2.2}$$

can easily be studied by representing it in a Jordan normal form $M$. We immediately get the following classification of linear vector fields:

**The case $M$ is hyperbolic** (All eigenvalues of $M$ are outside $i\mathbf{R}$): We can apply the theorem of Hartman-Grobman (see e.g. [Ro]), and conclude that $(X, 0)$ is locally $C^0$-conjugate to $(M, 0)$. Under $C^0$-conjugacy, the singularities can be subdivided into four classes, the topological type being determined by the "index" of $M$, i.e. the number of eigenvalues with negative real part.

*Note:* In the vector space of linear vector fields, the hyperbolic ones form an open and dense set; the non-hyperbolic ones form an algebraic variety of codimension 1. Hyperbolic vector fields are often called codimension-0 vector fields.

**The case $M$ is non-hyperbolic:** In that case $M$ does not determine the topological type of the singularity $(X, 0)$.

## 2.1 Stratification of the space of non-hyperbolic singularities depending on the linearized field

The case in which $M$ is not hyperbolic is of course the most interesting one to study, and we explore this case by making a distinction between the possible Jordan normal forms $M$ of $L$. In the vector space of linear vector fields, the following subsets are semi-algebraic sets:

$W_1$ (*codimension* 1): $M$ has exactly one eigenvalue on $i\mathbf{R}$, and this eigenvalue is 0.

$W_2$ (*codimension* 1): $M$ has two nonzero eigenvalues on $i\mathbf{R}$ and one real nonzero eigenvalue.

$W_3$ (*codimension* 2): $M$ has eigenvalues $(0, 0, \lambda)$, with $\lambda \neq 0$, and a nilpotent block.

$W_4$ (*codimension* 4): $M$ has eigenvalues $(0, 0, \lambda)$, with $\lambda \neq 0$, and is diagonal.

$W_5$ (*codimension* 2): $M$ has two nonzero eigenvalues on $i\mathbf{R}$ and one zero eigenvalue.

$W_6$ (*codimension* 3): $M$ has three zero eigenvalues and one Jordan-block.

There are two other cases with a triple zero eigenvalue: $M$ has two Jordan blocks or $M = 0$. In the former case the codimension is 5, while in the latter the codimension is 9.

We aim at studying all cases of codimension $\leq 4$, i.e. cases $W_1$–$W_6$. Because the codimension is $\leq 4$, these singularities show up in generic 4-parameter families of vector fields—this is due to Thom's transversality theorem (see e.g. [AR]). There are several reasons that motivate this study up to and including codimension 4. First, in the classification of the famous seven elementary catastrophes, four external parameters appear, hence our classification can be considered as a "non-elementary" analogue in $\mathbf{R}^3$ of the seven elementary catastrophes (see e.g. [Bro]). Let us however immediately stress that in the theory of elementary catastrophes the equivalence relation is a $C^\infty$ one. Second, some interesting phenomena appear in codimension 4 (see later). A third reason is the existence of a complete classification (for $C^0$-equivalence) of germs up to codimension 4 (see [DI2]).

Now, we will explain some basic techniques that are used in the study of $W_1$–$W_6$, and we will show how these techniques can be used.

## 2.2 Reduction to a center manifold

As a first technical tool let us recall the reduction to a center manifold.

**2.2 Theorem (Center manifold theorem [HPS])** *Let $X = \sum_{i=0}^{n} X_i \frac{\partial}{\partial x_i}$ be a $C^\infty$ vector field on $\mathbf{R}^n$ with $X(0) = 0$, such that $DX(0)$ has $c$ eigenvalues (counting multiplicity) on $i\mathbf{R}$, $c \in \mathbf{N}$. For any positive integer $\ell$ there exists a $c$-dimensional invariant manifold $W^c_X$ tangent to the eigenspace corresponding to the eigenvalues on $i\mathbf{R}$ of $DX(0)$.*

Given such an invariant manifold $W^c_X$, with local coordinates $(y_1, \ldots, y_c)$, we can define the restriction of $X$ to $W^c_X$:

$$\tilde{X} = \sum_{i=1}^{c} \tilde{X}_i(y_1, \ldots, y_c) \frac{\partial}{\partial y_i}.$$

**2.3 Proposition (Reduction to the center manifold [PS])** *Let $DX(0)$ have $m$ eigenvalues with positive real part (and $n-c-m$ eigenvalues with negative real part), and let $\tilde{X}$ be the restriction of $X$ to some center manifold $W_X^c$, then $(X,0)$ is $C^0$-equivalent to the germ of*

$$Y=\left(\sum_{i=1}^{c}\tilde{Y}_i(y_1,\ldots,y_c)\frac{\partial}{\partial y_i}\right)+\left(\sum_{i=c+1}^{c+m}y_i\frac{\partial}{\partial y_i}\right)+\left(\sum_{i=c+m+1}^{n}(-y_i)\frac{\partial}{\partial y_i}\right),$$

*where $\sum_{i=1}^{c}\tilde{Y}_i\frac{\partial}{\partial y_i}$ denotes any vector field $C^0$-equivalent to $(\tilde{X},0)$.*

The question of how the phase portrait of $(X,0)$ looks like can hence be reduced to the same question with respect to the phase portrait of the vector field reduced to a center manifold. Let us explain how this result can be used to handle the case $W_1$.

## 2.3 Singularities in the set $W_1$

The center manifolds are 1-dimensional here, and there is the following result related to 1-dimensional center manifolds:

**2.4 Proposition** *If $(X,0)$ has a one-dimensional center direction, and if $X$ satisfies a Łojasiewicz inequality (for example if $X$ is analytic), then all center manifolds of $(X,0)$ are $C^\infty$.*

(A vector field $(X,0)$ is said to satisfy a Łojasiewicz inequality if there exists a $k\in\mathbf{N}$ and a $c>0$ so that $\|X(x)\|\geq c\|x\|^k$ for all $x$ in some neighbourhood of 0.)

Using Propositions 2.3 and 2.4 we know that a singularity in $W_1$ is $C^0$-equivalent to

$$\varepsilon_1 x_1\frac{\partial}{\partial x_1}+\varepsilon_2 x_2\frac{\partial}{\partial x_2}+g(x_3)\frac{\partial}{\partial x_3},$$

with $\varepsilon_1=\pm1$, $\varepsilon_2=\pm1$, $g$ a $C^\infty$ function with $g(0)=g'(0)=0$. Writing

$$g(x_3)=a_2x_3^2+a_3x_3^3+a_4x_3^4+O(x_3^5),$$

we define the codimension of a $W_1$-singularity as $i-1$ if $i$ is the first nonzero coefficient in the expansion of $g(x_3)$. In order to get an idea of what the codimension means, we refer to the theorem of Tarski-Seidenberg (see e.g. [AR]) implying that for any $i\geq1$ the set of vector fields with a singularity in $W_1$ of codimension $i-1$ is a semi-algebraic set of codimension $i$ inside the vector space of $k$-jets, with $k\geq i$.

The dynamics of $W_1$-singularities are hence completely determined by $\tilde{X}=g(x_3)\frac{\partial}{\partial x_3}$:

**codimension 1-singularities** We have $g(x_3)=a_2x_3^2+O(x_3^3)$ with $a_2\neq0$. The topological type of $\tilde{X}$ is determined by the sign of $a_2$. In Fig. 1 the phase portraits of $\tilde{X}$ are drawn.



Figure 1: codimension 1: $a_2<0$ (left) and $a_2>0$ (right)

Figure 2: codimension 2: $a_3 < 0$ (left) and $a_3 > 0$ (right)

**codimension 2-singularities** We have $g(x_3) = a_3 x_3^3 + O(x_3^4)$ with $a_3 \neq 0$. Again, the topological type of $\tilde{X}$ is determined by the sign of $a_3$. See Fig. 2 for phase portraits of $\tilde{X}$.

## 2.4 Singularities in the sets $W_2$–$W_4$

In the sets $W_2$–$W_4$, the center manifolds are 2-dimensional, but they need not be $C^\infty$. In the case $W_2$ and if the vector field is analytic or if it satisfies some Łojasiewicz inequality one can prove the $C^\infty$ smoothness of the center manifolds (see [D2]).

By center manifold reduction, these singularities are $C^0$-equivalent to

$$g_1(x,y)\frac{\partial}{\partial x} + g_2(x,y)\frac{\partial}{\partial y} \pm z\frac{\partial}{\partial z},$$

with $g_1$ and $g_2$ some $C^\ell$ functions, and if $g = (g_1, g_2)$ then $g(0,0) = (0,0)$ and $Dg(0,0)$ has two eigenvalues on $i\mathbf{R}$.

The study of $W_2$–$W_4$ is hence reduced to the study of *planar* vector fields of codimension $\leq 4$, having a 1-jet linearly conjugate to

$$\begin{pmatrix} 0 & -\lambda \\ \lambda & 0 \end{pmatrix}, \quad \begin{pmatrix} 0 & 1 \\ 0 & 0 \end{pmatrix}, \quad \text{or} \quad \begin{pmatrix} 0 & 0 \\ 0 & 0 \end{pmatrix}$$

We refer to [D1] for an analysis of these types.

**Important remark** Different $C^0$-equivalence classes can be described by semi-algebraic regular manifolds of increasing codimension in certain jet spaces (at most 9-jet). This yields a semi-algebraic stratification of the $W_2$–$W_4$ singularities with codimension $\leq 4$.

We remark that it is possible to study more detailed behaviour than is necessary for the $C^0$ classification, such as the occurence of spirals, characteristic orbits and their mutual contact, etc.

## 2.5 Reduction to a normal form

The technique outlined in this section will be important in the preparation for further study and in the precise definition of "codimension".

Let $L = DX(0)$ and consider, at the level of homogeneous vector fields of degree $k$:

$$\mathrm{ad}_k\, L \colon H^k \to H^k \colon X_k \mapsto [L, X_k],$$

where $[L, X_k]$ stands for the Lie-bracket of $L$ and $X_k$. Let $B_k = \mathrm{Im}(\mathrm{ad}_k\, L)$ and take a complementary subspace $G_k$ (so that $B_k \oplus G_k = H_k$).

**2.5 Theorem (Normal form theorem)** *Let $X$ be a $C^r$ vector field, $r \in \mathbf{N}_*$, defined in a neighbourhood of $0 \in \mathbf{R}^n$ with $X(0) = 0$ and $DX(0) = L$. Let $B_k$ and $G_k$ be as above. Then there exists an analytic change of coordinates $\phi\colon (\mathbf{R}^n, 0) \to (\mathbf{R}^n, 0)$ in the neighbourhood of $0$ such that $X' = \phi_* X$ is of the form*

$$X'(y) = Ly + g_2(y) + \cdots + g_r(y) + O(\|y\|^{r+1}),$$

*with $g_i \in G_i$, for all $i \in \{2, \ldots, r\}$.*

As a possible choice of complementary subspaces $G_k$, one can always use $G_k = \ker(\mathrm{ad}_k\, L^{\mathrm{T}})$, where $L^{\mathrm{T}}$ denotes the transpose of $L$.

## 2.6 Singularities in the set $W_5$

Using Theorem 2.5 in combination with the theorem of Borel on the realization of formal power series by $C^\infty$ functions, we get that a $C^\infty$ coordinate change can bring any singularity in the set $W_5$ into the form

$$\lambda(1 + f(z, x^2 + y^2)) \left( x\frac{\partial}{\partial y} - y\frac{\partial}{\partial x} \right) + g(z, x^2 + y^2) \left( x\frac{\partial}{\partial x} + y\frac{\partial}{\partial y} \right) \\ + h(z, x^2 + y^2)\frac{\partial}{\partial z} + Y_1\frac{\partial}{\partial x} + Y_2\frac{\partial}{\partial y} + Y_3\frac{\partial}{\partial z},$$

with $f(0,0) = g(0,0) = h(0,0) = \frac{\partial h}{\partial z}(0,0) = 0$, and $j_\infty Y_j(0) = 0$ for $j = 1,2,3$.

Under mild conditions of Łojasiewicz type (e.g. $X$ analytic, see [D2]), one can even impose that the $z$-axis is invariant in this normal form.

The notion of $C^0$-equivalence allows us to multiply vector fields with positive functions: we can replace $\lambda(1 + f(z, x^2 + y^2))$ by $2\pi$, i.e. the $W_5$ vector field is locally $C^\infty$-equivalent to

$$2\pi \left( x\frac{\partial}{\partial y} - y\frac{\partial}{\partial x} \right) + g(z, x^2 + y^2) \left( x\frac{\partial}{\partial x} + y\frac{\partial}{\partial y} \right) \\ + h(z, x^2 + y^2)\frac{\partial}{\partial z} + Y_1\frac{\partial}{\partial x} + Y_2\frac{\partial}{\partial y} + Y_3\frac{\partial}{\partial z},$$

with $g(0,0) = h(0,0) = \frac{\partial h}{\partial z}(0,0) = 0$, and $j_\infty Y_j(0) = 0$ for $j = 1,2,3$, and with a condition on the $Y_i$ implying the invariance of the $z$-axis. In cylindrical coordinates, the above vector field can hence be written as

$$2\pi\frac{\partial}{\partial \theta} + X_R + Z(z, r, \theta),$$

with

$$X_R = g(z, r^2)r\frac{\partial}{\partial r} + h(z, r^2)\frac{\partial}{\partial z}. \tag{2.3}$$

and $j_\infty Z(0, 0, \theta) = 0, \forall \theta \in S^1$.

Consider now the plane $H = \{(x, y, z) \in \mathbf{R}^3 \colon y = 0, x > 0\}$ and the first return map of $X$. One shows (see [D2]) that this map can be extended to a $C^\infty$ diffeomorphism

$$P_X \colon \overline{H} \to \overline{H}.$$

As a consequence the study of $X$, up to $C^0$-equivalence, can be reduced to the study of $P_X$, up to $C^0$-conjugacy. By the construction we also have that $P_X$ is infinitely close (at $(0,0)$) to the (time 1)-map of the planar vector field $X_R$ in (2.3).

In [DRR] it is shown that the study of germs of diffeomorphisms for $C^0$ conjugacy can be reduced to the topological study of reduced vector fields $X_R$, at least under certain conditions. In [DI2] it has been checked that in all cases of codimension $\leq 4$, the necessary conditions are satisfied, and the results of [DRR] can be applied. (There was however one exception, but also in that case an extension of the results of [DRR] could lead to the result.) This elaboration reduces the whole problem to a study of the $C^0$-equivalence classes of $\mathbf{Z}_2$-equivariant ($(r,z) \mapsto (-r,z)$) vector fields up to codimension 4. This classification has been made in [DI3]. It is proven that this again gives a semi-algebraic stratification.

## 2.7 Blow up

The technique described here is the general quasi-homogeneous blow up. Consider

$$\begin{aligned} \Psi\colon S^{n-1} \times [0,\infty[ &\to \mathbf{R}^n \\ ((\overline{x}_1,\ldots,\overline{x}_n),r) &\mapsto (r^{\alpha_1}\overline{x}_1,\ldots,r^{\alpha_n}\overline{x}_n) \end{aligned}$$

(with nonzero weights $\alpha_i \in \mathbf{N}$). This transformation defines a $C^\infty$ vector field $\tilde{X}$ with $\Psi^*(\tilde{X}) = X$. An extra division by a factor $r^m$ leads to the "blown up vector field"

$$\overline{X} = \frac{1}{r^m}\tilde{X},$$

where $m$ is taken as large as possible, however keeping $S^{n-1} \times \{0\}$ invariant.

In calculations, one often uses charts on $S^{n-1}$, e.g. $\overline{x}_i = \pm 1$ for some $i \in \{1,\ldots,n\}$. The blow up construction restricted to a chart is called a directional blow up.

In the plane, a desingularization of singularities can be made by using successions of homogeneous blow up ($\alpha_1 = \alpha_2 = 1$), but the procedure to desingularize the vector field goes faster by choosing appropriate quasi-homogeneous blow ups based on a Newton diagram.

The use of blow up in the study of planar singularities has extensively been presented in [D].

Let us now apply this technique to singularities in the set $W_6$.

## 2.8 Singularities in the set $W_6$

The singularities have a 1-jet linearly conjugate to $y\frac{\partial}{\partial x} + z\frac{\partial}{\partial y}$, and the codimension is at least 3.

Normal form theory tells us that $(X,0)$ is $C^\infty$ conjugate to

$$y\frac{\partial}{\partial x} + z\frac{\partial}{\partial y} + f(x,y,z)\frac{\partial}{\partial z}, \tag{2.4}$$

with extra restrictions on $j_\infty f(0)$ (see [DI1]). We have

$$f(x,y,z) = ax^2 + bxy + cxz + dy^2 + O((x,y,z)^3).$$

**2.6 Theorem ([DI1])** *If $a \neq 0$, then the singularity (having codimension 3) is $C^0$-equivalent to*

$$y\frac{\partial}{\partial x} + z\frac{\partial}{\partial y} + x^2\frac{\partial}{\partial z}$$

*(or equivalently $\dddot{x} = x^2$).*

**Sketch of a proof** First, we can normalize the form of $f$ and set $a = 1$. We use a quasi-homogeneous blow up:

$$x = u^3\overline{x}, \qquad y = u^4\overline{y}, \qquad z = u^5\overline{z},$$

with $\overline{x}^2 + \overline{y}^2 + \overline{z}^2 = 1$ and $ut = \overline{t}$. This yields a $C^\infty$ vector field on $S^2 \times [0, \infty[$. To do further calculations, we use directional blow up. The most interesting behaviour can be seen in the $(\overline{y} = 1)$-direction; the blow up in this direction gives:

$$\overline{X}^y : \begin{cases} \dot{u} = \frac{1}{4}u\overline{z} \\ \dot{\overline{x}} = 1 - \frac{3}{4}\overline{x}\overline{z} \\ \dot{\overline{z}} = \overline{x}^2 - \frac{5}{4}\overline{z}^2 + O(u). \end{cases}$$

On $\{u = 0\}$ the vector field $\overline{X}^y$ has two singularities (hyperbolic ones), is symmetric with respect to $(\overline{x}, \overline{z}, \overline{t}) \mapsto (-\overline{x}, -\overline{z}, -\overline{t})$, and has interesting isoclines (see Fig. 3). Observe also that the divergence is negative on the positive quadrant.

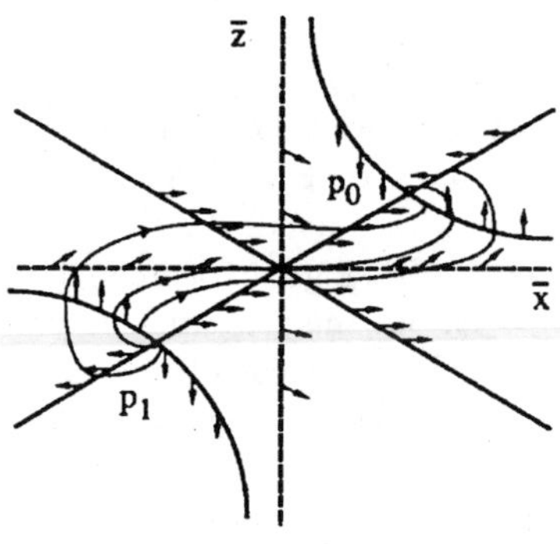


Figure 3: Blown-up vector field $\overline{X}^y$

Also in other charts, easy arguments can be applied to follow the orbits. One concludes that $\overline{X}|_{u=0}$ is a Morse-Smale system (of "North Pole-South Pole" type and it also is radially hyperbolic in the singularities). We refer to [DI1] for a full treatment. □

### 2.9 Singularities in the set $W_6$ of codimension 4

Assume the vector field is in the form (2.4), with

$$f(x, y, z) = ax^2 + bxy + cxz + dy^2 + e_1x^3 + \cdots$$

This time, we assume $a = 0$, $b \neq 0$ and $e_1 \neq 0$. After normalization, we can take $b = 2$ and $e_1 = 1$:

$$f(x, y, z) = 2xy + cxz + dy^2 + x^3 + \cdots$$

Changing $z$ to $z = \overline{z} + x^2$ (and dropping the bars) yields a vector field

$$X = y\frac{\partial}{\partial x} + (z + x^2)\frac{\partial}{\partial y} + (cxz + dy^2 + (1 + c)x^3 + e_2x^2z + e_3xz^2 + e_4x^2y + e_5xyz + e_6xy^2 + O((x, y, z)^4))\frac{\partial}{\partial z}.$$

In the study of equivalence classes, the parameters $(c, d)$ will play an important role, so let us write

$$X_{(c,d)} = X = y\frac{\partial}{\partial x} + (z + x^2)\frac{\partial}{\partial y} + (cxz + dy^2 + (1 + c)x^3 + \cdots)\frac{\partial}{\partial z}. \tag{2.5}$$

Again, we can use blow up:

$$x = u^2\overline{x}, \qquad y = u^3\overline{y}, \qquad z = u^4\overline{z}, \qquad ut = \overline{t},$$

with $\overline{x}^2 + \overline{y}^2 + \overline{z}^2 = 1$. This yields a blown up vector field $\overline{X}$. The most important information is found in the chart $\overline{z} = -1$, and there we find the vector field

$$X_z^- : \begin{cases} \dot{\overline{x}} = \overline{y} - 2u\overline{x}F(\overline{x}, \overline{y}) + O(u^2) \\ \dot{\overline{y}} = -1 + \overline{x}^2 - 3u\overline{y}F(\overline{x}, \overline{y}) + O(u^2) \\ \dot{u} = u^2F(\overline{x}, \overline{y}) + O(u^3), \end{cases}$$

with $F(\overline{x}, \overline{y}) = \frac{1}{4}(c\overline{x} - d\overline{y}^2 - (1 + c)\overline{x}^3)$.

A number of features do not facilitate the study of $X_z^-$:

(1) The vector field $X_z^-$ is nowhere hyperbolic in the radial direction.

(2) For $u = 0$ the vector field is a Hamiltonian vector field with Hamiltonian

$$H(\overline{x}, \overline{y}) = \frac{\overline{y}^2}{2} + \overline{x} - \frac{\overline{x}^3}{3}.$$

At $u = 0$ two singularities appear: $s = (1, 0)$ and $e = (-1, 0)$. Notice that $F(s) = -1$, which

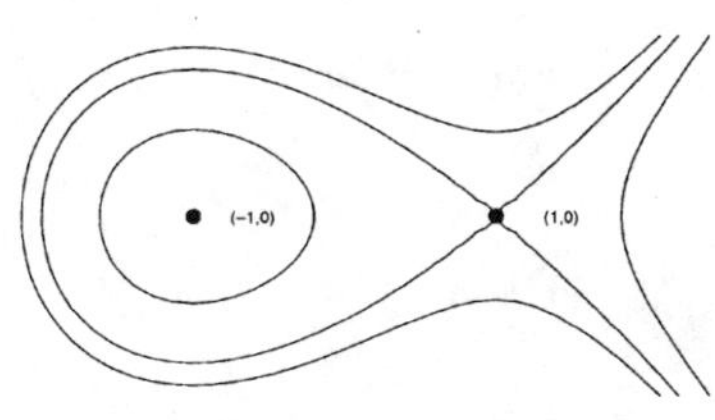


Figure 4: Compact components of level curves of $H$ for $h \in ]-2/3, 2/3[$

can be translated into an attracting center behaviour at $s$. Also $F(s) = +1$, which leads to a repelling center behaviour at $e$. The analysis of $\overline{X}$ at the singularities is far from being trivial and relies on the techniques introduced in [DRR]. For a thorough analysis, as well for this study as for other steps in our sketch of the proof, we refer to [DI1].

Another point we have to investigate is whether we can have invariant cylinders, hence blowing down to invariant cones. This resembles the traditional problem of studying limit cycles in planar Hamiltonian perturbations. We can try to use the same techniques, but of course we need to check carefully that they can really be applied. As such, in search for invariant cylinders, we consider the 1-form

$$\omega = F(\overline{x}, \overline{y})(2\overline{x}\, d\overline{y} - 3\overline{y}\, d\overline{x}).$$

($dH + u\omega$ is the 1-form associated to $j_1(X_z^-)$ at $u = 0$.) Then one proves that an invariant cylinder appears if and only if

$$\int_{\gamma_h} \omega = 0.$$

If such a cylinder appears, the behaviour on this cylinder is characterized by the function.

$$\int_0^{T(h)} F(\overline{x}_h(t), \overline{y}_h(t))\, dt.$$

The complete analysis leads to the following result:

**2.7 Proposition ([DI1])** *Let*

$$H(x, y) = \frac{y^2}{2} + x - \frac{x^3}{3}$$

*with $\gamma_h$ the compact component of $H = h$, for $h \in ]-2/3, 2/3[$. Let $(x_h, y_h)\colon [0, T(h)] \to \mathbf{R}^2$ be a solution of $X_H$ along $\gamma_h$, with $T(h)$ the period. Let $F(x, y) = \frac{1}{4}(cx - dy^2 - (1 + c)x^3)$ and $\omega = F(x, y)(2xdy - 3ydx)$.*

*Consider*

$$g_1(c, d) = -\frac{17}{7}c + \frac{12}{7}d - \frac{103}{28}, \qquad g_2(c, d) = c + \frac{11}{4},$$

*and*

$$g_3(h, c, d) = \int_0^{T(h)} F(x_h(t), y_h(t))\, dt.$$

*Let $S$ be the subset in the $(c, d)$-plane, where at least one of the following conditions is satisfied:*

(1) $g_1(c, d) = 0$;

(2) $g_2(c, d) = 0$;

(3) *there is a* $h \in ]-2/3, 2/3[$ *with* $g_3(h, c, d) = 0$ *and* $\int_{\gamma_h} \omega = 0$.

*Take $(c, d) \in \mathbf{R}^2 \setminus S$. Then for all $(c', d')$ sufficiently close to $(c, d)$ the vector field $X_{(c',d')}$ is $C^0$-equivalent to $X_{(c,d)}$ in (2.5).*

This means that in every region I–VI one $C^0$-equivalence class exists. The equivalent classes are mutually different except for the classes IV and V (see Fig. 5).

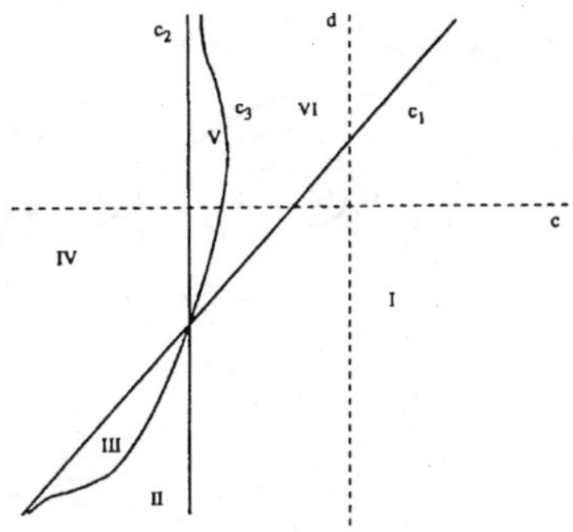


Figure 5: $C^0$-equivalence classes I–VI

**Description of Figure 5**

The curves $c_1\colon g_1(c,d) = 0$ and $c_2\colon g_2(c,d) = 0$ are lines, intersecting in $(-11/4,-7/4)$. These curves $c_2$ and $c_1$ correspond to a zero of the integral $\int_{\gamma_h}\omega$ at respectively $e$ and at the homoclinic loop. The third curve can be seen as a projection of the transverse intersection of two ruled surfaces. Let us describe it. We first consider in $(h,c,d)$-space the surface $g_3(h,c,d)=0$. One proves that

$$g_3(h,c,d) = 0 \quad \Longleftrightarrow \quad c - d + \frac{9}{4} - \frac{3}{2}\frac{hT(h)}{A(h)} = 0,$$

where $T$ is the period, $A$ is the area function and $\frac{d}{dh}\left(\frac{hT(H)}{A(h)}\right) > 0$. Hence, this surface is a ruled surface. On the other hand, the surface $\int_{\gamma_h}\omega = 0$ will also form a ruled surface: at $h = -2/3$ (corresponding to $e$) the integral is zero. At $h = +2/3$ (corresponding to the homoclinic connection of $s$) the integral is zero as well. Finally one can prove that the integral is zero at $\gamma_h$ along a line passing through $c_1 \cap c_2$ and rotating in a monotone way from $c_2$ to $c_1$ as $h$ increases. The intersection of the two surfaces is then projected onto the $(c,d)$-plane and forms the curve $c_3$.

**Description of the phase portraits**

In order to give an accurate, although simple description of the phase portraits in the different regions I–VI, we introduce vector fields whose blow up is of "Morse-Smale" type and $C^0$-equivalent to I–VI, respectively. In fact we do not give formulas for these vector fields but only describe how they look after homogeneous blow up.

Let $Y$ be a $C^\infty$ vector field with $j_{k-1}Y(0) = 0$, $j_kY(0) \neq 0$ for $k \geq 2$, and consider the blow up

$$\Psi\colon S^2 \times \mathbf{R} \to \mathbf{R}^3\colon (u,r) \mapsto r.u.$$

Write $\Psi_*(\hat{Y}) = Y$ and $\overline{Y} = (r^{k-1})^{-1}\hat{Y}$. We suppose that $\overline{Y}_0 := \overline{Y}|_{S^2\times\{0\}}$ is Morse-Smale and that $\overline{Y}$ is also hyperbolic in critical elements of $\overline{Y}_0$.

We suppose that $\overline{Y}_0$ has four singularities, at most one limit cycle and $\overline{Y}_0$ is like in I′–VI′ (see Fig. 6, where we remove from $S^2$ a sink $\tilde{p}$ in which $\overline{Y}$ is normally repelling).

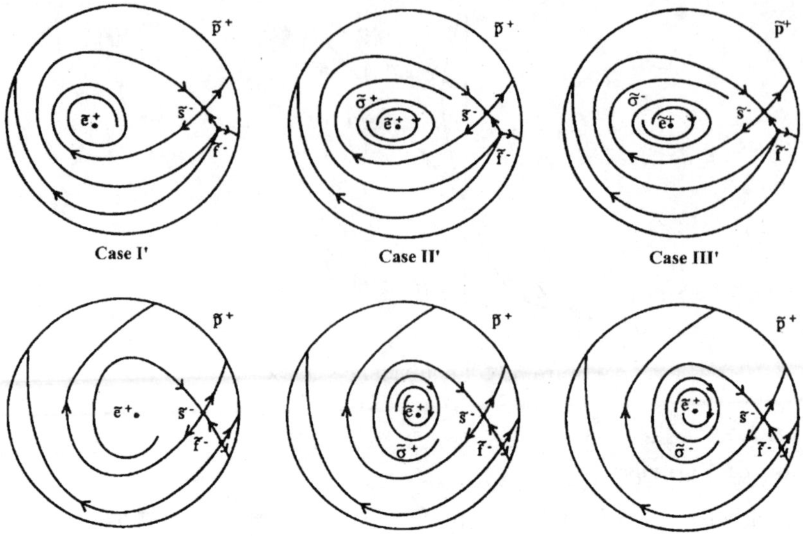


Figure 6: Blow up of Morse-Smale models

**2.8 Proposition** *If $Y$ is a $C^\infty$ vector field whose blow up satisfies one of the conditions I'–VI', then it is $C^0$-equivalent at 0 to any vector field $X$ satisfying the conditions I–VI, respectively. Such an equivalence is obtained by blowing down a "pinched $C^0$-equivalence" between $\overline{Y}$ and $\overline{X}$.*

**2.9 Definition** We call $\overline{Y}_{S^2\times[0,\infty[}$ and $\overline{X}_{S^2\times[0,\infty[}$ *pinched $C^0$-equivalent near $S^2 \times \{0\}$* if there exists a $C^0$-equivalence between $\overline{Y}_{S^2\times[0,\infty[}$ and $\overline{X}_{S^2\times[0,\infty[}$, respecting neighbourhoods of $S^2 \times \{0\}$.

The pinched equivalence at the blow up level permits to obtain Proposition 2.7 from Proposition 2.8 by means of standard techniques ([Ca, ULL]; The equivalence between IV and V is proved similarly).

### Non-algebraicity of the stratification

Here, we give a short description of the proof that the stratification is non-algebraic. More specifically, the curve $c_3$ (see Proposition 2.7) can be shown to be non-algebraic. For all details we refer to [DI2]. Let us recall that $c_3$ is defined in the $(c,d)$-plane by

$$\{\, g_3(h,c,d) = 0, \int_{\gamma_h} \omega = 0 \,\} \quad \text{for some } h$$

One can show that this is equivalent to

$$\begin{cases} c - d + \frac{9}{4} - \frac{3}{2}\left(\frac{\frac{15}{2}h^2 - 7hP(h)}{9h^2 - 4}\right) = 0 \\ -\frac{3}{2}(c - d + 1)h - \frac{1}{4}(-8c + 4d - 15)P(h) = 0, \end{cases}$$

where

$$P(h) = \frac{\int_{\gamma_h} yx\,dx}{\int_{\gamma_h} y\,dx}.$$

As such,

$$\begin{cases} c(h) = \dfrac{-30h + 24P(h) - 36h^2P(h) + 42hP(h)^2}{4P(h)(9h^2 - 4)} \\ d(h) = \dfrac{-30h - 12P(h) + 84hP(h)^2}{4P(h)(9h^2 - 4)} \end{cases}.$$

Hence, if $(c(h), d(h))$ is algebraic, then also $(h, P(h))$ is algebraic. However, $P$ is a solution of the Ricatti equation

$$(9h^2 - 4)P' = 7P^2 + 3hP - 5.$$

It represents an orbit going from a saddle to a resonant node for the associated planar vector field

$$\begin{cases} h' = 4 - 9h^2 \\ P' = 5 - 3hP - 7P^2 \end{cases}$$

Near the resonant node, the asymptotic development (in translated coordinates) is

$$v(u) = -\frac{5}{28}u \ln u + o(u).$$

This contradicts the assumption that $(c(h), d(h))$ is algebraic.

We can now end this paragraph by a number of interesting remarks, related to the results which we described.

(1) Also in $\mathbf{R}^2$ the stratification in $C^0$-equivalence classes is not algebraic (Il'yashenko [I] in the study of the center-focus problem). In the plane, the non-algebraicity starts at a much higher codimension.

(2) Up to codimension 4 the singularities in $\mathbf{R}^3$ are hence "$C^0$-determined" by some finite jet (Taylor approximation of finite order). In dimension 2, singularities of finite codimension are always $C^0$-determined by some finite jet; however, if the dimension is greater than 2, this is not always the case. There exist examples in $\mathbf{R}^3$ belonging to $W_5$ (see [D3]). In these examples, there exists an associated Poincaré map, exhibiting a hyperbolic sector in which it is possible to change the $C^0$-conjugacy class by means of infinitely flat terms. This changes the $C^0$-equivalence class of $X$ within the suspension of such a sector.

## 3 Local bifurcations—unfoldings of planar singularities

In this section we come back to the plane, but instead of studying the singularities of the vector fields we will describe their unfoldings. We will consider $C^\infty$ families of vector fields. The organizing center of such unfoldings is a singularity. Let us start by describing different classes of singularities that have to be considered.

Let $X$ be a vector field on $(\mathbf{R}^2, 0)$, with $X(0) = 0$. Let $L = DX(0)$, then a linear change of coordinates can put $L$ in Jordan normal form $M = A^{-1}LA$. As in Section 2 we make a distinction between the following possibilities:

(1) no eigenvalues on $i\mathbf{R}$ (codimension 0);

(2) one zero eigenvalue, 1 nonzero eigenvalue (codimension 1);

(3) a pair of purely imaginary eigenvalues (codimension 1);

(4) a nilpotent jet associated to two zero eigenvalues (codimension 2);

(5) a zero 1-jet (codimension 4).

The first class of vector fields is structurally stable. Vector fields in class 2 can be studied using a reduction to the 1-dimensional center manifold. For class 3, the use of adapted normal forms also reduces the study to a problem in dimension 1 (see [T1, D]). For both class 2 and 3 all generic (stable) $p$-parameter unfoldings are known. It has been shown that all bifurcation diagrams are cone-like from the origin, that all relevant information is described by zeros of families of polynomials in one real variable, and that a complete local bifurcation diagram can be detected by algebraic techniques.

For vector fields of class 5, the case where the 1-jet is zero, only very incomplete results are known. In the rest of this paragraph, we will now treat the case of a nilpotent jet with two zero eigenvalues.

So let $j_1X(0) = y\frac{\partial}{\partial x}$. Using normal form theory, one proves that the singularity can be given the infinite jet

$$j_\infty X(0) = y\frac{\partial}{\partial x} + ((a_2x^2 + a_3x^3 + \cdots) + y(b_1\dot{x} + b_2x^2 + b_3x^3 + \cdots))\frac{\partial}{\partial y}.$$

Any unfolding can be written as

$$X_\lambda := y\frac{\partial}{\partial x} + (F(x,\lambda) + yG(x,\lambda) + y^2Q(x,y,\lambda))\frac{\partial}{\partial y}$$

where $F$, $G$ and $Q$ $C^\infty$ are functions and $Q = O(\|(x,y,\lambda)\|^N)$, for $N$ as large as required. We assume that the singularity is put at $\lambda = 0$:

$$F(x,0) = a_2x^2 + a_3x^3 + \cdots, \qquad G(x,0) = b_1x + \cdots.$$

### 3.1 Nilpotent singularities of codimension 2

Assume $a_2b_1 \neq 0$. Then the family $X_\lambda$ can be brought into the form

$$y\frac{\partial}{\partial x} + \left(x^2 + \mu(\lambda) + y(\nu(\lambda) \pm x + x^2h(x,\lambda)) + y^2Q(x,y,\lambda)\right)\frac{\partial}{\partial y}.$$

In [B, T2], it is proved that all local phase portraits are described by the universal unfolding

$$y\frac{\partial}{\partial x} + (x^2 + \mu + y(\nu \pm x))\frac{\partial}{\partial y}.$$

This unfolding is called the Bogdanov-Takens bifurcation. The bifurcation diagram and the related phase portraits can be found in [D].

## 3.2 Nilpotent singularities of codimension 3

Here, we assume $a_2 b_1 = 0$, and we immediately distinguish two cases: $a_2 \neq 0$ and $a_2 = 0$. The first case is called the *cuspidal case.*

### 3.2.1 The cuspidal case

If $a_2 \neq 0$, $b_1 = 0$, then in a normal form also $b_2$ can be cancelled. So assume $b_2 = 0$, and $b_3 \neq 0$. In [DRS1], it is shown that a universal unfolding for this codimension-3 singularity is given by

$$y\frac{\partial}{\partial x} + (x^2 + \mu + y(\nu_0 + \nu_1 x \pm x^3))\frac{\partial}{\partial y}.$$

If $b_3 = 0$, and $b_4 \neq 0$, the codimension is 4, and in [LR] it is shown that

$$y\frac{\partial}{\partial x} + (x^2 + \mu + y(\nu_0 + \nu_1 x + \mu_2 x^3 \pm x^4))\frac{\partial}{\partial y}.$$

is a universal unfolding for this type of singularity. If also $b_4 = 0$, then the codimension is 5 or higher, and in [M] the following results have been obtained:

(1) The bifurcation diagrams are cone-like from the origin;

(2) all relevant information is described by zeros of families of polynomials in one real variable;

(3) a complete local bifurcation diagram can be detected by algebraic techniques.

The last observation means that calculations on a finite jet (depending on the codimension) permit to prove the presence of such an unfolding. It does not mean that the bifurcation diagram is an algebraic variety. In fact from codimension 3 on, the bifurcation diagram will not even be an analytic variety (see [DRS1]). Let us also mention that the cyclicity (maximum number of limit cycles) of the unfolding of a cuspidal singularity of codimension $n$ is $[\frac{2n+1}{3}]$.

### 3.2.2 The non-cuspidal case

Here, we assume $a_2 = 0$, but we restrict the study to the case where $a_3 \neq 0$. The cases $a_2 = a_3 = 0, a_4 \neq 0$ have codimension at least 4 and have not yet been studied. Assuming $b_1 \neq 0$, a $C^\infty$ normal form of a 3-parameter unfolding can be written as

$$y\frac{\partial}{\partial x} + (\varepsilon_1 x^3 + \mu_2 x + \mu_1 + y(\nu + b(\mu_1, \mu_2, \nu)x$$
$$+ \varepsilon_2 x^2 + x^3 h(x, \mu_1, \mu_2, \nu)) + y^2 Q(\mu_1, \mu_2, \nu))\frac{\partial}{\partial y},$$

with $(\mu_1, \mu_2, \nu)$ as independent parameters, $\varepsilon_1 = \pm 1$, $\varepsilon_2 = \pm 1$, all functions are $C^\infty$ and $b(0,0,0) = b > 0$. It is expected that a universal unfolding is given by

$$y\frac{\partial}{\partial x} + (\varepsilon_1 x^3 + \mu_2 x + \mu_1 + y(\nu + bx + x^2))\frac{\partial}{\partial y}, \tag{3.1}$$

with $\varepsilon_1 = \pm 1$, $\varepsilon_2 = \pm 1$ and $b > 0$. Three essentially different cases of codimension 3 can be considered, where the difference can already be seen in the phase portraits for $\mu_1 = \mu_2 = \nu = 0$:

(1) $\varepsilon_1 = 1$ (the saddle case);

(2) $\varepsilon_1 = -1$, $0 < b < 2\sqrt{2}$ (the focus case);

(3) $\varepsilon_1 = -1$, $b > 2\sqrt{2}$ (the elliptic case).

We refer to [D] for a description of the bifurcation diagrams for these cases.

The same paper provides steps in the proof that (3.1) indeed contains a universal unfolding.

Unfortunately in all cases the proof is incomplete. As usual in the study of planar vector fields the main problem deals with the limit cycles. Part of this study relies on studying zeros of Abelian integrals, and another part on making a global study of certain Liénard equations. The reduction to both is based on rescalings as can be seen in [D], and as will be shown in Section 3.3 for the study of the nilpotent saddle of codimension 4. However, the necessary treatment of the related Abelian integrals is missing. In fact, the treatment in [Z] contains a failure. For the saddle and the focus case it is the only missing point, while in the elliptic case there is also missing a part dealing with "large amplitude" limit cycles. For the latter however it is clear that, although the result is not yet available in the literature, it can be worked out using family blow up and calculations similar to the ones used in [DRS3]. Unfortunately it is not yet clear how to finish the study of the related Abelian integrals.

The main problem comes from the fact that $b > 0$. It turns out to be simpler to treat the case with $b = 0$. Seen the interest on its own, we will give an overview of results related to that case in Section 3.3. In Section 3.4 we will link the local study of unfoldings of nilpotent singularities to the global study of (generalized) Liénard equations, essentially paying attention to the limit cycles.

## 3.3 Nilpotent saddle of codimension 4

In connections with the problems about Abelian integrals related to the unfoldings of codimension 3 singularities and in combination with the bifurcation from elliptic Hamiltonians of degree 4 that we will treat in Section 3.5, let us consider the generic unfoldings of the nilpotent saddle of codimension 4.

Thus we consider

$$X_\lambda = y\frac{\partial}{\partial x} + (x^3 + \mu_2 x + \mu_1 + y(\nu + bx + \varepsilon x^2 + x^3 h(x,\lambda)) + y^2 Q(x,y,\lambda))\frac{\partial}{\partial y}, \tag{3.2}$$

where $\varepsilon = \pm 1$, $\lambda = (\mu_1, \mu_2, \nu, b)$ is a small parameter, and the functions $h$ and $Q$ are $C^\infty$, $Q = O(\|(x,y,\lambda)\|^N)$ for $N$ as large as required.

We can keep $\varepsilon = +1$, since the case $\varepsilon = -1$ can be reduced to it by the mapping $(x, y, \mu_1, \mu_2, \nu, b, t) \mapsto (-x, y, -\mu_1, \mu_2, -\nu, -b, -t)$.

We will now give an overview of the treatment made in [DFL]. For a precise statement of all results, as well as for the details of the discussion we refer to [DFL].

Exactly as in the study of the saddle of codimension 3, we make a first rescaling, called the "principal rescaling". It is given by

$$(x, y, \mu_1, \mu_2, \nu, b, t) = (\tau\overline{x}, \tau^2\overline{y}, \tau^3\overline{\mu}_1, \tau^2\overline{\mu}_2, \tau\overline{\nu}, \tau\overline{b}, \overline{t}/\tau),$$

and it changes (3.2) into

$$y\frac{\partial}{\partial x} + (x^3 + \overline{\mu}_2 x + \overline{\mu}_1 + \overline{\nu} y + \tau y(\overline{b}x + x^2) + O(\tau^2)y)\frac{\partial}{\partial y},$$

omitting the bars over $(x, y)$. For $\tau = 0$, this family becomes

$$y\frac{\partial}{\partial x} + (x^3 + \overline{\mu}_2 x + \overline{\mu}_1 + \overline{\nu} y)\frac{\partial}{\partial y}. \tag{3.3}$$

It is a family of cubic Liénard equations with constant damping. In the principal chart $\{\overline{\mu}_2 = -1\}$, which is the most interesting chart, the phase portraits are as shown in Fig. 7. Transversality arguments permit one to show that all phase portraits, except for those along

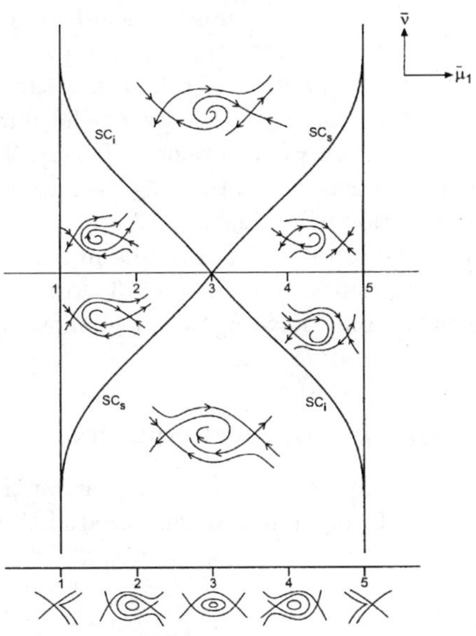


Figure 7: Phase portraits of the cubic Liénard equations with constant damping

$\{\overline{\nu} = 0\}$, persist for $\tau > 0$ sufficiently small. The Hamiltonian vector fields along the axis $\{\overline{\nu} = 0\}$, however, are not stable. In order to study how this line of vector fields changes for $\tau > 0$ we consider the following blow up in the parameter space:

$$(\overline{\nu}, \overline{\mu}_1, \overline{\mu}_2, \overline{b}, \tau) = (u\tilde{\nu}, \tilde{\mu}_1, \tilde{\mu}_2, \tilde{b}, ur)$$

with $\tilde{\nu}^2 + r^2 = 1$. For $r \sim 0$ we again recover the phase portraits of (3.3) for $\overline{\nu} \neq 0$. So it essentially remains to look at the chart $r = 1$, keeping $|\tilde{\nu}| \leq M_0$ for some $M_0 > 0$ sufficiently big. This is the same as applying the so-called "central rescaling",

$$(x, y, \mu_1, \mu_2, \nu, b, t) = (\delta\tilde{x}, \delta^2\tilde{y}, \delta^3\tilde{\mu}_1, \delta^2\tilde{\mu}_2, \delta^2\tilde{\nu}, \delta\tilde{b}, \tilde{t}/\delta),$$

to the original system (3.2). The system (3.2) gets transformed into $X_{\tilde{\lambda},\delta} + O(\delta^2)\frac{\partial}{\partial y}$, where $X_{\tilde{\lambda},\delta}$ stands for

$$y\frac{\partial}{\partial x} + (x^3 + \tilde{\mu}_2 x + \tilde{\mu}_1 + \delta(\tilde{\nu} + \tilde{b}x + x^2)y)\frac{\partial}{\partial y},$$

omitting the bars over $(x, y)$.

The really original part in the analysis now deals with the study of $X_{\tilde{\lambda},\delta}$ as perturbation of a family of Hamiltonian vector fields. An overview of the results will be given in Section 3.5.

This problem is now simpler than the analogous one in the study of the codimension 3-saddle (see Section 3.2), since in the latter the "central rescaling" does not lead to a Hamiltonian perturbation, but to a perturbation from a non-Hamiltonian integrable vector field. For the codimension 4-saddle the central rescaling unfortunately does not lead to a perturbation from a single Hamiltonian field but from a 1-parameter family. A complete study however has been made in [DL3], and for a precise statement of the results we refer to Section 3.5.

The study of the generic unfolding of the nilpotent saddle of codimension 4 has not been completed, since, among other things, it requires a full knowledge of the unfolding of the codimension 3-case. For a precise overview of the known results and remaining problems, we refer to [DFL]. It is expected that the cyclicity of the unfolding is 2, while for the precise conjectured bifurcation diagram we also refer to [DFL]. The description of the local bifurcations is complete, the most degenerate ones being the nilpotent saddle of codimension 3 and the Bogdanov-Takens bifurcation of codimension 3.

The bifurcation diagram of the codimension 4-saddle in fact describes the transition between two nilpotent saddle bifurcations of codimension 3, corresponding to $b > 0$ and $b < 0$, respectively. It also describes the link between two Bogdanov-Takens bifurcations of codimension 3, having an opposite orientation.

## 3.4 Limit cycles in polynomial Liénard equations

As already observed, the local study of unfoldings of singularities in the plane leads in a natural way to the global study of polynomial (generalized) Liénard equations. By this we mean systems of the form:

$$\begin{cases} \dot{x} = y \\ \dot{y} = -g(x) - yf(x) = -\sum\limits_{k=0}^{m} a_k x^k - y\left(\sum\limits_{k=0}^{n} b_k x^k\right) \end{cases} \tag{3.4}$$

with $a_m b_n \neq 0$. We will call such a system a Liénard equation of type $(m, n)$, i.e. of degree $m$ with damping of degree $n$, and the associated o.d.e. is

$$\ddot{x} + f(x)\dot{x} + g(x) = 0.$$

Some people call it a generalized Liénard equation if $m > 1$. We prefer to speak of the "classical Liénard equation" if $m = 1$. In (3.4) we can introduce the well-known Liénard transformation $Y = y + F(x)$, with $F'(x) = f(x)$, to come to the "standard" expression

$$\begin{cases} \dot{x} = Y - F(x) \\ \dot{Y} = -g(x) \end{cases} . \tag{3.5}$$

We still call it a Liénard vector field of type $(m, n)$. We warn the reader that in the literature, (3.5) is sometimes said to be of type $(m + 1, n)$ since $\deg F = m + 1$.

Define $H(m, n)$ to be the maximal number of limit cycles for Liénard vector fields of type $(m, n)$. The main question is to calculate $H(m, n)$ precisely. However, it is still not known that $H(m, n) < \infty$ for all fixed $(m, n)$. Let us recall some facts.

- $H(1,1) = 0$ (easy—see below).
- $H(1,2) = 1$ ([LMP]).
- $H(2,1) = 1$ ([Co]). This is the Bogdanov-Takens case.
- $H(3,1) = 1$ ([DRo, DL1]).
- $H(2,2) = 1$ ([DL2]).
- Conjecture ([LMP]) $H(1, n) = [n/2]$.

In fact, the calculation of $H(1, 3)$ is still an open problem, but it is expected that the bound is 1.

In all known cases with one limit cycle (i.e. $m + n \leq 4$ except for $(m, n) = (1, 3)$), it is known that the unique limit cycle is always hyperbolic. There also exists a classification of all phase portraits and all bifurcations.

The case $(m, n) = (1, 1)$ can e.g. be treated by the well-known technique of rotated vector fields, which can often be used in dealing with Liénard equations. If $m = n = 1$, one can put (3.4) in the form

$$X_\nu : \begin{cases} \dot{x} = y \\ \dot{y} = -x + (\nu + x)y \end{cases} .$$

For $\nu = 0$ the vector field is invariant under $(x, t) \mapsto (-x, -t)$, inducing a center behaviour around the origin. Hence, no isolated closed orbits exist. For $\nu$ varying, notice that

$$\begin{vmatrix} y & y \\ -x + (\nu + x)y & -x + (\nu' + x)y \end{vmatrix} = (\nu' - \nu)y^2,$$

which shows that this family is a family of rotating vector fields (for $y \neq 0$): at all points $(x, y)$ in the domain $\{y \neq 0\}$ the angle between the flow of $X_\nu$ and $X_0$ rotates in the same direction with respect to $\nu$. This means that if $\nu < 0$ (respectively $\nu > 0$), the flow of the vector field $X_\nu$ is directed everywhere inwards (or everywhere outwards) the closed orbits of $X_0$, and hence $X_\nu$ is a focus (global), and we cannot have closed orbits anymore. Of course, the non-occurrence of closed orbits for $\nu \neq 0$ also follows from the fact that $\operatorname{div} X_\nu = \nu$.

A full study of polynomial Liénard equations is out of reach for the moment. There are, however, a number of partial results of a rather general nature. We can e.g. mention the study of polynomial Liénard equations near infinity. It can be done by compactification, extending the polynomial vector field to an analytic one on a disc (half sphere). Such compactification can be done in a homogeneous way (using $(x, y) = (r^{-1}\cos\theta, r^{-1}\sin\theta)$ for $r \sim 0$) or in a quasi-homogeneous way (using $(x, y) = (r^{-p}\cos\theta, 1r^{-q}\sin\theta)$ for $r \sim 0$ and $p, q, \in \mathbf{N}_1$). A complete study for all $(m, n)$ can be found in [DH].

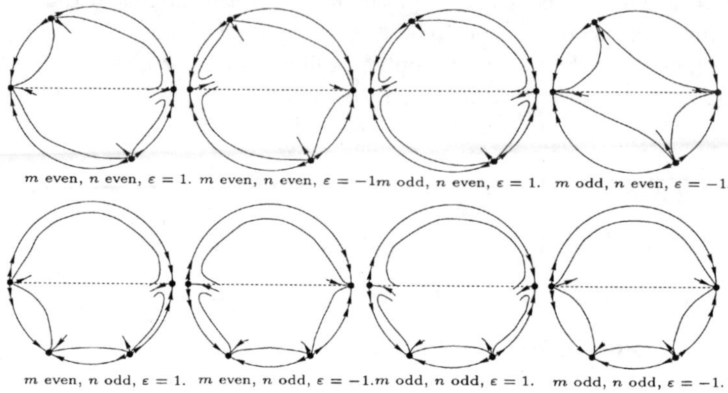


Figure 8: Behaviour near $\infty$ of Liénard equations of type $(m, n)$

In Fig. 8 we show the different possibilities for the Liénard equation of type $(m, n)$ with $m < 2n + 1$. The phase portrait is drawn on what is called a Poincaré-Lyapunov disc of type $(1, n + 1)$, i.e. a compactification obtained by using a quasi-homogeneous compactification with weights $(p, q) = (1, n + 1)$.

Another systematic study deals with the number of small amplitude limit cycles that can be bifurcated from a non-degenerate singularity of center-focus type. The following papers deal with it: [LL, CLl, CLy]. Although a complete list of these numbers (that we could denote by $H_\ell(m, n)$) is not yet known, we at least know the numbers $H_\ell(m, n)$ for $m \leq 2$ and $n \leq 2$: $H_\ell(n, 1) = H_\ell(1, n) = [n/2]$, while $H_\ell(n, 2) = H_\ell(2, n) = [(n + 1)/3]$. For $m \leq 2$ there are no examples available for which $H_\ell(m, n) < H(m, n)$. It is also true that $H_\ell(3, 1) = H(3, 1) = 1$. In Section 3.5 we will however see that $2 = H_\ell(3, 2) < H(3, 2)$, since we will show that $H(3, 2) \geq 5$.

A third systematic approach in getting partial results on Liénard equations of type $(m, n)$ deals with bifurcations from Hamiltonians. We deal with this approach in the next section.

## 3.5 Bifurcations from Hamiltonians

Consider

$$\begin{cases} \dot{x} = y \\ \dot{y} = -g(x) - y\delta f(x) = -\sum\limits_{k=0}^{m} a_k x^k - y\delta\left(\sum\limits_{k=0}^{n} b_k x^k\right) \end{cases} \tag{3.6}$$

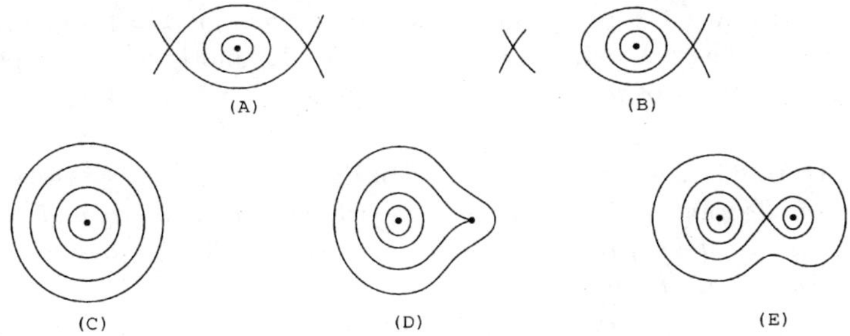


Figure 9: Phase portraits in cases (A)–(E)

with $a_m b_n \neq 0$ and $\delta$ close to 0. For $\delta = 0$, this is a family of Hamiltonian vector fields with Hamiltonians

$$H(x, y, (a_k)) = \frac{y^2}{2} + \sum_{k=0}^{m} \frac{a_k}{k+1} x^{k+1}.$$

We look for limit cycles in systems (3.6) with $\delta$ close to 0. The first order approximation of this problem is to look for isolated zeros of the Abelian integrals

$$I(h, (a_k), (b_\ell)) = \int_{\Gamma_{h,(a_k)}} y \left( \sum_{\ell=0}^{n} b_\ell x^\ell \right) dx$$

with $\Gamma_{h,(a_k)}$ the compact component of

$$\{ (x, y) \mid H(x, y, (a_k)) = h \}.$$

Let $\hat{H}(m, n)$ denote the maximum number of isolated zeros of $I(h, (a_k), (b_\ell))$. It has been shown by Khovanskii and Varchenko [K, V] that $\hat{H}(m, n) < \infty$. It has even been shown that for any fixed $m$, the sequence $\hat{H}(m, n)$ grows linearly in $n$. Unfortunately there is no effective algorithm to calculate $\hat{H}(m, n)$. It is rather easy to see that $\hat{H}(1, n) = [n/2]$, while Petrov has shown (see [P]) that $\hat{H}(2, n) = [(2n + 1)/3]$. Since a trivial calculation gives $\hat{H}(3, 1) = 1$, the next case to deal with is $(m, n) = (3, 2)$. The related Hamiltonians are elliptic of degree 4.

In four recent papers [DL3, DL4, DL5, DL6], Dumortier and Li have given a full description of the zeros of the Abelian integrals, proving among other things that $\hat{H}(3, 2) = 5$. This upper bound on the number of zeros can only be obtained for a specific class of Hamiltonians, while smaller upper bounds can be given for other classes of Hamiltonians. Let us now give an overview of the known results.

There are five different types shown in Fig. 9, and we will refer to them as cases (A)–(E). They are called (A): the case of the two saddle cycle; (B): the saddle loop case; (C): the global center case; (D): the cuspidal loop case; and (E): the figure eight-loop case. Note that (A) is a limiting case of (B), while (D) is a limiting case of (C) and (E).

### 3.5.1 Cases (A) and (B)

Up to a linear coordinate change there is a 1-parameter family

$$H_\lambda(x, y) := \frac{y^2}{2} - \frac{1}{4}x^4 - \frac{\lambda - 1}{3}x^3 + \frac{\lambda}{2}x^2 \tag{3.7}$$

of Hamiltonians, with $\lambda \geq 1$, representing a two-saddle-cycle for $\lambda = 1$ and a saddle loop for $\lambda > 1$. The case (A), where $\lambda = 1$, has been studied in [H] and is rather simple. It is shown that the maximum number of zeros is 1 for this case. For $\lambda > 1$, we state:

**3.1 Theorem ([DL3])** *If we integrate the 1-form $(x^2 + \beta x + \alpha)ydx$ over the compact level curves of the Hamiltonians (3.7) with $\lambda > 1$, then for all constants $\alpha$ and $\beta$ the maximum number of zeros is two, taking into account the multiplicity. The bifurcation diagram of the zeros is as represented in Fig. 10.*

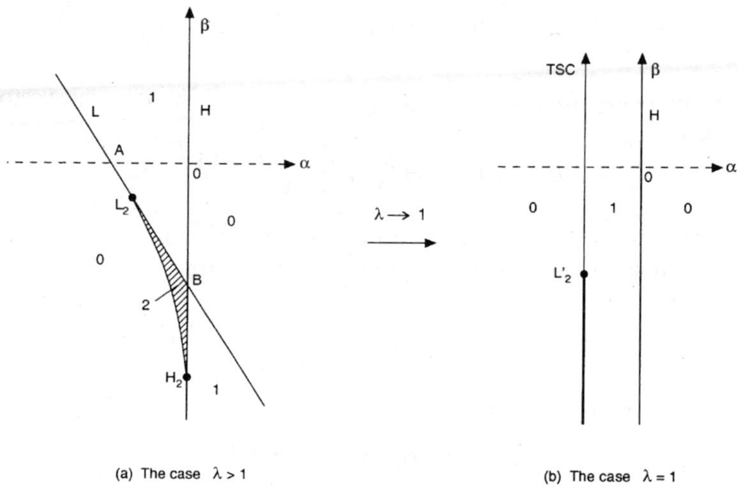


Figure 10: Bifurcation diagrams of zeros of Abelian integrals in cases (A) and (B)

In Fig. 10 we also show how the bifurcation diagram for case (B), represented in part (a) of the figure, degenerates into the bifurcation diagram of case (A), represented in part (b) of the figure.

### 3.5.2 Case (D)

A linear change of coordinates will lead to the Hamiltonian

$$H(x, y) = \frac{y^2}{2} + \frac{x^3}{3} + \frac{x^4}{4}. \tag{3.8}$$

The associated Hamiltonian vector field

$$X_H : \begin{cases} \dot{x} = y \\ \dot{y} = -x^2(1 + x) \end{cases}$$

has a non-degenerate center at $(-1, 0)$, a cusp point at $(0, 0)$ and a cuspidal loop $\Gamma_0$. Inside and outside of $\Gamma_0$, all orbits $\Gamma_h = H^{-1}(h)$ are closed, for $h \in (-1/12, 0) \cup (0, +\infty)$.

**3.2 Theorem ([DL4])** *If we integrate the 1-form $(x^2 + \beta x + \alpha)y\,dx$ over the compact level curves of the Hamiltonian (3.8), with $h \in (-1/12, \infty)$, then for all constants $\alpha$, $\beta$, the maximum number of zeros is 4, taking into account multiplicity. If we restrict to $h \in (-1/12, 0)$ (resp. $h \in ]0, \infty)$) representing closed curves inside (resp. outside) the cuspidal loop, then the maximum number of zeros is 2 (resp. 3). The bifurcation diagram of the different zeros is as represented in Fig. 11.*

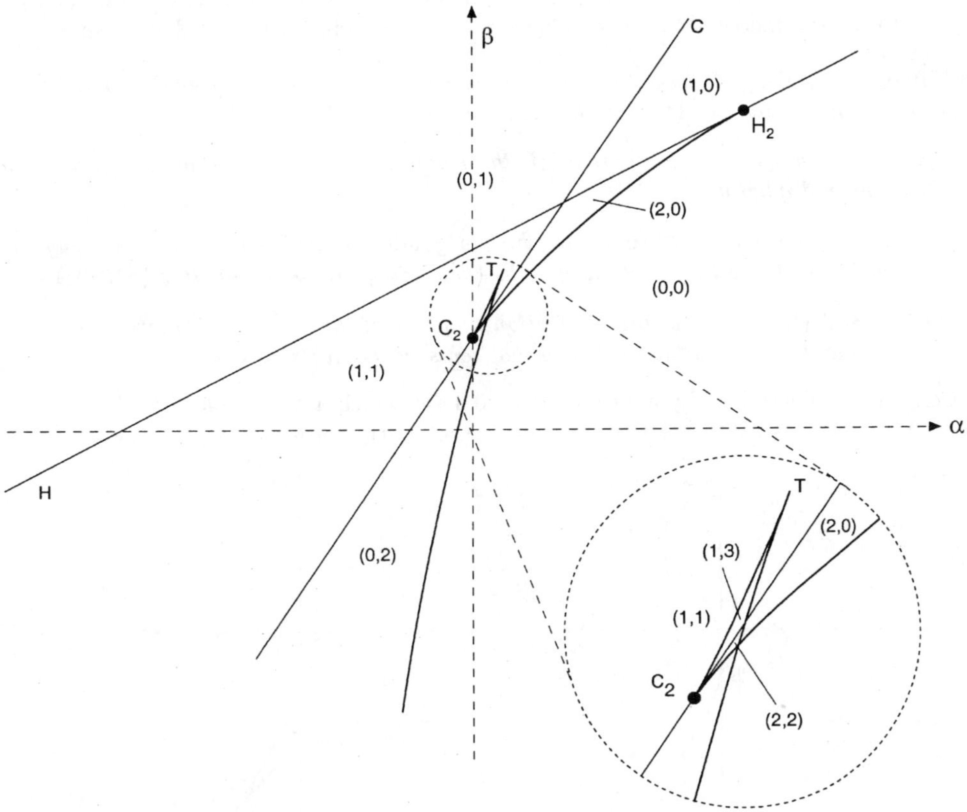


Figure 11: Bifurcation diagrams of zeros of Abelian integrals in case (D)

### 3.5.3 Case (C)

Up to a linear coordinate change, there is a 1-parameter family

$$H_\lambda(x, y) = \frac{y^2}{2} + \frac{1}{2}x^2 - \frac{2\lambda}{3}x^3 + \frac{1}{4}x^4 \tag{3.9}$$

of Hamiltonians, with $\lambda \in [0,1)$, representing a global center. For $\lambda \sim 1$ the family tends in a regular way to the cuspidal loop case. In this form, the perturbed family (3.6) yields

$$\begin{cases} \dot{x} = y \\ \dot{y} = -x(x^2 - 2\lambda x + 1) + \delta(\alpha + \beta x + x^2)y, \end{cases} \tag{3.10}$$

for some $\alpha, \beta \in \mathbf{R}$, some $\delta \sim 0$ (but positive) and $\lambda \in [0,1)$.

All orbits of (3.10) are closed, surrounding the center $(0,0)$. The system is symmetric for $\lambda = 0$. If $\lambda = 1$, then (3.9) has a cuspidal loop (case (D)) for $\delta = 0$. For $\lambda = 0$ it is easy to see, due to the symmetry, that one can have at most one zero for the Abelian integral.

**3.3 Theorem ([DL5])** *If we integrate the 1-form $(x^2 + \beta x + \alpha)y\,dx$ over the compact level curves of the Hamiltonian (3.9) with $h \in (0, \infty)$ and $\lambda \in (0,1)$ then:*

(1) *for all $\lambda$, and for all constants $\alpha$, $\beta$, the maximum number of zeros is 4, taking into account multiplicity.*

(2) *There exists some $\lambda^* \in (0,1)$ at which a quadruple zero shows up, occurring in a complete swallowtail-bifurcation of zeros (depending on the parameters $(\alpha, \beta, \lambda)$).*

(3) *For $\lambda \sim 0$ (resp. $\lambda \sim 1$) the bifurcation diagram of the zeros is as represented in the figure, the digit indicating the number of zeros in the different open regions.*

Conjecture: There is only one value $\lambda^* \in (0,1)$ at which a quadruple zero shows up; for $\lambda \in (0, \lambda^*)$ (resp. $\lambda \in (\lambda^*, 1)$) the bifurcation diagram of the zeros is as represented in Fig. 12.

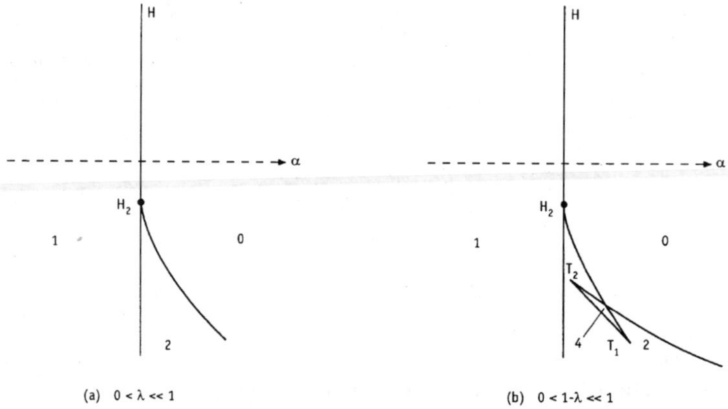


Figure 12: Bifurcation diagrams of zeros of Abelian integrals in case (C)

### 3.5.4 Case (E)

We can bring the system (3.6) in the form

$$\begin{cases} \dot{x} = y \\ \dot{y} = -x(x+1)(x-\lambda) + \delta(\alpha + \beta x + x^2)y, \end{cases}$$

with $\alpha, \beta \in \mathbf{R}$ and $\delta$ small. The corresponding Hamiltonian for $\delta = 0$ is given by the 1-parameter family

$$H_\lambda(x,y) = \frac{1}{2}y^2 - \frac{\lambda}{2}x^2 + \frac{1-\lambda}{3}x^3 + \frac{1}{4}x^4 \tag{3.11}$$

with $\lambda \in (0,1]$. If $\lambda \to 0$, then the vector field tends to the cuspidal loop case (D). If $\lambda = 1$ we get a symmetric case which was studied in [GH] and [CLW]. In the latter case the possible configurations of zeros, and the related bifurcation diagram with respect to $(\alpha, \beta)$, are represented in Fig. 15. $H_L$ (resp. $H_R$) stands for a zero at the left (resp. right) antisaddle, $S_L$ (resp. $S_R$) stands for a zero at the left (resp. right) saddle loop, $S_A$ (resp. $S_B$) stands for a zero that will give rise to a saddle connection above (resp. below) the saddle; the other curves represent double zeros.

**3.4 Theorem ([DL6])** *Let us consider Hamiltonians of the form* (3.11) *and the related elliptic integrals $I^{(i)}(h)$, for $i = 1,2,3$, obtained by integrating the 1-forms $(x^2 + \beta x + \alpha)ydx$ over connected components of level curves of $H_\lambda$, that surround respectively the left center ($i = 1$), the right center ($i = 2$) or all singularities ($i = 3$). Let $n_i$ denote the number of zeros of $I^{(i)}(h)$ for $i = 1,2,3$, depending on $(\alpha, \beta)$. Following statements hold:*

(1) *For all $\lambda \in (0,1)$ and for all $(\alpha, \beta)$, $n_1 + n_2 \le 2$, $n_3 \le 4$, $n_1 + n_2 + n_3 \le 5$.*

(2) *For $0 < \lambda \ll 1$ (resp. $0 < 1 - \lambda \ll 1$), the possible configurations $((n_1, n_2), n_3)$ and the related bifurcation diagram are as represented in Fig.* 13 *(resp. Fig.* 14*).*

(3) *For all $\lambda \in (0,1)$, the possible configurations $(n_1, n_2)$ and the related bifurcation diagram of the zeros $I^{(1)}$ and $I^{(2)}$ are as represented in Fig.* 13 *(and Fig.* 14*).*

(4) *There exists a value $\lambda^* \in (0,1)$, where $\lambda^* \sim 0.0549$, such that for $\lambda \in (0, \lambda^*)$ the bifurcation diagram of the zeros of $I^{(3)}(h)$ is as represented in Fig.* 16(a)*, and for $0 < \lambda - \lambda^* \ll 1$, the bifurcation diagram of the zeros of $I^{(3)}(h)$ is as represented in Fig.* 16(b). *For $(\alpha, \beta)$ in the triangular region where $I^{(3)}(h)$ has four zeros, the related $(n_1, n_2) = (1,0)$, implying the existence of a $((1,0),4)$-configuration of zeros.*

(5) *There exists a value $\lambda^0 \in (\lambda^*, 1)$ such that for $0 < \lambda^0 - \lambda \ll 1$ (resp. $0 < \lambda - \lambda^0 \ll 1$) the bifurcation diagram of the zeros of $I^{(3)}(h)$ is as represented in Fig.* 16(c) *(resp. Fig.* 16(d)*).*

# 4 Singular perturbations

The last section deals with singular perturbations and is meant to be a continuation of Chapter 5 in [D]. The problems we will deal with are planar, and our approach is of a "geometric singular perturbation" nature (see [Fe, J]). In these papers the study however is restricted to the "normally hyperbolic" situation, while, as in [D], we will essentially focus on the non-hyperbolic case.

Our main attention will still go to the "canard"-phenomenom in dimension 2, i.e. the possibility of passing from a "stable branch" of the slow manifold to an "unstable" one, when crossing a "turning"-point. In [D] this phenomenom has been presented for the van der Pol system

$$\varepsilon\ddot{x} + (x^2 + x)\dot{x} + x - a = 0.$$

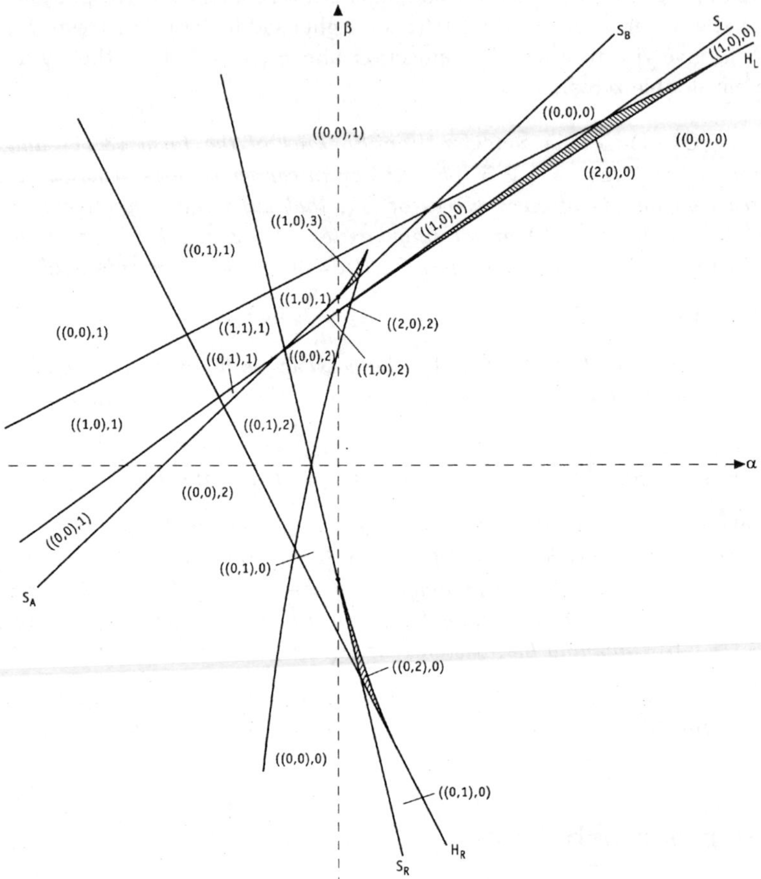


Figure 13: Case (E): Bifurcation diagram of zeros for $0 < \lambda \ll 1$

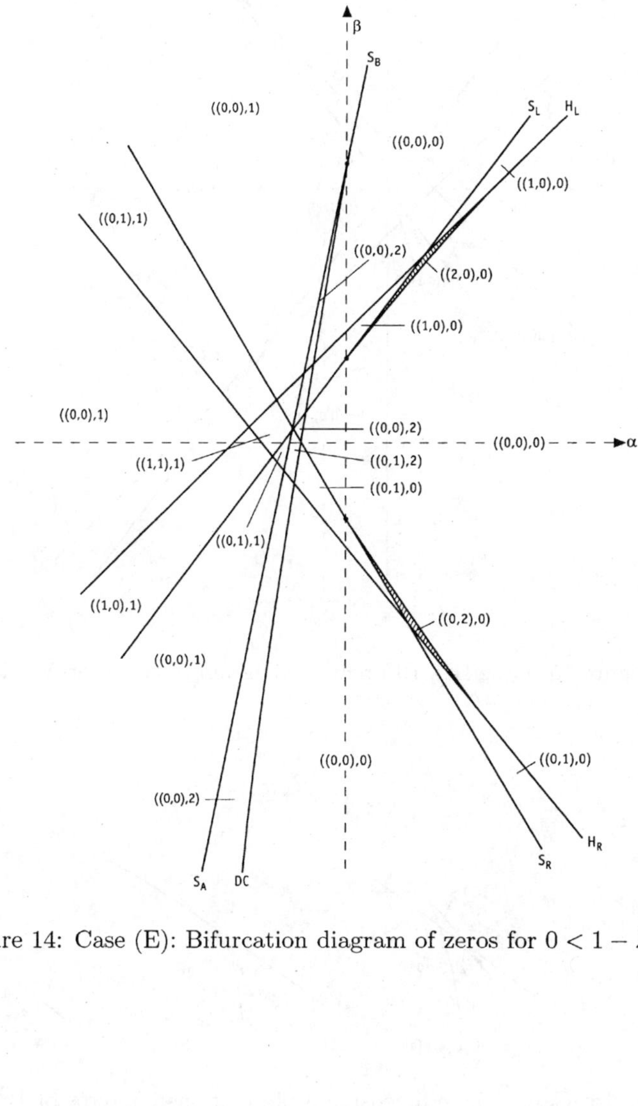


Figure 14: Case (E): Bifurcation diagram of zeros for $0 < 1 - \lambda \ll 1$

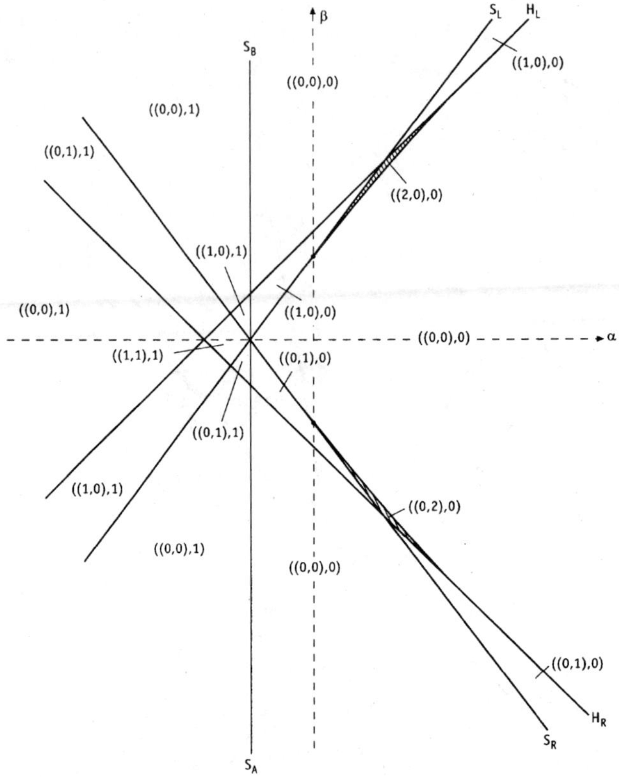


Figure 15: Case (E). Bifurcation diagram of zeros for $\lambda = 1$

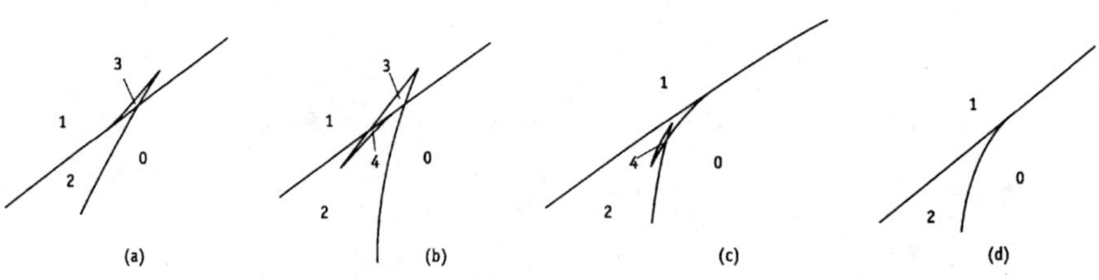


Figure 16: Case (E): codimension 3-bifurcations for zeros of $I^{(3)}(h)$

We now want to treat it in a more general context. In fact, we permit the phenomenom to occur on a 2-dimensional manifold. We will also not require the turning point to be "generic" as it is in the van der Pol case.

Section 4.1 will essentially be dedicated to the local study near such (general) turning points, while in Section 4.2 the turning point is again supposed to be "generic" but attention goes to a thorough study of multiple limit cycle bifurcations in such a singular perturbation. Because of its intrinsic importance and of its relation with Section 3.5 we will put the emphasis on applications to Liénard equations. Again, one of the main techniques is the (family) blow-up as already used in previous sections. In combination with the use of center manifolds it provides a firm base for a geometric explanation (based on the usual techniques from a dynamical systems approach) of the canard-phenomenom, including the fact that it occurs in exponentially small regions in parameter space. Again we will restrict to $C^\infty$ systems, not paying attention to extra properties in the analytic case, and not trying to generalize to $C^r$ situations with $r \in \mathbf{N}_1$.

## 4.1 Canard solutions near general turning points

### 4.1.1 Introduction

In this section, we describe a general framework in which the existence of canard solutions for singularly perturbed vector fields can be proved. To roughly state the ideas, consider the model vector field

$$X_{\varepsilon,a}\colon \begin{cases} \dot{x} = y - \dfrac{x^{2n}}{2n} + \varepsilon F(x,y,\varepsilon) \\ \dot{y} = \varepsilon(a - x^{2n-1}) + \varepsilon^2 G(x,y,\varepsilon) \end{cases}$$

with $\varepsilon$ small but positive, $a$ small and with $n$ an integer $\geq 2$. At $\varepsilon = 0$ the vector field degenerates in the sense that a whole curve $\gamma$ of singularities $y = x^{2n}/2n$ appears. The linear part of $X_{0,a}$ is:

$$DX_{0,a} = \begin{pmatrix} -x^{2n-1} & 1 \\ 0 & 0 \end{pmatrix}.$$

Apparently, at all points of $\gamma$ except at the origin the vector field $X_{0,a}$ is partially hyperbolic. One says that the vector field $X_{0,a}$ behaves *normally hyperbolic* at those points, and the center manifold is given by the curve of singularities $\gamma$.

A result of Fenichel [Fe] describes the dynamics near normally hyperbolic singularities of singularly perturbed vector fields. It is based on the center manifold reduction and use of appropiate normal forms. In the following sections, we repeat the result restricted to our specific 2-dimensional situation but giving a more detailed description of the $C^\infty$ character of the "slow manifolds".

The origin in the example is what is called a *turning point* (or alternatively, a contact point). At the origin, $X_{0,a}$ is no longer partially hyperbolic. A *canard phenomenom* occurs if for $\varepsilon \neq 0$ one encounters orbits that are close to both the stable and the unstable branch of the critical curve.

In this section, we describe a general framework in which the existence of canard solutions can be proved even if the turning point is non-generic, i.e. is a turning point with a higher but finite order of tangency. The result is not only valid on arbitrary 2-manifolds, but is also global in the sense that the constructed canard manifolds can satisfy arbitrary boundary conditions, provided some regularity assumptions are satisfied in these boundary conditions.

Finally, let us recall that in the literature, canard solutions are often called "*overstable solutions*". We prefer to make a distinction: in the above example, taking $F$ and $G$ analytic, the curve $\gamma$ can be complexified to form a curve over $\mathbf{C}$. An overstable solution is then a canard solution that can be complexified and that stays close to $\gamma$ in a full complex neighbourhood of the turning point, and not only along the real rays. In the complex setting, one does not speak about transition from attracting to repelling, instead in a full pinced neighbourhood of the origin, the nonzero eigenvalue $-x^{2n-1}$ is repelling in $2n-1$ complex sectors of the origin (these sectors are called mountains), and attracting in $2n-1$ complex sectors of the origin. For a complex orbit to stay near the curve $\gamma$ throughout all the mountain sectors is a far more restrictive condition than just demanding the orbit to stay near $\gamma$ on the real rays. This explains why canard solutions appear in significantly larger domains than overstable solutions. In the literature, some very interesting results have been proved with respect to overstable solutions, and a very complete local result on non-generic turning points can be found in [CRSS] using Gevrey-asymptotics techniques. We can also refer to [BFSW] for a number of results using non-standard analysis.

### 4.1.2 Non-generic turning points

We first give a description of what is called an "admissible turning point". The idea is to make the definition general enough to allow non-generic turning points, but to keep some constraints in order for the geometric techniques to work.

Let $X_\varepsilon$ be a $C^\infty$ family of vector fields on a 2-dimensional manifold $M$. Then $X_\varepsilon + 0\frac{\partial}{\partial\varepsilon}$ is a vector field on $M \times [0, \varepsilon_0[$.

**4.1 Definition** A *critical curve* of $X_\varepsilon$ is a curve of singularities of the vector field $X_0$, seen as a curve inside $M \times [0, \varepsilon_0[$. If the vector field depends on other parameters $\lambda$ we will still call it—by abuse of language—a critical curve.

In our study we will also use a special parameter—denoted by $a$, keeping $a \sim 0$—that will essentially be a parameter breaking the critical manifold in a regular way; a precise definition will follow. So we consider families $X_{\varepsilon,a,\lambda}$ on $M$, and the (trivial $\lambda$-family of) critical curve(s) is contained in $(a, \varepsilon) = (0, 0)$.

If $M = \mathbf{R}^2$, then one can write down corresponding differential equations, parametrizing $M$ by $(x, y)$:

$$X_{\varepsilon,a,\lambda} : \begin{cases} \dot{x} = f(x, y, \varepsilon, a, \lambda) \\ \dot{y} = \varepsilon g(x, y, \varepsilon, a, \lambda) \end{cases} . \tag{4.1}$$

In this notation, we assume that $\varepsilon \in [0, \varepsilon_0[$ for some small $\varepsilon_0 \in \mathbf{R}^+$, $\lambda \in \Lambda \subset \mathbf{R}^p$ is a $p$-dimensional parameter that is non-essential in our study. We assume that $a \in [-a_0, +a_0]$ is a parameter in a small compact interval, and that $f$ and $g$ are $C^\infty$ functions.

In fact, (4.1) is a representation of the singular perturbation problem in the so-called "fast time" $t$. We make a study near zeros of vector fields and the regular orbits describe the "fast movements". To study the "slow movements" one often prefers to rescale time to the so-called "slow time" $\tau = \varepsilon t$, changing (4.1) into

$$\begin{cases} \varepsilon \dot{x} = f(x, y, \varepsilon, a, \lambda) \\ \dot{y} = g(x, y, \varepsilon, a, \lambda) \end{cases} . \tag{4.2}$$

The critical curve is described by $f(x, y, 0, 0, \lambda) = 0$, and can be expressed as $y = F(x, 0, 0, \lambda)$ if, for example, we require $\frac{\partial f}{\partial y}(x, y, 0, 0, \lambda) \neq 0$ in the region under consideration. Important information on the slow movements is contained in the so-called "slow vector field" that we get by substituting $y = F(x, 0, 0, \lambda)$ into the second component of (4.2):

$$\frac{\partial F}{\partial x}(x, 0, 0, \lambda)\dot{x} = g(x, F(x, 0, 0, \lambda), 0, 0, \lambda).$$

We can also consider an equation which is equivalent to it outside the zeros of $\frac{\partial F}{\partial x}$ and possibly up to time reversal:

$$\dot{x} = \frac{\partial F}{\partial x}(x, 0, 0, \lambda)g(x, F(x, 0, 0, \lambda), 0, 0, \lambda).$$

It governs the "slow movements" at points where $g$ and $\frac{\partial F}{\partial x}$ are both different from zero. In other situations a more delicate elaboration is required. In this section our main concern will be focused on this relation between "slow and fast movements".

### 4.1.3 General assumptions (Existence of a critical curve)

The kind of vector fields that are of interest to us in this section have the following properties.

Assume that $X_{0,0,\lambda}$ has a curve of singularities $\gamma$. This curve may depend on $\lambda$, but we will not keep this dependence in the notation of $\gamma$. We require $\gamma = \gamma_\lambda$ to be a trivial $\lambda$-family of simple curves. We assume $\gamma$ contains a point $p_\gamma$ such that

- $p_\gamma$ divides $\gamma$ into two branches $\gamma_-$ and $\gamma_+$;
- $\gamma_- \cup \{p_\gamma\}$ and $\gamma_+ \cup \{p_\gamma\}$ are simple critical curves;
- $\gamma$ is oriented in such a way that $\gamma_+$ comes after $\gamma_-$;
- $X_0$ is normally attracting at all points of $\gamma_-$ and normally repelling at all points of $\gamma_+$.

The point $p_\gamma$ will be called a *turning point.*

**Assumption 1 (Admissible chart)** There is a (possibly $\lambda$-dependent) chart of $M$ in the neighbourhood of $p_\gamma$ so that in this chart $X_{\varepsilon,a,\lambda}$ has an expression like in (4.1), with the origin as turning point. With the notation as in (4.1), we assume

(1) $\gamma \subseteq \{\, (x, y)\ f(x, y, 0, 0, \lambda) = 0 \,\}$ is a graph over the real $x$-axis;

(2) the origin divides $\gamma$ in two pieces $\gamma_-$ and $\gamma_+$;

(3) along $\gamma \setminus \{(0, 0)\}$ both $\frac{\partial f}{\partial x}$ and $\frac{\partial f}{\partial y}$ are nonzero (at $\varepsilon = a = 0$);

(4) $\gamma_-$ is a graph over the negative real $x$-axis, and $\gamma_+$ is a graph over the positive real $x$-axis.

The main difficulty in the proof of the existence of canard solutions is of course the passage along the turning point $p_\gamma$. Nevertheless, one needs to study the passage along the normally hyperbolic branches of the critical curve $\gamma$ as well. The theory of Fenichel [Fe] can be used to obtain some results, but stronger results can be proved. In particular, we can give precise information on regularity aspects of "slow manifolds" and on their boundary conditions. Before stating such results let us impose a number of conditions.

Let $p$ be a point of $\gamma_-$ (or equally of $\gamma_+$), and consider the vector field $X = X_0 + 0\frac{\partial}{\partial\varepsilon}$ on $M \times [0, \varepsilon_0[$. Because the vector field $X_{0,0,\lambda}$ is normally attracting at $p$, there exists at least a 2-dimensional $C^1$ center manifold $W_p$. Consider such a center manifold $W_p$ (manifold with boundary if we keep $\varepsilon \in [0, \varepsilon_0[$).

Essentially, we want the dynamics on $W_p$ to be topologically equivalent to the dynamics of a model differential equation

$$\varepsilon\frac{\partial}{\partial x} + 0\frac{\partial}{\partial \varepsilon} \tag{4.3}$$

(so that for $\varepsilon = 0$ there is no passage, and for $\varepsilon > 0$ there is a regular flow from left to right in the $(x, \varepsilon)$-plane. More precisely:

There exists a $C^1$-embedding $\varphi\colon [0,1]^2 \to W_p\colon (x,\varepsilon) \mapsto \varphi(x,\varepsilon)$ so that

(1) $\varphi([0,1]^2)$ is a neighbourhood of $p$ inside $W_p$;

(2) $\varphi([0,1] \times \{0\}) \subset \{\varepsilon = 0\}$;

(3) $\varphi|_{\varepsilon=0}$ is orientation-preserving for the standard orientation on the $x$-axis and the chosen orientation on $\gamma$;

(4) $\varphi$ is a topological equivalence between $X|_{W_p}$ and the model vector field (4.3).

To ensure that this last assumption is satisfied, it suffices to assume the following:

**Assumption 2 (Normally hyperbolic passage)** For any point $p$ of $\gamma_- \cup \gamma_+$, there exists a sufficiently small neighbourhood $V$ of $p$ in $M \times [0, \varepsilon_0[$ so that in $V$ there are no singularities for $\varepsilon > 0$. Furthermore, inside center manifolds in $V$, the orientation of the orbits for $\varepsilon > 0$ must be compatible with the orientation on $\gamma$.

Our aim is now to prove the existence of $C^\infty$ invariant manifolds satisfying some specified boundary condition. The boundary condition is typically a smooth $\varepsilon$-graph, and the invariant manifold is obtained as a union of orbits through points of this boundary curve. The issue is the regularity of this construction as $\varepsilon \to 0$. To that end, let us introduce two definitions:

**4.2 Definition** Let $\Sigma$ be a smooth curve in $M \times [0, \varepsilon_0[$, possibly depending on some extra parameters. Assume that this curve is a $C^0$-graph in $\varepsilon \geq 0$. The *saturation* of $\Sigma$ is defined as the topological closure of the union of all orbits (w.r.t. the extended vector field $X_{\varepsilon,a,\lambda} + 0\frac{\partial}{\partial\varepsilon}$) of points of $\Sigma$. The need to take a topological closure becomes clear if one considers the limit point of $\Sigma$ as $\varepsilon \to 0$: the orbits become singular for this limit point.

**4.3 Definition** An *admissible entry boundary curve* $\Sigma_-$ in $M \times [0, \varepsilon_0[$-space (possibly $(a, \lambda)$-dependent) is a curve $(m(\varepsilon), \varepsilon)$ for which the "base point" $(m(0), 0)$ lies in the basin of attraction of $\gamma_-$. Furthermore, we assume that in any chart, $m$ is $C^\infty$ for $\varepsilon > 0$, and

$$\forall n \in \mathbf{N},\ \exists N \in \mathbf{N},\ \frac{\partial^n}{\partial \varepsilon^n} s_-(\varepsilon, a, \lambda) = O(\varepsilon^{-N}) \quad \text{as} \quad \varepsilon \to 0.$$

Similarly we define an *admissible exit boundary curve.*

The above definition allows $C^\infty$ curves $m$ to be chosen as entry boundary curves, but also curves that are parametrized by $\log \varepsilon$, $\varepsilon^{1/m}$, etc. In particular, entry boundary curves can have any finite order of contact with the manifold $M \times \{0\}$.

**4.4 Theorem** *Let Assumptions* 1 *and* 2 *be satisfied, and let* $\Sigma_-$ *be an admissible entry boundary curve. Through saturation in positive time one obtains an invariant manifold with boundary* $W_-$ *containing* $\Sigma_-$. *One has* $\partial W_- = F \cup S$, *where* $F$ *is an orbit of the fast flow from the base point* $\Sigma_- \cap \{\varepsilon = 0\}$ *up to and including its* $\omega$*-limit on* $\gamma_-$, *which we define as the corner point* $c_-$, *and where* $S$ *is a compact piece of the critical curve* $\gamma_-$, *from* $c_-$ *to any point* $p$ *(that comes after* $c_-$ *in the chosen orientation of* $\gamma$, *but before the turning point* $p_\gamma$*). Outside the boundary,* $W_-$ *is* $C^\infty$, *and* $W_-$ *is also* $C^\infty$ *along* $S \setminus \{c_-\}$. *Along* $F$, *the manifold is continuous.*

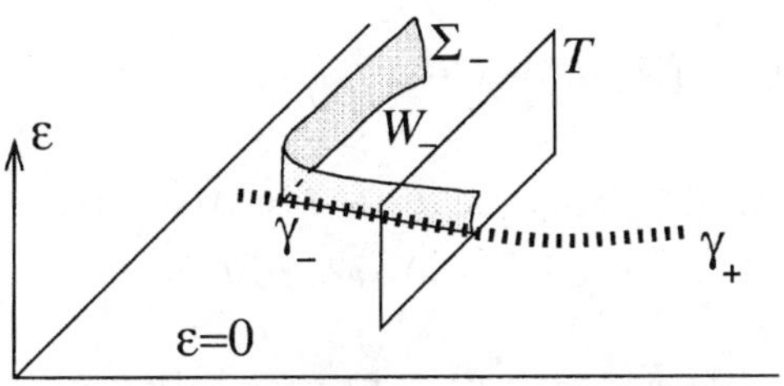


Figure 17: Saturation of an entry boundary curve along normally hyperbolic part of the critical curve

**Idea of the proof** First, observe that $C^\infty$ smoothness outside $\varepsilon = 0$ is trivial (there, the vector field is equivalent to a flow-box). Hence, if one proves that at all points $p$ of $S \setminus \{c_-\}$ the manifold is $C^k$-smooth, and this for all $k \in \mathbf{N}$, then the $C^\infty$ smoothness follows. This observation allows us to use $C^k$-normal forms, that could be obtained using the normal form theory of Takens, but also by using a result of Bonckaert [Bo], that we will recall below. Using these normal forms, one proves that the saturation of an initial boundary curve is of the form

$$y = y_0(\varepsilon, \lambda) \exp\left(-\int_{x_0(\varepsilon,\lambda)}^{x} \frac{ds}{g(s,\varepsilon,\lambda)}\right),$$

for some $C^k$-function $g$ that is positive, but zero at $\varepsilon = 0$. An analysis of this kind of expressions shows that the result is $C^k$ (even if $(x_0, y_0)$ only satisfy the mild conditions on boundary curves), as long as the piece of the critical curve $[x_0(0,\lambda), x] \subset \gamma$ is a $(C^0)$ Hausdorff limit of a regular orbit. This means that in the parameter space $[0, \varepsilon_0[\times \Lambda$, smoothness is obtained at all points where a passage is possible in the limit. □

**4.5 Remark** The intersection of a center manifold (of the critical manifold) with $\{\varepsilon = \varepsilon_0\}$, for small $\varepsilon_0 > 0$, is called a *slow manifold.* A slow manifold is not unique but different slow manifolds near the same compact piece of the critical manifold are very close to each other.

Let us now recall the normal form theorem that has been proved in [Bo], since it will also be of importance at a later stage.

**4.6 Theorem ([Bo])** *Let $X_\lambda(x, y, z, a)$ be a $C^\infty$ family of vector fields on $\mathbf{R}^4$ having the following properties:*

(1) *$X_\lambda$ is tangent to the foliation $da = 0$.*

(2) *$X_\lambda$ is tangent to the foliation $dF(y, z) = 0$ where $F(y, z) = y^p z^q$ for $(p, q) = (0, 1)$ or $p, q \in \mathbf{N}_0$ and relatively prime.*

(3) *$DX_\lambda(0, 0, 0, 0)$ has exactly one non-zero eigenvalue and the related eigenspace is given by $y = z = a = 0$.*

*Let $W$ be a $C^k$ center manifold of $X_\lambda$ at $(0, 0, 0, 0)$ with $k \in \mathbf{N}_0$.*

*Then there exists a local $C^k$ change of coordinates $\varphi$ of the form*

$$(x, y, z, a) \mapsto (\varphi_1(x, y, z, a, \lambda), \varphi_2(x, y, z, a, \lambda), \varphi_3(x, y, z, a, \lambda), a)$$

*with*

$$F(\varphi_2(x, y, z, a, \lambda), \varphi_3(x, y, z, a, \lambda)) = F(y, z),$$

*and a strictly positive $C^k$ function $h(x, y, z, a)$ such that*

$$[h.\varphi_* X](x, y, z, a) = \pm x \frac{\partial}{\partial x} + Y_\lambda(y, z, a),$$

*with $Y_\lambda$ of class $C^k$, $Y.a = 0$, $Y.F = 0$ and $\varphi(W) = \{x = 0\}$.*

### 4.1.4 Passing through turning points

Using Theorem 4.4, the passage along normally hyperbolic singular points can be studied. In order to study the passage near a turning point, we propose a study by means of family blow up. Some conditions are put in terms of the resulting blown up vector field. However, for an important class of singularly perturbed vector fields, the conditions are satisfied and simple assumptions are written down in a self-contained form (see Theorems 4.12 and 4.13).

If we have a 3-dimensional family of vector fields $X_{\varepsilon,\lambda} + 0\frac{\partial}{\partial \varepsilon}$, and we want to blow up the origin $(x, y, \varepsilon) = (0, 0, 0)$, then we use:

$$\begin{cases} x = u^p \bar{x} \\ y = u^q \bar{y} \\ \varepsilon = u^m \bar{\varepsilon} \end{cases} \tag{4.4}$$

with $(\bar{x}, \bar{y}, \bar{\varepsilon}) \in S^2$, $u \in \mathbf{R}^+$, and $p$, $q$, $m$ suitably chosen positive integers.

On the 2-sphere, we have the relation $\bar{x}^2 + \bar{y}^2 + \bar{\varepsilon}^2 = 1$, often implying the need of working in charts.

Suppose we want to look near the critical curve in these new coordinates, hence in the region $\varepsilon \sim 0$. By Assumption 1, in a well-chosen chart, $\gamma_-$ is a graph over the negative real axis, and hence $\gamma_-$ would be visible in the chart

$$\begin{cases} x = -u^p \\ y = u^q \overline{y} \\ \varepsilon = u^m \overline{\varepsilon} \end{cases} . \tag{4.5}$$

This chart is valid for $(\overline{y}, \overline{\varepsilon})$ small, and is called a *phase-directional rescaling chart*, or simply *the* $\{\overline{x} = -1\}$ *chart.* Notice that

$$d\varepsilon = 0 \quad \text{is replaced by} \quad d(u^m \overline{\varepsilon}) = 0.$$

With properly chosen weights the preimage of $\gamma_-$ under the blow up map has a limit for $u \to 0$. This is the point $P_-$.

**Assumption 3 (Regularity condition)** After blowing up at $(0,0,0)$ the $(x,y,\varepsilon)$-variables —in an admissible chart—of $X_{\varepsilon,0,\lambda}$, we get the following:

The preimages of $\gamma_-$ and $\gamma_+$ in the blow up space (including the endpoints of $\gamma_\pm$ on the blow up locus) are normally hyperbolic.

Define $P_\pm = \gamma_\pm \cap \Sigma$, where $\Sigma$ is the blow up locus, i.e. the preimage of $(x,y,\varepsilon) = (0,0,0)$ under the blow up map.

**Important remark** The regularity condition is restrictive in the sense that we require the blow up weights to be independent of $\lambda$, implying that the order of degeneracy of the critical curve must also not depend on $\lambda$.

We have in mind for example quadratic contacts where the second order angle changes:

$$\gamma \colon y = \lambda x^2 + O(x^3), \qquad \lambda > \lambda_0 > 0,$$

or contacts "of order $2n$", with a fixed $n$, i.e.

$$\gamma \colon y = \lambda x^{2n} + O(x^{2n+1}), \qquad \lambda > \lambda_0 > 0.$$

Again, the normal hyperbolicity at $p = P_-$ implies the existence of a 2-dimensional $C^1$ center manifold $W_p$ containing $P_-$. If we choose such a manifold $W_p$ near $P_-$, then we essentially want the vector field on $W_p$ to be $C^0$-equivalent to

$$-u\overline{\varepsilon}\frac{\partial}{\partial u} + \overline{\varepsilon}^2 \frac{\partial}{\partial \overline{\varepsilon}}. \tag{4.6}$$

More precisely:

Let $p = P_-$ be the endpoint of $\gamma_-$, in a blown up admissible chart, and let $W_p$ be a center manifold of $X$ at $p$. We require the existence of a $C^1$ embedding $\varphi \colon [0,1]^2 \to W_p \colon (u, \overline{\varepsilon}) \mapsto \varphi(u, \overline{\varepsilon})$ so that

(1) $\varphi([0,1]^2)$ is a neighbourhood of $P_-$ inside $W_p$;

(2) $\varphi([0,1] \times \{0\}) \subset \gamma_- \cup \{P_-\}$;

(3) $\varphi(\{0\} \times [0,1])$ is inside the blow up locus;

(4) $\varphi|_{[0,1]\times\{0\}}$ is orientation-preserving for the standard orientation on the $u$-axis and the negative orientation on $\gamma$;

(5) $\varphi$ is a topological equivalence between $X|_{W_p}$ and the model vector field (4.6).

This model behaviour will be ensured by Assumption 2 and the next assumption:

**Assumption 4 (Regular corner passage))** Let $p = P_-$ be the endpoint of $\gamma_-$ in a blown up admissible chart. Then there exists a sufficiently small neighbourhood $V$ of $p$ so that in $V$ there are no singularities for $\varepsilon = u^m\overline{\varepsilon} > 0$. A similar requirement is made for the endpoint $P_+$ of $\gamma_+$.

Assumption 4 implies the existence of a unique unstable separatrix of $P_-$ on the blow up locus $\{u = 0\}$ (separatrix between two hyperbolic sectors), and similarly the existence of a unique stable separatrix of $P_+$ on the blow up locus is implied. For a passage from orbits near $P_-$ to orbits near $P_+$ to be possible, the two separatrices must coincide. This will be visible in the family rescaling chart: continuing the orbits near $P_-$, $\overline{\varepsilon}$ keeps increasing, so that the orbits leave the chart $\{\overline{x} = -1\}$. We continue in "the chart $\{\overline{\varepsilon} = 1\}$", commonly known as the *family rescaling chart*:

$$\begin{cases} x = u^p\overline{x} \\ y = u^q\overline{y} \\ \varepsilon = u^m \end{cases} . \tag{4.7}$$

This chart is valid for $(\overline{x}, \overline{y})$ in a bounded set. Notice that

$$d\varepsilon = 0 \quad \text{is replaced by} \quad du = 0$$

This is the traditional chart where people do "rescaling".

**Assumption 5 (Connection condition)** Under the conditions of Assumption 3, there is a heteroclinic connection $\overline{\gamma}$ on the blow up locus $\Sigma$ connecting $P_-$ to $P_+$.

We assume that this connection consists of one orbit going from $P_-$ to $P_+$.

These conditions will suffice to saturate initial boundary curves to pass along the normally attracting singularities of $\gamma_-$, through the corner point $P_-$ in blow up space, and to pass along a compact piece of the regular connection $\overline{\gamma}$. Similarly, an initial exit boundary curve can be saturated in negative time to pass along points of $\gamma_+$, the corner point $P_+$ and compact pieces of the regular connection. In comparing the saturation of an entry boundary curve with the saturation of an exit boundary curve, we will use the following definition:

**4.7 Definition** A *center manifold* for the system (4.1) is the saturation of a local center manifold at a normally hyperbolic point on the critical curve, which may be obtained as the saturation of an entry/exit boundary curve in the basin of attraction/repulsion of the critical curve $\gamma$. The saturation defines an invariant manifold up to the turning point, or more specifically, up to a part of the blow up locus. We distinguish two classes of center manifolds:

- center manifolds along $\gamma_-$ (forward center manifolds, since they are obtained through saturation in positive time);

- center manifolds along $\gamma_+$ (backward center manifolds, since they are obtained through saturation in negative time).

Remark: these center manifolds may depend regularly on $(a, \lambda)$. (Nevertheless, at this stage, we have kept $a = 0$ in the blow up construction.)

One can compare forward and backward center manifolds by choosing a section in the family rescaling chart, transversally cutting the blow up locus. More precisely, in the family rescaling chart we can, for example, take a section $T$: $\{\overline{x} = 0\}$ and look at the intersection of the forward and backward center manifolds with $T$ (see Fig. 18). Choosing a coordinate $z$ on $\sigma := T \cap \{u = 0\}$, we can use $(z, u)$ as coordinates on $T$.

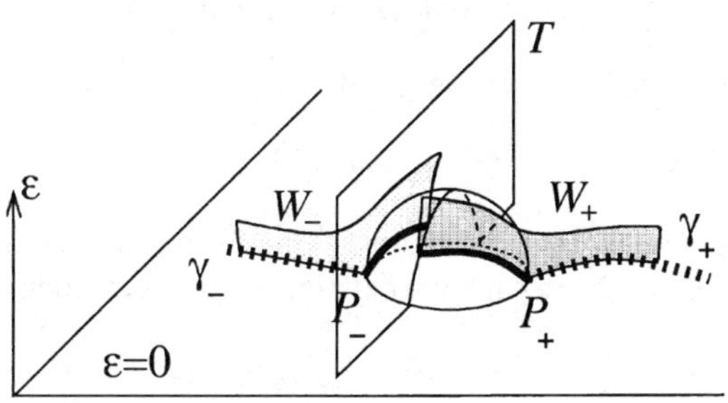


Figure 18: Forward and backward center manifolds intersecting a section $T$

The intersection of a pair of forward and backward manifolds with $T$ are hence graphs

$$z = \zeta_\pm(u, 0, \lambda). \tag{4.8}$$

Remember that at this stage we have kept $a = 0$ during the blow up construction and during the saturation of the manifolds. Of course, these graphs depend on the choice of center manifolds, but one proves that its $\infty$-jets are unique.

**Observation** In general, the two graphs *do not coincide*, instead, an angle appears between the two curves. Forgetting the presence of the special parameter $a$, one can calculate the 1-jet of $\zeta_+(u, 0, \lambda) - \zeta_-(u, 0, \lambda)$ by means of a Melnikov integral over the heteroclinic connection $\overline{\gamma}$, the generic case being the case where this integral is nonzero. In fact, the appearance of this angle is the reason why the parameter $a$ has been introduced. Let us explain this further. Because all backward center manifolds are infinitely tangent to $z_+(u, 0, \lambda)$, and all forward center manifolds to $z_-(u, 0, \lambda)$, the appearance of an angle (of finite order) between these two graphs implies that forward center manifolds cannot persist along the unstable branch of the critical curve $\gamma$. So one needs a condition on $(a, \varepsilon)$ to "match" the forward center manifolds to the backward center manifolds. In order to be able to apply the Implicit Function Theorem we will now impose that $a$ satisfies some condition of "regular breaking".

### 4.1.5 The role of the parameter $a$

We need to work in two steps, first starting by presenting the role of $a$ in case $\gamma$ remains a critical curve for $X_{\varepsilon,a,\lambda}$ with $a \neq 0$. In order to stress the specificity of this case, we write $a = A$. We have the following properties:

(1) The blow up can not only be performed for $A = 0$, but for all $A \sim 0$.

(2) Assumption 2 implies that the partially hyperbolic corner points $P_\pm$ persist for small $A \neq 0$.

(3) On the blow up locus the invariant separatrices at $P_-$ resp. $P_+$ persist for small $A \neq 0$.

(4) For $A = 0$ these invariant separatrices form a heteroclinic connection $\overline{\gamma}$, but for $A \neq 0$ this is not necessarily true.

(5) The graphs in (4.8) can be evaluated also for nonzero $A$.

Let us stress this last property, and define graphs

$$z = \zeta_\pm(u, A, \lambda) \tag{4.9}$$

that are obtained as the intersection of forward and backward center manifolds, this time depending regularly on $(A, \lambda)$. Note that the existence of the heteroclinic connection $\overline{\gamma}$ implies that

$$\zeta_+(0,0,\lambda) = \zeta_-(0,0,\lambda).$$

The matching of forward and backward center manifold, under a condition on $(A, \varepsilon)$ will be a consequence of the Implicit Function Theorem if we impose the following condition:

**4.8 Definition** We say that $A$ is a *regular breaking parameter* if

$$\rho(\lambda) := \frac{\partial}{\partial A}(\zeta_- - \zeta_+)(0,0,\lambda) \neq 0. \tag{4.10}$$

This definition does not depend on the choice of the regular coordinate $z$, nor on the transverse section $\sigma = T \cap \{u = 0\}$.

In order to describe the role of $a$ in the general case, we replace the family of vector fields $X_{\varepsilon,a,\lambda}$ by a subfamily $X_{\varepsilon,A\varepsilon^{n/m},\lambda}$; in this subfamily we know that the critical curve $\gamma$ of $X_{\varepsilon,0,\lambda}$ persists to a critical curve of $X_{\varepsilon,A\varepsilon^{n/m},\lambda}$ with $A \neq 0$, reducing the general case to the specific case discussed above. Remember that $m$ is the blow up weight of $\varepsilon$; the integer $n$ can be specified so as to obtain a regular breaking parameter:

**Assumption 6 (Breaking parameter)** Using the family rescaling chart expressed in (4.7), there exists some $n \in \mathbf{N}$ such that $A := a/u^n$ (hence $a = A\varepsilon^{n/m}$) is a regular breaking parameter.

In the search for canard solutions, we stay in a region in parameter space outside which no such solutions can exist. This is reflected in the rescaling

$$(a, \varepsilon) = (u^n A, u^m), \tag{4.11}$$

as used to express Assumption 6.

If, however, one wants to use family blow up to make a study in a complete neighbourhood of $(0,0)$ in the $(a,\varepsilon)$-plane, then one can perform a blow up in the parameter plane by writing

$$(a,\varepsilon) = (v^n A, v^m E), \tag{4.12}$$

for $A^2 + E^2 = 1$ (or equivalently working with charts $E = +1$ or $A = \pm 1$.) Thereafter one continues with blow up in the $(x,y,v)$-space:

$$(x,y,v) = (u^p \overline{x}, u^q \overline{y}, u\overline{\varepsilon}),$$

with $\overline{x}^2 + \overline{y}^2 + \overline{\varepsilon}^2 = 1$ (or equivalently working with charts). This of course leads to the same result as (4.11) under the condition $E = 1$.

To understand the exact significance of Assumption 6 let us treat an example: consider the van der Pol system

$$\begin{cases} \dot{x} = y - \dfrac{x^2}{2} - \dfrac{x^3}{3} \\ \dot{y} = \varepsilon(a - x) \end{cases}. \tag{4.13}$$

In this example, $a$ is not a regular breaking parameter; however one can show that $a/\sqrt{\varepsilon}$ is a regular breaking parameter. Hence, in (4.13) one can write

$$a = vA, \qquad \varepsilon = v^2 E,$$

with $A^2 + E^2 = 1$. Being interested in $A \sim 0$, we can consider the chart $E = 1$, and check the assumptions for the parameters $(v, A)$ instead of $(\varepsilon, a)$.

In any case, under all the above assumptions, a forward center manifold and a backward center manifold intersect the section $T$ in graphs $\zeta_\pm(u, A, \lambda)$ and

$$(\zeta_- - \zeta_+)(0,0,\lambda) = 0, \frac{\partial}{\partial A}(\zeta_- - \zeta_+)(0,0,\lambda) \neq 0.$$

The implicit function theorem now allows us to find a control curve $\mathcal{A}(u,\lambda)$ so that

$$(\zeta_- - \zeta_+)(u, \mathcal{A}(u,\lambda), \lambda) \equiv 0,$$

ensuring that the saturation of the forward center manifold persists as an invariant manifold along the unstable branch of the critical curve $\gamma$, hence forming a manifold of canard solutions:

**4.9 Theorem** *Let Assumptions* 1–6 *be satisfied for* (4.1). *Let* $\Sigma_\pm$ *be admissible entry/exit boundary curves.*

*Then for some* $m \in \mathbf{N}_1$ *and for* $\varepsilon \in [0, \varepsilon_0[$ *with* $\varepsilon_0 > 0$ *sufficiently small, there exists a unique smooth curve* $a = \mathcal{A}(\varepsilon^{1/m}, \lambda)$ *so that:*

- $\mathcal{A}(0,\lambda) = 0$.
- *The saturation of* $\Sigma_-$ *along* $X_{\varepsilon,\mathcal{A}(\varepsilon,\lambda),\lambda}$ *forms a manifold with boundary of canard solutions containing* $\Sigma_+$ *as well.*
- *The manifold with boundary is smooth in the blow up space for* $\varepsilon > 0$, *and at* $\varepsilon = 0$, *it is smooth everywhere on the blow up locus along the heteroclinic connection* $\overline{\gamma}$, *and also outside the blow up locus along the subset* $S$ *of the critical curve* $\gamma$ *between the two corner points (not including the corner points—see also the formulation of Theorem* 4.4).

• *The $\infty$-jet of $\mathcal{A}(u,\lambda)$ w.r.t. $u$ is independent of the chosen admissible entry/exit boundary curves.*

The $\infty$-jet of $\mathcal{A}$ being independent of the boundary curves, one is interested in the study of the formal expansion of $\mathcal{A}$. A formal study of the vector field sometimes yields results (see Theorem 4.11), but not always. In the general case, it is very hard to obtain an expansion for $\mathcal{A}$; even to obtain the first nonzero coefficient is not obvious:

**4.10 Theorem** *Under the conditions of Theorem 4.9, and with the notations introduced above, we have:*

(1) *The first nonzero coefficient of $\varepsilon^{1/m}$ in the expansion of $\mathcal{A}$ is related to the first nonzero coefficient in the expansion of $(\zeta_- - \zeta_+)(u,0,\lambda)$. In fact, they are proportional, and the ratio is precisely $-\rho(\lambda)$, with $\rho(\lambda)$ the nonzero coefficient in (4.10).*

(2) *The first nonzero coefficient of $\zeta_- - \zeta_+$ can be calculated, either using the previous item if having a preexisting knowledge of $\mathcal{A}$, or using a Melnikov integral along $\Gamma$ (the heteroclinic connection on the blow up locus from $P_-$ to $P_+$). The integrands can be obtained through a formal recursive process.*

Apart from the general result in Theorem 4.9, stating that any two admissible boundary curves can be joined by an invariant manifold, properly choosing a control curve $\mathcal{A}$, the next question to answer is a question regarding the regularity of the invariant manifolds at the turning point $p_\gamma$. In blow up coordinates, these manifolds with boundary are $C^\infty$ smooth, whereas in the original coordinates this need not be the case, as will be illustrated in the next example.

### 4.1.6 $C^1$-Canard manifolds

Consider

$$X_{\varepsilon,a}:\begin{cases}\dot{x} = y - \dfrac{x^4}{4} + \varepsilon x^2\\ \dot{y} = \varepsilon(a - x^3)\end{cases}$$

This vector field exhibits a turning point at the origin. One can check the validity of Assumptions 1–6 on this vector field, but one can also refer to Theorem 4.13, which will be discussed later. In any case, Theorem 4.9 can be applied to construct canard manifolds that are $C^\infty$ in blow up coordinates. The blow down of any such canard manifold is a graph $y = \phi(x,\varepsilon)$, and the control curve is given in this case by the $C^\infty$ graph $a = \mathcal{A}(\varepsilon^{1/4})$. The $C^k$-smoothness of $y = \phi(x,\varepsilon)$ implies the existence of a Taylor series up to order $k$, hence implying the existence of a formal object

$$y = \sum_{n=0}^{k} \phi_n(x)\varepsilon^n$$

that is a formal solution of

$$\left(y - \frac{x^4}{4} + \varepsilon x^2\right)\frac{dy}{dx} = \varepsilon(a - x^3).$$

A simple substitution of a candidate formal solution in the above equation leads to a contradiction at order $k = 2$.

This example shows that a formal study is not always possible: formal calculations do not always lead to the knowledge of an expansion for $\mathcal{A}$, although $\mathcal{A}$ is $C^\infty$ in $\varepsilon^{1/4}$. One can ask when do canard manifolds blow down to $C^\infty$ manifolds in the original coordinates. Clearly, the existence of a formal expansion satisfying the differential equations is a necessary condition. The following result states that it is the only extra restriction.

**4.11 Theorem** *Let Assumptions* 1–6 *be satisfied.*

*In admissible charts: manifolds from Theorem* 4.9 *are graphs* $y = \Psi(x, \varepsilon, \lambda)$ (*near turning point*).

*Assume that there exist formal power series*

$$\hat{a} = \sum_{n=0}^{\infty} a_n(\lambda)\varepsilon^n, \qquad \hat{y} = \sum_{n=0}^{\infty} y_n(x, \lambda)\varepsilon^n$$

*such that* $y_n$ *is smooth in a uniform neighbourhood of* $x = 0$, *and such that* $\hat{y}$ *is formally invariant under* $\hat{X}_{\varepsilon,\hat{a},\lambda}$.

*Then* $\Psi$ *will be smooth in a neighbourhood of the origin. The infinite jet* $j_\infty(\Psi)$ *will coincide with* $\hat{y}$, *and also* $j_\infty(\mathcal{A})$ *will coincide with* $\hat{a}$.

The proof of Theorems 4.10 and 4.11 as well as of the subsequent Theorems 4.12 and 4.13 can be found in [DD].

### 4.1.7 Two classes of vector fields

Here, we present two classes of vector fields where the Assumptions 1–6 have already been checked, so that a straightforward verification of the conditions on this theorem leads to an easy application of Theorem 4.9.

For $h = f$ or $g$, we introduce the notation

$$h_. := h(0,0,0,0,\lambda), \qquad h_x := \frac{\partial h}{\partial x}(0,0,0,0,\lambda),$$

and similar notations for other partial derivatives.

**4.12 Theorem** *If a family of vector field* $X_{\varepsilon,a,\lambda}$ *in the form* (4.1) *satisfies*

(1) $f_. = 0$, $f_y \neq 0$ (*existence of critical curve by means of implicit function theorem, at least locally around the origin*);

(2) $f_x = 0, f_{xx} \neq 0$ (*at the origin, normal hyperbolicity is lost in the most generic way*);

(3) $g_. = 0$, $g_x \neq 0$ (*connection condition*);

(4) $g_x \begin{vmatrix} f_y & f_a \\ f_{xy} & f_{xa} \end{vmatrix} - f_{xx} \begin{vmatrix} f_y & f_a \\ g_y & g_a \end{vmatrix} \neq 0$ (*breaking parameter condition*);

(5) *The product* $g_x f_y$ *is negative* (*transition from attracting to repelling*).

*Then, near* $(x, y) = (0, 0)$ *and for* $(a, \varepsilon)$ *sufficiently small, Assumptions* 1–5 *are satisfied, using the blow up* $(x, y, \varepsilon) = (u\overline{x}, u^2\overline{y}, u^2\overline{\varepsilon})$ *and writing* $a = uA$. *Also, the conditions of Theorem* 4.11 *are satisfied.*

In the above theorem dealing with a *generic* turning point, the conditions coincide entirely with the one-dimensional case treated in [CRSS], where, due to the presence of a formal expansion in original coordinates, one can control Gevrey estimates that lead to the same (but local) result. At more degenerate turning points, the formal expansion is in general not available:

**4.13 Theorem** *If a family of vector field $X_{\varepsilon,a,\lambda}$ in the form (4.1) satisfies*

(1) $f_{\cdot} = 0,\ f_y \neq 0$ *(existence of critical curve by means of implicit function theorem);*

(2) $f_x = 0, f_{xx} = 0, \ldots, f_{x^{2n-1}} = 0,\ f_{x^{2n}} \neq 0$ *(at the origin, normal hyperbolicity is lost);*

(3) $g_{\cdot} = 0, g_x = 0, \ldots, g_{x^{2n-2}} = 0,\ g_{x^{2n-1}} \neq 0$ *(connection condition);*

(4) $g_{x^{2n-1}} \begin{vmatrix} f_y & f_a \\ f_{xy} & f_{xa} \end{vmatrix} - f_{x^{2n}} \begin{vmatrix} f_y & f_a \\ g_y & g_a \end{vmatrix} \neq 0$ *(breaking parameter condition);*

(5) *The product $g_{x^{2n-1}} f_y$ is negative (ensures a transition from attracting to repelling).*

*Then, near $(x, y) = (0, 0)$ and for $(a, \varepsilon)$ sufficiently small, Assumptions 1–5 are satisfied, using the blow up $(x, y, \varepsilon) = (u\overline{x}, u^{2n}\overline{y}, u^{2n}\overline{\varepsilon})$ and writing $a = u^{2n-1}A$.*

Let us finally give also an example that does not fit in this class, but that also satisfies Assumptions 1–6, and let us illustrate briefly how to prove the assumptions.

### 4.1.8 Normal crossing of critical curves

Consider the scalar o.d.e.

$$\varepsilon \frac{dy}{dx} = a + x^{2n-1} y + \varepsilon F(x, y, \varepsilon, a, \lambda)$$

and the associated vector field

$$X_{\varepsilon,a,\lambda} : \begin{cases} \dot{x} = \varepsilon \\ \dot{y} = a + x^{2n-1} y + \varepsilon F(x, y, \varepsilon, a, \lambda), \end{cases}$$

with $F$ a $C^\infty$ function near $(x, y, \varepsilon, a) = (0, 0, 0, 0)$.

For $\varepsilon = a = 0$, we have a crossing of the lines of singularities: $x = 0$ and $y = 0$. Along $x = 0$, it would not be possible to satisfy the regular passage property, but along $y = 0$ it will be, so choose $y = 0$ as the critical curve.

Observe that along $y = 0$, for $x < 0$ the critical curve is attracting, and for $x > 0$ the critical curve is repelling. Given a point $p$ on the attracting part of the critical curve, then the regular passage assumption (Assumption 2) is trivially satisfied, since $\dot{x} = \varepsilon > 0$.

To check the remaining assumptions, we blow up the origin:

$$\varepsilon = v^{4n},\ a = v^{4n-1} A, \qquad x = u^2 \overline{x},\ y = u\overline{y},\ v = u\overline{v}.$$

In the phase-directional rescaling chart $\overline{x} = +1$, we find (after dividing by $u^{4n-2}$)

$$\begin{cases} \dot{u} = \dfrac{1}{2} u \overline{\varepsilon}^{4n} \\ \dot{\overline{v}} = -\dfrac{1}{2} \overline{v}^{4n+1} \\ \dot{\overline{y}} = \overline{v}^{4n-1} A + \overline{y} - \dfrac{1}{2} \overline{v}^{4n} \overline{y} + O(u) \end{cases} .$$

Clearly, Assumption 3 is satisfied, and due to the absence of singularities outside $u\overline{v} = 0$ Assumption 4 also holds. Look now in the family rescaling chart $\overline{v} = 1$:

$$\begin{cases} \dot{\overline{x}} = 1 \\ \dot{\overline{y}} = A + \overline{x}^{2n-1} \overline{y} + O(u). \end{cases}$$

Clearly, for $A = 0$ there is a connection $\overline{y} = 0$ connecting $P_-$ to $P_+$, and on this connection, no singularities appear. This shows that Assumption 5 is satisfied. Finally, in order to check Assumption 6, one can make use of an integrating factor. We refer to [DD] for the complete elaboration.

## 4.2 Multiple canard cycles

In the previous section we have studied canard solutions near turning points. As we have seen (Theorem 4.9) such canard solutions can link arbitrary given ("admissible") boundary curves $\Sigma_-$ and $\Sigma_+$. In some situations the basin of attraction of $\gamma_-$ coincides, entirely or partially, with the basin of repulsion of $\gamma_+$, so that we could choose $\Sigma_- = \Sigma_+$. In that case the canard solutions give rise to closed orbits, mostly limit cycles, so that we can talk of canard cycles.

In [D] in the treatment of the van der Pol equation one encoutered one (hyperbolic) limit cycle and no bifurcations on limit cycles. The technique of proof was not yet general enough to deal with multiple canard cycles and their bifurcations. In this section we will show how such a study can be performed: We will not only limit our presentation to the generic turning point, but we will also limit it to generalized Liénard equations. We will present an overview of the results that have been obtained in [DR2], together with some ideas of the proof.

Let us consider singular perturbation problems of the form

$$\begin{cases} \dot{x} = y - F(x, c) \\ \dot{y} = \varepsilon(a - x\overline{G}(x, c)), \end{cases} \tag{4.14}$$

with $F$ and $\overline{G}$ of class $C^\infty$ and $c \in C \subset \mathbf{R}^k$ with $C$ compact. We also write

$$a - x\overline{G}(x, c) = G(x, a, c). \tag{4.15}$$

We suppose

$$\overline{G}(x, c) = 1 + O(x). \tag{4.16}$$

In fact, it is possible to let $F$ and $G$ depend on $(x, c, a, \varepsilon)$, but will not do this in order to keep the notation and the presentation as simple as possible. There is also no need to limit oneself to (generalized) Liénard equations, but we prefer to leave it to the reader to make the necessary adaptions.

We impose that $F(x,c)$, for each $c$ separately, be a Morse function with a minimum as unique critical point on the domain under consideration. After the necessary translations and rescalings we can express this condition as

$$F(x,c) = \frac{1}{2}x^2 + o(x^2),$$

and we limit our attention to an appropiate domain as represented in Fig. 19. This means that we choose some $Y_1 > 0$ such that below $\{y = Y_1\}$, and for each $c$, the connected component of $\{y = F(x,c)\}$ containing $(0,0)$ represents a Morse function with 0 as a unique critical point. This property has to be verified for $c$ in the parameter range $C$ under consideration.

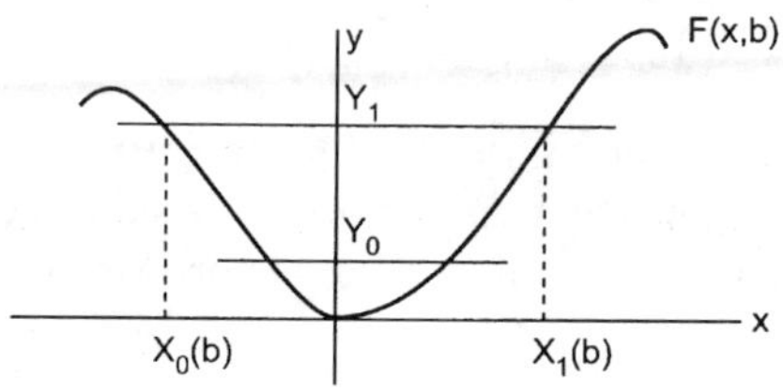


Figure 19: Domain on which $F$ has a minimum has a unique critical point

Along the critical manifold $\{y = F(x,c)\}$ for $\varepsilon = 0$ the 1-jet of the vector field is given by

$$\begin{pmatrix} -\frac{\partial F}{\partial x}(x, F(x,c)) & 1 \\ 0 & 0 \end{pmatrix}.$$

Its trace is clearly positive for $x \in [X_0(c), 0[$ and negative for $x \in ]0, X_1(c)]$; hence, as in Fig. 20, for $x < 0$ the critical manifold is normally hyperbolically repelling while for $x > 0$ it is normally hyperbolically attracting.

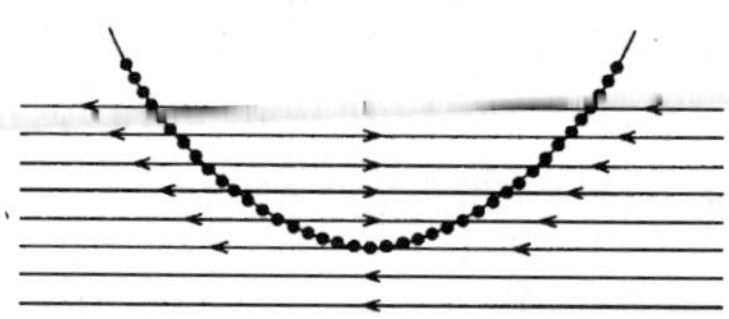

Figure 20: Phase portrait for $\varepsilon = 0$

In order to state a precise result let us recall:

**4.14 Definition** (of $C^\infty$-equivalence for functions) We say that two $C^\infty$ functions $A(y,\lambda)$ and $B(y,\lambda)$ are *left $C^\infty$ equivalent* if there exists some $C^\infty$ function $\psi(y,\lambda)$, everywhere different from zero, such that $A(y,\lambda) = \psi(y,\lambda) \cdot B(y,\lambda)$.

We introduce two $C^\infty$ functions

$$k_B(y,c) = \int_0^{x_0(y,c)} xh(x,c)f(x,c)^2\,dx, \tag{4.17}$$

$$k_F(y,c) = \int_0^{x_1(y,c)} xh(x,c)f(x,c)^2\,dx, \tag{4.18}$$

with

$$f(x,c) = \frac{1}{x}\frac{\partial F}{\partial x}(x,c) = 1 + O(x) > 0 \quad \text{and} \quad h(x,c) = -\frac{x}{G(x,0,c)} = 1 + O(x) > 0.$$

Let us also introduce the functions

$$L(c) = \frac{1}{2}\frac{\partial^2 G}{\partial x^2}(0,0,c), \tag{4.19}$$

$$\ell(c) = \frac{1}{3}\frac{\partial^3 F}{\partial x^3}(0,c). \tag{4.20}$$

**4.15 Theorem** *Considering system (4.14) with $F$ and $\overline{G}$ of class $C^\infty$, $c \in C \subset \mathbf{R}^k$ compact, with $a \sim 0$ and $\varepsilon \in [0,\infty[$, with $F$ having a unique generic minimum for $c \in C$, with $G$ as in (4.15) and (4.16), and using the notations introduced above, then for any $D > 0$ sufficiently small, there exists some neighbourhood $V$ of $0$ in $(a,\varepsilon)$-space such that the following properties hold:*

(1) *for $(a,\varepsilon)$ with $\varepsilon \leq a^2/D^2$, no closed orbit will cut the $y$-axis at $y$-values in $[Y_0, Y_1]$.*

(2) *If we write $(a,\varepsilon) = (uA, u^2)$, then there exists some $u_0 > 0$ and some $C^\infty$ function*

$$\Delta\colon [Y_0,Y_1] \times C \times [-D,D] \times [0,u_0] \to \mathbf{R},$$

*such that on the $y$-axis, for $y$-values in $[Y_0,Y_1]$ and $u > 0$, the position of the closed orbits is given by the equation*

$$\Delta(y,c,A,u) = 0.$$

*Moreover:*

(3) *The function $\frac{\partial \Delta}{\partial y}$, for $u > 0$, is left $C^\infty$-equivalent to*

$$-(I(y,c) + J(y,c,A,u)),$$

*where $J$ is $C^\infty$ and $J = O(u)$ and where*

$$I(y,c) = k_F(y,c) - k_B(y,c) = \int_{x_0(y,c)}^{x_1(y,c)} xh(x,c)f(x,c)^2\,dx. \tag{4.21}$$

(4) *There exists some manifold $\{a = c_0(c,\varepsilon)\}$ with $c_0$ of class $C^\infty$ and $c_0(c,\varepsilon) = -(L(c) + 3\ell(c))\varepsilon + O(\varepsilon^2)$, such that for any $K$,*

$$0 < K < \min\{\min(k_B(y,c), k_F(y,c)) \mid (y,c) \in [Y_0,Y_1] \times C\}$$

*and for any sequence $(\gamma_n)_n = ((y_n, c_n, a_n, \varepsilon_n))_n$ representing closed orbits of (4.14) with $(a_n,\varepsilon_n) \to 0$, there exists some $N \in \mathbf{N}$ such that for all $n \geq N$:*

$$|a_n - c_0(c_n,\varepsilon_n)| < e^{-K/\varepsilon_n}.$$

**Remark** The manifold $\{a = c_0(c, \varepsilon)\}$ can be called a "canard manifold" as a generalization of "canard curve" in the case of the van der Pol equation. It is in fact a $c$-family of curves in the $(a, \varepsilon)$-plane so that—in the spirit of Section 4.1—we could continue calling it a "canard curve" or "control curve". A "canard manifold" (or, by abuse of language, "canard curve") is not unique but its asymptotic development is. This situation is very comparable to what one usually encounters in working with center manifolds. For any $D > 0$ and $(a_0, \varepsilon)$ small enough, the canard curve(s) lie inside $\{\varepsilon \geq a^2/D^2\}$, while outside that set there are no closed orbits, cutting the $y$-axis in $[Y_0, Y_1]$.

Let us now give an idea of the proof together with some explanation concerning the functions $I$, $k_B$, $k_F$, $\ell$ and $L$. The proof still follows the main lines used in the treatment of the canard phenomenom in Van der Pol's equation (cf. [DR1, D]), and also used in Section 4.1 to study canard solutions in a more general context as in Theorem 4.9. In that section the attention first went to saturating admissible boundary curves, leading to $C^\infty$ invariant manifolds along the critical curve. The manifolds remained $C^\infty$ in the blow up locus of the turning point. In Section 4.1 we followed these saturations all the way until hitting a section $T$, defined in the family rescaling chart ($\bar{\varepsilon} = 1$), and chosen transverse to the orbits of the blown-up vector field. In this proof we consider the same construction but change the admissible curves $\Sigma_-$ and $\Sigma_+$ by the $(c, A)$-family of $C^\infty$ sections $[Y_0, Y_1] \times [0, u_0]$. Clearly saturation in forward time and in backward time until the section $T$, defines the respective $C^\infty$ mappings $\mathcal{F}$ and $\mathcal{B}$, expressing the passage from $[Y_0, Y_1] \times ]0, u_0]$ to $T$. Closed orbits for (4.14) correspond to zeros of

$$\Delta(Y, c, A, u) = \mathcal{F}(Y, c, A, u) - \mathcal{B}(Y, c, A, u).$$

For the calculation of the development of $\Delta$ it can be interesting to make a good choice of coordinates on $T$. We therefore consider again the blow up of (4.14) at $(x, y, a, \varepsilon) = (0, 0, 0, 0)$. The best way to proceed in view of explaining the whole construction is first to consider a rescaling in parameter space

$$(a, \varepsilon) = (sA, s^2). \tag{4.22}$$

It represents one specific chart ($E = 1$) in the blow up $(a, \varepsilon) = (sA, s^2E)$ with $A^2 + E^2 = 1$; it is, however, the most relevant chart since the other ones will merely be used to prove the non-occurrence of closed orbits. After performing the rescaling (4.22), we perform a family blow up

$$(x, y, s) = (u\bar{x}, u^2\bar{y}, uv) \tag{4.23}$$

with $\bar{x}^2 + \bar{y}^2 + v^2 = 1$, keeping $v \geq 0$ and $u \geq 0$. The section $T$ can be found in the domain of the chart $\{v = 1\}$, in which case (4.22) and (4.23) together can be written as

$$(x, y, a, \varepsilon) = (u\bar{x}, u^2\bar{y}, uA, u^2).$$

This operation, together with a time change by a factor $u$, changes (4.14) into

$$\begin{cases} \dot{\bar{x}} = \bar{y} - \dfrac{\bar{x}^2}{2} - \ell(c)u\bar{x}^3 + o(u) \\ \dot{\bar{y}} = A - \bar{x} + L(c)u\bar{x}^2 + o(u) \end{cases} \tag{4.24}$$

with $\ell(c)$ and $L(c)$ as defined in (4.20) and (4.19). For $(u, A) = (0, 0)$ this represents a time-reversible vector field $X_S$, having $e^{-\bar{y}}$ as integrating factor and

$$H(\bar{x}, \bar{y}) = -e^{-\bar{y}}\left(\bar{y} - \frac{\bar{x}^2}{2} + 1\right)$$

as related Hamiltonian. At the center $(0,0)$ we have $H(0,0) = -1$ and along the saddle connection $\overline{\gamma} = \{\overline{y} = \overline{x}^2/2 - 1\}$ we have $H(\overline{\gamma}) = 0$. As coordinates on $T$ it is appropiate to choose $(u,h)$, with $h$ being the value of the Hamiltonian $H$.

Along the closed orbits of $X_S$ the Poincaré mapping with respect to $T$ is given for $-1 < h < 0$ by

$$\begin{aligned} P_{u,A,c}(h) &= h + A\int_{\Gamma_h} e^{-\overline{y}}\, d\overline{x} + u\left(L(c)\int_{\Gamma_h} e^{-\overline{y}}\overline{x}^2\, d\overline{x} + \ell(c)\int_{\Gamma_h} e^{-\overline{y}}\overline{x}^3\, d\overline{y}\right) + o(|A,u|) \\ &= h + AI_1(h) + u(L(c) + 3\ell(c))I_2(h) + o(|A,u|), \end{aligned}$$

with $\Gamma_h = H^{-1}(h)$, $I_1(h) = \int_{\Gamma_h} e^{-\overline{y}}\, d\overline{x}$ and $I_2(h) = \frac{1}{3}\int_{\Gamma_h} e^{-\overline{y}}\overline{x}^3\, d\overline{y}$.

As $I_2(h)/I_1(h)$ has a strictly positive derivative with respect to $h$ on the interval $]-1,0[$ and $I_2(0)/I_1(0) = 1$, we see that the first order approximation of the canard curve is given by

$$A = -(L(c) + 3\ell(c))u.$$

For more details about these and other calculations we refer to [DR2].

If we choose some $Y_2 > Y_1$, inducing by saturation the $C^\infty$ mappings $\overline{\mathcal{F}}(u,c,A)$ and $\overline{\mathcal{B}}(u,c,A)$, then by using the appropiate normal forms along the critical curve, as well as along its blow up, as we did in Section 4.1, we can prove that

$$\overline{\mathcal{B}}(u,c,A) - \mathcal{B}(Y,u,c,A) = \exp\left(-\frac{k_B(Y,c)}{u^2}(1+O(u))\right); \tag{4.25}$$

$$\overline{\mathcal{F}}(u,c,A) - \mathcal{F}(Y,u,c,A) = \exp\left(-\frac{k_F(Y,c)}{u^2}(1+O(u))\right), \tag{4.26}$$

with $k_B$ and $k_F$ as in (4.17) and (4.18). The full proof can be found in [DR1, DR2]. It is based on the use of normal forms, inducing the exponentially small expression, and an analysis of the divergence integral leading to the simple expressions given in (4.17) and (4.18). The following heuristic argument indicates why the expressions (4.25) and (4.26) can be expected.

We write (4.14) as

$$X_{\varepsilon,a,c}\colon \begin{cases} \dot{x} = y - F(x,c) = y - \dfrac{x^2}{2} - x^3\overline{F}(x,c) \\ \dot{y} = \varepsilon G(x,a,c) = \varepsilon(a - x + x^2\tilde{G}(x,c)). \end{cases}$$

We know that the logarithm of the derivative of the passage map is essentially given by

$$\int \operatorname{div} X_{\varepsilon,a,c}\, dt = -\int \frac{\frac{\partial F}{\partial x}(x,c)}{\varepsilon G(x,c)}\, dy.$$

Under the parameter rescaling (4.22) and for $s \to 0$, the orbits, along which we integrate, tend to the critical curve given by $y = F(x,c)$. There, $dy = \frac{\partial F}{\partial x}(x,c)dx$.

As expected, one can show that for $s \to 0$, the integral $s^2\int \operatorname{div} X_{s^2,sA,c}\, dt$ tends to

$$\begin{aligned} \lim_{s\to 0} -\int \frac{\left(\frac{\partial F}{\partial x}(x,c)\right)^2}{sA - x + x^2\tilde{G}(x,c)}\, dx &= \int \frac{\left(\frac{\partial F}{\partial x}(x,c)\right)^2}{x - x^2\tilde{G}(x,c)}\, dx \\ &= \int \left(\frac{1}{x}\frac{\partial F}{\partial x}(x,c)\right)^2 x\left(-\frac{x}{G(x,0,c)}\right) dx \\ &= \int f(x,c)^2 x h(x,c)\, dx. \end{aligned}$$

For $\overline{\mathcal{B}}$ (resp. $\overline{\mathcal{F}}$) the integration has to be taken on respectively $[0, x_0(y,c)]$ and $[0, x_1(y,c)]$. A detailed analysis can be found in [DR1, DR2].

The relation with the Abelian integrals, calculated over the level curves of $H$, is given by:

$$\overline{\Delta}(u,c,A) = \overline{\mathcal{F}}(u,c,A) - \overline{\mathcal{B}}(u,c,A) = I_1(0)[A + (L(c) + 3\ell(c))u + o(|A,u|)],$$

and of course we have a similar statement for $\Delta$: we can write

$$\Delta(Y,u,c,A) = \overline{\Delta}(u,c,A) + \Phi(Y,u,c,A),$$

with all expressions being $C^\infty$. The Implicit Function Theorem now shows that there are $C^\infty$ functions $\overline{\mathcal{A}}(u,c)$ and $\mathcal{A}(Y,u,c)$ representing respectively $\{\overline{\Delta}(u,c,A) = 0\}$ and $\{\Delta(Y,u,c,A) = 0\}$.

By the construction we see that the $\infty$-jet of $\mathcal{A}$ with respect to $u$ does not depend on $Y$ and is equal to the $\infty$-jet of $\overline{\mathcal{A}}$. Since $(a,\varepsilon) = (uA, u^2)$, the expression of the control curve $\overline{\mathcal{A}}$ (and similarly for $\mathcal{A}$) in the $(a,\varepsilon)$-plane is given by

$$a = \sqrt{\varepsilon}\overline{\mathcal{A}}(\sqrt{\varepsilon}, c) = \overline{\mathcal{A}}_0(\sqrt{\varepsilon}, c),$$

with $\overline{\mathcal{A}}_0$ a $C^\infty$ function. That in this case we can write $\overline{\mathcal{A}}_0(\sqrt{\varepsilon}, c) = c_0(\varepsilon, c)$ for some $C^\infty$ function $c_0$ follows from the "symmetry" in the construction. For more details we refer to [DR2].

If we write

$$\Delta(Y,u,c,A) = \overline{\Delta}(u,c,A) + \exp\left(-\frac{k_B(Y,c)}{u^2}(1+O(u))\right) - \exp\left(-\frac{k_F(Y,c)}{u^2}(1+O(u))\right)$$

we get

$$\begin{aligned}\frac{\partial\Delta}{\partial Y}(Y,u,c,A) = &-\frac{1}{u^2}\frac{\partial k_B}{\partial Y}(Y,c)(1+O(u))\exp\left(-\frac{k_B(Y,c)}{u^2}(1+O(u))\right)\\ &+\frac{1}{u^2}\frac{\partial k_F}{\partial Y}(Y,c)(1+O(u))\exp\left(-\frac{k_F(Y,c)}{u^2}(1+O(u))\right)\end{aligned}$$

which can be written as

$$\frac{\partial\Delta}{\partial Y}(Y,u,c,A) = \frac{1}{u^2}\left(\exp\left(-\frac{k_F(Y,c)}{u^2}(1+O(u))\right) - \exp\left(-\frac{k_B(Y,c)}{u^2}(1+O(u))\right)\right),$$

since $\frac{\partial k_B}{\partial Y}$ and $\frac{\partial k_F}{\partial Y}$ are both strictly positive.

For $u > 0$ we see that $\frac{\partial\Delta}{\partial Y}(Y,u,c,A) = 0$ is given by

$$\exp\left(-\frac{k_F(Y,c)}{u^2}(1+O(u))\right) = \exp\left(-\frac{k_B(Y,c)}{u^2}(1+O(u))\right),$$

and hence by

$$I(Y,c) + J(Y,u,c,A) = 0,$$

for some $C^\infty$ function $J$ which is $O(u)$, and where $I$ is defined in (4.21).

It follows that $\frac{\partial\Delta}{\partial Y}$, for $u > 0$, is left $C^\infty$-equivalent to

$$-(I(Y,c) + J(Y,c,A,u)).$$

Here we end our survey of the proof of Theorem 4.15.

As a conclusion we see that the critical points of $\Delta$ are given by the zeros of $I(Y,c)+O(u)$. This set will be diffeomorphic to the one described by $I(Y,c)=0$ if it represents an elementary catastrophy. In that case, because of Rolle's theorem, we obtain a uniform upper bound on the number of limit cycles to be found in the set $[Y_0,Y_1]\times C\times[-D,D]\times[0,u_0]$. In order to get more precise information on the position and the bifurcation of the limit cycles, some extra work has to be done. We refer to [DR2] for precise results as well as for examples on which the theory can be applied.

Let us end by presenting one of the main theorems of [DR2].

We suppose that at some value $Y\in]Y_0,Y_1[$ and for some $c=c_0\in C$ we have

$$I(Y,c_0)=\frac{\partial}{\partial y}(Y,c_0)=\cdots=\frac{\partial^{r-1}I}{\partial y^{r-1}}(Y,c_0)=0,\ \frac{\partial^r I}{\partial y^r}(Y,c_0)\neq 0 \tag{4.27}$$

for some $r>1$. For simplicity in notation we will take $c_0=0$. On the $y$-axis we use the new variable $\tilde{y}=y-Y$. Using the Malgrange preparation theorem we can suppose that, for $\tilde{y}\sim 0$, and up to left $C^\infty$-equivalence (division by a strictly positive function $R(\tilde{y})$):

$$I(\tilde{y},c)=\tilde{y}^r+\sum_{i=0}^{r-1}\lambda_i(c)\tilde{y}^i, \tag{4.28}$$

with $\lambda_0(0)=\cdots=\lambda_{r-1}(0)=0$ and all $\lambda_i$ of class $C^\infty$. In case we deal with $I/R$ instead of $I$ we can also suppose that $R(0)=1$. As a generic condition on $I$ we require the mapping

$$c\mapsto(\lambda_0(c),\ldots,\lambda_{r-1}(c))$$

to be a submersion at $c=0$ and again for the sake of simplicity in notation, we restrict ourselves to the case $c\in C\subset\mathbf{R}^r$. By a reparametrization we can hence consider $(\lambda_0,\ldots,\lambda_{r-1})\in\mathbf{R}^r$ to be the new parameter $\lambda$ instead of $c$, at least in a neighbourhood of $c=0$, and write

$$I(\tilde{y},\lambda)=\tilde{y}^r+\sum_{i=0}^{r-1}\lambda_i\tilde{y}^i.$$

**4.16 Theorem** *Suppose system* (4.14) *to be like in Theorem* 4.15 *and suppose that at some point* $(y,c)=(Y,c_0)$ *the function* $I$ *fullfils condition* (4.27). *Suppose that there exists a* $C^\infty$ *reparametrization* $\lambda(c)$, *with* $\lambda(c_0)=0$, *such that the associated* $\frac{I(\tilde{y},\lambda)}{R(\tilde{y})}$ *is as in* (4.28) *for some* $C^\infty$ *function* $R(\tilde{y})>0$, *with* $y=Y+\tilde{y}$ *and* $R(0)=1$. *For suitably chosen functions* $\alpha$ *and* $\beta_i$ *with* $\tilde{\tilde{y}}=\tilde{y}(1+\alpha)$, $\tilde{\lambda}_i=\lambda_i+\beta_i$ *and introducing*

$$\tilde{\tilde{y}}=u^2\overline{y},\qquad \tilde{\lambda}_i=u^{2r-2i}\overline{\lambda}_i$$

*one can prove that for every sufficiently large ball* $L$ *in the parameter space* $(\overline{\lambda}_0,\ldots,\overline{\lambda}_{r-1})\in\mathbf{R}^r$, *for every* $\overline{Y}_0>0$ *and for every* $M_0>0$, *there exists* $u_0>0$ *sufficiently small such that for* $(\tilde{\tilde{y}},\tilde{\lambda}_0,\ldots,\tilde{\lambda}_{r-1})$ *inside*

$$\{(u^2\overline{y},u^{2r}\overline{\lambda}_0,\ldots,u^2\overline{\lambda}_{r-1})\mid 0<u<u_0, y\in[-\overline{Y}_0,\overline{Y}_0],(\overline{\lambda}_0,\ldots,\overline{\lambda}_{r-1})\in L\}$$

*and written in terms of* $(\overline{y}, \overline{\lambda}_0, \ldots, \overline{\lambda}_{r-1})$, *the function* $\Delta$ *is left* $C^\infty$*-equivalent to some* $C^\infty$ *function* $D(\overline{y}, \overline{\lambda}_i, u)$ *such that*

$$D(\overline{y}, \overline{\lambda}_i, u) = d_{r+1}(\overline{y}) + \sum_{i=0}^{r} \overline{\mu}_i(\overline{\lambda}_i) d_i(\overline{y}) + O(u)$$

*with* $d_0(\overline{y}) \equiv 1$, $d_i(\overline{y}) = \int_0^{\overline{y}} e^{-k_1 z} z^{i-1}\, dz$, *for* $i = 1, \ldots, r$ *and* $k_1 > 0$.

*Moreover,*

$$\overline{\mu}_i = \overline{\lambda}_{i-1} - \frac{i}{k_1}\overline{\lambda}_i, \quad \text{for} \quad i = 1, \ldots, r-1, \quad \text{and} \quad \overline{\mu}_r = \overline{\lambda}_{r-1} - \frac{r}{k_1}.$$

*The map* $(\overline{\lambda}_0, \ldots, \overline{\lambda}_{r-1}) \mapsto (\overline{\mu}_1, \ldots, \overline{\mu}_r)$ *is invertible. For any* $K > 0$, *if we keep* $|a - c_0(c, \varepsilon)| < K\varepsilon$, *then* $\overline{\mu}_0 = \overline{\mu}_0(a, \overline{\lambda}_0, \ldots, \overline{\lambda}_{r-1})$ *has a range which contains* $[-M_0, M_0]$ *for fixed* $(\overline{\lambda}_0, \ldots, \overline{\lambda}_{r-1})$ *and moreover* $\frac{\partial \overline{\mu}_0}{\partial a} > 0$.

The functions $(d_0(\overline{y}), \ldots, d_{r+1}(\overline{y}))$ form a so-called "complete Chebychev system" on $[-\overline{Y}_0, \overline{Y}_0]$, implying that they (at least) in a $C^0$-way exhibit an elementary catastrophy unfolding a multiple limit cycle of multiplicity $r+1$. For details and for applications we refer to [DR2].

## References


[AR] R. Abraham and J. Robbin, *Transversal Mappings and Flows*, Benjamin, New York-Amsterdam, 1967.

[B] R. I. Bogdanov, The versal deformation of a singular point of a vector field on the plane in the case of zero eigenvalues, *Trudy Sem. Petrovsk* **2** (1976), 37–65 (Russian); *Selecta. Math. Soviet.* **1** (1981), 389–421 (English).

[BFSW] E. Benoît, A. Fruchard, R. Schäfke, and G. Wallet, Solutions surstables des équations différentielles complexes lentes-rapides à point tournant, *Ann. Fac. Sci. Toulouse Math. (6)* **7** (4) (1998), 627–658.

[Bo] P. Bonckaert, Partially hyperbolic fixed points with constraints, *Trans. Amer. Math. Soc.* **348** (3) (1996), 997–1011.

[Bro] Th. Bröcker, *Differentiable Germs and Catastrophes*, London Math. Soc. Lecture Note Ser. **17** (1975).

[Ca] M. I. Camacho, A contribution to the topological classification of homogeneous vector fields in $\mathbf{R}^3$, *J. Differential Equations* **57** (2) (1985), 159–171.

[CLl] C. Christopher and N. G. Lloyd, Small-amplitude limit cycles in polynomial Liénard systems, *NoDEA Nonlinear Differential Equations Appl.* **3** (2) (1996), 183–190.

[CLW] S. Chow, C. Li, and D. Wang, *Normal Forms and Bifurcation of Planar Vector Fields*, Cambridge Univ. Press, Cambridge, 1994.

[CLy] C. Christopher and S. Lynch, Small-amplitude limit cycles bifurcations for Liénard equations with quadratic or cubic dampling or restoring forces, *Nonlinearity* **12** (4) (1999), 1099–1112.

[Co] W. Coppel, Some quadratic systems with at most one limit cycle, in: *Dynamics Reported*, Vol. 2, Dynam. Report. Ser. Dynam. Syst. Appl. **2**, Wiley, Chichester, 1989, 61–88.

[CRSS] M. Canalis-Durand, J.P. Ramis, R. Schäfke and Y. Sibuya, Gevrey solutions of singularly perturbed differential equations, *J. Reine Angew. Math.* **518** (2000), 95–129.

[D] F. Dumortier, Techniques in the theory of local bifurcations: blow-up, normal forms, nilpotent bifurcations, singular perturbations (notes written with B. Smits), in: *Bifurcations and Periodic Orbits of Vector Fields* (D. Schlomiuk, ed.), NATO ASI Ser. C Math. Phys. Sci. **408**, Kluwer Acad. Publ., Dordrecht, 1993, 19–73.

[D1] F. Dumortier, Singularities of vector fields on the plane, *J. Differential Equations* **23** (1) (1977), 53–106.

[D2] F. Dumortier, On the structure of germs of vector fields in $\mathbf{R}^3$ whose linear part generates rotations, in: *Dynamical Systems and Bifurcations (Groningen, 1984)* (B. L. J. Braaksma et al., eds.), Lecture Notes in Math. **1125**, Springer, Berlin, 1985, 15–46.

[D3] F. Dumortier, Nonstabilisable jets of diffeomorphisms in $R^2$ and of vector fields in $\mathbf{R}^3$, *Ann. Math. (2)* **124** (3) (1986), 405–440.

[DD] F. Dumortier and P. De Maesschalck, Canard solutions at non-generic turning points, preprint.

[DFL] F. Dumortier, P. Fiddelaers, and C. Li, Generic unfolding of the nilpotent saddle of codimension four, in: *Global Analysis of Dynamical Systems* (H. W. Broer, B. Krauskopf, and G. Vegter, eds.), IOP.

[DH] F. Dumortier and C. Herssens, Polynomial Liénard equations near infinity, *J. Differential Equations* **153** (1) (1999), 1–29.

[DI1] F. Dumortier and S. Ibáñez, Nilpotent singularities in generic 4-parameter families of 3-dimensional vector fields, *J. Differential Equations* **127** (2) (1996), 590–647.

[DI2] F. Dumortier and S. Ibáñez, Singularities of vector fields in $\mathbf{R}^3$, *Nonlinearity* **11** (4) (1998), 1037–1047.

[DI3] F. Dumortier and S. Ibáñez, Codimension 4 singularities of reflectionally symmetric planar vector fields, *Publ. Mat.* **43** (2) (1999), 501–533.

[DL1] F. Dumortier and C. Li, On the uniqueness of limit cycles surrounding one or more singularities for Liénard equations, *Nonlinearity* **9** (6) (1996), 1489–1500.

[DL2] F. Dumortier and C. Li, Quadratic Liénard equations with quadratic damping, *J. Differential Equations* **139** (1) (1997), 41–59.

[DL3] F. Dumortier and C. Li, Perturbations from an elliptic Hamiltonian of degree four. I. Saddle loop and two saddle cycle, *J. Differential Equations* **176** (1) (2001), 114–157.

[DL4] F. Dumortier and C. Li, Perturbations from an elliptic Hamiltonian of degree four. II. Cuspidal loop, *J. Differential Equations* **175** (2) (2001), 209–243.

[DL5] F. Dumortier and C. Li, Perturbation from an elliptic Hamiltonian of degree four. III. Global center, *J. Differential Equations*, to appear in 2003.

[DL6] F. Dumortier and C. Li, Perturbation from an elliptic Hamiltonian of degree four. IV. Figure Eight-Loop, *J. Differential Equations*, to appear in 2003.

[DR1] F. Dumortier and R. Roussarie, *Canard Cycles and Center Manifolds*, Mem. Amer. Math. Soc. **121** (577), 1996.

[DR2] F. Dumortier and R. Roussarie, Multiple canard cycles in generalized Liénard equations, *J. Differential Equations* **174** (1) (2001), 1–29.

[DRo] F. Dumortier and C. Rousseau, Cubic Liénard equations with linear damping, *Nonlinearity* **3** (4) (1990), 1015–1039.

[DRR] F. Dumortier, P. Rodrigues, and R. Roussarie, *Germs of Diffeomorphisms in the Plane*, Lecture Notes in Math. **902**, Springer-Verlag, Berlin-New York, 1981.

[DRS1] F. Dumortier, R. Roussarie, and J. Sotomayor, Generic 3-parameter families of vector fields on the plane, unfolding a singularity with nilpotent linear part. The cusp case of codimension 3, *Ergodic Theory Dynam. Systems* **7** (3) (1987), 375–413.

[DRS2] F. Dumortier, R. Roussarie, and J. Sotomayor, Generic three-parameter families of planar vector fields, unfoldings of saddle, focus and elliptic singularities with nilpotent linear parts, in: *Bifurcations of Planar Vector Fields: Nilpotent Singularities and Abelian Integrals* (F. Dumortier et al., eds.), Lecture Notes in Math. **1480**, Springer, Berlin, 1991, 1–164.

[DRS3] F. Dumortier, R. Roussarie, and J. Sotomayor, Bifurcations of cuspidal loops, *Nonlinearity* **10** (6) (1997), 1369–1408.

[Fe] N. Fenichel, Geometric singular perturbation theory for ordinary differential equations, *J. Differential Equations* **31** (1) (1979), 53–98.

[GH] J. Guckenheimer and P. Holmes, *Nonlinear oscillations, dynamical systems and bifurcations of vector fields*, Appl. Math. Sci. **42**, Springer, New York, 1983.

[H] E. Horozov, Versal deformations of equivariant vector fields in the case of symmetry of order 2 and 3, *Trudy Sem. Petrovsk.* (5) (1979), 163–192.

[HPS] M. Hirsch, C. C. Pugh, and M. Shub, *Invariant Manifolds*, Lecture Notes in Math. **583**, Springer, Berlin-New York, 1977.

[I] Y. Ilyashenko, Algebraic nonsolvability and almost algebraic solvability, *Funct. Anal. Appl.* (1972), 197–202.

[J] C. K. R. T. Jones, Geometric singular perturbation theory, in: *Dynamical Systems (Montecatini Terme, 1994)* (R. Johnson, ed.), Lecture Notes in Math. **1609**, Springer, Berlin, 1995, 44–118.

[K] A. Khovanskiĭ, Real analytic manifolds with the property of finiteness, and complex Abelian integrals, *Funktsional. Anal. i Prilozhen* **18** (2) (1984), 40–50.

[LL] N. G. Lloyd and S. Lynch, Small-amplitude limit cycles of certain Liénard systems, *Proc. Roy. Soc. London Ser. A* **418** (1854) (1988), 199–208.

[LMP] A. Lins, W. de Melo, and C.C. Pugh, On Liénard's equation, in: *Geometry and Topology (Proc. III Latin Amer. School of Math., IMPA, Rio de Janeiro, 1976)* (J. Palis and M. do Carmo, eds.), Lecture Notes in Math. **597**, Springer, Berlin, 1977, 335–357.

[LR] C. Li and C. Rousseau, A system with three limit cycles appearing in a Hopf bifurcation and dying in a homoclinic bifurcation, the cusp of order 4, *J. Differential Equations* **79** (1989), 132–167.

[M] P. Mardešić, Chebyshev Systems and the Versal Unfolding of the Cusps of Order $n$, Travaux en Cours **57**, Hermann, Paris, 1998.

[P] G. S. Petrov, Elliptic integrals and their nonoscillation, *Funktsional. Anal. i Prilozhen* **20** (1) (1986), 46–49, 96; The Chebyshev property of elliptic integrals, *Funktsional. Anal. i Prilozhen* **22** (1) (1988), 83–84; Nonoscillation of elliptic integrals, *Funktsional. Anal. i Prilozhen* **24** (3) (1990), 45–50, 96; On the nonoscillation of elliptic integrals, *Funktsional. Anal. i Prilozhen* **31** (4) (1997), 47–51, 95.

[PS] C. C. Pugh and M. Shub, Linearization of normally hyperbolic diffeomorphisms and dlows, *Invent. Math.* **10** (1970), 187–198.

[Ro] C. Robinson, *Dynamical Systems: Stability, Symbolic Dynamics, and Chaos*, CRC Press, New York, 1998.

[T1] F. Takens, Singularities of vector fields, *Inst. Hautes Études Sci. Publ. Math.* **43** (1974), 47–100.

[T2] F. Takens, Forced oscillations and bifurcations, in: *Applications of Global Analysis (Rijksuniversiteit Utrecht)* **1** (1974), 1–57; and in: *Global Analysis of Dynamical Systems*, IOP, Bristol, 2001 1–61.

[ULL] A. M. Urbina, M. León de la Barra, and G. Léon de la Barra, Stability of vector fields on $\mathbf{R}^3$, determined by the first non-vanishing jet, *J. Differential Equations* **79** (1) (1989), 178–187.

[V] A. N. Varchenko, Estimation of the number of zeros of an abelian integral depending on a parameter, and limit cycles, *Funktsional. Anal. i Prilozhen* **18** (2) (1984), 14–25.

[Z] H. Żołądek, Abelian integrals in unfoldings of codimension 3 singular planar vector fields, in: *Bifurcations of Planar Vector Fields: Nilpotent Singularities and Abelian Integrals* (F. Dumortier et al., eds.), Lecture Notes in Math. **1480**, Springer, Berlin-Heidelberg-New York, 1991, 165–224.

# Recent advances in the analysis of divergence and singularities

Jean ÉCALLE

*Laboratoire d'Analyse Harmonique*
*Département de Mathématiques*
*Université de Paris - Sud*
*91405 Orsay Cédex*
*France*

**Abstract**

We survey a number of related advances that have taken place over the last decade in the field of singularities, normal forms, ODEs, etc., as well as the analytic tools for tackling these problems, namely: resummation, resurgence, transseries, analysable functions. One such advance—the notion of well-behaved convolution average—has led to a simplification of the celebrated finiteness theorem for limit-cycles. Another one has clarified the (continuous) prenormalisation and (discontinuous) normalisation of local objects. Yet another—the notion of twisted resurgence monomials—has yielded a truly general method for canonical-explicit object synthesis (i.e. constructing local objects with prescribed analytic invariants). A fourth advance has shed new light on the classical KAM theorem about the survival of invariant tori. Lastly, a fifth development, which is arguably the most promising of all—the introduction of the new Lie algebra ARI—has led to a far-going elucidation of the arithmetics of MZV or "multiple zeta values".—Part of the results surveyed in this paper are joint work with F. Menous or B. Vallet, and mention is made of independent contributions by J. van der Hoeven.

## Contents





*Y. Ilyashenko and C. Rousseau (eds.),*
*Normal Forms, Bifurcations and Finiteness Problems in Differential Equations,* 87–186.

## General chapter-by-chapter overview

### 1 The Finiteness Theorem for limit-cycles and its resummation-theoretic proof twelve years on: review; simplification; aftermath

The proof in question, which relied on a constructive correspondence between a formal object (the *transseries* $\tilde{F}$) and its geometric counterpart (the *return map* $F$ ), has been significantly simplified, owing to the discovery of special *convolution-respecting averages* (in joint work with F. Menous). At the same time, the general theory of Transseries and Analysable Functions has grown and matured considerably, thanks mainly to the work of an Orsay-based mathematician, J. van der Hoeven. We shall attempt to cover some of the most exciting novelties in this fast-developing field.

### 2 Normal and prenormal forms

After reviewing the main classical results about *normal forms*, we shall introduce an altogether new type of *prenormalisation* (developed jointly with B. Vallet) and explain what it is good for.

### 3 Canonical object synthesis

We shall construct a new class of special functions that make possible a unified treatment of the *reverse classification problem*: finding 'canonical' local objects (differential systems, vector fields, diffeomorphisms, etc.) with prescribed invariants (formal or holomorphic); studying their iso-monodromic, or iso-Stokes, iso-Galois, etc., deformations.

### 4 An unexpected development in KAM theory: the non-existence of Supermultiple Small Denominators

Supermultiple Small Denominators or SSD for short (i.e. diophantine small denominators with an abnormally high multiplicity) were thought to be a central difficulty in KAM theory, and elaborate strategies had been evolved to account for their somewhat mysterious *'mutual compensation'*. But in 1994 it was discovered (jointly with B. Vallet) that, contrary to appearances, these much-dreaded SSD *simply do not exist*—they are a mirage conjured up by awkwardly conducted calculations. This startling claim, which at first was met with considerable skepticism, has now been fully vindicated. We shall explain the why and wherefore of this 'non-existence', and show what simplifications it brings to KAM theory, Floquet theory, etc.

### 5 A Tale of Three Structures

We shall try to present the outlines of an overarching structure which unifies three separate theories that *prima facie* would seem to be worlds apart:

(i) the Lie algebra ARI (which includes as sub-cases the Ihara braid algebra and the so-called renormalisation algebra);

(ii) the analysis of *'parametric divergence'* (an important type of divergence frequently encountered in physics);

(iii) the arithmetics of the *Multiple Zeta Values* or MZV: these constants tend to pop up everywhere (in Holomorphic Dynamics, Number Theory, Feynman Diagrams, Galois Theory, Knot Theory...) and have remained a focus of intense activity for more than a decade, but the main MZV-related conjectures stubbornly defied proving—that is, until the recent introduction of ARI.

# 1 Lesson One: The finiteness theorem for limit-cycles and its resummation-theoretic proof twelve years on: review; simplification; aftermath

## 1.1 Overall scheme

The finiteness theorem for limit cycles of analytic planar vector fields

$$X = P(x,y)\,\partial_x + Q(x,y)\,\partial_y \qquad (x, y \in \mathbb{R}) \tag{1.1}$$

states that *on compact subsets of the plane any such field has at most a finite number of limit-cycles, i.e. isolated periodic trajectories. In particular, a polynomial vector field has at most a finite number $N(X)$ of limit-cycles in the whole plane.*

By itself, the theorem doesn't solve Hilbert's celebrated 16th problem: it doesn't tell us what this number $N(X)$ is. It doesn't even settle the issue of whether this number is bounded for polynomials of a given degree $d$. At the moment, the question is still pending—even for degree 2—although in that case a (positive) answer appears to be close at hand.

Way back in the 1920s, Dulac gave a proof of the finiteness theorem, which alas was flawed and yet went unchallenged for more than fifty years. Eventually, two modern proofs appeared at the close of 1980s—one [I1, I2] due to Yu. S. Ilyashenko, geometric in spirit and non-constructive in nature; and another one [E6, E7] by myself. That latter proof relied on recent advances in resummation theory and fitted into a general program of "constructive formalisation" of local geometry, which stands on its own (it predated the current infatuation with polycycles and will hopefully outlive it).

## 1.2 Polycycles; local transit maps; global return map

Let $\mathcal{C}$ be a simple, closed curve on a real-analytic surface $\mathcal{S}$. Let $X$ be a real-analytic vector field, defined on a neighbourhood of $\mathcal{C}$ and such that $\mathcal{C}$ be a finite union of closed trajectories of $X$, with summits $S_1, S_2, \ldots, S_r$ corresponding to singular points of $X$, and $r$ analytic arcs $\mathcal{C}_i$ linking $S_{i-1}$ to $S_i$. On each $\mathcal{C}_i$ we fix an interior point $P_i$ and draw an analytic curve $\Gamma_i$ crossing $\mathcal{C}_i$ at $P_i$ at a right angle. We equip $\Gamma_i$ with an analytic abscissa $x_i$ that vanishes at $P_i$ and assumes positive values on the inside of the polycycle. Every interior orbit of $X$ close to $\mathcal{C}$ intersects $\Gamma_i$ at a point of abscissa $x_i = 1/z_i$ $(x_i \sim 0, z_i \sim +\infty)$ and $\Gamma_{i+1}$ at a point of abscissa $x_{i+1} = 1/z_{i+1}$ $(x_{i+1} \sim 0, z_{i+1} \sim +\infty)$. By setting $z_{i+1} = G_i(z_i)$ we define a self-mapping of $[\ldots, +\infty]$. The successive $G_i$ will be referred to as *transit maps* and their composition product $F := G_r \circ \cdots \circ G_2 \circ G_1$ as *return map* .

To sum up:

$$\text{i-th transit map:} \quad G_i : z_i = 1/x_i \mapsto z_{i+1} = 1/x_{i+1} \tag{1.2}$$

$$\text{(first) return map:} \quad F : z_1 = 1/x_1 \mapsto z_{r+1} = 1/x_{r+1} \tag{1.3}$$

$$F = G_r \circ \cdots \circ G_2 \circ G_1 \tag{1.4}$$

By a classical result on the reduction of planar singularities by means of successive blow-ups (see [Se]) it suffices to consider the case when each summit $S_i$ is either of *hyperbolic type* ($X$ has there two non-zero eigenvalues) or *semi-hyperbolic type* ($X$ has there one zero and one non-zero eigenvalue).

Actually, from the point of view of formalisation, three types of summits have to be distinguished:

$$\text{Type I:} \quad (\lambda', \lambda'') \quad \text{with} \quad \lambda := -\lambda'/\lambda'' \in \mathbb{R}^+ - \mathbb{Q}^+ \tag{1.5}$$

$$\text{Type II:} \quad (\lambda', \lambda'') \quad \text{with} \quad \lambda := -\lambda'/\lambda'' \in \mathbb{Q}^+ \tag{1.6}$$

$$\text{Type III:} \quad (\lambda', 0) \quad \text{or} \quad (0, \lambda'') \quad \text{with} \quad \lambda', \lambda'' \in \mathbb{R}^+ \tag{1.7}$$

For summits of type I or II (resp. III) the transit maps $G$ have formal counterparts $\tilde{G}$ which are either power series or elementary transseries (made up of two elementary series $\tilde{H}$ and $\tilde{K}$ at most) and of the form:

$$\begin{aligned} \text{Type I:} \quad \tilde{G} &= c z^{\lambda} \left\{ 1 + \sum_{\sigma \in \mathbb{N} + \lambda\mathbb{N}} c_\sigma z^{-\sigma} \right\} \\ &= \tilde{K} \circ P_\lambda \circ \tilde{H} \end{aligned} \tag{1.8}$$

$$\text{Type II:} \quad \tilde{G} = c z^{\lambda} \left\{ 1 + \sum_{m/\sigma \leq \text{const}} c_{\sigma,m} z^{-\sigma} (\log z)^m \right\} \tag{1.9}$$

$$\begin{aligned}\text{Type III}^+:\quad \tilde{G} &= e^{\lambda z}.\left\{c_0 + \sum_{m\geq 0, n\geq 0} c_{m,n} e^{-\lambda m z} z^{-n}\right\}\\ &= \tilde{K} \circ \exp \circ \tilde{H} \qquad (1.10)\end{aligned}$$

$$\begin{aligned}\text{Type III}^-:\quad \tilde{G} &= (\log z).\left\{c_0 + \sum_{m\geq 0, n\geq 0} c_{m,n} z^{-m} (\log z)^{-n}\right\}\\ &= \tilde{H} \circ \log \circ \tilde{K} \qquad (1.11)\end{aligned}$$

Only for summits of type I with a diophantine eigenvalue ratio $\lambda$ are the formal transit maps $G$ guaranteed to be convergent. In all other cases they are generically divergent, but with a quite specific type of divergence: *seriable* for type I with Liouvillian $\lambda$ and *resurgent* for types II and III. But they are always resummable under the standard Borel-Laplace procedure relative to a single "critical time class" $\{z_\star\}$[1]:

| | | |
|---|---|---|
| Type I and II: | critical time class | $\{z_\star\} = \{\log z\}$ |
| Type III$^+$: | critical time class | $\{z_\star\} = \{z^p\}$ |
| Type III$^-$: | critical time class | $\{z_\star\} = \{\log z\}$ |

## 1.3 The philosophy of "complete formalisation"

The methodological ideal—not only in this, but in a host of other problems—is to replace a local geometric object $F$ (say, a function germ) by a formal one $\tilde{F}$ (say, a series or transseries—the tilda signals formalness) without any loss of information ($F$ should be constructively recoverable from its idealisation $\tilde{F}$) and in such a way that all manipulations performed on $F$ (differentiation, integration, composition, etc.) should translate into purely formal— and therefore "mechanical", entirely perspicuous, and far simpler—operations on $\tilde{F}$. Graphically:

| Formal objects | biconstructive passage | Functions germs |
|---|---|---|
| $\tilde{F}$ | $\longleftrightarrow$ | $F$ |
| $\{+, \times, \partial, \circ, \prec\}$ | $\longleftrightarrow$ | $\{+, \times, \partial, \circ, \prec\}$ |

Here is a table that suggests the four levels of increasing difficulty that can be encountered in this process of biconstructive formalisation:

| convergent power series | straightforward summation | Weierstrass element defined in a full neighbourhood of $\infty$ |
|---|---|---|
| $\tilde{F}(z) = \sum a_n z^{-n}$ | $\longmapsto$ | $F(z)$ |

[1]Defined up to ordinary equivalence $\sim$.

| | | |
|---|---|---|
| divergent monocritical power series | Borel-Laplace summation | analytic germ defined in a sectorial neighbourhood of $+\infty$ |
| $\tilde{F}(z) = \sum a_n z^{-n}$ | $\longmapsto$ | $F(z)$ asymptotic to $\tilde{F}(z)$ |
| divergent polycritical power series | accelero-summation | analytic germ defined in a sectorial neighbourhood of $+\infty$ |
| $\tilde{F}(z) = \sum a_n z^{-n}$ | $\longmapsto$ | $F(z)$ asymptotic to $\tilde{F}(z)$ |
| divergent polycritical transseries | accelero-synthesis | analysable germ defined in a tapering neighbourhood of $+\infty$ |
| $\tilde{F}(z) = \sum a_\sigma A_\sigma(z)$ | $\longmapsto$ | $F(z)$ trans-asymptotic to $\tilde{F}(z)$ |

## 1.4 Transseries and analysable germs: some heuristics

**On the formal side,** since we demand complete closure under the direct and reverse operations $+, \times, \circ, \partial_z$, we should be prepared to tackle not only creatures like:

$$
\begin{array}{ccccccc}
z & \xrightarrow{\text{inversion}} & z^{-1} & \xrightarrow{\text{integration}} & \log z & \xrightarrow{\text{iteration}} & \log_n z := \log^{\circ n}(z) \\
\Big\uparrow \scriptstyle\text{integration} & & & & \Big\downarrow \scriptstyle\text{reciprocation} & & \Big\downarrow \scriptstyle\text{reciprocation} \\
1 & & & & \exp z & \xrightarrow{\text{iteration}} & \exp_n z := \exp^{\circ n}(z)
\end{array}
$$

but also hugely generalised "series" assembled from such elements. In other words, we should expect having to deal with *"transseries"* consisting of *well-ordered* sums of *"transmonomials"*, with a natural *transfinite indexation* on Cantor's interval $[0, \omega^\omega[$. But let's be a wee bit more specific.

(i) "Well-orderedness" means that any two transmonomials should be comparable (as large or small infinitesimals) and that each subtransseries $\tilde{H}$ obtained by truncating a given transseries $\tilde{F}$ should start with a "first" or "dominant" transmonomial $\tilde{A}_0$. This in turn is indispensable to ensure multiplicative invertibility:

$$\tilde{H} = a_0 \tilde{A}_0 + \tilde{K} \quad \longrightarrow \quad \tilde{H}^{-1} = \sum_{n \geq 1} \frac{(-1)^{n-1}}{a_0^n} \frac{\tilde{K}^{n-1}}{\tilde{A}_0^{\,n}} \qquad (a_0 \in \mathbb{R}^*)$$

(ii) Transseries will consist of an *asymptotic part*, with indexation on $[0, \omega[$, that can be derived elementarily from the corresponding germ , and a *trans-asymptotic part*, with indexation on $[\omega, \omega^\omega[$, whose geometric interpretation is going be more elusive, presupposing as it does a prior *resummation* of the asymptotic part.

(iii) The transmonomials themselves are "atomic" in the sense that they cannot be split into *sums* of simpler objects, but that does not prevent them from possessing a highly intricate arborescent *substructure* nor from often carrying a (countable) infinity of coefficients.

**On the analysis side,** we should be prepared to encounter *omnipresent divergence, of a severe but resummable sort.* Why so? Neither multiplication, nor composition, nor inversion, nor reciprocation, nor differentiation can of themselves cause divergence, though they may aggravate it. The primary source of divergence is *integration*, and of course the solving of functional equations (ODEs, PDEs, difference equations, etc.). Indeed, even quite simple transmonomials have divergent indefinite integral. For instance:

$$\int e^z z^\sigma dz \quad \longrightarrow \quad e^z \sum_{m\geq 0} a_m z^{1+\sigma-m} \qquad (\sigma \in \mathbb{R} \dot{-} \mathbb{N})$$

This actually holds true for most other transmonomials, including elementary ones[2] like:

$$e^{P(z)} z^\alpha, \qquad z^\alpha (\log z)^\beta, \qquad e^{P(z)} z^\alpha \prod_{i=1}^{r} (\log_i z)^{\beta_i}, \qquad \text{etc.}$$

In fact the only exceptions are $z^\sigma$, $e^{az}z^n$, $(\log z)^n$ $(n \in \mathbb{N})$ .

Thus, integration will produce divergence at every step, while the other operations, which on their own cannot generate divergence, must be counted upon to make it more severe when they encounter it in their way. Indeed, if $A, B, C$ be (large) transmonomials from different Archimedean scales[3], their indefinite integrals will generically exhibit monocritical divergence[4], but any operation that superposes or intermingles them, like $(\int A + \int B + \int C)^n$ or $\log_n(\int A + \int B + \int C)$, will produce *polycritical transseries* or, more baffling still, *polycritical transmonomials*, like $\exp_n(\int A + \int B + \int C)$ for $n$ large enough.

## 1.5 The algebra $\mathbb{R}[[[z]]]$ of transseries

We shall use the following suggestive notations:

| ...................... | infinitely large | infinitely small | unspecified |
|---|---|---|---|
| transmonomials | $\sqcap$ | $\sqcup$ | $\square$ |
| of exponential depth $m$ | ${}^m\sqcap$ | ${}^m\sqcup$ | ${}^m\square$ |
| transseries | $\sqcap\!\sqcap$ | $\sqcup\!\sqcup$ | $\square\!\square$ |
| of exponential depth $m$ | ${}^m\sqcap\!\sqcap$ | ${}^m\sqcup\!\sqcup$ | ${}^m\square\!\square$ |
| ...................... | .............. | .............. | .......... |

After naming our objects, let us proceed with their inductive definition.

[2]Elementary in the sense of carrying only a finite number of parameters or "internal coefficients".

[3]Meaning that their logarithms are non-equivalent.

[4]$\int A, \int B, \int C$ are going to be resurgent and resummable, each with a single critical time class (or more, if $A, B, C$ already carry internal divergence.

**Alogarithmic transmonomials and transseries of depth 0:**

*Alogarithmic transmonomials* of exponentiality 0 are ordinary monomials and *alogarithmic transseries* of exponentiality 0 are ordinary series

$$ {}^{0}\sqcap(z) = z^{\sigma} \qquad {}^{0}\sqcup(z) = z^{-\sigma} \qquad (\sigma > 0) $$
$$ {}^{0}\sqcap\!\sqcap(z) = \sum c_\alpha z^{\sigma_\alpha} \qquad {}^{0}\sqcup\!\sqcup(z) = \sum c_\alpha z^{-\sigma_\alpha} \qquad (\sigma_\alpha > 0) $$

with the finite generation axiom:

$$ {}^{0}\sqcap\!\sqcap = \text{finite sum}; \qquad {}^{0}\sqcup\!\sqcup \text{ in some } \mathbb{R}[[z^{-\sigma_1}, \ldots, z^{-\sigma_r}]] $$

and the natural order.

**Alogarithmic transmonomials and transseries of depth $m \geq 1$:**

An *alogarithmic transmonomial* of exponentiality $m$ is any expression:

$$ {}^{m}\sqcap(z) = z^{\sigma} \exp(\sqcap\!\sqcap(z)) \quad \text{or} \quad {}^{m}\sqcup(z) = z^{\sigma} \exp(-\sqcap\!\sqcap(z)) $$

with $\sigma \in \mathbb{R}$ and

$$ \sqcap\!\sqcap = {}^{m-1}\sqcap\!\sqcap + {}^{m-2}\sqcap\!\sqcap + \cdots + {}^{1}\sqcap\!\sqcap + {}^{0}\sqcap\!\sqcap > 0 \quad \text{(formally)} $$

The (total) order on such transmonomials is defined inductively by:

$$ \{\square_1 < \square_2\} \iff \{\square_1/\square_2 = \sqcup_3\} $$

A *homogeneous alogarithmic transseries* ${}^{m}\boxminus$ of exponentiality $m$ is any expression of the form:

$$ \sqcap\!\sqcap = \sum c_\alpha \, {}^{m}\sqcap_\alpha \quad \text{or} \quad \sqcup\!\sqcup = \sum c_\alpha \, {}^{m}\sqcup_\alpha $$

with a possibly transfinite, order-reflecting indexation $\alpha$, but with the finite generation axiom:

$$ {}^{m}\sqcap\!\sqcap / {}^{m}\sqcap_0 \quad \text{and} \quad {}^{m}\sqcup\!\sqcup \quad \text{in some} \quad \mathbb{R}[[\sqcup_1, \ldots, \sqcup_r]] $$

Lastly, a *general alogarithmic transseries* $\boxminus$ is any finite sum of homogeneous transseries.

**Unique additive decomposition:**

It is of the form:

$$ \begin{aligned} \boxminus &= \sqcap\!\sqcap + \text{const} + \sqcup\!\sqcup \\ &= {}^{m}\sqcap\!\sqcap + \cdots + {}^{1}\sqcap\!\sqcap + {}^{0}\sqcap\!\sqcap + \text{const} + {}^{0}\sqcup\!\sqcup + {}^{1}\sqcup\!\sqcup + \cdots + {}^{n}\sqcup\!\sqcup \\ &= \sum_{0 \leq \alpha < \omega_0 < \omega^{\omega}} c_\alpha \square_\alpha \end{aligned} $$

and can be obtained by repeated application of the "pull-down":

$$ \exp(\sqcap\!\sqcap + \text{const} + \sqcup\!\sqcup) = e^{\text{const}}.(\exp \sqcap\!\sqcap). \left(1 + \sum_{n \geq 1} \frac{1}{n!} \sqcup\!\sqcup^{n}\right) $$

under which all infinitesimals are expelled or "pulled down" from the exponentials, wherever (at whichever height) they may be found.

**General transmonomials and transseries of depth $m > -\infty$:**

They are obtained by postcomposing *alogarithmic* transmonomials or transseries by some finite iterate $\log_n := \log^{\circ n}$ of the logarithm. General transseries clearly inherit the unique additive expansion of their alogarithmic models.

**Example of general transseries:**

$$\begin{aligned} \boxed{\phantom{x}}\boxed{\phantom{x}} &= \\ \text{(larger)} \quad &+\sum \mathbf{a}.e^{e^{e^{3\,z}\sum \mathbf{c} \sum \mathbf{b} e^{-\mathbf{d} z} z^{n}}}.z^{-n} \\ \text{(large)} \quad &+\sum \mathbf{f}.e^{e^{z^2}\sum \mathbf{g} e^{-mz} z^{-n} (\log z)^p}.z^{-n^3} \\ \text{(small)} \quad &+\sum \mathbf{h}.e^{-z^2}.e^{mz}.z^{-m} \\ \text{(smaller)} \quad &+\sum \mathbf{k}.e^{-mz^2}.z^{-p}.(\log z)^q \end{aligned}$$

## 1.6 The algebra $\mathbb{R}\{\{\{z\}\}\}$ of analysable germs

To define the notion of *analysable germ*, which is the geometric counterpart of a (resummable) transseries, we require four main ingredients:

(1) the notion of *resurgent function* and *alien derivation*;

(2) the so-called *acceleration transforms*, which in some sense generalise the classical Laplace transform;

(3) the notion of *cohesive function* and cohesive continuation, which considerably extends that of analytic function and analytic continuation;

(4) the notion of *well-behaved average*, which makes it possible to turn multiform functions on $\mathbb{R}^+$ into uniform ones, in a way that is both agreeable to the convolution product and the acceleration transforms.

All these notions shall be briefly recalled in the next paragraphs (and in somewhat greater detail in the Appendix Section 6), but in order to convince ourselves of their necessity, and to show roughly where they fit into our scheme of things, we shall venture right away a (very sketchy) definition of *analysability*.

A real function germ $\varphi(z)$ at $+\infty$ is said to be *analysable* if it can be obtained from some transseries $\tilde{\varphi}(z)$ under a process known as *accelero-summation*, which involves a finite number of integral transforms (Borel, then several accelerations, then Laplace) and a finite number of critical time classes $\{z_i\} = \{h_i(z)\}$, with link-up functions $h_i(z)$ that are themselves required to be analysable, with their own finitely many secondary critical times $\{z_{i,j}\} = \{h_{i,j}(z)\}$, and so on, leading to a finite critical tree $\{z_{i_1,i_2,\dots,i_r}\}$.

Moreover, we may and often do demand (though this is by no means necessary) *stability under alien differentiation*: in other words, we may and often do ask that all alien derivatives $\mathbb{\Delta}_{\omega_s}^{\{z_{i_s}\}} \dots \mathbb{\Delta}_{\omega_2}^{\{z_{i_2}\}} \mathbb{\Delta}_{\omega_1}^{\{z_{i_1}\}} \varphi(z)$ be themselves analysable.

Here is a pictorial representation of this process of accelero-summation, relative to a sequence of $r$ critical times:

$$z_1 \prec z_2 \prec z_3 \prec \cdots \prec z_{r-1} \prec z_r \qquad (z_i \equiv h_i(z) \succ 1)$$

$$\begin{array}{cccccccc}
 & & & \text{Laplace} & & & & \\
\tilde{\varphi}_r(z_r) & z_r & \bullet & \longleftarrow & \bullet & \zeta_r & & \overset{\diamond}{\varphi}_r\,(\zeta_r) \\
 & & \downarrow & & \uparrow & \text{acceleration} & & \\
\varphi(z) & z & \bullet & & \bullet & \zeta_{r-1} & & \overset{\diamond}{\varphi}_{r-1}\,(\zeta_{r-1}) \\
 & & & & \uparrow & \text{acceleration} & & \\
 & & & & \cdots & \cdots & & \cdots \\
 & & & & \uparrow & \text{acceleration} & & \\
\tilde{\varphi}(z) & z & \bullet & & \bullet & \zeta_2 & & \overset{\diamond}{\varphi}_2\,(\zeta_2) \\
 & & \downarrow & & \uparrow & \text{acceleration} & & \\
\tilde{\varphi}_1(z_1) & z_1 & \bullet & \longrightarrow & \bullet & \zeta_1 & & \overset{\diamond}{\varphi}_1\,(\zeta_1) \\
 & & & \text{Borel} & & & &
\end{array}$$

Each acceleration $\hat{\varphi}_i(\zeta_i) \mapsto \hat{\varphi}_{i+1}(\zeta_{i+1})$ is actually three steps in one.

In step 1, the acceleration transform turns a uniform function $\hat{\varphi}_i(\zeta_i)$ defined *on* the whole of $\mathbb{R}^+$ into a function germ $\hat{\varphi}_{i+1}(\zeta_{i+1})$ defined at $+0$.

In step 2, the germ in question in extended, by analytic or cohesive continuation, into a global but multiform function, defined *over* the whole of $\mathbb{R}^+$.

In step 3, under a suitable process of averaging, this multiform $\hat{\varphi}_{i+1}(\zeta_{i+1})$ is turned once again into a uniform function defined *on* the whole of $\mathbb{R}^+$ , thus paving the way for the next three-stepped acceleration process.

Pictorially:

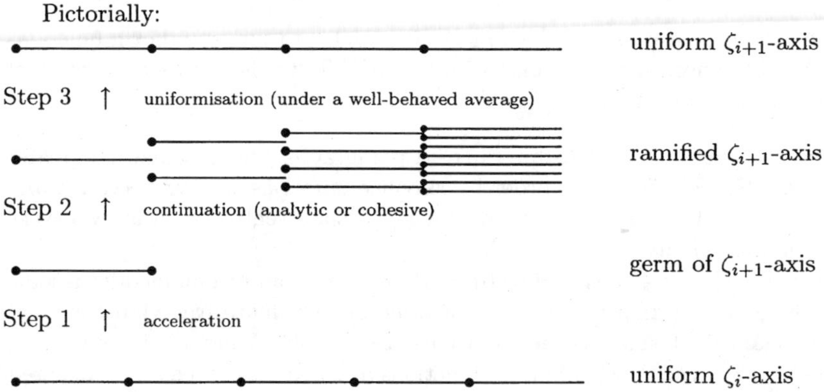


## 1.7 Resurgent functions and alien derivations

The algebra of *real* resurgent functions (this is the one that is relevant to the construction of the trigebra of analysable germs) consists of all (analytic or cohesive) function germs at

+0 that possess an *endless* (analytic or cohesive, but usually ramified) *forward* continuation *over* the whole of $\mathbb{R}^+$. They are subject to the convolution product, and there acts upon them an incredibly rich system of alien derivations.

More precisely, *locally* at 0, resurgent function may be thought of as microfunctions: they are pairs $\overset{\diamond}{\varphi}(\zeta) = \{\check{\varphi}(\zeta), \hat{\varphi}(\zeta)\}$ consisting of a major $\check{\varphi}$ defined upto a regular germ, and of a minor $\hat{\varphi}$, which is the "variation" of the major: see Section 6.2. The minor is exactly defined, but does not encapsulate the whole information about $\overset{\diamond}{\varphi}$, except in the important case of "integrable resurgent functions". For these the convolution product is defined by

$$(\hat{\varphi}_1 \star \hat{\varphi}_2)(\zeta) := \int_0^{\zeta} \hat{\varphi}_1(\zeta_1)\, \hat{\varphi}_2(\zeta - \zeta_1)\, d\zeta_1 \tag{1.12}$$

for $\zeta$ close to +0 and by analytic or cohesive continuation in the large. The convolution for general germs $\overset{\diamond}{\varphi}$ is defined in Section 6.2.

What really matters, however, is not the local or microfunction aspect, but the *global* properties of resurgent functions $\overset{\diamond}{\varphi}$, which come from their having endlessly continuable, but usually highly ramified minors $\hat{\varphi}$. This a source of many fascinating developments, chief amongst which is the existence of a system $\{\Delta_\omega,\ \omega \in \mathbb{R}^+\}$ of so-called alien derivations[5], which measure the singularities of the minor $\hat{\varphi}$ *over* the points $\omega$ and which, together, freely generate a infinite dimensional Lie algebra with an uncountable basis—despite their acting on functions of one single variable!

## 1.8 Basic transforms: Borel, Laplace, acceleration

**The Borel transform:**

It turns germs with subexponential growth at $\infty$ in the $z$-plane into germs at +0 in the $\zeta$-plane:

$$\varphi(z) \longmapsto \hat{\varphi}(\zeta) := \frac{1}{2\pi i} \int_{c-i\infty}^{c+i\infty} \varphi(z)\, e^{z\zeta}\, dz \tag{1.13}$$

**The Laplace transform:**

It reverses the Borel transform and turns uniform functions on $\mathbb{R}^+$ with (at most) exponential growth into germs at $\infty$ in the $z$-plane or half-plane:

$$\hat{\varphi}(\zeta) \longmapsto \varphi(z) := \int_{+0}^{+\infty} \hat{\varphi}(\zeta)\, e^{-z\zeta}\, d\zeta \tag{1.14}$$

**The acceleration transform:**

It turns global, uniform functions that are defined on a $\zeta_1$-axis and do not exceed a certain critical growth regime (the so-called *accelerable growth*, which depends on the acceleration but is always strictly faster-than-exponential) into germs that are defined at the origin of a $\zeta_2$-axis.

---

[5]Because they are indeed derivations, relative to the convolution product.

The acceleration of index $F = F_{1,2} = o(z)$ transmutes under Borel-Laplace the variable change $\varphi_1 \mapsto \varphi_2 := \varphi_1 \circ F_{1,2}$, with the "slower time" expressed in terms of the "faster time": $z_1 = F_{1,2}(z_2)$ and $z_1 \prec z_2$.

$$\begin{array}{ccccc} & & \text{Borel} & & \\ & \varphi_2(z_2) & \longmapsto & \hat{\varphi}_2(\zeta_2) & \\ \text{variable change} & \uparrow & & \uparrow & \text{acceleration} \\ & \varphi_1(z_1) & \longmapsto & \hat{\varphi}_1(\zeta_1) & \\ & & \text{Borel} & & \end{array}$$

It is an integral transform similar in form to the Laplace transform:

$$\hat{\varphi}_1(\zeta_1) \longmapsto \hat{\varphi}_2(\zeta_2) := \int_{+0}^{+\infty} C_F(\zeta_2, \zeta_1)\, \hat{\varphi}_1(\zeta_1)\, d\zeta \tag{1.15}$$

but with a kernel

$$C_F(\zeta_2, \zeta_1) := \frac{1}{2\pi i} \int_{c-i\infty}^{c+i\infty} e^{z_2\zeta_2 - z_1\zeta_1}\, d\zeta_1 \quad \text{with} \quad z_1 = F(z_2) \prec z_2 \tag{1.16}$$

that has always strictly faster-than-exponential decrease when $\zeta_1 \to \infty$ with fixed $\zeta_2 > 0$. This explains why all acceleration operators have larger domains of definition than the Laplace transform.

Moreover, for $F$ fixed, the closer $\zeta_2$ is to 0, the faster that rate of $\zeta_1$-decrease. This explains why accelerations usually yield only germs rather than (directly) global functions.

Lastly, and somewhat paradoxically, when $F$ gets closer and closer to the identity (i.e. when the gap between the two times becomes less), the kernel's rate of decrease as $\zeta_1 \to +\infty$ becomes fiercer. When $\log z_1 \prec \log z_2$, ie when the two critical times are not too close, the corresponding acceleration always yields analytic[6] germs $\hat{\varphi}_2(\zeta_2)$, irrespective of the nature of the accelerand $\hat{\varphi}_1(\zeta_1)$. However, for very close, logarithmically equivalent times (i.e. $\log z_1 \sim \log z_2$), the domain of definition of the corresponding acceleleration operator increases hugely, while the accelerates $\hat{\varphi}_2(\zeta_2)$ usually cease to be *analytic* and become merely *cohesive*—again, irrespective of the nature of the accelerands $\hat{\varphi}_1(\zeta_1)$.

**Algebra morphisms:**

- Borel turns multiplication into convolution.
- Accelerations respect convolution.
- Laplace turns convolution into multiplication.

Thus, each step of the accelero-summation process is an algebra morphism—which is absolutely essential, because we want the process to apply not just to linear, but also and above all to non-linear situations. In other words, if we calculate (by purely formal manipulations—inductive coefficient identification and the like) a formal transserial solution $\tilde{\varphi}$ to some non-linear problem (say, a differential equation $E$), we don't want to merely "resum" $\tilde{\varphi}$. We also insist on getting a sum $\varphi$ that actually satisfies the original equation $E$. This compels us to work with algebra morphisms only.

[6] Analytic, that is, right of 0, on some internal $]0, \ldots]$, but not necessarily at 0 itself.

## 1.9 Cohesive functions and cohesive continuation

Analytic functions on a closed real interval $I$ are characterised by the existence of uniform bounds on their derivatives:

$$|f^{(n)}(t)| \leq c_0 c_1^n n^n \qquad (\forall n \in \mathbb{N},\ \forall t \in I) \tag{1.17}$$

with constants $c_0, c_1$ which of course depend on $f$.

Similarly, the classical quasianalytic Denjoy classes ${}^{\alpha}DEN$ of finite order $\alpha \in \mathbb{N}$ are characterised by the uniform bounds:

$$|f^{(n)}(t)| \leq c_0(c_1/\log'_{1+\alpha}(n))^n \quad \Longleftrightarrow \tag{1.18}$$

$$|f^{(n)}(t)| \leq c_0(c_1.n.\log n \log_2 n \ldots \log_\alpha n)^n \tag{1.19}$$

But these classes are insufficient for our purposes. We also require "Denjoy classes" of transfinite order $\alpha < \omega^\omega$, defined by the same type of bounds:

$$|f^{(n)}(t)| \leq c_0(c_1/\log'_{1+\alpha} n)^n \tag{1.20}$$

but relatively to suitably defined *transfinite* iterates $\log_\alpha$ of the logarithm.

The importance of the algebra *COHES* of cohesive fonctions:

$$COHES = \bigcup_{\alpha<\omega^\omega} {}^{\alpha}DEN = \lim_{\alpha\to\omega^\omega} {}^{\alpha}DEN \tag{1.21}$$

comes:

(1) from the property of unique continuation, also known as quasianalyticity: a cohesive function defined on an interval $J$ vanishes on the whole of $J$ as soon as it vanishes on a given subinterval $I \subset J$ (this is I-quasianalyticity) or as soon as all its derivatives vanish at some point $t \in J$ (this is D-quasianalyticity).

(2) from its being vaster than the finite Denjoy classes: thus important functions like $\varphi(t) := \sum_{n\geq 1} 1/\exp_n(t)$ are in ${}^{\omega}DEN$ but in no finite ${}^{n}DEN$.

(3) from its stability: whereas the *union* of all quasianalytic Carleman classes is not quasianalytic, *COHES* is closed under $+, \times, \partial$ etc.

(4) from its close connection with the theory of acceleration: indeed, weak accelerates[7] are cohesive and, conversely, any cohesive function is a weak accelerate.

To fully appreciate the significance of the last point, let us recall that the whole process of recovering an analysable function $\varphi$ from its transseries $\tilde{\varphi}$ resolves into a succession of accelerations, each one of which decomposes into three substeps:

(1) the first substep turns a global, uniform function *on* $\mathbb{R}^+$ into a germ at the origin.

(2) the providential fact that this germ is either analytic or cohesive makes the second substep—unique forward continuation—possible and yields a global, but usually multiform function[8] *over* $\mathbb{R}^+$.

[7] I.e. *weak* means that $\log z_1 \sim \log z_2$.

[8] Strange though it may seem, *cohesive functions*, despite their being defined only on the real axis and nowhere else, may possess *cohesive singularities* and these may, in some non-obvious but precise and *constructive* sense, be "by-passed", either to the right or to the left, just like analytic singularities, thus giving rise to two different determinations beyond the singularity. See [E7, E11].

(3) the last substep consists in averaging this multiform function to make it uniform, so that it may be subjected to the next acceleration[9]. But serious difficulties lie in wait here, which we must now address.

## 1.10 The problem of uniformisation

### 1.10.1 The notion of "convolution-respecting average"

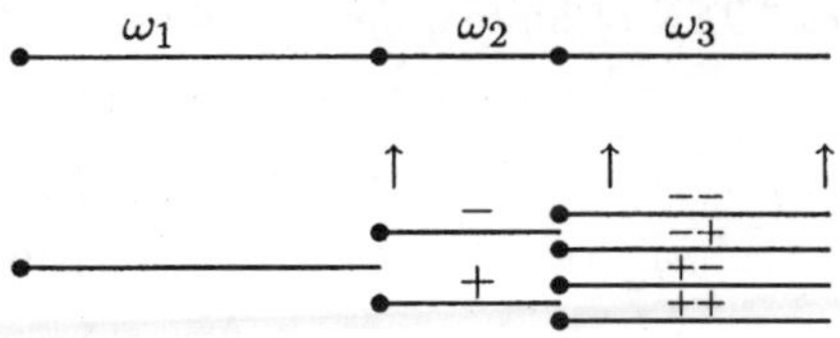


uniform function $(\mathbf{m}.\hat{\varphi})(\zeta)$

multiform function $\hat{\varphi}(\zeta)$

A uniformising average $\mathbf{m} : \hat{\varphi} \mapsto \mathbf{m}.\hat{\varphi}$ is defined by a system of weights $\mathbf{m}^{\binom{\epsilon_1,\dots,\epsilon_r}{\omega_1,\dots,\omega_r}}$ subject to the self-consistency relations:

$$\sum_{\epsilon_i \in \{+,-\}} \mathbf{m}^{\binom{\epsilon_1,\dots,\epsilon_i,\dots,\epsilon_r}{\omega_1,\dots,\omega_i,\dots,\omega_r}} = \mathbf{m}^{\binom{\epsilon_1,\dots,\epsilon_{i+1},\dots,\epsilon_r}{\omega_1,\dots,\omega_{i+1},\dots,\omega_r}} \qquad \forall i \tag{1.22}$$

and its action is as follows:

$$(\mathbf{m}.\hat{\varphi})(\zeta) := \sum_{\epsilon_i = \pm} \mathbf{m}^{\binom{\epsilon_1,\dots,\epsilon_r}{\omega_1,\dots,\omega_r}} \hat{\varphi}^{\binom{\epsilon_1,\dots,\epsilon_r}{\omega_1,\dots,\omega_r}}(\zeta) \tag{1.23}$$

with $\hat{\varphi}^{\binom{\epsilon_1,\dots,\epsilon_r}{\omega_1,\dots,\omega_r}}$ denoting the determination of $\hat{\varphi}$ on the branch of address $(\epsilon_1,\dots,\epsilon_r)$ over the interval $\zeta \in ]\omega_1 + \dots \omega_r, \omega_1 + \dots \omega_{r+1}[$ between two consecutive singularities.

To be really useful in the present context of accelero-summation, a uniformising average must fulfill three main conditions:

(C1) It must respect convolution[10], i.e. $\mathbf{m}.(\hat{\varphi}_1 \star \hat{\varphi}_2) = (\mathbf{m}.\hat{\varphi}_1) \star (\mathbf{m}.\hat{\varphi}_2)$.

(C2) It must respect realness, i.e. $\mathbf{m}.\hat{\varphi}(\zeta)$, as a global function, must be real whenever $\hat{\varphi}(\zeta)$, as a germ at $+0$, is real.

(C3) It must respect lateral growth, that is to say, $\mathbf{m}.\hat{\varphi}(\zeta)$ must not grow significantly faster than the two lateral determinations (right or left) of $\hat{\varphi}(\zeta)$ along the positive real axis.

Such averages will be declared "well-behaved".

- C1 is essential to get algebra morphisms.
- C2 is natural and, in many instances, indispensable.
- C3 ensures the convergence of the acceleration (or Laplace) integrals.

[9]Or to the Laplace transform, if we are at the very end of the resummation process.

[10]Observe that we are dealing here with two slightly different interpretations of the convolution product: in $\mathbf{m}.(\hat{\varphi}_1 \star \hat{\varphi}_2)$ we convolute two function germs near the origin, then use analytic (or cohesive) forward continuation to get a global ramified function, and lastly we uniformise it by means of $\mathbf{m}$, whereas in $(\mathbf{m}.\hat{\varphi}_1) \star (\mathbf{m}.\hat{\varphi}_2)$ we directly convolute two global, uniform functions.

C2's translation in terms of weights is straightforward: the weights should change into their complex conjugates when all signs $\epsilon_i$ are changed (if real, the weights should remain unchanged). As for C1 and C3 we shall see in a moment what they imply in terms of weights. But right now we must comment, however briefly, on this vexed issue of "lateral growth".

### 1.10.2 Central difficulty: faster-than-lateral growth

As pointed out, accelero-summation yields at every step $i$ a function $\hat{\varphi}_i(\zeta_i)$ which has precisely the right rate of growth, i.e. the one that makes the next acceleration (or Laplace) possible. But this applies only to singularity-free axes or, on singularity-carrying axes, to the two lateral determinations.[11] Most other determinations of $\hat{\varphi}_i(\zeta_i)$, especially the ones that correspond to oft-crossing paths, tend to display slightly faster-than-lateral growth.

For instance, if the lateral growth of $\hat{\varphi}(\zeta)$ is exponential (this is the growth that permits Laplace), the other determinations generally admit no better uniform bounds than $|\hat{\varphi}(\zeta)| \leq \gamma_0 e^{\gamma_1 |\zeta| |\log \zeta|}$ (so that they cannot be subjected to Laplace). Therefore, unless we resort to carefully honed averages $\mathbf{m}$, the averaged function $\mathbf{m}.\hat{\varphi}(\zeta)$ itself is going to display this slightly super-exponential growth.

This nuisance of faster-than-lateral growth is extremely common, generic almost. In the Dulac problem, it affects nearly all transit maps $G_i$ attached to summits of type III.

In order to show just how prevalent and inescapable the phenomenon of faster-than-lateral growth is, let us adduce the simplest conceivable instance of its occurence:

$$\tilde{A}(z) := \sum_{n\geq 1} n!\, z^{-n} \tag{1.24}$$

$$\hat{A}(\zeta) := \sum_{n\geq 0} \zeta^n = 1/(1-\zeta) \tag{1.25}$$

$$\tilde{B}(z) := \sum_{n\geq 1} A^n(z) = A(z)/(1-A(z)) \tag{1.26}$$

$$\hat{B}(\zeta) := \sum_{n\geq 1} \hat{A}^{\star n}(\zeta) \tag{1.27}$$

The divergent series $\tilde{A}(z)$ verifies the Euler equation $(1+\partial_z)\tilde{A}(z) = z^{-1}$ and its Borel transform $\hat{A}(\zeta)$, with its single pole at $\zeta = 1$, is a most elementary resurgent function. Yet a simple Möbius transform turns $\tilde{A}(z)$ into a series $\tilde{B}(z)$, also solution of a first-order differential equation, but with a Borel transform $\hat{B}(\zeta)$ that has singularities at every point $\zeta = n \in \mathbb{N}$, with simple poles as leading terms and logarithmic singularities as corrections:

$$\begin{aligned}\hat{B}(\zeta) = &+ R_B^{\epsilon_1,\ldots,\epsilon_{n-1},\star}.(\zeta-n)^{-1} && \text{(simple pole)}\\ &+ Reg_1^{\epsilon_1,\ldots,\epsilon_{n-1},\star}(\zeta-n).\log(\zeta-n) && \text{(logarithmic singularity)}\\ &+ Reg_0^{\epsilon_1,\ldots,\epsilon_{n-1},\star}(\zeta-n) && \text{(regular part)}\end{aligned}$$

[11] And that too even if we take care of choosing in the critical time class $\{z_i\}$ a suitably slow time $z_i$, which precaution has the effect of smoothing the singularities of $\hat{\varphi}_i(\zeta_i)$.

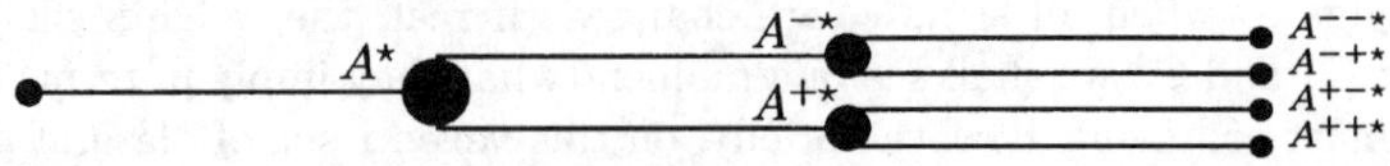


This even provides us with a discretised model of the phenomemenon of faster-than-lateral growth. Indeed, the residues $R_B^{\epsilon_1,\ldots,\epsilon_{n-1},\star}$ of address $\{\epsilon_i\}$ are calculable by a simple induction (see [E12]) which readily shows that they satisfy no better bounds than

$$|R_B^{\epsilon_1,\ldots,\epsilon_{n-1},\star}| \leq c_0 c_1^n n! \tag{1.28}$$

Therefore, although $\hat{B}$ has exponential growth (at most) on each singularity-free axis $\arg(\zeta) = \theta \neq 0$, on paths that criss-cross with the axis $\arg(\zeta) = 0$ (especially for constantly alternating $\epsilon_i$'s) it admits no better bounds than $|\hat{B}(\zeta)| \leq \gamma_0 e^{\gamma_1 |\zeta| |\log \zeta|}$.

### 1.10.3 Criteria for well-behaved convolution averages

To any uniformising average $\mathbf{m}$ we may attach the moulds[12]:

$$\mathbf{rem}^{\omega_1,\ldots,\omega_r} := (-1)^r \mathbf{m}^{\binom{+,\ldots,+}{\omega_1,\ldots,\omega_r}} \qquad (\text{"right-lateral mould"}) \tag{1.29}$$

$$\mathbf{lem}^{\omega_1,\ldots,\omega_r} := (-1)^r \mathbf{m}^{\binom{-,\ldots,-}{\omega_1,\ldots,\omega_r}} \qquad (\text{"left-lateral mould"}) \tag{1.30}$$

Due to the self-consistency relations, both the right- and left-lateral moulds encapsulate all the information about the entire weight system $\left\{\mathbf{m}^{\binom{\epsilon_1,\ldots,\epsilon_r}{\omega_1,\ldots,\omega_r}}\right\}$, and each one can be deduced from the other in a simple manner.

The three following conditions are equivalent:

(a) the uniformising average $\mathbf{m}$ *respects convolution*[13];

(b) its right-lateral mould $\mathbf{rem}^\bullet$ is *symmetrel*[14];

(c) its left-lateral mould $\mathbf{lem}^\bullet$ is *symmetrel.*

The two following conditions are equivalent:

(a) the uniformising average $\mathbf{m}$ *respects realness*[15];

(b) its moulds $\mathbf{rem}^\bullet$ and $\mathbf{lem}^\bullet$ are complex conjugate

The five following conditions are equivalent:

(d) the uniformising average $\mathbf{m}$ *respects lateral growth*[16];

(e) we have universal bounds $|\mathbf{rem}^{\omega^{\twoheadleftarrow}}| \leq C_1 (D_1)^{r(\omega^{\twoheadleftarrow})}$;

(f) we have universal bounds $|\mathbf{rem}^{\omega^{\rightarrowtail}}| \leq C_2 (D_2)^{r(\omega^{\rightarrowtail})}$;

[12]See Section 6.1.
[13]This is condition C1.
[14]See Section 6.1.
[15]This is condition C2.
[16]This is condition C3.

(g) we have universal bounds $|\mathbf{lem}^{\boldsymbol{\omega}^{\prec}}| \leq C_3(D_3)^{r(\boldsymbol{\omega}^{\prec})}$;

(h) we have universal bounds $|\mathbf{lem}^{\boldsymbol{\omega}^{\succ}}| \leq C_4(D_4)^{r(\boldsymbol{\omega}^{\succ})}$

for some constants $C_i, D_i$.

Here $M^{\bullet^{\prec}}$ (resp. $M^{\bullet^{\succ}}$ ) denotes the forward (resp backward) *contracting arborification* of the mould $M^\bullet$, whose definition is as follows:

$$M^{\boldsymbol{\omega}^{\prec}} := \sum_{\boldsymbol{\omega}^1 \gg \boldsymbol{\omega}^{\prec}} M^{\boldsymbol{\omega}^1}; \qquad M^{\boldsymbol{\omega}^{\succ}} := \sum_{\boldsymbol{\omega}^2 \gg \boldsymbol{\omega}^{\succ}} M^{\boldsymbol{\omega}^2} \tag{1.31}$$

The symbol $\boldsymbol{\omega}^{\prec}$ (resp. $\boldsymbol{\omega}^{\succ}$ ) denotes any sequence $\{\omega_i\}$ with an *arborescent* (resp. *anti-arborescent*) order on it, i.e. an order such that each element $\omega_i$ has at most one predecessor $\omega_{i_-}$ (resp. one successor $\omega_{i_+}$), whereas the sums on the right-hand side extend to all *totally ordered* sequences $\boldsymbol{\omega}^1$ (resp. $\boldsymbol{\omega}^2$) that can be obtained from $\boldsymbol{\omega}^{\prec}$ (resp. $\boldsymbol{\omega}^{\succ}$ ) with the possible *contraction* $\omega_i, \ldots, \omega_j \mapsto \omega_i + \cdots + \omega_j$ of several consecutive elements. For some details see Section 6.1 and for more details go to [E5, E12, EV3].

### 1.10.4 Examples of well-behaved averages

For any fixed $\tau \geq 0$, consider the multiplicative semigroup $\{{}^\tau g_\omega(\bullet),\ \omega > 0\}$ and its Fourier transform, the convolution semigroup $\{{}^\tau f_\omega(\bullet),\ \omega > 0\}$:

$${}^\tau g_\omega(y) := \exp(-\omega |y|^\tau) \qquad (y \in \mathbb{R}) \tag{1.32}$$

$${}^\tau f_\omega(x) := \frac{1}{2\pi i}\int_{\mathbb{R}} {}^\tau g_\omega(y) e^{ixy}\, dy \qquad (x \in \mathbb{R}) \tag{1.33}$$

$${}^\tau g_{\omega_1}(y)\, {}^\tau g_{\omega_2}(y) \equiv^\tau g_{\omega_1+\omega_2}(y) \tag{1.34}$$

$$({}^\tau f_{\omega_1} *^\tau f_{\omega_2})(x) := \int_{-\infty}^{+\infty} {}^\tau f_{\omega_1}(x_1)\, {}^\tau f_{\omega_2}(x - x_1)\, dx_1 \equiv^\tau f_{\omega_1+\omega_2}(x) \tag{1.35}$$

Each function ${}^\tau f_\omega$ may be viewed as representing the probability distribution at the time $t = \omega$, on the vertical axis $\omega + i\mathbb{R}$, of a particle starting from the origin at $t = 0$, moving along $\mathbb{R}^+$ at uniform horizontal speed, and diffusing randomly in the vertical direction.

We may then define ${}^\tau \mathbf{m}^{\binom{\epsilon_1, \ldots, \epsilon_r}{\omega_1, \ldots, \omega_r}}$ as the probability of our particle's successively crossing $\omega_1 + i\mathbb{R}^{\epsilon_1}, \omega_1 + \omega_2 + i\mathbb{R}^{\epsilon_2}, \ldots, \omega_1 + \omega_2 + \ldots \omega_r + i\mathbb{R}^{\epsilon_r}$.

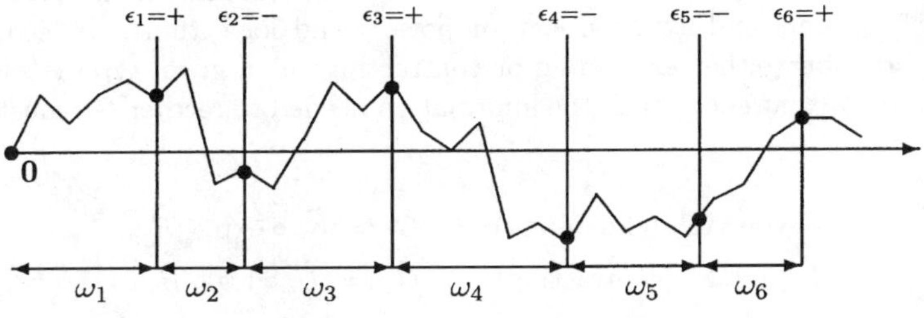


Since these numbers ${}^\tau \mathbf{m}^{\binom{\epsilon_1, \ldots, \epsilon_r}{\omega_1, \ldots, \epsilon_r}}$ satisfy the self-consistency relations (22), they may be regarded as *weights* defining a uniformising average ${}^\tau \mathbf{m}$. That average clearly respects *realness*

(condition C2) but also, less obviously, *convolution* and *lateral growth* (conditions C1 and C3). It is therefore well-behaved. Moreover, since the two parameters $\tau$ and $\omega$ essentially coalesce into one[17], ${}^\tau\mathbf{m}$ has the added distinction of being *scale-invariant*, i.e. invariant under a simultaneous, uniform dilation of all gaps $\omega_i$.

### 1.10.5 Limit-cases: the "standard" and "organic" averages

For $\tau \to +\infty$ the average ${}^\tau\mathbf{m}$ tends to the so-called *standard* (or *median*, or *uniform*) average, with weights:

$$\mathbf{m}^{\binom{\epsilon_1,\dots,\epsilon_r}{\omega_1,\dots,\omega_r}} := \frac{(2p)!\,(2q)!}{4^{p+q}\,(p+q)!\,p!\,q!} \qquad \begin{array}{l} p := \text{nb of } + \\ q := \text{nb of } - \end{array} \tag{1.36}$$

that do not depend on the gaps $\omega_i$ but only on the number $p, q$ of $\pm$ signs in the sequence $\{\epsilon_i\}$. It is actually the only convolution- and realness-respecting average with that property, but its simplicity is deceptive, since it does not respect lateral growth.

For $\tau \to +0$ the average ${}^\tau\mathbf{m}$ tends to the so-called *organic* average[18], whose weights are given by the following induction:

$$\mathbf{m}^{\binom{\epsilon_1,\dots,\epsilon_r}{\omega_1,\dots,\omega_r}} := \mathbf{m}^{\binom{\epsilon_1,\dots,\epsilon_r}{\omega_1,\dots,\omega_r}}.\left(1 - \frac{1}{2}\frac{\omega_r}{\omega_1 + \dots + \omega_r}\right) \quad \text{if} \quad \epsilon_{r-1} = \epsilon_r \tag{1.37}$$

$$\mathbf{m}^{\binom{\epsilon_1,\dots,\epsilon_r}{\omega_1,\dots,\omega_r}} := \mathbf{m}^{\binom{\epsilon_1,\dots,\epsilon_r}{\omega_1,\dots,\omega_r}}.\frac{1}{2}\frac{\omega_r}{\omega_1 + \dots + \omega_r} \quad \text{if} \quad \epsilon_{r-1} \neq \epsilon_r \tag{1.38}$$

and which, unlike the standard average, is *well-behaved*. A closer investigation reveals that it is, in some precise sense, the simplest of all such averages.

## 1.11 Application to Dulac's transit maps $G_i$

Let us now return to Dulac's problem and examine what simplifications the recourse to well-behaved averages brings about. For all three types of summit $S_i$ the formal maps $\tilde{G}_i$ may be divergent, but they are always resummable under the general resummation scheme of Section 1.6, with at most one critical time class for each summit[19]. Summits of type I or II (hyperbolic) offer no special difficulty, but those of type III (semi-hyperbolic) do, because for them the formal map is not a *bona fide* asymptotic series (with only one order of infinitesimals) but a strict *transseries* that mixes several orders of infinitesimals—in the present instance only two, namely powers and exponentials, or powers and logarithms. Indeed, depending on the transit direction (either expanding or contracting) at a given type III summit, any formalisation $\tilde{G}_i$ of $G_i$ that encodes all the information needed to recover $G_i$, must necessarily be of the form:

$$\begin{array}{ll} \text{Type III}^+ \text{ (expanding):} & \tilde{G}_i := \tilde{K}_i \circ \exp \circ \tilde{U}_i^\star \\ \text{Type III}^- \text{ (contracting):} & \tilde{G}_i :=^\star \tilde{U}_i \circ \log \circ \tilde{H}_i \end{array}$$

---

[17]Indeed ${}^\tau f_\omega(x) \equiv \omega^{-1/\tau}\;{}^\tau f_\omega(x/\omega^{1/\tau})$.

[18]Don't ask why it is dubbed "organic": it had to be given some name!

[19]So resummation here reduces to a simple Borel-Laplace shuttle.

with ordinary formal power series (of $z^{-1}$) at both ends and an exponential or logarithm as mid-factor. In the standard transseries expansion we get:

$$\tilde{G}_i(z) = \sum_{1\leq\alpha<\omega}^{\star} c_{i,\alpha}(z)A_{i,\alpha}(z) + \sum_{\omega\leq\alpha<\omega^2}^{\star\star} c_{i,\alpha}(z)A_{i,\alpha}(z) \tag{1.39}$$

i.e. we have a combination of an asymptotic part $\sum^{\star}$ with finite ordinals $\alpha$ as indices, and a transfinite part $\sum^{\star\star}$ with transfinite indices $\alpha$. As a consequence, *the very definition of the formal map $\tilde{G}_i$ becomes a non-trivial affair*, and involves three distinct steps, which we shall detail, for definiteness, in the expanding case (Type III$^+$):

**Step one:** we calculate the formal map $\tilde{U}_i^{\star}$ as the *asymptotic part* of the geometric map $\log\circ G_i$. This formal map $\tilde{U}_i^{\star}$ turns out to be the normalising map of an identity-tangent, unitary map[20] $U_i$, which is none other than the holonomy map of the field $X$ at $S_i$. Therefore the Borel transform of $\tilde{U}_i^{\star}$ is convergent, with singularities over $\mathbb{Z}$.

**Step two:** we resum $\tilde{U}_i^{\star}$ to a true map germ $U_i^{\star}$ by Borel-Laplace, *relative to some convolution preserving average of our choosing, but which has to be the same for all summits.*

**Step three:** we calculate $\tilde{K}_i$ as the asymptotic part of the germ $G_i\circ {}^{\star}U_i\circ\log$, where ${}^{\star}U_i$ is of course the germ reciprocal to $U_i^{\star}$.

## 1.12 Application to Dulac's return map $F$

If we now turn to the global Dulac problem, i.e. to the investigation of the return map $F$ and its formalisation $\tilde{F}$:

$$\tilde{F} := \tilde{G}_r\circ\cdots\circ\tilde{G}_2\circ\tilde{G}_1(z) = \sum_{0\leq\alpha<\gamma<\omega^{\omega}} c_\alpha A_\alpha(z \tag{1.40}$$

($c_\alpha$ scalar, $A_\alpha(z)$ transmonomial) we find that this latitude in the choice of the convolution-respecting average $\mathbf{m}$ can lead to three different methods: *crude*; *smarter*; *smartest.*

### First method (crude): with the lateral average

We may select the trivial *right-* or *left-lateral* average (either). Then of course we have no problem preserving lateral growth, but we are saddled with sums $U_i^{\star}$ that carry *imaginary parts.* The other factor, namely $\tilde{K}_i$, will be convergent. Both $\tilde{K}_i$ and its trivial sum $K_i$ will have their own imaginary parts, which will cancel out the imaginary parts in $U_i^{\star}$, so that the composition product $K_i\circ\exp\circ U_i^{\star}$ will indeed yield the *real germ* $G_i$.

Still, the procedure introduces imaginary parts in the (strictly) trans-asymptotic coefficients $c_\alpha$ of the transseries $\tilde{F}$ and, even worse, inside some of its transmonomials $\tilde{A}$—namely "upstairs", inside the towers of piled-up exponentials. This is a severe drawback for two reasons. First, imaginary parts are *personae non gratae* in the formalisation of an inherently *real object* such as $F$. Second, the imaginary numbers tucked away upstairs inside the exponential towers might create oscillations in the *sums* $A_\alpha$ of some of the transmonomials and so in

[20] I.e. a map of the form $z\mapsto U_i(z) = z(1+\sum a_{i,n}z^{-n})$ with $\bar{U}_i\circ U_i = id$.

$F(z)-z$ itself. By a careful inductive reasoning, we can show *that this in fact is not the case*, because the imaginary parts sitting upstairs are always neutralised by larger infinitesimals which are *purely real*. Nonetheless, the presence of imaginary parts is an aesthetic irritant and a practical nuisance. It robs the non-oscillation of $F(z)-z$ of the intuitive obviousness which it ought to possess, and which is restored with the second and third methods.

### Second method (smarter): with the standard average

We now select the *standard* or *median* average[21]. This does away with all imaginary parts, but at the cost of introducing faster-than-lateral growth in the *uniformised* or *averaged* Borel transform $\mathbf{m}\hat{U}_i^\star$. This is offset, fortunately, by the phenomenon of *emanated resurgence* which has been analysed at great length in [E7] and which induces both *divergence* and *resurgence* inside the factor $\tilde{K}_i$. This time, both $\tilde{U}_i^\star$ and $\tilde{K}_i$ are *real* and *divergent-resurgent*, and the faster-than-exponential growth *cancels out* in the (uniformised) *joint Borel transform* of $\tilde{K}_i \circ \exp \circ \tilde{U}_i^\star$, so that all $\tilde{G}_1, \ldots, \tilde{G}_r$ and $\tilde{F}$ can be *exactly* accelero-summed to $G_1, \ldots, G_r$ and $F$.

There does, however, remain a slight flaw: unlike the transseries $\tilde{G}_i$ and $\tilde{F}$ taken as a whole, *some partial sums of these transseries may not always be resummed exactly, but only up to arbitrarily small germ ideals.* Due once again to *emanation resurgence*[22] these ideals can be rendered as small as $1/\exp_n(z)$ for any given iteration order $n$, that is to say, *smaller than any term present in a given transseries.* This is sufficient for all intents and purposes, and in particular more than sufficient for proving the non-oscillation of $F$. But the impossibility of resumming *exactly* (rather than modulo smaller and smaller ideals) certain sub-transseries of our transseries is slightly irksome. This last remaining imperfection vanishes with the third method.

### Third method (smartest): with the organic average

We select a *well-behaved average* $\mathbf{m}$, like the organic average of Section 1.10.5 or the diffusion-induced averages of Section 1.10.4 Then the formal factors $\tilde{K}_i$ will automatically be *convergent*, like in the first method, but also *real*, like in the second method. There will be no faster-than-lateral growth to worry about, nor any need for any compensation of any sort. And not only will all our transseries be exactly resummable to their correct sums, but so too will all their sub-transseries.

Let us summarise the main differences between the three methods in the following table. There $\tilde{G}_i^\#$ denotes a strict sub-transseries of $\tilde{G}_i$, while $G_i^\#$ and $G_i$ stand for the corresponding sums. Whereas $G_i^\#$ depends on the method, $G_i$ obviouly doesn't. Lastly, when applied to sums, *nearly real* means *real upto an exponentially small part* $\exp(-c_0 z)$, and *almost exact* means *defined up to transexponential accuracy*, ie up to infinitesimals of the form $1/\exp_n(z)$ for any finite $n$.

---

[21] See Section 1.10.5 supra. In [E7] it was denoted by **med** and in [EM] by **mun**.

[22] See [E7, pp. 78–82].

| | Method I | Method II | Method III |
|---|---|---|---|
| .... | .................. | .................. | .................. |
| $\tilde{U}_i^\star$ | real<br>divergent, resurgent | real<br>divergent, resurgent | real<br>divergent, resurgent |
| $\tilde{K}_i$ | complex<br>convergent | real<br>divergent, resurgent | real<br>convergent |
| $\tilde{G}_i$ | complex<br>divergent, resurgent | real<br>divergent, resurgent | real<br>divergent, resurgent |
| $\tilde{G}_i^\#$ | complex<br>divergent, resurgent | real<br>divergent, resurgent | real<br>divergent, resurgent |
| .... | .................. | .................. | .................. |
| $U_i^\star$ | nearly real<br>exact | real<br>almost exact | real<br>exact |
| $K_i$ | nearly real<br>exact | real<br>almost exact | real<br>exact |
| $G_i$ | real<br>exact | real<br>exact | real<br>exact |
| $G_i^\#$ | nearly real<br>exact | real<br>almost exact | real<br>exact |

## 1.13 Recapitulation. What has been gained?

Let us enumerate some of the main gains that accrue come from working with a *well-behaved average*—preferably the *organic average*—first in the general theory of analysabilty, then in the particular application most relevant to this Summer School: Dulac's problem and the non-oscillation of the return map.

### Exact resummation of all sub-transseries and sub-transmonomials

This applies not only to the ones 'downstairs' but also to those 'upstairs'.

### Neat representation of the tree of all "critical time classes"

Each critical time class can now be represented by an (infinitely large) transmonomial $A(z)$, which is the common leading term shared by all representatives of that class.

### Neat, canonical choice of slow critical times

Namely $B(z) := A(z) - A(z)/\log_n(A(z))$ for $n$ large enough.

**Accelero-synthesis without parasitical divergence**

The process, which leads to the gradual actualisation of all subtransseries of a given transseries (in each model $\zeta_i$ only those parts formally exponential or subexponential in $z_i$ can be actualised), has been described in [E7, E10]. It still applies, but now we get rid of the parasitical resurgence, since the latter stemmed from the so-called emanation resurgence, which in turn was a consequence of working with the median average (not well-behaved).

**Decelero-analysis and the "trans-formalisation" algorithm**

This is the reverse process: going from $F$ to $\tilde{F}$. In classical asymptotics it is essentially trivial, but in transasymptotics it ceases to be so. There are indeed several ways, such as ${}^{\tau_1}\tilde{F}$ and ${}^{\tau_2}\tilde{F}$, of formalising one and the same analysable germ, depending on the w.-b. average we select, say ${}^{\tau_1}\mathbf{m}$ or ${}^{\tau_2}\mathbf{m}$, but there also exists an object even more general than the transseries, namely the *display* (it carries the *pseudo-variables* attached to *all critical times*, see [E13, ch. 1]) which contains at once, in explicit fashion, *all* the information about *all* the possible formalisations, and also shows how to "trans-formalise", i.e. how to go from one formalisation to the other.

**Dulac's problem: the all-convergent case**

In the all-convergent case (ie when all $\tilde{G}_i$'s are convergent) we have the commutative diagram:

$$\begin{array}{ccccc} & \{G_r,\dots,G_2,G_1\} & \overset{\text{compose}}{\longrightarrow} & G_r\circ\cdots\circ G_2\circ G_1 =: F & \\ \text{sum} & \uparrow\ \dots\ \uparrow\ \uparrow & & \uparrow & \text{sum} \\ & \{\tilde{G}_r,\dots,\tilde{G}_2,\tilde{G}_1\} & \overset{\text{compose}}{\longrightarrow} & \tilde{G}_r\circ\cdots\circ \tilde{G}_2\circ \tilde{G}_1 =: \tilde{F} & \end{array}$$

which tranforms the formal dichotomy:

$$\text{Either } \tilde{F}(z)\equiv z \text{ or } \tilde{F}(z)\equiv z + c_0\tilde{\Box}_0(z)+\tilde{o}(\tilde{\Box}_0(z)) \quad \text{with} \quad c_0\neq 0$$

into the geometric or 'effective' dichotomy:

$$\text{Either } F(z)\equiv \mathrm{z} \text{ or } F(z)\equiv z + c_0\Box_0(z)+o(\Box_0(z)) \quad \text{with} \quad c_0\neq 0$$

**Dulac's problem: the general case**

$$z_1 \prec z_2 \prec z_3 \prec \cdots \prec z_{r-1} \prec z_r \qquad (z_i \equiv h_i(z) \succ 1)$$

$$
\begin{array}{ccccccc}
 & & & \text{Laplace} & & & \\
\tilde{F}_r(z_r) & z_r & \bullet & \longleftarrow & \bullet & \zeta_r & \overset{\diamond}{F}_r\,(\zeta_r) \\
 & & \downarrow & & \uparrow & \text{acceleration} & \\
F(z) & z & \bullet & & \bullet & \zeta_{r-1} & \overset{\diamond}{F}_{r-1}\,(\zeta_{r-1}) \\
 & & & & \uparrow & \text{acceleration} & \\
 & & & & \cdots & \cdots & \cdots \\
 & & & & \uparrow & \text{acceleration} & \\
\tilde{F}(z) & z & \bullet & & \bullet & \zeta_2 & \overset{\diamond}{F}_2\,(\zeta_2) \\
 & & \downarrow & & \uparrow & \text{acceleration} & \\
\tilde{F}_1(z_1) & z_1 & \bullet & \longrightarrow & \bullet & \zeta_1 & \overset{\diamond}{F}_1\,(\zeta_1) \\
 & & & \text{Borel} & & &
\end{array}
$$

Here, we have as many commutative diagrams as there are steps (even substeps) in the accelero-summation process, but the conclusion is the same as in the all-convergent case. We still have a formal dichotomy:

$$\text{Either } \tilde{F}(z) \equiv z \text{ or } \tilde{F}(z) \equiv z + c_0\tilde{\Box}_0(z) + \tilde{o}(\tilde{\Box}_0(z)) \quad \text{with} \quad c_0 \neq 0$$

that automatically induces a geometric or 'effective' one:

$$\text{Either } F(z) \equiv z \text{ or } F(z) \equiv z + c_0\Box_0(z) + o(\Box_0(z)) \quad \textit{with} \quad c_0 \neq 0$$

**An aside: the case when $F = \mathrm{id}$**

In that case there seem to be extremely strong constraints that imply the "pairing" of divergent summits. But results here are still far from complete.

## 1.14 Is the finiteness proof credible? Q & A

**Q:** *Formalisation—in the present instance, replacing $F$ by $\tilde{F}$—may be highly efficient, but does it not also bring about an impoverishment? Does it not rob problem-solving of all spontaneity and personal initiative, turning everything into a drab, mechanical, mindless routine?*

**A:** In a sense, maybe. But that's only part of the picture. I agree: the excitement of having to grope in the dark, the attractions of trial-and-error, etc. may get eroded, but the "beauty" and "substance" are not gone: rather, *they have been absorbed into the apparatus on which formalisation relies.* Think of all these enchanting structures—resurgence, alien calculus, mould calculus, accelero-summation, cohesiveness, transseries, well-behaved averages, arborification, etc.—which formalisation has unearthed, polished, furbished, and brought under the clear light of day! Can you dispute that the result is a neat gain, not only in terms of problem-cracking power, but even from a purely aesthetic point of view?

**Q:** *Still, you must be aware that Yu. S. Ilyashenko has come up with an alternative proof of the finiteness theorem, which is almost entirely geometric in nature. It sticks to what you would call the 'multiplicative plane', never bothers with the 'Borel plane', and makes practically no use of resummation theory, at least not in your sense. Does it not undermine your case? Does it not weaken your claims about the merits of 'formalisation'?*

**A:** It does not. For one thing, there exist degrees in the accuracy of the various descriptions of $F$, and non-oscillation is but one feature among many. Then we should not forget that the return map $F$ is a very special type of analysable germ—namely a composition product of fairly elementary transit maps $G_i$. It is a far cry from "general analysability". Due to its idiosyncracies, $F$ may indeed yield to several alternative approaches, including purely geometric ones. But faced with a truly generic instance of analysability, the most that 'geometry' may hope to achieve, by going to the fringes of the $z$-sectors of regularity, is to capture *something* of the *lateral* behaviour (right and left of a singularity-carrying axis) in the $\zeta$-plane (see picture), whereas the real difficulties (faster-than-lateral growth, etc.) arise precisely on oft-crossing paths, which are completely off-limits for 'geometry' (it cannot even *see* them). Really, I cannot even begin to imagine how 'geometry' could possibly duplicate such things as *well-behaved derivations, w.-b. averages, w.-b. resurgence monomials*, etc., which, at some stage, become simply indispensible for getting a full grasp of analysability.

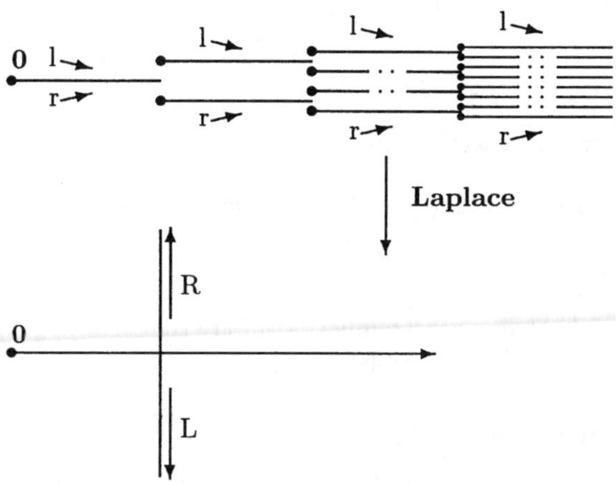


**Q:** *Did not Professor Zołądek pass disparaging comments on your proof, while declaring himself "convinced" by Ilyashenko's?*

**A:** He did. A colleague of mine has brought Zołądek's ten-line 'assessment' to my attention. What can I say? Z. does not give the slightest indication of having read, let alone understood what he quaintly calls the "French proof". So please forgive me for not commenting on his comments. But I must say I was slightly puzzled by the enthusiasm he expressed there for the "Russian proof". So I asked my colleague what she made of this, and she gave me a quite interesting answer: in her estimate, what Z. did really absorb was not the complete finiteness proof as set forth in [I2], but the 100 pages long proof in [I1] of the special case when summits of type $\mathrm{III}^+$ alternate with summits of type $\mathrm{III}^-$ (with any number of type I or II summits in between—they make no difference). Now, this is an *interesting, non-trivial, but still rather elementary* special case. And if one goes for comparisons, then one should compare comparable things. In my scheme of things, that case offers few difficulties (it presents no full-blown transmonomials, no exponential towers, no cohesive steps, no critical time tree, and requires only elementary accelerations) and it can be dealt with quite expeditiously [E7, p. 202].

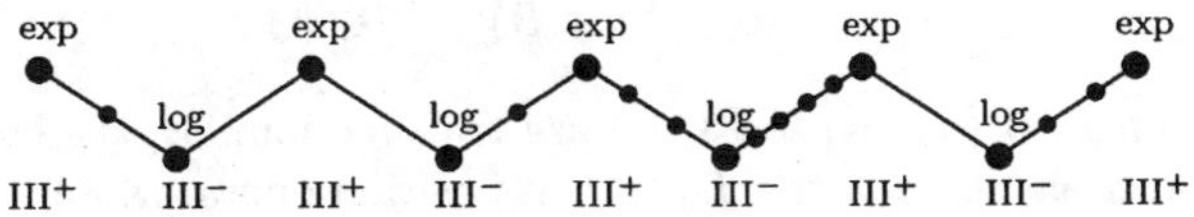


**Q:** *Going back to the general case and its two long-winded proofs: What trust can we repose in them? Have they been checked from A to Z by serious mathematicians? Is there going to be certainty/agreement any time soon?*

**A:** I can only speak for myself. I don't think my book has been read and checked, from cover to cover, by anyone really intent on verifying the proof. But I do know that its *methods* have come under close scrutiny, and that too from people who desired to *apply* them, including in limit situations, and also to *extend* them. *These methods have not been found wanting.* And let us not forget that we are speaking here about a *constructive* proof. Experience teaches that such proofs have a self-righting quality, a resilience all their own. The notion that, due to some gap or mistake lurking somewhere in the meanders of the proofs, the whole thing might suddenly collapse—that notion, I think, can be discounted rightaway. But you asked: when will there be general acceptance? Well, I would imagine that it will come in due course, but gradually, and not as a result of *someone actually checking all the minutiae of the proof* (a thankless undertaking!), but due to *a growing familiarity with, and confidence in, the methods on which it rests.* Possibly, 15 or 20 years from now, *analysability* and all the wherewithal will have become so much part of the mathematical landscape that people, when faced with an object like the return map, will say, after a brief mental evaluation: "Hm, hm! This $F$ is a composition product of clearly analysable factors $G_i$. But everyone knows that composition preserves analysability. QED!" In the meantime, however, there are bound to remain doubters alongside the believers, and that's how it should be!

## 1.15 Complement 1: The trans-Lagrange inversion formula

To convey something of the flavour of the theory of transseries, let us mention the analogue of the classical Lagrange formula for calculating the functional inverse. Everything being formal here, we drop all tildas.

**Framework** We consider mutually reciprocal transseries $f, g : f \circ g = id$ with bounded logarithmic depth (only those make full transasymptotic sense). To further simplify, we apply a "flattening" change of variables, i.e. we precompose by $\log_q := \log^{\circ q}$ and postcompose by $\exp_q := \exp^{\circ q}$, with $q$ large enough to ensure that $f(x) = x + o(1), g(x) = x + o(1)$.

**Notations** For any transseries $\varphi$ and any transmonomial $A$, let $Proj_A(\varphi)$ denote the scalar coefficient $c_A \in \mathbb{R}$ of $A$ in the canonical expansion $\varphi = \sum c_A A$. In particular, let $Proj_1(\varphi)$ denote the constant term in that same expansion (it separates the infinitely large transmonomials to its left from those infinitely small to its right.) Lastly, let $\varphi'$ (resp. $'\varphi$) denote the derivative of $\varphi$ (resp. its indefinite integral *without constant term*).

**Matrix elements** For any two transseries $A, B$ and any $f \succ 1$, we set

$$f_{[A,B]} := \mathrm{Proj}_1(A \circ f.B) \qquad (\in \mathbb{R}) \tag{1.41}$$

In the special case when $A, B$ (and so too $1/B$) are transmonomials, the coefficients $f_{[A,1/B]}$ may be viewed as matrix elements of the $f$-postcomposition operator. Indeed we have:

$$\begin{aligned} A \circ f &\equiv \sum_B f_{[A,1/B]}.B \qquad \text{(transfinite sum)} \\ (f_1 \circ f_2)_{[A,1/C]} &\equiv \sum_B f_{1[A,1/B]}.f_{2[B,1/C]} \qquad \text{(finite sum)} \end{aligned} \tag{1.42}$$

$$(f_1, f_2 \text{ transseries} \succ 1, \quad A, B, C \text{ transmonomials}) \tag{1.43}$$

**The trans-Lagrange inversion formula**

**1.1 Lemma** *Mutually reciprocal transseries $f, g$, flattened so as to satisfy $f(x) = x + o(1)$, $g(x) = x + o(1)$, always have their matrix elements linked by the simple involutive formula:*

$$g_{[A,B]} = -f_{[A^\star,B^\star]} \quad \textit{with} \quad A^\star = '(B.\log_q'), \quad B^\star = A'/\log_q' \tag{1.44}$$

*valid for any $q$ larger than the logarithmic depth of $A, B, f, g$.*

**The Lagrange inversion formula** If $f, g$ are ordinary power series:

$$f(x) = x(1 + \sum a_n x^{-n}) \qquad (x \sim +\infty) \tag{1.45}$$

$$g(x) = x(1 + \sum b_n x^{-n}) \qquad (x \sim +\infty) \tag{1.46}$$

and if we take $(A, B)$ to be ordinary monomials $(x^m, x^n)$, we may choose $q = 1$ in the above formula, and then we get:

$$f_{[x^m,x^n]} = \frac{m}{n}\, g_{[x^n,x^m]} \qquad (m, n \in \mathbb{Z}^\star) \tag{1.47}$$

which is just an unusual presentation of the classical Lagrange inversion formula.

## 1.16 Complement 2: Beyond the exponential range

The *indiscernibility theorem*, very roughly, asserts that *from a purely asymptotic viewpoint there exists no canonical notion of fractional or transfinite iteration for the maps* log *or* exp. More precisely, *for any strictly fractional $\alpha \in \mathbb{Q}^+ - \mathbb{N}$ or any strictly transfinite ordinal $\alpha \in [\omega, \omega^\omega[$, no purely asymptotic criterion at $+\infty$, based on the behaviour (sign) of expressions finitely constructed from* $\log_\alpha$ *or* $\exp_\alpha$*, plus the direct/inverse operations $+, \times, \partial, \circ$, plus a* *finite number of analysable germs $\varphi_i$, can enable us to isolate a privileged representative in the class of all possible iterates of the same order $\alpha$—no matter what degree of regularity*[23] *we choose to impose.*

This and a few other results of the same ilk suggest that, in some sense, the pair of trigebras

$$\mathbb{R}[[[z]]] \text{ (transseries)} \quad \text{and} \quad \mathbb{R}\{\{\{z\}\}\} \text{ (analysable germs)}$$

are the "largest of their kind". If we insist on enlarging them, some essential properties are bound to give way. Fortunately, these two provide a framework capacious and flexible enough for most problems of smooth asymptotics. It is worth noting, in particular, that differential equations can never generate an asymptotic behaviour even remotely like that of $\log_\alpha$ or $\exp_\alpha$ for a fractional or transfinite $\alpha$ (unless this behaviour is already there in the data).

## 1.17 Complement 3: van der Hoeven's intermediate value theorem

**1.2 Theorem (J. van der Hoeven)** *Let $P$ be a differential polynomial with transserial coefficients, i.e. with coefficients in the trigebra $\mathbb{R}[[[z]]]$. Then, given any pair of transseries $\tilde{F} < \tilde{G}$ in $\mathbb{R}[[[z]]]$ such that $P(\tilde{F}) < 0 < P(\tilde{G})$ there exists $\tilde{H}$ in $\mathbb{R}[[[z]]]$ such that $\tilde{F} < \tilde{H} < \tilde{G}$ and $P(\tilde{H}) = 0$. Furthermore, there exists a fully algorithmic procedure for constructing such an "intermediate solution" $\tilde{H}$.*

## 1.18 Complement 4: van der Hoeven's complex transseries as a tool for solving algebraic differential equations

The difficulty with complexifying transseries is that the presence of imaginary parts "high up" in the transmonomials (inside exponentials) creates oscillations, so that we can no longer compare, much less totally order, these complex transmonomials. But there is a way round this difficulty: we may equip $\mathbb{C}$ with an addition-respecting (total) order, regard our complex transmonomials as pure symbols, and then proceed with the inductive construction of complex transseries exactly as in the real case. There are as many addition-respecting orders on $\mathbb{C}$ as there are "positive half-planes" $\mathbb{H}^+$ containing $\mathbb{R}^+$ and excluding $\mathbb{R}^-$, with an additional convention for including half the boundary of $\mathbb{H}^+$ and excluding the other half. Different choices lead to different constructions $\mathbb{C}^{\mathbb{H}^+}[[[z]]]$, but these are all isomorphic. So it is permissible to speak of *the* trigebra $\mathbb{C}[[[z]]]$ of complex transseries.

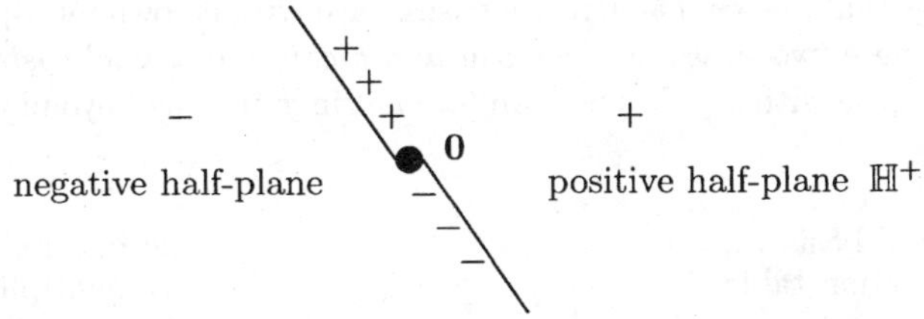


**1.3 Theorem (J. van der Hoeven)** *Any algebraic differential equation $P(y) = 0$ polynomial in $y, y', y'', \ldots, y^{(d)}$ and with arbitrary transserial coefficients in $\mathbb{C}[[[x]]]$ admits solutions in $\mathbb{C}[[[x]]]$.*

[23] Whether cohesivity, analyticity, softness (see [E7, p. 287]) or any combination of these.

However, there doesn't always exist a full $d$-parameter family of solutions, so that not *all* solutions are caught in our net—which is another way of saying that $\mathbb{C}[[[x]]]$ is not algebraically-differentially closed.

The true challenge, however, would be to resum as many complex transseries as possible—beginning with those of natural origin, like the algebraic-differential elements of $\mathbb{C}[[[x]]]$. Encouraging forays have been made in this direction, but there are serious obstacles, mainly because it is not at all clear what should take the place of accelero-summation in the new context.

## 1.19 Complement 5: Finitary transseries and the dimorphic core $\mathbb{N}\mathrm{a}$ of $\mathbb{R}$

Let us now return to the real transseries. If in our inductive construction of $\mathbb{R}[[[x]]]$ in Section 1.5 we replace $\mathbb{R}$ by the field $\mathbb{A}$ of algebraic numbers, and the rings of power series $\mathbb{R}[[z^{-\sigma_1},\ldots,z^{-\sigma_1}]]$ and $\mathbb{R}[[\sqcup_1,\ldots,\sqcup_r]]$ by the polynomial rings $\mathbb{A}[z^{-\sigma_1},\ldots,z^{-\sigma_1}]$ and $\mathbb{A}[\sqcup_1,\ldots,\sqcup_r]$, we get the trigebra $\mathbb{A}^{\mathrm{fin}}[[[x]]]$ of *finitary transseries*, so-called because their definition (or construction) draws only on a finite number of parameters.

Two huge sets of exotic derivations—the *alien derivations* $\Delta_\omega^{\{z_i\}}$ and the *foreign derivations* $\nabla_\omega^{\{z_i\}}$, each relative to a specific index $\omega \in \mathbb{R}^+$ and a specific time class $\{z_i\}$—act on these finitary transseries and produce transcendental constants, the so-called *naturals*, with (unlike the over-publicised *periods*) come equipped with a natural indexation. The *naturals* order themselves spontaneously into a hierarchy of *dimorphic rings* $\mathbb{D}_1,\mathbb{D}_2,\mathbb{D}_3,\mathbb{D}_4,\ldots$:

| | | | | |
|---|---|---|---|---|
| $\mathbb{Q}$ | = | rationals | | |
| $\mathbb{A}$ | = | algebraic numbers | $\Longleftarrow$ | Galois theory |
| $\mathbb{D}_1$ | = | *multizetas* | $\Longleftarrow$ | ARI/GARI |
| $\mathbb{D}_2$ | = | hyperlogarithmic constants | $\Longleftarrow$ | ARI/GARI? |
| $\mathbb{D}_3$ | = | linear alg. differ. numbers | $\Longleftarrow$ | ?? |
| $\mathbb{D}_4$ | = | non linear alg. differ. numbers | $\Longleftarrow$ | ??? |

that is two say, rings that possess two natural bases, each with its own multiplication rule, and, half-way between these two bases, a *canonical and highly non-trivial system of irreducibles.* That at least is the case with $\mathbb{D}_1$. See Section 5. The rings that lie beyond are still very much uncharted territory:

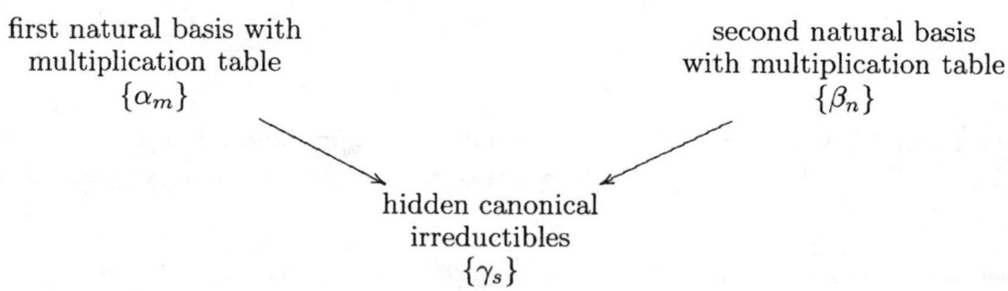

# 2 Lesson Two: Normal and prenormal forms

## 2.1 Local objects and their homogeneous components. Resonance, quasiresonance, nihilence

By *local analytic object* we shall mean, primarily:

(1) germs of singular analytic vector fields at 0 on $\mathbb{C}^\nu$, often referred to as just *fields* for short;

(2) germs of analytic diffeomorphisms of $\mathbb{C}^\nu$ into itself, with 0 as fixed point, or *diffeos* for short;

and, secondarily, all those equations or systems (differential, difference, functional, etc.) which may, in a standard manner, be rephrased in terms of fields or diffeos.

Fields will be denoted

$$X = \sum_{1 \le i \le \nu} X_i(x)\, \partial_{x_i}; \qquad X_i(x) \in \mathbb{C}\{x_1, \ldots, x_\nu\} \tag{2.1}$$

but instead of diffeos proper:

$$f : x_i \mapsto f_i(x), \qquad i = 1, \ldots, \nu; \quad f_i(x) \in \mathbb{C}\{x_1, \ldots, x_\nu\} \tag{2.2}$$

it will be more convenient to handle the corresponding substitution operators $F$ (same symbols, but capitalised):

$$(F\varphi)(x) := \varphi(f(x)); \qquad \forall \varphi(x) \in \mathbb{C}\{x_1, \ldots, x_\nu\} \tag{2.3}$$

Working with *operators* makes it easier to isolate a local object's *homogeneous components*:

$$X = X^{\text{lin}} + \sum_n \mathbb{B}_n \qquad \text{(for a field)} \tag{2.4}$$

$$F = (1 + \sum_n \mathbb{B}_n) F^{\text{lin}} \qquad \text{(for a diffeo)} \tag{2.5}$$

$$\text{with } \mathbb{B}_n : x^m.\mathbb{C} \to x^{m+n}.\mathbb{C} \qquad (m, n \text{ multiintegers}) \tag{2.6}$$

For simplicity, we shall assume that the linear part is diagonalisable, and work in an analytic chart where it is actually diagonal:

$$X^{\text{lin}} = \sum_i \lambda_i x_i\, \partial_{x_i} \tag{2.7}$$

$$F^{\text{lin}} : \varphi(x_1, \ldots, x_r) \mapsto \varphi(l_1 x_1, \ldots, l_r x_r) \quad \text{with} \quad l_i \in \mathbb{C}^\star \tag{2.8}$$

The discussion hinges on the nature of the object's spectrum, i.e. the eigenvalues of its linear part: $\lambda = (\lambda_1, \ldots, \lambda_\nu)$ for a field; $l = (l_1, \ldots, l_\nu)$ for a diffeo. If the spectrum is "generic", then the object is analytically conjugate to its linear part—and that ends the matter, at least from the local point of view.

Difficulties arise only in the case of

- *resonance*: relations of type $0 = \lambda_i - \sum m_i \lambda_j$ or $0 = l_i - \prod l_j^{m_j}$ with non-negative integers $m_j$.
- *quasiresonance*: i.e. when Bryuno's well-known diophantine condition, which minorises the above expressions in terms of $\|m\|$, is not fulfilled[24].
- *nihilence*: this complication, which presupposes resonance but bears on coefficients of all orders, occurs mostly, though not only, in symplectic or volume-preserving objects[25].

The more 'complicated' an object, the larger its set of *invariants* tends to be. Alongside the *formal* and *analytic invariants* (i.e. relative to formal or analytic coordinate changes) we have the notion of *holomorphic invariants*—i.e. invariants that depend holomorphically on the object **Ob** (or, in practical terms, its Taylor coefficients), at least when **Ob** remains within a fixed formal conjugacy class[26].

Resonance generates *formal invariants* (other than the spectrum itself), of which there may be an infinite number[27] if the resonance degree is $\geq 2$.

Each of the above-mentioned complications—resonance, quasiresonance, and nihilence—whether in isolation or in combination, gives rise to analytic invariants (strictly analytic, i.e. non formal). Moreover, when resonance alone is at work, there tend to exist[28] complete systems of analytic-cum-holomorphic invariants $\{\mathbf{A}_\omega\}$.

## 2.2 Continuous prenormalisation versus discontinuous normalisation

A resonant object $X$ or $F$ usually cannot be fully linearised—not even formally—but, under a formal change of coordinates corresponding to a substitution operator $\Theta_{\text{pre}}$, it can always be brought into a so-called *prenormal form* $X^{\text{pre}}$ or $F^{\text{pre}}$

$$X = \Theta_{\text{pre}} X^{\text{pre}} \Theta_{\text{pre}}^{-1} \tag{2.9}$$
$$F = \Theta_{\text{pre}} F^{\text{pre}} \Theta_{\text{pre}}^{-1} \tag{2.10}$$

that is to say, a form which commutes with the linear part of $X$:

$$[X^{\text{lin}}, X^{\text{pre}}] = 0; \qquad [F^{\text{lin}}, F^{\text{pre}}] = 0 \tag{2.11}$$

and which therefore involves only *resonant* homogeneous components $B_n^\star$:

$$X^{\text{pre}} = X^{\text{lin}} + \sum \mathbb{B}_n^\star \quad \text{with} \quad \langle n, \lambda \rangle = 0 \tag{2.12}$$
$$F^{\text{pre}} = (1 + \sum \mathbb{B}_n^\star) F^{\text{lin}} \quad \text{with} \quad l^n = 1 \tag{2.13}$$

If the number of (non-vanishing) resonant components $\mathbb{B}_n^\star$ is minimal (which implies that their coefficients are formal invariants of the object), then $X^{\text{pre}}$ or $F^{\text{pre}}$ deserves to be regarded as a *normal form* and will be noted $X^{\text{nor}}$ or $F^{\text{nor}}$. For objects with one (resp. several) degrees of resonance, there exist finitely (resp. infinitely) many independent formal invariants.

---

[24] See conditions (4.32)+(4.33) in Lesson 4

[25] See Lesson 4.

[26] Thus, when applied to *invariants*, the words *analytic* and *holomorphic* assume quite different meanings: analytic invariants are not necessarily holomorphic.

[27] More precisely, an infinite number of *independent* formal invariants.

[28] With slight qualifications, see e.g. [E5].

Though simplest in terms of outward shape, the normal forms $X^{\text{nor}}$ or $F^{\text{nor}}$ have quite a few drawbacks. One is the unavoidably non-continuous nature of the maps $X \mapsto X^{\text{nor}}$ or $F \mapsto F^{\text{nor}}$, even when we keep the linear parts $X^{\text{lin}}$ or $F^{\text{lin}}$ fixed. Another is the absence, thus far, of general and truly algorithmic procedures for determining them, especially in the case of highly resonant spectra. A third drawback (manifest even in the case of *simple* resonance) is their unsuitability for mechanical computation: the exact shape of the normal forms always depends on one or several discrete invariants (such as the "levels", see for example [E3]), whose exact values, in turn, depend on whether certain real or complex numbers (which depend polynomially on the Taylor coefficients of the object) do or do not vanish—a matter which a computer usually cannot decide, for want of algorithmic tests.

So, for many purposes, it is preferable to work with *continuous prenormal forms* or rather, to be quite precise, with *prenormal forms that depend continuously on the homogeneous components* $\mathbb{B}_n$ *of the object, while the linear part is kept fixed*[29].

In concrete terms, a *continuous prenormalisation* is a universal correspondence of the form:

$$X = X^{\text{lin}} + \sum \mathbb{B}_n \mapsto X^{\text{pre}} = X^{\text{lin}} + \sum \text{pran}^\bullet \mathbb{B}_\bullet \tag{2.14}$$

$$= X^{\text{lin}} + \sum \text{pran}^{\omega_1,\ldots,\omega_r} \mathbb{B}_{n_1,\ldots,n_r} \tag{2.15}$$

$$F = (1 + \sum \mathbb{B}_n) F^{\text{lin}} \mapsto F^{\text{pre}} = (1 + \sum \text{pren}^\bullet \mathbb{B}_\bullet) F^{\text{lin}} \tag{2.16}$$

$$= (1 + \sum \text{pren}^{\omega_1,\ldots,\omega_r} \mathbb{B}_{n_1,\ldots,n_r}) F^{\text{lin}} \tag{2.17}$$

$$\text{with} \quad \mathbb{B}_{n_1,n_2,\ldots,n_r} := \mathbb{B}_{n_r} \ldots \mathbb{B}_{n_2} \mathbb{B}_{n_1} \tag{2.18}$$

which is entirely determined by a given system $\{\text{pran}^\bullet\}$ or $\{\text{pren}^\bullet\}$ of scalar coefficients[30] independent of the object and indexed by sequences $\omega$ constructed from the 'frequencies' $\omega_i := \langle n_i, \lambda \rangle$.

## 2.3 Examples of prenormalisations: 'regal' and 'royal'

Let us concentrate on *vector fields* for simplicity. In practice, the construction of the (alternal) *prenormal* mould $\text{pran}^\bullet$ goes hand in hand with the construction of the (symmetral) *prenormalising* mould *mathrmspran*$^\bullet$ that yields the prenormalising substitution operator:

$$\Theta_{\text{pre}} := 1 + \sum \text{spran}^\bullet \mathbb{B}_\bullet = 1 + \sum \text{spran}^{\omega_1,\ldots,\omega_r} \mathbb{B}_{n_1,\ldots,n_r}.$$

Here are two examples—the simplest of all, as it will turn out.

### Regal prenormalisation:

The moulds are characterised by

$$\Box\, \text{spran}^\bullet = \text{pran}^\bullet \times \text{spran}^\bullet - \text{spran}^\bullet \times \text{I}^\bullet \tag{2.19}$$

$$\text{lang} \,.\, \text{spran}^\bullet = \text{copran}^\bullet \times \text{spran}^\bullet \tag{2.20}$$

[29]Continuous dependence on the *linear part* would be an obviously impossible demand.

[30]The vowel *a* (resp. *e*) alludes to the symmetry type—symmetral or alternal (resp. symmetrel or alternel)—of the moulds $\text{pran}^\bullet$ etc. (resp. $\text{pren}^\bullet$ etc.) that are used in the case of fields (resp. diffeos).

together with the initial conditions

$$\text{spran}^{\emptyset} = 1 \qquad \text{(due to symmetrality)} \tag{2.21}$$

$$\text{pran}^{\emptyset} = \text{copran}^{\emptyset} = 0 \qquad \text{(due to alternality)} \tag{2.22}$$

and the "complementarity" conditions

$$\text{pran}^{\boldsymbol{\omega}} = 0 \qquad (\text{if } \|\boldsymbol{\omega}\| \neq 0) \tag{2.23}$$

$$\text{copran}^{\boldsymbol{\omega}} = 0 \qquad (\text{if } \|\boldsymbol{\omega}\| = 0\,) \tag{2.24}$$

The two mould operators $\square$ and lang occuring in the above system multiply moulds by, respectively, the sum $\|\boldsymbol{\omega}\|$ or the length $r(\boldsymbol{\omega})$ of their indexing sequences $\boldsymbol{\omega}$, and the obvious fact of their being *mould derivations*:

$$\square\,(A^{\bullet} \times B^{\bullet}) \equiv (\square\, A^{\bullet}) \times B^{\bullet} + A^{\bullet} \times (\square\, B^{\bullet})$$

$$\text{lang}(A^{\bullet} \times B^{\bullet}) \equiv (\text{lang}\, A^{\bullet}) \times B^{\bullet} + A^{\bullet} \times (\text{lang}\, B^{\bullet})$$

ensures that the (clearly unique) solution of the above system consists of a mould pair $\text{spran}^{\bullet}/\text{pran}^{\bullet}$ of *symmetral/alternal* type.

**Royal prenormalisation:**

It is defined by the same system of mould equations as the *regal form*, except that we replace the mould derivation lang by the mould derivation lan, which mutiplies a mould, not by its length $r(\boldsymbol{\omega})$ (total number of indices), but its reduced length $r^{\star}(\boldsymbol{\omega})$, i.e. the number of its non-zero indices $\omega_i$. So the new system reads:

$$\square\,\text{spran}^{\bullet} = \text{pran}^{\bullet} \times \text{spran}^{\bullet} - \text{spran}^{\bullet} \times \text{I}^{\bullet} \tag{2.25}$$

$$\text{lan}\,.\,\text{spran}^{\bullet} = \text{copran}^{\bullet} \times \text{spran}^{\bullet} \tag{2.26}$$

with the same initial conditions as above and also the same "complementarity" conditions.

This innocuous-looking change brings with it considerable simplifications, and in fact yields what must be adjudged the *simplest* and most regular of all prenormalisations—hence its name *royal.*

## 2.4 Mould amplification

To show that the *royal prenormalisation* cannot be simplified or improved upon, and also to establish its analytic properties (namely, generic divergence + resurgence), we require a mould transform—the so-called amplification—that preserves a mould's symmetry type but isolates the contribution of its vanishing indices $\omega_i = 0$. Indeed, if a mould $M^{\bullet}$ is alternal (resp. symmetral), then the amplified mould $\underline{M}^{\bullet}$, which is defined by

$$\underline{M}^{\varpi_1,\ldots,\varpi_r} := \sum_{n_i \geq 0} M^{\omega_1,0^{[n_1]},\ldots,\omega_r,0^{[n_r]}} (\check{a}_1)^{n_1} (\check{a}_2)^{n_2} \ldots (\check{a}_r)^{n_r} \tag{2.27}$$

$$\varpi_i := \begin{pmatrix} \omega_i \\ a_i \end{pmatrix} \in \begin{pmatrix} \mathbb{C}^{\star} \\ \mathbb{C} \end{pmatrix}; \qquad \check{a}_i := a_1 + a_2 + \cdots + a_i; \qquad 0^{[n_i]} := (0, \ldots, 0)\ (n_i \text{ times})$$

is also alternal (resp. symmetral). Amplification "almost" respects mould multiplication (see [EV1, p. 16]), and has a simple effect on most other mould operations. For instance, the three mould derivations that enter the definition of the *regal* and *royal* prenormalisations become after amplification:

$$\underline{\square}\,.\,\underline{M}^{\varpi} := \|\varpi\|\underline{M}^{\varpi} := (\|a\| + \|\omega\|)\underline{M}^{\varpi} \tag{2.28}$$

$$\underline{\mathrm{lang}}\,.\,\underline{\mathrm{M}}^{\varpi} = (r(\omega) + \sum a_i\, \partial_{a_i})\underline{\mathrm{M}}^{\varpi} \tag{2.29}$$

$$\underline{\mathrm{lan}}\,.\,\underline{\mathrm{M}}^{\varpi} = r^{\star}(\omega)\underline{\mathrm{M}}^{\varpi} \tag{2.30}$$

Thus, one easily finds that the first non-trivial values of $pran^{\varpi}$, corresponding to a length $r = 2$ and $\omega_1 + \omega_2 = 0$, $\omega_i \neq 0$, are as follows:

$$\underline{\mathrm{pran}}^{\varpi_1,\varpi_2} = \frac{1}{a_1 + a_2}\log\left(\frac{1 + a_2/\omega_2}{1 + a_1/\omega_1}\right); \qquad \text{“regal”} \tag{2.31}$$

$$\underline{\mathrm{pran}}^{\varpi_1,\varpi_2} = \frac{1}{2}\left(\frac{1}{\omega_2 + a_2} - \frac{1}{\omega_1 + a_1}\right); \qquad \text{“royal”} \tag{2.32}$$

More generally, the moulds $\mathrm{pran}^{\bullet}$ and $\mathrm{spran}^{\bullet}$ associated with the royal (resp. regal) prenormalisation are rational (resp. hyperlogarithmic) functions of their two sets of indices ($a_i$ and $\omega_i$).

Let us ponder a moment the above formulas.

(1) Though the regal and royal $\mathrm{pran}^{\bullet}$ differ, they are seen to coincide for $a_1 + a_2 = 0$ (recall that $\omega_1 + \omega_2 = 0$ anyhow).

(2) For fixed indices $\omega_i$ both are regular functions of the $a_i$-variables at the origin, but with a (common) singular locus away from the origin, and no natural analytic boundaries.

Those two observations, still valid beyond $r = 2$, are highly significant, since the presence of singularities in the $a_i$ variables stands in close relation to the *divergence* of the corresponding prenormalisation, while the absence of obstacles to endless analytic continuation (around these singularities) is related to the *resurgence* properties of that same prenormalisation.

This immediately raises the question: are these singularities in the $a_i$ variables peculiar to the two prenormalisations just examined, or are they absolutely unavoidable? As it happens, the latter is the case, but to establish this we require a mould $\mathrm{hard}^{\bullet}$ that describes "all that is common" (singularities included) to *all* (not just the *regal* or *royal*) *prenormalisations.*

## 2.5 The 'hard core' common to all prenormalisations

Taking our clue from the above remark, we define $\underline{\mathrm{prann}}^{\varpi}$ as being the restriction of $\underline{\mathrm{pran}}^{\varpi}$ on the hyperplane $\|a\| = 0$.[31] We then define the 'hard core' $\mathrm{hard}^{\bullet}$ in this way: for any set $\varpi^1, \varpi^2, \ldots, \varpi^m$ of unbreakable, zero-sum sequences[32]. and any natural integer $n$ no greater than $m$ we set

$$\mathrm{hard}_n^{\varpi^1;\ldots;\varpi^m} := \sum_{\varpi^{(1)}\ldots\varpi^{(n)}\in\mathrm{circ}(\varpi^1;\ldots;\varpi^m)} \underline{\mathrm{prann}}^{\varpi^{(1)}}\ldots\underline{\mathrm{prann}}^{\varpi^{(n)}} \tag{2.33}$$

[31] Of course, for $\underline{\mathrm{pran}}^{\varpi}$ and thus $\underline{\mathrm{prann}}^{\varpi}$ to be $\neq 0$, we must also have $\|\omega\| = 0$.

[32] I.e. not a product of zero-sum subsequences.

with sequences $\varpi^{(i)}$ made up of one or several sequences $\varpi^j$ and with a sum $\sum$ ranging over all decompositions of the form:

$$\varpi^{(1)}\varpi^{(2)}\ldots\varpi^{(n)} \equiv \varpi^i\varpi^{i+1}\ldots\varpi^m\varpi^1\ldots\varpi^{(i-2)}\varpi^{(i-1)} \qquad \text{(for some } i\text{)}$$

Thus for $m = 1, 2, 3$ we get:

$$\begin{aligned}
\mathrm{hard}_1^{\varpi^1} &:= \underline{\mathrm{prann}^{\varpi^1}}\\
\mathrm{hard}_1^{\varpi^1;\varpi^2} &:= \underline{\mathrm{prann}^{\varpi^1\varpi^2}} + \underline{\mathrm{prann}^{\varpi^2\varpi^1}}\\
\mathrm{hard}_2^{\varpi^1;\varpi^2} &:= \underline{\mathrm{prann}^{\varpi^1}}\,\underline{\mathrm{prann}^{\varpi^2}}\\
\mathrm{hard}_1^{\varpi^1;\varpi^2;\varpi^3} &:= \underline{\mathrm{prann}^{\varpi^1\varpi^2\varpi^3}} + \underline{\mathrm{prann}^{\varpi^2\varpi^3\varpi^1}} + \underline{\mathrm{prann}^{\varpi^3\varpi^1\varpi^2}}\\
\mathrm{hard}_2^{\varpi^1;\varpi^2;\varpi^3} &:= \underline{\mathrm{prann}^{\varpi^1\varpi^2}}\,\underline{\mathrm{prann}^{\varpi^3}} + \underline{\mathrm{prann}^{\varpi^2\varpi^3}}\,\underline{\mathrm{prann}^{\varpi^1}} + \underline{\mathrm{prann}^{\varpi^3\varpi^1}}\,\underline{\mathrm{prann}^{\varpi^2}}\\
\mathrm{hard}_3^{\varpi^1;\varpi^2;\varpi^3} &:= \underline{\mathrm{prann}^{\varpi^1}}\,\underline{\mathrm{prann}^{\varpi^2}}\,\underline{\mathrm{prann}^{\varpi^3}}
\end{aligned}$$

Note that we must take care to avoid repetitions in the permutations: thus, for $n = 2, m = 3$ we get three terms on the right-hand side, not six. More generally, the number of terms on the right-hand side is exactly $\sum \frac{m!}{m_1!\ldots m_n!}$, with a sum ranging over $1 \le m_1 \le m_2 \le \cdots \le m_n$ and $m_1 + \cdots + m_n = m$.

**Universality of the 'hard core' hard$^\bullet$**

*For any fixed set of zero-sum, unbreakable sequences $\varpi^1, \varpi^2, \ldots, \varpi^m$ and any integer $n \le m$, the complex number $\mathrm{hard}_n^{\varpi^1;\ldots;\varpi^m}$ does not depend on the actual choice of prenormalisation.*

**Analytic expression of the 'hard core' hard$^\bullet$**

*Again, for any set of $m$ zero-sum unbreakable sequences $\varpi^i$ and $n \le m$, the universal 'hard core' is given by the formula:*

$$\mathrm{hard}_n^{\varpi^1;\ldots;\varpi^m} = \frac{\mathrm{daa}^{m-n}}{(m-n)!}(\mathrm{Taa}^{\varpi^1}\ldots\mathrm{Taa}^{\varpi^1}) \quad \textit{with} \tag{2.34}$$

$$\mathrm{Taa}^{\varpi} := \frac{(-1)^{r-1}}{\breve{\varpi}_1\ldots\breve{\varpi}_{r-1}} = \frac{1}{\hat{\varpi}_2\ldots\hat{\varpi}_r} \quad \textit{and}$$

$$\frac{\mathrm{daa}^s}{s!}\mathrm{Taa}^{\varpi} := \mathrm{Taa}^{\varpi}\sum_{1\le i_1\cdots\le i_s\le r-1}\frac{1}{\breve{\varpi}_{i_1}\ldots\breve{\varpi}_{i_s}} = \mathrm{Taa}^{\varpi}\sum_{2\le i_1\le\cdots\le i_s\le r}\frac{(-1)^s}{\hat{\varpi}_{i_1}\ldots\hat{\varpi}_{i_s}}$$

*with the predictable abbreviations $\breve{\varpi}_i := \breve{\omega}_i + \breve{a}_i$, $\hat{\varpi}_i := \hat{\omega}_i + \hat{a}_i$ and with* daa *viewed as a symbolic derivation that acts according to the Leibniz rule on the product on the right-hand side of* (2.34).

The proofs, as well as the parallel formulas for diffeos (with alternel/symmetrel moulds in place of alternal/ symmetral ones for vector fields) may be found in [EV1].

The above expression of the "hard core" and the presence in it of unavoidable singularities on the hyperplanes

$$(\omega_i + \omega_{i+1} + \cdots + \omega_j) + (a_i + a_{i+1} + \cdots + a_j) = 0$$

confirms what we were suggesting a moment ago:

- any universal prenormalisation necessarily involves divergence (which is unfortunate) but of resurgent type (which is fortunate), not only in the prenormalising transformations (which was predictable enough) but also in the prenormal form itself (which is more surprising)
- of all prenormalisations, the "royal" one stands out as the simplest choice:
  (i) it displays no parasitical singularities and is acted upon non-trivially by a minimal number of alien derivations (see Section 2.6 *infra*);
  (ii) its resurgence equations, though complex enough, are 'simplest';
  (iii) the associated moulds $\mathrm{pran}^\bullet$ and $\mathrm{spran}^\bullet$ are simplest too: namely, rational in the $\boldsymbol{\omega}$ variables, and with amplifications $\underline{\mathrm{pran}}^\bullet, \underline{\mathrm{spran}}^\bullet$ also rational in both the $\boldsymbol{\omega}$ and $\boldsymbol{a}$ variables.

## 2.6 Normalisation and prenomalisation resurgence

We first recall the normalisation and prenormalisation equations for a resonant (for definiteness, simply resonant) vector field:

$$X = \Theta_{\mathrm{nor}} X^{\mathrm{nor}} \Theta_{\mathrm{nor}}^{-1} \qquad (X^{\mathrm{nor}} = \text{normal form})$$
$$X = \Theta_{\mathrm{pre}} X^{\mathrm{pre}} \Theta_{\mathrm{pre}}^{-1} \qquad (X^{\mathrm{pre}} = \text{prenormal form})$$

Both transformations give rise to resurgence, but of a rather different sort—even when we choose, as we shall in the sequel, the *royal prenormalisation*, which is closest to normalisation.

This difference merely reflects that between the moulds $\mathrm{va}^\bullet(z)$ and $\mathrm{wa}^\bullet(z)$ which enter the expansions of the normaliser and prenormaliser:

$$\Theta_{\mathrm{nor}} = \sum \mathrm{va}^\bullet(z)\mathbb{B}_\bullet \qquad \text{(normaliser)}$$
$$\Theta_{\mathrm{pre}} = \sum \mathrm{wa}^\bullet(z)\mathbb{B}_\bullet \qquad \text{(royal prenormaliser)}$$

As usual, we work in a chart $z, u_1, \ldots, u_{\nu-1}$ so chosen as to 'unload' the whole divergence and resurgence on one single variable $z$, the remaining $u_i$ being inert parameters. As for the $\mathbb{B}_\bullet$, they are ordinary differential operators elementarily constructed from the Taylor coefficients of the field $X$. For numerous examples, see e.g. [E3, E5]. Both moulds $\mathrm{va}^\bullet(z)$ and $\mathrm{wa}^\bullet(z)$ are symmetral, and both have components that are divergent-resurgent in $z$. But there is a marked difference in complexity. Whereas $\mathrm{va}^\bullet(z)$ is given by the simple induction:

$$(\partial_z + \|\omega\|)\mathrm{va}^{\eta_1,\ldots,\eta_r}(z) = -\mathrm{va}^{\eta_1,\ldots,\eta_{r-1}}(z) z^{-\sigma_r-1} \quad \text{with} \quad \eta_i := \binom{\omega_i}{\sigma_i}$$

or more explicitely:

$$(\partial_z + \omega_1 + \cdots + \omega_r)\mathrm{va}^{\binom{\omega_1,\ldots,\omega_r}{\sigma_1,\ldots,\sigma_r}}(z) = -\mathrm{va}^{\binom{\omega_1,\ldots,\omega_{r-1}}{\sigma_1,\ldots,\sigma_{r-1}}}(z) z^{-\sigma_r-1} \tag{2.35}$$

the definition of $\mathrm{wa}^\bullet(z)$ is more involved:

$$\mathrm{wa}^{\boldsymbol{\eta}}(z) := \sum_{\substack{\boldsymbol{\eta}=\boldsymbol{\eta}^1\boldsymbol{\eta}^2\ldots\boldsymbol{\eta}^{s-1}\boldsymbol{\eta}^s\boldsymbol{\eta}^* \\ \|\boldsymbol{\omega}^1\|=\cdots=\|\boldsymbol{\omega}^s\|=0,\ \|\boldsymbol{\omega}^*\|\neq 0}} (-1)^s \frac{\mathrm{lava}^{\boldsymbol{\eta}^1}(z)\,\mathrm{lava}^{\boldsymbol{\eta}^2}(z)\ldots\mathrm{lava}^{\boldsymbol{\eta}^s}(z)\mathrm{va}^{\boldsymbol{\eta}^*}(z)}{r_1.(r_1+r_2)\ldots(r_1+\cdots+r_s)}$$

with $r_i :=$ length of $\boldsymbol{\eta}^i$ and

$$\mathrm{lava}^\bullet(z) := (\mathrm{lan}\,.\,\mathrm{va}^\bullet(z)) \times (\mathrm{va}^\bullet(z))^{-1} \qquad (\mathrm{lava}^\bullet \text{ alternal})$$
$$\mathrm{lan}\,.\,\mathrm{va}^{\boldsymbol{\eta}}(z) := (\sum_{\omega_i \neq 0} 1)\,.\,\mathrm{va}^{\boldsymbol{\eta}}(z) \qquad (\mathrm{lan}^\bullet = \text{mould derivation})$$

The main difference, however, is in the Borel transforms $\widehat{\mathrm{va}}^\bullet(\zeta)$ and $\widehat{\mathrm{wa}}^\bullet(\zeta)$. Assume for simplicity that all $\sigma_i$ are 0. Then $\widehat{\mathrm{va}}^\bullet(\zeta)$ is regular *at* the origin $\zeta = 0$, but usually not *over* the origin, i.e. at the points that lie over 0 on the other Riemann leaves, whereas $\widehat{\mathrm{wa}}^\bullet(\zeta)$ is always regular both *at* and *over* the origin.

The shortest way to check this is via the resurgence equations of $\mathrm{va}^\bullet$ and those of the alternal mould $\mathrm{lava}^\bullet$ which, alongside $\mathrm{va}^\bullet$, is the main building block of $\mathrm{wa}^\bullet$. These resurgence equations involve one and the same constant (i.e. $z$-independent) and alternal mould $\mathrm{ka}^\bullet$, but again there is a significant gap in complexity:

$$\Delta\, \mathrm{va}^\bullet(z) = \mathrm{ka}^\bullet \times \mathrm{va}^\bullet(z) \tag{2.36}$$
$$\Delta\, \mathrm{lava}^\bullet(z) = \mathrm{lan}\,.\,\mathrm{ka}^\bullet + \mathrm{ka}^\bullet \times \mathrm{lava}^\bullet(z) - \mathrm{lava}^\bullet(z) \times \mathrm{ka}^\bullet \tag{2.37}$$

From here it is but a short step to derive the resurgence equations for our main objects. For the (direct/inverse) normaliser they read:

$$[\boldsymbol{\Delta}_\omega, \Theta_{\mathrm{nor}}] = -\Theta_{\mathrm{nor}}\mathbb{A}_\omega \qquad \text{(B.E. for direct normaliser)} \tag{2.38}$$
$$[\boldsymbol{\Delta}_\omega, \Theta_{\mathrm{nor}}^{-1}] = +\mathbb{A}_\omega\Theta_{\mathrm{nor}}^{-1} \qquad \text{(B.E. for inverse normaliser)} \tag{2.39}$$

and are none other than the Bridge Equation, which is an amazingly general and flexible tool for extracting the object's *analytic invariants* $\mathbb{A}_\omega$ and even, if we so wish, for expressing these $\mathbb{A}_\omega$ in terms of the two basic ingredients:

- the operators $\mathbb{B}_n$ which represent the field $X$ via its Taylor coefficients;
- the mould $\mathrm{ka}^\bullet$ which represents the "universal-transcendent" part.

Similar, but more complex, resurgence equations hold for the prenormalising transformation $\Theta_{\mathrm{pre}}$ as well as the premormalised field $X^{\mathrm{pre}}$ (which, unlike $X^{\mathrm{nor}}$, is itself resurgent). Writing down these equations—they are entirely deducible from the system (2.36)–(2.37)—is a rewarding exercise, which we highly recommend.

# 3 Lesson Three: Canonical object synthesis

## 3.1 Local objects and analytic invariants

We revert to the notations and conventions of Lesson Two. Here too, by *local analytic object* we shall mean essentially:

(1) germs of singular analytic vector fields at 0 on $\mathbb{C}^\nu$ or *fields* for short:

$$X = \sum_{1 \le i \le \nu} X_i(x)\, \partial_{x_i}; \qquad X_i(x) \in \mathbb{C}\{x_1, \dots, x_\nu\} \tag{3.1}$$

(2) germs of analytic diffeomorphisms of $\mathbb{C}^\nu$ into itself or *diffeos* for short:

$$f : x_i \mapsto f_i(x); \qquad i = 1, \ldots, \nu, \quad f_i(x) \in \mathbb{C}\{x_1, \ldots, x_\nu\} \tag{3.2}$$

with the corresponding substitution operators $F$ denoted by a capital letter:

$$(F\varphi)(x) := \varphi(f(x)); \qquad \forall \varphi(x) \in \mathbb{C}\{x_1, \ldots, x_\nu\} \tag{3.3}$$

We recall that everything depends on the nature of the object's spectrum, i.e. the eigenvalues of its linear part: $\lambda = (\lambda_1, \ldots, \lambda_\nu)$ for a field or $l = (l_1, \ldots, l_\nu)$ for a diffeo. If the spectrum is 'generic', then the object is analytically conjugate to its linear part. Difficulties arise only in case of

- *resonance* of the spectrum (see Section 2.1);
- *quasiresonance* of the spectrum (see Section 2.1);
- *nihilence* of the object (see Section 2.1).

As already pointed out, the more 'complicated' an object, the larger its set of *invariants* tends to be. Alongside the *formal* and *analytic invariants* (i.e. relative to formal or analytic coordinate changes) we have the notion of *holomorphic invariants*—i.e. those invariants that depend holomorphically on the object **Ob** (or, in practical terms, its Taylor coefficients), at least when **Ob** remains within a fixed formal conjugacy class.

Resonance generates *formal invariants*: finitely many for one degree of resonance, infinitely many for several degrees.

Each of the afore-mentioned complications—resonance, quasiresonance, nihilence—also gives rise to analytic invariants[33]. These complications may combine, but when resonance alone is at work, there tend to exist[34] complete systems of analytic-cum-holomorphic invariants $\{\mathbb{A}_\omega\}$.

*We shall in this Third Lesson deal with the purely resonant case.* This covers in particular such important objects as:

- identity-tangent diffeomorphisms;
- vector fields with (one or several) vanishing eigenvalues;
- most singular differential equations or systems[35].

*"Object analysis"* starts from some object **Ob** and is concerned with finding its invariants. For resonant objects, which alone matter to us here, there is a method of sweeping generality—the Bridge Equation (see *infra*)—for conctructing complete systems $\{\mathbb{A}_\omega\}$ of analytic-cum-holomorphic invariants, in the form of specific differential operators $\mathbb{A}_\omega$, with indices $\omega$ running through a countable set $\Omega$ generated by the object's spectrum. Moreover, relative to a special class of "nice" bases $\{\Delta_\omega^{\text{nice}}\}$ of the algebra ALIEN of alien derivations, technically known as *well-behaved* bases, the Bridge Equation yields systems $\{\mathbb{A}_\omega^{\text{nice}}\}$ that can

[33]Strictly analytic, i.e. not deducible from formal invariants.

[34]With slight qualifications, see e.g. [E3, E5].

[35]After their standard rephrasing as time-autonomous systems or, equivalently, as vector fields.

be characterised by means of simple, transparent growth conditions on the invariants $\mathbb{A}_\omega^{\text{nice}}$ as $\omega$ increases.

*"Object synthesis"* is the converse problem: starting from a prescribed system $\{\mathbb{A}_\omega\}$ with the proper growth pattern, find an object **Ob** whose invariants coincide with that system. There are actually four degrees:

- *existence:* showing that such an object **Ob** does exist;
- *constructiveness:* producing an effective procedure for constructing it;
- *expliciteness:* expanding the object **Ob**, in a manner both explicit and universal, by means of elementary special functions, the so-called resurgence monomials, that are not constructed *ad hoc*, but given once and for all;
- *canonicity:* examining whether perchance there exists a "canonical" choice for **Ob** and also (since we don't want to forego *expliciteness*) a corresponding system of "canonical" resurgence monomials.

We won't recall (see [E2, E3, E15]) the basic facts about *existential, constructive, explicit* synthesis, nor shall we bother with the various strategies, some going back to the late 70s, for establishing the related results. Our concern here is with the dream-goal of *explicit-canonical* synthesis, particularly for non-linear problems. The earliest attempts in this direction were based on the notion of *hyperlogarithmic monomials.* We shall show why these attempts, while interesting in their own way and insightful, were doomed to partial failure. We shall then mention the existence of a whole new class of *resurgence monomials,* based on "prodiffusions" and which on account of their nice growth properties, *do permit explicit synthesis* in all cases. Lastly, we shall show that there exists a particular subclass, the so-called *paralogarithmic* or *spherical monomials,* which unquestionably stand out as 'canonical' and which can be harnessed to synthesise objects **Ob** that inherit their 'canonicity'.

## 3.2 Object Analysis: the Bridge Equation

Let **Ob** be some (purely) resonant object —field or diffeo—expressed in a particular analytic chart $x = \{x_1, \ldots, x_\nu\}$ that diagonalises the object's linear part. The object's complete linearisation is usually impossible, even formally, and what takes its place is *formal normalisation,* which removes all but a few resonant monomials, or the more radical step of *formal trivialisation,* which forfeits *entireness*[36] but reduces the object to the simplest conceivable form, namely $\partial_z$ for as field and $z \mapsto z+1$ for a diffeo.

Let $y = \{y_1, \ldots, y_\nu\}$ be a formal normal chart, and consider the formal-entire coordinate changes $y_i = \theta_i(x)$ and $x_i = \theta_i^{-1}(y)$ with the substitution operators $\Theta$ and $\Theta^{-1}$ that go with them: $\Theta^{\pm}\varphi := \varphi \circ \theta^{\pm}$

Also consider the "trivial chart" $\{z, u\} = \{z, u_1, \ldots, u_{\nu-1}\}$. Expressing the given coordinates $x = \{x_i\}$ in terms of the trivial coordinates $\{z, u_i\}$, we get the so-called *formal integral* $x(z, u) = \{x_1(z, u), \ldots, x_\nu(z, u)\}$, which satisfies:

$$\partial_z x_i(z,u) \equiv X_i(x(z,u)) \quad \forall i \quad \text{for a field} \tag{3.4}$$

$$f_i(x(z,u)) \equiv x_i(z+1,u) \quad \forall i \quad \text{for a diffeo} \tag{3.5}$$

[36] It no longer relies on entire power series.

The Bridge Equation (B.E.) is a powerful and versatile tool for extracting the object's *invariants* from the divergence-resurgence of the *trivialising* or (direct/inverse) *normalising* transformations. Here are its three main forms:

$$\mathbf{\Delta}_\omega\, x(z,u) = \mathbb{A}_\omega x(z,u) \qquad \text{(B.E. for the formal integral)} \tag{3.6}$$

$$[\mathbf{\Delta}_\omega, \Theta] = -\Theta\mathbb{A}_\omega \qquad \text{(B.E. for the direct normaliser)} \tag{3.7}$$

$$[\mathbf{\Delta}_\omega, \Theta^{-1}] = +\mathbb{A}_\omega\Theta^{-1} \qquad \text{(B.E. for the inverse normaliser)} \tag{3.8}$$

The indices $\omega$ on both sides of the Bridge Equation range through a countable set $\Omega$ spanned by the object's multipliers, i.e. the $\lambda_j$ in the case of a field, and the $\log l_j$ (to which one must add the universal multiplier $\lambda_0 := 2\pi i$) in the case of a diffeo.

The $\mathbf{\Delta}_\omega$ on the left-hand side denotes the alien derivation relative to the variable $z$ and index $\omega$ but with a built-in exponential factor $\mathbf{\Delta}_\omega := e^{-\omega z}\Delta_\omega$ that makes it commute with $\partial_z$ and ensures the invariance rule under non-divergent changes of equivalent variables $z \mapsto z_\star$ with $z \sim z_\star$[37].

$$\mathbf{\Delta}_\omega^{(z)}\varphi(z) \equiv \mathbf{\Delta}_\omega^{(z_\star)}\varphi_\star(z_\star) \quad \text{if} \quad \varphi(z) \equiv \varphi_\star(z_\star) \quad \text{and} \quad z \sim z_\star \tag{3.9}$$

The alien-differentiation variable $z$, also known as *critical variable*[38], is always $\sim \infty$. In (3.6) it is simply the $z$ inside the formal integral. In (3.7) or (3.8) it is the inverse of some *resonant monomial*, i.e. $z := 1/x^m$ or $z := 1/y^m$. Due to the afore-mentioned invariance property of alien differentiation[39], the *critical variable* is actually defined up to equivalence $\sim$ and so the proper intrinsic notion is in fact that of *critical class*.

The $\mathbb{A}_\omega$ on the right-hand side are ordinary differential operators—in the variables $(z, u_i)$ or $(x_i)$ or $(y_i)$ respectively. They are constructively determined, even overdetermined, by the requirement of equality in the Bridge Equation—whichever of its variants we choose to work with, and whichever critical variable we pick (within the critical class) for alien differentiation. Each single $\mathbb{A}_\omega$ is an invariant of the object **Ob**, and the total collection $\{\mathbb{A}_\omega, \omega \in \Omega\}$ constitutes a set, both complete and free, of analytic-cum-holomorphic invariants.

All these claims, as sketchy as they are sweeping, clearly cry for explanations and qualifications, which cannot be supplied here but are available in the literature [E2, E3, E5, E10]. We recalled these statements simply as a general backdrop for the twin problems of object analysis and synthesis—but to illustrate the method we shall focus on just four typical examples.

## 3.3 Examples of local objects

**Example 1:** One-dimensional identity-tangent diffeo.

$$f^{\text{nor}} : z \longrightarrow z + 2\pi i \tag{3.10}$$

$$f : z \longrightarrow z + 2\pi i + \sum_{2 \le n} a_n z^{-n} \tag{3.11}$$

Remark: We might of course have chosen the *unit shift* as our normal form, but choosing the $2\pi i$-shift has the advantage of placing the singularities over $\mathbb{Z}$ in the Borel-plane, and of rendering the parallel with Example 2 (*infra*) more obvious.

[37] E.g. $z_\star := z(1 + \sum a_n z^{-n})$ or $z_\star := z(1 + \sum a_n (\log z)^{-n})$ with convergent series $\sum$.

[38] Or *critical time*.

[39] It mirrors the invariance rule $\varphi'(z)\, dz \equiv \varphi'_\star(z_\star)\, dz_\star$ for ordinary differentiation.

**Example 2:** Singular, non-linear differential equation[40].

$$d_z y^{\text{nor}} = y^{\text{nor}} \tag{3.12}$$

$$d_z y = y + \sum_{1+n\geq 0} b_n(z) y^{1+n} \qquad b_n(z) \in \mathbb{C}\{z^{-1}\} \tag{3.13}$$

**Example 3:** Singular linear differential system.

$$d_z y_i^{\text{nor}} = \lambda_i y_i^{\text{nor}} \qquad (i = 1, \ldots, \nu;\ \lambda_i \neq \lambda_j \text{ if } i \neq j) \tag{3.14}$$

$$d_z y_i = \lambda_i y_i + \sum_{1\leq j\leq \nu} b_{i,j}(z) y_j \qquad b_{i,j}(z) \in \mathbb{C}\{z^{-1}\} \tag{3.15}$$

**Example 4:** Singular non-linear differential system.

$$d_z y_i^{\text{nor}} = \lambda_i\, y_i^{\text{nor}} \qquad (1 \leq i \leq \nu;\ \lambda \text{ not res. nor quasi. res.}) \tag{3.16}$$

$$d_z y_i = \lambda_i y_i + \sum_{\substack{1+n_i\geq 0 \\ n_j\geq 0 \text{ if } j\neq i}} b_{i,n}(z) y_i y^n \qquad b_{i,n} \in \mathbb{C}\{z^{-1}\} \tag{3.17}$$

### 3.4 Resurgence equations and analytic invariants

**Example 1:** One-dimensional identity-tangent diffeo. We may work with the formal normalising map $f^\star$ or its inverse ${}^\star f$:

$$f^\star \circ f \equiv f^{\text{nor}} \circ f^\star \quad \text{i.e.} \quad f^\star(f(z)) \equiv 2\pi i + f^\star(z) \tag{3.18}$$

$$f \circ {}^\star f \equiv {}^\star f \circ f^{\text{nor}} \quad \text{i.e.} \quad f({}^\star f(z)) \equiv {}^\star f(z + 2\pi i) \tag{3.19}$$

Both are generically divergent but always resurgent. They satisfy the resurgence equations:

$$\mathbf{\Delta}_n f^\star(z) \equiv -A_n \exp(-n f^\star(z)) \qquad (\forall n \in \mathbb{Z}^\star) \tag{3.20}$$

$$\mathbf{\Delta}_n {}^\star f(z) \equiv +A_n e^{-nz} \partial_z {}^\star f(z) =: \mathbb{A}_n {}^\star f(z) \qquad (\forall n \in \mathbb{Z}^\star) \tag{3.21}$$

which in turn yield the complete and free system of analytic invariants:

$$\mathbb{A} = \{\mathbb{A}_n := A_n e^{-nz} \partial_z;\ n \in \mathbb{Z}^\star,\ A_n \in \mathbb{C}\} \tag{3.22}$$

**Example 2:** Singular, non-linear differential equation. We may work with the formal integral

$$y(z,u) \in \mathbb{C}[[z^{-1}, ue^z]] \qquad (u = \text{integration parameter}) \tag{3.23}$$

which is generically divergent (in $z$) but always resurgent (again, in $z$) and satisfies the Bridge Equation:

$$\mathbf{\Delta}_n y(z,u) \equiv A_n u^{n+1}\, \partial_u y(z,u) =: \mathbb{A}_n y(z,u) \qquad (n = -1, 1, 2, 3, \ldots) \tag{3.24}$$

yielding the complete and free system of analytic invariants:

$$\mathbb{A} = \{\mathbb{A}_n := A_n u^{n+1}\, \partial_u;\ n \in \{-1\} \cup \mathbb{N}^\star,\ A_n \in \mathbb{C}\} \tag{3.25}$$

[40]Under addition of a variable $x := z^{-1}$ it translates into a two-dimensional vector field, local (at $0 \in \mathbb{C}^2$), singular, and resonant (with one vanishing and one non-vanishing multiplier).

**Example 3:** Singular linear differential system. Here the formal integral reduces to

$$y(z,u) = \sum_{1\le i\le \nu} b_i(z) e^{\lambda_i z} u_i \quad \text{with} \quad b_i(z) \in \mathbb{C}[[z^{-1}]] \tag{3.26}$$

The Bridge Equation reads:

$$\mathbb{\Delta}_{\lambda_i-\lambda_j} y(z,u) \equiv A_{\lambda_i-\lambda_j}\, u_j\, \partial_{u_i} y(z,u) =: \mathbb{A}_{\lambda_i-\lambda_j} y(z,u) \qquad (i \neq j) \tag{3.27}$$

and once again yields a complete and free, but this time *finite*, system of analytic invariants:

$$\{\mathbb{A}_{\lambda_i-\lambda_j} := A_{\lambda_i-\lambda_j} u_j\, \partial_{u_i}; \qquad 1 \le i \neq j \le \nu\} \tag{3.28}$$

**Example 4:** Singular non-linear differential system. The formal integral involves $\nu$ integration parameters $u_i$, each with its accompanying exponential factor:

$$y(z,u) \in \mathbb{C}[[z^{-1}, u_1 e^{\lambda_1 z}, \ldots, u_\nu e^{\lambda_\nu z}]] \qquad (\text{with } \mathbb{Q}\text{-independent } \lambda_i\text{'s}) \tag{3.29}$$

The Bridge Equation reads:

$$\mathbb{\Delta}_\omega y(z,u) \equiv \mathbb{A}_\omega y(z,u) \qquad (\forall \omega = \mathbf{\Omega}) \tag{3.30}$$

with indices $\omega$ running through a set:

$$\mathbf{\Omega} = \left\{ \omega;\ \omega = \sum_{1\le i\le\nu} m_i \lambda_i,\ m_i \ge -1,\ \sum_{m_i=-1} 1 = 0 \text{ or } 1 \right\} \tag{3.31}$$

and with differential operators of the form:

$$\mathbb{A}_\omega := u_1^{m_1} \ldots u_\nu^{m_\nu} \sum_{1\le i\le\nu} A^i_\omega u_i\, \partial_{u_i} \quad \text{if} \quad \omega = \sum m_i \lambda_i \qquad (A^i_\omega \in \mathbb{C}) \tag{3.32}$$

which, together, constitute a complete and free system $\{\mathbb{A}_\omega;\ \omega = \mathbf{\Omega}\}$ of analytic invariants.

## 3.5 Alien derivations: lateral, standard or organic

Alien derivations $\Delta_\omega$ are determined by systems of weights $\mathbf{d}^{\binom{\epsilon_1,\ldots,\epsilon_i,\ldots,\epsilon_r}{\omega_1,\ldots,\omega_i,\ldots,\omega_r}}$ subject to the self-consistency relations:

$$\sum_{\epsilon_i \in \{+,-\}} \mathbf{d}^{\binom{\epsilon_1,\ldots,\epsilon_i,\ldots,\epsilon_r}{\omega_1,\ldots,\omega_i,\ldots,\omega_r}} := \mathbf{d}^{\binom{\epsilon_1,\ldots,\epsilon_{i+1},\ldots,\epsilon_r}{\omega_1,\ldots,\omega_i+\omega_{i+1},\ldots,\omega_r}} \qquad \forall i \tag{3.33}$$

and their action in the *convolutive model* is given by:

$$\Delta_\omega \hat\varphi(\zeta) := \sum_{\epsilon_i=\pm} \mathbf{d}^{\binom{\epsilon_1,\ldots,\epsilon_r}{\omega_1,\ldots,\omega_r}}\, \hat\varphi^{\binom{\epsilon_1,\ldots,\epsilon_r}{\omega_1,\ldots,\omega_r}}(\zeta+\omega) \tag{3.34}$$

for $\zeta$ close to $+0$ and by analytic continuation in the large. There being no scope for confusion, we also use the same symbols to denote the alien derivations acting in the *multplicative models (formal or germinal)*, i.e. the pull-backs by Borel-Laplace of the operators $\Delta_\omega$ as defined by (3.34).

### 3.5.1 Lateral alien operators $\Delta_\omega^+$ and $\Delta_\omega^-$

The right-lateral operators $\Delta_\omega^+$ are defined by:

$$\mathbf{d}^{\binom{\epsilon_1,\dots,\epsilon_r}{\omega_1,\dots,\omega_r}} := \begin{cases} \epsilon_r 1 & \text{if } (\epsilon_1,\dots,\epsilon_{r-1}) = (+,\dots,+) \\ 0 & \text{otherwise} \end{cases}$$

and the left-lateral operators $\Delta_\omega^-$ are defined by:

$$\mathbf{d}^{\binom{\epsilon_1,\dots,\epsilon_r}{\omega_1,\dots,\omega_r}} := \begin{cases} -\epsilon_r 1 & \text{if } (\epsilon_1,\dots,\epsilon_{r-1}) = (-,\dots,-) \\ 0 & \text{otherwise} \end{cases}$$

Their simplicity is deceptive because they are not *first-order* derivations. Instead of satisfying (in the multiplicative models) the Leibniz rule:

$$\Delta_\omega(\varphi_1\varphi_2) \equiv (\Delta_\omega\varphi_1)\varphi_2 + \varphi_1(\Delta_\omega\varphi_2)$$

they satisfy messier relations:

$$\Delta_\omega^\pm(\varphi_1\varphi_2) \equiv (\Delta_\omega^\pm\varphi_1)\varphi_2 + \varphi_1(\Delta_\omega^\pm\varphi_2) + \sum_{\omega_1+\omega_2=\omega} (\Delta_{\omega_1}^\pm\varphi_1)(\Delta_{\omega_2}^\pm\varphi_2)$$

with a sum extending to all $\omega_1, \omega_2$ colinear with $\omega$. For any given pair of test functions $\varphi_1, \varphi_2$ the above sum makes sense, since it can never involve more than a finite number of non-zero terms.

### 3.5.2 Standard alien derivations $\Delta_\omega = \Delta_\omega^{\text{stan}}$

They correspond to the weights:

$$\mathbf{d}^{\binom{\epsilon_1,\dots,\epsilon_r}{\omega_1,\dots,\omega_r}} := \epsilon_r \frac{p!\,q!}{(p+q+1)!} = \epsilon_r \frac{p!\,q!}{r!} \qquad \begin{aligned} p &:= \#\{1 \le i < r;\ \epsilon_i = +\} \\ q &:= \#\{1 \le i < r;\ \epsilon_i = -\} \end{aligned}$$

They are first-order alien derivations (and thus an impovement on the $\Delta_\omega^\pm$) but they are not *well-behaved.*

### 3.5.3 Organic alien derivations $\Delta_\omega = \Delta_\omega^{\text{org}}$

They correspond to the weights:

$$\mathbf{d}^{\binom{\epsilon_1,\dots,\epsilon_r}{\omega_1,\dots,\omega_r}} := \begin{cases} \frac{\epsilon_r}{2} \frac{\omega_{p+1}}{\omega_1+\dots+\omega_r} & \text{if } (\epsilon_1,\dots,\epsilon_r) = ((+)^p, (-)^q, \epsilon_r) \\ \frac{\epsilon_r}{2} \frac{\omega_{q+1}}{\omega_1+\dots+\omega_r} & \text{if } (\epsilon_1,\dots,\epsilon_r) = ((-)^q, (+)^p, \epsilon_r) \\ 0 & \text{otherwise} \end{cases}$$

They are *first-order and well-behaved* alien derivations. In fact, at the moment, they are the simplest extant system of such derivations.

### 3.5.4 Conversion rule

Each system $\{\Delta_\omega^{\text{org}}\}$ or $\{\Delta_\omega^{\text{stan}}\}$ freely generates the algebra ALIEN of alien derivations, and we have the simple conversion rules:

$$\Delta_{\omega_0}^{\text{org}} \equiv \Delta_{\omega_0}^{\text{stan}} + \sum_{3 \leq r \text{ odd}} H_{\omega_0}^{\omega_1,\ldots,\omega_r} [[\Delta_{\omega_r}^{\text{stan}}, \ldots, [\Delta_{\omega_2}^{\text{stan}}, \Delta_{\omega_1}^{\text{stan}}]] \tag{3.35}$$

$$\Delta_{\omega_0}^{\text{stan}} \equiv \Delta_{\omega_0}^{\text{org}} + \sum_{3 \leq r \text{ odd}} K_{\omega_0}^{\omega_1,\ldots,\omega_r} [[\Delta_{\omega_r}^{\text{org}}, \ldots, [\Delta_{\omega_2}^{\text{org}}, \Delta_{\omega_1}^{\text{org}}]] \tag{3.36}$$

with scalar coefficients $H_{\omega_0}^{\omega_1,\ldots,\omega_r}$ and $K_{\omega_0}^{\omega_1,\ldots,\omega_r}$ which of course vanish unless $\omega_0 \neq \sum \omega_i$. For instance:

$$H_{\omega_0}^{\omega_1,\ldots,\omega_{2r+1}} := \frac{1}{2r+1} \sum_{0 \leq i \leq 2r} \frac{(-1)^i (2r)!}{i!\,(2r-i)!} \frac{\omega_{i+1}}{\omega_0} \quad \text{if} \quad \omega_0 = \sum \omega_i \tag{3.37}$$

### 3.5.5 Characterisation of well-behaved alien derivations

To each system of alien derivations we may associate the moulds:

$$\mathbf{red}^{\omega_1,\ldots,\omega_r} := (-1)^r \mathbf{d}^{\binom{+,\ldots,+}{\omega_1,\ldots,\omega_r}} \qquad \text{(right-lateral mould)} \tag{3.38}$$

$$\mathbf{led}^{\omega_1,\ldots,\omega_r} := (-1)^r \mathbf{d}^{\binom{-,\ldots,-}{\omega_1,\ldots,\omega_r}} \qquad \text{(left-lateral mould)} \tag{3.39}$$

Due to the self-consistency relations, both the right- and left-lateral moulds encapsulate all the information about the entire weight system $\left\{\mathbf{d}^{\binom{\epsilon_1,\ldots,\epsilon_r}{\omega_1,\ldots,\omega_r}}\right\}$, and each one can be deduced from the other in a simple manner.

The three following conditions are equivalent:

(a) the alien operators $\Delta_\omega$ in (3.34) are (first-order) *alien derivations*;

(b) the right-lateral mould $\mathbf{red}^\bullet$ is *alternel*;

(c) the left-lateral mould $\mathbf{led}^\bullet$ is *alternel.*

The five following conditions are equivalent:

(d) the system $\Delta_\omega$ in (3.34) of (first-order) alien derivations is *well-behaved*;

(e) we have universal bounds $|\mathbf{red}^{\omega^{\twoheadleftarrow}}| \leq C_1 (D_1)^{r(\omega^{\twoheadleftarrow})}$;

(f) we have universal bounds $|\mathbf{red}^{\omega^{\twoheadrightarrow}}| \leq C_2 (D_2)^{r(\omega^{\twoheadrightarrow})}$;

(g) we have universal bounds $|\mathbf{led}^{\omega^{\twoheadleftarrow}}| \leq C_3 (D_3)^{r(\omega^{\twoheadleftarrow})}$;

(h) we have universal bounds $|\mathbf{led}^{\omega^{\twoheadrightarrow}}| \leq C_4 (D_4)^{r(\omega^{\twoheadrightarrow})}$.

## 3.6 Well-behaved resurgence monomials

### 3.6.1 Multiplicative systems of resurgence monomials

They are basically systems $\{\mathcal{U}^{\boldsymbol{\omega}}(z) = \mathcal{U}^{\omega_1,\dots,\omega_r}(z)\}$ of elementary resurgent functions which:

(1) behave simply under multiplication[41];

(2) behave simply under alien differentiation;

(3) are "complete" in the sense that they should enable us to expand (or approximate) any given resurgent function $\varphi$:

$$\varphi(z) \text{``}=\text{''} \sum_{\boldsymbol{\omega}} c_{\boldsymbol{\omega}}(z)\mathcal{U}^{\boldsymbol{\omega}}(z) := \sum_{r\geq 0}\sum_{\omega_i} c_{\omega_1,\dots,\omega_r}(z)\mathcal{U}^{\omega_1,\dots,\omega_r}(z) \tag{3.40}$$

with coefficients $c_{\boldsymbol{\omega}}(z)$ that are either ordinary constants or "resurgence constants", that is to say, functions with only vanishing alien derivatives: $\Delta_{\omega_0} c_{\boldsymbol{\omega}}(z) \equiv 0$, $\forall \omega_0$.

Condition (2) is relative to a given basis $\{\Delta_\omega\}$ of the algebra ALIEN of alien derivations. In concrete terms the condition stipulates that:

$$\Delta_{\omega_0}\mathcal{U}^{\omega_1,\dots,\omega_r} \equiv \begin{cases} \Delta_{\omega_0}\mathcal{U}^{\omega_2,\dots,\omega_r} & \text{if } \omega_0 = \omega_1 \\ 0 & \text{if } \omega_0 \neq \omega_1. \end{cases} \tag{3.41}$$

Condition (1) means that:

$$\mathcal{U}^{\boldsymbol{\omega}'}\mathcal{U}^{\boldsymbol{\omega}''} \equiv \sum_{\boldsymbol{\omega}\in \mathbf{shu}(\boldsymbol{\omega}',\, \boldsymbol{\omega}'')} \mathcal{U}^{\boldsymbol{\omega}} \tag{3.42}$$

with a sum extending to all sequences $\boldsymbol{\omega}$ obtained by *shuffling* the two factor sequences $\boldsymbol{\omega}', \boldsymbol{\omega}''$. In other words, the mould $\mathcal{U}^{\bullet}$ should be *symmetral*[42].

Clearly, there exist infinitely many multiplicative systems of resurgence monomials. Indeed, if $\mathcal{U}^{\bullet}(z)$ is one such system, so will be the system $\mathcal{U}^{\bullet}_{\mathcal{C}}(z) := \mathcal{U}^{\bullet}(z) \times \mathcal{C}^{\bullet}(z)$ derived therefrom by postmultiplication[43] by any symmetral, resurgence-constant mould $\mathcal{C}^{\bullet}(z)$.

### 3.6.2 Integral alien calculus

Multiplicative systems of resurgence monomials are extremely useful to solve resurgence equations, or systems of such equations, and to express their solutions in the form of expansions of type (3.40), often with constant coefficients $c_{\boldsymbol{\omega}}$. Thus, if we revert to Example 1 and try to solve the system of resurgence equations (3.20) that characterise the normalising transformation $f^\star$, we find:

$$f^\star(z) := z - \sum_r \sum_{n_i} A_{n_1} \dots A_{n_1} \Gamma_{n_1,\dots,n_r} \mathcal{U}^{n_1,\dots,n_r}(z) \tag{3.43}$$

$$\text{with} \quad \Gamma_{n_1,\dots,n_r} := (n_1)(n_1+n_2)\dots(n_1+n_2+\dots+n_{r-1}) \tag{3.44}$$

[41] Or convolution, depending on the model.

[42] See Section 6.1.

[43] In the sense of moulds.

But the real issue of course is convergence. We might try to solve it on an *ad hoc* basis, i.e. by choosing our resurgence monomials differently for each problem. But we are more ambitious: we want resurgence monomials that work in *all* cases[44]. That may seem a tall order, but it is feasibble! The answer lies in the notion of *well-behaved* systems of resurgence monomials. And not only do such systems exist, but there is a canonical choice!

### 3.6.3 Characterisation of well-behaved resurgence monomials

To any given system of resurgence monomials we may associate a *right-lateral* mould ${}^{re}\mathcal{U}^\bullet(z)$ and a *left-lateral* mould ${}^{le}\mathcal{U}^\bullet(z)$ characterised by the orthogonality conditions:

$$\{{}^{re}\mathcal{U}^{\omega_1,\dots,\omega_r}\} \quad \text{orthogonal to} \quad \{\Delta^+_{\omega_1,\dots,\omega_r} := \Delta^+_{\omega_r}\dots\Delta^+_{\omega_1}\} \tag{3.45}$$

$$\{{}^{le}\mathcal{U}^{\omega_1,\dots,\omega_r}\} \quad \text{orthogonal to} \quad \{\Delta^-_{\omega_1,\dots,\omega_r} := \Delta^-_{\omega_r}\dots\Delta^-_{\omega_1}\} \tag{3.46}$$

The three following conditions are equivalent:

(a) a system of resurgence monomials is *multiplicative*;

(b) its right-lateral mould ${}^{re}\mathcal{U}^\bullet$ is *symmetrel*[45];

(c) its left-lateral mould ${}^{le}\mathcal{U}^\bullet$ is *symmetrel*;

Observe that the criterion here is *symmetrel* and not *symmetral* as in Section 3.6.1 This is because the lateral alien operators $\Delta^\pm_\omega$ are not first-order alien derivations (see Section 3.5.3).

The five following conditions are equivalent:

(d) a system of multiplicative resurgence monomials is *well-behaved*;

(e) we have universal bounds $\|{}^{re}\mathcal{U}^{\omega^{\leftarrowtail}}\| \le C_1(D_1)^{r(\omega^{\leftarrowtail})}$;

(f) we have universal bounds $\|{}^{re}\mathcal{U}^{\omega^{\rightarrowtail}}\| \le C_2(D_2)^{r(\omega^{\rightarrowtail})}$;

(g) we have universal bounds $\|{}^{le}\mathcal{U}^{\omega^{\leftarrowtail}}\| \le C_3(D_3)^{r(\omega^{\leftarrowtail})}$;

(h) we have universal bounds $\|{}^{le}\mathcal{U}^{\omega^{\rightarrowtail}}\| \le C_4(D_4)^{r(\omega^{\rightarrowtail})}$.

for a suitable norm $\|.\|$ (see [E15]).

Resorting to so-called 'prodiffusion integrals' (rather similar to the diffusion integrals which we used in Lesson One, Section 1.10.4, to construct our well-behaved averages, but with Borel-Laplace replacing the Fourier transform) one may produce a large variety of well-behaved systems of resurgence monomials. Furthermore, there exits a canonical choice, the only remaining latitude being in the determination of a single real parameter $c$.

[44]I.e. for all systems of resurgence equations that do admit solutions.

[45]See Section 6.1.

### 3.6.4 The canonical choice: "spherical" or "twisted" monomials

As just mentioned, they depend on a positive parameter $c > 0$ (the "twist") and are defined by the absolutely convergent integrals:

$$\mathcal{U}a_c^{\omega_1,\dots,\omega_r}(z) := \frac{\text{S.P.A.}}{(2\pi i)^r}\int_0^\infty \frac{e^{-\sum_1^r(\omega_i y_i + c^2\bar{\omega}_i y_i^{-1})}\,dy_1\cdots dy_r}{(y_r - y_{r-1})\dots(y_2-y_1)(y_1-z)} \tag{3.47}$$

$$\mathcal{U}e_c^{\omega_1,\dots,\omega_r}(z) := \mathcal{U}a_c^{\omega_1,\dots,\omega_r}(z)e^{(\sum_1^r\omega_i)z+(\sum_1^r\bar{\omega}_i)c^2z^{-1}} \tag{3.48}$$

$$\mathcal{U}_c^{\omega_1,\dots,\omega_r}(z) := \mathcal{U}a_c^{\omega_1,\dots,\omega_r}(z)e^{(\sum_1^r\bar{\omega}_i)c^2z^{-1}} \tag{3.49}$$

with integration along the rays $\arg(\omega_i y_i) = \arg(\bar{\omega}_i/y_i) = 0$.

The $\mathcal{U}a_c^{\boldsymbol{\omega}}$ are auxiliary expressions. The resurgence monomials proper are the $\mathcal{U}_c^{\boldsymbol{\omega}}$ (orthogonal to the ordinary alien derivations $\Delta_\omega$) and the exponential-carrying $\mathcal{U}e_c^{\boldsymbol{\omega}}$ (orthogonal to the exponential-carrying alien derivations $\boldsymbol{\Delta}_\omega$).

#### Interpretation of S.P.A.

S.P.A. in front of the integral means *suitable path average.* If we integrate first in $y_1$, then $y_2$, etc., the question arises as to how (i.e. on which side) $y_i$ should bypass the next (yet unused) variable $y_{i+1}$. If to the right, we set $\epsilon_i := +$. If to the left, we set $\epsilon_i := -$. To each choice $\{\epsilon_1, \epsilon_2, \dots\}$ there corresponds a different integration path, and S.P.A. means that one should take a precise average of all such paths, depending on which system $\Delta_\omega$ of alien derivations one wishes the $\mathcal{U}^{\boldsymbol{\omega}}$ to be orthogonal to. But for the right- or left-lateral moulds (characterised by orthogonality to $\Delta_\omega^{\pm}$) the S.P.A. average reduces to one single path, with all $\epsilon_i$ identical (either $+$ or $-$).

#### Interpretation of $1/(y_1 - z)$

The integral $(\dots)$ defines $\mathcal{U}^{\boldsymbol{\omega}}$ in all three models (formal, geometric, convolutive) at one stroke, depending on how we construe $1/(y_1 - z)$:

- as a power series in $z^{-1}$;
- or as a function germ at $\infty$;
- or again as its own Borel tranform.

| | | |
|---|---|---|
| Formal model: | $\frac{1}{y_1-z} \to$ | $-\sum_0^\infty z^{-n-1}y_1^n$ |
| $\Longrightarrow$ | $\mathcal{U}_c^{\boldsymbol{\omega}}(z)$ | as a formal power series |
| Geometric model: | $\frac{1}{y_1-z} \to$ | $z$-germ at $\infty$ |
| $\Longrightarrow$ | $\mathcal{U}_c^{\boldsymbol{\omega}}(z)$ | as a sectorial $z$-germ at $\infty$ |
| Convolutive model: | $\frac{1}{y_1-z} \to$ | $-\exp(y_1\zeta)$ |
| $\Longrightarrow$ | $\hat{\mathcal{U}}_c^{\boldsymbol{\omega}}(\zeta)$ | as a full $\zeta$-germ at $0$ |

**Main result**

*For positive values $c > 0$ of the twist, $\mathcal{U}^\bullet_c$ constitutes a well-behaved, multiplicative system of resurgence monomials.*

*Gist of the proof:* The difficult bit of course is *well-behaved.* We use the criteria of Section 3.6.3 and the fact (see Section 6.1) that the moulds $\mathrm{tas}^\bullet_{a,\infty}$, $\mathrm{tas}^\bullet_{\infty,b}$ as well as the moulds $\mathrm{sofo}^\bullet$, $\mathrm{sefo}^\bullet$ (which are essentially their "Fourier transforms") *conserve their form* under either arborification or anti-arborification, and as a consequence do not register any significant increase in component size[46].

**The limit-case $c = 0$**

In the limit-case $c = 0$ the integrals (3.47) remain well-defined and we still have a multiplicative system $\mathcal{U}^\bullet_0$ of resurgence monomials, but it is no longer *well-behaved.* In fact, it coincides with the much more ancient system $\mathcal{U}^\bullet$ of hyperlogarithmic monomials, so-called because their dependence in the $\omega_i$'s is indeed of hyperlogarithmic type. In contradistinction, the $\mathcal{U}^\bullet_c$ and the host of special functions attached to them (see Section 6.7) are called *paralogarithmic.*

**Why "twisted" and why "spherical"?**

The presence of a free parameter $c$ slightly detracts from the "canonicity" of our system, but this cannot be helped: no system of well-behaved resurgence monomials can suffice for all problems unless there is at least one free parameter that can be adjusted from case to case. The miracle is rather that *one* parameter should be enough! So much for the twist. As for "spherical", it refers to the striking symmetry of behaviour which our monomials $\mathcal{U}^\bullet_c$ exhibit at the antipodes 0 and $\infty$ of the Riemann sphere when $c > 0$, and which, remarkably enough, disappears when we "untwist" them, i.e. when $c = 0$.

## 3.7 Canonical object synthesis at work

Basically, with the twisted monomials at our disposal, Object Synthesis becomes a purely mechanical affair. This is precisely what we had set out to achieve: to reduce the whole process to a succession of formal manipulations. But here we must be content with outlining the six main steps:

**Step 1:** Select a formal class of local analytic objects, characterised by a formal normal form $\mathbf{Ob}^{\mathrm{nor}}$, and start from any given *admissible* system of analytic invariants $\mathbb{A} = \{\mathbb{A}_\omega;\ \omega \in \mathbf{\Omega}\}$.

**Step 2:** Choose a *well-behaved system of alien derivations*, preferably the *organic* system $\Delta^{\mathrm{org}} = \{\Delta^{\mathrm{org}}_\omega;\ \omega \in \mathbb{C}^\star\}$, and express the analytic invariants in the corresponding basis $\mathbb{A}^{\mathrm{org}} = \{\mathbb{A}^{\mathrm{org}}_\omega;\ \omega \in \mathbf{\Omega}\}$.

**Step 3:** Solve *"mechanically"*[47] the system of resurgence equations that characterise the direct or inverse normaliser $\Theta^{\pm 1}$. For instance, in the case of simply resonant fields we find

[46] See towards the end of Section 6.1 and also [E15, EV3].

[47] I.e. without worrying about convergence. Mark the choice of words: *mechanically*, i.e. by means of expansions into series of abstract resurgence monomials, rather than *formally*, which would suggest solving the problem in the ring of formal power series.

these expansions[48]:

$$\Theta \overset{\text{always}}{:=} 1 + \sum_{1\leq r} \sum_{\omega_i \in \Omega} (-1)^r \mathcal{U}e^{\omega_1,\dots,\omega_r}(z) \mathbb{A}_{\omega_r} \dots \mathbb{A}_{\omega_1} \tag{3.50}$$

$$\Theta^{-1} \overset{\text{conditionally}}{:=} 1 + \sum_{1\leq r} \sum_{\omega_i \in \Omega} \mathcal{U}e^{\omega_r,\dots,\omega_1}(z) \mathbb{A}_{\omega_r} \dots \mathbb{A}_{\omega_1} \tag{3.51}$$

**Step 4:** Replace in that "mechanical" solution the abstract monomials $\mathcal{U}^{\omega}_{\text{org}}(z)$ by the *twisted* or *spherical* monomials $\mathcal{U}^{\omega}_{\text{c,org}}(z)$ for a large enough twist $c$.

**Step 5:** *Arborify* the expansions according to the rules given in Section 6.1 so as to render the previously obtained expansions for $\Theta^{\pm 1}$ *convergent* in the space of resurgent functions.

**Step 6:** Construct the sought-after analytic object **Ob** from its normaliser by using $\mathbf{Ob} = \Theta \mathbf{Ob}^{\text{nor}} \Theta^{-1}$.

The reader may easily work this out in the case of our four Examples. For details and more examples, he may turn to [E15].

### 3.7.1 Remarks and complements

**Remark 1: Antipodal involution**

As already pointed out, our twisted monomials have much the same behaviour at both poles of the Riemann sphere. The exact correspondence is described in Section 6.7 using the so-called *antipodal involution*:

$$\text{pod} : z \mapsto c^2 z^{-1} \quad \text{and} \quad (\omega_i, c^2\bar{\omega}_i) \mapsto (\bar{\omega}_i, c^2\omega_i) \tag{3.52}$$

In terms of the objects being produced, this means that canonical object synthesis automatically generates two objects for the price of one: the 'true' object, local at $\infty$ and with exactly the prescribed invariants, and a 'mirror reflection', local at 0 and with closely related invariants. Depending on the nature of the problem (linear/non-linear, etc.) and of the invariants, these two objects may or may not link up under analytic continuation on the Riemann sphere.

**Remark 2: Analogy with $q$-equations**

Authors like Sauloy recently observed that $q$-difference equations are in some sense easier to tackle than difference or differential equations, due to dilations $z \mapsto qz$ having two fixed points 0 and $\infty$, whereas shifts $z \mapsto a + z$ have only one, namely $\infty$. It is certainly no coincidence that the simplest resurgence monomials that permit object synthesis are precisely the twisted ones ($c \neq 0$), for whom the antipodal symmetry is restored, and that the twistless monomials ($c = 0$), though apparently more simple, turn out to be inadequate for this particular purpose.

[48] The second expansion, for the reverse normaliser, is valid only if all the invariants $\mathbb{A}_\omega$ have no $\partial_z$-component and so commute with the resurgence monomials $\mathcal{U}e^{\omega}(z)$. When this is not the case, the expansion (3.51) should be slightly modified, but one can also be content to work with (3.50), which is always valid, and then derive $\Theta^{-1}$ by straightforward inversion of $\Theta$.

**Remark 3: Necessity of a one-parameter freedom**

The necessity of having at least one degree of freedom in object synthesis has been known since the 1980s at least. This holds even for such elementary objects as linear systems (Example 3). Indeed, in most cases, the twist $c$ must exceed a certain lower bound $c_{\min}$ that depends on the invariants $\{\mathbb{A}_\omega\}$. There exists, however, an important exception: the so-called *unilateral classes*, when for instance all non-vanishing $\mathbb{A}_\omega$ have their indices on the same half-line. This applies in particular to Example 2 when $\mathbb{A}_{-1} = 0$.

**Remark 4: Iso-invariant deformations**

There exists a closed system of formulae (see Section 6.7 and [E15]) to describe the exact dependence (partial derivatives, asymptotics, etc.) of our canonical resurgence monomials as functions of their variable $z$, twist $c$ and indices $\omega_i, \bar{\omega}_i$. As a result, one may write down the—often unexpectedly simple—partial differential equations which govern the sundry deformations (iso-invariant, iso-monodromic, iso-resurgent, iso-Galoisian, etc.) of our synthesised objects.

**Remark 5: WB derivations and WB monomials: unequal status**

Working with well-behaved alien derivations is merely convenient, whereas the recourse to well-behaved resurgence monomials is truly indispensible. There is a subtle difference here, which should be well understood. Indeed, the choice of this or that system of WB alien derivations *does not affect the result*: it simply gives us a comfortable basis of ALIEN to work with. Besides, there is always the lazy option of working with the lateral alien operators $\Delta_\omega^\pm$, the only drawback being that the corresponding invariants $\mathbb{A}_\omega^\pm$ cease to be first-order differential operators. In complete contrast, the synthesised object very much depends on the choice of the system of WB monomials. And in the absence of well-behaved monomials, canonical synthesis would founder altogether.

**Remark 6: Non-canonical synthesis**

*Linear* object synthesis (Example 3) is of course a very old subject. As for *non-linear* object synthesis, adequate methods (especially for situations like in our Examples 1, 2, and 4) were evolved in the late 1970s (see [E2, E3]) but they were not always explicit, much less canonical, except in a few special instances.

# 4 Lesson Four: A surprise development in KAM theory: the non-existence of Supermultiple Small Denominators

## 4.1 A few landmarks

• *Poincaré* leaned (circa 1880) towards the view that the small denominators occuring in Hamiltonian systems must generically create unsurmountable divergence in the so-called *Lindstedt series* (the formally quasiperiodic Fourier series which describe the motion).

• The first truly general convergence results in the presence of small denominators were established circa 1940 by *Siegel* [Si1, Si2], but under unnecessarily strong diophantine assumptions, and moreover in non-hamiltonian situations, where the risk of SSD (*supermultiple small denominators* , i.e. small denominators aggravated by *resonance*) does not arise.

• Using successive approximations, *Kolmogorov* proved in 1954 his epoch-making result about the survival of invariant tori near integrability (under Siegel's condition), thus indirectly establishing the convergence of the Lindstedt series—and proving Poincaré wrong. This set off an extremely active line of research (*Arnold*, *Moser*, etc.), known as KAM theory[49].

• *Cherry* before World War II and (far more systematically) *Bryuno* and *Rüssmann* in the the 70s began proving small denominator results under a diophantine condition weaker than Siegel's and probably optimal (i.e. minimal), known as Bryuno's condition.

• Towards the close of the 80s *Yoccoz* and in his wake *Perez-Marco*, using an original renormalisation approach, proved the optimality of Bryuno's condition for a number of non-linear problems (in low dimensions).

• About the same time *Eliasson* re-proved the convergence result for the Lindstedt series near integrability, but *directly*, by a super-meticulous examination of their coefficients. This was a *premiere* of sorts, but it used a forbidding amount of analysis (Eliasson's famed method of "sign compensations") and it somehow entrenched the impression that the SSD (supermultiple small multipliers) were "really there". Numerous epigones (mainly Italians, *Gallavotti* etc., also *Krikorian*) took up and developed the method, which came to be known as the "direct method" in KAM theory.

In the course of this Fourth Lesson, I shall draw on joint work by B. Vallet and myself, and show, by purely algebraic or rather combinatorial arguments, that the SSD actually *do not exist*. We shall focus on three closely related problems, of slightly increasing difficulty:

- the so-called "correction" of resonant vector fields;
- Floquet theory (differential equations with quasi-periodic coefficients);
- the survival of tori for near-integrable Hamitonian systems.

## 4.2 cor and pre: differences/similarities

Let $X$ be a local, singular, analytic vector field with diagonal linear part $X^{\mathrm{lin}}$ and homogeneous components $\mathbb{B}_{\mathbf{n}}$:

$$X = X^{\mathrm{lin}} + \mathbb{B} = X^{\mathrm{lin}} + \sum_{\mathbf{n}} \mathbb{B}_{\mathbf{n}} \quad \text{with} \tag{4.1}$$

$$X^{\mathrm{lin}} = \sum_{1\le i\le \nu} \lambda_i x_i\, \partial_{x_i} \quad \text{and} \quad \mathbb{B}_{\mathbf{n}} = x^{\mathbf{n}} \sum_{1\le i\le \nu} b_{\mathbf{n},i} x_i\, \partial_{x_i}$$

In the non-resonant case, $X$ is formally linearisable, even analytically so if there is no quasiresonance. In the resonant case, however, the simplification cannot be so thorough. The most we can hope for is to remove all non-resonant terms, and allow resonant terms on one side of the conjugation equation only. If we allow them on the left-hand side, we get familiar objects,

[49] An acronym for the three main contributors: Kolmogorov, Arnold, Moser.

the so-called normal or prenormal forms $X^{\text{nor}}, X^{\text{pre}}$, which we studied at length in Lesson Two. But if we allow them on the right-hand side, we encounter another type of object, the so-called *correction* $X^{\text{cor}}$, with altogether different properties, both formal and analytic.

$$X \overset{\text{conj.}}{\sim} X^{\text{lin}} \qquad \text{(non-resonant case)} \tag{4.2}$$

$$X \overset{\text{conj.}}{\sim} X^{\text{lin}} + X^{\text{pre}} \qquad (\text{resonant case, } X^{\text{pre}} \in \text{Reson}) \tag{4.3}$$

$$X - X^{\text{cor}} \overset{\text{conj.}}{\sim} X^{\text{lin}} \qquad (\text{resonant case, } X^{\text{cor}} \in \text{Reson}) \tag{4.4}$$

The main differences are these:

| $X^{\text{cor}}$ ("correction") | $X^{\text{pre}}$ ("prenormal form") |
|---|---|
| unique (in any given chart) | non-unique (even after chart choice)<br>but unique mod[Reson, Reson] |
| generically convergent | generically divergent-resurgent |
| interpretation: recondite<br>(Gallavotti's 'Wick Invariant') | interpretation: simple<br>(geometrical-dynamical) |

## 4.3 The correction of resonant vector fields

The corrected form $X - X^{\text{cor}}$ and its linearising transformation $\Theta_{\text{cor}}$:

$$X \mapsto X^{\text{corrd}} = X - X^{\text{cor}} \qquad (X^{\text{cor}} \text{ unique}) \tag{4.5}$$

$$X - X^{\text{cor}} = \Theta_{\text{cor}} X^{\text{lin}} \Theta_{\text{cor}}^{-1} \qquad (\Theta_{\text{cor}} \text{ non-unique}) \tag{4.6}$$

$$[X^{\text{lin}}, X^{\text{corrd}}] = [X^{\text{lin}}, X^{\text{cor}}] = 0 \tag{4.7}$$

both admit mould expansions:

$$X^{\text{cor}} = \sum \text{Carr}^{\bullet} \mathbb{B}_{\bullet} = \sum_{r \geq 1} \sum \text{Carr}^{\omega_1,\ldots,\omega_r} \mathbb{B}_{n_r} \ldots \mathbb{B}_{n_1} \tag{4.8}$$

$$\Theta_{\text{cor}} = \sum \text{Scarr}^{\bullet} \mathbb{B}_{\bullet} = \sum_{r \geq 1} \sum \text{Scarr}^{\omega_1,\ldots,\omega_r} \mathbb{B}_{n_r} \ldots \mathbb{B}_{n_1} \tag{4.9}$$

with "frequencies" $\omega_i := \langle \lambda, n_i \rangle$ and with moulds $\text{Carr}^{\bullet}$ (alternal) and $\text{Scarr}^{\bullet}$ (symmetral) that depend rationally on these frequencies. Here, we shall leave $\text{Scarr}^{\bullet}$ alone and concentrate on the mould $\text{Carr}^{\bullet}$. It is calculable from the following induction:

$$\text{var}_i \text{Carr}^{\boldsymbol{\omega}} = \sum_{\mathbf{a}\omega_i\mathbf{bc}=\boldsymbol{\omega}} \text{Carr}^{\mathbf{a}\omega_i\mathbf{c}} \text{Carr}^{\mathbf{b}} - \sum_{\mathbf{ab}\omega_i\mathbf{c}=\boldsymbol{\omega}} \text{Carr}^{\mathbf{b}} \text{Carr}^{\mathbf{a}\omega_i\mathbf{c}} \tag{4.10}$$

with the initial conditions:

$$\text{Carr}^{\emptyset} = 0; \quad \text{Carr}^{0} = 1; \quad \text{Carr}^{\omega_1} = 0 \quad \text{if} \quad \omega_1 \neq 0 \tag{4.11}$$

and with 'variation operators'[50] $\text{var}_i$ that act like this:

$$\text{var}_i M^{\omega_1,\ldots,\omega_r} := \omega_i M^{\omega_1,\ldots,\omega_r} + M^{\omega_1,\ldots,\omega_i+\omega_{i+1},\ldots,\omega_r} - M^{\omega_1,\ldots,\omega_{i-1}+\omega_i,\ldots,\omega_r} \tag{4.12}$$

[50] For their geometric interpretation, see [EV2].

**Key lemma: Non-repetition of denominators in** $\mathrm{Carr}^{\boldsymbol{\omega}}$

*For sequences* $\boldsymbol{\omega}$ *of a given length* $r$ *and a fixed degeneracy pattern of order* $d$ *(see below) the correction coefficient* $\mathrm{Carr}^{\boldsymbol{\omega}}$*, as a rational function of its* $r-d$ *independent variables* $\omega_i$*, has only poles of the form* $\eta^{-\mu}$*, with linear combinations* $\eta$ *of the* $\omega_i$ *obtained by splitting unbreakable zero-sum sub-sequences* $\boldsymbol{\omega}^\star$ *of* $\boldsymbol{\omega}$*:*

$$\pm\eta = \|\boldsymbol{\omega}'\| = -\|\boldsymbol{\omega}''\| \qquad (\boldsymbol{\omega}'\boldsymbol{\omega}'' = \boldsymbol{\omega}^\star = (\omega_i, \ldots, \omega_j)) \tag{4.13}$$

*and with a multiplicity* $\mu$ *no larger than the number of unbreakable, zero-sum sequences* $\boldsymbol{\omega}^\star$ *which, when split, can produce* $\eta$*. Moreover, although there is in general no canonical "best way" of decomposing* $\mathrm{Carr}^{\boldsymbol{\omega}}$*, there always exist decompositions of the form*

$$\mathrm{Carr}^{\boldsymbol{\omega}} = \sum m_P \prod_{i=1\ldots s_P} (\eta_{P,i})^{-\mu_{P,i}} \quad \textit{with} \quad \sum_i \mu_{P,i} \equiv r-1 \tag{4.14}$$

*which involve only effective poles* $\eta$*, with a multiplicity never exceeding the intrinsic multiplicity* $\mu$ *and with integral coefficients* $m_P$ *bounded by:*

$$\sum_P |m_P| \leq \frac{(2r-2)!}{(r-1)!\,r!} \leq 4^r \qquad (r = r(\boldsymbol{\omega});\ m_p \in \mathbb{Z}^\star) \tag{4.15}$$

**Comments and proof**

A fixed *degeneracy pattern* of order $d$ is a set of $d$ pairs $(i,j)$ satisfying $1 \leq i \leq j \leq r$ and such that the sequence $\boldsymbol{\omega}^\star = (\omega_i, \ldots, \omega_j)$ be unbreakable with zero-sum. "Unbreakability", we recall, rules out non-trivial factorisations $\boldsymbol{\omega}^\star = \boldsymbol{\omega}'\boldsymbol{\omega}''$ with either $\|\boldsymbol{\omega}'\| = 0$ or $\|\boldsymbol{\omega}''\| = 0$, but it does not rule out non-trivial factorisations of the form $\boldsymbol{\omega}^\star = \boldsymbol{\omega}'\boldsymbol{\omega}''\boldsymbol{\omega}'''$ with a zero-sum middle factor $\|\boldsymbol{\omega}''\| = 0$.

What the above lemma tells us about the multiplicities $\mu$ of the poles $\eta^{-\mu}$ amounts to this: *the denominators* $\eta$ *of* $\mathrm{Carr}^{\boldsymbol{\omega}}$ *undergo no repetitions unless they are already repeated within the sequence* $\boldsymbol{\omega}$*, these repetitions being induced by the degeneracy pattern itself.* This is in sharp contrast to the behaviour of most other moulds, such as $\mathrm{sa}^\bullet$, $\mathrm{musa}^\bullet$, etc., and all prenormalisation moulds $\mathrm{pran}^\bullet$.

Thus, if $\boldsymbol{\omega}$ has even length $r = 2r'$ and the following degeneracy pattern:

$$0 = \omega_1 + \omega_r; \qquad 0 = \omega_2 + \omega_3 = \omega_4 + \omega_5 = \cdots = \omega_{r-2} + \omega_{r-1} \tag{4.16}$$

(which leaves as independent variables $\omega_1$ and $\omega_2, \omega_4, \omega_6, \omega_{r-2}$) we have on the one hand, for all coefficients $\mathrm{pran}^{\boldsymbol{\omega}}$[51]:

$$\mathrm{pran}^{\boldsymbol{\omega}} = -(\omega_1)^{-r'} \prod_{i=1}^{r'-1} (\omega_1 + \omega_{2i})^{-1} \tag{4.17}$$

with a "huge" multiplicity (either $r'$ or $1+r'$) for $\omega_1$; and on the other hand, calculating $\mathrm{Carr}^{\boldsymbol{\omega}}$ by the procedure which we shall spell out in a moment, we find:

$$\mathrm{Carr}^{\boldsymbol{\omega}} = (-1)^{r'-1}(\omega_1)^{-1} \prod_{1\leq i\leq r'} (\omega_{2i})^{-1}(\omega_1 + \omega_{2i})^{-1} \tag{4.18}$$

---

[51] More accurately, for all prenormalisations that are free of parasitical singularities, like the *royal* or *regal* one, but for the others the phenomenon of "supermultiplicity" is either just as bad or worse.

in full agreement with the above lemma, and with no unwarranted repetition of poles.

We say that a ZUS (short for "zero-sum, unbreakable sequence") $\boldsymbol{\omega}^\star$ is adjacent to a component $\omega_i$ if $\omega_i$ either *initiates* or *terminates* $\boldsymbol{\omega}^\star$ (as its first or last element) or if it immediately *precedes* or *follows* $\boldsymbol{\omega}^\star$. Thus, any $\omega_i$ has at most four distinct adjacent ZUS. If $\omega_i$ initiates a (necessarily unique) ZUS $\boldsymbol{\omega}^+$, we set $r_i^+ = r(\boldsymbol{\omega}^+) =$ length of $\boldsymbol{\omega}^+$ and, if not, we set $r_i^+ = 0$. Similarly, if $\omega_i$ terminates a (necessarily unique) ZUS $\boldsymbol{\omega}^-$, we set $r_i^- = r(\boldsymbol{\omega}^-) =$ length of $\boldsymbol{\omega}^-$ and, if not, we set $r_i^- = 0$.

We can now state the two selection rules for the index $i$ of $var_i$. They read:

$$[C_1] \qquad \{0 < r_i^+ + r_i^-\}$$
$$[C_2] \qquad \{0 < r_i^+ + r_{i+1}^+ \text{ or } 0 = r_{i+1}^+\} \quad \text{and} \quad \{0 < r_i^- + r_{i-1}^- \text{ or } 0 = r_{i-1}^-\}$$

and can be interpreted as follows:

$[C_1]$ says that $\omega_i$ should be the first or last component of some ZUS, or both.

$[C_2]$ says that $\omega_i$ should not be squeezed between two adjacent ZUS $\boldsymbol{\omega}^\star$ and $\boldsymbol{\omega}^{\star\star}$ such that $\boldsymbol{\omega}^\star \subset \boldsymbol{\omega}^{\star\star}$. It says, too, that if $\omega_i$ is *externally adjacent* to some ZUS $\boldsymbol{\omega}^\star$, it should also be *internally adjacent* to some other ZUS $\boldsymbol{\omega}^{\star\star}$ that overlaps with $\boldsymbol{\omega}^\star$, but doesn't contain it.

There clearly exist $25 = 5^2$ distinct *adjacency types* $T_1, T_2, \ldots, T_{25}$ but a careful check shows that only $8 = 3^2 - 1$ of them, namely $T_1, T_2, \ldots, T_8$ are allowed under the selection rule $[C_1] + [C_2]$. All licit and illicit adjacency types are listed in the table below, where all sub-sequences $\lceil \ldots \rceil$ or $\lfloor \ldots \rfloor$ squeezed between two opposite upper or lower brackets (facing one another) are assumed to be of ZUS type.

| licit | | illicit | | illicit | |
|---|---|---|---|---|---|
| $T_1$ | $\lfloor \omega_i \mathbf{c} \rfloor$ | $T_9$ | $\omega_i$ | $T_{17}$ | $\lfloor \mathbf{b} \rfloor \lfloor \omega_i \lceil \mathbf{c} \rfloor \mathbf{d} \rceil$ |
| $T_2$ | $\lceil \mathbf{b} \omega_i \rceil$ | $T_{10}$ | $\omega_i \lceil \mathbf{c} \rceil$ | $T_{18}$ | $\lfloor \mathbf{b} \rfloor \lfloor \omega_i \lceil \mathbf{c} \rceil \mathbf{d} \rfloor$ |
| $T_3$ | $\lceil \mathbf{b} \lfloor \omega_i \rceil \mathbf{c} \rfloor$ | $T_{11}$ | $\lfloor \mathbf{b} \rfloor \omega_i$ | $T_{19}$ | $\lceil \mathbf{b} \lfloor \omega_i \rceil \lceil \mathbf{c} \rceil \mathbf{d} \rfloor$ |
| $T_4$ | $\lfloor \omega_i \lceil \mathbf{c} \rfloor \mathbf{d} \rceil$ | $T_{12}$ | $\lfloor \mathbf{b} \rfloor \omega_i \lceil \mathbf{c} \rceil$ | $T_{20}$ | $\lfloor \mathbf{a} \lceil \mathbf{b} \rfloor \omega_i \rceil \lceil \mathbf{c} \rceil$ |
| $T_5$ | $\lfloor \mathbf{a} \lceil \mathbf{b} \rfloor \omega_i \rceil$ | $T_{13}$ | $\lceil \mathbf{b} \omega_i \rceil \lfloor \mathbf{c} \rfloor$ | $T_{21}$ | $\lceil \mathbf{a} \lfloor \mathbf{b} \rfloor \omega_i \rceil \lceil \mathbf{c} \rceil$ |
| $T_6$ | $\lceil \mathbf{b} \lfloor \omega_i \rceil \lceil \mathbf{c} \rfloor \mathbf{d} \rceil$ | $T_{14}$ | $\lfloor \mathbf{b} \rfloor \lfloor \omega_i \mathbf{c} \rfloor$ | $T_{22}$ | $\lceil \mathbf{a} \lfloor \mathbf{b} \rfloor \lfloor \omega_i \rceil \mathbf{c} \rfloor$ |
| $T_7$ | $\lfloor \mathbf{a} \lceil \mathbf{b} \rfloor \lfloor \omega_i \rceil \mathbf{c} \rfloor$ | $T_{15}$ | $\lfloor \omega_i \lceil \mathbf{c} \rceil \mathbf{d} \rfloor$ | $T_{23}$ | $\lfloor \mathbf{a} \lceil \mathbf{b} \rfloor \lfloor \omega_i \rceil \lceil \mathbf{c} \rceil \mathbf{d} \rfloor$ |
| $T_8$ | $\lfloor \mathbf{a} \lceil \mathbf{b} \rfloor \lfloor \omega_i \rceil \lceil \mathbf{c} \rfloor \mathbf{d} \rceil$ | $T_{16}$ | $\lceil \mathbf{a} \lfloor \mathbf{b} \rfloor \omega_i \rceil$ | $T_{24}$ | $\lceil \mathbf{a} \lfloor \mathbf{b} \rfloor \lfloor \omega_i \rceil \lceil \mathbf{c} \rfloor \mathbf{d} \rceil$ |
| | | | | $T_{25}$ | $\lceil \mathbf{a} \lfloor \mathbf{b} \rfloor \lfloor \omega_i \rceil \lceil \mathbf{c} \rceil \mathbf{d} \rfloor$, |

The next step is to prove that any $\boldsymbol{\omega}$, whatever its degeneracy type, has at least *one* licit component $\omega_i$. In fact, it always has *two*, and often much more: see [EV2].

A linear combination $\eta = \omega_p + \cdots + \omega_q$ is said to be a *formal pole* of the sequence $\boldsymbol{\omega}$ if it can be obtained by breaking up some ZUS. The *formal multiplicity* of $[{}^{\eta}_{\boldsymbol{\omega}}]$ of $\eta$ in $\boldsymbol{\omega}$ is defined as being the number of ZUS which, when split, can produce $\eta$.

Collecting all the above results, rewriting the induction rule as follows:

$$\mathrm{Carr}^{\boldsymbol{\omega}} = \frac{1}{\omega_i}\Big\{+\mathrm{Carr}^{\boldsymbol{\omega}^1} - \mathrm{Carr}^{\boldsymbol{\omega}^2} + \sum \mathrm{Carr}^{\boldsymbol{\omega}^3}\mathrm{Carr}^{\boldsymbol{\omega}^4} - \sum \mathrm{Carr}^{\boldsymbol{\omega}^5}\mathrm{Carr}^{\boldsymbol{\omega}^6}\Big\}$$

with

$$\begin{aligned} &\boldsymbol{\omega}^1 = (\ldots, \omega_{i-1}+\omega_i, \ldots); && \boldsymbol{\omega}^2 = (\ldots, \omega_i+\omega_{i+1}, \ldots) \\ &\boldsymbol{\omega}^3 = \boldsymbol{\omega}^{34}\omega_i\boldsymbol{\omega}^{43} && \text{for} \quad \boldsymbol{\omega} = \boldsymbol{\omega}^{34}\omega_i\boldsymbol{\omega}^4\boldsymbol{\omega}^{43} \\ &\boldsymbol{\omega}^6 = \boldsymbol{\omega}^{65}\omega_i\boldsymbol{\omega}^{56} && \text{for} \quad \boldsymbol{\omega} = \boldsymbol{\omega}^{65}\boldsymbol{\omega}^5\omega_i\boldsymbol{\omega}^{56} \end{aligned}$$

and taking care of applying this rewritten induction rule for licit $\omega_i$'s only, we can easily verify, by going through all the 8 licit cases, that the *actual multiplicity of* $\eta$ *as pole of the rational function* $\mathrm{Carr}^{\boldsymbol{\omega}}$ *is no larger than the formal multiplicity of* $\eta$ *in the sequence* $\boldsymbol{\omega}$. QED

### Arborification

The above argument shows the absence of SSD (supermultiple small denominators) but is not enough to establish the convergence of $X^{\mathrm{cor}}$ under Bryuno's condition. Indeed, the straightforward mould expansion (4.8) is usually divergent. But it can easily be rendered convergent by resorting to the usual trick of arborification-coarborification (see Section 6.1):

$$X^{\mathrm{cor}} = \sum \mathrm{Carr}^{\bullet^{\prec}}\mathbb{B}_{\bullet^{\prec}} = \sum_{r\geq 1}\sum \mathrm{Carr}^{\omega_1,\ldots,\omega_r{}^{\prec}}\mathbb{B}_{n_1,\ldots,n_r{}^{\prec}} \tag{4.19}$$

and by dutifully 'arborifying' the whole argument that we went through. Thus the induction rule becomes:

$$\mathrm{var}_i\mathrm{Carr}^{\boldsymbol{\omega}^{\prec}} = \sum_{\substack{\omega_i\in\mathbf{a}^{\prec}\\ \omega_i<\mathbf{b}^{\prec}}} \mathrm{Carr}^{\mathbf{a}^{\prec}}\mathrm{Carr}^{\mathbf{b}^{\prec}} - \sum_{\substack{\omega_i\in\mathbf{c}^{\prec}\\ \omega_i>\mathbf{b}^{\prec}}} \mathrm{Carr}^{\mathbf{b}^{\prec}}\mathrm{Carr}^{\mathbf{c}^{\prec}} \tag{4.20}$$

The sums now extend to all connected sub-trees $\mathbf{b}^{\prec}$ of $\boldsymbol{\omega}^{\prec}$ either directly following or preceding $\omega_i$ and $\mathbf{a}^{\prec}$ (resp. $\mathbf{c}^{\prec}$) denotes the remaining part of $\boldsymbol{\omega}^{\prec}$, with the arborescent order inherited from the parent sequence $\boldsymbol{\omega}^{\prec}$. The variance operator $var_i$ also must be arborified. The transposition goes like this:

$$\mathrm{var}_i\mathrm{Carr}^{\boldsymbol{\omega}^{\prec}} = \omega_i\mathrm{Carr}^{\boldsymbol{\omega}^{\prec}} - \mathrm{Carr}^{\boldsymbol{\eta}^{\prec}} + \sum \mathrm{Carr}^{\boldsymbol{\sigma}^{\prec}} \tag{4.21}$$

where $\boldsymbol{\eta}^{\prec}$ denotes the unique arborified sequence obtained by contracting $\omega_i$ with its (necessarily unique) immediate predecessor $\omega_{i\_}$; and the sum $\sum$ extends to all arborescent sequences $\boldsymbol{\sigma}^{\prec}$ obtainable by contracting $\omega_i$ with any one of its immediate successors $\omega_j$ (i.e. all $\omega_j$ such that $\omega_{j\_} = \omega_i$). For details and numerous examples see [EV2].

The upshot is that not only *is the correction* $X^{\mathrm{cor}}$ *always analytic under Bryuno's diophantine condition* but also that *its analyticity owes nothing to any so-called compensation of supermultiple small denominators, given that these SSD quite simply do not exist.*

## 4.4 Application to Floquet theory

Floquet theory deals with differential equations whose coefficients are q.p. (quasiperiodic) functions of the variable. The central difficulty, as with the correction or KAM theory, comes from the *small denominators*, which integration creates at every step. Let us consider a simple but fairly typical case, that of a homogeneous linear differential system:

$$\partial_t X(t) = U(t)X(t) \tag{4.22}$$

for a q.p. matrix $U(\bullet)$ with frequencies in a fixed set $\Omega$:

$$U(t) := \sum_{\omega\in\Omega} e^{i\omega t} U_\omega \qquad (U_\omega = \text{const},\ t \in \mathbb{R}) \tag{4.23}$$

$$\omega \in \Omega := \lambda_1\mathbb{Z} + \lambda_2\mathbb{Z} + \cdots + \lambda_\nu\mathbb{Z} \tag{4.24}$$

An important, very popular sub-case is:

$$U(t) := lA + \epsilon B(t) = lA + \epsilon\left(\sum e^{i\omega t} B_\omega\right) \tag{4.25}$$

$$(A,\ B_\omega \text{ constant matrices}, 1 \ll l,\ \epsilon \ll 1) \tag{4.26}$$

**Some notations**

The method is to reduce (4.22) to a trivial equation

$$\partial_t Y(t) = VY(t) \quad \text{with} \quad V = \text{const} \tag{4.27}$$

by means of a change of unknown

$$X(t) = \Theta(t)Y(t) \tag{4.28}$$

The problem thus reduces to finding a q.p. matrix $\Theta(\bullet)$ verifying:

$$V + \Theta^{-1}(t)\,\partial_t\Theta(t) = \Theta^{-1}U(t)\Theta(t) \tag{4.29}$$

On the space of q.p. matrices with frequencies in $\Omega$, i.e. of the form (4.23), let there be defined the following operators:

$$\overline{V} := (\partial_t - \text{ad}(V)); \qquad \underline{V} := (\partial_t - \text{ad}(V))^{-1} \tag{4.30}$$

The 'derivation' $\overline{V}$ acts on all components $e^{i\omega t}U_\omega$. The 'integration' $\underline{V}$ acts unambiguously only on those components that are non-constant ($\omega \neq 0$). For instance, if we take for $U_\omega$ the general matrix $[a_{j,k}]$ and for $V$ the diagonal matrix with spectrum $\{iv_j\}$, we get for each $\omega \neq 0$:

$$\underline{V}(e^{i\omega t}[a_{j,k}]) = e^{i\omega t}[b_{j,k}] \quad \text{with} \quad b_{j,k} = \frac{a_{j,k}}{i(\omega - v_{j,k})};\ v_{j,k} := v_j - v_k \tag{4.31}$$

A similar formula holds for a general input $V$ and, in all cases, the only denominators to appear are the $(\omega - v_{j,k})$, which of course includes the $\omega$ themselves (for $j = k$). We call such denominators *intrinsic*[52].

[52] As opposed to the *extrinsic* denominators which bedevil all approximation methods.

The occurrence of (potentially small) denominators other that the $\omega$ makes it necessary to impose a modified form of Bryuno's classical diophantine condition:

$$S(\lambda) := \sum_k 2^{-k} \log(\frac{1}{\varpi(2^k)}) < \infty \tag{4.32}$$

with

$$\varpi(K) := \inf \left|\sum m_i \lambda_i\right| \quad \text{for} \quad \sum m_i \lambda_i \neq 0 \quad \text{and} \quad \sum |m_i| \leq K \tag{4.33}$$

Here, we must replace $S(\lambda)$ by $S_{\boldsymbol{v}}(\lambda)$, with the same definition as in (4.32), but relative the diophantine gauge $\varpi_{\boldsymbol{v}}(K)$:

$$\varpi_{\boldsymbol{v}}(K) := \inf \left|\left(\sum m_i \lambda_i\right) - (v_j - v_k)\right| \tag{4.34}$$

Everything hinges on the arithmetics of the spectrum $i\boldsymbol{v} = \{iv_1, \ldots, iv_n\}$ of the constant matrix $V$. The parameters $v_j$ are the so-called Floquet exponents[53]. They are defined only modulo $\Omega$ (or, exceptionnaly, modulo a slightly larger group), but they possess a "principal determination" [54], and the way to calculate the latter is as follows: start from the fundamental *matricial* solution $X(t)$ of (4.22), with initial conditions $X(0) = I$ and of course $\det X(t) \neq 0$; then follows a continuous determination of $(1/t)\log X(t)$; lastly take the limit of its spectrum as $t \to \pm\infty$.

Since the frequencies $\lambda_i$ that span $\Omega$ were assumed to satisfy the Bryuno-type condition (4.32)+(4.34), we can pick $\epsilon \leq \epsilon_0(B)$ small enough to ensure the convergence not only of the familiar-looking series:

$$C = \sum_r \epsilon^r \sum_{\omega_i} \mathrm{Carr}^{\omega_1,\ldots,\omega_r} B_{\omega_r} \ldots B_{\omega_2} B_{\omega_1} \tag{4.35}$$

$$\Theta = 1 + \sum_r \epsilon^r \sum_{\omega_i} e^{i(\sum \omega_j)t} \mathrm{Scarr}^{\omega_1,\ldots,\omega_r} B_{\omega_r} \ldots B_{\omega_2} B_{\omega_1} \tag{4.36}$$

but also of these two new series:

$$C = \sum_r \epsilon^r \sum_{\omega_i} \mathrm{Carr}^{B_{\omega_1},\ldots,B_{\omega_r}} \tag{4.37}$$

$$\Theta = 1 + \sum_r \epsilon^r \sum_{\omega_i} e^{i(\sum \omega_j)t} \mathrm{Scarr}^{B_{\omega_1},\ldots,B_{\omega_r}} \tag{4.38}$$

The first pair of series is clearly patterned on the expansions which we found for the "correction" of a resonant vector field (see Section 4.3 *supra*). The second pair is derived from the first by applying a simple transposition rule which can be easily grasped from the

[53]For the *cognoscenti*: in the 'hyperbolic' case, the $v_j$ may possess an imaginary part, which creates in the $iv_j$ a real part that corresponds to the so-called Lyapunov exponents.

[54]Which in the case of $\mathrm{sl}(2,\mathbf{R})$ reduces to a classical (but generally complex) 'rotation number'.

following example, for a length $r = 3$:

$$\begin{aligned}
\mathrm{Carr}^{\omega_1,\omega_2,\omega_3} B_{\omega_3} B_{\omega_2} B_{\omega_1} :=& \frac{1}{3}\left(\frac{1}{\omega_1(\omega_1+\omega_2)} - \frac{1}{\omega_1\omega_3} + \frac{1}{\omega_3(\omega_2+\omega_3)}\right) B_{\omega_3} B_{\omega_2} B_{\omega_1} \\
\longrightarrow \quad \mathrm{Carr}^{B_{\omega_1},B_{\omega_2},B_{\omega_3}} :=& + \frac{(-1)^3}{3}((e^{i\omega_3}B_{\omega_3})(\underline{V}((e^{i\omega_2}B_{\omega_2})(\underline{V}e^{i\omega_1}B_{\omega_1})))\\
&- \frac{(-1)^3}{3}((\underline{V}e^{i\omega_3}B_{\omega_3})(e^{i\omega_2}B_{\omega_2})(\underline{V}e^{i\omega_1}B_{\omega_1})))\\
&+ \frac{(-1)^3}{3}((\underline{V}((\underline{V}e^{i\omega_3}B_{\omega_3})(e^{i\omega_2}B_{\omega_2})))(e^{i\omega_1}B_{\omega_1}))
\end{aligned}$$

Since the coefficients $\mathrm{Carr}^\bullet$ (unlike $\mathrm{Scarr}^\bullet$) vanish for $\omega_1 + \cdots + \omega_r \neq 0$, the series $C$ (unlike $\Theta(t)$) is $t$-constant, as indeed it should be. Nonetheless, in order to properly carry out the above transposition, one should take care of adopting not *any* expression of $\mathrm{Carr}^\bullet$ and $\mathrm{Scarr}^\bullet$, but a *suitably symmetrised* one. Even so, the expression is not unique, but this does not affect the one-to-one character of the transposition, since to each algebraic identity of type:

$$\frac{1}{\omega_1(\omega_1+\omega_2)} + \frac{1}{\omega_2(\omega_1+\omega_2)} = \frac{1}{\omega_1\omega_2} \tag{4.39}$$

there corresponds an non-commutative integration-by-parts identity:

$$\begin{aligned}
&\underline{V}((e^{i\omega_2}B_{\omega_2})(\underline{V}(e^{i\omega_1}B_{\omega_1}))) + \underline{V}(\underline{V}(e^{i\omega_2}B_{\omega_2}))(e^{i\omega_1}B_{\omega_1}))) \\
&\qquad = (\underline{V}(e^{i\omega_2}B_{\omega_2}))(\underline{V}(e^{i\omega_1}B_{\omega_1}))
\end{aligned} \tag{4.40}$$

For $V = 0$, the series (4.37)–(4.38) coincide with the series (4.35)–(4.36). So by construction they are guaranteed to converge, since at the outset the perturbation $\epsilon B(\bullet)$ was assumed to be small enough.

For $V \neq 0$ on the other hand, the series (4.37)–(4.38) converge only if $S_{\boldsymbol{v}}(\lambda)$ is close enough to $S(\lambda)$. If this already holds for the 'principal determination' $\boldsymbol{v} := \boldsymbol{u}$, so much the better, for that means that our construction is at an end: $U(\bullet)$ is indeed of the form $V - C + \epsilon B(\bullet)$ for some matrix $V$ with spectrum $i\boldsymbol{u}$ (we may even, if we so wish, exactly calculate that matrix) and the normalising matrix $\Theta(\bullet)$ is explicitly given by (4.38).

When applied to a Floquet equation such as (4.22)+(4.25), the above considerations immediately establish the reducibility theorem[55] *for a set $l$ with positive measure.* In fact, a slight refinement of the argument[56] yields reducibility for a set $l$ of *full measure.* See also [El1, El2].

## 4.5 Application to KAM theory

We shall perforce remain extremely sketchy. We work under the classical (analytic) KAM assumptions, ie we perturb an integrable hamiltonian $h$:

$$h(y) = \langle \lambda, y \rangle + \langle y, Q, y \rangle = \sum \lambda_i y_i + \sum Q_{i,j} y_i y_j \tag{4.41}$$

[55] I.e. the existence of a q.p. non-singular matrix $\Theta(t)$ that reduces the given equation (4.22) to the trivial form (4.27).

[56] By playing on the indeterminacy of the Floquet exponents, which enables us to change $v_i$ into $v_i + \omega_i$ with $\omega_i$ in $\Omega$ and large enough, so as to make $S_{\boldsymbol{v}}(\lambda)$ as close to $S(\lambda)$ as necessary.

(with $\mathbb{Q}$-independent basic frequencies $\lambda_i$) into a non-integrable $H$:

$$H(x,y) = h(y) + \epsilon b(x,y) \qquad (x \in \mathbb{T}^\nu,\ y \in \mathbb{R}^\nu_0) \tag{4.42}$$

$$= \langle \lambda, y \rangle + \sum_{m,n} H_{m,n}(x,y) \tag{4.43}$$

and we wish to establish, under Bryuno's (not Siegel's) diophantine assumption on the $\lambda_i$'s, the convergence, for $y = 0$ and for small enough values of the perturbation parameter $\epsilon$, of the "uncorrected" Lindstedt series:

$$\sum H_{m,n}(x,y) = \sum c_{m,n}(\epsilon) e^{2\pi i \langle x, m \rangle} y^n = \sum c_{m,n}(\epsilon) e^{2\pi i \langle \lambda, m \rangle t} y^n \tag{4.44}$$

$$\omega := \langle m, \lambda \rangle = \text{“frequency”} \qquad (m \in \mathbb{Z}^\nu) \tag{4.45}$$

$$\eta := |n| - 1 = \sum n_i - 1 = \text{“grade”} \geq -1 \qquad (n \in \mathbb{N}^\nu) \tag{4.46}$$

We prefer to work with fields rather than their potentials. The thing is to partially "correct" and partially "normalise"[57] our field $X^H$:

$$X^H - X^{\mathrm{cor}} \overset{\mathrm{conj}}{\sim} X^{\mathrm{lin}} + X^{\mathrm{nor}} \tag{4.47}$$

$$\mathrm{frequency}(X^{\mathrm{cor}}) = 0, \quad \mathrm{frequency}(X^{\mathrm{nor}}) = 0 \tag{4.48}$$

$$\mathrm{grade}(X^{\mathrm{cor}}) = 0, \quad \mathrm{grade}(X^{\mathrm{nor}}) \neq 0 \tag{4.49}$$

by allowing only terms of zero (resp. non-zero) grade on the left- (resp. right-hand) side of (4.47). As in Section 4.3 the correction still possesses a mould expansion of type:

$$X^{\mathrm{cor}} = \sum_{r \geq 1} \sum \mathrm{Bicarr}^{\binom{\omega_1, \ldots, \omega_r}{\eta_1, \ldots, \eta_r}} X^{H_{m_r,n_r}} \ldots X^{H_{m_2,n_2}} X^{H_{m_1,n_1}} \tag{4.50}$$

with frequencies $\omega_i := \langle m_i, \lambda \rangle$ and grades $\eta_i := -1 + \sum n_i$. The normal part $X^{\mathrm{nor}}$ also has a similar mould expansion, but we need not worry about it, since it vanishes for $y = 0$ and so does not contribute to the Lindstedt series.

The alternal mould $\mathrm{Bicarr}^\bullet$ is more complex than, but essentially similar to, the mould $\mathrm{Carr}^\bullet$ of Section 4.3. In fact, $\mathrm{Bicarr}^\bullet$ reduces to $\mathrm{Carr}^\bullet$ when all the grades $\eta_i$ are 0 or, more generally, when to each vanishing partial sum $\omega_i + \cdots + \omega_j = 0$ there corresponds a vanishing partial sum $\eta_i + \cdots + \eta_j = 0$.

To cut a long story short: we can duplicate in this case all the steps of Section 4.3 and prove, once again, the non-occurence of *supermultiple small denominators*, except that now the *formal multiplicity* of a divisor is exactly *twice* what it was in Section 4.3. That apart, precious little changes. We still must *arborify* to get the convergence of $X^{\mathrm{cor}}$. This establishes, for a small enough pertubation parameter $\epsilon$, the convergence of the Lindstedt series for the *corrected hamiltonian*. But a standard argument going back to Poincaré (known as "killing the constants" and using the possibility of changing the integration constants) readily yields the convergence of the Lindstedt series for the *given* hamiltonian itself.

[57] In the sense of Section 2. Note that in this case, due to symplecticity, there is no difference between *normalising* and *prenormalising* : all prenormalisations coincide with one another and with the normalisation.

## 4.6 KAM theory "in three theorems"

Since we are on this subject of KAM theory, I cannot resist mentioning that the formalisation approach applies with equal success to three distinct cases, which between themselves cover almost the whole spectrum of possibilities:

### (1) The general case: no assumption on the $\lambda$ whatsoever

The proper tool here is the (genuine) theory of compensation[58] and its *compensators.* We get special, divergent but resummable expansions of the movement, valid for nearly optimal time intervals $-t_1(\epsilon) < t < t_2(\epsilon)$.

### (2) Maximal resonance: all $\lambda_i$'s commensurate

There are no small denominators at all. There is divergence, but of a perfectly manageable sort: resurgent and exactly resummable. We therefore get a good description of the movement, including the tendency towards escape (the so-called Arnold diffusion).

### (3) The generic case: $\lambda$ diophantine

This is the case we have just discussed: we have a Cantor of surviving invariant tori near integrability. We cannot beat Kolmogorov's original proof for brevity or elegance, but we dispense with the non-intrinsic approximation process inherent in that proof. We also get behind the "true reasons" for the convergence of the Lindstedt series, by a direct examination of their coefficients, and unravel the mechanism that inhibits the occurence of SSD.

## 4.7 Genuine versus illusory compensation

To preclude any terminological confusion, let us point out that the notion of "sign compensation", as used by the "Italo-Nordic KAM school" (Eliasson, Gallavotti, etc.) bears no relation whatsoever to the genuine notion of compensation as introduced and used by us in the late 80s [E9, E10]. The latter belongs to an altogether different context, namely that of local objects with a Liouvillian (ie non-diophantine) spectrum. Even when such objects can be formally linearised, they generically resist analytical linearisation. However, after ramifying *one*, *two*, or at the utmost *three* coordinates (i.e. by considering expansions involving non-entire powers of these coordinates), it is always possible to construct a *ramified* and yet *effective* (as opposed to formal) linearisation of these Liouvillian objects[59]. The real compensation at work here takes place within the so-called "compensators", the simplest of which have the form:

$$z^{\sigma_0,\sigma_1,\ldots,\sigma_r} := \sum_{0\leq i\leq r} z^{\sigma_i} \prod_{j\neq i}(\sigma_i-\sigma_j)^{-1} \qquad (z\in\mathbb{C}_\bullet,\ \sigma_i\in\mathbb{R}^+) \tag{4.51}$$

When the exponents $\sigma_i$ are pair-wise different, but very close to each other, the coefficient in front of the power $z^{\sigma_i}$ (which has a genuine, individualised existence) becomes very large

[58] See Section 4.7 *infra* and [E9].

[59] Inside ramified neighbourhoods of the origin with spiral-like boundaries. Compensation (in our sense) also comes into its own in KAM theory, but in statements like Section 4.6, case (1), and not at all for establishing the convergence of the Lindstedt series.

and yet, due to a true phenomenon of compensation, the finite sum on the right-hand side of (4.51) remains bounded.

To recap, let us illustrate the difference on a trivial, yet telling example:

$$\begin{array}{lllll} \text{Bogus compensation:} & f(\omega_1,\omega_2) & := & \frac{1}{\omega_1(\omega_1+\omega_2)}+\frac{1}{\omega_2(\omega_1+\omega_2)} & \equiv \frac{1}{\omega_1\omega_2} \\ \text{Genuine compensation:} & g_t(\sigma_1,\sigma_2) & := & \frac{t^{\sigma_1}}{\sigma_1-\sigma_2}+\frac{t^{\sigma_2}}{\sigma_2-\sigma_1} & \end{array}$$

# 5 Lesson Five: A Tale of Three Structures: Singular perturbations; multizetas; ARI/GARI

## 5.1 Dimorphy: functional or numerical

Strange to say, but the notion of dimorphy is itself 'dimorphous':

(i) For a space $\mathbb{D}$ of *functions*, dimorphy means closure under two *distinct* products: usually point-wise multiplication and some form or other of convolution[60].

(ii) For a space $\mathbb{D}$ of *numbers*, dimorphy means being a countable $\mathbb{Q}$-ring and possessing two distinct, *natural* bases $\{\alpha_m\}$ and $\{\beta_n\}$, each with its countable indexation $m$ and $n$[61], with finite conversion laws:

$$\alpha_m \equiv \sum_n H_m^n \beta_n; \qquad \beta_n \equiv \sum_m K_n^m \alpha_m \tag{5.1}$$

and with *two distinct ways* of calculating the *one and only* product on $\mathbb{D}$, which is ordinary number multiplication:

$$\alpha_m \alpha_n \equiv \sum_r A_{m,n}^r \alpha_r; \qquad \beta_m \beta_n \equiv \sum_r B_{m,n}^r \beta_r \tag{5.2}$$

All four sums have to be finite, with rational constants $H, K, A, B$.

Clearly, since one may always concoct artificial bases $\{\alpha_m\}$, $\{\beta_n\}$ to meet the above conditions, the whole emphasis in this notion of numerical dimorphy must lie on the *naturalness* of the two bases. This may seem a rather shaky foundation for a mathematical definition, but we venture to suggest that in fact it is not: in all known instances of dimorphy, there is no scope for hesitation ; the two bases $\{\alpha_m\}$, $\{\beta_m\}$ are clearly *there for all to see*, unmistakably *nature-given*, whereas it often takes a considerable amount of toil to *extract the hidden core* of $\mathbb{D}$, which usually is an algebraically free system $\{\gamma_r\}$ of irreducibles. This third set $\{\gamma_r\}$, typically, lies buried deep below the surface and, at least when taken in canonical form, tends to be found exactly mid-way between the two 'emerging' sets $\{\alpha_m\}$, $\{\beta_n\}$. So, even though it may be argued that numerical dimorphy is, ultimately, pure *māyā*, it is the sort of *māyā* that you must work hard to dispel...

In any case, functional and numerical dimorphy go hand in hand, and the proper framework for their joint investigation would seem to be, not the so-called 'theory of periods'[62], but the twin systems $\mathcal{N}\!a, \mathbb{N}a$, whose construction, very roughly, goes like this:

---

[60] See for example the algebra of *biresurgent functions* in [E13, Section 4].

[61] $m$ an $n$ usually do not range through $\mathbb{N}$, but through more complex sets, like the monoid generated by $\mathbb{N}$.

[62] Despite all the hullabaloo about 'periods', the constants there are given pell-mell, with no natural indexation, and all the symmetries central to dimorphy are broken, beginning with the sum/integral symmetry.

(i) We produce the function germs $f$ (of $z$, at $\infty$) in $\mathcal{N}\!a$ 'out of nothing', i.e. from $f(z) \equiv 1$, by taking larger and larger closures under the (direct and reverse) operations $+, \times, \partial_z, \circ$, with $\mathbb{Q}$ or $\mathbb{A}$ as scalar field. Very early on in this enlargement process, chronic *divergence* appears in the formal series or transseries $\tilde{f}$ which 'expand' at $\infty$ the germs $f$ in $\mathcal{N}\!a$.

(ii) We carefully refrain from introducing artificial derivations on $\mathcal{N}\!a$, for fear of compromising the natural character of the construction. Rather, we ask: are there—*already, without our doing*—exotic derivations (i.e. derivations not generated by $\partial_z$) that act on $\mathcal{N}\!a$ and respect its natural topology? And we find that there is indeed a teeming profusion of them—two systems in fact, the alien derivations $\Delta_\omega$ and the foreign derivations $\nabla_\omega$. The reason for this plethora of exotic derivations is the omnipresence of *divergence* in $\mathcal{N}\!a$: to analyse this divergence, qualitatively and quantitatively, suitable operators are called for, which are precisely the exotic derivations.

(iii) To produce $\mathbb{N}\mathbf{a}$ from $\mathcal{N}\!a$ (i.e. numbers from function germs, monics from monomials) we *do not* evaluate our germs at given points[63]. Rather, we let the exotic derivations act on these germs, and it turns out that the exotic derivatives of our monomials are expressible as sums of 'simpler' or 'earlier' monomials, with well-defined, generically transcendental scalar coefficients.

(iv) We harvest all these coefficients, declare them to be 'monics', and call $\mathbb{N}\mathbf{a}$ the ring generated by them.

The advantage of *exotic derivation* over *pointwise evaluation* lies not at all in the nature of the constants being produced (they are much the same with both methods) but rather in the orderliness of the procedure, which turns out monics directly in *mould form*, automatically gives them the *right type of indexation*, and tells us to which fundamental *symmetry type* they belong (as moulds).

There are at least four major 'domains of dimorphy', of increasing size, in $\mathbb{N}\mathbf{a}$. They comprise, respectively:

(i) the multizetas;

(ii) the general hyperlogarithmic monics (see Section 4);

(iii) the monics associated with monomials that verify *affine* differential equations, with coefficients in $\mathbb{Q}[z]$ or $\mathbb{A}[z]$;

(iv) the monics associated with monomials that verify *'bipolynomial'*[64] differential equations, again with coefficients in $\mathbb{Q}[z]$ or $\mathbb{A}[z]$.

The third and (especially) fourth domain of dimorphy are incredibly large and would seem to encompass more or less all constants encountered in 'real life'. In fact, dimorphy appears to extend as far as the sight reaches: the whole of $\mathbb{R}$'s explorable-constructive part seems to be 'dimorphic' to the core.

Needless to say, these constructions certainly admit many variations, and their exhaustive investigation (for instance extending to Domains 2, 3, 4 the whole algebraic apparatus developped for Domain 1) would require huge efforts—and might not repay them. Yet, strangely,

[63] Say, rational or algebraic points.

[64] 'Bipolynomial' means that the differential equation may involve not only ordinary products of $f$, $f'$, $f''$, etc., but also convolution products of type $\ast$.

the central fact about $\mathbb{Na}$, namely *numerical dimorphy*, is easy enough to establish, at least for these four domains. It directly mirrors the fact of *functional dimorphy*, which follows from the stability of $\mathcal{Na}$ under the generalised 'Borel-Laplace' transform, which itself is but an adaptation of the Fourier transform. So the least we can say is that dimorphy is 'well-connected'! It is nothing of an accident or localised oddity.

**P.S.** This section owes much to discussions I had with Joris van der Hoeven.

### 5.1.1 Are there exotic derivations acting on numbers?

Dimorphy is by no means the end of the story. After identifying the phenomenon and acknowledging its scope, we must raise another question: does dimorphy really exhaust the arithmetico-algebraic structure of $\mathbb{Na}$, the 'explorable part' of $\mathbb{R}$? For instance, narrowing the focus to Domain 1: do the 'quadratic relations' of Section 5.3.1 exhaust the set of algebraic constraints on multizetas? One would assume the answer to be yes, but at the moment the tool-kit of transcendence theory seems woefully inadequate to tackle such questions[65]. That might change, however, if we had at our disposal, for *numbers*, the sort of high-powered machinery that we have for *functions*, namely: exotic derivations. We might then go about disproving the existence of 'undesirable relations' $R(\alpha, \beta, \dots) = 0$ for *numbers* by subjecting them to exotic differentiation, in search of a contradiction. The scheme works wonders with the resurgent or analysable *functions* and the *alien derivations* that operate on them. So do there exist numerical derivations, non-elementary and useful, which annihilate $\mathbb{Q}$ and $\mathbb{A}$ but act non-trivially on some countable ring $\mathbb{D}$ containing $\mathbb{A}$ (:= the field of algebraic numbers)? Well, it is too early to say, but our canonical decomposition for the (formal) multizetas does strongly suggest a positive answer. See Section 5.3.6 *infra* and [E16].

## 5.2 The overarching structure: ARI/GARI

### 5.2.1 Bimoulds. Swap/Push. Contractions

A few basic facts about the *mould* formalism have been recalled in Section 6.1. *Bimoulds* are moulds that depend on double sequences:

$$A^{\bullet} = A^{\mathbf{w}} = A^{w_1,\dots,w_r} = A^{\binom{u_1,\dots,u_r}{v_1,\dots,v_r}}$$

and, more crucially, that are subjected to operations which mix up intimately the two sequences[66].

One such operation is the basic involution *swap*:

$$A_*^{\bullet} = \mathrm{swap}(A^{\bullet}) \quad \Longleftrightarrow \quad A_*^{\binom{u_1,\dots,u_r}{v_1,\dots,v_r}} = A^{\binom{v_r,\,v_{r-1:r},\dots,v_{2:3},\,v_{1:2}}{u_{1\dots r},\,u_{1\dots r-1},\dots,u_{1\,2},\,u_1}} \tag{5.3}$$

Another operation is the *push*:

$$A_*^{\bullet} = \mathrm{push}(A^{\bullet}) \quad \Longleftrightarrow \quad A_*^{\binom{u_1,\dots,u_r}{v_1,\dots,v_r}} = A^{\binom{-u_{1\dots r},\,u_1,\,u_2,\dots,u_{r-1}}{-v_r,\,v_{1:r},\,v_{2:r},\dots,v_{r-1:r}}} \tag{5.4}$$

[65] Despite spectacular but localised break-throughs by Apéry and, quite recently, Rivoal.

[66] Otherwise they ought to be regarded as moulds whose indices $w_i = \binom{u_i}{v_i}$ simply happen to be in $\mathbb{C}^2$ rather than in $\mathbb{C}$.

(We make constant use of the shorthand $u_{12} := u_1 + u_2 v_{1:2} := v_1 - v_2$ etc.).

It is often convenient to represent bimoulds in the so-called 'augmented notation', which consists in adding to any given sequence $\mathbf{w}$ a *redundant* initial term $w_0 = \binom{u_0}{v_0}$. The **u**-variables are then constrained by the condition $u_0 + u_1 + \cdots + u_r = 0$ and, dually, the **v**-variables are defined upto addition of a common constant. Thus:

$$A^{\binom{u_1,\ldots,u_r}{v_1,\ldots,v_r}} \equiv \operatorname{aug} A^{\binom{u_0,u_1,\ldots,u_r}{v_0,v_1,\ldots,v_r}} \quad \text{with} \quad u_0 := -u_{1\ldots r};\ v_0 := 0. \tag{5.5}$$

With the augmented notations, for instance, the *push* reduces to a unit shift on the sequence **w**.

More operations on bimoulds shall be defined in the sequel, but nearly all of them involve four specific types of *sequence contractions*, denoted by the symbols $\rceil, \lceil, \rfloor, \lfloor$. These are always relative to some given *factorisation* $\mathbf{w} = \mathbf{w}^1\mathbf{w}^2 \ldots \mathbf{w}^s$ of the total sequence. The contraction rules are immediately apparent from the following example. Relative to the factorisation:

$$\mathbf{w} = \ldots\mathbf{a}.\mathbf{b}\ldots = \ldots\begin{pmatrix} u_3, u_4, u_5 \\ v_3, v_4, v_5 \end{pmatrix}\begin{pmatrix} u_6, u_7, u_8, u_9 \\ v_6, v_7, v_8, v_9 \end{pmatrix}\ldots$$

the symbols $\rceil, \lceil, \rfloor, \lfloor$ signal the following changes:

$$\mathbf{a}\rceil := \begin{pmatrix} u_3, u_4, u_{56789} \\ v_3, v_4, v_5 \end{pmatrix} \qquad \lceil\mathbf{b} := \begin{pmatrix} u_{3456}, u_7, u_8, u_9 \\ v_6, v_7, v_8, v_9 \end{pmatrix} \tag{5.6}$$

$$\mathbf{a}\rfloor := \begin{pmatrix} u_3, u_4, u_5 \\ v_{3:6}, v_{4:6}, v_{5:6} \end{pmatrix} \qquad \lfloor\mathbf{b} := \begin{pmatrix} u_6, u_7, u_8, u_9 \\ v_{6:5}, v_{7:5}, v_{8:5}, v_{9:5} \end{pmatrix} \tag{5.7}$$

with the usual abbreviations for sums and differences.

Thus we see that the contractor $\rceil$ *adds* to *the* upper-right element of **a** *all* upper elements of neighbouring **b**, whereas the contractor $\rfloor$ *subtracts* from *all* lower elements of **a** *the* lower-left element of neighbouring **b**. And *vice versa* for $\lceil$ and $\lfloor$. Indeed, the **u**-variables are meant to be added together, and the **v**-variables to be substracted from one another.

### 5.2.2 The Lie algebra ARI

Consider the bilinear product *ari*:

$$\begin{aligned} C^\bullet = \operatorname{ari}(A^\bullet, B^\bullet) \iff C^{\mathbf{w}} &= \sum_{\mathbf{w}=\mathbf{b}.\mathbf{c}} (A^{\mathbf{b}}B^{\mathbf{c}} - B^{\mathbf{b}}A^{\mathbf{c}}) \\ &+ \sum_{\mathbf{w}=\mathbf{b}.\mathbf{c}.\mathbf{d}} (A^{\lfloor\mathbf{c}}B^{\mathbf{b}\rceil\mathbf{d}} - B^{\lfloor\mathbf{c}}A^{\mathbf{b}\rceil\mathbf{d}}) \\ &+ \sum_{\mathbf{w}=\mathbf{a}.\mathbf{b}.\mathbf{c}} (A^{\mathbf{a}\lceil\mathbf{c}}B^{\mathbf{b}\rfloor} - B^{\mathbf{a}\lceil\mathbf{c}}A^{\mathbf{b}\rfloor}) \end{aligned} \tag{5.8}$$

with $\mathbf{b} \neq \emptyset, \mathbf{c} \neq \emptyset$ in all three sums (but **a** and **d** may be empty).

*The* ari-*bracket is anti-commutative, satisfies the Jacobi identity, and turns the space of all bimoulds such that* $A^\emptyset = 0$ *into a Lie algebra, known as* ARI.

### 5.2.3 The Lie group GARI

Consider the binary law *gari*:

$$
\begin{aligned}
C^\bullet &= \mathrm{gari}(A^\bullet, B^\bullet) \iff \\
C^{\mathbf{w}} &= \sum_{\mathbf{w}=\mathbf{a}^1.\mathbf{b}^1.\mathbf{c}^1\ldots\mathbf{a}^s.\mathbf{b}^s.\mathbf{c}^s.\mathbf{a}^{s+1}} A^{\lceil\mathbf{b}^1\rceil\ldots\lceil\mathbf{b}^s\rceil} B^{\mathbf{a}^1\rfloor}\ldots B^{\mathbf{a}^s\rfloor} B^{\mathbf{a}^{s+1}\rfloor} B_\star^{\lfloor\mathbf{c}^1}\ldots B_\star^{\lfloor\mathbf{c}^s}
\end{aligned} \tag{5.9}
$$

with summation over all $s \geq 1$ and with factor sequences subject only to $\mathbf{b}^i \neq \emptyset$ and $\mathbf{c}^i.\mathbf{a}^{i+1} \neq \emptyset$ (but consecutive factors $\mathbf{c}^i$ and $\mathbf{a}^{i+1}$ may be empty *separately* and the extreme factors $\mathbf{a}^1, \mathbf{c}^s, \mathbf{a}^{s+1}$ and even the product $\mathbf{c}^s.\mathbf{a}^{s+1}$ may also be empty, *separetly or simultaneously*). Here $B_\star^\bullet$ denotes the inverse $\mathrm{invmu}(B^\bullet)$ of $B^\bullet$ relative to the ordinary (associative, non-commutative) product (mu or $\times$) on moulds:

$$
C^\bullet = \mathrm{mu}(A^\bullet, B^\bullet) = A^\bullet \times B^\bullet \iff C^{\mathbf{w}} = \sum_{\mathbf{w}=\mathbf{w}^1.\mathbf{w}^2} A^{\mathbf{w}^1} B^{\mathbf{w}^2} \tag{5.10}
$$

This *gari*-product is clearly affine in $A^\bullet$ but severely non-linear in $B^\bullet$.

*It is also associative, and turns the set of all bimoulds such that* $A^\emptyset = 1$ *into a Lie group, known as* GARI, *whose Lie algebra is* ARI.

### 5.2.4 Some properties of ARI/GARI. Allied structures

Like ordinary moulds, most interesting bimoulds fall into a few basic *symmetry types*. The definition for *symmetral/alternal* and *symmetrel/alternel* is exactly the same as for ordinary moulds, but in the case *symmetril/alternil* the contraction $w_i \otimes w_j$ should be interpreted as:

$$
A^{\ldots,w_i\otimes w_j,\ldots} := P(v_{i:j})A^{\ldots,\binom{u_{ij}}{v_i},\ldots} + P(v_{j:i})A^{\ldots,\binom{u_{ij}}{v_j},\ldots} \quad \text{with} \quad P(t) := 1/t \tag{5.11}
$$

Thus, for a *symmetral* bimould $A^\bullet$ and factor sequences of length 1 and 2 we get:

$$
A^{\binom{u_1}{v_1}} A^{\binom{u_2,u_3}{v_2,v_3}} \equiv A^{\binom{u_1,u_2,u_3}{v_1,v_2,v_3}} + A^{\binom{u_2,u_1,u_3}{v_2,v_1,v_3}} + A^{\binom{u_2,u_3,u_1}{v_2,v_3,v_1}}
$$

but if $A^\bullet$ is *symmetrel* (resp. *symmetril*) we get additional, 'contracted' terms on the right-hand side, namely $A^{\binom{u_{12},u_3}{v_{12},v_3}} + A^{\binom{u_2,u_{13}}{v_2,v_{13}}}$ resp.

$$
P(v_{1:2})A^{\binom{u_{12},u_3}{v_1,v_3}} + P(v_{2:1})A^{\binom{u_{12},u_3}{v_2,v_3}} + P(v_{1:3})A^{\binom{u_2,u_{13}}{v_2,v_1}} + P(v_{3:1})A^{\binom{u_2,u_{13}}{v_2,v_{13}}}
$$

*The set of all alternal bimoulds is a subalgebra of* ARI *. That of all symmetral bimoulds is a subgroup of* GARI.

These are closure properties for moulds with a simple symmetry. But ARI/GARI is specially well-suited for the study of bimoulds with a double symmetry:

*The set* $\mathrm{ARI}_{\underline{\mathrm{al/al}}}$ *of bialternal even bimoulds (i.e. bimoulds that are alternal and whose* swappee *is also alternal) constitute an important subalgebra of* ARI, *and similarly the set* $\mathrm{GARI}_{\underline{\mathrm{as/as}}}$ *of bisymmetral even bimoulds is an important subgroup of* GARI.

Here, "even" means that, for any given length $r$, the component $A^{w_1,\ldots,w_r}$ is an even functions of $\mathbf{w}$. Actually, 'evenness' is *almost* a consequence of the double symmetry: thus, it may be shown that a bialternal bimould automatically has even components for all lengths

$r$, except at most for $r = 1$. But to ensure stability under the ARI-bracket, length-one components also have to be even. This subsidiary parity condition is signalled by underlining: e.g. $\underline{\mathrm{al/al}}$ and $\underline{\mathrm{as/as}}$.

Even more important for our purpose is *the subalgebra* $\mathrm{ARI}_{\underline{\mathrm{al/il}}}$ *of alternal bimoulds with an alternil swappee, and the subgroup* $\mathrm{GARI}_{\underline{\mathrm{as/is}}}$ *of symmetral bimoulds with an symmetril swappee.* Here also, the parity condition implies the evenness of length-one components, but is slightly more technical for $r \geq 2$.

The double symmetry has other consequences: it implies invariance under some form or other of idem-potent transformation, like

(i) the *push* for $\mathrm{ARI}_{\underline{\mathrm{al/al}}}$;

(ii) the *spush* for $\mathrm{GARI}_{\underline{\mathrm{as/as}}}$;

(iii) variants of these for $\mathrm{ARI}_{\underline{\mathrm{al/il}}}$ and $\mathrm{GARI}_{\underline{\mathrm{as/is}}}$ (see [E14, E16, E17]).

It also ensures the existence of an *involutive (or group) automorphism*: thus, the involution *swap*, which is no algebra automorphism on ARI as a whole, becomes one when restricted to $\mathrm{ARI}_{\underline{\mathrm{al/al}}}$.

### 5.2.5 Some remarkable elements of ARI

Bimoulds with a double symmetry *do matter*—and in more ways than one. But they are rather thin on the ground, and not so easy to construct. So it comes as a relief to know that *most of them*, and in some important cases *all of them*, can be derived from a small set of rather elementary bimoulds, the so-called bielementals $\mathrm{belam}_r^\bullet/\mathrm{belim}_r^\bullet$. These depend only on the component length $r$ and on a 2-variable function $\mathrm{xaxi}(w_1) := \mathrm{xa}(u_1)\mathrm{xi}(v_1)$[67], or rather the even part of xaxi. All components of $\mathrm{belam}_r^\bullet/\mathrm{belim}_r^\bullet$ are $\equiv 0$, except the component of length $r$, which reduces to a simple superposition:

$$\begin{aligned}\mathrm{belam}_{r,\mathrm{xaxi}}^{w_1,\ldots,w_r} &= \mathrm{belam}_{r,\mathrm{xaxi}}^{\binom{u_1,\ldots,u_r}{v_1,\ldots,v_r}} \\ &:= \sum_{\substack{i,j,m,n\in\mathbb{Z}_{r+1}\\ \cdots<i\leq m<j\leq n<\cdots}} \mathrm{bel}_r^{i,j;m,n}\mathrm{xa}(u_i+u_{i+1}+\cdots+u_{j-1})\mathrm{xi}(v_m-v_n) \\ &\equiv \sum_{\substack{i,j,m,n\in\mathbb{Z}_{r+1}\\ \cdots<i\leq m<j\leq n<\cdots}} \frac{1}{2}\mathrm{bel}_r^{i,j;m,n}(\mathrm{xa}(u_{i\ldots j-1})\mathrm{xi}(v_{m:n})+\mathrm{xa}(u_{j\ldots i-1})\mathrm{xi}(v_{n:m}))\end{aligned} \tag{5.12}$$

with a *swappee*

$$\mathrm{belim}_{r,\mathrm{xaxi}}^\bullet := \mathrm{swap}(\mathrm{belam}_{r,\mathrm{xaxi}}^\bullet) \equiv \mathrm{belam}_{r,\mathrm{xixa}}^\bullet \tag{5.13}$$

and with integer coefficients

$$\mathrm{bel}_r^{i,j;m,n} \equiv \mathrm{bel}_r^{j,i;n,m} := \frac{(-1)^{[m-i]_r+[n-j]_r}[r-1]_r!}{[m-i]_r!\,[n-j]_r!\,[j-m-1]_r!\,[i-n-1]_r!} \tag{5.14}$$

This calls for a few comments:

[67]With variables $u_1, v_1$ in two (possibly different) abelian groups.

The above formulas use the cyclic *augmented notation*: we index the variable $u_i, v_i$ of the $r$-th component of a bimould on $\mathbb{Z}_{r+1} := \mathbb{Z}/(r+1)\mathbb{Z}$ after adding the two 'redundant' variables $u_0 := -u_{1\ldots r}$ et $v_0 := 0$ . The inequalities under the $\sum$ sign are of course relative to the cyclic order on $\mathbb{Z}_{r+1}$ and, for any $k \in \mathbb{Z}_{r+1}$, $[k]_r$ denotes the representative of $k$ in $\{0, 1, \ldots r\}$.

Formula (5.13) shows that the involution *swap* leaves bielementals unchanged, apart from swapping xa and xi. But the main facts are these:

(i) all bimoulds $\mathrm{belam}^{\bullet}_{r,\mathrm{xaxi}}$ are *bialternal*;

(ii) they vanish for *odd* or (iff $r \geq 2$) *semi-constant* functions xaxi;

(iii) they are non-zero for *even* functions xaxi constant in neither variable;

(iv) they generate *most other* bialternals under the *ari*-bracket.

### 5.2.6 Further remarkable elements of ARI

As usual, we set: $P(t) := 1/t$ and $Q_c(t) := c/\tan(ct)$ for some $c \in \mathbb{C}$.

The identities

$$\mathrm{pa}_1^{w_1} := P(u_1); \qquad \mathrm{pa}_r^{w_1,\ldots,w_r} := P(u_{12\ldots r})(\mathrm{pa}_{r-1}^{w_1,\ldots,w_{r-1}} - \mathrm{pa}_{r-1}^{w_2,\ldots,w_r}) \tag{5.15}$$

$$\mathrm{pi}_r^{w_1,\ldots,w_r} := (v_1 + v_2 + \ldots v_r)P(v_1)P(v_{1:2})P(v_{2:3})\ldots P(v_{r-1:r})P(v_r) \tag{5.16}$$

define (the former by induction, the latter directly) two series of rather peculiar bimoulds, the $\mathrm{pa}^{\bullet}_r$ and $\mathrm{pi}^{\bullet}_r$, which depend each on one set of variables—the $u_i$ or $v_i$, and have only one non-zero component, that of length $r$. The $\mathrm{pa}^{\bullet}_r$ and $\mathrm{pi}^{\bullet}_r$ are *alternal*, and although not *bialternal*, they still possess a double symmetry of sorts, since they are exchanged, not under the involution swap, but under another important, if less general involution: the *slap*[68]. Moreover they self-reproduce under the ARI-bracket:

$$\mathrm{ari}(\mathrm{pa}^{\bullet}_{r_1}, \mathrm{pa}^{\bullet}_{r_2}) = (r_1 - r_2)\mathrm{pa}_{r_1+r_2} \tag{5.17}$$

$$\mathrm{ari}(\mathrm{pi}^{\bullet}_{r_1}, \mathrm{pi}^{\bullet}_{r_2}) = (r_1 - r_2)\mathrm{pi}_{r_1+r_2} \tag{5.18}$$

which means that the subalgebras $\mathrm{ARI}_{\mathrm{pa}}$ and $\mathrm{ARI}_{\mathrm{pi}}$ of ARI generated by the bimoulds $\mathrm{pa}^{\bullet}_r$ or $\mathrm{pi}^{\bullet}_r$ are each isomorphic to the algebra $\mathrm{Diff}_t$ spanned by the differential operators $t^{n+1}\partial_t$.

### 5.2.7 Some remarkable elements of GARI

By Lie exponentiation, the algebra isomorphisms just mentioned induce group isomorphisms between each of the subgroups:

(i) $\mathrm{GARI}_{\mathrm{pa}} := \mathrm{expari}(\mathrm{ARI}_{\mathrm{pa}})$;

(ii) $\mathrm{GARI}_{\mathrm{pi}} := \mathrm{expari}(\mathrm{ARI}_{\mathrm{pi}})$;

and the group:

---

[68] Which acts as an automorphism, but only on the subalgebra $\mathrm{ARI}_{\mathrm{eupol}} \subset \mathrm{ARI}$ consisting of so-called *eupolar* bimoulds, which are particular rational functions of $\mathbf{u}$ or $\mathbf{v}$.

(iii) $\text{Diffeo}_t := \exp(\text{Diff}_t)$ of formal, identity-tangent diffeomorphisms $t \mapsto t + O(t^2)$ of $\mathbb{C}_{,0}$ unto itself.

Of special interest are the images $\text{par}^\bullet \in \text{GARI}_{\text{pa}}$ and $\text{pil}^\bullet \in \text{GARI}_{\text{pi}}$ of the diffeomorphism $f \in \text{Diffeo}_t$ defined by $f(t) := 1 - \exp(-t)$. Like all bimoulds in $\text{GARI}_{\text{pa}}$ and $\text{GARI}_{\text{pi}}$, $\text{par}^\bullet$ and $\text{pil}^\bullet$ are symmetral, but the remarkable and unexpected thing is that their *swappees* $\text{pir}^\bullet := \text{swap}(\text{par}^\bullet)$ and $\text{pal}^\bullet := \text{swap}(\text{pil}^\bullet)$ are symmetral too.

*The bisymmetral pairs* $\text{pal}^\bullet/\text{pil}^\bullet$ *and* $\text{par}^\bullet/\text{pir}^\bullet$ *thus defined are central to the theory.* They *do not* fulfill the parity condition and so *do not* belong to $\text{GARI}_{\underline{\text{as/as}}}$. Indeed, upto rescaling and under suitable additional conditions ("eupolarity"), they are the only bisymmetral (bi)moulds[69] that depend on one set of variables only ($\mathbf{u}$ or $\mathbf{v}$) and whose $r$-th component is homogeneous of degree $-r$.

Of the two pairs, $\text{pal}^\bullet/\text{pil}^\bullet$ is the more important by far. It has a 'eutrigonometric' counterpart $\text{tal}^\bullet/\text{til}^\bullet$, obtained by replacing $P$ by $Q_c$ and adding suitable corrective terms that involve only even powers of $c$—see [E14].

These bisymmetral bimoulds enjoy an incredible number of properties and sit at the hub of a galaxy of some sixty 'special bimoulds', which are investigated in [E17] and whose applications far outstrip multizeta theory.

### 5.2.8 Further remarkable elements of GARI

The *scramble* is a general bimould transform defined by:

$$A^\bullet \mapsto B^\bullet = \text{scramble}(A^\bullet) \quad \text{with} \quad B^\bullet := \sum_{\mathbf{w}^* \in \text{scram}(\mathbf{w})} .\epsilon(\mathbf{w}, \mathbf{w}^*) A^{\mathbf{w}^*} \tag{5.19}$$

where $\text{scram}(\mathbf{w})$ is the set of all sequences $\mathbf{w}^* = \binom{u_1^*,\ldots,u_r^*}{v_1^*,\ldots,v_r^*}$ which have the same length $r$ as $\mathbf{w} = \binom{u_1,\ldots,u_r}{v_1,\ldots,v_r}$ and are characterised by the property that for each $j \in \{1, \ldots, r\}$:

$$u_1^* v_1^* + u_2^* v_2^* + \ldots u_j^* v_j^* = \sum_{1 \le i \le j} \left( \sum_{p_{j,i-1} < p \le p_{j,i}} u_p \right) v_{q_{j,i}} \tag{5.20}$$

for some pair $\{p_{j,k}\}, \{q_{j,k}\}$ of intertwined sequences:

$$0 = p_{j,0} < q_{j,1} \le p_{j,1} < q_{j,2} \le p_{j,2} < \cdots < q_{j,j} \le p_{j,j} < r$$

There are exactly $r!! := 1.3\ldots(2r-1)$ such sequences $\mathbf{w}^*$. Each $u_j^*$ is a sum of one or several consecutive $u_i$, and each $v_j^*$ is either of the form $v_{j_\star}$, in which case we set $\epsilon(\mathbf{w}, \mathbf{w}^*, j) := 1$, or of the form $v_{j_\star} - v_{j_{\star\star}}$, in which case we set $\epsilon(\mathbf{w}, \mathbf{w}^*, j) := \text{sign}(j_{\star\star} - j_\star)$. (Mark the inversion). Multiplied together, these signs define the global sign factor $\epsilon(\mathbf{w}, \mathbf{w}^*) := \prod_{j=1}^{j=r} \epsilon(\mathbf{w}, \mathbf{w}^*, j)$ in the definition of the scramble transform.

In the above definition (5.19), $A^\bullet$ was assumed to be a bimould, but it could just as well be a mere mould, in which case $A^{\mathbf{w}^*}$ should be interpreted as $A^{u_1^* v_1^*, \ldots, u_r^* v_r^*}$. Thus, the *scramble* turns moulds and bimoulds alike into bimoulds.

[69] Being constant in one series of variables, bimoulds like $\text{pal}^\bullet/\text{pil}^\bullet$, $\text{par}^\bullet/\text{pir}^\bullet$ etc. are often referred to as "moulds".

One of the reasons behind the importance of the *scramble* is that it preserves the two basic symmetry types: *if the mould or bimould $A^\bullet$ is alternal (resp. symmetral), so is the bimould* $B^\bullet = \text{scramble}(A^\bullet)$.

Remarkable (bi)moulds tend to have remarkable *'scramblees'*. Thus the symmetral mould $\mathcal{V}^\bullet(z)$, (see Section 6.7.7) which is central to *equational resurgence* and *Singular Systems*, yields the bimould $S^\bullet(x) := \text{scramble}(\mathcal{V}^\bullet(x))$, which is central to the theory of *co-equational resurgence* and *Singularly Perturbed Systems*. This theory is *one of the three structures* referred to in the title of Lesson Five. In this hasty Survey, it must be given short shrift, but a detailed treatment is available in [E8].

Closely linked to the symmetral resurgence monomials $\mathcal{V}^\bullet(z)$ are the alternal hyperlogarithmic monics $V^\bullet = V^\bullet_{\omega_0}$ featuring[70] in the resurgence equations:

$$\Delta_{\omega_0}\mathcal{V}^\bullet(z) = V^\bullet_{\omega_0} \times \mathcal{V}^\bullet(z) \tag{5.21}$$

When scrambled, that mould yields the so-called *tesselation* mould $\text{tes}^\bullet := \text{scramble}(V^\bullet)$ [E8], which dominates the geometry of co-equational resurgence in the Borel planes, and possesses many arresting features, like *being locally constant in its two series of variables, the $u_i$ as well as the $v_i$* (although $\text{tes}^\bullet$ is a superposition of several highly complex functions). Thus for $r = 2$, the tesselation coefficient:

$$\text{tes}^{w_1,w_2} := V^{u_1 v_1, u_2 v_2} + V^{u_{12} v_2, u_1 v_{1:2}} - V^{u_{12} v_1, u_2 v_{2:1}}$$

is, contrary to appearances, locally constant on $\mathbb{C}^4$ and assumes only three distinct values there, namely 0 and $\pm 1$.

### 5.2.9 Basic complexity of ARI/GARI

The basic complexity of ARI/GARI (as reflected in its main operations, associated structures, fundamental bimoulds, etc.) is quite high. Thus, for a given component length $r$, the inversion *invgari* in GARI or the Lie exponential *expari* of ARI into GARI resolves into a sum of a fast increasing number (marked # in the table below) of terms, each of which fills upto half a line, or more, of small print:

| length $r$ | 1 | 2 | 3 | 4 | 5 | 6 | 7 | 8 | ... |
|---|---|---|---|---|---|---|---|---|---|
| #(invgari) | 1 | 4 | 20 | 112 | 672 | 4,224 | 27,459 | 183,040 | ... |
| #(expari) | 1 | 4 | 21 | 126 | 818 | 5,594 | 39,693 | 289,510 | ... |

For $r = 8$, we already get six digit numbers, and spelling out the corresponding formulas in full would take about one hundred pages. This means that one must often rely heavily on automatic computation when *exploring* the fringes and by-paths of ARI/GARI. Fortunately, however, the whole field is so strongly structured, and so harmonious too, offering so many hints and props to intuition, that facts and formulas are easy to guess and, once guessed, quickly yield to rigorous proving. Writing down all these proofs is of course another matter, due to the sheer mass of the facts already unearthed or yet to emerge!

[70]The lower index is actually redundant and may be dropped, since $V^{\omega_1,\ldots,\omega_r}_{\omega_0} \equiv 0$ unless $\omega_1 + \ldots \omega_r = \omega_0$.

## 5.3 The arithmetics of multizetas

### 5.3.1 Formal multizetas $\mathrm{za}^\bullet/\mathrm{ze}^\bullet$

In the *first encoding*, the *generalised* or *modulated* multizetas are defined by:

$$\mathrm{Ze}^{\binom{\epsilon_1,\ldots,\epsilon_r}{s_1,\ldots,s_r}} := \sum_{n_1>\ldots n_r>0} n_1^{-s_1}\ldots n_r^{-s_r} e_1^{n_1}\ldots e_r^{n_r} \tag{5.22}$$

with $s_j \in \mathbb{N}^*$, $\epsilon_j \in \mathbb{Q}/\mathbb{Z}$, $e_j := \exp(2\pi i \epsilon_j)$.

The *second encoding* may be directly defined, via the polylogarithms, but it is more expeditious to derive it from the first encoding:

$$\mathrm{Za}^{e_1,0^{(s_1-1)},\ldots,e_r,0^{(s_r-1)}} \overset{\text{essentially}}{:=} \mathrm{Ze}^{\binom{\epsilon_r,\epsilon_{r-1:r},\ldots,\epsilon_{1:2}}{s_r,s_{r-1},\ldots,s_1}} \tag{5.23}$$

Setting $\epsilon_j \equiv 0$, $e_j \equiv 1$, we get the *usual* or *plain* multizetas.

On its obvious domain of convergence $\Re(s_1) > 1, \Re(s_{1\,2}) > 2, \ldots$, the series (2.28) defines a holomorphic function *with a meromorphic extension to the whole of* $\mathbb{C}^r$ which in turn possesses:

(i) a remarkable singular locus;

(ii) remarkably simple multipoles described by the *Bernoully mould*;

(iii) a *'parity property'* reminiscent of the *reflexion property* of the Riemann zeta function[71].

From the arithmetical point of view, however, the $\mathbb{Q}$-ring generated by the values of the multizeta function on $\mathbb{Z}^r$ (at regular points or even at singular ones, after canonical removal of the multipole) is no larger than the $\mathbb{Q}$-ring generated (in fact: spanned) by its values on $\mathbb{N}^r$. So we may restrict our attention to the latter.

The multizetas, whether plain or modulated, are eminently 'dimorphic' entities: they are doubly closed under multiplication, since to the two encodings there correspond two distinct ways of calculating their products. These are the two classical systems of *quadratic relations*, which can be derived in any number of ways. In pithy mould language, with the conventions of Section 6.1, they can be enuntiated as follows:

(i) *The mould* $\mathrm{Ze}^\bullet$*, where defined, is symmetrel*[72]*, and there is a unique extension to the divergent case that keeps it symmetrel and gives* $\mathrm{Ze}^{\binom{0}{1}} = 0$.

(ii) *The mould* $\mathrm{Za}^\bullet$*, where defined, is symmetral and there is a unique extension to the divergent case that keeps it symmetral and gives* $\mathrm{Za}^0 = \mathrm{Za}^1 = 0$.

But these two extensions do not exactly coincide. There is a slight discrepancy, which calls for some simple corrective terms [E14] in the conversion formula (5.23). Hence the mention "essentially" in the middle of (5.23).

All the indications, numerical and theoretical, are that the two sets of 'quadratic relations' do express the totality of algebraic constraints on multizetas. So we may confidently replace

[71]Somewhat confusingly known as 'functional equation'.

[72]Since the mould $Ze^\bullet$ has two-storeyed indices $\omega_i = \binom{\epsilon_i}{s_i}$, the contractions $\omega_i + \omega_j$ must of course be interpreted as $\binom{\epsilon_i+\epsilon_j}{s_i+s_j}$.

the *true* multizetas $\mathrm{Ze}^\bullet, \mathrm{Za}^\bullet$, which at the moment are still largely beyond the reach of arithmetics[73], by their *formal* or *symbolic* counterparts $\mathrm{ze}^\bullet, \mathrm{za}^\bullet$, written in lower-case letters and subject only to

(i) the symmetrelity of $\mathrm{ze}^\bullet$;

(ii) the symmetrality of $\mathrm{za}^\bullet$;

(iii) the conversion rules (5.23),

and address the problem of unravelling all the algebraic consequences.

### 5.3.2 Generating functions $\mathrm{zag}^\bullet/\mathrm{zig}^\bullet$

In *scalar* form, the multizetas are rather unwieldy, and it is convenient to replace them by *generating series*, so tailored as to preserve the simplicity of the two symmetries and the transparency of the conversion rule. The proper definitions are:

$$\mathrm{zig}^{\binom{\epsilon_1,\dots,\epsilon_r}{v_1,\dots,v_r}} := \sum_{1\le s_j} \mathrm{ze}^{\binom{\epsilon_1,\dots,\epsilon_r}{s_1,\dots,s_r}} v_1^{s_1-1}\dots v_r^{s_r-1} \tag{5.24}$$

$$\mathrm{zag}^{\binom{u_1,\dots,u_r}{\epsilon_1,\dots,\epsilon_r}} := \sum_{1\le s_j} \mathrm{za}^{e_1,0^{(s_1-1)},\dots,e_r,0^{(s_r-1)}} u_1^{s_1-1} u_{12}^{s_2-1}\dots u_{12\dots r}^{s_r-1} \tag{5.25}$$

In the *formal* case the components of the two new moulds $\mathrm{zag}^\bullet/\mathrm{zig}^\bullet$ are mere power series, but in the *genuine* case, i.e. for the moulds $\mathrm{Zag}^\bullet/\mathrm{Zig}^\bullet$ built from the *numerical* multizetas, these power series sum up to meromorphic functions with interesting properties, such as satisfying simple difference equations [E14].

Moreover, we have the implications:

$$\mathrm{ze}^\bullet \text{ symmetrel} \iff \mathrm{zig}^\bullet \text{ symmetril}$$
$$\mathrm{za}^\bullet \text{ symmetral} \iff \mathrm{zag}^\bullet \text{ symmetral}$$

and the conversion rule (5.23) translates into:

$$\mathrm{swap}(\mathrm{zig}^\bullet) \overset{\text{exactly}}{=} \mathrm{mu}(\mathrm{zag}^\bullet, \mathrm{mono}^\bullet) \tag{5.26}$$

Here, $\mathrm{mono}^\bullet$ is an elementary, constant-valued mould: up to rational factors, its values are 'monozetas' $\zeta(s)$, hence its name. So for the generating functions, the conversion rule essentially reduces to the involution *swap*. Remark that formula (5.26) uses the primary mould product *mu* (see Section 6.1). But due to the elementary nature of $\mathrm{mono}^\bullet$, the right-hand side of (5.26) may also be written as an (exceptionnally commutative) product in GARI. Indeed:

$$\mathrm{mu}(\mathrm{zag}^\bullet, \mathrm{mono}^\bullet) \equiv \mathrm{gari}(\mathrm{zag}^\bullet, \mathrm{mono}^\bullet) \equiv \mathrm{gari}(\mathrm{mono}^\bullet, \mathrm{zag}^\bullet)$$

Thus, studying the formal multizetas boils down to *finding and describing all the symmetral/symmetril pairs of (essential) swappees* $\mathrm{zag}^\bullet/\mathrm{zig}^\bullet$ *with values in the ring of formal power series.*

[73] Despite the trail-blazing work of R. Apéry and T. Rivoal.

### 5.3.3 Immediate bipartion and arduous tripartition

As a natural element of GARI, the mould $\mathrm{zag}^\bullet$ splits into two and even three factors:

$$\mathrm{zag}^\bullet = \mathrm{gari}(\mathrm{zag}^\bullet_{\mathrm{I+II}}, \mathrm{zag}^\bullet_{\mathrm{III}}) \qquad (\mathrm{zag}^\bullet_{\mathrm{III}} \in \mathrm{GARI}^{\mathrm{o.l.}}_{\underline{\mathrm{as/is}}}) \tag{5.27}$$

$$\mathrm{zag}^\bullet = \mathrm{gari}(\mathrm{zag}^\bullet_{\mathrm{I}}, \mathrm{zag}^\bullet_{\mathrm{II}}, \mathrm{zag}^\bullet_{\mathrm{III}}) \qquad (\mathrm{zag}^\bullet_{\mathrm{II}} \in \mathrm{GARI}^{\mathrm{e.l.}}_{\underline{\mathrm{as/is}}}) \tag{5.28}$$

The factors $\mathrm{zag}^\bullet_{\mathrm{I}}$, $\mathrm{zag}^\bullet_{\mathrm{II}}$, $\mathrm{zag}^\bullet_{\mathrm{I+II}}$ are of type "e.l.", meaning that their components of even/odd length are even/odd functions of $\mathbf{w}$. The factor $\mathrm{zag}^\bullet_{\mathrm{III}}$ on the other hand is of type "o.l.", meaning that its components of even/odd length are odd/even functions of $\mathbf{w}$. Under the algebra isomorphisms of Section 5.3.4 *infra*, bimoulds of type "e.l." (resp. "o.l.") correspond to bimoulds that are (in both cases) even functions of $\mathbf{w}$ but whose only non-zero components have even (resp. odd) lengths[74]. Hence the abbreviations e.l. (even-lengthed) and o.l. (odd-lengthed).

But there is a major difference between (5.27) and (5.28). The first factorisation is elementary, immediate, and indisputably canonical, with the $\mathrm{zag}^\bullet_{\mathrm{III}}$ factor given by:

$$\mathrm{gari}(\mathrm{zag}^\bullet_{\mathrm{III}}, \mathrm{zag}^\bullet_{\mathrm{III}}) = \mathrm{gari}(\mathrm{imne}(\mathrm{invgari}(\mathrm{zag}^\bullet)), \mathrm{zag}^\bullet) \tag{5.29}$$

where *imne* denotes the elementary ARI/GARI automorphism:

$$\mathrm{imne} := \mathrm{impar} \circ \mathrm{neg} : A^{\binom{u_1,\dots,u_r}{v_1,\dots,v_r}} \mapsto (-1)^r A^{\binom{-u_1,\dots,-u_r}{-v_1,\dots,-v_r}} \tag{5.30}$$

Since all elements of GARI have exactly one square root, (5.29) determines $\mathrm{zag}^\bullet_{\mathrm{III}}$ and then (5.27) determines $\mathrm{zag}^\bullet_{\mathrm{I+II}}$ by division.

The difficulty with the second, more precise factorisation (5.28), which consists in disentangling the factors $\mathrm{zag}^\bullet_{\mathrm{I}}$ and $\mathrm{zag}^\bullet_{\mathrm{II}}$, is not its existence, which again is quite straightforward, but its canonicity: there are infinitely many ways of detaching the 'polar-trigonometric' factor $\mathrm{zag}^\bullet_{\mathrm{I}}$, which carries the $\pi^2$-dependence, from $\mathrm{zag}^\bullet_{\mathrm{I+II}}$, and the 'right' choice will be sketched in Section 5.3.6.

These factorisations hold not only in the *formal*, but also in the *genuine* case, and thus lead to two canonical splittings of the $\mathbb{Q}$-ring of multizetas, one immediate, the other more recondite:

$$\mathbb{Z}\mathrm{eta} = \mathbb{Z}\mathrm{eta}_{\mathrm{I+II}} \otimes \mathbb{Z}\mathrm{eta}_{\mathrm{III}} \tag{5.31}$$

$$\mathbb{Z}\mathrm{eta} = \mathbb{Z}\mathrm{eta}_{\mathrm{I}} \otimes \mathbb{Z}\mathrm{eta}_{\mathrm{II}} \otimes \mathbb{Z}\mathrm{eta}_{\mathrm{III}} \quad \text{with} \quad \mathbb{Z}\mathrm{eta}_{\mathrm{I}} = \mathbb{Q}[\pi^2] \tag{5.32}$$

The ring $\mathbb{Z}\mathrm{eta}_{\mathrm{III}}$ (resp. $\mathbb{Z}\mathrm{eta}_{\mathrm{I+II}}$) is generated by all *irreducibles* of odd length (resp. by those of even length, plus the odd man out $\zeta(2) = \pi^2/6$ ).

### 5.3.4 The free generation theorem

*The* $\mathbb{Q}$*-ring* $\mathbb{Z}$eta *of formal multizetas, as well as the three factor-rings* I, II, III, *are polynomial rings, that is to say, they are freely generated on* $\mathbb{Q}$ *by a countable system of 'irreducibles'. This holds equally for the* plain *and the more general* modulated *multizetas.*

---

[74] This applies only to the factors $\mathrm{zag}^\bullet_{\mathrm{I}}$ and $\mathrm{zag}^\bullet_{\mathrm{II}}$.

Although it has been open for the better part of the nineties, this *free generation theorem* is a very simple affair. The only difficulty is to establish the closure under the *ari*-bracket of the space $\mathrm{ARI}_{\underline{\mathrm{al/il}}}$ of all alternal/alternil bimoulds (with the subsidiary 'parity' condition). The neatest proof consists in observing that the space $\mathrm{ARI}_{\underline{\mathrm{al/al}}}$ of bialternal bimoulds is (trivially) a subalgebra of ARI, and in using either of the two explicit isomorphisms:

$$\mathrm{adari}(\mathrm{pal}^\bullet) \quad \text{or} \quad \mathrm{adari}(\mathrm{par}^\bullet) : \mathrm{ARI}_{\underline{\mathrm{al/al}}} \Rightarrow \mathrm{ARI}_{\underline{\mathrm{al/il}}} \tag{5.33}$$

where $\mathrm{adari}(\mathrm{pal}^\bullet)$ (resp. $\mathrm{adari}(\mathrm{par}^\bullet)$) denotes the adjoint action in ARI of the bisymmetral mould $\mathrm{pal}^\bullet$ (resp. $\mathrm{par}^\bullet$ ) constructed in Section 5.2.6.

Observe that, while these isomophisms make it certain that $\mathrm{ARI}^{\mathrm{ent}}_{\underline{\mathrm{al/il}}}$ *is* a subalgebra, they do not exchange the subalgebras $\mathrm{ARI}^{\mathrm{ent}}_{\underline{\mathrm{al/al}}}$ and $\mathrm{ARI}^{\mathrm{ent}}_{\underline{\mathrm{al/il}}}$ of 'entire-valued' bimoulds (i.e. bimoulds with values in the ring of formal power series). In fact, these two subalgebras *are not* isomorphic.

The *general entire-valued*, symmetral/symmetril pair of swappees $\mathrm{zag}^\bullet/\mathrm{zig}^\bullet$ is then obtained by postcomposition in GARI of a *particular* $\mathrm{zag}^\bullet$ (e.g. the 'genuine', 'numerical' $\mathrm{Zag}^\bullet$ or its first factor $\mathrm{Zag}^\bullet_I$) by the *general* element of $\mathrm{GARI}^{\mathrm{ent}/\#}_{\underline{\mathrm{as/is}}} = \mathrm{expari}(\mathrm{ARI}^{\mathrm{ent}/\#}_{\underline{\mathrm{al/il}}})$. Here, *ent* means *entire-valued* as usual, and # denotes an additional condition which depends on the ring of multizetas that is being considered: thus for the *plain* multizetas, # simply means *constant in the v-variables.* For the *modulated* multizetas, see Section 5.3.5 below.

The problem has thus been completely linearised, and the *free generators* of $\mathbb{Z}$eta, or *irreducibles* (other than $\pi^2$), are seen to be in one-to-one correspondence with the generators of $\mathrm{ARI}^{\mathrm{ent}/\#}_{\underline{\mathrm{al/il}}}$ *as a vector space.*

But *as a Lie algebra*, $\mathrm{ARI}^{\mathrm{ent}/\#}_{\underline{\mathrm{al/il}}}$ has far more structure on it than the $\mathbb{Q}$-ring $\mathbb{Z}$eta of formal multizetas, and its *linear generators* may be further analysed—down to *Lie generators.* This is where the exciting work on multizetas actually *begins*!

### 5.3.5 Generators and dimensions

#### The Broadhurst-Kreimer conjectures

Let us consider jointly the *plain multizetas* (without unit roots) and the *Eulerian multizetas* (modulated by the unit roots $\pm 1$)—the former because of their obvious importance; the latter because, contrary to appearences, they are actually *simpler.* Let $P_{s,r}$ (resp. $E_{s,r}$) be the smallest number of irreducibles of length $r$ and weight $s$ ($:= s_1 + \cdots + s_r$) needed to produce, jointly, a complete system of irreducibles for the plain (resp. Eulerian) multizetas.

Relying on extensive numerical computations and some inspired guess-work, Broadhurst and Kreimer have conjectured that the dimensions $P_{s,r}$ and $E_{s,r}$ for the 'genuine', as opposed to 'formal', multizetas could be read off the generating functions:

$$\prod_{s\geq 3, r\geq 1} (1 - x^s y^r)^{P_{s,r}} \stackrel{?}{=} 1 - \frac{x^3 y}{1 - x^2} + \frac{x^{12} y^2 (1 - y^2)}{(1 - x^4)(1 - x^6)} \tag{5.34}$$

$$\prod_{s\geq 3, r\geq 1} (1 - x^s y^r)^{E_{s,r}} \stackrel{?}{=} 1 - \frac{x^3 y}{(1 - x^2)(1 - xy)} \tag{5.35}$$

**The two series of relevant algebras**

For any $p \geq 1$ let $\mathbb{Z}_p$ be the subgroup $\{0, 1/p, \ldots, (p-1)/p\}$ of $\mathbb{Q}/\mathbb{Z}$ and let $\mathrm{ARI}^{\mathrm{ent}/\mathbb{Z}_p}$ denote the subalgebra of ARI (it is one!) which regroups all bimoulds $A^\bullet$:

(i) with **u**-variables ranging through $\mathbb{C}$;

(ii) with **v**-variables ranging through $\mathbb{Z}_p$;

(iii) with values in the ring of formal power series in **u**;

(iv) with the self-correlation constraints:

$$A^{\binom{u_1,\ldots,u_r}{qv_1,\ldots,qv_r}} \equiv \sum_{qv_i^*=qv_i} A^{\binom{qu_1,\ldots,qu_r}{v_1^*,\ldots,v_r^*}} \qquad (\forall q \mid p) \tag{5.36}$$

and in particular (for $q = p$):

$$A^{\binom{u_1,\ldots,u_r}{0,\ldots,0}} \equiv \sum_{v_i^*\in\mathbb{Z}_p} A^{\binom{qu_1,\ldots,qu_r}{v_1^*,\ldots,v_r^*}} \tag{5.37}$$

The subalgebras directly relevant to the study of the multizetas modulated by unit roots of order $p$ are $\mathrm{ARI}^{\mathrm{ent}/\mathbb{Z}_p}_{\underline{\mathrm{al/al}}}$ and $\mathrm{ARI}^{\mathrm{ent}/\mathbb{Z}_p}_{\underline{\mathrm{al/il}}}$. For $p = 1$ their elements are simply **v**-constant bimoulds.

**Eulerian multizetas and generators of $\mathrm{ARI}^{\mathrm{ent}/\mathbb{Z}_2}_{\underline{\mathrm{al/al}}}$**

For $r = 1$ and $d$ even we set:

$$\mathrm{bela}^{w_1}_{1,d} = u_1^d \text{ (resp. } (2^{-d}-1)u_1^d)) \quad \text{if} \quad v_1 = 0 \text{ (resp. } v_1 = 1/2) \tag{5.38}$$

and for $r \geq 1$ and $d$ even we set:

$$\mathrm{bela}^\bullet_{r,d} := \mathrm{belam}^\bullet_{r,\mathrm{xaxi}} \quad \text{with} \quad \mathrm{xa}(t) := t^d;\ \mathrm{xi}(0) := 0;\ \mathrm{xi}(1/2) := 1 \tag{5.39}$$

(i) *All bimoulds* $\mathrm{bela}^\bullet_{r,d}$ *(for* $r = 1, 2, 3, \ldots$ *and* $d = 2, 4, 6, \ldots$*) are non-zero, bialternal, and self-correlated.*

(ii) *They freely generate a sub-algebra* $\underline{\mathrm{ARI}}^{\mathrm{ent}/\mathbb{Z}_2}_{\underline{\mathrm{al/al}}}$ *the* $\mathrm{ARI}^{\mathrm{ent}/\mathbb{Z}_2}_{\underline{\mathrm{al/al}}}$ *consisting of all bimoulds of type* $\underline{\mathrm{al/al}}$ *that possess an extension of type* $\underline{\mathrm{al/il}}$.

**Eulerian multizetas and generators of $\mathrm{ARI}^{\mathrm{ent}/\mathbb{Z}_2}_{\underline{\mathrm{al/il}}}$**

(i) *Each bialternal* $\mathrm{bela}^\bullet_{r,d}$ *has a canonical counterpart or 'extension'* $\mathrm{bema}^\bullet_{r,d}$*, of alternal/alternil type, self-correlated, and with a first non-zero component (i.e. the one of length* $r$*) equal to the single non-zero component of* $\mathrm{bela}^\bullet_{r,d}$.

(ii) *These* $\mathrm{bema}^\bullet_{r,d}$ *freely generate the algebra* $\mathrm{ARI}^{\mathrm{ent}/\mathbb{Z}_2}_{\underline{\mathrm{al/il}}}$.

(iii) *The Eulerian irreducibles correspond one-to-one to the bialternals spanning* $\underline{\mathrm{ARI}}^{\mathrm{ent}/\mathbb{Z}_2}_{\underline{\mathrm{al/al}}}$. *More precisely, the number* $E_{s,r}$ *of independent irreducibles of weight s and length r coincides with the dimension of the cell of* $\mathrm{ARI}^{\mathrm{ent}/\mathbb{Z}_2}_{\underline{\mathrm{al/al}}}$ *consisting of bimoulds of length r and total degree* $d = s - r$.

(iv) *This establishes the Broadhurst-Kreimer conjecture* (5.35) *for the* formal *Eulerian multizetas*[75].

### Plain multizetas and generators of $\mathrm{ARI}^{\mathrm{ent}/\mathbb{Z}_1}_{\underline{\mathrm{al/il}}}$

(i) *For each even d there is a canonical pair of alternal/alternil swappees* $\mathrm{ma}^\bullet_d/\mathrm{mi}^\bullet_d$ *with initial components* $\mathrm{ma}^{w_1} := u_1^d$, $\mathrm{ma}^{w_1} := v_1^d$ *and with a total of d non-zero components.*

(ii) *It is conjectured that the* $\mathrm{ma}^\bullet_d$ *freely generate* $\mathrm{ARI}^{\mathrm{ent}/\mathbb{Z}_1}_{\underline{\mathrm{al/il}}}$.

This conjecture (formulated in a different, less flexible framework) has been around for quite a while now, but it is rather crude and can be considerably sharpened, by reasoning on bialternals. Indeed:

### Plain multizetas and generators of $\mathrm{ARI}^{\mathrm{ent}/\mathbb{Z}_1}_{\underline{\mathrm{al/al}}}$

We require three integer sequences $\alpha, \beta, \gamma$ (with $\alpha(d) \equiv \beta(d) + \gamma(d-2)$):

$$\sum \alpha(d)x^d := x^6(1-x^2)^{-1}(1-x^4)^{-1} \tag{5.40}$$

$$\sum \beta(d)x^d := x^6(1-x^2)^{-1}(1-x^6)^{-1} \tag{5.41}$$

$$\sum \gamma(d)x^d := x^8(1-x^4)^{-1}(1-x^6)^{-1} \tag{5.42}$$

and three series of bialternals:

$$\begin{array}{rl} \mathrm{ekma}^\bullet_d/\mathrm{ekmi}^\bullet_d & d \text{ even} \geq 2 \\ \mathrm{doma}^\bullet_{d,b}/\mathrm{domi}^\bullet_{d,b} & d \text{ even} \geq 10,\ 1 \leq b \leq \beta(d) \\ \mathrm{carma}^\bullet_{d,c}/\mathrm{carmi}^\bullet_{d,c} & d \text{ even} \geq 8,\ 1 \leq c \leq \gamma(d) \end{array}$$

of total degree $d$ and with a single non-zero component of length respectively $1, 2, 4$. The definition of the first two pairs is straightforward:

$$\mathrm{ekma}^{w_1}_d := u_1^d; \qquad \mathrm{ekmi}^{w_1}_d := v_1^d \tag{5.43}$$

$$\mathrm{doma}^{w_1,w_2}_{d,b} := \mathrm{fa}(u_1,u_2)(\mathrm{ga}(u_1,u_2))^{b-1}(\mathrm{ha}(u_1,u_2))^{d/2-3b} \tag{5.44}$$

$$\mathrm{domi}^{w_1,w_2}_{d,b} := \mathrm{fi}(v_1,v_2)(\mathrm{gi}(v_1,v_2))^{b-1}(\mathrm{hi}(v_1,v_2))^{d/2-3b} \tag{5.45}$$

with

$$\mathrm{fa}(u_1,u_2) := u_1u_2(u_1-u_2)(u_1+u_2)(2u_1+u_2)(2u_2+u_1)$$

$$\mathrm{ga}(u_1,u_2) := (u_1+u_2)^2u_1^2u_2^2; \qquad \mathrm{ha}(u_1,u_2) := u_1^2+u_1u_2+u_2^2$$

$$\mathrm{fi}(v_1,v_2) := v_1v_2(v_1-v_2)(v_1+v_2)(2v_1-v_2)(2v_2-v_1)$$

$$\mathrm{gi}(v_1,v_2) := (v_1-v_2)^2v_1^2v_2^2; \qquad \mathrm{hi}(v_1,v_2) := v_1^2-v_1v_2+v_2^2$$

[75] Instead of Eulerian multizetas, Broadhurst and Kreimer speak of 'Euler sums'.

The definition of the last pair, carma$^\bullet$/carmi$^\bullet$, is more roundabout. Observe first that the ekma$^\bullet_d$ are not *free* in ARI, but *bound* (for each degree $d$) by exactly $\gamma(d)$ independent relations of the form:

$$\sum_{d_1+d_2=d+2} R_c^{d_1,d_2}[\mathrm{ekma}^\bullet_{d_1},\mathrm{ekma}^\bullet_{d_2}] = 0^\bullet \qquad (1\le c\le \gamma(d),\ R_c^{d_1,d_2}\in\mathbb{Q}) \tag{5.46}$$

which result from the expansions:

$$[\mathrm{ekma}^\bullet_{d_1},\mathrm{ekma}^\bullet_{d_2}] = \sum_{1\le b\le\beta(d_1+d_2)} K^b_{d_1,d_2}\mathrm{doma}^\bullet_{d_1+d_2,b} \qquad (K^b_{d_1,d_2}\in\mathbb{Q}) \tag{5.47}$$

Next, consider the moulds:

$$\mathrm{vima}^\bullet_{d,c} := \sum_{d_1+d_2=d+2} R_c^{d_1,d_2}[\mathrm{ma}^\bullet_{d_1},\mathrm{ma}^\bullet_{d_2}] \neq 0^\bullet \tag{5.48}$$

with $R_c^{d_1,d_2}$ as in (3.13). By construction:

(a) vima$^\bullet_{d,c}$ *is of alternal/alternil type*;

(b) *its components of length* 1, 2, 3 *vanish*;

(c) *its (non-vanishing) component of length* 4 *defines a bialternal mould, which is precisely the sought-after mould* carma$^\bullet_{d,c}$

Now, the crude conjecture at the end of the last para can be replaced by the much sharper, but also more tractable statements:

(i) *The moulds* ekma$^\bullet_d$ *are free under the ari-bracket upto the contraints* (3.13). *More precisely, the number* $P^*_{s,r}$ *of linearly independent bialternals of length* $r$ , *degree* $d$ (*and weight* $s := d + r$) *generated by the* ekma$^\bullet_d$ *is given by the BK-like formula:*

$$\prod_{s\ge3,r\ge1}(1-x^s y^r)^{P^*_{s,r}} = 1-\frac{x^3y}{1-x^2}+\frac{x^{12}y^2}{(1-x^4)(1-x^6)} \tag{5.49}$$

(ii) *The moulds* carma$^\bullet_{d,k}$ *are free under the ari-bracket. As a consequence, the number* $P^{**}_{s,r}$ *of linearly independent bialternals of length* $r$, *degree* $d$ (*and weight* $s := d+r$) *generated by the* carma$^\bullet_{d,k}$ *is given by the BK-like formula:*

$$\prod_{s\ge3,r\ge1}(1-x^s y^r)^{P^{**}_{s,r}} = 1-\frac{x^{12}y^4}{(1-x^4)(1-x^6)} \tag{5.50}$$

(iii) *In combination, the* ekma$^\bullet_d$ *and* carma$^\bullet_{d,k}$ *generate the bialternal algebra* $\mathrm{ARI}^{\mathrm{ent}/\mathbb{Z}_1}_{\mathrm{al/al}}$, *freely upto the sole constraints* (3.13). *As a consequence, the total number* $P_{s,r}$ *of linearly independent bialternals of length* $r$, *degree* $d$ (*and weight* $s := d + r$) *is given by the BK formula* (5.34).

Unlike in the Eulerian case, the three above statements have been established only upto length $r = 7$ ($\forall d$) and remain *conjectural* beyond that. But the supporting evidence is overwhelming [E16] and and in any case they have the merit of completely removing the weirdness of the artificial-looking corrective term $x^{12}y^2(1-y^2)/((1-x^4)(1-x^6))$ in the BK-formula: the explanation is simply that to each 'missing' bialternal of length 2 (there just aren't 'enough' doma$^\bullet$/domi$^\bullet$ around) there answers, under the transparent mechanism (3.13)–(3.15), a 'stop-gap' bialternal of length 4 (namely the 'unexpected' carma$^\bullet$/carmi$^\bullet$).

### 5.3.6 Canonical-explicit decomposition into irreducibles

Let $\delta^n$ be the upper dilation operator on bimoulds and let $\tau^s$ denote the homogeneous projector that retains only the $\boldsymbol{u}$-homogeneous part of total $\boldsymbol{u}$-degree $s-r$:

$$\delta^n M^{\binom{u_1,\ldots,u_r}{v_1,\ldots,v_r}} := n^{-r} M^{\binom{u_1/n,\ldots,u_r/n}{v_1,\ldots,v_r}} \tag{5.51}$$

$$\tau^s M^{\binom{u_1,\ldots,u_r}{v_1,\ldots,v_r}} := M^{\binom{u_1,\ldots,u_r}{v_1,\ldots,v_r}} \|_{\boldsymbol{u}\text{-part of degree } s-r} \tag{5.52}$$

There exists an explicit bimould $\mathrm{loma}^\bullet/\mathrm{lomi}^\bullet$:

- of alternal/alternil type;
- with components depending on the $u_i$'s alone and regular at the origin;
- carrying only rational coefficients, ie with components in $\mathbb{Q}[[u_1,\ldots,u_r]]$;
- with a length-1 component equal to $1/(1-u_1^2)$;
- with a length-2 component that is a rational function of $u_1, u_2$
- with length-$r$ components that are meromorphic functions of $u_1,\ldots,u_r$ and possess a *'minimal'* and yet (for $r \geq 3$) *infinite* number of elementary multipoles over the multi-integers.

This bimould $\mathrm{loma}^\bullet/\mathrm{lomi}^\bullet$ possesses innumerable properties. A whole book, mostly devoted to it, is currently being written. It (the bimould!) holds the key to the arcanes of multizeta aritmetics, mainly because the "numerical" $\mathrm{Zag}^\bullet$ and its factors can be reconstituded from it, by means of the following relations, which involve the anti-action arit of ARI on BIMU and the anti-action garit of GARI on BIMU:

$$\mathrm{garit}(\mathrm{Zag}^\bullet_{\mathrm{II}}) := \sum_{r \text{ even}} B_\star^{n_1,\ldots,n_r} \mathrm{arit}(\delta^{n_r}\mathrm{loma}^\bullet)\ldots\mathrm{arit}(\delta^{n_1}\mathrm{loma}^\bullet) \tag{5.53}$$

$$\mathrm{garit}(\mathrm{Zag}^\bullet_{\mathrm{III}}) := \sum_{r \text{ odd}} C_\star^{n_1,\ldots,n_r} \mathrm{arit}(\delta^{n_r}\mathrm{loma}^\bullet)\ldots\mathrm{arit}(\delta^{n_1}\mathrm{loma}^\bullet) \tag{5.54}$$

$$\mathrm{garit}(\mathrm{Zag}^\bullet_{\mathrm{II+III}}) := \sum_{r\geq 1} D_\star^{n_1,\ldots,n_r} \mathrm{arit}(\delta^{n_r}\mathrm{loma}^\bullet)\ldots\mathrm{arit}(\delta^{n_1}\mathrm{loma}^\bullet) \tag{5.55}$$

$$\mathrm{garit}(\tau^s\mathrm{Zag}^\bullet_{\mathrm{II}}) := \sum_{r \text{ even}} B^{s_1,\ldots,s_r} \mathrm{arit}(\tau^{s_r}\mathrm{loma}^\bullet)\ldots\mathrm{arit}(\tau^{s_1}\mathrm{loma}^\bullet) \tag{5.56}$$

$$\mathrm{garit}(\tau^s\mathrm{Zag}^\bullet_{\mathrm{III}}) := \sum_{r \text{ odd}} C^{s_1,\ldots,s_r} \mathrm{arit}(\tau^{s_r}\mathrm{loma}^\bullet)\ldots\mathrm{arit}(\tau^{s_1}\mathrm{loma}^\bullet) \tag{5.57}$$

$$\mathrm{garit}(\tau^s\mathrm{Zag}^\bullet_{\mathrm{II+III}}) := \sum_{r\geq 1} D^{s_1,\ldots,s_r} \mathrm{arit}(\tau^{s_r}\mathrm{loma}^\bullet)\ldots\mathrm{arit}(\tau^{s_1}\mathrm{loma}^\bullet) \tag{5.58}$$

In the first system of relations the summation extends to all integers $n_i \geq 1$ but the sums are absolutely convergent. All three moulds $B_\star^\bullet, C_\star^\bullet, D_\star^\bullet$ are *symmetral*, of *perinomal type*[76], and assume *rational* values.

[76] This is a new class of functions, which are omnipresent in multizeta arithmetics. Their properties are reminiscent of *polynomial, periodic, and modular* functions all at once!

In the second system of relations the summation extends to all *odd* integers $s_i \geq 3$ but the sums are finite since $s_1 + \cdots + s_r = s$. The three new moulds $B^\bullet, C^\bullet, D^\bullet$ are once again *symmetral* and of *perinomal type*, but *a priori* transcendental. The third mould $D^\bullet$ in particular constitutes (together with $\pi^2$ which comes from the factor $I$) a *complete and free*[77] *system of irreducibles* for the formal multizetas. This also leads to the construction of *abstract numerical derivations*—by all accounts an extremely promising development [E16, E17].

# 6 Reminders and complements

## 6.1 Moulds/bimoulds/comoulds. Basic symmetries and operations

### 6.1.1 Main operations on moulds and bimoulds

Moulds are *functions of a variable number of variables*: they depend on sequences $\boldsymbol{\omega} := (\omega_1, \ldots, \omega_r)$ of arbitrary length $r = r(\boldsymbol{\omega})$. The sum $\|\boldsymbol{\omega}\|$ of a sequence is simply $\sum_1^r \omega_i$. Sequences are systematically written in boldface, with upper indexation when such is called for, and with the product denoting concatenation: e.g. $\boldsymbol{\omega} = \boldsymbol{\omega}^1.\boldsymbol{\omega}^2$. The elements $\omega_i$ which make up these sequences are written in normal print, with lower indexation. The sequences themselves are affixed to the moulds as upper indices $A^\bullet = \{A^{\boldsymbol{\omega}}\}$, since moulds are meant to be contracted

$$A^\bullet, B_\bullet \mapsto \langle A^\bullet, B_\bullet \rangle := \sum A^{\boldsymbol{\omega}} B_{\boldsymbol{\omega}}$$

with dual objects (often differential operators or elements of an associative algebra), the so-called co-moulds $B_\bullet = \{B_{\boldsymbol{\omega}}\}$, which carry their own indices in lower position.

**Basic mould operations:**

Moulds may be *added, multiplied, composed.*

Mould addition is what you expect: components of equal length get added.

Mould multiplication (*mu* or $\times$) is associative, but non-commutative:

$$C^\bullet = A^\bullet \times B^\bullet \iff C^{\boldsymbol{\omega}} = \sum_{\boldsymbol{\omega} = \boldsymbol{\omega}^1.\boldsymbol{\omega}^2} A^{\boldsymbol{\omega}^1} B^{\boldsymbol{\omega}^2} \tag{6.1}$$

(This includes the trivial decompositions $\boldsymbol{\omega} = \boldsymbol{\omega}.\emptyset$ and $\boldsymbol{\omega} = \emptyset.\boldsymbol{\omega}$).

Mould composition ($\circ$) too is associative and non-commutative:

$$C^\bullet = (A^\bullet \circ B^\bullet) \iff C^{\boldsymbol{\omega}} = \sum_{\boldsymbol{\omega} = \boldsymbol{\omega}^1 \ldots \boldsymbol{\omega}^s} A^{\|\boldsymbol{\omega}^1\|, \ldots, \|\boldsymbol{\omega}^s\|} B^{\boldsymbol{\omega}^1} \ldots B^{\boldsymbol{\omega}^s} \tag{6.2}$$

with a sum extending to all possible decompositions of $\boldsymbol{\omega}$ into $s \leq r(\boldsymbol{\omega})$ non-empty factor sequences $\boldsymbol{\omega}^i$.

The operations $(+, \times, \circ)$ on moulds interact in exactly the same way as their namesakes for power series. Thus $(A^\bullet \times B^\bullet) \circ C^\bullet \equiv (A^\bullet \circ C^\bullet) \times (B^\bullet \circ C^\bullet)$.

[77]Free, of course, *up to the symmetrality relations*, but one can most easily derive from it a system of absolute irreducibles.

**Basic bimould operations:**

Bimoulds are moulds with indices of the form $w_i = \binom{u_i}{v_i}$. Some twenty-odd operations are defined on them. All are defined via four types of contractions $\mathbf{w} \mapsto \lceil \mathbf{w}, \lfloor \mathbf{w}, \mathbf{w} \rceil, \mathbf{w} \rfloor$ (see Section 5.1) which *add* the $u_i$'s and *subtract* the $v_i$'s in a "symplectic" way, i.e. respecting both $\sum u_i v_i$ and $\sum du_i \wedge dv_i$. We always use the following short-hand for **u**-sums and **v**-differences: $u_{12} = u_1 + u_2, u_{123} = u_1 + u_2 + u_3, \ldots, v_{1:2} := v_1 - v_2$, etc.

The main operations on bimoulds are

- the ordinary mould multiplication mu, giving rise to the algebra BIMU;
- the Lie bracket ari, giving rise to the Lie algebra ARI;
- the associative, non-commutative law gari, giving rise to the group GARI;
- the anti-action arit of ARI in BIMU [78];
- the anti-action garit of GARI in BIMU[79].

But there also exist no less than five sets of operations that run parallel to ari, gari, arit, garit. There is also an interesting and useful structure of Lie super-algebra SUARI which mirrors that of ARI. These are neither artefacts nor idle constructions. They all play an important part in unravelling the arithmetics of multizetas, and more generally in investigating *dimorphy*.

### 6.1.2 Main symmetries for moulds and bimoulds

Most useful moulds fall into a few basic *symmetry types*.

A mould $A^\bullet$ is said to be *symmetral* (resp. *alternal*) iff:

$$\sum_{\omega \in \mathrm{sha}(\boldsymbol{\omega}^1, \boldsymbol{\omega}^2)} A^{\omega} = A^{\boldsymbol{\omega}^1} A^{\boldsymbol{\omega}^2} \quad (\text{resp. } 0) \qquad \forall \boldsymbol{\omega}^1 \neq \emptyset,\ \forall \boldsymbol{\omega}^2 \neq \emptyset \tag{6.3}$$

A mould $A^\bullet$ is said to be *symmetrel* (resp. *alternel*) iff:

$$\sum_{\omega \in \mathrm{she}(\boldsymbol{\omega}^1, \boldsymbol{\omega}^2)} A^{\omega} = A^{\boldsymbol{\omega}^1} A^{\boldsymbol{\omega}^2} \quad (\text{resp. } 0) \qquad \forall \boldsymbol{\omega}^1 \neq \emptyset,\ \forall \boldsymbol{\omega}^2 \neq \emptyset \tag{6.4}$$

Here $\mathrm{sha}(\boldsymbol{\omega}^1, \boldsymbol{\omega}^2)$ (resp. $\mathrm{she}(\boldsymbol{\omega}^1, \boldsymbol{\omega}^2)$) denotes the set of all sequences $\boldsymbol{\omega}$ obtained from $\boldsymbol{\omega}^1$ and $\boldsymbol{\omega}^2$ under ordinary (resp. contracting) shuffling. In a contracting shuffle, two adjacent indices $\omega_i$ and $\omega_j$ stemming from $\boldsymbol{\omega}^1$ and $\boldsymbol{\omega}^2$ respectively may coalesce into $\omega_{ij} := \omega_i + \omega_j$.

The definition of *symmetral/alternal* carries over, unchanged, to the case of bimoulds. The definition of *symmetril/alternil* for bimoulds resembles that of symmetrel/alternel for moulds except that the contractions $\omega_i + \omega_j$ get replaced by $w_i \otimes w_j$ with:

$$A^{\ldots, w_i \otimes w_j, \ldots} := v_{i:j}^{-1} A^{\ldots, w_{i/j}, \ldots} + v_{j:i}^{-1} A^{\ldots, w_{j/i}, \ldots}$$

$$\text{and} \quad v_{i:j} := v_i - v_j, \quad w_{i/j} := \begin{pmatrix} u_i + u_j \\ v_i \end{pmatrix}$$

[78] It should be carefully distinguished from the *adjoint action* adari of ARI on itself.

[79] It should be carefully distinguished from the *adjoint action adgari* of GARI on ARI.

### 6.1.3 Arborification and co-arborification

Straightforward mould expansions $\sum_{\omega} A^{\omega} B_{\omega}$, which typically pair a symmetral or alternal mould $A^{\bullet}$ with a cosymmetral comould $B_{\bullet}$ (or a symmetrel or alternel mould $A^{\bullet}$ with a cosymmetrel comould $B_{\bullet}$) often fail to converge absolutely, i.e. $\sum_{\omega} \|A^{\omega} B_{\omega}\| = +\infty$, although the underlying series may well be convergent. Fortunately, there is an extremely general method for restoring convergence. In essence, it replaces expansions indexed by *totally ordered* sequences $\boldsymbol{\omega}$ by others whose indices are *arborescent sequences* $\boldsymbol{\omega}^{\prec}$ or $\boldsymbol{\omega}^{\nprec}$, like this:

$$\sum_{\omega} A^{\omega} B_{\omega} \mapsto \sum_{\omega^{\prec}} A^{\omega^{\prec}} B_{\omega^{\prec}} \qquad \text{(ordinary arborification)} \qquad (6.5)$$

$$\sum_{\omega} A^{\omega} B_{\omega} \mapsto \sum_{\omega^{\nprec}} A^{\omega^{\nprec}} B_{\omega^{\nprec}} \qquad \text{(contracting arborification)} \qquad (6.6)$$

The dual arborification/coarborification transforms:

arborification $\Longrightarrow$

ordinary: $\quad A^{\omega^{\prec}} := \sum_{\omega^{\prec} < \omega} A^{\omega} \qquad$ (complete definition)

contracting: $\quad A^{\omega^{\nprec}} := \sum_{\omega^{\nprec} \ll \omega} A^{\omega} \qquad$ (complete definition)

coarborification $\Longrightarrow$

ordinary: $\quad B_{\omega} := \sum_{\omega^{\prec} < \omega} B_{\omega^{\prec}} \qquad$ (mere constraint)

contracting: $\quad B_{\omega} := \sum_{\omega^{\nprec} \ll \omega} B_{\omega^{\nprec}} \qquad$ (mere constraint)

are defined in such a way as to:

(1) leave the expansions formally unchanged: they amount to a simple redistribution of terms;

(2) drastically reduce the size of the comould part: it typically gets divided by a factor of order $r! := r(\omega)!$;

(3) prevent a concomitant increase of the mould part: it typically retains the same order of magnitude, despite being changed into a sum of almost $r!$ similar terms!

But whereas the reduction (2) is automatic and universal, the non-increase (3) relies on specific identities, of an algebraic or combinatorial nature, which can never be taken for granted, and yet tend to take place, with providential regularity, whenever we require them!

For the *coaborification* rule, we refer to [E5] or [EV3].

The ordinary (resp. contracting) *arborification* rule boils down to summing all the terms $A^{\boldsymbol{\omega}}$ with totally ordered sequences $\boldsymbol{\omega}$ whose order is compatible with the arborecent order of $\boldsymbol{\omega}^{\prec}$ (resp. with that of $\boldsymbol{\omega}^{\prec\!\!\!\prec}$, but allowing contractions of consecutive elements $\omega_i$). The following example should make this amply clear.Assume:

$$\begin{array}{ccccc} & & & & \omega_4 \\ & & & \nearrow & \\ \boldsymbol{\omega}^{\prec} \text{ (or } \boldsymbol{\omega}^{\prec\!\!\!\prec}\text{ )} := & \omega_1 & \longrightarrow & \omega_2 & \longrightarrow \ \omega_3 \end{array}$$

Then the arborification rules means:

$$\begin{aligned} A^{\boldsymbol{\omega}^{\prec}} &:= A^{\omega_1,\omega_2,\omega_3,\omega_4} + A^{\omega_1,\omega_2,\omega_4,\omega_3} + A^{\omega_1,\omega_4,\omega_2,\omega_3} \\ A^{\boldsymbol{\omega}^{\prec\!\!\!\prec}} &:= A^{\omega_1,\omega_2,\omega_3,\omega_4} + A^{\omega_1,\omega_2,\omega_4,\omega_3} + A^{\omega_1,\omega_4,\omega_2,\omega_3} + A^{\omega_1,\omega_2,\omega_3+\omega_4} + A^{\omega_1,\omega_2+\omega_4,\omega_3} \\ &:= A^{\boldsymbol{\omega}^{\prec}} + A^{\omega_1,\omega_2,\omega_3+\omega_4} + A^{\omega_1,\omega_2+\omega_4,\omega_3} \end{aligned}$$

### 6.1.4 Some elementary yet useful moulds

#### Constant-type moulds

| mould | value | symmetry | associated series |
|---|---|---|---|
| $\mathbf{1}^\bullet$ | 1 if $r=0$ (0 otherwise) | symmetral | 1 |
| $\mathrm{I}^\bullet$ | 1 if $r=1$ (0 otherwise) | alternal | $x$ |
| $\log^\bullet$ | $\frac{(-1)^{r-1}}{r}$ | alternel | $\log(1+x)$ |
| $\exp_a^\bullet$ | $\frac{a^r}{r!}$ | symmetral | $e^{ax}$ |
| $\mathrm{tu}_a^\bullet$ | $\frac{(-1)^r}{r!}\frac{\Gamma(r-a)}{\Gamma(-a)}$ | symmetrel | $(1+x)^a$ |

#### Difference-type flat moulds

$$\begin{aligned} \mathrm{sad}^{\emptyset} &:= 1 \\ \mathrm{sad}^{t_1,\ldots,t_r} &:= 1 \quad \text{if} \quad t_1 < t_2 < \cdots < t_r \\ \mathrm{sad}^{t_1,\ldots,t_r} &:= 0 \quad \text{otherwise} \end{aligned}$$

$$\begin{aligned} \mathrm{lad}^{\emptyset} &:= 0 \\ \mathrm{lad}^{t_1,\ldots,t_r} &:= (-1)^q \frac{p!\,q!}{(p+q+1)!} = (-1)^q \frac{p!\,q!}{r!} \\ \text{with} \quad p &:= \sum_{t_i<t_{i+1}} 1 \quad \text{and} \quad q := \sum_{t_i>t_{i+1}} 1 \end{aligned}$$

**Difference-type polar moulds**

Relevant to the construction of w.-b. averages and w.-b. monomials.

$$\begin{aligned}
\mathrm{tas}^{\emptyset}_{a,b} &:= 1 \\
\mathrm{tas}^{t_1}_{a,b} &:= \frac{a-b}{(a-t_1)(t_1-b)} \\
\mathrm{tas}^{t_1,\ldots,t_r}_{a,b} &:= \frac{a-b}{(a-t_1)(t_1-t_2)\ldots(t_{r-1}-t_r)(t_r-b)} \\
\mathrm{tas}^{\emptyset}_{\star} &:= 0 \\
\mathrm{tas}^{t_1}_{\star} &:= \frac{1}{(-t_1)(t_1)} \\
\mathrm{tas}^{t_1,\ldots,t_r}_{\star} &:= \frac{1}{(-t_1)(t_1-t_2)\ldots(t_{r-1}-t_r)(t_r)}
\end{aligned}$$

Identities:

$$\begin{aligned}
\mathrm{tas}^{\bullet}_{a,b} \times \mathrm{tas}^{\bullet}_{b,c} &= \mathrm{tas}^{\bullet}_{a,c} \\
\mathrm{tas}^{\bullet}_{a,b} \times \mathrm{tas}^{\bullet}_{b,a} &= \mathbf{1}^{\bullet}
\end{aligned}$$

**Sum-type flat moulds**

We use the auxiliary notations:

$$\begin{aligned}
\boldsymbol{x} &:= (x_1,\ldots,x_r) \\
\check{x}_i &:= x_1+\cdots+x_i \\
\hat{x}_i &:= x_i+\cdots+x_r \\
\|\boldsymbol{x}\| &:= x_1+\cdots+x_r = \hat{x}_1 = \check{x}_r \\
\sigma(x) &:= +1 \text{ if } x>0 \quad (\text{resp. } -1 \text{ if } x<0) \\
\delta &:= \text{dirac}
\end{aligned}$$

$$\begin{aligned}
\mathrm{sofo}^{\boldsymbol{x}}_{\pm} &:= (-1)^r \sigma_{\pm}(\check{x}_1)\ldots\sigma_{\pm}(\check{x}_r) \\
\mathrm{antisofo}^{\boldsymbol{x}}_{a} &:= (-1)^r \sigma_{\pm}(\check{x}_1)\ldots\sigma_{\pm}(\check{x}_r) \\
\mathrm{sefo}^{\boldsymbol{x}}_{\pm} &:= (-1)^{r-1} \sigma_{\pm}(\check{x}_1)\ldots\sigma_{\pm}(\check{x}_{r-1})\,\sigma_{\mp}(\check{x}_r) \\
\mathrm{antisefo}^{\boldsymbol{x}}_{\pm} &:= (-1)^{r-1} \sigma_{\mp}(\check{x}_1)\sigma_{\pm}(\check{x}_{r-1})\ldots\sigma_{\pm}(\check{x}_r) \\
\mathrm{lefo}^{\boldsymbol{x}}_{\pm} &:= (-1)^r \sigma_{\pm}(\check{x}_1)\ldots\sigma_{\pm}(\check{x}_{r-1})\delta(\check{x}_r) \\
\mathrm{antilefo}^{\boldsymbol{x}}_{\pm} &:= (-1)^r \delta(\check{x}_1)\sigma_{\pm}(\check{x}_{r-1})\ldots\sigma_{\pm}(\check{x}_r)
\end{aligned}$$

**Sum-type polar moulds**

Relevant to the construction of the organic averages and alien derivations. As usual $\breve{\omega}_i := \omega_1 + \cdots + \omega_i$ and $\hat{\omega}_i := \omega_i + \cdots + \omega_r$.

$$\mathrm{sa}^{\boldsymbol{\omega}} := \prod_{i=1}^{i=r} \frac{\omega_i}{\breve{\omega}_i} \qquad \mathrm{musa}^{\boldsymbol{\omega}} := (-1)^r \prod_{i=1}^{i=r} \frac{\omega_i}{\hat{\omega}_i}$$

$$\mathrm{romo}_a^{\boldsymbol{\omega}} := \prod_{i=1}^{i=r} \left( a\frac{\omega_i}{\breve{\omega}_i} - 1 \right) \qquad \mathrm{antiromo}_a^{\boldsymbol{\omega}} := \prod_{i=1}^{i=r} \left( a\frac{\omega_i}{\hat{\omega}_i} - 1 \right)$$

$$\mathrm{remo}_a^{\boldsymbol{\omega}} := a\frac{\omega_r}{\breve{\omega}_r} \prod_{i=1}^{i=r-1} \left( a\frac{\omega_i}{\breve{\omega}_i} - 1 \right) \qquad \mathrm{antiremo}_a^{\boldsymbol{\omega}} := a\frac{\omega_1}{\hat{\omega}_1} \prod_{i=2}^{i=r} \left( a\frac{\omega_i}{\hat{\omega}_i} - 1 \right)$$

$$\mathrm{somo}^{\bullet}_{a,b} := \mathrm{remo}^{\bullet}_a \times \mathrm{antiromo}^{\bullet}_{1-b} \tag{6.7}$$

$$:= \mathrm{romo}^{\bullet}_a \times \mathrm{antiremo}^{\bullet}_{1-b} \tag{6.8}$$

$$= \mathrm{romo}^{\bullet}_{a/b} \times \mathrm{remo}^{\bullet}_b \tag{6.9}$$

$$\mathrm{somo}^{\bullet}_{\left[\begin{smallmatrix} a & b \\ c & d \end{smallmatrix}\right]} := \mathrm{somo}^{\bullet}_{\frac{c-b}{d-b}, \frac{a-b}{d-b}} \tag{6.10}$$

$$\mathrm{somo}^{\bullet}_{\left[\begin{smallmatrix} b & 0 \\ a & 1 \end{smallmatrix}\right]} := \mathrm{somo}^{\bullet}_{a,b} \tag{6.11}$$

Identities:

$$\mathrm{remo}^{\bullet}_a \times \mathrm{antiromo}^{\bullet}_{1-a} = \mathbf{1}^{\bullet} \tag{6.12}$$

$$\mathrm{romo}^{\bullet}_a \times \mathrm{antiremo}^{\bullet}_{1-a} = \mathbf{1}^{\bullet} \tag{6.13}$$

| | | | | |
|---|---|---|---|---|
| multiplicative inverse: | $\mathrm{somo}^{\bullet}_{\left[\begin{smallmatrix} a & b \\ c & d \end{smallmatrix}\right]}$ | $\leftrightarrow$ | $\mathrm{somo}^{\bullet}_{\left[\begin{smallmatrix} c & b \\ a & d \end{smallmatrix}\right]}$ | ($a, c$ exchanged) |
| composition inverse: | $\mathrm{somo}^{\bullet}_{\left[\begin{smallmatrix} a & b \\ c & d \end{smallmatrix}\right]}$ | $\leftrightarrow$ | $\mathrm{somo}^{\bullet}_{\left[\begin{smallmatrix} b & a \\ d & c \end{smallmatrix}\right]}$ | (columns exchanged) |
| sequence reversal: | $\mathrm{somo}^{\bullet}_{\left[\begin{smallmatrix} a & b \\ c & d \end{smallmatrix}\right]}$ | $\overset{\mathbf{anti}}{\leftrightarrow}$ | $\mathrm{somo}^{\bullet}_{\left[\begin{smallmatrix} c & d \\ a & b \end{smallmatrix}\right]}$ | (rows exchanged) |

multiplication: $\mathrm{somo}^{\bullet}_{a_1,a_2} \times \mathrm{somo}^{\bullet}_{a_2,a_3} = \mathrm{somo}^{\bullet}_{a_1,a_3}$

composition: $\mathrm{somo}^{\bullet}_{a_1,b_1} \circ \mathrm{somo}^{\bullet}_{a_2,b_2} = \mathrm{somo}^{\bullet}_{(a_2-b_2)a_1+b_2,(a_2-b_2)b_1+b_2}$

multiplication: $\mathrm{somo}^{\bullet}_{\left[\begin{smallmatrix} a_0 & b_1 \\ a_2 & b_2 \end{smallmatrix}\right]} \times \mathrm{somo}^{\bullet}_{\left[\begin{smallmatrix} a_1 & b_1 \\ a_0 & b_2 \end{smallmatrix}\right]} = \mathrm{somo}^{\bullet}_{\left[\begin{smallmatrix} a_1 & b_1 \\ a_2 & b_2 \end{smallmatrix}\right]}$

composition: $\mathrm{somo}^{\bullet}_{\left[\begin{smallmatrix} a_1 & b_1 \\ a_2 & b_2 \end{smallmatrix}\right]} \circ \mathrm{somo}^{\bullet}_{\left[\begin{smallmatrix} b_1 & c_1 \\ b_2 & c_2 \end{smallmatrix}\right]} = \mathrm{somo}^{\bullet}_{\left[\begin{smallmatrix} a_1 & c_1 \\ a_2 & c_2 \end{smallmatrix}\right]}$

$$\begin{aligned} \mathrm{redo}^{\bullet}_a &:= -\square^{-1}(\mathrm{romo}^{\bullet}_{1-a} \times \square \mathrm{I}^{\bullet} \times \mathrm{antiromo}^{\bullet}_a) \\ &= -\square^{-1}(\mathrm{romo}^{\bullet}_{1-a} \times \square \mathrm{I}^{\bullet} \times (\mathbf{1}^{\bullet} + \mathrm{I}^{\bullet})^{-1} \times (\mathrm{romo}^{\bullet}_{1-a})^{-1}) \end{aligned}$$

**Symmetry types**

All the above moulds fall into one or the other of the main symmetry types.

*Alternal:* $\mathrm{lad}^\bullet$, $\mathrm{tas}^\bullet_\star$

*Symmetral:* $\mathrm{sad}^\bullet$, $\mathrm{tas}^\bullet_{a,b}$, $\mathrm{sa}^\bullet$, $\mathrm{musa}^\bullet$

*Alternel:* $\mathrm{lefo}^\bullet_\pm$, $\mathrm{redo}^\bullet_\pm$

*Symmetrel:* $\mathrm{tu}^\bullet_a$, $\mathrm{sofo}^\bullet_\pm$, $\mathrm{sefo}^\bullet_\pm$, $\mathrm{romo}^\bullet_a$, $\mathrm{remo}^\bullet_a$

Moulds $\mathrm{something}^\bullet$ and $\mathrm{antisomething}^\bullet$ have the same symmetry type.

**Smooth arborification**

All the above moulds possess the property of smooth arborification (meaning that their arborified variants admit essentially the same type of bounds) the only exception being the mould $\mathrm{tu}^\bullet_a$ for $a \notin \mathbb{Z}$ and in particular for $a = 1/2$. This is in relation to the fact that the *standard* or *median* average (which admits $\mathrm{tu}^\bullet_{1/2}$ as its right- *and* left-lateral mould) is not *well-behaved.*

Of course, for alternal or symmetral (resp. alternel or symmetrel) moulds, one should take the ordinary (resp. contracting) form of arborification.

**Form-preserving arborification**

All the sum-type moulds listed above, ie all those moulds whose definition involves forward sums $\hat{x}_i$ or $\hat{\omega}_i$ (resp. backward sums $\check{x}_i$ or $\check{\omega}_i$) have the stronger and very useful property of form-preserving arborification. This means that they retain their *outward analytical expression,* except that the sums $\hat{x}_i$ or $\hat{\omega}_i$ (resp. $\check{x}_i$ or $\check{\omega}_i$) are now relative to the arborescent (resp. antiarborescent) order. The same holds for the difference-type moulds $\mathrm{tas}^\bullet_{a,\infty}$ and $\mathrm{tas}^\bullet_{\infty,b}$.

## 6.2 Resurgent functions

**Minors/majors:**

Real-majors and natural-majors

$$\hat{\varphi}(\zeta) = -\frac{1}{2\pi i}(\check{\varphi}_{\mathrm{real}}(e^{\pi i}\zeta) - \check{\varphi}_{\mathrm{real}}(e^{-\pi i}\zeta)) \tag{6.14}$$

$$\hat{\varphi}(\zeta) = \check{\varphi}_{\mathrm{nat}}(\zeta) - \check{\varphi}_{\mathrm{nat}}(e^{-2\pi i}\zeta) \tag{6.15}$$

$$\check{\varphi}_{\mathrm{real}}(\zeta) \equiv 2\pi i\check{\varphi}_{\mathrm{nat}}(e^{-\pi i}\zeta) \tag{6.16}$$

The formulae below use real-majors.

**Standard Borel transform:**

$\varphi(z) \to \overset{\diamond}{\varphi}(\zeta) = (\hat{\varphi}(\zeta), \check{\varphi}(\zeta))$

$$\hat{\varphi}(\zeta) = \frac{1}{2\pi i}\int_{c-i\infty}^{c+i\infty} \exp(z\zeta)\varphi(z)\,dz \qquad (1 \ll c;\ \arg\zeta = 0) \tag{6.17}$$

$$\check{\varphi}(\zeta) = \int_{+u}^{+\infty} \exp(-z\zeta)\varphi(z)\,dz \qquad (1 \ll u;\ |\arg\zeta| \leq \pi) \tag{6.18}$$

**Standard Laplace transform:**

$\overset{\diamond}{\varphi}(\zeta) = (\hat{\varphi}(\zeta), \check{\varphi}(\zeta)) \to \varphi(z)$

$$\varphi(z) = \int_{+0}^{+\infty} \exp(-z\zeta)\hat{\varphi}(\zeta)\,d\zeta \qquad (\text{for } \overset{\diamond}{\varphi} \text{ integrable at } 0_\bullet) \tag{6.19}$$

$$\varphi(z) = \frac{1}{2\pi i}\int_{e^{-\pi i}\infty}^{e^{\pi i}\infty} \exp(z\zeta)\check{\varphi}(\zeta)\,d\zeta \qquad (\text{for any } \overset{\diamond}{\varphi};\ \arg z = 0) \tag{6.20}$$

**Elementary (standard) Borel/Laplace transforms:**

$$\begin{aligned} &\varphi(z) = z^{-\sigma} \qquad (\text{for } \sigma \in \mathbb{C} - \mathbb{N}^\star)\\ &\hat{\varphi}(\zeta) = \zeta^{\sigma-1}/\Gamma(\sigma)\\ &\check{\varphi}(\zeta) = \zeta^{\sigma-1}\Gamma(1-\sigma) \end{aligned} \tag{6.21}$$

$$\begin{aligned} &\varphi(z) = z^{-n} \qquad (\text{for } n \in \mathbb{N}^\star)\\ &\hat{\varphi}(\zeta) = \zeta^{n-1}/\Gamma(n)\\ &\check{\varphi}(\zeta) = (-1)^n\zeta^{n-1}\log\zeta/\Gamma(n) \end{aligned} \tag{6.22}$$

## 6.3 Alien derivations

$$\Delta_\omega\hat{\varphi}(\zeta) := \sum_{\epsilon_i=\pm} \mathbf{d}^{\binom{\epsilon_1,\ldots,\epsilon_r}{\omega_1,\ldots,\omega_r}}\hat{\varphi}^{\binom{\epsilon_1,\ldots,\epsilon_r}{\omega_1,\ldots,\omega_r}}(\zeta+\omega) \tag{6.23}$$

Standard alien derivations (weights $\mathbf{d} = \mathbf{dun}$):

$$\mathbf{d}^{\binom{\epsilon_1,\ldots,\epsilon_r}{\omega_1,\ldots,\omega_r}} := \epsilon_r \frac{p!\,q!}{(p+q+1)!} \qquad \begin{aligned} p &:= \#\{1 \leq i < r;\ \epsilon_i = +\}\\ q &:= \#\{1 \leq i < r;\ \epsilon_i = -\}\end{aligned} \tag{6.24}$$

Organic alien derivations (weights $\mathbf{d} = \mathbf{dom}$):

$$\begin{aligned} &\mathbf{dun}^{\binom{\epsilon_1,\ldots,\epsilon_r}{\omega_1,\ldots,\omega_r}} := \tfrac{\epsilon_r}{2\pi i}\tfrac{p!\,q!}{r!} \quad \text{with} \quad p := \#\{\epsilon_i = +,\ i \leq r-1\}\\ &\qquad\qquad\qquad\qquad\qquad \text{and} \quad q := \#\{\epsilon_i = -,\ i \leq r-1\}\\ &\mathbf{dom}^{\binom{\epsilon_1,\ldots,\epsilon_r}{\omega_1,\ldots,\omega_r}} := \begin{cases} \frac{\epsilon_r}{2}\frac{\omega_{p+1}}{\omega_1+\cdots+\omega_r} & \text{if } (\epsilon_1,\ldots,\epsilon_r) = ((+)^p,(-)^q,\epsilon_r)\\ \frac{\epsilon_r}{2}\frac{\omega_{q+1}}{\omega_1+\cdots+\omega_r} & \text{if } (\epsilon_1,\ldots,\epsilon_r) = ((-)^q,(+)^p,\epsilon_r)\\ 0 & \text{otherwise} \end{cases} \end{aligned} \tag{6.25}$$

## 6.4 Convolution averages

$$(\mathbf{m}.\hat{\varphi})(\zeta) := \sum_{\epsilon_i=\pm} \mathbf{m}^{\binom{\epsilon_1,\ldots,\epsilon_r}{\omega_1,\ldots,\omega_r}} \hat{\varphi}^{\binom{\epsilon_1,\ldots,\epsilon_r}{\omega_1,\ldots,\omega_r}}(\zeta) \tag{6.26}$$

Standard convolution average (weights $\mathbf{m} = \mathbf{mun}$):

$$\mathbf{m}^{\binom{\epsilon_1,\ldots,\epsilon_r}{\omega_1,\ldots,\omega_r}} := \frac{(2p)!\,(2q)!}{4^{p+q}(p+q)!\,p!\,q!} \qquad (p := \text{nb of } +;\ q := \text{nb of } -) \tag{6.27}$$

Organic convolution average (weights $\mathbf{m} = \mathbf{mon}$):

$$\mathbf{m}^{\binom{\epsilon_1,\ldots,\epsilon_r}{\omega_1,\ldots,\omega_r}} := \mathbf{m}^{\binom{\epsilon_1,\ldots,\epsilon_{r-1}}{\omega_1,\ldots,\omega_{r-1}}}.\left(1 - \frac{1}{2}\frac{\omega_r}{\omega_1+\cdots+\omega_r}\right) \quad \text{if} \quad \epsilon_{r-1} = \epsilon_r \tag{6.28}$$

$$\mathbf{m}^{\binom{\epsilon_1,\ldots,\epsilon_r}{\omega_1,\ldots,\omega_r}} := \mathbf{m}^{\binom{\epsilon_1,\ldots,\epsilon_{r-1}}{\omega_1,\ldots,\omega_{r-1}}}.\frac{1}{2}\frac{\omega_r}{\omega_1+\cdots+\omega_r} \quad \text{if} \quad \epsilon_{r-1} \neq \epsilon_r \tag{6.29}$$

## 6.5 Comparative tables

$$\mathbf{m}^{\bar{\epsilon}_1,\ldots,\bar{\epsilon}_r} \equiv +\mathbf{m}^{\epsilon_1,\ldots,\epsilon_r}; \qquad \mathbf{d}^{\bar{\epsilon}_1,\ldots,\bar{\epsilon}_r} \equiv -\mathbf{d}^{\epsilon_1,\ldots,\epsilon_r} \tag{6.30}$$

| object | average | average | derivation | derivation | derivation |
|---|---|---|---|---|---|
| species | uniform | organic | uniform | organic | organic |
| nature | "bad" | "good" | "bad" | "good" | "good" |
| name | **mun** | **mun** | **dun** | **don** | **dom** |
| (+) | 1/2 | 1/2 | 1 | 1 | 1 |
| sums | **1/2** | **1/2** | **1** | **1** | **1** |
| (+,+) | 3/8 | 3/8 | 1/2 | 1/2 | 1/2 |
| (−,+) * | 1/8 | 1/8 | 1/2 | 1/2 | 1/2 |
| sums | **1/2** | **1/2** | **1** | **1** | **1** |
| (+,+,+) | 5/16 | 5/16 | 1/3 | 1/3 | 1/3 |
| (−,+,+) | 1/16 | 5/48 | 1/6 | 1/6 | 1/6 |
| (+,−,+) * | 1/16 | 1/48 | 1/6 | 1/6 | 1/6 |
| (−,−,+) | 1/16 | 1/16 | 1/3 | 1/3 | 1/3 |
| sums | **1/2** | **1/2** | **1** | **1** | **1** |
| (+,+,+,+) | 35/128 | 35/128 | 1/4 | 1/4 | 1/4 |
| (−,+,+,+) | 5/128 | 35/384 | 1/12 | 5/48 | 1/8 |
| (+,−,+,+) | 5/128 | 7/384 | 1/12 | 1/24 | 0 |
| (−,−,+,+) | 3/128 | 7/128 | 1/12 | 5/48 | 1/8 |
| (+,+,−,+) | 5/128 | 1/128 | 1/12 | 5/48 | 1/8 |
| (−,+,−,+) * | 3/128 | 1/384 | 1/12 | 1/24 | 0 |
| (+,−,−,+) | 3/128 | 5/384 | 1/12 | 5/48 | 1/8 |
| (−,−,−,+) | 5/128 | 5/128 | 1/4 | 1/4 | 1/4 |
| sums | **1/2** | **1/2** | **1** | **1** | **1** |

| name | **mun** | **mun** | **dun** | **don** | **dom** |
|---|---|---|---|---|---|
| $(+,+,+,+,+)$ | 63/256 | 63/256 | 1/5 | 1/5 | 1/5 |
| $(-,+,+,+,+)$ | 7/256 | 21/256 | 1/20 | 37/480 | 1/10 |
| $(+,-,+,+,+)$ | 7/256 | 21/1280 | 1/20 | 11/480 | 0 |
| $(-,-,+,+,+)$ | 3/256 | 63/1280 | 1/30 | 1/16 | 1/10 |
| $(+,+,-,+,+)$ | 7/256 | 9/1280 | 1/20 | 11/480 | 0 |
| $(-,+,-,+,+)$ | 3/256 | 3/1280 | 1/30 | 1/120 | 0 |
| $(+,-,-,+,+)$ | 3/256 | 3/256 | 1/30 | 7/240 | 0 |
| $(-,-,-,+,+)$ | 3/256 | 9/256 | 1/20 | 37/480 | 1/10 |
| $(+,+,+,-,+)$ | 7/256 | 1/256 | 1/20 | 37/480 | 1/10 |
| $(-,+,+,-,+)$ | 3/256 | 1/768 | 1/30 | 7/240 | 0 |
| $(+,-,+,-,+)$ * | 3/256 | 1/3840 | 1/30 | 1/120 | 0 |
| $(-,-,+,-,+)$ | 3/256 | 1/1280 | 1/20 | 11/480 | 0 |
| $(+,+,-,-,+)$ | 3/256 | 7/1280 | 1/30 | 1/16 | 1/10 |
| $(-,+,-,-,+)$ | 3/256 | 7/3840 | 1/20 | 11/480 | 0 |
| $(+,-,-,-,+)$ | 3/256 | 7/768 | 1/20 | 37/480 | 1/10 |
| $(-,-,-,-,+)$ | 7/256 | 7/256 | 1/5 | 1/5 | 1/5 |
| sums | **1/2** | **1/2** | **1** | **1** | **1** |
| $(+,+,+,+,+,+)$ | 231/1024 | 231/1024 | 1/6 | 1/6 | 1/6 |
| $(-,+,+,+,+,+)$ | 21/1024 | 77/1024 | 1/30 | 71/1152 | 1/12 |
| $(+,-,+,+,+,+)$ | 21/1024 | 77/5120 | 1/30 | 23/1440 | 0 |
| $(-,-,+,+,+,+)$ | 7/1024 | 231/5120 | 1/60 | 29/640 | 1/12 |
| $(+,+,-,+,+,+)$ | 21/1024 | 33/5120 | 1/30 | 11/960 | 0 |
| $(-,+,-,+,+,+)$ | 7/1024 | 11/5120 | 1/60 | 23/5760 | 0 |
| $(+,-,-,+,+,+)$ | 7/1024 | 11/1024 | 1/60 | 47/2880 | 0 |
| $(-,-,-,+,+,+)$ | 5/1024 | 33/1024 | 1/60 | 29/640 | 1/12 |
| $(+,+,+,-,+,+)$ | 21/1024 | 11/3072 | 1/30 | 23/1440 | 0 |
| $(-,+,+,-,+,+)$ | 7/1024 | 11/9216 | 1/60 | 11/1920 | 0 |
| $(+,-,+,-,+,+)$ | 7/1024 | 11/46080 | 1/60 | 1/720 | 0 |
| $(-,-,+,-,+,+)$ | 5/1024 | 11/15360 | 1/60 | 23/5760 | 0 |
| $(+,+,-,-,+,+)$ | 7/1024 | 77/15360 | 1/60 | 47/2880 | 0 |
| $(-,+,-,-,+,+)$ | 5/1024 | 77/46080 | 1/60 | 11/1920 | 0 |
| $(+,-,-,-,+,+)$ | 5/1024 | 77/9216 | 1/60 | 13/576 | 0 |
| $(-,-,-,-,+,+)$ | 7/1024 | 77/3072 | 1/30 | 71/1152 | 1/12 |
| $(+,+,+,+,-,+)$ | 21/1024 | 7/3072 | 1/30 | 71/1152 | 1/12 |
| $(-,+,+,+,-,+)$ | 7/1024 | 7/9216 | 1/60 | 13/576 | 0 |
| $(+,-,+,+,-,+)$ | 7/1024 | 7/46080 | 1/60 | 11/1920 | 0 |
| $(-,-,+,+,-,+)$ | 5/1024 | 7/15360 | 1/60 | 47/2880 | 0 |
| $(+,+,-,+,-,+)$ | 7/1024 | 1/15360 | 1/60 | 23/5760 | 0 |
| $(-,+,-,+,-,+)$ * | 5/1024 | 1/46080 | 1/60 | 1/720 | 0 |
| $(+,-,-,+,-,+)$ | 5/1024 | 1/9216 | 1/60 | 11/1920 | 0 |
| $(-,-,-,+,-,+)$ | 7/1024 | 1/3072 | 1/30 | 23/1440 | 0 |
| $(+,+,+,-,-,+)$ | 7/1024 | 3/1024 | 1/60 | 29/640 | 1/12 |

| | | | | | |
|---|---|---|---|---|---|
| $(-,+,+,-,-,+)$ | 5/1024 | 1/1024 | 1/60 | 47/2880 | 0 |
| $(+,-,+,-,-,+)$ | 5/1024 | 1/5120 | 1/60 | 23/5760 | 0 |
| $(-,-,+,-,-,+)$ | 7/1024 | 3/5120 | 1/30 | 11/960 | 0 |
| $(+,+,-,-,-,+)$ | 5/1024 | 21/5120 | 1/60 | 29/640 | 1/12 |
| $(-,+,-,-,-,+)$ | 7/1024 | 7/5120 | 1/30 | 23/1440 | 0 |
| $(+,-,-,-,-,+)$ | 7/1024 | 7/1024 | 1/30 | 71/1152 | 1/12 |
| $(-,-,-,-,-,+)$ | 21/1024 | 21/1024 | 1/6 | 1/6 | 1/6 |
| sums | **1/2** | **1/2** | **1** | **1** | **1** |

## 6.6 Difference between "good" and "bad"

We first set some notations:

$$\text{cnb}(r_1, r_2) := \sum_{k=0}^{k=r_2} (-1)^k \frac{r_2!}{k!\,(r_2-k)!} (r_2-k)^{r_1} \tag{6.31}$$

In particular:

$$\text{cnb}(r,1) = 1 \tag{6.32}$$

$$\text{cnb}(r,r) = r! \tag{6.33}$$

$$\text{cnb}(r,r_\star) \geq 2 \quad \text{if} \quad 1 < r_\star < r \tag{6.34}$$

$$\text{cnb}(r,r_\star) = 0 \quad \text{if} \quad r < r_\star \tag{6.35}$$

Then we set:

$$\text{f}(r_0, r_1) = \sum_{r_2=1}^{r_2=r_1} \text{f}(r_0+r_2)\text{cnb}(r_1, r_2) \tag{6.36}$$

Next, let $\boldsymbol{\omega}^{\leftarrowtail}_{r_0,r_1}$ (resp. $\boldsymbol{\omega}^{\rightarrowtail}_{r_0,r_1}$) be the arborescent (resp. antiarborescent) sequence obtained by *suffixing* (resp. *prefixing*) the *totally non-ordered* sequence $(\omega'_1,\ldots,\omega'_{r_1})$ to the *totally ordered* sequence $(\omega_1,\ldots,\omega_{r_0})$.

Assume now that $F^\bullet$ is some *constant-type mould* like $\text{tu}^\bullet_a$ (Section 6.1.2 *supra*), i.e. a mould whose values depend solely on the sequence length $r$, so that $F^{\omega_1,\ldots,\omega_r} \equiv f(r)$. In view of what precedes, it is clear that after a contracting arborification or antiarborification we get:

$$\text{F}^{\boldsymbol{\omega}^{\leftarrowtail}_{r_0,r_1}} \equiv \text{F}^{\boldsymbol{\omega}^{\rightarrowtail}_{r_0,r_1}} \equiv \text{f}(r_0,r_1) \quad \text{with} \quad \text{f}(r_0,r_1) \text{ as in (6.36)} \tag{6.37}$$

If we take $F^\bullet := \text{tu}^\bullet_a$ with $a \in \mathbb{Z}$, then $\text{tu}^\bullet_a$ is well-behaved and indeed we can see (trivially for $a < 0$, less so for $a > 0$) that:

$$\limsup_{r_1 \to +\infty} \left( \frac{\log |\text{f}(r_0,r_1)|}{r_0 + r_1} \right) < +\infty \qquad (\forall r_0 \text{ fixed and } \geq 1) \tag{6.38}$$

But if $a \notin \mathbb{Z}$, then $\text{tu}^\bullet_a$ is not well-behaved and we can show that:

$$\limsup_{r_1 = +\infty} \left( \frac{\log |\text{f}(r_0,r_1)|}{r_0 + r_1} \right) = +\infty \qquad (\forall r_0 \text{ fixed and } \geq 1) \tag{6.39}$$

and in fact:

$$\limsup_{r_1 \to +\infty} \left( \frac{\log |\text{f}(r_0,r_1)|}{(r_0+r_1)\log(r_0+r_1)} \right) > 0 \qquad (\forall r_0 \text{ fixed and } \geq 1) \tag{6.40}$$

## 6.7 Resurgence monomials

### 6.7.1 Hyperlogarithms, perilogarithms, paralogarithms

*Perilogarithms* have indices $\varpi_i = (\omega_i, \omega_i^\star) \in (\mathbb{C}^\star, \mathbb{C}^\star)$.
*Hyperlogarithms* have indices $\omega_i \in \mathbb{C}^\star$.
*Perilogarithms* have indices $\varpi_i = (\omega_i, \omega_i^\star) \in (\mathbb{C}^\star, \mathbb{C}^\star)$ with $\omega_i \omega_i^\star \in \mathbb{R}^+$.
*Perilogarithms* have indices $\varpi_i = (\omega_i, c^2 \bar{\omega}_i) \in (\mathbb{C}^\star, \mathbb{C}^\star)$.
Usually $c$ is fixed, so that only $\omega_i$ is mentioned.

### 6.7.2 $D$- or $\Delta$-friendly monomials and monics

*Monics* depend only on the indices $\omega_i$ or $\varpi_i$.
*Monomials* depend on a variable $z$ as well.
$D$-friendly monomials behave simply under ordinary $z$-differentiation, but less so under alien $z$-differentiation: their alien derivatives necessarily involve a number of so-called $D$-friendly monics.
$\Delta$-friendly monomials behave simply under alien $z$-differentiation, but less so under ordinary $z$-differentiation: their ordinary derivatives necessarily involve a number of so-called $\Delta$-friendly monics.
$D$-friendly (resp $\Delta$-friendly) monomials all carry a calligraphic $\mathcal{V}$ (resp. $\mathcal{U}$) in their names while the corresponding monics carry an upper-case $V$ (resp. $U$), sometimes supplemented by a suitable string of pre- or suffixes.

### 6.7.3 Total closure

The monomials, as functions of $z$, are acted upon by *one* ordinary derivation $D := \partial_z$ but by *infinitely many* independent alien derivations $\Delta_{\omega_i}$ or their exponential-carrying variants $\mathbf{\Delta}_\omega := e^{-\omega z} \Delta_\omega$. The latter have the advantage of commuting with the ordinary derivation $D$, but at the cost of introducing an exponential factor and thus ceasing to act internally on the ring of formal power series of $z^{-1}$.

To highlight the $D \leftrightarrow \Delta$ duality, it is sometimes convenient to (formally) regroup all alien derivations under one single symbol:

$$\Delta := \sum \Delta_\omega; \qquad \mathbf{\Delta} := \sum \mathbf{\Delta}_\omega$$

We also require the following mould derivations:

$$\begin{aligned}
\square A^{\varpi} &:= \|\omega\| \, A^{\varpi} := \left(\sum \omega_i\right) A^{\varpi} \\
\square^\star A^{\varpi} &:= \|\omega^\star\| \, A^{\varpi} := \left(\sum \omega_i^\star\right) A^{\varpi} \\
\square_{\omega_i} &:= [\partial_{\omega_i}, \square] \\
\square^\star_{\omega_i} &:= [\partial_{\omega_i^\star}, \square^\star]
\end{aligned}$$

### 6.7.4 Hyperlogarithmic monomials and monics

| | | |
|---|---|---|
| $\Delta$-friendly monomials | $\mathcal{U}^\bullet(z), \mathcal{U}e^\bullet(z)$ | symmetral |
| $\Delta$-friendly monics | $U^\bullet$ | alternal |
| $\Delta$-friendly monics | $US^\bullet, SU^\bullet$ | symmetral |
| $D$-friendly monomials | $\mathcal{V}^\bullet(z), \mathcal{V}e^\bullet(z)$ | symmetral |
| $D$-friendly monics | $V^\bullet$ | alternal |
| $D$-friendly monics | $VS^\bullet, SV^\bullet$ | symmetral |

### 6.7.5 Basic relations

$$\mathcal{U}e^\bullet(z) = \exp(z\,\Box)\,.\,\mathcal{U}^\bullet(z)$$
$$\mathcal{V}e^\bullet(z) = \exp(z\,\Box)\,.\,\mathcal{V}^\bullet(z)$$

$$\mathcal{U}^\bullet(z) = \mathcal{V}^\bullet(z) \circ U^\bullet$$
$$\mathcal{V}^\bullet(z) = \mathcal{U}^\bullet(z) \circ V^\bullet$$

$$\mathbf{1}^\bullet = U^\bullet \circ V^\bullet = V^\bullet \circ U^\bullet$$
$$\mathbf{1}^\bullet = US^\bullet \times SU^\bullet = VS^\bullet \times SV^\bullet$$

$$U^\bullet = US^\bullet \times I^\bullet \times SU^\bullet \quad \text{if all} \quad \omega_i \in \mathbb{R}^+$$
$$V^\bullet = VS^\bullet \times I^\bullet \times SV^\bullet \quad \text{if all} \quad \omega_i \in \mathbb{R}^+$$

### 6.7.6 More relations

$$\partial_{\omega_i}\mathcal{U}^\bullet(z) = -\mathcal{U}^\bullet(z) \times \left(\frac{\Box_i}{\Box}U^\bullet\right) - z\,\Box_i\,\mathcal{U}^\bullet(z) \tag{6.41}$$
$$z\,\partial_z\mathcal{U}^\bullet(z) = -z\,\Box\,\mathcal{U}^\bullet(z) - \mathcal{U}^\bullet(z) \times U^\bullet \tag{6.42}$$

$$\partial_{\omega_i}\mathcal{U}e^\bullet(z) = -\mathcal{U}e^\bullet(z) \times \left(\exp(z\,\Box)\,.\,\frac{\Box_i}{\Box}\,.\,U^\bullet\right) \tag{6.43}$$
$$z\,\partial_z\mathcal{U}e^\bullet(z) = -\mathcal{U}^\bullet(z) \times (\exp(z\,\Box)\,.\,U^\bullet) \tag{6.44}$$

$$\partial_{\omega_i}U^\bullet = +\left(\frac{\Box_i}{\Box}U^\bullet\right) \times U^\bullet - U^\bullet \times \left(\frac{\Box_i}{\Box}U^\bullet\right) \tag{6.45}$$
$$\partial_{\omega_i}US^\bullet = +\left(\frac{\Box_i}{\Box}U^\bullet\right) \times US^\bullet \tag{6.46}$$
$$\partial_{\omega_i}SU^\bullet = -U^\bullet \times \left(\frac{\Box_i}{\Box}SU^\bullet\right) \tag{6.47}$$

$\partial_\omega := \sum \omega_i \partial_{\omega_i}$

$$\partial_\omega U^\bullet = 0 \tag{6.48}$$
$$\partial_\omega US^\bullet = +U^\bullet \times US^\bullet \tag{6.49}$$
$$\partial_\omega SU^\bullet = -SU^\bullet \times U^\bullet \tag{6.50}$$

### 6.7.7 Ordinary and alien differentiation. The $D \leftrightarrow \Delta$ duality

$$(z\,\partial_z + z\,\Box)\mathcal{U}^\bullet(z) = -\mathcal{U}^\bullet(z) \times U^\bullet \tag{6.51}$$
$$(z\,\partial_z + z\,\Box)\mathcal{V}^\bullet(z) = -\mathcal{V}^\bullet(z) \times I^\bullet \tag{6.52}$$
$$\Delta\mathcal{U}^\bullet(z) = I^\bullet \times \mathcal{U}^\bullet(z) \tag{6.53}$$
$$\Delta\mathcal{V}^\bullet(z) = V^\bullet \times \mathcal{V}^\bullet(z) \tag{6.54}$$

with $\Box$ standing as usual for multiplication by $\| \bullet \| = \sum \omega_i$ and

$$I^{\omega_1} := 1; \qquad I^{\omega_1,\dots,\omega_r} := 0 \quad \text{if} \quad r \neq 1 \tag{6.55}$$

### 6.7.8 Perilogarithmic monomials and monics

*Perilogarithms* have indices $\boldsymbol{\varpi} := (\varpi_1, \dots, \varpi_r)$ with $\varpi_i = (\omega_i, \omega_i^\star) \in (\mathbb{C}^\star, \mathbb{C}^\star)$ but with products $\omega_i \omega_i^\star \in \mathbb{R}^+$.
Antipodal involution:

$$\# : \quad {}^{\#}M^{\varpi_1,\dots,\varpi_r} := M^{\varpi_r^\star,\dots,\varpi_1^\star} \quad \text{with} \quad \varpi^\star = (\omega_i^\star, \omega_i) \quad \text{if} \quad \varpi = (\omega_i, \omega_i^\star)$$

$\Delta$-frienly perilogarithms:

| | primary | secundary | type |
|---|---|---|---|
| monomials | $\mathcal{U}a^\bullet$ | $\mathcal{U}^\bullet, \mathcal{U}e^\bullet$ | symmetral |
| monics | $U^\bullet$ | $UR^\bullet, UL^\bullet$ | alternal |
| monics | $USS^\bullet, SSU^\bullet$ | $US^\bullet, SU^\bullet$ | symmetral |

$D$-frienly perilogarithms:

| | primary | secundary | type |
|---|---|---|---|
| monomials | $\mathcal{V}a^\bullet$ | $\mathcal{V}^\bullet, \mathcal{V}e^\bullet$ | symmetral |
| monics | $V^\bullet$ | $VR^\bullet, VL^\bullet$ | alternal |
| monics | $VSS^\bullet, SSV^\bullet$ | $VS^\bullet, SV^\bullet$ | symmetral |

### 6.7.9 Basic relations

$$\mathcal{U}e^\bullet(z) = \exp(z\,\Box + z^{-1}\,\Box^\star)\,.\,\mathcal{U}a^\bullet(z) \tag{6.56}$$
$$\mathcal{U}^\bullet(z) = \exp(z^{-1}\,\Box^\star)\,.\,\mathcal{U}a^\bullet(z) \tag{6.57}$$
$$\mathbf{1}^\bullet = USS^\bullet \times SSU^\bullet = US^\bullet \times SU^\bullet \tag{6.58}$$
$$UR^\bullet = USS^\bullet \times (\Box^\star\, SSU^\bullet) \tag{6.59}$$
$$UL^\bullet = USS^\bullet \times (\Box\, SSU^\bullet) \tag{6.60}$$
$$USS^\bullet = US^\bullet \times {}^{\#}US^\bullet \tag{6.61}$$
$$SSU^\bullet = {}^{\#}SU^\bullet \times SU^\bullet \tag{6.62}$$
$$SU^\bullet = \mathcal{U}a^\bullet(1) \tag{6.63}$$

### 6.7.10 Integral formulae for the Δ-friendly monomials and monics

The main ingredients of the integral formulae are the SPA or "Standard Path Averaging" (see Section 3.6.4) and the CCI or "Common Core Integrand":

$$\mathrm{CCI} := \frac{1}{(2\pi i)^r} \frac{\exp(-\sum \omega_i t_i - \sum \omega_i^\star / t_i)}{(t_r - t_{r-1}) \dots (t_3 - t_2)(t_2 - t_1)} \tag{6.64}$$

and

**Monomials:**

$$\begin{aligned}
\mathcal{U}a^{\varpi}(z) &= \mathrm{SPA} \int_0^\infty \mathrm{CCI}(t_1 - z)^{-1}\, dt_1 \dots dt_r \\
\mathcal{U}e^{\varpi}(z) &= e^{\|\omega\| z + \|\omega^\star\| z^{-1}} \mathcal{U}a^{\varpi}(z) \\
\mathcal{U}^{\varpi}(z) &= e^{\|\omega^\star\| z^{-1}} \mathcal{U}a^{\varpi}(z)
\end{aligned}$$

**Monics:**

$$\begin{aligned}
U^{\varpi} &= \mathrm{SPA} \int_0^\infty \mathrm{CCI} \left(\sum \omega_i\right) dt_1 \dots dt_r \\
&= \mathrm{SPA} \int_0^\infty \mathrm{CCI} \left(\sum \omega_i^\star / t_i^2\right) dt_1 \dots dt_r \\
UL^{\varpi} &= \mathrm{SPA} \int_0^\infty \mathrm{CCI} \left(\sum \omega_i / t_i\right) dt_1 \dots dt_r \\
UR^{\varpi} &= \mathrm{SPA} \int_0^\infty \mathrm{CCI} \left(\sum \omega_i^\star / t_i\right) dt_1 \dots dt_r \\
SU^{\varpi} &= \mathrm{SPA} \int_0^\infty \mathrm{CCI}(t_1 - 1)^{-1}\, dt_1 \dots dt_r \\
US^{\varpi} &= \mathrm{SPA} \int_0^\infty \mathrm{CCI}(1 - t_r)^{-1}\, dt_1 \dots dt_r \\
SSU^{\varpi} &= \mathrm{SPA} \int_0^\infty \mathrm{CCI}(1/t_1)\, dt_1 \dots dy_r \\
USS^{\varpi} &= \mathrm{SPA} \int_0^\infty \mathrm{CCI}(-1/t_r)\, dt_1 \dots dt_r
\end{aligned}$$

### 6.7.11 More relations for the Δ-friendly perilogarithms

**Monomials:**

$$\partial_{\omega_i} \mathcal{U}a^\bullet(z) = -\mathcal{U}a^\bullet(z) \times \frac{\square_i}{\square} U^\bullet - z\, \square_i \mathcal{U}a^\bullet(z) \tag{6.65}$$

$$\partial_{\omega_i^\star} \mathcal{U}a^\bullet(z) = +z^{-1} \mathcal{U}a^\bullet(z) \times USS^\bullet \times \square_i^\star SSU^\bullet - z^{-1}\, \square_i^\star \mathcal{U}a^\bullet(z) \tag{6.66}$$

$$z\, \partial_z \mathcal{U}a^\bullet(z) = (-z\, \square + z^{-1}\, \square^\star) \mathcal{U}a^\bullet(z) - \mathcal{U}a^\bullet(z) \times (U^\bullet + z^{-1} UR^\bullet) \tag{6.67}$$

**Monics:**

$$\partial_{\omega_i} U^\bullet = +\left(\frac{\square_i}{\square} U^\bullet\right) \times U^\bullet - U^\bullet \times \left(\frac{\square_i}{\square} U^\bullet\right) - \square_i \, UR^\bullet \tag{6.68}$$
$$\partial_{\omega_i^\star} U^\bullet = +\square((\square_i^\star \, USS^\bullet) \times SSU^\bullet) \tag{6.69}$$

$$\partial_{\omega_i} USS^\bullet = +\left(\frac{\square_i}{\square} U^\bullet\right) \times USS^\bullet \tag{6.70}$$
$$\partial_{\omega_i^\star} USS^\bullet = +USS^\bullet \times \left(\frac{\square_i^\star}{\square^\star} \, {}^\# U^\bullet\right) \tag{6.71}$$
$$\partial_{\omega_i} SSU^\bullet = -SSU^\bullet \times \left(\frac{\square_i}{\square} U^\bullet\right) \tag{6.72}$$
$$\partial_{\omega_i^\star} SSU^\bullet = -\left(\frac{\square_i^\star}{\square^\star} \, {}^\# U^\bullet\right) \times SSU^\bullet \tag{6.73}$$

$$\partial_{\omega_i} US^\bullet = +\left(\frac{\square_i}{\square} U^\bullet\right) \times US^\bullet - \square_i \, US^\bullet \tag{6.74}$$
$$\partial_{\omega_i^\star} US^\bullet = +US^\bullet \times (\square_i^\star \, {}^\# US^\bullet) \times SU^\bullet \tag{6.75}$$
$$\partial_{\omega_i} {}^\# US^\bullet = +SU^\bullet \times (\square_i \, US^\bullet) \times {}^\# US^\bullet \tag{6.76}$$
$$\partial_{\omega_i^\star} {}^\# US^\bullet = +{}^\# US^\bullet \times \left(\frac{\square_i^\star}{\square^\star} \, {}^\# U^\bullet\right) - \square_i^\star \, {}^\# US^\bullet \tag{6.77}$$

$$\partial_{\omega_i} SU^\bullet = -SU^\bullet \times \left(\frac{\square_i}{\square} U^\bullet\right) - \square_i \, SU^\bullet \tag{6.78}$$
$$\partial_{\omega_i^\star} SU^\bullet = +{}^\# US^\bullet \times (\square_i^\star \, {}^\# SU^\bullet) \times SU^\bullet \tag{6.79}$$
$$\partial_{\omega_i} {}^\# SU^\bullet = +{}^\# SU^\bullet \times (\square_i \, SU^\bullet) \times US^\bullet \tag{6.80}$$
$$\partial_{\omega_i^\star} {}^\# SU^\bullet = -\left(\frac{\square_i^\star}{\square^\star} \, {}^\# U^\bullet\right) \times {}^\# SU^\bullet - \square_i^\star \, {}^\# SU^\bullet \tag{6.81}$$

### 6.7.12 Yet more relations for the Δ-friendly perilogarithms

The partial differentiation rules relative to

$$\partial_\omega := \sum \omega_i \partial_{\omega_i} \quad \text{and} \quad \partial_{\omega^\star} := \sum \omega_i^\star \partial_{\omega_i^\star}$$

though deducible from the above, are also worth mentioning.

$$\partial_\omega \mathcal{U}a^\bullet(z) = -\mathcal{U}a^\bullet(z) \times U^\bullet - z \, \square \, \mathcal{U}a^\bullet(z) \tag{6.82}$$
$$\partial_{\omega^\star} \mathcal{U}a^\bullet(z) = +z^{-1} \mathcal{U}a^\bullet(z) \times UR^\bullet - z^{-1} \, \square \, \mathcal{U}a^\bullet(z) \tag{6.83}$$

$$\begin{aligned}
\partial_\omega U^\bullet = \partial_{\omega^\star} U^\bullet &= \square((\square^\star \, USS^\bullet) \times SSU^\bullet &&= -\square \, UR^\bullet \\
\partial_\omega USS^\bullet = \partial_{\omega^\star} USS^\bullet &= +U^\bullet \times USS^\bullet &&= USS^\bullet \times {}^\# U^\bullet \\
\partial_\omega SSU^\bullet = \partial_{\omega^\star} SSU^\bullet &= -SSU^\bullet \times U^\bullet &&= -\, {}^\# U^\bullet SSU^\bullet
\end{aligned}$$

$$\partial_\omega US^\bullet = +U^\bullet \times US^\bullet - \square US^\bullet \tag{6.84}$$

$$\partial_{\omega^\star} US^\bullet = +US^\bullet \times (\square^\star\, {}^{\#}US^\bullet) \times {}^{\#}SU^\bullet \tag{6.85}$$

$$\partial_\omega SU^\bullet = -SU^\bullet \times U^\bullet - \square SU^\bullet \tag{6.86}$$

$$\partial_{\omega^\star} SU^\bullet = +\,{}^{\#}US^\bullet \times (\square^\star\, {}^{\#}SU^\bullet) \times SU^\bullet \tag{6.87}$$

### 6.7.13 A glimpse of the $D$-friendly perilogarithms

Their inductive definition is:

$$(z\,\partial_z + z\,\square - z^{-1}\,\square^\star)\mathcal{V}a^\bullet(z) = -\mathcal{V}a^\bullet(z) \times \mathrm{Ja}^\bullet(z) \tag{6.88}$$

with an elementary, one-component mould $\mathrm{Ja}^\bullet$:

$$\mathrm{Ja}^{\underline{\varpi}_1}(z) = \mathrm{Ja}^{\binom{\varpi_1}{\tau_1}}(z) := \begin{cases} 1 & \text{if } \tau_1 = 0 \\ cz^{-1} & \text{if } \tau_1 = 1 \end{cases}$$

$$\mathrm{Ja}^{\underline{\varpi}_1,\dots,\underline{\varpi}_r}(z) = \mathrm{Ja}^{\binom{\varpi_1,\dots,\varpi_r}{\tau_1,\dots,\tau_r}}(z) := 0 \quad \text{if} \quad r \neq 1$$

### 6.7.14 From $D$- to $\Delta$-friendly

$$\mathcal{U}a^\bullet(z) = \mathcal{V}a^\bullet \circ U^\bullet \tag{6.89}$$

which is short-hand for

$$\mathcal{U}a^{\varpi}(z) = \sum_{\left\{\substack{\varpi^1\dots\varpi^s=\varpi \\ s\geq 1,\ \tau_i\in\{0,1\}}\right\}} \mathcal{V}a^{\binom{\|\varpi^1\|,\dots,\|\varpi^s\|}{\tau_1,\dots,\tau_s}}(z)U^{\binom{\varpi^1}{\tau_1}}\dots U^{\binom{\varpi^s}{\tau_s}} \tag{6.90}$$

with

$$U^{\underline{\varpi}} = U^{\binom{\varpi}{\tau_0}} := \begin{cases} U^{\varpi} & \text{if } \tau_0 = 0 \\ UR^{\varpi} & \text{if } \tau_0 = 1 \end{cases}$$

### 6.7.15 Resurgence equations

$$\Delta\mathcal{U}^\bullet = I^\bullet \times \mathcal{U}^\bullet \qquad \text{(with indices } \varpi_i) \tag{6.91}$$

$$\Delta\mathcal{V}^\bullet = V^\bullet \times \mathcal{V}^\bullet \qquad \text{(with indices } \underline{\varpi}_i = \tbinom{\varpi_i}{\tau_i}) \tag{6.92}$$

The endearingly simple relation $I^\bullet = V^\bullet \circ U^\bullet$ connecting the $\Delta$- and $D$-friendly hyperlogarithmic monics carries over to the perilogarithmic monics, but with doubled storeyed indices $\underline{\omega}_i = \binom{\omega_i}{\tau_i}$, with a double-storeyed $U^\bullet$ and a double-storeyed mould composition $\circ$ defined as above.

### 6.7.16 Paralogarithmic monomials and monics

We now replace tha antipodal involution:

$$({}^{\#}M)^{\varpi_1,\dots,\varpi_r} := M^{c^2\bar{\varpi}_r,\dots,c^2\bar{\varpi}_1} \tag{6.93}$$

by the more convenient variant:

$$({}^{\sharp}M)^{\varpi_1,\dots,\varpi_r} := M^{\bar{\varpi}_r,\dots,\bar{\varpi}_1} \tag{6.94}$$

and we get:

$$^{\sharp}SSU_c^{\bullet} = SSU_c^{\bullet} \tag{6.95}$$
$$^{\sharp}USS_c^{\bullet} = USS_c^{\bullet} \tag{6.96}$$
$$^{\sharp}U_c^{\bullet} = SSU_c^{\bullet} \times U_c^{\bullet} \times USS_c^{\bullet} \tag{6.97}$$
$$^{\sharp}UR_c^{\bullet} = SSU_c^{\bullet} \times UL_c^{\bullet} \times USS_c^{\bullet} \tag{6.98}$$
$$^{\sharp}UL_c^{\bullet} = SSU_c^{\bullet} \times UR_c^{\bullet} \times US\S_c^{\bullet} \tag{6.99}$$

$$({}^{\sharp}\mathcal{U}a)_c^{\bullet}(z) \times (\mathcal{U}a)_c^{\bullet}(c^2/z) \equiv SSU_c^{\bullet} \tag{6.100}$$

The integral formulae remain unchanged, except that the extreme factors $(y_1 - 1)^{-1}$ and $(1 - y_r)^{-1}$ become $(y_1 - c)^{-1}$ and $(c - y_r)^{-1}$. One should always integrate along the axes $\mathrm{Arg}(\omega_i y_i) = \mathrm{Arg}(\bar{\omega_i}/y_i) = 0$ and heed the "SPA" rules of mutual bypassing whenever several consecutive $\mathrm{Arg}(\omega_i)$ coincide. The partial differentiation rules for the perilogarithms particularise to the paralogarithms.

## 6.8 Acceleration operators and acceleration kernels

**Acceleration/pseudoacceleration kernels:**

Acceleration $z_1 \to z_2$ with $z_1 \ll z_2$ and $z_1 \equiv F(z_2)$:

$$C_F(\zeta_2,\zeta_1) := \frac{1}{2\pi i}\int_{c-i\infty}^{c+\infty} e^{z_2\zeta_2 - z_1\zeta_1}\,dz_2$$
$$C^F(\zeta_2,\zeta_1) := \int_{+\infty}^{+u} e^{-z_2\zeta_2 + z_1\zeta_1}\,dz_2 \quad \text{with} \quad u \gg 1$$

Pseudocceleration $z_1 \to z_0$ with $z_0 = z_1 + F(z_1)$ and $1 \ll F(z) \ll z$:

$$C_{I+F}(\zeta_1,\zeta_0) := C_F(\zeta_1 - \zeta 0, \zeta_0) \qquad (0 < \zeta_0 < \zeta_1$$
$$C^{I+F}(\zeta_1,\zeta_0) := C^F(\zeta_1 - \zeta 0, \zeta_0) \qquad (0 < \zeta_0 < \zeta_1)$$

**Acceleration integrals:**

Acceleration of minors : infinite interval

$$\hat{\varphi}_2(\zeta_2) = \int_{+0}^{+\infty} C_F(\zeta_2, \zeta_1)\hat{\varphi}_1(\zeta_1)\, d\zeta_1$$

Acceleration of majors : infinite loop

$$\check{\varphi}_2(\zeta_2) = \frac{1}{2\pi i}\int_{\infty_1}^{\infty_2} C^F(\zeta_2, \zeta_1)\check{\varphi}_1(\zeta_1)\, d\zeta_1$$

Deceleration of minors : finite loop

$$\zeta_1\hat{\varphi}_1(\zeta_1) = \frac{1}{2\pi i}\int_{0_1}^{0_2} \zeta_2\hat{\varphi}_2(\zeta_2)C^F(\zeta_2, \zeta_1)\, d\zeta_2$$

Deceleration of majors : finite interval

$$\zeta_1\check{\varphi}_1(\zeta_1) = \int_{+0}^{+v} \zeta_2\check{\varphi}_2(\zeta_2)C_F(\zeta_2, \zeta_1)\, d\zeta_2$$

**Pseudo-acceleration integrals**

Pseudodeceler. of minors : finite interval

$$\begin{aligned}\hat{\varphi}_1(\zeta_1) &= \int_0^{\zeta_1} C_{I+F}(\zeta_1, \zeta_0)\hat{\varphi}_0(\zeta_0)\, d\zeta_0 \\ &= \int_0^{\zeta_1} C_F(\zeta_1 - \zeta_0, \zeta_0)\hat{\varphi}_0(\zeta_0)\, d\zeta_0\end{aligned}$$

Pseudodeceler. of majors : finite loop

$$\begin{aligned}\check{\varphi}_1(\zeta_1) &= \frac{1}{2\pi i}\int_{v_1}^{v_2} C^{I+F}(\zeta_1, \zeta_0)\check{\varphi}_0(\zeta_0)\, d\zeta_0 \\ &= \frac{1}{2\pi i}\int_{v_1}^{v_2} C^F(\zeta_1 - \zeta_0, \zeta_0)\check{\varphi}_0(\zeta_0)\, d\zeta_0\end{aligned}$$

Pseudoacceler. of minors : finite loop

$$\begin{aligned}\zeta_0\hat{\varphi}_0(\zeta_0) &= \frac{1}{2\pi i}\int_{0_1}^{0_2} \zeta_1\hat{\varphi}_1(\zeta_1)C^{I+F}(\zeta_1, \zeta_0)\, d\zeta_1 \\ &= \frac{1}{2\pi i}\int_{0_1}^{0_2} \zeta_1\hat{\varphi}_1(\zeta_1)C^F(\zeta_1 - \zeta_0, \zeta_0) d\zeta_1\end{aligned}$$

Pseudoacceler. of majors : finite interval

$$\begin{aligned}\zeta_0\check{\varphi}_0(\zeta_0) &= \int_{\zeta_0}^{v} \zeta_1\check{\varphi}_1(\zeta_1)C_{I+F}(\zeta_1, \zeta_0)\, d\zeta_1 \\ &= \int_{\zeta_0}^{v} \zeta_1\check{\varphi}_1(\zeta_1)C_F(\zeta_1 - \zeta_0, \zeta_0)\, d\zeta_1\end{aligned}$$

## References


[Bra] B. Braaksma, Multisummability of formal power series solutions of non-linear meromorphic differential equations, *Ann. Inst. Fourier* **42** (1992), 517–540.

[Bro] Broadhurst, Conjectured enumeration of irreducible multiple Zeta values, from knots and Feynman diagrams, preprint MK76AA, Open University Milton Keynes, November 1996.

[Bry] A. D. Bryuno, Analytical forms of differential equations, I, II, *Trudy Mosk. Mat. Ob.* **25** (1971), 119–262, **26** (1972), 199–239; English translation in *Trans. Mosc. Soc.* **25** (1971), **26** (1972)

[E1] J. Écalle, *Les fonctions résurgentes, Vol. 1, Algèbres de fonctions résurgentes*, Publ. Math. Orsay, 1981.

[E2] J. Écalle, *Les fonctions résurgentes, Vol. 2, Les fonctions résurgentes appliquées à l'itération*, Publ. Math. Orsay, 1981.

[E3] J. Écalle, *Les fonctions résurgentes, Vol. 3, L'équation du pont et la classification analytique des objets locaux*, Publ. Math. Orsay, 1985.

[E4] J. Écalle, *The acceleration operators and their applications*, Proc. Internat. Cong. Math., Kyoto, 1990, vol. 2, Springer, Tokyo, 1991.

[E5] J. Écalle, Singularités non abordables par la géometrie, *Ann. Inst. Fourier* **42** (1992), 73–164.

[E6] J. Écalle, Finitude des cycles-limites et accéléro-sommation de l'application de retour, Prépubl. Math. Orsay, 1990.

[E7] J. Écalle, *Introduction aux fonctions analysables et preuve constructive de la conjecture de Dulac*, Actual. Math., Hermann, Paris, 1992.

[E8] J. Écalle, Weighted products and parametric resurgence, in: *Méthodes résurgentes*, Travaux en Cours **47**, L. Boutet de Monvel, 1994, 7–49.

[E9] J. Écalle, Compensation of small denominators and ramified linearisation of local objects, in: *Complex Analytic Methods in Dynamical Systems*, IMPA, Aster., 1994, 135–199.

[E10] J. Écalle, Six lectures on transseries, analysable functions and the constructive proof of Dulac's conjecture, in: *Bifurcations and Periodic Orbits of Vector Fields* (D. Schlomiuk, ed.), Kluwer, 1993, 75–184.

[E11] J. Écalle, Cohesive functions and weak accelerations, *J. Analyse Math.*, Szolem Mandelbrojt Memorial Volume Vol. 60, 1993.

[E12] J. Écalle, Annexe C: Discretised resurgence, in: [M], 161–218.

[E13] J. Écalle, A tale of three structures: the arithmetics of multizetas, the analysis of singularities, the Lie algebra ARI, in: *Differential Equations and the Stokes Phenomenon* (B. L. J. Braaksma et al., eds.), World Scient. Publ., 2002, 89–145.

[E14] J. Écalle, ARI/GARI, la dimorphie et l'arithmétique des multizêtas: un premier bilan, Prépubl. Math. Orsay, 2002, to appear in: *Ann. Math. Bordeaux.*

[E15] J. Écalle, Twisted resurgence monomials and canonical synthesis of local objects, to appear in: *Proc. of the June 2002 Edinburgh conference on Analysable Functions* (O. Costin, ed.), World Scientific.

[E16] J. Écalle, Canonical-explicit decomposition of multizetas into irreducibles, to appear in: *Ann. M. Toulouse.*

[E17] J. Écalle, *Multizeta Arithmetics and the Underlying Structures*, Publ. Math. Orsay, 2004.

[EM] J. Écalle and F. Menous, Well-behaved convolution averages, in: *The Stokes Phenomenon and Hilbert's 16th Problem* (B. L. J. Braaksma et al., eds.), World Scient. Publ., 71–101.

[EV1] J. Écalle and B. Vallet, Prenormalisazation, correction, and linearization of resonant vector fields or diffeomorphisms, Prépubl. Orsay, 1995.

[EV2] J. Écalle and B. Vallet, Correction and linearization of resonant vector fields or diffeomorphisms, *Math. Z.* **229** (1998), 249–318.

[EV3] J. Écalle and B. Vallet, The arborication-coarborification transform: analytic and algebraic aspects, to appear in: *Ann. M. Toulouse.*

[El1] L. H. Eliasson, Generalization of an estimate of small divisors by Siegel, in: *Analysis et cetera (sic)* (Rabinowitz and Zehnder, eds.), Academic Press, 1990.

[El2] L. H. Eliasson, Absolutely convergent series expansions for quasiperiodic motions, preprint 2, Univ. of Stockholm, 1988.

[El3] L. H. Eliasson, Hamiltonian systems with linear form near an invariant torus, in: *Non-linear Dynamics* (Turchetti, ed.), World Scientific, Singapore, 1989.

[El4] L. H. Eliasson, Floquet solutions for the one-dimensional quasi-periodic Schrödinger equation, *Comm. Math. Phys.* **146** (1992), 447–482.

[G] A. B. Goncharov, Polylogarithms in arithmetic and geometry, *Proc. ICM-94* (1995), 374–387.

[H1] J. van der Hoeven, *Automatic Asymptotics*, Ph.D. Thesis, École Polytechnique, 1997.

[H2] J. van der Hoeven, A differential intermediate value theorem, Prépub. Math. Orsay, 2001.

[H3] J. van der Hoeven, Complex transseries solutions to algebraic differential equations, Prépub. Math. Orsay, 2001.

[I1] Yu. S. Ilyashenko, Finiteness theorems for limit cycles, English translation in: *Russian Math. Surveys* **45** (1990).

[I2] Yu. S. Ilyashenko, *Finiteness for Limit Cycles*, Transl. Math. Monogr. **94**, Amer. Math. Soc., Providence, R.I.

[M] F. Menous, *Les bonnes moyennes uniformisantes et leurs applications à la resommation réelle*, Ph.D. Thesis, Orsay, 1996.

[PM] R. Perez-Marco, Solution complète au problème de Siegel de linéarisation d' une application holomorphe au voisinage d'un point fixe, d'après J.-C. Yoccoz, *Séminaire Bourbaki 1991–92* **753** (1992).

[Se] A. Seidenberg, Reduction of singularities of the differential equation $A\,dy = B\,dx$, *Amer. J. Math.* **90** (1968), 248–269.

[Si1] C. L. Siegel, Iteration of analytic functions, *Ann. Math.* **43** (1942), 607–612.

[Si2] C. L. Siegel, Über die Normalform analytischer Differentialgleichungen in der Nähe einer Gleichgewichtslösung, *Nachr. Akad. Wiss. Göttingen Math. Phys. Abt.* (1952).

[Ri] T. Rivoal, *Propriétés diophantiennes des valeurs de la fonction zêta de Riemann aux entiers impairs*, Ph.D. Thesis, Caen, 2001.

[Ru] Rüssmann, Über die Iteration analytischer Funktionen., *J. Math. Mech.* **17** (1967), 523–532.

[Z] D. Zagier, *Values of Zeta Functions and their Applications*, First European Congress of Mathematics, Vol. 2, Birkhäuser, Boston, 1994, 427–512.

# Local bifurcations of limit cycles, Abel equations and Liénard systems

J.-P. FRANÇOISE

*Université de Paris VI, Site Chevaleret*
*175 Rue du Chevaleret*
*75013 Paris*
*France*

Notes taken by Hur SEOK

**Abstract**

The class of Liénard equations is introduced and motivated with several examples from modeling physiology. The location and number of limit cycles are first obtained in the limit of fast-slow dynamics. The general case is then discussed in the framework of a quantitative version of Bautin's theory. Abel equations are related to the center-focus problem and a formula is provided for all the successive derivatives of the associated return map.

## 1 Introduction

In this series of lectures, a summary of recent contributions to bifurcation theory of differential systems is provided. We present the tools necessary to a mathematics student to grasp the latest developments around the local Hilbert's 16th-problem on limit cycles in the direction of Abel equations and algebraic moments. This direction of research was initiated jointly with M. Briskin and Y. Yomdin [3, 4] and was followed by several contributions; among them those of C. Christopher [6], F. Pakovich [32]. One motivation to study this field of research may be the variety of the methods employed which includes monodromy groups, Galois extensions and Wermer's theory of Riemann surfaces. But there is certainly another motivation which relates to applications. The theory of limit cycles initiated by H. Poincaré found a strong development with the famous contributions of Van der Pol [35–37] to the study of electrical circuits. This line of research became extremely active with its applications to electrophysiology. In Section 2 we try to provide an intuitive idea of the notion of limit cycles with examples of fast-slow dynamics from modeling physiology. The basic problem posed by Hilbert, to count and locate the limit cycles and to study their deformation when the parameters of the equation change, becomes quite meaningful for Liénard systems in the limit of fast-slow dynamics (which encompass all the examples from physiology). New complex analytic techniques added to the classical approach of Bautin allow to obtain bounds for Liénard systems in general (not only in the fast-slow limit). The most general approach, due to Y. Yomdin and the author [15, 19, 20], follows the same techniques but needs deeper



*Y. Ilyashenko and C. Rousseau (eds.),*
*Normal Forms, Bifurcations and Finiteness Problems in Differential Equations,* 187–209.

tools such as the Hironaka polynomial division theorem. This approach is then completed with the line of research initiated by M. Briskin, Y. Yomdin and the author [3,4] to find the Bautin ideal in the case of homogeneous perturbations via the Cherkas transform and Abel equations. A final new computation is given in this article: all the successive derivatives of the converse of the return mapping of Abel equations are computed and related to iterated moments. This should lead to future developments of research in the field.

It is a pleasure for me to thank C. Rousseau for the invitation to give this course. I express my hearty thanks to J. Demongeot for many inspiring discussions on mathematical biology.

# 2 Van der Pol systems, FitzHugh-Nagumo systems, singular Hopf bifurcation, fast-slow limits of Liénard equations, bursting oscillations

## 2.1 Van der Pol systems, FitzHugh-Nagumo dynamics, singular Hopf bifurcation

The harmonic oscillator:

$$\ddot{x} + x = 0, \tag{2.1a}$$

has all its orbits periodic of equal period. The origin of the plane is stable of center type. If we introduce a dissipative term and change the equation into:

$$\ddot{x} - A\dot{x} + x = 0; \tag{2.1b}$$

the origin is not a center any more but a focus (stable or unstable, depending of the sign of $A$). In his famous articles on electrical circuits [35–37], Van der Pol considered the case where the friction term (resistance) is no longer a constant but depends of the variable $x$ and is an even function of $x$. The simplest case is

$$\ddot{x} - A(1 - x^2)\dot{x} + x = 0. \tag{2.1c}$$

Assume that $A$ is positive, hence the system has a resistance which for a small amplitude is negative. Van der Pol observed that there is still an oscillation which for small values of $A$ closely resembles sinusoidal oscillations. When $A$ is large instead, the solution does not look sinusoidal any more and displays fast, then slow, alternating phases. This new type of oscillation was named relaxation oscillations by Van der Pol. How to understand intuitively the occurrence and the pattern of this oscillation when $A$ is large? Write the system as follows:

$$\begin{aligned} \dot{x} &= y \\ \dot{y} &= -x + A(1 - x^2)y, \end{aligned} \tag{2.2}$$

then change $y$ into $Y$ such that $y = Y + AF(x)$, with $F(x) = x - x^3/3$. This yields:

$$\begin{aligned} \dot{x} &= Y + AF(x) \\ \dot{Y} &= -x. \end{aligned} \tag{2.3}$$

Now change $Y$ into $AY$ and time $t$ into $At$, and obtain:

$$\begin{aligned} \epsilon\dot{x} &= Y + F(x) \\ \dot{y} &= -x \end{aligned} \tag{2.4}$$

with $\epsilon = 1/A^2$. This is an example of fast-slow dynamics. Let $-F(c_0)$ be the value of $Y$ corresponding to the local minimum of the cubic nullcline $Y = -F(X)$. Consider an initial point $(x_0, y_0)$ such that $y_0 < -F(c_0)$ which is not on the nullcline. The orbit through this point starts almost parallel to the $x$-axis going fast to the left till it touches the nullcline. There, under the influence of the slow dynamics, it goes slowly upwards along the cubic till it reaches the local maximum. Then it goes fast to the right almost parallel to the $x$-axis till it reaches the nullcline and then goes down. This is elementary evidence of the existence of an attractive limit cycle for $\epsilon \neq 0$, and of an attractive "limit periodic" set (in the sense of [20]) for $\epsilon = 0$.

The most general so-called FitzHugh-Nagumo dynamics:

$$\begin{aligned} \epsilon\dot{x} &= y - F(x) \\ \dot{y} &= \alpha x + \beta y + \gamma, \end{aligned} \tag{2.5}$$

where $F(x)$ is a cubic polynomial, is used to modeling in physiology. It was derived initially by FitzHugh [12, 13] as a natural limit of the Hodgkin-Huxley equations which model the propagation of the nerve impulse along an axon of the squid. The same type of equations were written later for many different physiological systems such as the electrical activity of the cardiac cells of the lobster, the production of insulin by the $\beta$-cells of the pancreas, thalamic neurons of cats, etc. (see for instance Keener-Sneyd, Murray [23, 30]). Let us focus now on the 1-parameter family:

$$\begin{aligned} \epsilon\dot{x} &= y - F(x) \\ \dot{y} &= x - c. \end{aligned} \tag{2.6}$$

Denote by $c_0$ the local minimum of the cubic function $F(x)$. If $c < c_0$, there is an attractive focus, and for some initial data near the equilibrium position and above the cubic nullcline the flow is quickly attracted to the focus. Instead, for some other initial data still near the equilibrium position but below the cubic nullcline, the flow undergoes a rather long path before going back to the attractive focus. This special feature has little to do with the qualitative dynamics but is strongly related to the singular perturbation analysis and the so-called excitability of the stable equilibrium position.

In the case $c > c_0$, the equilibrium position becomes unstable. Under the influence of the fast-slow dynamics, the point jumps till it intersects the cubic nullcline. Then it goes up slowly along the cubic nullcline till it reaches the local maxima. There, under the influence of the fast dynamics, it is pushed back to the other branch of the cubic. It then goes back slowly near the equilibrium position and then starts again. This shows that an attractive limit cycle has been born in the bifurcation. The bifurcation is thus a supercritical Hopf bifurcation. The classical phase plane analysis has to be adapted to the singular perturbation context (cf. F. Dumortier, R. Roussarie [10]).

The general class formed by the equations

$$\begin{aligned} \epsilon\dot{x} &= y - F(x) \\ \dot{y} &= \alpha x + \beta y + \gamma, \end{aligned} \tag{2.7}$$

where now $F(x)$ is any polynomial (which vanishes at $x = 0$) is named slow-fast Liénard dynamics. We can easily extend the preceding notions to this more general context and relate the number of limit-periodic sets to simple topological invariants of the nullcline $y = F(x)$.

## 2.2 On a class of generic Liénard equations and the notions of excitability and multi-excitability

Consider the system:

$$\begin{aligned} \epsilon\dot{x} &= y - F(x) \\ \dot{y} &= x - c, \end{aligned}$$

where $F(x)$ is a polynomial of degree $2n + 1$ (or $2n$). Assume that $F$ has $2n$ (or $2n - 1$) distinct critical points and that the corresponding critical values are all distinct. Assume furthermore that $c$ is not a critical point.

The dynamics of the system is encoded in the parameter $c$; the minima of $F$: $c_0, c_1, \ldots, c_{n-1}$, the maxima of $F$: $d_0, \ldots, d_{n-1}$ (in case the degree of $F$ is $2n+1$; with an easy change of notation in case degree of $F$ is $2n$). It depends only on the ordering of the minima $F(c_i)$ and of the maxima $F(d_j)$ and the relative position of $c$ among these values. The only singular point is generic because $c$ is not a critical value.

Let us focus on the more specific example:

$$\begin{aligned} \epsilon\dot{x} &= y - F(x) \\ \dot{y} &= x - c, \end{aligned} \tag{2.8}$$

where $F(x) = -x^2(x-\alpha)(x-\beta)(x-\gamma)$ is a quintic polynomial, $\alpha < \beta < \gamma$ are positive constants, and $0 < c < \alpha$. This system has just one singular point which is unstable. The two local minima of the nullcline $y = F(x)$ are $c_0 = 0$ and $c_1$. Assume that $F(c_1) < 0$. Denote by $d_0 < d_1$ the two local maxima of $y = F(x)$ and assume that $F(d_1) > F(d_0)$.

**2.1 Theorem** *The system displays an attractive limit cycle (relaxation oscillation) which is excitable. The corresponding threshold is* $F(c_1)$.

Consider now the same example but with $c < 0$. Then the singular point is stable and excitable. But there are indeed two different scales of length of excursion before going back to the attractor. We refer to this situation as a case of multi-excitability (bi-excitability more precisely in this case). The notion of multi-excitability relates to multi-stability; A. Goldebeter gave several examples of biochemical systems which behave like this (Goldebeter [22]).

To conclude this part: We have motivated with several examples the interest of the class of Liénard systems. In the limit of a fast-slow dynamics, the number and the location of the limit cycles is easily deduced from the topology of the nullcline. There is a bound for the number of attractive limit cycles, which is $n$ in case the degree of the polynomial $F(x)$ is $2n$ or $2n + 1$.

## 2.3 Bursting oscillations with fast dynamics in normal form

Bifurcations of limit cycles in the framework of fast-slow dynamics became an essential ingredient to modeling the so-called bursting oscillations. There is a now standard approach known as dissection of bursting (Rinzel [33]). The silent phase corresponds to an equilibrium of the fast dynamics, and the repetitive spiking corresponds to a stable limit cycle of the fast dynamics. The slow dynamics acts as a driver of the parameters which makes the fast dynamics bifurcate.

A quite simple example can be indeed realized, which involves three time scales, with the FitzHugh-Nagumo system as the fast system and a slow periodic dynamics driving periodically the parameter $c$ back and forth symmetrically to $c_0$. This reads:

$$\begin{aligned} \epsilon\dot{x} &= x - F(x) \\ \delta\dot{y} &= x - c \\ \ddot{c} &= -\omega^2 c. \end{aligned} \tag{2.9}$$

Systems of this type and the qualitative aspects of the corresponding "wave form" for the solution $x$ have been discussed in the recent article (Doss-Bachelet, Françoise, Piquet [8]). We include here two other examples built with a fast dynamics in normal form.

The simplest example of a supercritical Hopf bifurcation is given by the normal form:

$$\begin{aligned} \dot{x} &= y + cx - x(x^2 + y^2) \\ \dot{y} &= -x + cy - y(x^2 + y^2), \end{aligned} \tag{2.10a}$$

which can be conveniently written in the complex notation $z = x + \mathrm{i}y$:

$$\dot{z} = (\mathrm{i} + c)z - z(z\bar{z}). \tag{2.10b}$$

When $c < 0$, the origin is globally stable. At the bifurcation point $c = 0$, the origin loses its stability and a stable limit cycle $z\bar{z} = c$ is born. The following fast-slow dynamics displays bursting oscillations for properly chosen initial data:

$$\begin{aligned} \epsilon\dot{z} &= (\mathrm{i} + c)z - z(z\bar{z}) \\ \ddot{c} &= -\omega^2 c. \end{aligned} \tag{2.11}$$

In this example, the slow dynamics is two-dimensional and periodic. A drawback of this example is that the amplitude of the oscillations decreases (as $1/t$) when the time $t$ increases.

The second example is obtained from a subcritical Hopf bifurcation, also given by a normal form:

$$\dot{z} = (\mathrm{i} + c)z + 2z(z\bar{z}) - z(z\bar{z})^2. \tag{2.12}$$

When $c < 0$, the origin is stable. When $c = -1$, a double limit cycle $z\bar{z} = 1$ appears and it splits into a stable limit cycle and an unstable limit cycle. When $c = 0$, the unstable limit cycle disappears at the origin which becomes unstable. The following fast-slow dynamics displays bursting oscillations:

$$\begin{aligned} \epsilon\dot{z}(\mathrm{i} + c)z &+ 2z(z\bar{z}) - z(z\bar{z})^2 \\ \dot{c} &= c - 3 + 5z\bar{z}. \end{aligned} \tag{2.13}$$

The slow dynamics is one-dimensional and yields an hysteresis mechanism on the fast dynamics. If an initial data $(x_0, y_0, c_0)$ is chosen such that $c_0 > -3$ and $(c_0, z_0\bar{z}_0)$ below the linear nullcline: $c - 3 + 5z(z\bar{z}) = 0$, the origin is stable and quickly attracts the orbit (silent phase). But the parameter $c$ increases and when $c = 0$, the origin becomes unstable and the orbit jumps at the stable limit cycle. Doing so, the orbit goes above the nullcline and then $c$ decreases. The point undergoes now rapid oscillations (spikes) along the stable limit cycle. But as $c$ decreases, when it reaches the value $c = -1$, the limit cycle dissapears into a saddle-node of periodics. Then, the origin being now the only attractor, the orbits goes down to the the axis $z\bar{z} = 0$ (silent phase), and then it starts again.

# 3 Homogeneous perturbations of a rotation and Bautin's approach to Hilbert's 16th-problem

## 3.1 Construction of the return mapping by Bautin's method

We consider a polynomial plane vector field $X$ of type:

$$X = x\partial/\partial y - y\partial/\partial x + \sum_{i,j/i+j=d} [a_{i,j}x^i y^j \partial/\partial x + b_{i,j}x^i y^j \partial/\partial y]. \tag{3.1}$$

and the associated flow, solution of the system:

$$\begin{aligned} \dot{x} &= -y + \sum_{i,j/i+j=d} a_{i,j}x^i y^j = -y + P(x,y), \\ \dot{y} &= x + \sum_{i,j/i+j=d} b_{i,j}x^i y^j = -y + Q(x,y). \end{aligned} \tag{3.2}$$

The parameters of the vector field $(a, b)$ can take any real values and thus $(a, b)$ should be considered as a point in the vector space $\mathbb{R}^{2(d+1)}$. We first recall Bautin's approach to find the return mapping of (3.1) near the origin. Write (3.2) in polar coordinates $(r, \theta)$:

$$x = r\cos(\theta), \qquad y = r\sin(\theta). \tag{3.3}$$

This leads to:

$$2r\dot{r} = 2(x\dot{x} + y\dot{y}), \qquad r\dot{r} = xP + yQ = r^{d+1}A(\theta), \tag{3.4}$$

$$\dot{\theta} = (x\dot{y} - y\dot{x})/(x^2 + y^2) = 1 + r^{d-1}B(\theta), \tag{3.5}$$

where $A(\theta)$ and $B(\theta)$ are two trigonometric polynomials (in $\cos(\theta), \sin(\theta)$) linear in the parameters $(a, b)$.

This yields:

$$dr/d\theta = r^d A(\theta)/[1 + r^{d-1}B(\theta)], \tag{3.6}$$

and thus

$$dr/d\theta = \sum_{k=0}^{\infty} (-1)^k r^{k(d-1)+d} A(\theta) B(\theta)^k. \tag{3.7}$$

This equation (3.7) may be rewritten:

$$dr/d\theta = \sum_{k \geq d} r^k R_k(\theta), \tag{3.8}$$

where the coefficients $R_k(\theta)$ are trigonometric polynomials in $(\cos(\theta), \sin(\theta))$ and polynomials in the parameters $(a, b)$. To simplify the notation, the dependence on the parameters $(a, b)$ is not made explicit. Bautin looks for a solution of (3.8), $r = r(\theta)$, so that $r(0) = r_0$, given as an expansion:

$$r = r_0 + v_2(\theta)r_0^2 + \cdots + v_k(\theta)r_0^k + \cdots \tag{3.9}$$

Comparison between (3.8) and (3.9) yields:

$$v_2(\theta) = \cdots = v_{d-1}(\theta) = 0, \tag{3.10}$$

$$v_d'(\theta) = R_d(\theta), \tag{3.11}$$

$$v_k'(\theta) = \sum_{i=2}^{k} B_{ik}[v_d(\theta), \ldots, v_{k-1}(\theta)]R_i(\theta), \quad k \geq d+1. \tag{3.12}$$

The polynomial $B_{ik}[a_d, \ldots, a_{k-1}]$ has integer coefficients and is the coefficient of $X^{k-i}$ in $(X + a_dX^d + \cdots + a_pX^p + \cdots)^i$. The relation (3.12) allows one to determine inductively the functions $v_k(\theta)$:

$$v_k(\theta) = \int_0^\theta \left[\sum_{i=2}^{k} B_{ik}[v_d(\phi), \ldots, v_{k-1}(\phi)]R_i(\phi)\right] d\phi. \tag{3.13}$$

Two facts can easily be derived from this construction:

(i) $v_k(\theta)$ is polynomial in $\theta$ (of degree less than $k$) and in $(\sin(\theta), \cos(\theta))$.

(ii) $v_k(\theta)$ is polynomial in the parameters $(a, b)$ of the vector field. Thus in particular the coefficients $v_k(2\pi)$ of the return mapping are polynomials in $(a, b)$.

**3.1 Definition** The vector field $X$ *displays a center at the origin* $(0 \in \mathbb{R}^2)$ (or is said *to be a center*) if and only if $r(2\pi) = r(0)$ for all values of $r(0)$ close to 0.

Let $f_\lambda(x) = \sum a_k(\lambda)x^k$ be an analytic series in $x$ with polynomial coefficients in the parameters $\lambda = (\lambda_0, \ldots, \lambda_D)$. Denote by $\mid a_k \mid$ (norm of the polynomial $a_k(\lambda)$) the sum of the absolute value of the coefficients.

**3.2 Definition** The series $f_\lambda$ is called an $A_0$*-series* if the following two conditions are satisfied:

There are positive constants $K_1, K_2, K_3, K_4$ such that

(1) $\deg(a_k) \leq K_1k + K_2$;

(2) $\mid a_k \mid\leq K_3K_4^k$.

The $A_0$-series form a subring of the ring of formal power series in $x$ with polynomial coefficients in $\lambda$. All the usual analytic operations, like substitution to a given analytic function, composition, inversion, etc., transform $A_0$-series into $A_0$-series. (The proof is rather straightforward and is not provided here). The $A_0$-series have been introduced (in the subject) by M. Briskin and Y. Yomdin [4]. The proof of the following easy lemma is left to the reader:

**3.3 Lemma** *An $A_0$-series $f_\lambda(x)$ converges in the disc $D(0, R)$ of radius*

$$R = [1/(K_4(1+ \mid \lambda \mid))^{K_1}].$$

In the following, we also denote by $f_\lambda$ the complex analytic function defined for all $\lambda \in C^D$ on the disc $D(0,R)$ by the $A_0$-series.

**3.4 Proposition** *For all $\theta$, the series $\sum_{k\geq d} x^k R_k(\theta)$, is an $A_0$-series with $K_1 = 1/(d-1)$, $K_2 = -1/(d-1)$, $K_3 = 1/[2(d+1)]^{1/d-1}$, $K_4 = [2(d+1)]^{1/d-1}$.*

**3.5 Theorem** *For all $\theta$, the series $\sum_{k\geq d} x^k v_k(\theta)$, is an $A_0$-series with*

$$\begin{aligned} K_1' &= 1/(d-1), \\ K_2' &= 0, \\ K_3' &= [2\pi K_3/4K_4^2][K_4 - 2C + ((K_4-2C)^2 - K_4^2)]^2, \\ K_4' &= [2K_4^2]/(K_4 - 2C + (K_4-2C)^2 - K_4^2), \\ C &= K_4 + 2\pi K_3 K_4^2. \end{aligned}$$

## 3.2 Definitions of the center set, the Bautin ideal, and the Bautin index

**3.6 Definition** The *center set* $C$ is the set of values of the parameters $(a,b)$ so that the corresponding vector field $X$ has a center at the origin.

**3.7 Proposition** *The center set $C$ is a real algebraic manifold.*

**Proof** The center set $C$ is given as the zero set of the coefficients $v_k(2\pi)$, which are polynomials in the parameters $(a,b)$. □

From now on, it is appropriate to change the notation and denote by $L_k(a,b) = v_k(2\pi)$ the coefficients of the return mapping to emphasize their dependence in terms of the parameters $(a,b)$. We denote by

$$r \mapsto L(r) = r + L_d(a,b)r^d + \cdots + L_k(a,b)r^k + \cdots \tag{3.14}$$

the return mapping defined in (3.9) for $\theta = 2\pi$.

**3.8 Definition** The *Bautin ideal* is the ideal generated in the ring $R[a,b]$ by the coefficients $L_k(a,b)$.

**3.9 Definition** The *Bautin index* is the first integer $k_0$ so that the polynomials $L_d(a,b),\ldots,$ $L_{k_0}(a,b)$ generate the Bautin ideal.

The local Hilbert's 16th problem is to find a bound depending only on $d$ for the number of isolated periodic orbits of $X$ in a neighborhood of the origin. Isolated periodic orbits of $X$ defined in the domain of definition of the return mapping correspond exactly to the isolated solutions of the equation:

$$L(r) - r = 0. \tag{3.15}$$

We gradually change into notations more pertinent to the general algebraic geometry setting. Let $\Phi(x,\lambda)$ be an analytic series in $x$ with polynomial coefficients in the parameter $\lambda = (\lambda_1,\ldots,\lambda_D)$.

**3.10 Definition** The *Bautin ideal* of the series $\Phi(x, \lambda)$ is the ideal generated in the ring $R[\lambda]$ by the coefficients $\Phi_k(x, \lambda)$. The *center set* of the series $\Phi(x, \lambda)$ is the zero set of its Bautin ideal. The *Bautin index* of the series $\Phi(x, \lambda)$ is the minimal integer $d$ such that the coefficients $\Phi_1(x, \lambda), \dots, \Phi_d(x, \lambda)$ generate the Bautin ideal of the series $\Phi(x, \lambda)$.

## 3.3 A bound for the number of real fixed points

We recall here Bautin's proof for the existence of a bound for the number of real fixed points.

Let $\Phi(x, \lambda)$ be an analytic series in $x$ with polynomial coefficients in the parameters $\lambda$ and with Bautin index $d$. Assume that $\Phi(0, \lambda) = 0$ and $\Phi(x, 0) = 0$.

**3.11 Theorem** *There is a ball $B \subset \mathbb{R}^D$, centered at 0, in the space $\mathbb{R}^D$ and an interval $I$ containing 0 such that for all $\lambda \in B$, the number of zeros of $\Phi(x, \lambda)$ contained in $I$ is less than or equal to $d$.*

**Proof** Using the definition of the Bautin index $d$, we write:

$$\Phi(x, \lambda) = \sum_{i=1}^{d} \Phi_i(\lambda)[1 + \Psi_i(x, \lambda)]x^i, \tag{3.16}$$

with $\Psi_i(x, \lambda)$ analytic in $x$, and with polynomial coefficients in $\lambda$ such that $\Psi_i(0, \lambda) = \Psi_i(x, 0) = 0$. Assume that $B$ and $I$ are small enough so that (for instance) $\mid \Psi_i(x, \lambda) \mid \leq 1/2$ on $I \times B$. Then divide $\Phi(x, \lambda)$ by $[1 + \Psi_d(x, \lambda)]$ and rewrite:

$$[\Phi(x, \lambda)]/[1 + \Psi_d(x, \lambda)] = \Phi_1(\lambda) + \Phi_2(\lambda)[1 + \Psi_i'(x, \lambda)]x^2 + \cdots . \tag{3.17}$$

Then it follows from Rolle's lemma that the number of zeros of $\Phi(x, \lambda)$ is less than 1+ number of zeros of the derivative $[\Phi(x, \lambda)]/[1 + \Psi_d(x, \lambda)]'$. Write this derivative as

$$\Phi_2(\lambda)[1 + \Psi_i^{(2)}(x, \lambda)]x + \cdots + \Phi_{d-1}(\lambda)[1 + \Psi_i^{(2)}(x, \lambda)]x^{d-1}. \tag{3.18}$$

Then repeat the process (sometimes referred to as the division-derivation algorithm). We obtain the result by an easy induction. □

It is clear that one cannot go far with this proof if we want an explicit control of the size of the domain (either of $I$ or of $B$). For instance, we know that the coefficients $\Phi_i(\lambda)$ are combinations of the first $d$ coefficients:

$$\Phi_i(\lambda) = P_{(i,1)}(\lambda)\Phi_1(\lambda) + \cdots + P_{(i,d)}(\lambda)\Phi_d(\lambda), \tag{3.19}$$

but how to control the size of the "quotients" $P_{(i,k)}(\lambda)$ which enter in the construction of the $\Psi_i(x, \lambda)$? We will see in the next chapter how to circumvent these difficulties using complex analytic methods such as the polynomial Hironaka division theorem.

## 3.4 Formulation of the problem in terms of projection of analytic sets

Let $\Phi : \mathbb{R} \times \mathbb{R}^D \mapsto \mathbb{R}$ be an analytic series with polynomial coefficients:

$$\Phi(x, \lambda) = x + \Phi_2(\lambda)x^2 + \cdots + \Phi_k(\lambda)x^k + \cdots . \tag{3.20}$$

We consider the subset $\Sigma \subset \mathbb{R} \times \mathbb{R}^D$ defined as the zero-set of $\Phi(x, \lambda) - x$:

$$\Sigma = [(x, \lambda)/\Phi(x, \lambda) - x = 0]. \tag{3.21}$$

Let $\pi : \Sigma \mapsto \mathbb{R}^D$ be the restriction to $\Sigma$ of the natural projection $\pi : \mathbb{R} \times \mathbb{R}^D \mapsto \mathbb{R}^D$. The center set associated to the analytic series $\Phi(x, \lambda) - x$ defined in (3.20) is the set $C \subset \mathbb{R}^D$ of parameters $\lambda$ such that the fiber of the projection, $\pi^{-1}(\lambda)$, is contained in the set $\Sigma$. The next theorem displays the following geometric interpretation:

**3.12 Theorem** *There is a neighborhood $I \times B$ of $(0, 0)$ in $\mathbb{R} \times \mathbb{R}^D$ such that for all points $\lambda$ of $B$ the number of isolated points of the fibers $\pi^{-1}(\lambda)$ restricted to $\Sigma$ is less than the Bautin index $d$ of the analytic series $\Phi(x, \lambda) - x$.*

# 4 A quantitative version of Bautin's approach. The example of Liénard systems

Bautin's method as presented in the preceding chapter does not allow one to produce a precise estimate on the size of the domain on which the number of zeros of an $A_0$-series (such as the displacement function) is controlled. It is not surprising that to get such an estimate, complex analytic methods are more appropriate.

## 4.1 Finiteness properties, Bernstein inequality

The closed disc centered at $0 \in \mathbb{C}$ of radius $R$ is denoted by $\bar{D}_R$. This paragraph relies entirely on [9, 10].

**4.1 Definition** Let $R > 0$, $0 < \alpha < 1$ and $K > 0$, and let $f$ be holomorphic in a neighborhood of $\bar{D}_R$. The function $f$ belongs to the *Bernstein class* $B^1_{R,\alpha,K}$ if and only if

$$\max[| f(z) |, \ z \in \bar{D}_R]/\max[| f(z) |, \ z \in \bar{D}_{\alpha R}] \leq K. \tag{4.1}$$

**4.2 Definition** Let $N$ be an integer, $R > 0$ and $c > 0$, and let $f(z) = \sum_{i \geq 0} a_i z^i$ be an analytic function on a neighborhood of $0 \in \mathbb{C}$. The function $f$ belongs to the *Bernstein class* $B^2_{N,R,c}$ if and only if for all $j \geq N$:

$$| a_j | R^j \leq c \max_{0 \leq i \leq N} | a_i | R^i. \tag{4.2}$$

**4.3 Theorem** *Let $f$ be an element of $B^2_{N,R,c}$. Then $f$ is analytic on the open disc $D_R$ and belongs to $B^1_{R',\alpha,K}$ for all $R' < R$, $\alpha < 1$, and*

$$K = \alpha^{-N}[1 + \alpha(1 - \alpha^N)/(1 - \alpha) + c\beta/(1 - \beta)], \quad \textit{where} \quad \beta = R'/R.$$

*Conversely if $f$ belongs to $B^1_{R,\alpha,c}$, then $f \in B^2_{N,R,c}$ with*

$$N = [(\log_2(K) - \log_2(1 - \alpha) + 1)/(\log_2(1/\alpha))],$$
$$c = [K(2K + 1)/(1 - \alpha)^2].$$

The number of zeros of a function which belongs to a Bernstein class is bounded according to what follows.

**4.4 Theorem** *If $f \in B^1_{R,\alpha,K}$, then the number of zeros of $f$ in the disc $\bar{D}_{\alpha R}$ is less than*

$$\log_2(K)/[\log_2[(1+\alpha^2)/2\alpha]]. \tag{4.3}$$

**4.5 Lemma** *Let $f \in B^2_{N,R,c}$ and $R'' = R/[2^{3N}\max(c,2)]$. Then the number of zeros of $f$ in the disc $\bar{D}_{R''}$ is less than $N$.*

## 4.2 The Hironaka polynomial division theorem

The precise statement of the following theorem was provided by P. Milman. The terminology "Hironaka polynomial division theorem" is also due to him. This technical result allows one to clarify the presentation of the "Quantitative Bautin's theorem" first proved in [9].

**4.6 Theorem** *Let $I$ be an ideal of $\mathbb{C}[\lambda_1,\dots,\lambda_D]$. There is a system of generators $g_1,\dots,g_s$ of the ideal $I$ and constants $C$ and $C_1$ such that for all elements $f \in I$ of degree $k$, there is a decomposition*

$$f(\lambda) = \sum_{i=1}^{s} \phi_i(\lambda) g_i(\lambda), \tag{4.4}$$

*with*

$$\deg(\phi_i) \le \deg(f) = k, \tag{4.5a}$$

*and*

$$|\phi_i| \le CC_1^k \,|f|. \tag{4.5b}$$

## 4.3 The main theorem

**4.7 Theorem** *Let $f_\lambda(x) = \sum_{k\ge 1} f_k(\lambda)x^k$ be an $A_0$-series. For any $\lambda \in \mathbb{C}^D$, the function $f_\lambda$ belongs to the Bernstein class $B^2_{d-1,R,c}$ where $d$ is the Bautin index of the series $f_\lambda(x)$,*

$$R = [(C_1\bar{\lambda})^{K_1}K_4]^{-1}, \qquad \bar{\lambda} = \max(1, |\lambda|)$$

$$c = \begin{cases} [MCK_3(C_1\bar{\lambda})^{K_2}]/R^d, & \text{if } R \le 1, \\ [MCK_3(C_1\bar{\lambda})^{K_2}]/R, & \text{if } R \ge 1. \end{cases}$$

*The constants $K_1,\dots,K_4$ are those which appear in the definition of an $A_0$-series (Definition 3.2). The constants $C$ and $C_1$ appear in the "Hironaka polynomial division theorem", and $M$ is the norm of the matrix of the change of basis $(f_k)$ to a Gröbner basis $(g_l)$ of the Bautin ideal of the series $f_\lambda(x)$.*

Theorem 4.7 and Lemma 4.5 now imply the following:

**4.8 Theorem** *The number of zeros of $f_\lambda(x)$ in the disc $\bar{D}_{R''}$ is less than $d-1$ with*

$$R'' = R/[2^{3(d-1)}\max(c,2)]. \tag{4.6}$$

## 4.4 Computation of the Bautin ideal for Liénard systems

This part deals with Liénard equations whose particular interest was motivated in the first paragraph. Liénard equations play also a key role in Hilbert's 16th problem, as suggested for instance in [20, 21], because of the topological simplicity of the return mapping . Limit cycles encircle the origin and are necessarily contained in the domain of existence of the return mapping. This domain of existence may of course be different from to the domain of convergence of the analytic series which defines the return mapping in a neighborhood of the origin. Nevertheless it is interesting to produce an estimate of the size of this domain of convergence. Many contributions have previously been made to Liénard equations, and particularly by N. G. Lloyd and his co-workers [1, 5, 15–18]. But the approach via the Lyapunoff series which is used in all these references does not entail information on the domain of convergence of the first return mapping itself. A direct computation of the return mapping (indeed of its converse) is possible.

This provides a bound to the number of complex limit cycles in a fixed neighborhood of the origin.

Examples produced with perturbation theory (first-order) of Hamiltonians by A. Lins Neto, W. de Melo and C. C. Pugh [13] show that this bound is optimal

This paragraph is thus concerned with the polynomial vector fields $X$ of following type

$$X = -x\partial/\partial y + y\partial/\partial x + \sum_{i=1}^{d}[\lambda_i x^i]y\partial/\partial y, \tag{4.7}$$

and the associated flow, solution of the system:

$$\dot{x} = y, \tag{4.8a}$$

$$\dot{y} = -x + \sum_{i=1}^{d}[\lambda_i x^i]y = -x + p(x)y. \tag{4.8b}$$

The equations of the flow yield a second order differential equation classically called the Liénard equation. Note that this differential system is sometimes written:

$$\dot{x} = Y - P(x), \tag{4.8c}$$

$$\dot{Y} = -x, \tag{4.8d}$$

which is easily changed into the preceding ones with $y = Y - P(x)$, if $P'(x) = -p(x)$.

Writing (4.8a)–(4.8d) in polar coordinates $(r, \theta)$:

$$x = r\cos\theta, \qquad y = r\sin\theta, \tag{4.9}$$

one obtains

$$2r\dot{r} = 2(x\dot{x} + y\dot{y}), \qquad r\dot{r} = r^2 p(r\cos\theta)\sin^2\theta \tag{4.10}$$

$$\dot{\theta} = (x\dot{y} - y\dot{x})/(x^2 + y^2) = -1 + \sin\theta\cos\theta p(r\cos\theta). \tag{4.11}$$

This yields:

$$dr/d\theta = rp(r\cos\theta)\sin^2\theta/[-1 + \sin\theta\cos\theta p(r\cos\theta)]. \tag{4.12}$$

Bautin's approach is based on the study of solutions of (4.12) $r = r(\theta)$ so that $r(0) = r_0$, given as an expansion:

$$r_0 = r + v_2(\theta)r^2 + \cdots + v_k(\theta)r^k + \cdots \tag{4.13}$$

Comparison between (4.12) and (4.13) yields:

$$\sum_{k\geq 1} v_k'(\theta)r^k[-1 + \sin\theta\cos\theta p(r\cos\theta)] + \sum_{k\geq 1}[kv_k(\theta)p(r\cos\theta)\sin^2\theta]r^k = 0. \tag{4.14}$$

From this one obtains the following recurrency relation on the coefficients $v_k(\theta)$:

$$v_k'(\theta) = \sum_{l=1}^{d} \lambda_l(\cos\theta)^l\sin\theta[\cos\theta v_{k-l}'(\theta) + (k-l)\sin\theta v_{k-l}(\theta)]. \tag{4.15}$$

The following result was proved in [14]:

**4.9 Theorem** *Let $2n$ be the largest even integer such that $2n \leq d$. The Bautin ideal of the differential system* (4.8a)–(4.8d) *is generated by $(\lambda_2, \ldots, \lambda_{2n})$, and its Bautin index is equal to $2n+1$.*

The proof tells slightly more. Consider the increasing sequence of ideals generated by the coefficients:

$$(v_2(2\pi), \ldots, v_k(2\pi)). \tag{4.16}$$

The first coefficient $v_2(2\pi)$ is zero, hence the first one which contributes to the Bautin ideal is $v_3(2\pi)$, which is proportional to $\lambda_2$. Consider now increasing values of the integer $k$. The sequence of ideals does not increase at a step $v_k(2\pi)$ with even $k$. When $k$ is odd, $k = 2k_0 + 1$, then $v_{2k_0+1}$ contributes with a term $c_{2k_0+1}\lambda_{2k_0}$ with $c_{2k_0+1} \neq 0$, hence the ideal strictly increases and becomes equal to:

$$(\lambda_2, \ldots, \lambda_{2k_0}). \tag{4.17}$$

This proceeds till

$$k_0 = n = [d/2],$$

and later the ideal does not increase any more. The Bautin ideal is thus equal to:

$$(\lambda_2, \ldots, \lambda_{2n}), \tag{4.18}$$

and the increasing sequence of ideals (4.16) stabilizes at the step $2k_0 + 1 = 2n + 1$.

The classical Bautin approach entails the following:

**4.10 Theorem** *For all values of $\lambda$, there is a neighborhood of the origin on which the number of real limit cycles is less than $n-1$ ($n$ if we consider the perturbation of a focus).*

Note that this bound is in agreement with the bound foreseen in the Lins Neto-de Melo-Pugh conjecture (cf. [13]). Of course this computation does not provide any control of the size of the domain on which the number of limit cycles is less than $n$ in terms of the coefficients of the perturbation. This is the reason to develop further analysis based on the complexification of the return mapping. Let us consider again the famous example of the van der Pol equation:

$$\begin{aligned} \dot{x} &= y, \\ \dot{y} &= -x + p(x)y, \qquad p(x) = \lambda(1 - x^2). \end{aligned} \tag{4.19}$$

In that case, there is a perturbation of the linear part into a focus, $d = 2$, and thus $n = 1$ and we obtain a bound of 1 for the number of limit cycles. This is a well known result and it is also known that in the van der Pol example there is effectively a limit cycle which is born from a supercritical Hopf bifurcation. Note that the existence of a subcritical Hopf bifurcation giving birth to two limit cycles needs a polynomial $p(x)$ of degree $d = 4$.

## 4.5 Estimates of the radius of convergence of the first return mapping in the Liénard case

In the Liénard case, the following holds:

**4.11 Proposition** *The series* (4.13) *is an* $A_0$*-series for all values of* $\theta$*, with:*

$$K_1 = 1, \qquad K_2 = -1, \qquad K_3 = \pi/2, \qquad K_4 = 2, \tag{4.20}$$

*if* $|\lambda| < 1$

**4.12 Theorem** *The return mapping of the vector field* $X$ *converges at least on the disc* $D(0, R(\rho))$ *of radius* $R(\rho) = \rho/2$*, where* $\rho$ *is the unique positive real number such that*

$$\rho^d \mid \lambda_d \mid + \cdots + \rho \mid \lambda_1 \mid = 1. \tag{4.21}$$

This last estimate improves (4.6) for small $|\lambda|$ but not for large ones.

**Proof** Change of coordinates $(x, y)$ into $(\rho x, \rho y)$ transforms

$$\begin{aligned} \dot{x} &= y \\ \dot{y} &= -x + p(x)y, \end{aligned} \tag{4.22a}$$

into

$$\begin{aligned} \dot{x} &= y \\ \dot{y} &= -x + \sum_{i=1}^{d} (\lambda_i \rho^i) x^i y. \end{aligned} \tag{4.22b}$$

This means that any result obtained on the disc $D(0, R)$ for the equation (4.22a) is valid on the disc $D(0, \rho R)$ for the new equation $\lambda_i \mapsto \lambda_i' = \rho^i \lambda_i$. □

## 4.6 Estimates of the number of complex limit cycles on a fixed neighborhood of the origin for Liénard's equations

Note now that in comparison with the "polynomial Hironaka division theorem" that we need to use in the general situation, the division step is quite easy.

In the case of Liénard equations, $(\lambda_2, \ldots, \lambda_{2n})$ defines clearly a Hironaka basis (standard basis or Gröbner basis).

Following the techniques of [9,10], we prove now that the first return mapping belongs to a Bernstein class. It is convenient at this point to change notation and write

$$f(\lambda, r) = r + v_2(2\pi)r^2 + \cdots + v_k(2\pi)r^k + \cdots \tag{4.23a}$$

for the (converse of the) first return mapping, and $f_k(\lambda) = v_k(2\pi)$ for its coefficients. Indeed, we consider the analytic extension of the series to the complex domain and denote now $x$ as the complex variable in place of the real variable $r$.

$$f(\lambda, x) = x + f_2(\lambda)x^2 + \cdots + f_k(\lambda)x^k + \cdots . \tag{4.23b}$$

**4.13 Proposition** *The analytic series* (4.23a)–(4.23b) *belongs to the Bernstein class* $B^2_{(B,R,c)}$ *with*

$$R = K_4(1+ \mid \lambda \mid), \tag{4.24a}$$

$$c = \frac{nK_3^2 K_4^{2n-1}}{c_{2n+1}^n}. \tag{4.24b}$$

The following holds:

**4.14 Theorem** *The number of zeros of* $f_\lambda(x)$ *in the disc* $D(0, R')$ *is less than* $B - 1 = 2n$ *with*

$$R' = \frac{c_{2n+1}^n}{2^{10n-1} n\pi^2 (1+ \mid \lambda \mid)^{4n+1}}. \tag{4.25}$$

For small values of $\mid \lambda \mid$, this estimate can be improved with the scaling property presented previously:

**4.15 Theorem** *The vector field* $X$ *has less than* $2n$ *complex limit cycles on* $D(0, R'(\rho))$, *the disc of radius*

$$R'(\rho) = \frac{\rho c_{2n+1}^n}{\pi^2 n 2^{14n}}, \tag{4.26a}$$

*where* $\rho$ *is the unique positive real number such that*

$$\rho^d \mid \lambda_d \mid + \cdots + \rho \mid \lambda_1 \mid = 1. \tag{4.26b}$$

# 5 Abel equation, algebraic moments

## 5.1 Center problem for plane polynomial vector fields. Reduction to the Abel equation

Consider a system on the plane

$$\begin{aligned} \dot{x} &= -y + F(x,y) \\ \dot{y} &= x + G(x,y) \end{aligned} \tag{5.1}$$

with $F$ and $G$ homogeneous polynomials of degree $d$. The classical Center-Focus problem, posed by Poincaré in the 1880's, concerns the possibility to write explicit conditions for an isolated singularity of the system to be a center (all orbits near the origin are closed).

This problem can be reduced to the center problem for an Abel equation $y' = p(x)y^2 + q(x)y^3$ (with $p$ and $q$ polynomials in $\sin x$, $\cos x$), as follows.

Write (5.1) in polar coordinates $x = r\cos\theta$, $y = r\sin\theta$. We obtain

$$\dot{x} = \dot{r}\cos\theta - r\sin\theta\cdot\dot{\theta} = -r\sin\theta + F(r\cos\theta, r\sin\theta), \tag{5.2a}$$

$$\dot{y} = \dot{r}\sin\theta + r\cos\theta\cdot\dot{\theta} = r\cos\theta + G(r\cos\theta, r\sin\theta). \tag{5.2b}$$

Multiplying the first equation by $\cos\theta$, the second by $\sin\theta$ and adding, we get

$$\begin{aligned} \dot{r} &= F(r\cos\theta, r\ \sin\theta)\cos\theta + G(r\cos\theta, r\sin\theta)\sin\theta \\ &= r^d(F(\cos\theta,\sin\theta)\cos\theta + G(\cos\theta,\sin\theta)\sin\theta) \\ &= r^d f(\theta). \end{aligned} \tag{5.3a}$$

In a similar way,

$$\dot{\theta} = 1 + r^{d-1}(-F(\cos\theta,\sin\theta)\sin\theta + G(\cos\theta,\sin\theta)\cos\theta) = 1 + r^{d-1}g(\theta). \tag{5.3b}$$

Finally, we get

$$\frac{dr}{d\theta} = \frac{\dot{r}}{\dot{\theta}} = \frac{r^d f(\theta)}{1 + r^{d-1}g(\theta)}. \tag{5.4}$$

Then we apply a transformation, suggested by L. Cherkas:

$$\rho = \frac{r^{d-1}}{1 + r^{d-1}g(\theta)}r^{d-1} = \frac{\rho}{1 - \rho g(\theta)}. \tag{5.5}$$

Notice that for $d = 2$ this transformation and its inverse are regular at $r = \rho = 0$. For $d > 2$, Cherkas' transformation, considered over complex numbers, ramifies at zero, but for small real $r > 0$, $\rho > 0$, it is monotone.

Denote $1 + r^{d-1}g(\theta)$ by $A$. We have

$$\begin{aligned} \frac{d\rho}{d\theta} &= \frac{1}{A^2}\left((d-1)r^{d-2}\cdot\frac{dr}{d\theta}\cdot A - r^{d-1}\left((d-1)r^{d-2}\frac{dr}{d\theta}g(\theta) + r^{d-1}g'(\theta)\right)\right) \\ &= \frac{r^{d-1}}{A^2}[((d-1)f(\theta) - g'(\theta))r^{d-1} - (d-1)f(\theta)g(\theta)r^{2d-2}A] \\ &= ((d-1)f(\theta) - g'(\theta))\rho^2 - (d-1)f(\theta)g(\theta)\rho^3. \end{aligned} \tag{5.6}$$

Thus we obtain

$$\frac{d\rho}{d\theta} = p(\theta)\rho^2 + q(\theta)\rho^3, \tag{5.7}$$

where $p(\theta) = (d-1)f(\theta) + g'(\theta)$ and $q(\theta) = -(d-1)f(\theta)g(\theta)$ are polynomials in $\sin\theta$, $\cos\theta$ of degrees $d+1$ and $2d+2$, respectively.

Clearly, the trajectory of (5.1) near the origin is closed if and only if the corresponding solution of (5.7) satisfies $\rho(2\pi) = \rho(0)$ (and hence is periodic). Therefore the center-focus problem for (5.1) is translated into the problem of finding conditions on $p$ and $q$, under which all the solutions of (5.7) are periodic. In reference [3], we constructed for the Abel equation (5.7) the Bautin ideal, whose zeros correspond to the coefficients of $p$ and $q$, for which all the solutions are periodic, and discussed the relation of this property with the existence of the integrating factor for (5.2a)–(5.2b).

If we concentrate on a polynomial Abel equation

$$y' = p(x)y^2 + q(x)y^3 \tag{5.8}$$

with $p$ and $q$ polynomials with complex coefficients in a complex variable $x$, the corresponding center problem is to find conditions for $y(0) \equiv y(1)$ for all the solutions of (5.8).

This setting does not correspond directly to any question about polynomial systems on the plane (although in many special situations similar equations arise.) However, it is believed to provide essentially the same difficulties as the original problem, while bringing possible algebraic simplifications.

## 5.2 Iterated moments of Abel equations

Iterated moments naturally appear with the computation of the successive derivatives of the solution of Abel equations

$$\frac{dy}{dx} = p(x)y^2 + \epsilon q(x)y^3 \tag{5.9}$$

relative to the parameter $\epsilon$.

We computed in [1], the first derivative relative to $\epsilon$ of the solution $y(1,\epsilon,c)$ with initial data $y(0,\epsilon) = c$. We give here a formula for the successive derivatives at any order.

For $\epsilon = 0$, the equation becomes

$$\frac{dy}{dx} = p(x)y^2, \tag{5.10}$$

which yields

$$y(x) = \frac{c}{1 - cP(x)}, \tag{5.11}$$

with $P(x) = \int_0^x p(t)\,dt$. For $\epsilon \neq 0$, the solution is defined by

$$dy - p(x)y^2\,dx - \epsilon q(x)y^3\,dx = 0, \tag{5.12}$$

or equivalently,

$$dH + \epsilon q(x)y\,dx = 0, \tag{5.13}$$

with

$$H(x,y) = -\frac{1}{y} + P(x).$$

Note that $H$ and $c$ are essentially equivalent ($H = 1/c$). Assume further that $P(1) = 0$.

Let $\omega$ be a 1-form:

$$\omega = a(x, y)\, dx + b(x, y)\, dy, \tag{5.14}$$

and let $\gamma_0(c)$ be the arc of solution of the Abel equation (5.10):

$$\gamma_0(c) = (x, y), \qquad 0 \leq x \leq 1, \qquad y = \frac{c}{1 - cP(x)}.$$

Denote more generally by $\gamma_\epsilon(c)$ the arc of solution of the Abel equation (5.8) such that $0 \leq x \leq 1$.

Restriction of the 1-form $\omega$ to the curve $\gamma_0(c)$ yields:

$$\left[ a\left(x, \frac{c}{1 - cP(x)}\right) + b\left(x, \frac{c}{1 - cP(x)}\right) \frac{c^2 P'(x)}{(1 - cP(x))^2} \right] dx. \tag{5.15}$$

Let $F(x, c)$ be the unique function so that

$$\frac{\partial F}{\partial x} = \left[ a\left(x, \frac{c}{1 - cP(x)}\right) + b\left(x, \frac{c}{1 - cP(x)}\right) \frac{c^2 P'(x)}{(1 - cP(x))^2} \right], \tag{5.16}$$

and

$$F(0, c) \equiv 0.$$

We have indeed shown the following:

**5.1 Proposition** *Any 1-form $\omega = a(x, y)dx + b(x, y)dy$ can be decomposed as follows:*

$$\omega = dF + g(x, y)\, dc,$$

*where*

$$\begin{aligned} g(x, y) &= \overline{g}(x, c)\, |_{c = y/(1 + yP(x))}, \\ \overline{g}(x, c) &= b\left(x, \frac{c}{1 - cP(x)}\right) \frac{1}{(1 - cP(x))^2} - \frac{\partial F}{\partial c}(x, c). \end{aligned} \tag{5.17}$$

This proposition is obviously quite close to the relative cohomology result shown in (Françoise [16]) for the function $H = c$. The difference here is that there is a decomposition even if the integral of the 1-form does not vanish on $\gamma_0(c)$. So in some sense the result is closer to the decomposition obtained recently in (Françoise-Pelletier [17]). Now we have the following:

**5.2 Corollary** *For any 1-form $\omega = a(x, y)\, dx + b(x, y)\, dy$, the vanishing (for all $c$)*

$$\int_{\gamma_0(c)} \omega \equiv 0,$$

*is equivalent to*

$$F(0, c) \equiv F(1, c).$$

Our aim is to compute the first non-zero derivatives of $y(1,\epsilon,c)$ relative to the parameter $\epsilon$. We follow the algorithm of the successive derivatives of (Françoise [16]). This algorithm was developped for the Poincaré center-focus problem. It is quite interesting to compare the same algorithm used now for polynomial Abel equations because there are noticeable differences. The algorithm provides exactly the first return mapping. In the previous studies related to the center-focus problem, it was only possible to compute the first non-zero derivative (relative to the parameter $\epsilon$) of the return mapping. The situation would be different in the case of trigonometric Abel equations (to be discussed in a separate article). The main idea of the algorithm relates to integrable 1-parameter deformations of exact forms. A 1-parameter deformation

$$dH + \epsilon\omega$$

of the exact form $dH$ has an integrating factor

$$\left(1 + \sum_{k=1}^{\infty} \epsilon^k g_k\right)$$

if there is a series

$$H + \sum_{k=1}^{\infty} \epsilon^k R_k,$$

such that

$$\left(1 + \sum_{k=1}^{\infty} \epsilon^k g_k\right)(dH + \epsilon\omega) = d\left(H + \sum_{k=1}^{\infty} \epsilon^k R_k\right). \tag{5.18}$$

This yields:

$$g_{k-1}\omega + g_k\, dH = dR_k. \tag{5.19}$$

We have:

$$\int_{\gamma_\epsilon(c)} [dH + \epsilon\omega] \equiv 0, \tag{5.20}$$

with

$$\omega = q(x)y\, dx.$$

Write

$$y(1,\epsilon,c) = c + \epsilon y_1(c) + \cdots + \epsilon^k y_k(c) + \cdots.$$

Now define inductively the sequence of 1-forms $\omega_k$ and of functions $R_k$ by

$$\begin{aligned} \omega_1 = \omega &= dF + g(x,y)\, dc, \\ R_1(x,c) &= F(x,c), \end{aligned} \tag{5.21}$$

and

$$\omega_k = dR_k - g_k(x,y)\, dc = g_{k-1}\omega. \tag{5.22}$$

Denote

$$R(x,y,\epsilon) = H + \sum_{k=1}^{\infty} \epsilon^k R_k. \tag{5.23}$$

The formula

$$\int_{\gamma_\epsilon(c)} (1+\epsilon g_1+\cdots+\epsilon^k g_k+\cdots)\,[dH+\epsilon\omega]\equiv 0 \tag{5.24}$$

implies

$$R(1,y(1,\epsilon),\epsilon)=0.$$

The function $R(1,c,\epsilon)$ is thus the converse of the first return mapping $y(1,c,\epsilon)$. Obviously, finding the successive coefficients of one series is equivalent to the determination of the coefficients of the other, and the Bautin ideals of the two series are the same.

The general framework can be applied to

$$\omega = q(x)y\,dx.$$

This yields:

$$F(x,c)=F_1(x,c)=\int_0^x q(t)\frac{c}{1-cP(t)}. \tag{5.25}$$

The vanishing of the first derivative is thus equivalent to the vanishing of all the moments:

$$\begin{gathered} F(1,c)\equiv 0,\\ \int_0^1 q(t)P(t)^n\,dt=0,\quad n=0,1,\ldots, \end{gathered} \tag{5.26}$$

as was shown in (Briskin-Françoise-Yomdin [4]). The first function $R_1(x,c)=F(x,c)$ displays:

$$F(x,c)=\int_0^x q(t)\frac{c}{1-cP(t)}.$$

The function $g_1$ is just

$$g_1=\int_0^x q(t)\frac{\partial}{\partial c}\cdot\frac{c}{1-cP(t)},$$

thus the second derivative is equal to

$$\int_0^1 q(x_1)\frac{c}{1-cP(x_1)}\int_0^{x_1} q(x_2)\frac{\partial}{\partial c}\cdot\frac{c}{1-cP(x_2)}\,dx_2\,dx_1. \tag{5.27}$$

And more generally, the $k$-th derivative has the form

$$\int_{0\le x_k\le x_{k-1}\le\cdots\le x_1\le 1} q(x_1)\cdots q(x_k)\left(\frac{c}{1-cP(x_1)}\frac{\partial}{\partial c}\right)\cdot\left(\frac{c}{1-cP(x_2)}\frac{\partial}{\partial c}\right)\cdot\cdots\cdot\left(\frac{c}{1-cP(x_k)}\right). \tag{5.28}$$

The coefficients of (5.28) when it is expanded as a series in $c$ are of the form

$$\int_{0\le x_k\le x_{k-1}\le\cdots\le x_1\le 1} q(x_1)\cdots q(x_k)P(x_1)^{l_1}\cdots P(x_k)^{l_k}. \tag{5.29}$$

## References

[1] N. N. Bautin, On the number of limit cycles which appear with the variation of coefficients from an equilibrium position of focus or center type, *Mat. Sb.* **30** (1952), 181–196; *Amer. Math. Soc. Transl.* **5** (1962), 396–413.

[2] T. R. Blows and N. G. Lloyd, The number of small-amplitude limit cycles of Liénard equations, *Math. Proc. Cambridge Philos. Soc.* **95** (1984), 359–366.

[3] M. Briskin, J.-P. Françoise, and Y. Yomdin, The Bautin ideal of the Abel equation, *Nonlinearity* **11** (1998), 431–443.

[4] M. Briskin, J.-P. Françoise, and Y. Yomdin, Center conditions, composition of polynomials and moments on algebraic curves, *Ergodic Theory Dynam. Systems* **19** (1999), 1201–1220.

[5] M. Briskin and Y. Yomdin, Algebraic families of analytic functions, *J. Differential Equations* **136** (2) (1997), 248–267.

[6] C. Christopher, Abel equations: composition conjectures and the model problem, *Bull. London Math. Soc.* **32** (2000), 332–338.

[7] C. J. Christopher and N. G. Lloyd, Small-amplitude limit cycles in Liénard systems, *NoDEA Non-linear Differential Equations Appl.* **3** (1996), 183–190.

[8] C. Doss-Bachelet, J.-P. Françoise, and C. Piquet, Bursting oscillations in two coupled Fitztlugh-Nagumo systems, to appear in *Complex Us.*

[9] F. Dumortier, Techniques in the theory of local bifurcations: blow-up, normal form, nilpotent bifurcations, singular perturbations, in: *Bifurcations and Periodic Orbits of Vector Fields* (D. Schlomiuk, ed.), NATO ASI series C **408**, Kluwer, Dordrecht, 1993.

[10] F. Dumortier and R. Roussarie, Canard cycles and center manifolds, Mem. Amer. Math. Soc. **577**, 1–97.

[11] F. Dumortier and C. Rousseau, Cubic Liénard equations with linear damping, *Nonlinearity* **9** (1990), 1489–1500.

[12] R. FitzHugh, Thresholds and plateaus in the Hodgkin-Huxley nerve equations, *J. Gen. Physiol.* **43** (1960), 867–896.

[13] R. FitzHugh, Mathematical models of excitation and propagation in nerves, in: *Biological Engineering* (H. P. Schwan, ed.), McGraw-Hill, New York, 1969, 1–85.

[14] J.-P. Françoise, Analytic properties of the return mapping of Liénard equations, *Math. Res. Lett.* **9** (2002), 1–12.

[15] J.-P. Françoise, On the bifurcation of periodic orbits, *Comput. Appl. Math.* **20** (12) (2001), 91–119.

[16] J.-P. Françoise, Successive derivatives of a first-return map, application to quadratic vector fields, *Ergodic Theory Dynam. Systems* **16** (1996), 87-96.

[17] J.-P. Françoise and M. Pelletier, Higher order Melnikov functions. In preparation.

[18] J.-P. Françoise and R. Pons, Computer algebra methods and the stability of differential systems, *Random Comput. Dynam.* **3** (4) (1995), 265–287.

[19] J.-P. Françoise and C. C. Pugh, Keeping track of limit cycles, *J. Differential Equations* **65** (1986), 139–157.

[20] J.-P. Françoise and Y. Yomdin, Bernstein inequality and applications to analytic geometry and differential equations, *J. Funct. Anal.* **146** (1997), 185–205.

[21] J.-P. Françoise and Y. Yomdin, Projection of analytic sets and Bernstein inequalities, in: *Singularities Symposium-Lojasiewicz 70* (B. Jacubczyk et al., eds.), Banach Center Publ. **44**, Warszawa, 1998, 103–108.

[22] A. Gasull and J. Torregrosa, Small-amplitude limit cycles in Liénard systems via multiplicity, *J. Differential Equations* **159** (1999), 168–211.

[23] A. Goldbeter, *Biochemical Oscillations and Cellular Rythms*, Cambridge University Press, 1996.

[24] J. Keener and J. Sneyd, *Mathematical Physiology*, Interdiscip. Appl. Math. **8**, Springer-Verlag, New York, 1998.

[25] A. I. Khibnik, B. Krauskopf, and C. Rousseau, Global study of a family of cubic Liénard equations, *Nonlinearity* **11** (1998), 1505–1519.

[26] A. Lins Neto, W. de Melo, and C. C. Pugh, On Liénard equations, in: *Geometry and Topology (Rio de Janeiro, 1976)*, Lecture Notes in Math. **597**, Springer-Verlag, New York, 1977, 335–357.

[27] N. G.Lloyd, Liénard systems with several limit cycles, *Math. Proc. Cambridge Philos. Soc.* **102** (1987), 565-572.

[28] N. G. Lloyd and S. Lynch, Small-amplitude limit cycles of certain Liénard systems, *Proc. Roy. Soc. London Ser. A* **418** (1988), 199–208.

[29] S. Lynch, Small-amplitude limit cycles of Liénard systems, *Calcolo* **27** (1990), 1–32.

[30] S. Lynch, Liénard systems and the second part of Hilbert's 16th problem, *Nonlinear Anal.* **30** (1997), 1395-1403.

[31] J. D. Murray, *Mathematical Biology*, Biomathematics **19**, Springer-Verlag, Berlin, 1989.

[32] F. Pakovich, A counterexample to the composition conjecture, *Arxiv. Math.* (2002).

[33] J. Rinzel, Bursting oscillations in an excitable membrane model, in: *Ordinary and Partial Differential Equations* (B. D. Sleeman and R. J. Jarvis, eds.), Springer-Verlag, New-York, 1985.

[34] S. Smale, Dynamics retrospective: great problems, attempts that failed, *Phys. D* **51** (1991), 267–273.

[35] S. Smale, Mathematical problems for the next century, *Math. Intelligencer* **20** (1998), 7–15.

[36] B. van der Pol, On relaxation-oscillations, *Philos. Mag.* **3** (7) (1926), 978–992.

[37] B. van der Pol and J. van der Mark, The heartbeat considered as a relaxation oscillation, and an electrical model of the heart, *Philos. Mag.* **6** (7) (1928), 763–775.

[38] B. van der Pol, Oscillations sinusoidales et de relaxation, *L'Onde électrique* (1931), 245–256.

[39] Y. Yomdin, Global finiteness properties of analytic families and algebra of their Taylor coefficients, in: *The Arnol'd Fest (Toronto, Ontario, 1997)*, Fields Inst. Commun. **24**, Amer. Math. Soc., Providence, RI, 1999, 527–555.

[40] H. Żołądek, Quadratic systems with center and their perturbations, *J. Differential Equations* **109** (1994), 223–273.

# Complexity of computations with Pfaffian and Noetherian functions

Andrei GABRIELOV

*Department of Mathematics*
*Purdue University*
*West Lafayette, IN 47907*
*USA*

Nicolai VOROBJOV

*Department of Computer Science*
*University of Bath*
*Bath, BA2 7AY, England*
*UK*

**Abstract**

This paper is a survey of the upper bounds on the complexity of basic algebraic and geometric operations with Pfaffian and Noetherian functions, and with sets definable by these functions. Among other results, we consider bounds on Betti numbers of sub-Pfaffian sets, multiplicities of Pfaffian intersections, effective Łojasiewicz inequality for Pfaffian functions, computing the frontier and closure of restricted semi-Pfaffian sets, constructing smooth stratifications and cylindrical cell decompositions (including an effective version of the complement theorem for restricted sub-Pfaffian sets), relative closures of non-restricted semi-Pfaffian sets and bounds on the number of their connected components, bounds on multiplicities of isolated solutions of systems of Noetherian equations.

## 1 Introduction

Pfaffian functions, introduced by Khovanskii [24, 25] at the end of the 1970s, are analytic functions satisfying triangular systems of Pfaffian (first order partial differential) equations with polynomial coefficients. Over $\mathbb{R}$, these functions, and the corresponding semi- and sub-Pfaffian sets, are characterized by global finiteness properties similar to the properties of polynomials and semialgebraic sets. This allows one to establish efficient upper bounds on the complexity of different algebraic and geometric operations with these functions and sets.

One of the important applications of the Pfaffian theory is in the real algebraic geometry of *fewnomials* — polynomials defined by simple formulas, possibly of a high degree. The complexity of operations with fewnomials in many cases allows upper bounds in terms of the complexity of the defining formulas, independent of their degree.



*Y. Ilyashenko and C. Rousseau (eds.),*
*Normal Forms, Bifurcations and Finiteness Problems in Differential Equations,* 211–250.

Over $\mathbb{C}$, Pfaffian functions, and more general Noetherian functions (satisfying the same kind of equations but without the triangularity condition) are characterized by local finiteness properties, sufficient for upper bounds on the complexity of stratification, frontier and closure.

This paper is a survey of results about the upper bounds for operations on Pfaffian and Noetherian functions, and on sets definable by these functions. All bounds are functions of a finite set, called *format*, of some natural parameters (like the degree or the number of variables) associated with a Pfaffian or Noetherian function or a definable set. The goal is to obtain upper bounds that are as low as possible.

The content of the paper is as follows. We start with definitions and examples of Pfaffian functions (including fewnomials), semi- and sub-Pfaffian sets. We discuss Khovanskii's bound on the number of isolated solutions of a system of Pfaffian equations, and its extensions for Betti numbers of semi- and sub-Pfaffian sets.

Next, we derive an upper bound for the multiplicity of a Pfaffian intersection. This bound is then used to obtain an effective version of the Łojasiewicz inequality for a Pfaffian function, and an algorithm for constructing the frontier and closure (in its domain of definition) of a semi-Pfaffian set. The bound for Pfaffian multiplicities is also used to construct an algorithm for a weak stratification of a semi-Pfaffian set, i.e., for a representation of the set as a disjoint union of smooth manifolds. The complexity (running time) of the stratification algorithm is explicitly estimated in terms of the format of the input semi-Pfaffian set. This also implies explicit upper bounds on the number of strata and their formats. Stratification is used for an effective proof of the following *complement theorem*: the complement of a projection of a restricted (relatively compact in its domain of definition) semi-Pfaffian set to a subspace is again a projection of a semi-Pfaffian set. The proof of the theorem uses an algorithm for a *cylindrical cell decomposition* of a restricted semi-Pfaffian set, a construction which is important in its own right.

For general (not necessarily restricted) semi-Pfaffian sets, the complement theorem is not known to be true. We describe a wider category of sets, called *limit sets*, which is a Boolean algebra and is also closed under projections to subspaces. We describe an explicit upper bound on the number of connected components of a limit set in terms of the format of the set.

Noetherian functions do not generally satisfy the global finiteness properties of Pfaffian functions. However, it is possible to establish some *local* finiteness properties. We describe in some detail a proof of an explicit upper bound on the multiplicity of an isolated solution of a system of Noetherian equations. This proof involves several stages. First, we give a brief introduction to integration over Euler characteristics. Next, we consider the univariate case, which implies, in particular, an upper bound on the vanishing order of a multivariate Noetherian function. As another application of the univariate result, we derive an upper bound for the degree of nonholonomy of a system of polynomial vector fields. The last stage of the proof for the multivariate case involves a lower bound on the codimension of the set of intersections of high multiplicity.

## 2 Pfaffian functions and sub-Pfaffian sets

Pfaffian functions, introduced by Khovanskii at the end of the 1970s, are real or complex analytic functions satisfying triangular systems of Pfaffian (first order partial differential) equations with polynomial coefficients. We use the notation $\mathbb{K}^n$, where $\mathbb{K}$ is either $\mathbb{R}$ or $\mathbb{C}$, in the statements relevant to both cases.

**2.1 Definition ([24, 25, 18])** A *Pfaffian chain* of the order $r \geq 0$ and degree $\alpha \geq 1$ in an open domain $G \subset \mathbb{K}^n$ is a sequence of analytic functions $f_1, \dots, f_r$ in $G$ satisfying differential equations

$$df_j(\mathbf{x}) = \sum_{1 \leq i \leq n} g_{ij}(\mathbf{x}, f_1(\mathbf{x}), \dots, f_j(\mathbf{x}))\, dx_i \tag{2.1}$$

for $1 \leq j \leq r$. Here $g_{ij}(\mathbf{x}, y_1, \dots, y_j)$ are polynomials in $\mathbf{x} = (x_1, \dots, x_n)$, $y_1, \dots, y_j$ of degrees not exceeding $\alpha$. A function $f(\mathbf{x}) = P(\mathbf{x}, f_1(\mathbf{x}), \dots, f_r(\mathbf{x}))$, where $P(\mathbf{x}, y_1, \dots, y_r)$ is a polynomial of a degree not exceeding $\beta \geq 1$, is called a *Pfaffian function* of order $r$ and degree $(\alpha, \beta)$. Note that the Pfaffian function $f$ is defined only in the domain $G$ where all functions $f_1, \dots, f_r$ are analytic, even if $f$ itself can be extended as an analytic function to a larger domain.

**2.2 Remark** This definition is more restrictive than the definition from [25], where the Pfaffian chains are defined as sequences of nested integral manifolds of polynomial 1-forms. Both definitions lead to essentially the same class of Pfaffian functions, although the orders and degrees of Pfaffian chains for the same Pfaffian function can be different according to these two definitions. We found Definition 2.1 to be more convenient to trace the behaviour of parameters of Pfaffian functions under different operations.

More general definitions of Pfaffian functions, where the coefficients of (2.1) are not necessarily polynomial, were considered in [29, 32]. Most of our constructions can be adjusted to this more general definition, however upper bounds on the complexity may not be efficient enough in this case.

**2.3 Examples** (a) Pfaffian functions of order 0 and degree $(1, \beta)$ are polynomials of degrees not exceeding $\beta$.

(b) The exponential function $f(x) = e^{ax}$ is a Pfaffian function of order 1 and degree $(1,1)$ in $\mathbb{R}$, due to the equation $df(x) = af(x)\, dx$. More generally, for $i = 1, 2, \dots, r$, let $E_i(x) := e^{E_{i-1}(x)}$, $E_0(x) = ax$. Then $E_r(x)$ is a Pfaffian function of order $r$ and degree $(r, 1)$, since $dE_r(x) = aE_1(x) \cdots E_r(x)\, dx$.

(c) The function $f(x) = 1/x$ is a Pfaffian function of order 1 and degree $(2,1)$ in the domain $\{x \in \mathbb{R} \mid x \neq 0\}$, due to the equation $df(x) = -f^2(x)\, dx$.

(d) The logarithmic function $f(x) = \ln(|x|)$ is a Pfaffian function of order 2 and degree $(2,1)$ in the domain $\{x \in \mathbb{R} \mid x \neq 0\}$, due to the equations $df(x) = g(x)\, dx$ and $dg(x) = -g^2(x)\, dx$, where $g(x) = 1/x$.

(e) The polynomial $f(x) = x^m$ can be viewed as a Pfaffian function of order 2 and degree $(2,1)$ in the domain $\{x \in \mathbb{R} \mid x \neq 0\}$ (but not in $\mathbb{R}$), due to the equations $df(x) = mf(x)g(x)\, dx$ and $dg(x) = -g^2(x)\, dx$, where $g(x) = 1/x$. In some cases a better way to deal with $x^m$ is to change the variable $x = e^u$ reducing this case to (b).

(f) The function $f(x) = \tan(x)$ is a Pfaffian function of order 1 and degree $(2,1)$ in the domain $\bigcap_{k \in \mathbb{Z}} \{x \in \mathbb{R} \mid x \neq \pi/2 + k\pi\}$, due to the equation $df(x) = (1 + f^2(x))\, dx$.

(g) The function $f(x) = \arctan(x)$ is a Pfaffian function in $\mathbb{R}$ of order 2 and degree $(3,1)$, due to the equations $df(x) = g(x)\, dx$ and $dg(x) = -2xg^2(x)\, dx$, where $g(x) = (x^2+1)^{-1}$.

(h) The function $\cos(x)$ is a Pfaffian function of order 2 and degree $(2,1)$ in the domain $\bigcap_{k\in\mathbb{Z}}\{x \in \mathbb{R} \mid x \neq \pi + 2k\pi\}$, due to the equations $\cos(x) = 2f(x) - 1$, $df(x) = -f(x)g(x)\,dx$, and $dg(x) = ((1+g^2(x))/2)\,dx$, where $f(x) = \cos^2(x/2)$ and $g(x) = \tan(x/2)$. Also, since $\cos(x)$ is a polynomial of degree $m$ of $\cos(x/m)$, the function $\cos(x)$ is Pfaffian of order 2 and degree $(2,m)$ in the domain $\bigcap_{k\in\mathbb{Z}}\{x \in \mathbb{R} \mid x \neq m\pi + 2km\pi\}$. The same is true, of course, for any shift of this domain by a multiple of $\pi$. However, $\cos(x)$ is not a Pfaffian function in the whole real line.

**2.4 Lemma** *The sum (resp. product) of two Pfaffian functions $f_1$ and $f_2$ of orders $r_1$ and $r_2$ and degrees $(\alpha_1,\beta_1)$ and $(\alpha_2,\beta_2)$ respectively, is a Pfaffian function of order $r_1+r_2$ and degree $(\alpha, \max\{\beta_1,\beta_2\})$ (resp. $(\alpha, \beta_1+\beta_2)$), where $\alpha = \max\{\alpha_1,\alpha_2\}$. If the two functions are defined by the same Pfaffian chain of order $r$, then the orders of the sum and of the product are both equal to $r$.*

**Proof** Combine Pfaffian chains for $f_1$ and $f_2$ into a Pfaffian chain for $f_1+f_2$ and $f_1f_2$. If a Pfaffian chain is common for the two functions, then it is also a Pfaffian chain for their sum and product. □

**2.5 Lemma** *A partial derivative of a Pfaffian function of order $r$ and degree $(\alpha,\beta)$ is a Pfaffian function having the same Pfaffian chain of order $r$ and degree $(\alpha, \alpha+\beta-1)$.*

**Proof** is straightforward. □

**2.6 Example** (*Fewnomials*) Generalizing Example 2.3 (e), we can view a polynomial $f \in \mathbb{R}[\mathbf{x}] = \mathbb{R}[x_1,\ldots,x_n]$ as a Pfaffian function in the following sense. Each monomial

$$f_{i_1\cdots i_n} := a_{i_1\cdots i_n} x_1^{i_1}\cdots x_n^{i_n}$$

of $f$ with $a_{i_1\cdots i_n} \neq 0$ is a Pfaffian function in the domain $G := \{\mathbf{x} \in \mathbb{K}^n \mid x_1\cdots x_n \neq 0\}$, of order $n+1$ and degree $(2,1)$, due to the equations

$$df_{i_1\cdots i_n} = \sum_{1\le j\le n} i_j f_{i_1\cdots i_n} g_j\,dx_j,$$
$$dg_j = -g_j^2\,dx_j,$$

where $g_j = 1/x_j$. According to Lemma 2.4, $f$ is a Pfaffian function in $G$ of order $n+m$ and degree $(2,1)$, where $m$ is the number of all monomials in $f$ (with non-zero coefficients). Let $\mathcal{K}$ be a set of all monomials of $f$. Then $f$ is called a *fewnomial* with the support $\mathcal{K}$.

A polynomial $F = P(x_1,\ldots,x_n,u_1,\ldots,u_m)$ of degree $\beta$ in $x_1,\ldots,x_n,u_1,\ldots,u_m$, where $x_1,\ldots,x_n$ are variables and $u_1,\ldots,u_m \in \mathcal{K}$ are monomials, is called a *fewnomial of pseudodegree* $\beta$ with support $\mathcal{K}$. Obviously $F$ is a Pfaffian function of order $n+m$ and of degree $(2,\beta)$. Note that $\beta$ may be different from the degree $d$ of the polynomial $P$ after the substitution of the monomials $u_j$. We call $d$ the *degree* of $F$.

In the sequel we will reserve the term "polynomial" for Pfaffian functions of order 0 and degree $(1,\beta)$ (see Example 2.3 (a)).

We define specializations for the class of Pfaffian functions over $\mathbb{R}$ of more general concepts of semi- and subanalytic sets (see, e.g., [4]).

**2.7 Definition** (*Semi-Pfaffian sets*) A set $X \subset \mathbb{R}^n$ is called *semi-Pfaffian* in an open domain $G \subset \mathbb{R}^n$ if it consists of points in $G$ satisfying a Boolean combination $\mathcal{F}$ of some atomic equations and inequalities $f = 0$, $g > 0$, where $f, g$ are Pfaffian functions having a common Pfaffian chain defined in $G$. We will write $X = \{\mathcal{F}\}$. A semi-Pfaffian set $X$ is *restricted in $G$* if its topological closure lies in $G$. A semi-Pfaffian set is called *basic* if the Boolean combination is just a conjunction of equations and strict inequalities.

**2.8 Definition** (*Sub-Pfaffian sets*) A set $X \subset \mathbb{R}^n$ is called *sub-Pfaffian* in an open domain $G \subset \mathbb{R}^n$ if it is an image of a semi-Pfaffian set under a projection into a subspace.

Our main object of study will be a following subclass of sub-Pfaffian sets.

**2.9 Definition** (*Restricted sub-Pfaffian sets*) Consider the closed cube $\mathcal{I}^{m+n} := [-1, 1]^{m+n}$ in an open domain $G \subset \mathbb{R}^{m+n}$ and the projection map $\pi : \mathbb{R}^{m+n} \to \mathbb{R}^n$. A subset $Y \subset \mathcal{I}^n$ is called *restricted sub-Pfaffian* if $Y = \pi(X)$ for a restricted semi-Pfaffian set $X \subset \mathcal{I}^{m+n}$.

A restricted sub-Pfaffian set need not be semi-Pfaffian as the following example, due to Osgood [35], shows. This fact is the most significant difference between the theories of semi- and sub-Pfaffian sets on the one hand, and semialgebraic sets on the other.

**2.10 Example** Let

$$Y := \{(x, y, z) \in \mathcal{I}^3 \mid \exists u \in [0, 1]\ (y = xu, z = xe^u)\} \subset \mathbb{R}^3.$$

Then $Y$ is a two-dimensional restricted sub-Pfaffian set such that any real analytic function vanishing on $Y$ in the neighbourhood of the origin is identically zero. Hence $Y$ is not semi-Pfaffian.

Restricted sub-Pfaffian sets form a Boolean algebra. Finite unions and intersections of arbitrary sub-Pfaffian sets are clearly sub-Pfaffian. The fact that *complement* of a restricted sub-Pfaffian set in $\mathcal{I}^n$ is also restricted sub-Pfaffian is a particular case of Gabrielov's *complement theorem* [13]. We will consider an algorithmic version of this theorem in Section 7.

**2.11 Definition** (*Format*) Consider a semi-Pfaffian set

$$X := \bigcup_{1 \le i \le M} \{\mathbf{x} \in \mathbb{R}^s \mid f_{i1} = \cdots = f_{iI_i}, g_{i1} > 0, \ldots, g_{iJ_i} > 0\} \subset G, \tag{2.2}$$

where $f_{ij}$, $g_{ij}$ are Pfaffian functions with a common Pfaffian chain of order $r$ and degree $(\alpha, \beta)$, defined in an open domain $G$. Its *format* is a tuple $(r, N, \alpha, \beta, s)$, where $N \ge \sum_{1 \le i \le M}(I_i + J_i)$. For $s = m + n$ and a sub-Pfaffian set $Y \subset \mathbb{R}^n$ such that $Y = \pi(X)$, its *format* is the format of $X$.

We will refer to the representation of a semi-Pfaffian set in the form (2.2) as to *disjunctive normal form* (*DNF*).

**2.12 Remark** In this paper we are concerned with upper bounds of various characteristics of semi- and sub-Pfaffian sets and complexities of computations, as functions of the format. In fact, these characteristics and complexities also depend on the domain $G$ in which the corresponding Pfaffian chains are defined. Note that Definition 2.1 imposes no restrictions on an open set $G$, thus allowing it to be arbitrarily complex and induce this complexity on the corresponding semi- and sub-Pfaffian sets. To avoid this, we will always assume in the sequel (unless explicitly stated otherwise) that $G$ is "simple", like $\mathbb{R}^n$, $\mathcal{I}^n$, $\{\mathbf{x} \mid x_1 > 0, \dots, x_n > 0\}$, or $\{\mathbf{x} \mid \|\mathbf{x}\|^2 < 1\}$. A more general approach allows $G$ to be a semi-Pfaffian set defined by Pfaffian functions in a larger domain $G' \supset G$, which in turn is defined by Pfaffian functions in some $G'' \supset G'$, and so on (see details in [16]).

# 3 Betti numbers of sub-Pfaffian sets

## 3.1 Topological complexity of semi-Pfaffian sets

We start with the following fundamental result of Khovanskii which can be considered as an analog of Bezout's theorem for Pfaffian functions.

**3.1 Theorem ([24, 25])** *Consider a system of equations $f_1 = \cdots = f_n = 0$, where $f_i$, $1 \le i \le n$, are Pfaffian functions in a domain $G \subset \mathbb{R}^n$, having a common Pfaffian chain of order $r$ and degrees $(\alpha, \beta_i)$, respectively. Then the number of non-degenerate solutions of this system does not exceed*

$$\mathcal{M}(n, r, \alpha, \beta_1, \dots, \beta_n) := 2^{r(r-1)/2} \beta_1 \cdots \beta_n (\min\{n, r\}\alpha + \beta_1 + \cdots + \beta_n - n + 1)^r. \tag{3.1}$$

Applying directly this result to fewnomials (see Example 2.6) $f_1, \dots, f_n$ in the domain $\{\mathbf{x} \mid x_1 > 0, \dots, x_n > 0\} \subset \mathbb{R}^n$, we get the upper bound

$$2^{(n+m)(n+m-1)/2}(2n+1)^{n+m}$$

on the number of non-degenerate solutions of the system $f_1 = \cdots = f_n = 0$. Here $m$ is the number of different monomials occurring in at least one of polynomials $f_i$. There is, however, a better upper bound.

**3.2 Corollary** *The number of non-degenerate solutions of a system of polynomial equations $f_1 = \cdots = f_n = 0$ belonging to the octant $\{\mathbf{x} \mid x_1 > 0, \dots, x_n > 0\}$ does not exceed*

$$2^{m(m-1)/2}(n+1)^m,$$

*where $m$ is the number of different monomials occurring in at least one of the polynomials $f_i$.*

**Proof** Making a change of variables $x_j = e^{y_j}$ we reduce the system $f_1 = \cdots = f_n = 0$ to a system of linear equations in $m$ exponential functions of the kind $e^{i_1 y_1 + \cdots + i_n y_n}$. The left-hand sides of these equations are Pfaffian functions in $\mathbb{R}^n$ having a common Pfaffian chain of order $m$ and degrees equal to $(1, 1)$. Now the bound follows directly from Theorem 3.1. □

Theorem 3.1 implies various upper bounds on the topological complexity of a semi-Pfaffian set as functions of the format by applying almost without change some well-developed techniques for semialgebraic sets. Here is an example.

**3.3 Corollary** *Consider a system of equations $f_1 = \cdots = f_k = 0$, where $f_i$, $1 \leq i \leq k$, are Pfaffian functions in a domain $G \subset \mathbb{R}^n$, having a common Pfaffian chain of order $r$ and degrees $(\alpha, \beta_i)$, respectively. Then the number of connected components of $X := \{f_1 = \cdots = f_k = 0\}$ does not exceed*

$$2^{r(r-1)/2+1}\beta(\alpha + 2\beta - 1)^{n-1}((2n-1)(\alpha+\beta) - 2n + 2)^r, \tag{3.2}$$

*where $\beta := \max_{1 \leq i \leq k}\{\beta_i\}$.*

**Proof** (cf. [33]) Choose a large enough positive $R \in \mathbb{R}$ such that $\{\mathbf{x} \in X \mid \|\mathbf{x}\|^2 \leq R\}$ has the same number of connected components as $X$. For a sufficiently small real $\varepsilon > 0$ the number of connected components of $X$ does not exceed the number of connected components of the smooth compact hypersurface $X_\varepsilon := \{F = 0\}$, where $F := f_1^2 + \cdots + f_k^2 + \varepsilon(\|\mathbf{x}\|^2 - R)$.

After a generic rotation of coordinates in $\mathbb{R}^n$, the projection of $X_\varepsilon$ on any of them is a Morse function. In particular, all solutions of the system of equations

$$F = \frac{\partial F}{\partial x_2} = \cdots = \frac{\partial F}{\partial x_n} = 0 \tag{3.3}$$

are non-degenerate and thus, by Theorem 3.1 and Lemma 2.5, their number does not exceed (3.2). Because $X_\varepsilon$ is compact, each of its connected components contains a solution of system (3.3). This concludes the proof. □

The most general (to our knowledge) upper bound on the topological complexity of semi-Pfaffian sets is provided by the following theorem (see [49]).

**3.4 Theorem** *Consider a semi-Pfaffian set $X := \{\mathcal{F}\} \subset G \subset \mathbb{R}^n$, where $G$ is an open domain, $\mathcal{F}$ is either a conjunction of equations and strict inequalities or a Boolean combination (with no negations) of non-strict inequalities. Let $\mathcal{F}$ contain equations or inequalities of the kind $f * 0$, where $* \in \{=, >, \geq\}$, and there are $s$ different Pfaffian functions $f$ in $G$ having a common Pfaffian chain of order $r$ and degree $(\alpha, \beta)$. Then the sum of the Betti numbers of $X$ does not exceed*

$$s^n 2^{r(r-1)/2} O(n\beta + \min\{n, r\}\alpha)^{n+r}. \tag{3.4}$$

This theorem is a direct analog of the Basu [2] refinement of the Petrovskii-Oleinik-Thom-Milnor [37, 34, 42, 33] bounds for semialgebraic sets, and can be proved in a similar way replacing references to Bezout's theorem by references to Khovanskii's Theorem 3.1.

**3.5 Remark** Corollary 3.3, Theorem 3.4, and most of the other upper bounds for Pfaffian functions appearing in the sequel can be reformulated for a particular case of fewnomials similar to how it was done for Theorem 3.1. These specifications are straightforward and will be omitted.

## 3.2 Number of consistent sign assignments

**3.6 Definition** For a given finite family $h_1, \ldots, h_k$ of Pfaffian functions $h_i$ in an open domain $G$ define its *consistent sign assignment* as a non-empty semi-Pfaffian set in $G$ of the kind

$$\{\mathbf{x} \in G \mid h_{i_1} = \cdots = h_{i_{k_1}} = 0,\ h_{i_{k_1}+1} > 0, \ldots, h_{i_{k_2}} > 0,\ h_{i_{k_2}+1} < 0, \ldots, h_{i_k} < 0\},$$

where $i_1, \ldots, i_{k_1}, \ldots, i_{k_2}, \ldots, i_k$ is a permutation of $1, \ldots, k$.

**3.7 Theorem ([18])** *Let $h_1, \dots, h_k$ be Pfaffian functions in $G$ having a common Pfaffian chain of order $r$ and degrees $(\alpha, \beta_1), \dots, (\alpha, \beta_k)$, respectively. Then the number of distinct consistent sign assignments for $h_1, \dots, h_k$ does not exceed*

$$\min\{3^k, 2^{r(r-1)/2+1}(2n+1)^r(\alpha+8k\beta)^{n+r+1}\},$$

*where $\beta := \max_{1 \le i \le k}\{\beta_i\}$.*

**Proof** The bound $3^k$ is trivial. Choose in every consistent sign assignment one arbitrary point. Let $\Lambda$ be the set of all chosen points. There exists a positive $\varepsilon \in \mathbb{R}$ such that for every $\mathbf{x} \in \Lambda$ and every $i = 1, \dots, k$ the inequality $h_i(\mathbf{x}) > 0$ implies $h_i(\mathbf{x}) > \varepsilon$, and $h_i(\mathbf{x}) < 0$ implies $h_i(\mathbf{x}) < -\varepsilon$. It is easy to prove that the number of consistent sign assignments does not exceed the number of connected components of the semi-Pfaffian set

$$S := \left\{ \mathbf{x} \in G \mid h := \prod_{1 \le i \le k} (h_i + \varepsilon)^2 (h_i - \varepsilon)^2 > 0 \right\}.$$

For a small enough positive $\delta < \varepsilon$, the number of connected components of $S$ does not exceed the number of connected components of $\{\mathbf{x} \in G \mid h = \delta\}$. It remains to apply Corollary 3.3 to $\{\mathbf{x} \in G \mid h = \delta\}$. □

## 3.3 Sub-Pfaffian sets defined by formulae with quantifiers

We now address the problem of estimating the topological complexity of restricted *sub-Pfaffian* sets. Until recently this question did not have a satisfactory solution even in the particular case of projections of *semialgebraic* sets defined by Boolean formulae. In the Pfaffian category the situation is further complicated by the fact that the quantifier elimination process cannot be used.

In this section we describe a reduction of estimating of Betti numbers of sets defined by formulae with quantifiers to a similar problem for sets defined by quantifier-free formulae. More precisely, let $X$ be a subset in $\mathcal{I}^{n_0} = [-1, 1]^{n_0} \subset \mathbb{R}^{n_0}$ defined by a formula

$$X = \{\mathbf{x}_0 \mid Q_1\mathbf{x}_1 Q_2\mathbf{x}_2 \cdots Q_\nu\mathbf{x}_\nu((\mathbf{x}_0, \mathbf{x}_1, \dots, \mathbf{x}_\nu) \in X_\nu)\}, \tag{3.5}$$

where $Q_i \in \{\exists, \forall\}$, $Q_i \ne Q_{i+1}$, $\mathbf{x}_i \in \mathbb{R}^{n_i}$, and let $X_\nu$ be either an open or a closed set in $\mathcal{I}^{n_0+\cdots+n_\nu}$ being a difference between a finite $CW$-complex and one of its subcomplexes. For instance, if $\nu = 1$ and $Q_1 = \exists$, then $X$ is the projection of $X_\nu$.

We express an upper bound on each Betti number of $X$ via a sum of Betti numbers of some sets defined by quantifier-free formulae involving $X_\nu$. In conjunction with Theorem 3.4 this implies an upper bound for restricted sub-Pfaffian sets defined by formulae with quantifiers.

Throughout this section each topological space is assumed to be a difference between a finite $CW$-complex and one of its subcomplexes.

**3.8 Example** The closure $X$ of the interior of a compact set $Y \subset \mathcal{I}^n$ is homotopy equivalent to

$$X_{\varepsilon,\delta} = \{\mathbf{x} \mid \exists \mathbf{y}(\|\mathbf{x} - \mathbf{y}\| \le \delta)\ \forall \mathbf{z}(\|\mathbf{y} - \mathbf{z}\| < \varepsilon)\ (\mathbf{z} \in Y)\}$$

for small enough $\delta, \varepsilon > 0$ such that $\delta \gg \varepsilon$. Representing $X_{\varepsilon,\delta}$ in the form (3.5), we conclude that $X$ is homotopy equivalent to $X_{\varepsilon,\delta} = \{\mathbf{x} \mid \exists \mathbf{y}\ \forall \mathbf{z}\ X_2\}$, where

$$X_2 = \{(\mathbf{x}, \mathbf{y}, \mathbf{z}) \mid (\|\mathbf{x} - \mathbf{y}\| \leq \delta \wedge (\|\mathbf{y} - \mathbf{z}\| \geq \varepsilon \vee \mathbf{z} \in Y))\}$$

is a closed set in $\mathcal{I}^{3n}$. Our results allow to bound from above the Betti numbers of $X$ in terms of the Betti numbers of some sets defined by quantifier-free formulae involving $X_2$.

## 3.4 A spectral sequence associated with a surjective map

**3.9 Definition** A continuous map $f : X \to Y$ is *locally split* if for any $y \in Y$ there is an open neighbourhood $U$ of $y$ and a section $s : U \to X$ of $f$ (i.e., $s$ is continuous and $fs = \mathrm{Id}$). In particular, a projection of an open set in $\mathbb{R}^n$ on a subspace of $\mathbb{R}^n$ is always locally split.

For any two continuous surjective maps $f_1 : X_1 \to Y$ and $f_2 : X_2 \to Y$, define the operation $\times_Y$:

$$X_1 \times_Y X_2 := \bigcup_{\mathbf{y} \in Y} f_1^{-1}(\mathbf{y}) \times f_2^{-1}(\mathbf{y}).$$

Note that if $Y$ is a singleton, then $X_1 \times_Y X_2$ coincides with the usual Cartesian product $X_1 \times X_2$.

**3.10 Theorem ([20])** *Let $f : X \to Y$ be a surjective cellular map. Assume that $f$ is either closed or locally split. Then for any Abelian group $G$, there exists a spectral sequence $E^r_{p,q}$ converging to $H_*(Y, G)$ with*

$$E^1_{p,q} = H_q(W_p, G), \tag{3.6}$$

*where*

$$W_p = \underbrace{X \times_Y \cdots \times_Y X}_{p+1 \text{ times}}. \tag{3.7}$$

*In particular,*

$$\dim H_k(Y, G) \leq \sum_{p+q=k} \dim H_q(W_p, G) \tag{3.8}$$

*for all $k$.*

**3.11 Remark** Let $X, Y \subset \mathbb{R}^n$ and $f$ a surjective cellular map satisfying the following property. For any convergent sequence in $Y$ there is an infinite subsequence which is an $f$-image of a convergent sequence in $X$. This condition includes both the closed and the locally split cases and may be more convenient for applications. For such an $f$, Theorem 3.10 is also true.

## 3.5 Upper bounds for Betti numbers of sub-Pfaffian sets

Let $X = X_0 \subset \mathbb{R}^{n_0}$ be a sub-Pfaffian set defined by a formula

$$Q_1 \mathbf{x}_1 Q_2 \mathbf{x}_2 \cdots Q_\nu \mathbf{x}_\nu \mathcal{F}(\mathbf{x}_0, \mathbf{x}_1, \ldots, \mathbf{x}_\nu), \tag{3.9}$$

where $\mathcal{F}$ is a quantifier-free Boolean formula with no negations having $s$ atoms of the kind $f > 0$. Let all $f$'s be Pfaffian functions in an open domain $G$ having a common Pfaffian chain of order $r$ and degrees at most $(\alpha, \beta)$. Assume for definiteness that $X$ is closed (the case of an open set is similar).

Consider first the case of a single quantifier $Q_1 = \exists$. Then $\nu = 1$ and (3.9) reduces to $\exists \mathbf{x}_1 X_1$, where $X_1 = \{\mathcal{F}(\mathbf{x}_0, \mathbf{x}_1)\}$. Note that in this case $X = f(X_1)$, where $f$ is the projection map onto a subspace. According to Theorem 3.10,

$$\mathrm{b}_{q_0}(X) \leq \sum_{p_1+q_1=q_0} \mathrm{b}_{q_1}(\underbrace{X_1 \times_X \cdots \times_X X_1}_{p_1+1 \text{ times}}), \tag{3.10}$$

where $\mathrm{b}_i$ stands for the $i$th Betti number. Observe that $X_1 \times_X \cdots \times_X X_1$ is a closed set definable by a Boolean combination with no negations of $(p_1+1)s$ atoms of the kind $g > 0$, where the $g$'s are Pfaffian functions in an open domain $G \subset \mathbb{R}^{n_0+(p_1+1)n_1}$ having a common Pfaffian chain of order $(p_1+1)r$, degrees $(\alpha, \beta)$, and $n_0 + (p_1+1)n_1$ variables. Let $t_k := n_0 + (k+1)n_1$. According to Theorem 3.4, for any $q_1 \leq \dim(X)$,

$$\mathrm{b}_{q_1}(X_1 \times_X \cdots \times_X X_1) \leq ((p_1+1)s)^{t_{p_1}} 2^{(p_1+1)r((p_1+1)r-1)/2} O(t_{p_1}\beta + \min\{p_1 r, t_{p_1}\}\alpha)^{t_{p_1}+(p_1+1)r}.$$

Then, due to (3.10), for any $k \leq \dim(X) \leq n_0$,

$$\begin{aligned}\mathrm{b}_k(X) &\leq \sum_{p_1+q_1=k} \mathrm{b}_{q_1}(X_1 \times_X \cdots \times_X X_1)\\ &\leq k((k+1)s)^{t_k} 2^{(k+1)r((k+1)r-1)/2} O(t_k\beta + \min\{kr, t_k\}\alpha)^{t_k+(k+1)r}.\end{aligned}$$

Relaxing the obtained bound, we get

$$\mathrm{b}_k(X) \leq (ks)^{O(t_k)} 2^{(O(kr))^2} (t_k(\alpha+\beta))^{O(t_k+kr)}.$$

In [20] this bound is generalized for formulae with $\nu$ quantifiers. More precisely, if $X \subset \mathbb{R}^{n_0}$ is defined by (3.9), then for any $k \leq \dim(X) \leq n_0$,

$$\mathrm{b}_k(X) \leq s^{O(u_\nu)} 2^{O(\nu u_\nu + r^2 v_\nu^2)} (u_\nu(\alpha+\beta))^{O(u_\nu + r v_\nu)},$$

where

$$u_\nu := 2^\nu n_0 n_1 \cdots n_\nu, \qquad v_\nu := 2^{2\nu} n_0^2 n_1^2 \cdots n_{\nu-2}^2 n_{\nu-1}.$$

## 4 Multiplicities of Pfaffian intersections

**4.1 Definition** A *deformation* of a Pfaffian function $f(\mathbf{x})$ in $G \subset \mathbb{K}^n$ is an analytic function $\theta(\mathbf{x}, \varepsilon)$ in a domain $G' \subset \mathbb{K}^{n+1}$ such that $G = G' \cap \{\varepsilon = 0\}$, $\theta(\mathbf{x}, 0) = f(\mathbf{x})$, and, for a fixed $\varepsilon$, the function $\theta(\mathbf{x}, \varepsilon)$ is Pfaffian having the same Pfaffian chain and the same degree as $f(\mathbf{x})$.

**4.2 Definition** Let $f_1(\mathbf{x}), \ldots, f_n(\mathbf{x})$ be Pfaffian functions in $G \subset \mathbb{K}^n$. The *multiplicity* at $\mathbf{y} \in G$ of the *Pfaffian intersection* $f_1 = \cdots = f_n = 0$ is the maximal number of isolated *complex* solutions, for a fixed $\varepsilon \neq 0$, of the system of equations $\theta_1(\mathbf{x}, \varepsilon) = \cdots = \theta_n(\mathbf{x}, \varepsilon) = 0$ converging to $\mathbf{y}$ as $\varepsilon \to 0$. Here $\theta_i(\mathbf{x}, \varepsilon)$ is any deformation of $f_i(\mathbf{x})$ for all $1 \leq i \leq n$.

**4.3 Theorem ([11])** *Let $f_1(\mathbf{x}), \ldots, f_n(\mathbf{x})$ be Pfaffian functions in $G \subset \mathbb{K}^n$ having a common Pfaffian chain of order $r$ and degrees $(\alpha, \beta_1), \ldots, (\alpha, \beta_n)$, respectively. Then the multiplicity of the Pfaffian intersection $f_1 = \cdots = f_n = 0$ at any point $\mathbf{y} \in G$ does not exceed* (3.1).

**4.4 Corollary (Pfaffian Łojasiewicz inequality [11])** *Let $f$ be a Pfaffian function in an open domain $G \subset \mathbb{R}^n$ of order $r$ and degree $(\alpha, \beta)$. Then there is a neighbourhood $U$ of $\{f = 0\}$ in $G$ such that for any $\mathbf{x} \in U$,*

$$|f(\mathbf{x})| \geq C(\text{dist}(\mathbf{x}, \{f = 0\}))^q,$$

*for a real $C > 0$ and a positive integer*

$$\begin{aligned} q &\leq 2^{r(r-1)+1} 4^{n-1} \beta(\alpha + \beta - 1)^{n-1} (\min\{n, r\}\alpha + (n-1)(4\alpha + 3\beta - 5) + \beta)^r \\ &\leq 2^{r(r-1)+1} O(n)^r O(\alpha + \beta)^{n+r}. \end{aligned}$$

The following corollary will be needed in Section 5 for proving upper bounds on the frontier and closure of a semi-Pfaffian set and for an algorithm which computes frontier and closure.

Let

$$\mathcal{K}(m, n, r, \alpha, \beta) := \mathcal{M}(n, r, \alpha, \underbrace{\beta, \ldots, \beta}_{m+1}, \underbrace{\tau_m, \ldots, \tau_m}_{n-m-1}),$$

where $\mathcal{M}(n, r, \alpha, \beta_1, \ldots, \beta_n)$ is defined in (3.1) and $\tau_m = (m+2)(\alpha + \beta - 1)$. Let

$$\mathcal{K}(n, r, \alpha, \beta) := \max_{0 \leq m < n} \mathcal{K}(m, n, r, \alpha, \beta).$$

**4.5 Corollary ([14])** *Let $f_1, \ldots, f_I, g_1, \ldots, g_J$ be Pfaffian functions in a domain $G \subset \mathbb{R}^n$ having a common Pfaffian chain of order $r$ and degrees $(\alpha, \beta)$ each. Let $\kappa > \mathcal{K}(n, r, \alpha, \beta)$ be an integer. For a point $\mathbf{y} \in G$, let $F_i(\mathbf{x}, \mathbf{y})$ be the Taylor expansion of $f_i(\mathbf{x})$ at $\mathbf{y}$ of order $\kappa^2$, and let $G_j(\mathbf{x}, \mathbf{y})$ be the Taylor expansion of $g_j(\mathbf{x})$ at $\mathbf{y}$ of order $\kappa$. Then the closure of the semi-Pfaffian set*

$$X := \{\mathbf{x} \in G \mid f_1(\mathbf{x}) = \cdots = f_I(\mathbf{x}) = 0,\ g_1(\mathbf{x}) > 0, \ldots, g_J(\mathbf{x}) > 0\}$$

*contains $\mathbf{y}$ if and only if the closure of the following* semialgebraic *set $X_{\mathbf{y}}$ contains $\mathbf{y}$:*

$$X_{\mathbf{y}} := \{\mathbf{x} \in G \mid F_i(\mathbf{x}, \mathbf{y}) \leq |\mathbf{x} - \mathbf{y}|^{\kappa^2},\ \textit{for } i = 1, \ldots, I,\ G_j(\mathbf{x}, \mathbf{y}) > |\mathbf{x} - \mathbf{y}|^{\kappa},\ \textit{for } j = 1, \ldots, J\}.$$

The next corollary will be needed in Section 6 for a stratification algorithm.

**4.6 Definition** For a set of differentiable functions $\mathbf{h} = (h_1, \ldots, h_k)$, a set of distinct indices $\mathbf{i} = (i_1, \ldots, i_k)$ with $1 \leq i_\nu \leq n$, and an index $j$, $1 \leq j \leq n$, different from all $i_\nu$ we define a partial differential operator

$$\partial_{\mathbf{h},\mathbf{i},j} := \det \begin{pmatrix} \partial h_1/\partial x_{i_1} & \ldots & \partial h_1/\partial x_{i_k} & \partial h_1/\partial x_j \\ \cdots & \cdots & \cdots & \cdots \\ \partial h_k/\partial x_{i_1} & \ldots & \partial h_k/\partial x_{i_k} & \partial h_k/\partial x_j \\ \partial/\partial x_{i_1} & \ldots & \partial/\partial x_{i_k} & \partial/\partial x_j \end{pmatrix}.$$

When $k = 0$, the corresponding operator is simply $\partial_j := \partial/\partial x_j$. For $m \geq 0$ we define $\partial^m_{\mathbf{h},\mathbf{i},j}$ (resp. $\partial^m_j$) as the $m$th iteration of $\partial_{\mathbf{h},\mathbf{i},j}$ (resp. $\partial_j$).

**4.7 Corollary** *Let* $\mathbf{i} = (i_1, \ldots, i_k)$ *be a set of distinct indices,* $1 \leq i_\nu \leq n$. *Let* $f$ *be a Pfaffian function in an open neighbourhood* $G$ *of a point* $\mathbf{x} \in \mathbb{R}^n$ *of order* $r$ *and degree* $(\alpha, \beta_{k+1})$. *Let* $\mathbf{h} = (h_1, \ldots, h_k)$ *be a set of Pfaffian functions in* $G$ *of order* $r$ *and degrees* $(\alpha, \beta_1), \ldots, (\alpha, \beta_k)$ *respectively, each having the same Pfaffian chain as* $f$, *and such that* $h_1(\mathbf{x}) = \cdots = h_k(\mathbf{x}) = 0$,

$$\det \begin{pmatrix} \partial h_1/\partial x_{i_1} & \ldots & \partial h_1/\partial x_{i_k} \\ \cdots & \cdots & \cdots \\ \partial h_k/\partial x_{i_1} & \ldots & \partial h_k/\partial x_{i_k} \end{pmatrix} (\mathbf{x}) \neq 0.$$

*Suppose that* $\partial^{m_1}_{\mathbf{h},\mathbf{i},1} \cdots \partial^{m_n}_{\mathbf{h},\mathbf{i},n} f(\mathbf{x}) = 0$ *for* $0 \leq m_1 + \cdots + m_n \leq \mathcal{M}(k+1, r, \alpha, \beta_1, \ldots, \beta_{k+1})$, $m_{i_1} = \cdots = m_{i_k} = 0$. *Then* $f$ *vanishes identically on* $\{\mathbf{y} \in \mathbb{R}^n \mid h_1(\mathbf{y}) = \cdots = h_k(\mathbf{y}) = 0\}$ *in the neighbourhood of* $\mathbf{x}$.

# 5 Frontier and closure of a semi-Pfaffian set

## 5.1 Bounds on formats of frontier and closure

**5.1 Definition** The *closure* $\bar{X}$ of a set $X$ in an open domain $G$ is the intersection with $G$ of the usual topological closure of $X$:

$$\bar{X} := \{\mathbf{x} \in G \mid \forall \varepsilon > 0\ \exists \mathbf{y} \in X\ (|\mathbf{x} - \mathbf{y}| < \varepsilon)\}.$$

The *frontier* $\partial X$ of $X$ in $G$ is $\partial X := \bar{X} \setminus X$.

From the definition one could hope to infer that the closure and frontier of a semi-Pfaffian set are sub-Pfaffian. It turns out that a much stronger statement is true: closure and frontier are actually semi-Pfaffian.

**5.2 Theorem** ([14]) *Consider a semi-Pfaffian set in disjunctive normal form (DNF)*

$$X := \bigcup_{1 \leq i \leq M} \{\mathbf{x} \in G \mid f_{i1} = \cdots = f_{iI_i} = 0,\ g_{i1} > 0, \ldots, g_{iJ_i} > 0\} \subset \mathbb{R}^n$$

*having format* $(r, N, \alpha, \beta, n)$. *Then the closure* $\bar{X}$ *and frontier* $\partial X$ *of* $X$ *in* $G$ *are semi-Pfaffian sets which can be represented in DNF with the format of* $\bar{X}$ *being* $(r, N', \alpha, \beta', n)$, *where*

$$N' = (ND)^{(n+r+1)O(n)},$$

$\beta' = D^{O(n)}$, $D = \beta + \alpha(\mathcal{K}(n, r, \alpha, \beta) + 1)^2$, *and* $\mathcal{K}(n, r, \alpha, \beta)$ *is as defined in Section 4. The format of* $\partial X$ *is* $(r, N'', \alpha, \beta', n)$, *where*

$$N'' = (ND)^{(n+r+1)^2 O(n)}.$$

**Proof** The idea is to reduce the problem of describing the closure to the *semialgebraic* case using Corollary 4.5.

Since the closure of the union of sets equals to the union of closures, it is sufficient to consider just the case of a basic semi-Pfaffian set

$$X := \{\mathbf{x} \in G \mid f_1 = \cdots = f_I = 0,\ g_1 > 0, \ldots, g_J > 0\}$$

with $I + J = N$. We let $\kappa := \mathcal{K}(n, r, \alpha, \beta) + 1$ and use the notations from Corollary 4.5. According to Corollary 4.5, the closure $\bar{X}$ contains $\mathbf{y} \in G$ if and only if $\mathbf{y}$ belongs to the closure $\bar{X}_{\mathbf{y}}$. Let $h_1(\mathbf{x}), \ldots, h_r(\mathbf{x})$ be the common Pfaffian chain for $f_i$, $g_j$, ($i = 1, \ldots, I$, $j = 1, \ldots, J$). Direct calculation shows that

$$F_i(\mathbf{x}, \mathbf{y}) = \Phi_i(\mathbf{x}, \mathbf{y}, h_1(\mathbf{y}), \ldots, h_r(\mathbf{y})) \quad \text{and} \quad G_j(\mathbf{x}, \mathbf{y}) = \Psi_j(\mathbf{x}, \mathbf{y}, h_1(\mathbf{y}), \ldots, h_r(\mathbf{y})),$$

where $\Phi_i$ and $\Psi_j$ are polynomials in $\mathbf{x}$, $\mathbf{y}$, and $\mathbf{h} = (h_1, \ldots, h_r)$, of degrees not exceeding $\beta + \alpha\kappa^2$ and $\beta + \alpha\kappa$, respectively.

Thus, $\mathbf{y} \in \bar{X}$ if and only if

$$\begin{aligned}(\mathbf{y}, h_1(\mathbf{y}), \ldots, h_r(\mathbf{y})) \in \{(\mathbf{z}, h_1, \ldots, h_r) \in G \times \mathbb{R}^r \mid \forall \varepsilon > 0\ \exists \mathbf{x}\ (|\mathbf{x} - \mathbf{z}| \leq \varepsilon,\\ \Phi_i(\mathbf{x}, \mathbf{z}, \mathbf{h}) \leq |\mathbf{x} - \mathbf{z}|^{\kappa^2} \text{ for } i = 1, \ldots, I,\ \Psi_j(\mathbf{x}, \mathbf{z}, \mathbf{h}) > |\mathbf{x} - \mathbf{z}|^{\kappa} \text{ for } j = 1, \ldots, J)\}, \quad (5.1)\end{aligned}$$

where $\mathbf{z} = (z_1, \ldots, z_n)$ are variables. Formula

$$\begin{aligned}&\forall \varepsilon > 0\ \exists \mathbf{x}\ (|\mathbf{x} - \mathbf{z}| \leq \varepsilon,\\ &\quad \Phi_i(\mathbf{x}, \mathbf{z}, \mathbf{h}) \leq |\mathbf{x} - \mathbf{z}|^{\kappa^2} \text{ for } i = 1, \ldots, I,\ \Psi_j(\mathbf{x}, \mathbf{z}, \mathbf{h}) > |\mathbf{x} - \mathbf{z}|^{\kappa} \text{ for } j = 1, \ldots, J) \quad (5.2)\end{aligned}$$

of the first order theory of $\mathbb{R}$ contains two blocks of quantifiers of sizes 1 (for $\varepsilon$) and $n$ (for $x$), $n+r$ free variables, and $N$ polynomials of degrees at most $\beta + \alpha\kappa^2$. According to the efficient quantifier elimination algorithm [38] (see also [3]), there is an equivalent (defining the same set in $\mathbb{R}^{n+r}$) quantifier-free formula in DNF with format $(r, N', \alpha, \beta', n)$. Substituting $h_k(\mathbf{y})$ for $h_k$ into this formula, we obtain a semi-Pfaffian set in DNF with the properties required in the theorem.

The statement for $\partial X = \bar{X} \setminus X$ follows from the statement for $\bar{X}$, using Theorem 3.7 to represent the difference of two sets in DNF as a set in DNF. □

## 5.2 Complexity of computing frontier and closure

The proof of Theorem 5.2 shows that closure and frontier of a semi-Pfaffian set can be efficiently computed. Indeed, for given input functions $f_{i1}, \ldots, f_{iI_i}, g_{i1}, \ldots, g_{iJ_i}$, $i = 1, \ldots, M$, we can write out an explicit formula (5.2) with concrete polynomials $\Phi_i(\mathbf{x}, \mathbf{z}, \mathbf{h})$, $\Psi_j(\mathbf{x}, \mathbf{z}, \mathbf{h})$, and concrete integer $\kappa$. Then the quantifier elimination algorithm from [38, 3] is applied, which represents the closure $\bar{X}$ as a quantifier-free formula in DNF. To find the frontier $\partial X$, the algorithm first lists all consistent sign assignments for the family of all Pfaffian functions involved in DNF formulae for $X$ and $\bar{X}$. Then it selects all the assignments $A$ which lie in $\bar{X}$ and do not lie in $X$, by checking whether $A \cap \bar{X} \neq \emptyset$ and $A \cap X = \emptyset$. The union of the selected assignments coincides with $\partial X$.

In order to estimate the "efficiency" of a computation we need to specify more precisely a *model of computation*. As such we use a *real numbers machine* which is an analogy of a classical Turing machine but allows the exact arithmetic and comparisons on real numbers. Since we are interested only in upper complexity bounds for algorithms, we have no need in a formal definition of this model of computation (it can be found in [5]). In most of our computational problems we will need to modify the standard real numbers machine by equipping it with an *oracle* for deciding feasibility of any system of Pfaffian equations and inequalities. An oracle is a subroutine which can be used by a given algorithm any time the latter needs to check feasibility. We assume that this procedure always gives a correct answer ("true" or "false") though we do not specify how it actually works[1]. An *elementary step* of a real numbers machine is either an arithmetic operation, or a comparison (branching) operation, or an oracle call. The *complexity* of a real numbers machine is the number of elementary steps it makes in the worst case until termination, as a function of the format of the input.

Using the complexity upper bound for the quantifier elimination procedure from [38, 3], we obtain the following statement.

**5.3 Theorem** *There are two algorithms which for an input semi-Pfaffian set $X$ defined as in Theorem 5.2 produce the closure $\bar{X}$ and the frontier $\partial X$, respectively, representing them as semi-Pfaffian sets in DNF with formats described in Theorem 5.2. The algorithm for $\bar{X}$ does not use the oracle, its complexity does not exceed $(ND)^{(n+r+1)O(n)}$. The algorithm for $\partial X$ uses at most $(ND)^{(n+r+1)O(n)}$ calls of the oracle, its complexity does not exceed $(ND)^{(n+r+1)^2O(n)}$.*

**5.4 Remark** An analysis of the proofs of Theorems 5.2 and 5.3 easily shows that they are also true in a parametric form. More precisely, consider a set $X(\mathbf{t})$ defined by a formula in DNF, where all atomic Pfaffian functions $f_{ij}, g_{ij}$ depend on variables $\mathbf{x} \in \mathbb{R}^n$ and $\mathbf{t} \in \mathbb{R}^s$. Then there is an algorithm which computes formulae $\Gamma(\mathbf{x}, \mathbf{t})$ and $\Delta(\mathbf{x}, \mathbf{t})$ in DNF such that for any fixed $\mathbf{t}_0 \in \mathbb{R}^s$ we have $\{\mathbf{x} \in \mathbb{R}^n \mid \Gamma(\mathbf{x}, \mathbf{t}_0)\} = \overline{X(\mathbf{t}_0)}$ and $\{\mathbf{x} \in \mathbb{R}^n \mid \Delta(\mathbf{x}, \mathbf{t}_0)\} = \partial X(\mathbf{t}_0)$. Upper bounds on the formats of $\Gamma(\mathbf{x}, \mathbf{t})$, $\Delta(\mathbf{x}, \mathbf{t})$, and on the complexity of the algorithm are similar to the analogous bounds from Theorems 5.2 and 5.3.

## 5.3 Infinitesimal quantifiers

Theorems 5.2 and 5.3 can be interpreted using a language of "infinitesimal quantifiers" [39].

**5.5 Definition** Let $\mathcal{F}(\mathbf{x}, \mathbf{y})$ be a Boolean combination of some atomic equations and inequalities $f = 0$, $g > 0$, where $f, g$ are Pfaffian functions in variable vectors $\mathbf{x} \in \mathbb{R}^n$, $\mathbf{y} \in \mathbb{R}^m$ in an open domain $G \subset \mathbb{R}^{n+m}$. Then

- $(\exists \mathbf{y} \sim 0)\mathcal{F}$ stands for $\forall \varepsilon > 0\ \exists \mathbf{y}$ ($|\mathbf{y}| < \varepsilon$ and $\mathcal{F}$),
- $(\forall \mathbf{y} \sim 0)\mathcal{F}$ stands for $\exists \varepsilon > 0\ \forall \mathbf{y}$ (if $|\mathbf{y}| < \varepsilon$, then $\mathcal{F}$).

The operators $(\exists \mathbf{y} \sim 0)$ and $(\forall \mathbf{y} \sim 0)$ are called *infinitesimal quantifiers*, and can be read "there exists an arbitrarily small $\mathbf{y}$ such that $\mathcal{F}$ is true" and "for all sufficiently small $\mathbf{y}$, $\mathcal{F}$ is true", respectively.

It is easy to see that

$$\neg((\exists \mathbf{y} \sim 0)\mathcal{F}) \equiv (\forall \mathbf{y} \sim 0)\neg\mathcal{F} \quad \text{and} \quad \neg((\forall \mathbf{y} \sim 0)\mathcal{F}) \equiv (\exists \mathbf{y} \sim 0)\neg\mathcal{F}. \tag{5.3}$$

Infinitesimal quantifiers are convenient for describing some $\varepsilon/\delta$-constructions.

[1]For some classes of Pfaffian functions the feasibility problem is decidable on standard real numbers machines or Turing machines. Apart from polynomials, such a class is formed, for example, by terms of the kind $P(e^{x_1}, x_1, x_2, \dots, x_n)$, where $P$ is a polynomial in the variables $x_0, x_1, \dots, x_n$ (see [45]). For such classes the oracle can be replaced by a deciding procedure, and we get an algorithm in the usual sense.

**5.6 Example** A point $\mathbf{x}$ is a local maximum of a Pfaffian function $f$ if and only if

$$(\forall \mathbf{y} \sim 0)((\mathbf{y} \neq 0) \to (f(\mathbf{x}+\mathbf{y}) \leq f(\mathbf{x}))).$$

It is not immediately obvious that the set of all local maxima of $f$ is semi-Pfaffian.

Unlike ordinary quantifiers, the infinitesimal ones *can be eliminated* in the restricted case. Due to (5.3), it is sufficient to prove this just for the existential quantifier.

Consider in an open domain $G$ a semi-Pfaffian set $X := \{(\mathbf{x}, \mathbf{y}) \in G \mid \mathcal{F}(\mathbf{x}, \mathbf{y})\} \subset \mathcal{I}^{n+m} \subset G$ and the projection map $\pi : X \to \mathbb{R}^n$ on the subspace of the $\mathbf{x}$-coordinates. Let $X(\mathbf{x}) := \pi^{-1}(\mathbf{x}) \cap X$.

**5.7 Lemma** *The set* $\{\mathbf{x} \in \mathcal{I}^n \mid (\exists \mathbf{y} \sim 0)\mathcal{F}(\mathbf{x}, \mathbf{y})\}$ *coincides with* $\{\mathbf{x} \in \mathcal{I}^n \mid (\mathbf{x}, \mathbf{0}) \in \overline{X(\mathbf{x})}\}$, *where* $\mathbf{0}$ *is the origin in* $\mathbb{R}^m$.

**Proof** If a semi-Pfaffian set $Z$ is contained in the closed cube $\mathcal{I}^m$, then its topological closure coincides with $\bar{Z}$. Then the statement $(\exists \mathbf{y} \sim 0)(\mathbf{y} \in Z)$ is equivalent to $\mathbf{0} \in \bar{Z}$. It follows that for any fixed $\mathbf{x} \in \mathcal{I}^n$ the statement $(\exists \mathbf{y} \sim 0)\mathcal{F}(\mathbf{x}, \mathbf{y})$ is equivalent to $(\mathbf{x}, \mathbf{0}) \in \overline{X(\mathbf{x})}$. □

Due to Remark 5.4, the set $\{\mathbf{x} \in \mathcal{I}^n \mid (\mathbf{x}, \mathbf{0}) \in \overline{X(\mathbf{x})}\}$ is semi-Pfaffian, with explicit upper bounds on the format and on the complexity of the algorithm for computing this set. Lemma 5.7 implies that the same is true for the set $\{\mathbf{x} \in \mathcal{I}^n \mid (\exists \mathbf{y} \sim 0)\mathcal{F}(\mathbf{x}, \mathbf{y})\}$.

In particular, the set of all points of local maxima (Example 5.6) is semi-Pfaffian. In [39] the singular locus of a semi-Pfaffian set in $\mathcal{I}^n$ is defined by a formula using only infinitesimal quantifiers, thus the singular locus is semi-Pfaffian with an explicit upper bound on the format.

# 6 Stratification of a semi-Pfaffian set

In [46] Whitney proved that an algebraic set in $\mathbb{R}^n$ can be represented as a finite disjoint union of smooth manifolds which are semialgebraic sets. Łojasiewicz [30, 31] extended Whitney's theorem to the class of real semianalytic sets. Later Gabrielov [13] showed, as a part of an elementary proof of his *complement theorem*, that the smooth strata of a semianalytic set $X$ can be defined by functions belonging to the smallest extension of a family of functions defining $X$ which is closed under addition, multiplication and taking partial derivatives. Important classes sharing this property are Pfaffian functions and their special subclasses: polynomials, fewnomials, exponential polynomials. Combined with estimates on multiplicities of Pfaffian intersection from Section 4, this result allows to construct an algorithm which produces a smooth stratification of a semi-Pfaffian set, to estimate its complexity, and to bound formats of the resulting strata [18]. Similar results are true for sets in $\mathbb{C}^n$ defined by Boolean combinations of atomic formulae of the kind $f = 0$ and $f \neq 0$, where $f$ is a Pfaffian function over $\mathbb{C}$.

**6.1 Definition** A *weak stratification* of a semi-Pfaffian set $X$ is a partition of $X$ into a disjoint union of smooth (i.e., nonsingular), not necessarily connected, possibly empty, semi-Pfaffian subsets $X_i$ called *strata*. A stratification is *basic* if all strata are basic semi-Pfaffian sets which are *effectively nonsingular*, i.e., the system of equations and inequalities for each stratum $X_i$ of codimension $k$ includes a set of $k$ Pfaffian functions $h_{i1}, \ldots, h_{ik}$ such that the restriction $h_{ij}|_{X_i} \equiv 0$ for $j = 1, \dots, k$, and $dh_{i1} \wedge \dots \wedge dh_{ik} \neq 0$ at every point of $X_i$. Note that we do not require the boundary of a stratum to coincide with a union of some other strata, let alone any regularity conditions.

Consider a semi-Pfaffian set $X$ defined by (2.2) having a format $(r, N, \alpha, \beta, n)$. Let

$$\mathcal{B} := (\alpha + \beta + 1)^{(r+1)^{O(n)}}.$$

**6.2 Theorem ([18])** *There is an algorithm (without an oracle) which produces a finite basic weak stratification of a semi-Pfaffian set $X$. The number of strata (some of which may be empty) is $N^{n+r}\mathcal{B}$, each having the format $(r, N\mathcal{B}, \alpha, \mathcal{B}, n)$. The complexity of the algorithm does not exceed $3^N N^{n+1}\mathcal{B}$. If the oracle is allowed, the algorithm produces only non-empty strata and its complexity does not exceed $N^{n+r}\mathcal{B}$.*

**Outline of a proof** (inspired by [46, 13]). Let $X = \{\mathbf{x} \in \mathbb{R}^2 \mid f(\mathbf{x}) = 0, g(\mathbf{x}) > 0\}$ be a basic semi-Pfaffian set (curve) in $\mathbb{R}^2$, and let zero be a regular value of $f$. Then the sets $X^1 := \{\mathbf{x} \in X \mid (\partial_1 f, \partial_2 f)(\mathbf{x}) \neq 0\}$ and $X^2 := \{\mathbf{x} \in X \mid (\partial_1 f, \partial_2 f)(\mathbf{x}) = 0\}$ form a stratification of $X$. If zero is a singular value of $f$, then to define a stratification we will need to consider partial derivatives of $f$ of higher orders, but, due to Corollary 4.7, not higher than $\mathcal{M}(1, r, \alpha, \beta)$. The proof of the theorem is a far-reaching generalization of this idea.

Assume that the use of the oracle is allowed. Note that Theorem 3.7 can be turned into an algorithm (with oracle) for listing all consistent sign assignments for a given family of Pfaffian functions. Thus, we can assume that $X$ is a *basic* semi-Pfaffian set defined by a system of equations and strict inequalities:

$$X := \{\mathbf{x} \mid f_1(\mathbf{x}) = \dots = f_\lambda(\mathbf{x}) = 0,\ g_1(\mathbf{x}) > 0, \dots, g_\mu(\mathbf{x}) > 0\}.$$

We now employ the notation for partial differential operators from Definition 4.6. Additionally, let $\mathcal{M}_1 := \mathcal{M}(1, r, \alpha, \beta)$, $\beta_k := \beta + (\mathcal{M}_k - 1)((\alpha - 1)k + \beta_1 + \dots + \beta_{k-1})$ and $\mathcal{M}_{k+1} := \mathcal{M}(k+1, r, \beta_k, \beta_1, \dots, \beta_k)$.

Consider a list of all partial derivatives $\partial_n^{q_n} \dots \partial_1^{q_1}$ of functions $f_j$ with $q_1 + \dots + q_n \leq \mathcal{M}(0, r, \alpha, \beta)$, ordered lexicographically in $(q_n, \dots, q_1, j)$. Let $X^0 \subset X$ be a subset where all these derivatives vanish. According to Corollary 4.7, all the functions $f_j$ are identically zero in a neighbourhood of each $\mathbf{x} \in X^0$, hence $X^0$, if nonempty, is a smooth open set in $\mathbb{R}^n$ coinciding with $X$.

Suppose now that $X^0 = \emptyset$, thus there can be found a function $f_j \not\equiv 0$. Then any $\mathbf{x} \in X$ belongs to one and only one of the sets

$$Z^1_{j_1, \mathbf{m}^1} \subset X, \quad \mathbf{m}^1 := (m^1_{i_1}, \dots, m^1_1), \quad m^1_1 + \dots + m^1_{i_1} \leq \mathcal{M}_1, \quad m^1_{i_1} > 0,$$

at whose points the derivative $h_1' := \partial_{i_1}^{m^1_{i_1}} \dots \partial_1^{m^1_1} f_{j_1}$ is different from 0, while all derivatives in the lexicographically ordered list preceding $h_1'$ vanish. Let $h_1 := \partial_{i_1}^{m^1_{i_1}-1} \partial_{i_1}^{m^1_{i_1}-1} \dots \partial_1^{m^1_1} f_{j_1}$, so that $h_1' = \partial_{i_1} h_1$, and consider a smooth manifold $Y^1 := \{\mathbf{x} \in X \mid h_1(\mathbf{x}) = 0, h_1'(\mathbf{x}) \neq 0\} \supset Z^1_{j_1, \mathbf{m}^1}$ of codimension 1. Denote by $F^1$ the set of all functions that appear in the equations defining $Z^1_{j_1, \mathbf{m}^1}$. Note that for any $i < i_1$ all functions from $F^1$, including $h_1$, do not depend on $x_i$, due to Corollary 4.7.

We now consider partial derivatives

$$\hat{\partial}_n^{q_n} \cdots \hat{\partial}_{i_1+1}^{q_{i_1+1}} := \partial_{h_1,i_1,n}^{q_n} \cdots \partial_{h_1,i_1,i_1+1}^{q_{i_1+1}}$$

of functions $f_\nu \in F^1$, $\nu = (q_{i_1}, \ldots, q_1, j)$, "along" the manifold $Y^1 \cap \{x_i = 0 \mid i < i_1\}$ with $q_{i_1+1} + \cdots + q_n \leq \mathcal{M}_2$, ordered lexicographically in $(q_n, \ldots, q_{i_1+1}, \nu)$. Let $X^1_{j_1,\mathbf{m}^1} \subset Y^1 \cap \{x_i = 0 \mid i < i_1\}$ be a subset of $X$ where all these derivatives vanish. According to Corollary 4.7, all the functions $f_\nu$ are identically zero on $Y^1 \cap \{x_i = 0 \mid i < i_1\}$ in a neighbourhood of each $\mathbf{x} \in X^1_{j_1,\mathbf{m}^1}$, hence $X^1_{j_1,\mathbf{m}^1}$, if nonempty, is an open submanifold of $Y^1 \cap \{x_i = 0 \mid i < i_1\}$. Since all the functions in the equations defining $X^1_{j_1,\mathbf{m}^1}$ and $Y^1$ do not depend on $x_i$, for $i < i_1$, this implies that $X^1_{j_1,\mathbf{m}^1}$ is a smooth manifold of codimension 1.

If $\mathbf{x} \notin X^1_{j_1,\mathbf{m}^1}$, then $\mathbf{x}$ belongs to one and only one of the sets

$$Z^2_{j_1,j_2,\mathbf{m}^1,\mathbf{m}^2} \subset Z^1_{j_1,\mathbf{m}^1}, \quad \mathbf{m}^2 = (m^2_{i_2}, \ldots, m^2_1), \quad m^2_{i_1+1} + \cdots + m^2_{i_2} \leq \mathcal{M}_2,$$
$$m^2_{i_2} > 0, \quad (m^2_{i_1}, \ldots, m^2_1, j_2) \prec (m^1_{i_1}, \ldots, m^1_1, j_1),$$

where $\prec$ is the lexicographic order, at whose points the derivative $h_2' := \hat{\partial}_{i_2}^{m^2_{i_2}} \cdots \hat{\partial}_1^{m^2_{i_2}} f_{j_2}$ (where $\hat{\partial}_i := \partial_{h_1,i_1,i}$) is different from 0, while all derivatives in the lexicographically ordered list preceding $h_2'$ vanish. Let

$$h_2 := \hat{\partial}_{i_2}^{m^2_{i_2}-1} \partial_{i_2}^{m^2_{i_2-1}} \cdots \partial_1^{m^2_1} f_{j_2},$$

so that

$$h_2' = \hat{\partial}_{i_2} h_2 = \partial_{i_1} h_1(\mathbf{x}) \partial_{i_2} h_2(\mathbf{x}) - \partial_{i_2} h_1(\mathbf{x}) \partial_{i_1} h_2(\mathbf{x}),$$

and consider a smooth manifold

$$Y^2 := \{\mathbf{x} \in X \mid h_1(\mathbf{x}) = h_2(\mathbf{x}) = 0, h_2'(\mathbf{x}) \neq 0\} \supset Z^2_{j_1,j_2,\mathbf{m}^1,\mathbf{m}^2}$$

of codimension 2.

The continuation of this procedure for $k = 2, \ldots, n$ leads to the consecutive definition of the sets $X^k_{j_1,\ldots,j_k,\mathbf{m}^1,\ldots,\mathbf{m}^k}$, where

$$1 \leq j_t \leq \lambda, \quad 0 \leq m^t_1 + \cdots + m^t_{i_1} \leq \mathcal{M}_1, \ldots, 0 \leq m^t_{i_{t-1}+1} + \cdots + m^t_{i_t} \leq \mathcal{M}_t, \quad m^t_{i_t} > 0,$$
$$(m^t_{i_s}, \ldots, m^t_1, j_t) \prec (m^s_{i_s}, \ldots, m^s_1, j_s) \quad \text{for} \quad 1 \leq s \leq t.$$

The same arguments as above show that the sets $X^k_{j_1,\ldots,j_k,\mathbf{m}^1,\ldots,\mathbf{m}^k}$ form a stratification of $X$, i.e., they are disjoint smooth manifolds, and their union is $X$. The number of strata, their formats, and complexity of producing them can be estimated from the process of generating the sets $X^k_{j_1,\ldots,j_k,\mathbf{m}^1,\ldots,\mathbf{m}^k}$. □

## 7 Cylindrical decompositions of sub-Pfaffian sets

In [10] it was proved that the complement of any subanalytic set in a cube $\mathcal{I}^n$ is also subanalytic. This *complement theorem* plays a key role in real analytic geometry (see [4, 8]) and in the model-theoretic study of o-minimality [9, 47].

The complement theorem immediately follows from the existence of a *cylindrical cell decomposition* of the ambient space compatible with a given subanalytic set. The existence was proved in [13] by means of a quasi-constructive process of manipulation with symbols of real analytic functions and their derivatives. In [19] the method from [13] was modified, so that being applied to a sub-Pfaffian set it yields an algorithm producing a cylindrical cell decomposition of this set. There is also an alternative algorithm for a cylindrical decomposition with a slightly better complexity bound [36].

For the special case of *semialgebraic sets* similar complexity results are known for a cylindrical cell decomposition problem [7, 48], and significantly better results are known for the *quantifier elimination* problem (the latter is stronger than the complement theorem).

**7.1 Definition** A *cylindrical cell* is defined by induction as follows.

(1) A cylindrical 0-cell in $\mathbb{R}^n$ is an isolated point.

(2) A cylindrical 1-cell in $\mathbb{R}$ is an open interval $(a, b) \subset \mathbb{R}$.

(3) For $n \geq 2$ and $0 \leq k < n$, a cylindrical $(k+1)$-cell in $\mathbb{R}^n$ is either a graph of a continuous bounded function $f : C \to \mathbb{R}$, where $C$ is a cylindrical $(k+1)$-cell in $\mathbb{R}^{n-1}$, or else a set of the form

$$\{(x_1, \dots, x_n) \in \mathbb{R}^n \mid (x_1, \dots, x_{n-1}) \in C \text{ and } f(x_1, \dots, x_{n-1}) < x_n < g(x_1, \dots, x_{n-1})\},$$

where $C$ is a cylindrical $k$-cell in $\mathbb{R}^{n-1}$, and $f, g : C \to \mathbb{R}$ are continuous bounded functions such that $f(x_1, \dots, x_{n-1}) < g(x_1, \dots, x_{n-1})$ for all points $(x_1, \dots, x_{n-1}) \in C$.

The definition implies that any $k$-cell is homeomorphic to an open $k$-dimensional ball.

**7.2 Definition** A *cylindrical cell decomposition* $\mathcal{D}$ of a subset $A \subset \mathbb{R}^n$ is defined by induction as follows.

(1) If $n = 1$, then $\mathcal{D}$ is a finite family of pair-wise disjoint cylindrical cells (i.e., isolated points and intervals) whose union is $A$.

(2) If $n \geq 2$, then $\mathcal{D}$ is a finite family of pair-wise disjoint cylindrical cells in $\mathbb{R}^n$ whose union is $A$ and there is a cylindrical cell decomposition of $\pi(A)$ such that $\pi(C)$ is its cell for each $C \in \mathcal{D}$, where $\pi : \mathbb{R}^n \to \mathbb{R}^{n-1}$ is the projection map onto the coordinate subspace of $x_1, \dots, x_{n-1}$.

**7.3 Definition** Let $B \subset A \subset \mathbb{R}^n$ and $\mathcal{D}$ a cylindrical cell decomposition of $A$. Then $\mathcal{D}$ is *compatible* with $B$ if for any $C \in \mathcal{D}$ we have either $C \subset B$ or $C \cap B = \emptyset$ (i.e., some subset $\mathcal{D}' \subset \mathcal{D}$ is a cylindrical cell decomposition of $B$).

**7.4 Theorem ([19])** *Let $X$ be a semi-Pfaffian set in an open domain $G \subset \mathbb{R}^{m+n}$ defined by (2.2) with $s = m + n$, format $(r, N, \alpha, \beta, m + n)$, and contained in an open cube $\hat{\mathcal{I}}^{m+n} := (-1, 1)^{m+n}$ such that the closure $\mathcal{I}^{m+n} \subset G$. Let $\pi : \mathbb{R}^{m+n} \to \mathbb{R}^n$ be the projection function, $Y := \pi(X)$, and $d := \dim(Y)$. Then there is an algorithm (with oracle) producing a cylindrical cell decomposition $\mathcal{D}$ of the image of $\hat{\mathcal{I}}^n = \pi(\hat{\mathcal{I}}^{m+n})$ under a linear coordinate change such that $\mathcal{D}$ is compatible with the image of $Y$.*

*Each cell is described as a projection of a semi-Pfaffian set in DNF, i.e., by a formula of the type*

$$\pi'\left(\bigcup_{1\le i\le M}\bigcap_{1\le j\le M_i}\{h_{ij} *_{ij} 0\}\right),$$

*where $h_{ij}$ are Pfaffian functions in $n' \geq m+n$ variables, $\pi' : \mathbb{R}^{n'} \to \mathbb{R}^n$ is the projection function, $*_{ij} \in \{=,>\}$, and $M, M_i$ $(i = 1, \dots, M)$ are certain integers.*

*The number of cells in the decomposition is less than*

$$\mathcal{N} := N^{(d!)^2(m+2n)^d(r+m+2n)^d}(\alpha+\beta)^{r^{O(d(m+dn))}},$$

*the format of each cell is*

$$(r, \mathcal{N}, (\alpha+\beta)^{r^{O(d(m+dn))}}, (\alpha+\beta)^{r^{O(d(m+dn))}}, \mathcal{N}).$$

*The complexity of the algorithm is*

$$N^{(r+m+n)^{O(d)}}(\alpha+\beta)^{(r+m+n)^{O(d(m+dn))}}.$$

**7.5 Corollary** *Under the assumptions of Theorem 7.4 the complement $\widetilde{Y} := \hat{\mathcal{I}}^n \setminus Y$ is a sub-Pfaffian set. There is an algorithm for computing $\widetilde{Y}$ having the same complexity as the algorithm from Theorem 7.4. The complement $\widetilde{Y}$ is represented by the algorithm as a union of some cells of the cylindrical cell decomposition a linear image of $\hat{\mathcal{I}}^n$ described in Theorem 7.4.*

**Outline of a proof of Theorem 7.4** Assume that the points in $\mathbb{R}^{m+n}$ are of the kind $(\mathbf{y}, \mathbf{x}) = (x_1, \dots, x_n, x_{n+1}, \dots, x_{m+n})$ and $\pi(\mathbf{y}, \mathbf{x}) = \mathbf{y}$.

### Computing the dimension $d = \dim(Y)$

The algorithm computes a weak stratification of $X$ using Theorem 6.2. Each stratum $X_\lambda = \{(\mathbf{y}, \mathbf{x}) \in \mathbb{R}^{m+n} \mid \mathbf{f} = 0, \mathbf{g} > 0\}$, where $\mathbf{f}, \mathbf{g}$ are vectors of Pfaffian functions and relations $=, >$ are understood component-wise, is effectively nonsingular (see Definition 6.1). In particular, the dimension $d'_\lambda := \dim(X_\lambda)$ is computed. To find $d_\lambda := \dim(\pi(X_\lambda))$, observe that by Sard's theorem, $\operatorname{rank}(\partial\mathbf{f}/\partial\mathbf{x})(\mathbf{y}, \mathbf{x}) \leq m - d'_\lambda + d_\lambda$ for any $(\mathbf{y}, \mathbf{x}) \in X_\lambda$, while the equality is attained for *almost* any $(\mathbf{y}, \mathbf{x}) \in X_\lambda$. Using the oracle, choose the maximal $a$ such that

$$\{(\mathbf{y}, \mathbf{x}) \in X_\lambda \mid \operatorname{rank}(\partial\mathbf{f}/\partial\mathbf{x})(\mathbf{y}, \mathbf{x}) = m - d'_\lambda + a\} \neq \emptyset,$$

then $a = d_\lambda$ and $d = \max_\lambda\{d_\lambda\}$.

### An example

To illustrate the idea of the algorithm, consider

$$X = \{\mathbf{x} = (x_1, x_2, x_3) \mid f := x_1^2 + x_2^2 + x_3^2 - 1/2 = 0\}, \qquad Y = \{\mathbf{y} = (x_1, x_2) \mid x_1^2 + x_2^2 \leq 1/2\},$$

hence $n = d = 2$, $m = 1$.

The algorithm consists of two recursive procedures which we call *down* and *up*. Starting the first step of the *down* procedure observe that $X$ is effectively non-singular. Let

$$\begin{aligned} X' &:= \{(x_1,x_2,x_3) \in X \mid \partial f/\partial x_3 \neq 0\}, \\ V' &:= \{(x_1,x_2,x_3) \in X \mid \partial f/\partial x_3 = 0\} \\ &= \{(x_1,x_2,x_3) \in X \mid \hat{f}_1 := x_3 = 0,\ \hat{f}_2 := x_1^2 + x_2^2 - 1/2 = 0\} \end{aligned}$$

be semi-Pfaffian sets of all regular and of all singular points, respectively, of the restriction of the projection map $\pi : (x_1,x_2,x_3) \mapsto (x_1,x_2)$ on $X$. Introduce the new notation: $X_2 := V'$, $Y_2 := \pi(V')$, $d_2 := \dim(Y_2) = 1$. We have described the first recursive step of *down*.

Now we start the second step with $X_2$, $Y_2$ playing the role of $X$, $Y$. Observe that $X_2$ is effectively non-singular and all points are regular for the restriction $\pi|_{X_2}$. Consider the projection map $\rho_2 : (x_1,x_2) \mapsto x_1$. Let

$$\begin{aligned} S_2 &:= \left\{(x_1,x_2,x_3) \in X_2 \mid \det \begin{pmatrix} \partial\hat{f}_1/\partial x_2 & \partial\hat{f}_1/\partial x_3 \\ \partial\hat{f}_2/\partial x_2 & \partial\hat{f}_2/\partial x_3 \end{pmatrix} = 0\right\} \\ &= \{(x_1,x_2,x_3) \in \mathbb{R}^3 \mid x_1^2 = 1/2,\ x_2 = x_3 = 0\} \end{aligned}$$

be the set of all critical points of $\rho_2\pi|_{X_2}$. Let $Z_2 := \rho_2\pi(S_2)$, then $\dim(Z_2) = \dim(Y_2 \cap \rho_2^{-1}(Z_2)) = 1 < d = 2$. Let $X_3 := S_2$, $Y_3 := Y_2 \cap \rho_2^{-1}(Z_2)$, $d_3 := \dim(Y_3) = 0$ We have completed the second recursive step of *down*.

In the last (degenerate) recursive step, $X_3$ is effectively non-singular and all points are regular for the restriction $\pi|_{X_3}$. For the projection map $\rho_3 : (x_1,x_2) \mapsto 0$, the set $S_3$ of all critical points of $\rho_3\pi|_{X_3}$ is empty, thus the projection $Z_3 := \rho_2\pi(S_3) = \emptyset$. The *down* procedure is completed.

Now the algorithm starts the recursive *up* procedure. In the first recursive step consider the sub-Pfaffian set $Y_3$ constructed on the last step of *down*. Since $Y_3$ consists of just two points, $(1/\sqrt{2},0)$ and $(-1/\sqrt{2},0)$, the construction of a cylindrical cell decomposition $\mathcal{D}_3$ of $\hat{\mathcal{I}}^2$ compatible with $Y_3$ is trivial.

In the second recursive step consider sub-Pfaffian sets $Y_2$ and $Z_2$. The decomposition $\mathcal{D}_3$, being cylindrical, induces a cell decomposition $\mathcal{D}_3'$ of $\hat{\mathcal{I}}^1 = \{x_1 \mid -1 < x_1 < 1\}$ into five cells:

- $C_1 := \{x_1 \mid -1 < x_1 < -1/\sqrt{2}\}$,
- $C_2 := \{x_1 \mid x_1 = -1/\sqrt{2}\}$,
- $C_3 := \{x_1 \mid -1/\sqrt{2} < x_1 < 1/\sqrt{2}\}$,
- $C_4 := \{x_1 \mid x_1 = 1/\sqrt{2}\}$,
- $C_5 := \{x_1 \mid 1/\sqrt{2} < x_1 < 1\}$.

For any $z \in C_3$ the cardinality of the fibre $\rho_2^{-1}(z) \cap Y_2$ is constant ($= 2$). Moreover, the following five cells form a cylindrical cell decomposition of $\rho_2^{-1}(C_3) \cap \hat{\mathcal{I}}^2$ compatible with $\rho_2^{-1}(C_3) \cap Y$:

- $\{(x_1,x_2) \in \rho_2^{-1}(C_3) \cap \hat{\mathcal{I}}^2 |\ \exists (y_1,y_2) \in Y\ \exists (y_1',y_2') \in Y\ (y_1 = y_1', y_2 < y_2' < x_2)\}$
- $\{(x_1,x_2) \in \rho_2^{-1}(C_3) \cap \hat{\mathcal{I}}^2 |\ \exists (y_1,y_2) \in Y\ \exists (y_1',y_2') \in Y\ (y_1 = y_1', y_2 < y_2' = x_2)\}$

- $\{(x_1, x_2) \in \rho_2^{-1}(C_3) \cap \hat{\mathcal{I}}^2 |\ \exists (y_1, y_2) \in Y\ \exists (y_1', y_2') \in Y\ (y_1 = y_1', y_2 < x_2 < y_2')\}$
- $\{(x_1, x_2) \in \rho_2^{-1}(C_3) \cap \hat{\mathcal{I}}^2 |\ \exists (y_1, y_2) \in Y\ \exists (y_1', y_2') \in Y\ (y_1 = y_1', y_2 = x_2 < y_2')\}$
- $\{(x_1, x_2) \in \rho_2^{-1}(C_3) \cap \hat{\mathcal{I}}^2 |\ \exists (y_1, y_2) \in Y\ \exists (y_1', y_2') \in Y\ (y_1 = y_1', x_2 < y_2 < y_2')\}$.

Similar cylindrical cell decompositions of $\rho_2^{-1}(C_i) \cap \hat{\mathcal{I}}^2$ can be constructed for all other cells $C_i$, but in fact such decompositions have already been produced as parts of the cell decomposition $\mathcal{D}_3$. Combining all cell decomposition for $\rho_2^{-1}(C_i) \cap \hat{\mathcal{I}}^2$ with $\mathcal{D}_3$ we get a cylindrical cell decomposition of $\hat{\mathcal{I}}^2$ compatible with $Y$. This completes the *up* procedure and the whole construction.

### General algorithm: the *down* procedure

After computing $d$ the algorithm uses one after another two recursive procedures, *down* and *up*. We start with *down* by conducting a descending recursion on $d$. We will describe in some detail only the first (typical) step of the recursion. For each stratum $X_\lambda$ with $\dim(X_\lambda) \geq d$ the algorithm finds the semi-Pfaffian set $X_\lambda'$ of all regular and the semi-Pfaffian set $V_\lambda'$ of all critical points of the projection map $\pi|_{X_\lambda}$. Note that $X_\lambda'$ and $V_\lambda'$ can be described by explicit quantifier-free formulae and any of them may be empty. By Sard's theorem, $\dim(\pi(V_\lambda')) < d$. For any $y \in Y \setminus \pi(V_\lambda')$ the intersection $\pi^{-1}(y) \cap X_\lambda'$ is smooth.

Now the algorithm works with $X_\lambda'$. The next step is to select in this smooth manifold a semi-Pfaffian subset of dimension $d$ having the same $\pi$-projection as $X_\lambda'$. The algorithm finds a function $g:\ \mathbb{R}^{m+n} \to \mathbb{R}$ and a semi-Pfaffian set $V_\lambda'' \subset X_\lambda'$ such that $\dim(\pi(V_\lambda'')) < d$, and for any $y \in Y \setminus \pi(V_\lambda' \cup V_\lambda'')$ the critical points of $g|_{\pi^{-1}(y) \cap X_\lambda'}$ are non-degenerate, in particular isolated. The function $g$ can be defined by an expression

$$\left( \prod_j h_j \right) (1 + (\mathbf{c}, \mathbf{x})),$$

where $\mathbf{c}$ is a vector of integers, and the product of a zero number of factors is assumed to be equal to 1 (see details in [19]). The set $X_\lambda''$ such that for any $y \in Y \setminus \pi(V_\lambda' \cup V_\lambda'')$ the intersection $X_\lambda'' \cap \pi^{-1}(y)$ is a finite set of all critical points of $g|_{\pi^{-1}(y) \cap X_\lambda'}$, can be described by an explicit quantifier-free formula. If $X_\lambda'' \neq \emptyset$, then $\dim(X_\lambda'') = d$.

The algorithm computes a weak stratification of $X_\lambda''$ using Theorem 6.2. For each stratum of $X_{\lambda\mu}$ the (maximal) dimension $d$ the algorithm finds the semi-Pfaffian set $X_{\lambda\mu}'$ of all regular and the semi-Pfaffian set $V_{\lambda\mu}'$ of all critical points of the projection map $\pi|_{X_{\lambda\mu}}$. These sets can be described by explicit quantifier-free formulae. By Sard's theorem, $\dim(\pi(V_{\lambda\mu}'')) < d$. Let $V_{\lambda\mu} := V_\lambda' \cup V_\lambda'' \cup V_{\lambda\mu}'$, and let $T$ be the union of all strata of $X$ of dimension less than $d$.

If $d = n$, then define

$$X_{\text{new}} := \bigcup_{\lambda\mu} (\partial X_{\lambda\mu}' \cup V_{\lambda\mu}) \cup T,$$

$$Y_{\text{new}} := \bigcup_{\lambda,\mu} \pi(\partial X_{\lambda\mu}' \cup V_{\lambda\mu}) \cup \pi(T),$$

where the frontier $\partial X_{\lambda\mu}'$ can be computed using Theorem 5.3. Observe that $\dim(Y_{\text{new}}) < d$. The algorithm goes to the next recursive step with $X_{\text{new}}$ and $Y_{\text{new}}$ replacing $X$ and $Y$, respectively.

If $d < n$, then the algorithm continues the current recursive step. Consider the projection function $\rho : \mathbb{R}^n \to \mathbb{R}^d$, where $\mathbb{R}^n$ is equipped with coordinates $\mathbf{y}$. The algorithm performs a linear transformation of coordinates $\mathbf{y}$ such that in the new coordinates for each $\mathbf{y}$ in the closure $\bar{Y}$ the set $\rho^{-1}(\rho(\mathbf{y}))$ is finite[2]. In the sequel we assume that this condition is satisfied.

Consider the set $S_{\lambda\mu}$ of all critical points of the restriction of the composition $\rho\pi : \mathbb{R}^{m+n} \to \mathbb{R}^d$ to $X'_{\lambda\mu}$. This set can be described by an explicit quantifier-free formula. Observe that $\dim(\rho\pi(S_{\lambda\mu})) < d$ by Sard's theorem, hence $\dim(\pi(S_{\lambda\mu})) < d$ by the choice of the linear transformation.

The aim of the next action of the algorithm is to identify a subset in $\mathbb{R}^d$ of positive codimension such that within each connected component of the complement of this set any two points have 0-dimensional $\rho$-fibers with the same lexicographic order. Introduce the following sets:

$$\begin{aligned} W_{i,\varepsilon} := \; & \{(\mathbf{y}, \varepsilon) \mid \mathbf{y} = (z_1, \ldots, z_{n-d}, y_{n-d+1}, \ldots, y_n) \in Y,\ \varepsilon \in \mathbb{R}, \\ & \exists \mathbf{y}' = (z'_1, \ldots, z'_{n-d}, y_{n-d+1}, \ldots, y_n) \in Y,\ \rho(\mathbf{y}') = \rho(\mathbf{y}), \\ & z'_1 = z_1, \ldots, z'_{i-1} = z_{i-1},\ z'_i \neq z_i,\ |z'_i - z_i| < \varepsilon\} \subset \mathbb{R}^{n+1}; \\ W_i := \; & \overline{W}_{i,\varepsilon} \cap \{(\mathbf{y}, \varepsilon) \mid \varepsilon = 0\} \subset \mathbb{R}^n; \\ W := \; & \bigcup_{1 \le i \le n-d} W_i \subset \mathbb{R}^n; \\ Z := \; & \rho\pi\left(\bigcup_{\lambda,\mu} (\partial X'_{\lambda\mu} \cup V_{\lambda\mu} \cup S_{\lambda\mu})\right) \cup \rho(W) \cup \rho\pi(T) \subset \mathbb{R}^d. \end{aligned}$$

Then $\dim(Z) = \dim(Y \cap \rho^{-1}(Z)) < d$. Observe that $W$ is a sub-Pfaffian set, more precisely, there exists an integer $n_{\text{new}}$ such that $n + m \le n_{\text{new}} \le 2n + m$, and a semi-Pfaffian set $U \subset \mathbb{R}^{n_{\text{new}}}$ such that $\pi_{\text{new}}(U) = W$ for the projection map $\pi_{\text{new}} : \mathbb{R}^{n_{\text{new}}} \to \mathbb{R}^n$. There is an explicit quantifier-free formulae defining $U$.

Let $U'$ denote the semi-Pfaffian set defined in $\mathbb{R}^{n_{\text{new}}}$ by the same quantifier-free formula as

$$\bigcup_{\lambda\mu} (\partial X'_{\lambda\mu} \cup V_{\lambda\mu} \cup S_{\lambda\mu}) \cup T \subset \mathbb{R}^{m+n}.$$

Let $Y_{\text{new}} := Y \cap \rho^{-1}(Z) = \pi_{\text{new}}(U \cup U')$, $X_{\text{new}} := \pi_{\text{new}}^{-1}(Y \cap \rho^{-1}(Z)) = U \cup U'$. Observe that $X_{\text{new}}$ is defined by an explicit quantifier-free formula with Pfaffian functions in $n_{\text{new}}$ variables. The algorithm determines $d_{\text{new}} := \dim(Y_{\text{new}}) = \dim(Z) < d$, using the method described at the beginning of the proof.

This completes the description of the recursive step. In the last recursive step (let it have number $l \le d$) the dimension $\dim(Y_{\text{new}}) = 0$ and $Z = \emptyset$.

### General algorithm: the *up* procedure

The algorithm then starts the *up* recursion procedure. An input data of the $r$th recursive step consists of:

---

[2]According to the Koopman-Brown theorem such a transformation exists, for details on how to find it see [19].

- a pair $Y, Z \subset \mathbb{R}^n$ of sub-Pfaffian sets constructed in steps $l-r$ and $l-r+1$, respectively, of the "down" procedure, and
- a cylindrical cell decomposition $\mathcal{D}$ of $\hat{\mathcal{I}}^n$ compatible with $Y \cap \rho^{-1}(Z)$.

The decomposition $\mathcal{D}$ induces a cylindrical cell decomposition $\mathcal{D}'$ of $\rho(\hat{\mathcal{I}}^n) = \hat{\mathcal{I}}^d$ compatible with $Z$, namely, the elements of $\mathcal{D}'$ are exactly the $\rho$-projections of the elements of $\mathcal{D}$. By the definition of $Z$, for any $d$-dimensional cell $C \in \mathcal{D}'$ and any $\mathbf{z} \in C$, the cardinality of $\rho^{-1}(\mathbf{z}) \cap Y$ is a constant, say $\mathcal{L}$. Moreover, the union

$$\bigcup_{1 \leq \nu \leq \mathcal{L}+1} \{\mathbf{y} \in \rho^{-1}(C) \cap \hat{\mathcal{I}}^n \mid \exists \mathbf{y}_1 \in Y \cdots \exists \mathbf{y}_{\mathcal{L}} \in Y\ (\mathbf{y}_i \neq \mathbf{y}_j \text{ for all } i, j, i \neq j,$$
$$\mathbf{y}_1 \prec \cdots \prec \mathbf{y}_{\nu-1} \prec \mathbf{y} \prec \mathbf{y}_\nu \prec \cdots \prec \mathbf{y}_{\mathcal{L}}, \rho(\mathbf{y}_1) = \cdots = \rho(\mathbf{y}_{\mathcal{L}}) = \rho(\mathbf{y}))\},$$

where the relation $\mathbf{u} \prec \mathbf{v}$ for $\mathbf{u} = (u_1, \ldots, u_n), \mathbf{v} = (v_1, \ldots, v_n) \in \mathbb{R}^n$ stands for the disjunction

$$\bigvee_{d+1 \leq i \leq n+1} (u_1 = v_1, \ldots, u_{i-1} = v_{i-1},\ u_i < v_i),$$

represents a cylindrical cell decomposition of $\rho^{-1}(C) \cap \hat{\mathcal{I}}^n$ compatible with $\rho^{-1}(C) \cap Y$.[3]

Note that Corollary 3.3 provides an upper bound for $\mathcal{L}$ as an explicit function, say $\mathcal{M}$, of the format of $Y$. The algorithm finds $\mathcal{L}$ as the maximal $l$, $1 \leq l \leq \mathcal{M}$, such that the statement

$$\exists \mathbf{y}_1 \in \rho^{-1}(C) \cdots \exists \mathbf{y}_l \in \rho^{-1}(C)\ (\mathbf{y}_i \neq \mathbf{y}_j \text{ for all } i, j,\ i \neq j,\ \rho(\mathbf{y}_1) = \cdots = \rho(\mathbf{y}_l))$$

is true. Then the algorithm computes the cell decomposition of $\rho^{-1}(C) \cap \hat{\mathcal{I}}^n$.

Combining the cylindrical cell decompositions for $\rho^{-1}(C) \cap \hat{\mathcal{I}}^n$ for all $d$-dimensional cells $C$ of $\mathcal{D}'$, with the cell decomposition $\mathcal{D}$, the algorithm gets a cylindrical cell decomposition of $\hat{\mathcal{I}}^n$ compatible with $Y$. This finishes the description of the *up* procedure.

The formats of cells in the resulting cylindrical decomposition and the complexity of the algorithm can be estimated from the description of the algorithm using the upper bounds discussed in previous sections (see [19]). □

## 8 Limit sets

In this section we remove the condition on semi-Pfaffian sets to be *restricted.*

For arbitrary (including non-restricted) semi-Pfaffian sets a "closure at infinity" operation was introduced in [6] and [47]. The main theorem from [6, 47] (see also [23, 29, 41]) implies that the sets constructed from semi-Pfaffian sets by a finite sequence of projections on subspaces and closures at infinity constitute an o-minimal structure.

In [16], Gabrielov introduced the "relative closure" operation for a one-parameter family of semi-Pfaffian sets, and the concept of a "limit set" as a finite union of relative closures of semi-Pfaffian families. Every semi-Pfaffian set is a limit set. The main results of [16] state that limit sets constitute an *effectively* o-minimal structure, i.e., any first-order formula

[3] A more detailed description of this decomposition can be found in [19].

with limit sets defines a limit set which admits an upper complexity bound in terms of the complexity of the formula. In [21] an explicit complexity bound was obtained for the number of connected components of a limit set. We now proceed to a more detailed description of these results.

## 8.1 Exponential Łojasiewicz inequality

We start with another version of the Łojasiewicz inequality. Unlike the inequality from Corollary 4.4, it describes the rate of growth of a Pfaffian function not only in a neighbourhood of a point in the domain $G$ but also in a neighbourhood of a point on the boundary of $G$. The price paid for this extension is a much weaker lower bound.

Let $\mathrm{cl}(Z)$ denote the topological closure of a set $Z$. Introduce $\mathrm{fr}(Z) := \mathrm{cl}(Z) \setminus Z$.[4] We assume that the closure points of $X$ at infinity are included in $\mathrm{cl}(Z)$ and $\mathrm{fr}(Z)$. To avoid the separate treatment of infinity, we assume that $\mathbb{R}^n$ is embedded in a projective space and all constructions are performed in an affine chart such that $Z$ is relatively compact in that chart. To achieve this it may be necessary to subdivide $Z$ into smaller pieces, each of them relatively compact in its own chart.

**8.1 Theorem (Exponential Łojasiewicz inequality [16, 22, 27, 28])** *Let $X$ be a semi-Pfaffian set in a domain $G \subset \mathbb{R}^n$ defined by a formula with Pfaffian functions of order $r$, and let $f(\mathbf{x})$ be a Pfaffian function in $\mathbb{R}^n$. Suppose that $\mathbf{0} \in cl(X \cap \{\mathbf{x} \in \mathbb{R}^n \mid f(\mathbf{x}) > 0\})$. Then*

$$\mathbf{0} \in \mathrm{cl}(\{\mathbf{x} \in X \mid f(\mathbf{x}) \geq 1/E_r(\|\mathbf{x}\|^{-q})\}),$$

*for some integer $q > 0$. Here $E_r$ is the iterated exponential function from Example* 2.3 (b).

A proof of this theorem can be found in [16], it uses an iterated exponential upper bound from [40] on the asymptotic growth of a function in a Hardy field .

## 8.2 Relative closure and limit sets

Let the space $\mathbb{R}^n \times \mathbb{R}$ have coordinates $(\mathbf{x}, \lambda)$. For a set $X \subset \mathbb{R}^n \times \mathbb{R}$ we define: $X_+ := X \cap \{\lambda > 0\}$, $X_\lambda := X \cap \{\lambda = \mathrm{const}\}$, and $\check{X} := \mathrm{cl}(X_+) \cap \{\lambda = 0\}$. The coordinate $\lambda$ is considered as a *parameter*, and $X$ is considered as a *family* of sets $X_\lambda$ in $\mathbb{R}^n$.

**8.2 Definition** (*Semi-Pfaffian family*) A semi-Pfaffian set $X$ in DNF in a domain $G$ is called a *semi-Pfaffian family* if for any $\varepsilon > 0$ the intersection $X \cap \{\lambda > \varepsilon\}$ is restricted in $G$. The *format* of $X$ is defined as the format of a semi-Pfaffian set $X_\lambda$ for a small $\lambda > 0$.

**8.3 Lemma** *Let $X$ be a semi-Pfaffian family. Then* $\mathrm{cl}(X)_+$ *and* $\mathrm{fr}(X)_+$ *are semi-Pfaffian families. The formats of these families admit upper bounds in terms of the format of $X$.*

**Proof** The set $\mathrm{cl}(X)_+$ is contained in $G$ since $X \cap \{\lambda > \varepsilon\}$ is restricted in $G$ for any $\varepsilon > 0$. Hence, also $\mathrm{fr}(X)_+ \subset G$. According to Theorem 5.2, the sets $\mathrm{cl}(X)_+$ and $\mathrm{fr}(X)_+$ are semi-Pfaffian in $G$. The sets $\mathrm{cl}(X)_+ \cap \{\lambda > \varepsilon\}$ and $\mathrm{fr}(X)_+ \cap \{\lambda > \varepsilon\}$ are restricted in $G$ for any $\varepsilon > 0$, since this is true for $X$.

[4] For a semi-Pfaffian set $Z$ in a domain $G$ the topological closure $\mathrm{cl}(Z)$ may be different from the closure $\bar{Z} = \mathrm{cl}(Z) \cap G$, and $\mathrm{fr}(Z)$ may be different from $\partial Z$ (see Definition 5.1).

The statement on formats follows from Theorem 5.2, since $\mathrm{cl}(X)_\lambda = \mathrm{cl}(X_\lambda)$ and $\mathrm{fr}(X)_\lambda = \mathrm{fr}(X_\lambda)$ for a generic $\lambda > 0$. These equalities can be derived from the existence of a weak stratification (Theorem 6.2), Sard's theorem, and the finiteness properties of semi-Pfaffian sets. □

**8.4 Definition** (*Semi-Pfaffian couple*) Two semi-Pfaffian families $X$ and $Y$ with a common domain $G \subset \mathbb{R}^n \times \mathbb{R}$ form a *semi-Pfaffian couple* $(X, Y)$ in $G$ if the set $Y$ is relatively closed in $\{\lambda > 0\}$ (i.e., $\mathrm{cl}(Y)_+ = Y_+$) and contains $\mathrm{fr}(X)_+$. The format of a couple $(X, Y)$ is defined as the component-wise maximum of the formats of $X$ and $Y$.

**8.5 Definition** (*Relative closure*) Let $(X, Y)$ be a semi-Pfaffian couple in $G \subset \mathbb{R}^n \times \mathbb{R}$. The *relative closure* of $(X, Y)$ is defined as

$$(X, Y)_0 := \check{X} \setminus \check{Y} \subset \check{G} \subset \mathbb{R}^n.$$

The format of $(X, Y)_0$ is defined as the format of the couple $(X, Y)$.

**8.6 Definition** (*Limit set*) A *limit set* in $\Omega \subset \mathbb{R}^n$ is a finite union of relative closures $(X_i, Y_i)_0$ of semi-Pfaffian couples $(X_i, Y_i)$ in $G_i \subset \mathbb{R}^n \times \mathbb{R}$, such that $\check{G}_i = \Omega$ for all $i$. The format of a limit set is defined as $(K, r, N, \alpha, \beta, s)$, where $(r, N, \alpha, \beta, s)$ is the component-wise maximum of the formats of couples $(X_i, Y_i)$, and $K$ is the number of these couples.

**8.7 Example** Any (not necessarily restricted) semi-Pfaffian set is a limit set. If is sufficient to show that a basic set

$$X := \{\mathbf{x} \in G \mid f_1(\mathbf{x}) = \cdots = f_I(\mathbf{x}) = 0,\ g_1(\mathbf{x}) > 0, \ldots, g_J(\mathbf{x}) > 0\}$$

in a domain $G := \{\mathbf{x} \in \mathbb{R}^n \mid h_1(\mathbf{x}) > 0, \ldots, h_\ell(\mathbf{x}) > 0\}$ (see Remark 2.12) is the relative closure of a semi-Pfaffian couple. Let $g := g_1 \cdots g_J$, $h := h_1 \cdots h_\ell$. Define sets

$$\begin{aligned}
W &:= \{(\mathbf{x}, \lambda) \in X \times (0,1] \mid h(\mathbf{x}) > \lambda,\ \|\mathbf{x}\| < \lambda^{-1}\};\\
Y_1 &:= \{(\mathbf{x}, \lambda) \in G \times (0,1] \mid f_1(\mathbf{x}) = \cdots = f_I(\mathbf{x}) = 0,\ g(\mathbf{x}) = 0, h(\mathbf{x}) \geq \lambda,\ \|\mathbf{x}\| \leq \lambda^{-1}\};\\
Y_2 &:= \{(\mathbf{x}, \lambda) \in G \times (0,1] \mid f_1(\mathbf{x}) = \cdots = f_I(\mathbf{x}) = 0,\ h(\mathbf{x}) = \lambda,\ \|\mathbf{x}\| \leq \lambda^{-1}\};\\
Y_3 &:= \{(\mathbf{x}, \lambda) \in G \times (0,1] \mid f_1(\mathbf{x}) = \cdots = f_I(\mathbf{x}) = 0,\ h(\mathbf{x}) \geq \lambda,\ \|\mathbf{x}\| = \lambda^{-1}\}.
\end{aligned}$$

Observe that $(W, Y_1 \cup Y_2 \cup Y_3)$ is a semi-Pfaffian couple. Its relative closure is $X$.

## 8.3 Boolean and projection operations over limit sets

**8.8 Lemma** *Let $(X, Y)$ be a semi-Pfaffian couple in a domain $G \subset \mathbb{R}^n \times \mathbb{R}$. Then the complement $\check{G} \setminus (X, Y)_0$ of $(X, Y)_0$ in $\check{G}$ is a limit set with the format admitting an upper bound in terms of the format of $(X, Y)$.*

**Proof** Let $G = \{(\mathbf{x}, \lambda) \in \mathbb{R}^n \times \mathbb{R} \mid h_1(\mathbf{x}, \lambda) > 0, \ldots, h_\ell(\mathbf{x}, \lambda) > 0\}$ (see Remark 2.12) and introduce $h := h_1 \cdots h_\ell$. Then $h$ is positive in $G$ and vanishes on $\mathrm{fr}(G)$. Let

$$G' := \{(\mathbf{x}, \lambda) \in G \mid \lambda > 0,\ h(\mathbf{x}, \lambda) \geq 1/E_r(\lambda^{-q})\},$$

where $r$ is the order of Pfaffian functions in the formula defining $X$ and $q$ is a positive integer. Let $Z := G \setminus X$ and $Z' := Z \cap G'$. It is clear that $Z'$ is a semi-Pfaffian family in $G$ (while $Z$ may not be). It follows from the exponential Łojasiewicz inequality (Theorem 8.1) that $\check{Z} = \check{Z}'$ for large $q$. We now prove that

$$\check{G} \setminus (X, Y)_0 = (Z', \mathrm{cl}(X)_+)_0 \cup (Y, \emptyset)_0. \tag{8.1}$$

Indeed, by the definition of the relative closure, the right-hand side of (8.1) coincides with

$$(\check{Z}' \setminus \check{X}) \cup \check{Y} = (\check{Z} \setminus \check{X}) \cup \check{Y}.$$

Since $(X, Y)_0 \cap (\check{Z} \setminus \check{X}) = \emptyset$ and $(X, Y)_0 \cap \check{Y} = \emptyset$, the left-hand side of (8.1) contains its right-hand side. Let now $\mathbf{x} \in \check{G} \setminus (X, Y)_0$. Note that $x \in \check{X} \cup \check{Z}$. If $x \in \check{X}$, then $x \in \check{Y}$, else $x \in \check{Z} \setminus \check{X}$. Thus, the right-hand side of (8.1) contains its left-hand side.

The right-hand side of (8.1) is a limit set. The proof of the statement about its format is straightforward. □

**8.9 Lemma** *Let $(X, Y)$ and $(X', Y')$ be two semi-Pfaffian couples in domains $G \subset \mathbb{R}^n \times \mathbb{R}$ and $G' \subset \mathbb{R}^n \times \mathbb{R}$ respectively. Then $(X, Y)_0 \cap (X', Y')_0$ is a limit set with the format admitting an upper bound in terms of formats of $(X, Y)$ and $(X', Y')$.*

A proof of this lemma, technically similar to the proof of Lemma 8.8 above, can be found in [16].

**8.10 Theorem** *Limit sets form a Boolean algebra. The format of a limit set defined by a Boolean combination of limit sets $X_1, \ldots, X_\ell$ admits an upper bound in terms of the complexity of the formula and the formats of $X_1, \ldots, X_\ell$.*

**Proof** This immediately follows from Lemmas 8.8 and 8.9. □

**8.11 Theorem** *Let $(X, Y)$ be a semi-Pfaffian couple in a domain $G \subset \mathbb{R}^{m+n} \times \mathbb{R}$, and let $\pi : \mathbb{R}^{m+n} \to \mathbb{R}^n$ be the projection map. Then $\pi((X, Y)_0)$ is a limit set in $\pi(\check{G}) \subset \mathbb{R}^n$, and its format admits an upper bound in terms of the format of $(X, Y)$.*

A proof of this theorem can be found in [16] and is difficult.

**8.12 Remark** Theorems 8.10 and 8.11 imply that limit sets form an *o-minimal structure* [9]. Moreover, it is an *effectively* o-minimal structure, i.e., the formats of the results of Boolean and projection operations over limit sets admit upper bounds in terms of the formats of these sets. However, no explicit expressions for the bounds have yet been obtained. This is a promising direction for future research.

## 8.4 Connected components of limit sets

We now establish an explicit upper bound on the number of connected components of the relative closure of a semi-Pfaffian couple $(X, Y)$.

Note that if $Y = \emptyset$, then $\mathrm{fr}(X)_+ = \emptyset$, i.e., $X_\lambda$ is compact for any $\lambda > 0$. In this case the relative closure $(X, Y)_0 = \check{X}$ is also compact, and therefore the number of its connected components does not exceed the number of the connected components of $X_\lambda$, for all sufficiently small $\lambda > 0$. An upper bound for $X_\lambda$ follows from Theorem 3.4.

Suppose now that $Y \neq \emptyset$. For $\mathbf{x} \in \mathbb{R}^n$ and $\lambda > 0$, let $\Psi_\lambda(\mathbf{x}) := \min_{\mathbf{y} \in Y_\lambda} \|\mathbf{x} - \mathbf{y}\|^2$ be the (squared) distance from $\mathbf{x}$ to $Y_\lambda$, let $\Psi(\mathbf{x}) := \min_{\mathbf{y} \in \check{Y}} \|\mathbf{x} - \mathbf{y}\|^2$ be the distance from $\mathbf{x}$ to $\check{Y}$. Note that these two functions are well defined (minima exist) since $Y_\lambda$ and $\check{Y}$ are both closed. Let $Z_\lambda$ be the set of local maxima of $\Psi_\lambda(\mathbf{x})$ on $X_\lambda$.

**8.13 Lemma** *The number of connected components of $(X, Y)_0$ does not exceed the number of connected components of $Z_\lambda$ for all sufficiently small $\lambda > 0$.*

**Proof** Let $C$ be a connected component of $(X, Y)_0$. By the definition of relative closure, if $\mathbf{x} \in C$, then $\mathbf{x} \notin \check{Y}$. Thus, $\Psi(\mathbf{x}) > 0$. Since $\mathrm{fr}(C) \subset \check{Y}$, the function $\Psi$ vanishes on $\mathrm{fr}(C)$. It follows that $\Psi$ has a local maximum, say $\mathbf{x}_0$, in $C$. There exists $\mathbf{x}_\lambda \in X_\lambda$ such that $\|\mathbf{x}_\lambda - \mathbf{x}_0\| \to 0$ as $\lambda \searrow 0$. It follows that $\lim_{\lambda \searrow 0} \Psi_\lambda(\mathbf{x}_\lambda) = \Psi(\mathbf{x}_0) > 0$. In particular, there exists a constant $\varepsilon > 0$ such that $\Psi_\lambda(\mathbf{x}_\lambda) > \varepsilon$ for all sufficiently small $\lambda > 0$. Let $W_{\lambda,\varepsilon} := \{\mathbf{x} \in X_\lambda \mid \Psi_\lambda(\mathbf{x}) > \varepsilon\}$ and let $C_\lambda$ be the connected component of $W_{\lambda,\varepsilon}$ which contains $\mathbf{x}_\lambda$. Since $\Psi_\lambda(\mathbf{x}) > \varepsilon$ for any $\mathbf{x} \in C_\lambda$, the sets $C_\lambda$ are "close" to $C$ for sufficiently small $\lambda > 0$, i.e., $\mathrm{cl}(\bigcup_{\lambda > 0} C_\lambda) \cap \{(\mathbf{x}, \lambda) \mid \lambda = 0\}$ is the connected subset of $(X, Y)_0$ containing $\mathbf{x}_0$, hence a subset of $C$. From the definition of $C_\lambda$, there exists a local maximum $\mathbf{z}_\lambda$ of $\Psi_\lambda$ on $C_\lambda$, and a connected component $V_\lambda$ of $Z_\lambda$ containing $\mathbf{z}_\lambda$ lies in $C_\lambda$. It follows that $V_\lambda$ is "close" to $C$ for all sufficiently small $\lambda > 0$. This implies the lemma. □

**8.14 Theorem** *Let $(X, Y)$ be a semi-Pfaffian couple. The number of connected components of $(X, Y)_0$ is finite and admits an explicit upper bound in terms of the format of $(X, Y)$.*

**Proof** According to Lemma 8.13, it is sufficient to bound from above the number of connected components of $Z_\lambda$. Since $Z_\lambda$ is a *restricted sub-Pfaffian* set, the number of its connected components is finite. Moreover, due to Theorem 7.4, $Z_\lambda$ admits a cylindrical cell decomposition with the number of cells explicitly bounded from above in terms of the format of $(X, Y)$. The number of connected components of $Z_\lambda$ does not exceed the number of cells in the decomposition, which implies the second statement of the theorem. □

The upper bound (via a cylindrical cell decomposition) which can be extracted from the proof of Theorem 8.14 is doubly exponential in the number $n$ of variables, and is not best possible. A better bound can be obtained by applying the results of Section 3.5 to a formula with quantifiers describing $Z_\lambda$. In [21] a much more specialized method was used to prove the following, currently the best, upper bound.

**8.15 Theorem ([21])** *Let the format of a semi-Pfaffian couple $(X, Y)$ be $(r, N, \alpha, \beta, n)$. Then the number of connected components of $(X, Y)_0$ does not exceed*

$$N^2 2^{O(n^2 r^2)} (n(\alpha + \beta))^{O(n^2 + nr)}.$$

A proof of the theorem and a slightly more precise bound (not using the $O$-notation) can be found in [21]. Further results on upper bounds for higher Betti numbers of limit sets were recently obtained in [50].

# 9 Noetherian functions

Noetherian functions are analytic functions in $G \subset \mathbb{K}^n$ defined by equations similar to (2.1) but without triangularity condition. Given a *Noetherian chain*, i.e., a sequence of functions $f_1(\mathbf{x}), \dots, f_r(\mathbf{x})$ analytic in $G$, such that

$$df_j(\mathbf{x}) = \sum_{1 \le i \le n} g_{ij}(\mathbf{x}, f_1(\mathbf{x}), \dots, f_r(\mathbf{x}))\, dx_i \quad \text{for all} \quad j, \tag{9.1}$$

with $g_{ij}$ polynomials in $n + r$ variables of degree at most $\alpha$, a *Noetherian function* of order $r$ and degree $(\alpha, \beta)$ is a function $\phi(\mathbf{x}) = P(\mathbf{x}, f_1(\mathbf{x}), \dots, f_r(\mathbf{x}))$, where $P(\mathbf{x}, \mathbf{y})$ is a polynomial of degree $\beta$ in $n + r$ variables.

Alternatively, a Noetherian chain can be defined as an integral manifold $\Lambda = \{(z_1 = f_1(\mathbf{x}), \dots, z_r = f_r(\mathbf{x}))\}$ of a $n$-dimensional distribution in $\mathbb{K}^{n+r}$:

$$dy_j = \sum_{i=1}^{n} g_{ij}(\mathbf{x}, \mathbf{y})\, dx_i, \quad \text{for} \quad j = 1, \dots, r, \tag{9.2}$$

and a Noetherian function as a restriction of $P(\mathbf{x}, \mathbf{y})$ to $\Lambda$.

All Noetherian functions defined by the same chain constitute a subring $\mathfrak{R}$ of the ring of analytic functions in $G$, finitely generated over the polynomial ring $\mathbb{K}[\mathbf{x}]$ and closed under differentiation, i.e., for any $\phi \in \mathfrak{R}$, all partial derivatives $\partial\phi/\partial x_i$ are in $\mathfrak{R}$. Conversely, any such ring is a ring of Noetherian functions. Any set of its generators can be taken as a Noetherian chain. The name "Noetherian function" introduced by Tougeron [43] refers to the Noetherian property of that ring.

For the univariate case $\mathbf{x} = t \in \mathbb{K}$, Noetherian functions are simply polynomials restricted to a solution $y_j = f_j(t)$ of a system $dy_j/dt = g_j(t, y_1, \dots, y_r)$ of ordinary differential equations with polynomial coefficients.

The simplest example of a Noetherian chain, $f_1 = \sin t, f_2 = \cos t$, in $\mathbb{R}$ shows that the global finiteness properties of Pfaffian functions do not hold for Noetherian functions. In the early eighties, Khovanskii conjectured that Noetherian functions (also in the complex domain) satisfy *local* finiteness properties. Assume $0 \in G$, and consider a germ $X_0$ at 0 of a set $X$ defined by equations and inequalities between Noetherian functions (a *semi-Noetherian set*), or an intersection $X_\delta$ of the set $X$ with a ball of radius $\delta$ centered at 0. Khovanskii's conjecture states that the topological and geometric complexity of $X_0$ (or of $X_\delta$, for a small $\delta > 0$) can be bounded from above by an explicit function of its *format* (see (2.11)).

Considerable progress towards proving this conjecture was made in [17]:

**9.1 Theorem** *Let $\phi_1, \dots, \phi_n$ be Noetherian functions of order r and degree $(\alpha, \beta)$ in $G \subset \mathbb{C}^n$, with the same Noetherian chain. Then the multiplicity of any isolated solution of the system of equations $\phi_1 = \dots = \phi_n = 0$ does not exceed the maximum of the following two numbers:*

$$\begin{gathered} \frac{1}{2}Q((r+1)(\alpha-1)[2\alpha(n+r+2)-2r-2]^{2r+2}+2\alpha(n+2)-2)^{2(r+n)}, \\ \frac{1}{2}Q(2(Q+n)^n(\beta+Q(\alpha-1)))^{2(r+n)}, \quad \text{where} \quad Q = en\left(\frac{e(n+r)}{\sqrt{n}}\right)^{\ln n+1}\left(\frac{n}{e^2}\right)^n. \end{gathered} \tag{9.3}$$

**9.2 Remark** One can show (see [43]) that, for any given integers $n$, $r$, $\alpha$, and $\beta$, there exists an integer $M(n,r,\alpha,\beta)$ such that the multiplicity $\mu$ of any isolated intersection $\phi_1 = \cdots = \phi_n = 0$, where $\phi_j(\mathbf{y}) = P_j(\mathbf{y}, f_1(\mathbf{y}), \ldots, f_r(\mathbf{y}))$ are Noetherian functions in $G \subset \mathbb{C}^n$ of degree at most $\beta$, with a Noetherian chain $f_1, \ldots, f_r$ of order $r$ and degree $\alpha$, does not exceed $M(n,r,\alpha,\beta)$.

To prove this, recall first that the condition $\mu \geq M$, for any analytic functions $\phi_1, \ldots, \phi_n$, can be formulated as a system of polynomial equations on the values of the functions $\phi_j$ and their partial derivatives of order not exceeding $M$. For Noetherian functions, the values of their partial derivatives can be expressed as polynomials of the variables $y_i$, the values of $\phi_j$, and the coefficients of the polynomials $g_{ij}$ and $P_j$ in their definition. Consider now the ring $\mathfrak{S}$ of polynomials in all these variables. The condition $\mu \geq M$ is represented by an ideal $I_M$ in $\mathfrak{S}$. As $I_M$ is an increasing sequence and $\mathfrak{S}$ is a Noetherian ring, the sequence $I_M$ stabilizes at some $M = M(n,r,\alpha,\beta)$. This means that any intersection with multiplicity $\mu \geq M(n,r,\alpha,\beta)$ has infinite multiplicity. Theorem 9.1 can be interpreted as an explicit upper bound for the number $M(n,r,\alpha,\beta)$

A sketch of the proof of Theorem 9.1 is given below, based on several preliminary results. First, we give a brief introduction to integration over Euler characteristics (see [44]). Next, we consider the univariate case $n = 1$. The univariate result implies, in particular, an upper bound on the vanishing order of a multivariate Noetherian function. As an application of the univariate result, we derive an upper bound for the degree of nonholonomy of a system of polynomial vector fields. In the multivariate case, we have to take care of possible non-integrability of the distribution (9.2). Finally, we need a lower bound on the codimension of the set of intersections of high multiplicity.

## 9.1 Integration over Euler characteristics

Assume that we are working in a category of "tame" sets (such as semialgebraic or global subanalytic sets). This means that our category is closed under Boolean operations and projections, and each set is homotopy equivalent to a finite simplicial complex. In the language of model theory, this is called an *o-minimal structure* (see [9] for the definitions and properties of sets definable in an o-minimal structure). In particular, the Betti numbers $\mathrm{b}_i(X)$ and the Euler characteristic $\chi(X) = \sum_i (-1)^i \mathrm{b}_i(X)$ are finite for any set $X$. The key properties of the Euler characteristic of compact sets are its additivity and multiplicativity:

$$\chi(X \cup Y) + \chi(X \cap Y) = \chi(X) + \chi(Y) \tag{9.4}$$

$$\chi(X \times Y) = \chi(X)\,\chi(Y). \tag{9.5}$$

These properties allow one to extend $\chi$ as an additive and multiplicative function to all (not necessarily compact) sets. Of course, $\chi(X)$ for a non-compact set may be different from the topological Euler characteristic. For example, if $B^n$ is a closed ball in $\mathbb{R}^n$ and $X = B^n \setminus S^{n-1}$ is an open ball, then $\chi(X) = \chi(B^n) - \chi(S^{n-1})$ is $(-1)^n$.

A *constructible* function $f(\mathbf{x})$ in $\mathbb{K}^n$ is a function with finitely many values $y_i$ such that all its level sets $X_i = \{\mathbf{x} : f(\mathbf{x}) = y_i\}$ are "tame". Its *integral over Euler characteristic* over $U \subset \mathbb{K}^n$ is defined as

$$\int_U f(\mathbf{x})\, d\chi := \sum_i y_i\, \chi(X_i \cap U). \tag{9.6}$$

The properties (9.4) and (9.5) imply finite additivity and the Fubini theorem for this integral:

**9.3 Theorem** *For two constructible functions, $f$ and $g$,*

$$\int_U f(\mathbf{x})\,d\chi + \int_U g(\mathbf{x})\,d\chi = \int_U (f(\mathbf{x}) + g(\mathbf{x}))\,d\chi.$$

**9.4 Theorem** *Let $f(\mathbf{x}, \mathbf{y})$ be a constructible function in $\mathbb{K}^{n+m}$. For $U \subset \mathbb{K}^n$ and $V \subset \mathbb{K}^m$, the function*

$$g(\mathbf{x}) = \int_V f(\mathbf{x}, \mathbf{y})|_{\mathbf{x}=\text{const}}\,d\chi$$

*is constructible and*

$$\int_U g(\mathbf{x})\,d\chi = \int_{U\times V} f(\mathbf{x}, \mathbf{y})\,d\chi.$$

## 9.2 Univariate case

When $n = 1$, Noetherian chains are trajectories of vector fields with polynomial coefficients, and Noetherian functions are polynomials restricted to trajectories of such vector fields.

Let $t \in \mathbb{C}$, $\mathbf{y} = (y_1, \ldots, y_r) \in \mathbb{C}^r$, and let $\gamma = \{\mathbf{y} = \mathbf{y}(t)\}$ be a germ of a trajectory through $0 \in \mathbb{C}^{r+1}$ of a vector field $\xi = g_0(t, \mathbf{y})\partial/\partial t + \sum_i g_i(t, \mathbf{y})\partial/\partial y_i$, where $g_i$ are germs of analytic functions at $0 \in \mathbb{C}^{r+1}$, $g_0(0) \neq 0$. Let $P(t, \mathbf{y})$ be a germ of an analytic function at $0 \in \mathbb{C}^{r+1}$, and let $\phi(t) = P(t, \mathbf{y}(t))$ be the restriction of $P(t, \mathbf{y})$ to $\gamma$. Suppose that $\phi(t) \not\equiv 0$, and let $\mu$ be the order of a zero of $\phi$ at $t = 0$. Let $S(t, \mathbf{y}, \varepsilon)$ be a one-parameter deformation of $P$, i.e., a germ of an analytic function at $0 \in \mathbb{C}^{r+2}$ such that $S(t, \mathbf{y}, 0) = P(t, \mathbf{y})$. We write $S_\varepsilon(t, \mathbf{y})$ for $S(t, \mathbf{y}, \varepsilon)|_{\varepsilon=\text{const}}$ considered as a function in $\mathbb{C}^{r+1}$.

**9.5 Definition** For a positive integer $q$, the *Milnor fiber* $Z_q(\xi, S)$ of the deformation $S$ with respect to a vector field $\xi$ is the intersection of a ball $\|(t, \mathbf{y})\| \leq \delta$ in $\mathbb{C}^{r+1}$ with a set $S_\varepsilon = \xi S_\varepsilon = \cdots = \xi^{q-1} S_\varepsilon = 0$, for a small positive $\delta$ and a complex nonzero $\varepsilon$ much smaller than $\delta$.

One can show (see [26]) that the homotopy type of $Z_q(\xi, S)$ does not depend on $\varepsilon$ and $\delta$, as long as $|\varepsilon| \ll \delta \ll 1$. Unless $P$ has an isolated singularity, this homotopy type does depend on the deformation $S$.

**9.6 Theorem** *Let $S$ be a one-parameter deformation of an analytic function $P$. For a positive integer $q$, let $Z_q = Z_q(\xi, S)$ be the Milnor fibers of $S$ with respect to an analytic vector field $\xi$, and let $\chi(Z_q)$ be the Euler characteristic of $Z_q$. Suppose that $P$ restricted to a trajectory of $\xi$ through $0$ has a zero of order $\mu < \infty$ at $0$. Let $Q := \max\{q : Z_q \neq \emptyset\}$. Then*

$$\mu = \sum_{q=1}^{Q} \chi(Z_q). \tag{9.7}$$

**Proof** One can assume, after a change of coordinates in $(\mathbb{C}^{r+1}, 0)$, that $\xi = \partial/\partial t$. It follows from [26] that the homotopy type of $Z_q$ does not depend on the coordinate system. Let $\pi$ be the projection $\mathbb{C}^{r+1} \to \mathbb{C}^r$ along the $t$-axis. Let $B_\eta = \{\|\mathbf{y}\| \leq \eta\}$ be a ball of radius $\eta$ in $\mathbb{C}^r$. We can replace the ball $\{\|(t, \mathbf{y})\| \leq \delta\}$ in Definition 9.5 by $D_{\delta,\eta} = \{|t| \leq \delta,\ \mathbf{y} \in B_\eta\}$, where

$0 < \eta \ll \delta$, so that the projection $\pi : \{P = 0\} \cap D_{\delta,\eta} \to B_\eta$ is a finite $\mu$-fold ramified covering (counting the multiplicities). This also would not change the homotopy type of $Z_q$, as long as $|\varepsilon| \ll \eta$.

For $\mathbf{y} \in B_\eta$, each set $\pi^{-1}(\mathbf{y}) \cap Z_q$ is finite, and its Euler characteristic $\zeta_q(\mathbf{y})$ equals the number of points in it (not counting multiplicities). From Theorem 9.4,

$$\int_{B_\eta} \zeta_q(\mathbf{y})\, d\chi = \chi(Z_q). \tag{9.8}$$

Note that each point $(t, \mathbf{y}) \in \{S_\varepsilon = 0\} \cap D_{\delta,\eta}$ belongs to exactly $k$ sets $Z_1, \ldots, Z_k$, where $k$ equals the multiplicity of $t$ in $\pi^{-1}(\mathbf{y}) \cap \{S_\varepsilon = 0\}$. Hence $\sum_{q=1}^{Q} \zeta_q(\mathbf{y}) \equiv \mu$ does not depend on $\mathbf{y}$. From (9.8) and Theorem 9.3,

$$\sum_{q=1}^{Q} \chi(Z_q) = \int_{B_\eta} \sum_{q=1}^{Q} \zeta_q(\mathbf{y})\, d\chi = \int_{B_\eta} \mu\, d\chi = \mu\chi(B_\eta) = \mu.$$

□

The following is a special case of Thom's transversality theorem (see Lemma 1 in [15]).

**9.7 Lemma** *Let $\xi = g_0(t, \mathbf{y})\partial/\partial t + \sum_{i=1}^{r} g_i(t, \mathbf{y})\partial/\partial y_i$ be a germ at $0 \in \mathbb{K}^{r+1}$ of an analytic vector field, $g_0(0) \neq 0$, and $P(t, \mathbf{y})$ a germ of an analytic function. Let $c = (c_0, \ldots, c_r) \in \mathbb{K}^{r+1}$ and*

$$S_c(t, \mathbf{y}, \varepsilon) = P(t, \mathbf{y}) + \varepsilon \sum_{i=0}^{r} c_i t^i. \tag{9.9}$$

*For a generic $c$, the sets $Z_q = Z_q(\xi, S_c)$ are nonsingular of codimension $q$ for $q = 1, \ldots, r+1$, and empty for $q > r + 1$.*

**9.8 Corollary** *Let $\xi$ and $P$ be as in Lemma 9.7, and let $\mu < \infty$ be the multiplicity of $P$ on the trajectory of $\xi$ through $0$. Let $Z_q = Z_q(\xi, S_c)$ be the Milnor fibers of (9.9) with respect to $\xi$. For a generic $c$,*

$$\mu = \chi(Z_1) + \cdots + \chi(Z_{r+1}). \tag{9.10}$$

**9.9 Theorem** *Let $\xi = g_0(t, \mathbf{y})\partial/\partial t + \sum_i g_i(t, \mathbf{y})\partial/\partial y_i$, where $g_i$ are polynomials of degree not exceeding $\alpha \geq 1$, $g_0(0) \neq 0$, and let $P$ be a polynomial of degree not exceeding $\beta \geq r$. Let $\mu < \infty$ be the multiplicity of $P$ on the trajectory of $\xi$ through $0$. Then $\mu$ does not exceed*

$$\frac{1}{2} \sum_{k=0}^{r} [2\beta + 2k(\alpha - 1)]^{2r+2}. \tag{9.11}$$

**Proof** This follows from (9.10) and from an estimate [33] of the Euler characteristic of a set $Z_q$ defined by polynomial equations of degree not exceeding $\beta + (q-1)(\alpha - 1)$. □

**9.10 Corollary** *Let $\phi \not\equiv 0$ be a Noetherian function of order $r$ and degree $(\alpha, \beta)$ in a neighborhood of $0 \subset \mathbb{K}^n$. Then the vanishing order of $\phi$ at $0$ does not exceed* (9.11).

**Proof** This follows from Theorem 9.9 after restricting $\phi$ to a generic line through $0$. □

## 9.3 Degree of nonholonomy of a system of vector fields

**9.11 Definition** Let $\Xi$ be a system of analytic vector fields $\xi_i$ in $G \subset \mathbb{K}^n$. Let $\mathcal{L}_1(\Xi)$ be the space of all linear combinations of $\xi_i$ with coefficients in $\mathbb{K}$. For $k \geq 2$, define $\mathcal{L}_k(\Xi) := \mathcal{L}_{k-1}(\Xi) + [\mathcal{L}_1(\Xi), \mathcal{L}_{k-1}(\Xi)]$. Then $\mathcal{L}(\Xi) := \bigcup_k \mathcal{L}_k(\Xi)$ is the Lie algebra generated by the vector fields $\xi_i$. For $z \in \mathbb{K}^n$, let $d_k(z)$ be the dimension of the subspace generated by the values at $z$ of the vector fields from $\mathcal{L}_k(\Xi)$, and let $d(\Xi, z) := \max_k d_k(z)$ be the dimension of the subspace generated by the values at $z$ of the vector fields from $\mathcal{L}(\Xi)$. In particular, when $d(\Xi, z) = n$, the system $\Xi$ is called totally nonholonomic (controllable). The minimal $k$ such that $d_k(z) = d(\Xi, z)$ is called the *degree of nonholonomy* of $\Xi$ at $z$. It is easy to check that the values $d_k$, and the degree of nonholonomy, do not change if we allow linear combinations of vector fields with analytic (instead of constant) coefficients.

It is shown in [12] that an upper bound for the multiplicity of a zero of a polynomial on a trajectory of a polynomial vector field implies an upper bound on the degree of nonholonomy of a system of polynomial vector fields. In particular, the upper bound in Theorem 9.9 implies the following upper bound for the degree of nonholonomy:

**9.12 Theorem** *Let $\Xi := \{\xi_i\}$ be a system of vector fields in $\mathbb{K}^n$ with polynomial coefficients of degree not exceeding $p \geq 1$. For $z \in \mathbb{K}^n$, let $d = d(\Xi, z) > 1$. Then the degree of nonholonomy of $\Xi$ at $z$ does not exceed*

$$F_d + \frac{F_{d-1}}{2} \sum_{k=0}^{n-1} [2p(F_{d+2} - 1) + 2k(pF_d - 1)]^{2n} \tag{9.12}$$

*where $F_i$ are the Fibonacci numbers.*

**Proof** According to Proposition 1 of [12], there exist vector fields $\chi_0, \dots, \chi_{d-1}$ such that (a) $\chi_0$ and $\chi_1$ are some of $\xi_i$, and $\chi_0(z) \neq 0$; (b) for $k > 1$, $\chi_k$ is either one of $\xi_i$ or a linear combination of brackets $[\chi_i, f\chi_j]$ where $i, j < k$ and $f$ is a linear function; (c) for a generic small $\varepsilon = (\varepsilon_1, \dots, \varepsilon_{d-2})$, $Q = \chi_0 \wedge \cdots \wedge \chi_{d-1}$ does not vanish identically on a trajectory $\gamma$ of $\chi_\varepsilon = \chi_0 + \varepsilon_1\chi_1 + \cdots + \varepsilon_{d-2}\chi_{d-2}$ through $z$. Taking into account that $[\chi, \chi] = 0$ for any vector field $\chi$, the arguments in the proof of Proposition 1 of [12] can be modified to replace (b) by (b$'$) for $k > 1$, $\chi_k$ is either one of $\xi_i$ or a linear combination of brackets $[\chi_i, f\chi_j]$, where $j < i < k$ and $f$ is a linear function. In particular, each $\chi_k$ is a vector field with polynomial coefficients of degree not exceeding $pF_{k+1}$, where $F_i$ are the Fibonacci numbers: $F_1 = 1,\ F_2 = 1,\ F_{i+1} = F_i + F_{i-1}$.

Let $x_1, \dots, x_n$ be linear coordinates in $\mathbb{K}^n$. Then

$$Q = \sum_{i_1 < \cdots < i_d} Q_{i_1 \dots i_d} \frac{\partial}{\partial x_{i_1}} \wedge \cdots \wedge \frac{\partial}{\partial x_{i_d}},$$

where $Q_{i_1 \dots i_d}$ are polynomials of degrees not exceeding $\beta = p(F_1 + \cdots + F_d) = p(F_{d+2} - 1)$. Due to (c), some of these polynomials do not vanish identically on the trajectory $\gamma$ of a vector field $\chi_\varepsilon$ with polynomial coefficients of degree not exceeding $\alpha = pF_d$. Due to Theorem 9.9, the multiplicity $\mu$ of a zero of such a polynomial restricted to $\gamma$ does not exceed

$$\frac{1}{2} \sum_{k=0}^{n-1} [2\beta + 2k(\alpha - 1)]^{2n} = \frac{1}{2} \sum_{k=0}^{n-1} [2p(F_{d+2} - 1) + 2k(pF_d - 1)]^{2n}.$$

Each derivation of $Q$ along $\chi_\varepsilon$ decreases this multiplicity by 1. Hence the result of $\mu$ consecutive derivations of $Q$ along $\chi_\varepsilon$ does not vanish at $z$. From (b′), each $\chi_k$ is a linear combination with polynomial coefficients of brackets of $\xi_i$ of order at most $F_{k+1}$, and $\chi_\varepsilon$ is a combination of brackets of $\xi_i$ of order at most $F_{d-1}$. Taking into account a formula for a derivation along $\chi_\varepsilon$:

$$\partial_{\chi_\varepsilon}(\chi_0 \wedge \cdots \wedge \chi_{d-1}) = \sum_{i=0}^{d-1} \chi_0 \wedge \cdots \wedge [\chi_\varepsilon, \chi_i] \wedge \cdots \wedge \chi_{d-1},$$

we see that the result of $\mu$ derivations of $Q$ along $\chi_\varepsilon$ is a linear combination, with polynomial coefficients, of wedge-products of vector fields which are brackets of $\xi_i$ of order not exceeding $F_d + F_{d-1}\mu$, which is equal to (9.12). Since the result of $\mu$ derivations of $Q$ along $\chi_\varepsilon$ does not vanish at $z$, there exist $d = d(\Xi, z)$ brackets of $\xi_i$ of order not exceeding (9.12) which are linearly independent at $z$. □

## 9.4 Multivariate case: integrability

For a system (9.1) of partial differential equations in $\mathbb{K}^n$, when $n > 1$, analytic solutions $f_1(\mathbf{x}), \dots, f_r(\mathbf{x})$ do not exist, in general, due to non-integrability. When such solutions do exist, their union is an analytic set (Lemma 9.13 below). When the coefficients $g_{ij}$ are polynomials, this set is algebraic.

**9.13 Lemma** *Let* $\mathbf{x} \in \mathbb{K}^n$, $\mathbf{y} \in \mathbb{K}^r$, *and let* $g_{ij}(\mathbf{x}, \mathbf{y})$ *be analytic functions in* $U \subset \mathbb{K}^{n+r}$, *for* $i = 1, \dots, n$ *and* $j = 1, \dots, r$. *The union of all integral manifolds of* (9.2) *is an analytic subset of* $U$.

**Proof** The statement can be proved locally, in a neighborhood of a point $(\mathbf{x}, \mathbf{y}) \in U$. Let $\xi_i$, $i = 1, \dots, n$, be the following vector fields tangent to the distribution (9.2):

$$\xi_i = \frac{\partial}{\partial x_i} + \sum_{j=1}^{r} g_{ij}(\mathbf{x}, \mathbf{y}) \frac{\partial}{\partial y_j}. \tag{9.13}$$

For a nonzero $c = (c_1, \dots, c_n) \in \mathbb{K}^n$, let $\xi_c = c_1\xi_1 + \cdots + c_n\xi_n$, and let $\gamma_c$ be a germ of a trajectory of $\xi_c$ through $(\mathbf{x}, \mathbf{y})$. Let $\Lambda$ be the union of all $\gamma_c$. Then $\Lambda$ is a germ of an $n$-dimensional manifold.

If there exists an integral manifold of (9.2) through $(\mathbf{x}, \mathbf{y})$, it contains all trajectories $\gamma_c$ and its germ at $(\mathbf{x}, \mathbf{y})$ coincides with $\Lambda$. Since all vector fields $\xi_i$ are tangent to $\Lambda$, any bracket $[\xi_j, \xi_k]$ at each point of $\Lambda$ belongs to the subspace generated by $\xi_i$.

Due to Lemmas 4 and 5 in [12], the converse is also true: if any bracket $[\xi_j, \xi_k]$ at each point of $\Lambda$ belongs to the subspace generated by $\xi_i$, then $\Lambda$ is an integral manifold of (9.2).

Let $Q_{jk} = [\xi_j, \xi_k] \wedge \xi_1 \cdots \wedge \xi_n$. If $\Lambda$ is an integral manifold of (9.2) then all $Q_{jk}$ vanish identically on $\gamma_c$. In particular,

$$\xi_c^\nu Q_{jk}(\mathbf{x}, \mathbf{y}), \tag{9.14}$$

the value at $(\mathbf{x}, \mathbf{y})$ of the $\nu$-th derivative of $Q_{jk}$ along $\xi_c$, is zero for each $j$, $k$, and $\nu$. In the opposite case, when $Q_{jk}$ does not vanish identically on $\gamma_c$ for some $j$, $k$, and $c$, there exists a nonzero derivative (9.14).

Hence, $(\mathbf{x}, \mathbf{y})$ belongs to an integral manifold of (9.2) if and only if all derivatives (9.14) vanish. As these derivatives are analytic in $(\mathbf{x}, \mathbf{y})$, their common zeros constitute an analytic set. □

**9.14 Proposition** *Let $g_{ij}$ in (9.2) be polynomials of degree not exceeding $\alpha \geq 1$, and let $Y$ be the union of all integral manifolds of (9.2). Then $Y$ can be defined by a system of algebraic equations of degree not exceeding*

$$d_Y(n, r, \alpha) = \frac{1}{2}(r+1)(\alpha-1)[2\alpha(n+r+2) - 2r - 2]^{2r+2} + 2\alpha(n+2) - 2. \tag{9.15}$$

**Proof** From the proof of Lemma 9.13, the set $Y$ is defined by zeros of derivatives (9.14) of $Q_{ij}$ along $\xi_c$, for all $i$, $j$, and $c$. Here $\xi_c$ is a linear combination of the vector fields $\xi_i$ defined in (9.13), and $Q_{jk} = [\xi_j, \xi_k] \wedge \xi_1 \cdots \wedge \xi_n$.

As the coefficients of $\xi_i$ are polynomials of degree not exceeding $\alpha$, the coefficients of $Q_{jk}$ are polynomials of degree not exceeding $B = \alpha(n+2) - 1$. If $(\mathbf{x}, \mathbf{y})$ does not belong to $Y$, some of these coefficients do not vanish identically on the trajectory $\gamma_c$ of $\xi_c$ through $(\mathbf{x}, \mathbf{y})$. Due to Theorem 9.9, the multiplicity of zero at $(\mathbf{x}, \mathbf{y})$ of these coefficients restricted to $\gamma_c$ cannot exceed

$$N = \frac{1}{2}\sum_{k=0}^{r}[2B + 2k(\alpha-1)]^{2r+2} = \frac{1}{2}\sum_{k=0}^{r}[2\alpha(n+k+2) - 2k - 2]^{2r+2}.$$

This means that $Y$ can be defined by zeros of derivatives (9.14) with $\nu \leq N$. But these derivatives are polynomials in $(\mathbf{x}, \mathbf{y})$ of degree at most $B + N(\alpha - 1)$. Hence $Y$ is defined by a system of algebraic equations of degree at most

$$\frac{\alpha-1}{2}\sum_{k=0}^{r}[2\alpha(n+k+2) - 2k - 2]^{2r+2} + \alpha(n+2) - 1.$$

As $\alpha \geq 1$, this does not exceed the right hand side of (9.15). □

## 9.5 Multivariate case: Milnor fibers

Let $g_{ij}(\mathbf{x}, \mathbf{y})$ be germs of analytic functions at $0 \in \mathbb{C}^{n+r}$, and let $\{\mathbf{y} = f(\mathbf{x})\}$ be a germ of an integral manifold of (9.2) through 0. Let $P(\mathbf{x}, \mathbf{y}) = (P_1(\mathbf{x}, \mathbf{y}), \ldots, P_n(\mathbf{x}, \mathbf{y}))$ be a germ of an analytic vector-function at 0, and let $\phi_i(\mathbf{x}) = P_i(\mathbf{x}, f(\mathbf{x}))$. Let $S(\mathbf{x}, \mathbf{y}, \varepsilon)$ be a one-parameter deformation of $P(\mathbf{x}, \mathbf{y})$, i.e., a germ of an analytic vector-function at $0 \in \mathbb{C}^{n+r+1}$ such that $S(\mathbf{x}, \mathbf{y}, 0) = P(\mathbf{x}, \mathbf{y})$. For a fixed $\varepsilon$, we write $S_\varepsilon(\mathbf{x}, \mathbf{y})$ for $S(\mathbf{x}, \mathbf{y}, \varepsilon)$ considered as a function in $\mathbb{C}^{n+r}$.

Let $Y$ be the union of all integral manifolds of (9.2). Due to Lemma 9.13, $Y$ is a germ of an analytic set. For $(\mathbf{x}, \mathbf{y}) \in Y$, let $\mu_\varepsilon(\mathbf{x}, \mathbf{y})$ be the multiplicity of the intersection $S_{1,\varepsilon}|_\Lambda = \cdots = S_{n,\varepsilon}|_\Lambda = 0$ at $(\mathbf{x}, \mathbf{y})$, where $\Lambda$ is an integral manifold of (9.2) through $(\mathbf{x}, \mathbf{y})$. Let $W_q$ be the set of those $(\mathbf{x}, \mathbf{y}, \varepsilon)$ where $(\mathbf{x}, \mathbf{y}) \in Y$ and $\mu_\varepsilon(\mathbf{x}, \mathbf{y}) \geq q$.

**9.15 Lemma** *$W_q$ is an analytic set.*

**Proof** One can choose a system of coordinates $(\mathbf{x}, \mathbf{z})$ in the neighborhood of $0 \in \mathbb{C}^{n+r}$ so that each integral manifold of (9.2) is defined by $\mathbf{z} = \text{const}$.

Due to Lemma 5.5 in [1], the condition $\mu_\varepsilon(\mathbf{x}_0, \mathbf{z}_0) \geq q$ depends only on the Taylor expansion $\check{S}_i$ in $\mathbf{x}$ of $S_i$ at $(\mathbf{x}_0, \mathbf{z}_0, \varepsilon)$ of order $q-1$. The coefficients of $\check{S}_i$ are

$$\frac{\partial^{|\nu|} S_i}{\partial x^\nu}(\mathbf{x}_0, \mathbf{z}_0, \varepsilon). \tag{9.16}$$

Let $K = \binom{q+n-1}{n}$ be the number of monomials in $n$ variables of degree less than $q$. For fixed $\mathbf{z}_0$ and $\varepsilon$, consider $\check{S}_i$ as a vector in $\mathbb{C}^K$. For any multi-index $\nu = (\nu_1, \ldots, \nu_n)$, with $|\nu| = \nu_1 + \cdots + \nu_n < q$, consider $(\mathbf{x} - \mathbf{x}_0)^\nu \check{S}_i$ as a vector in $\mathbb{C}^K$, disregarding terms of order $q$ and higher in $\mathbf{x} - \mathbf{x}_0$. The condition $\mu_\varepsilon(\mathbf{x}_0, \mathbf{z}_0) \geq q$ means that rank of the set of $Kn$ vectors $(\mathbf{x} - \mathbf{x}_0)^\nu \check{S}_i$ in $\mathbb{C}^K$ is at most $K - q$. This implies the vanishing of all $(K-q+1)$-minors of a $(K \times Kn)$-matrix composed of these vectors. As the elements of this matrix are the partial derivatives (9.16), which are analytic in $(\mathbf{x}_0, \mathbf{z}_0, \varepsilon)$, this, in combination with the equations for $Y$, provides a system of analytic equations for $W_q$. □

**9.16 Definition** For a positive integer $q$, the *Milnor fiber* $Z_q$ of the deformation $S$ with respect to the distribution (9.2) is the intersection of $W_q$ with a closed ball $\{\|(\mathbf{x}, \mathbf{y})\| \leq \delta\}$ in the space $\{\varepsilon = \text{const}\} \subset \mathbb{C}^{n+r+1}$, for a small positive $\delta$ and a complex nonzero $\varepsilon$ much smaller than $\delta$. According to [26], the homotopy type of $Z_q$ does not depend on $\varepsilon$ and $\delta$. Let $\chi(Z_q)$ be the Euler characteristic of $Z_q$.

**9.17 Theorem** *Let $P = (P_1(\mathbf{x}, \mathbf{y}), \ldots, P_n(\mathbf{x}, \mathbf{y}))$ be a germ of an analytic function at 0 in $\mathbb{C}^{n+r}$. Let $f = (f_1(\mathbf{x}), \ldots, f_r(\mathbf{x}))$ be a germ of an analytic function at $0 \in \mathbb{C}^n$ satisfying (9.1) and let $\phi_i(\mathbf{x}) = P_i(\mathbf{x}, f(\mathbf{x}))$. Suppose that the intersection $\phi_1(\mathbf{x}) = \cdots = \phi_n(\mathbf{x}) = 0$ is isolated at $\mathbf{x} = 0$, with multiplicity $\mu$. Let $S = S(\mathbf{x}, \mathbf{y}, \varepsilon)$ be a one-parameter deformation of $P$, let $Z_q$ be the Milnor fibers of $S$ with respect to the distribution (9.2), and let $Q = \max\{q : Z_q \neq \emptyset\}$. Then*

$$\mu = \sum_{q=1}^{Q} \chi(Z_q). \tag{9.17}$$

**Proof** The arguments, based on integration over Euler characteristic, essentially repeat those in the proof of Theorem 9.6, except everything should be restricted to the set $Y$ of all integral manifolds of (9.2).

Let us choose a system of coordinates $(\mathbf{x}, \mathbf{z})$ in $\mathbb{C}^{n+r}$, as in the proof of Lemma 9.15, so that each integral manifold of (9.2) through $(\mathbf{x}_0, \mathbf{z}_0) \in Y$ is defined by $\mathbf{z} = \mathbf{z}_0$. Let $\pi$ be projection $\mathbb{C}^{n+r} \to \mathbb{C}^r$. Let $B_\eta = \{\|\mathbf{z}\| \leq \eta\}$ be a ball of radius $\eta$ in $\mathbb{C}^r$. We can replace the ball $\{\|(\mathbf{x}, \mathbf{y})\| \leq \delta\}$ in Definition 9.16 by $D_{\delta,\eta} = \{\|\mathbf{x}\| \leq \delta,\ \mathbf{z} \in B_\eta\}$, where $0 < \eta \ll \delta$, so that the projection $\pi : \{P = 0\} \cap D_{\delta,\eta} \to B_\eta$ is a finite covering. This would not change the homotopy type of $Z_q$, as long as $|\varepsilon| \ll \eta$. Then $\pi : \{D_{\delta,\eta} = 0\} \cap Y \to B_\eta \cap \pi(Y)$ is a finite $\mu$-fold ramified covering (counting the multiplicities). Let $\zeta_q(\mathbf{z}) = \chi(\pi^{-1}(\mathbf{z}) \cap Z_q)$ be the number of preimages of a point $\mathbf{z}$ in $Z_q$, not counting multiplicities. Then $\sum_{q=1}^{Q} \zeta_q(\mathbf{z}) \equiv \mu$ does not depend on $\mathbf{z}$. We have

$$\int_{B_\eta \cap \pi(Y)} \zeta_q \, d\chi = \chi(Z_q), \quad \text{and} \quad \int_{B_\eta \cap \pi(Y)} \sum_{q=1}^{Q} \zeta_q \, d\chi = \int_{B_\eta \cap \pi(Y)} \mu \, d\chi = \mu.$$

The last equality holds because $B_\eta \cap \pi(Y)$ is a compact contractible set, hence its Euler characteristic equals 1. □

**9.18 Proposition** *Let $g_{ij}$ in the expression* (9.2) *be polynomials of degree not exceeding $\alpha$, and let $S_i(\mathbf{x}, \mathbf{y}, \varepsilon)$ be polynomials of degree at most $\beta$ in $(\mathbf{x}, \mathbf{y})$, with coefficients analytic in $\varepsilon$. Then each Milnor fiber $Z_q$ of $S$ with respect to* (9.2) *can be defined by polynomial equations of degree not exceeding maximum of* (9.15) *and*

$$d(n, q, \alpha, \beta) = (K - q + 1)[\beta + (q - 1)(\alpha - 1)], \quad \text{where} \quad K = \binom{q + n - 1}{n}. \tag{9.18}$$

**Proof** Let us fix a small nonzero $\varepsilon$. According to the arguments in the proof of Lemma 9.15, the condition $(\mathbf{x}_0, \mathbf{y}_0) \in Z_q$ is equivalent to the vanishing of all $(K - q + 1)$-minors of a matrix composed of the partial derivatives (9.16) of $S_\varepsilon$ of order $\nu < q$, in a system of coordinates $(\mathbf{x}, \mathbf{z})$ where the integral manifolds of (9.2) are rectified.

Let us define a germ of the integral manifold $\Lambda$ of (9.2) through $(\mathbf{x}_0, \mathbf{y}_0)$ by a function $f(\mathbf{x}) = (f_1(\mathbf{x}), \ldots, f_r(\mathbf{x}))$ satisfying (9.1). The equations (9.1) allow one to represent a partial derivative (9.16) as a polynomial in $(\mathbf{x}_0, \mathbf{y}_0)$ of degree not exceeding $\beta + \nu(\alpha - 1)$. Hence the elements of our matrix are polynomials in $(\mathbf{x}_0, \mathbf{y}_0)$ of degree not exceeding $\beta + (q-1)(\alpha-1)$, and its $(K-q+1)$-minors are polynomials in $(\mathbf{x}_0, \mathbf{y}_0)$ of degree not exceeding $(K-q+1)[\beta + (q-1)(\alpha-1)$. These polynomials, in combination with the equations for $Y$, provide a system of equations for $Z_q$. □

**9.19 Corollary** *Under conditions of Proposition* 9.18, *the absolute value of the Euler characteristic of $Z_q$ does not exceed*

$$\frac{1}{2} \max(2d_Y(n, r, \alpha), 2d(n, q, \alpha, \beta))^{2(n+r)}, \tag{9.19}$$

*where $d_Y(n, r, \alpha)$ and $d(n, q, \alpha, \beta)$ are defined in* (9.15) *and* (9.18), *respectively.*

**Proof** This follows from an estimate [33] of the Euler characteristic of the set $Z_q$ defined by polynomial equations of degree not exceeding maximum of (9.15) and (9.18). □

## 9.6 Proof of Theorem 9.1

To complete the proof of Theorem 9.1, we need to find a low-degree deformation $S$ of a polynomial $P$ such that the Milnor fibers $Z_q$ with large $q$ would be empty, i.e., to derive an upper bound on the value of $Q$ in Theorem 9.17. Such an upper bound can be derived from the following theorem (see Theorem 7 in [17]):

**9.20 Theorem** *Let $\nu = (\nu_1, \ldots, \nu_n)$ be a sequence of nonnegative integers, $|\nu| = \nu_1 + \cdots + \nu_n$, and $\mathbf{x}^\nu = x_1^{\nu_1} \cdots x_n^{\nu_n}$. For a nonnegative integer $r$, an analytic map $P = (P_1, \ldots, P_n) : \mathbb{C}^n \to \mathbb{C}^n$, and a vector $c = (c_1, \ldots, c_n)$ where $c_j = c_{j,\nu}$, $|\nu| \le r$, let*

$$P^c = (P_1^c, \ldots, P_n^c) \quad \text{where} \quad P_j^c(\mathbf{x}) = P_j(\mathbf{x}) + \sum_{\nu : |\nu| \le r} c_{\nu,j} x^\nu. \tag{9.20}$$

*The set of those* $(\mathbf{x}, c)$ *where the multiplicity of* $P^c$ *at* $\mathbf{x}$ *exceeds*

$$Q(n,r) = \left(\frac{n+r}{1+\cdots+1/n}\right)^{1+\cdots+1/n} \prod_{k=1}^{n} \left(\frac{(k-1)!}{k}\right)^{1/k}. \tag{9.21}$$

*has codimension greater than* $n+r$.

**9.21 Corollary** *Let* $P_{\mathbf{y}}(\mathbf{x}) = (P_{\mathbf{y},1}(\mathbf{x}), \ldots, P_{\mathbf{y},n}(\mathbf{x}))$ *be a generic family of analytic maps* $\mathbb{C}^n \to \mathbb{C}^n$ *depending on parameters* $\mathbf{y} \in \mathbb{C}^r$. *Then the multiplicity of* $P_{\mathbf{y}}$, *at any point* $\mathbf{x} \in \mathbb{C}^n$ *and for any* $\mathbf{y}$, *is less than* (9.21).

Consider now a deformation $S(\mathbf{x}, \mathbf{y}, \varepsilon)$ of the polynomial $P(\mathbf{x}, \mathbf{y})$ defined by

$$S_j(\mathbf{x}, \mathbf{y}, \varepsilon) = P_j(\mathbf{x}, \mathbf{y}) + \varepsilon \sum_{\nu:|\nu|\le r} c_{\nu,j} x^{\nu},$$

where $c_{\nu,j}$ are generic complex numbers. From Theorem 9.20, the Milnor fibers $Z_q$ of this deformation are empty for $q \ge Q(n,r)$, where $Q(n,r)$ is defined in (9.20).

According to (9.17),

$$\mu = \sum_{q=1}^{Q(n,r)} \chi(Z_q) \le Q(n,r) \max_{q \le Q(n,r)} |\chi(Z_q)|. \tag{9.22}$$

From (9.19), the right hand side of (9.22) does not exceed

$$\frac{1}{2} Q(n,r) \max(2d_Y(n,r,\alpha), 2d(n, Q(n,r), \alpha, \beta))^{2(m+n)},$$

where $d_Y$ and $d$ are defined in (9.15) and (9.18), respectively. The value of $d(n,q,\alpha,\beta)$ in (9.18) does not exceed

$$(q+n-1)^n(\beta + (q-1)(\alpha-1)) < (q+n)^n(\beta + q(\alpha-1)).$$

The statement of Theorem 9.1 follows now from the following estimate for $Q(n,r)$ (see Proposition 1 in [17]):

**9.22 Proposition** *The value of* $Q(n,r)$ *in* (9.20) *does not exceed*

$$en \left(\frac{e(n+r)}{\sqrt{n}}\right)^{\ln n + 1} \left(\frac{n}{e^2}\right)^n.$$

## References


[1] V. I. Arnold, S. M. Gusein-Zade, and A. N. Varchenko, *Singularities of Differentiable Maps*, Vol. I, Birkhäuser, Basel, 1986.

[2] S. Basu, On bounding Betti numbers and computing Euler characteristic of semi-algebraic sets, *Discrete Comput. Geom.* **22** (1999), 1–18.

[3] S. Basu, R. Pollack, and M.-F. Roy, On the combinatorial and algebraic complexity of quantifier elimination, *J. ACM* **43** (1996), 1002–1045.

[4] E. Bierstone and P. Milman, Geometric and differential properties of subanalytic sets, *Bull. Amer. Math. Soc.* **25** (1991), 385–393.

[5] L. Blum, F. Cucker, M. Shub, and S. Smale, *Complexity and Real Computation*, Springer, New York, 1997.

[6] J.-Y. Charbonnel, Sur certains sous-ensembles de l'espace euclidien, *Ann. Inst. Fourier (Grenoble)* **41** (1991), 679–717.

[7] G. E. Collins, Quantifier elimination for real closed fields by cylindrical algebraic decomposition, Lecture Notes in Comput. Sci. **33**, Springer, 1975, 134–183.

[8] J. Denef and L. van den Dries, $p$-adic and real subanalytic sets, *Ann. of Math. (2)* **128** (1988), 79–138.

[9] L. van den Dries, *Tame Topology and o-Minimal Structures*, London Math. Soc. Lecture Note Ser. **248**, Cambridge University Press, Cambridge, 1998.

[10] A. Gabrielov, Projections of semianalytic sets, *Funct. Anal. Appl.* **2** (1968), 282–291.

[11] A. Gabrielov, Multiplicities of Pfaffian intersections and the Łojasiewicz inequality, *Selecta Math. (N.S.)* **1** (1995), 113–127.

[12] A. Gabrielov, Multiplicities of zeros of polynomials on trajectories of polynomial vector fields and bounds on degree of nonholonomy, *Math. Res. Lett.* **2** (1995), 437–451.

[13] A. Gabrielov, Complements of subanalytic sets and existential formulas for analytic functions, *Invent. Math.* **125** (1996), 1–12.

[14] A. Gabrielov, Frontier and closure of a semi-Pfaffian set, *Discrete Comput. Geom.* **19** (1998), 605–617.

[15] A. Gabrielov, Multiplicity of a zero of an analytic function on a trajectory of a vector field, in: *The Arnoldfest*, Fields Inst. Commun. **24**, Amer. Math. Soc., Providence, RI, 1999, 191–200.

[16] A. Gabrielov, Relative closure and complexity of Pfaffian elimination, to appear in: *Discrete and Computational Geometry (Goodman-Pollack Festschrift)* (B. Aronov et al., eds.), Springer, 2003. Available at `http://www.math.purdue.edu/~agabriel`

[17] A. Gabrielov and A. Khovanskii, Multiplicity of a Noetherian intersection, in: *Geometry of Differential Equations* (A. Khovanskii et al., eds.), Amer. Math. Soc. Transl. Ser. 2 **186**, Amer. Math. Soc., Providence, RI, 1998, 119–131.

[18] A. Gabrielov and N. Vorobjov, Complexity of stratifications of semi-Pfaffian sets, *Discrete Comput. Geom.* **14** (1995), 71–91.

[19] A. Gabrielov and N. Vorobjov, Complexity of cylindrical decompositions of sub-Pfaffian sets, *J. Pure Appl. Algebra* **164** (2001), 179–197.

[20] A. Gabrielov, N. Vorobjov, and T. Zell, Betti numbers of semialgebraic and sub-Pfaffian sets, to appear in *J. London Math. Soc.*, 2003. Available at `http://www.math.purdue.edu/~agabriel`

[21] A. Gabrielov and T. Zell, On the number of connected components of the relative closure of a semi-Pfaffian family, to appear in: *Algorithmic and Quantitative Real Algebraic Geometry*, DIMACS Ser. Theoret. Math. and Theoret. Comput. Sci., Amer. Math. Soc., Providence, RI, 2003. Available at `http://www.math.purdue.edu/~agabriel`

[22] D. Grigoriev, Deviation theorems for Pfaffian sigmoids, *St. Petersburg Math. J.* **6** (1995), 107–111.

[23] M. Karpinski and A. Macintyre, A generalization of Wilkie's theorem of the complement, and an application to Pfaffian closure, *Selecta Math. (N.S.)* **5** (1999), 507–515.

[24] A. Khovanskii, On a class of systems of transcendental equations, *Soviet Math. Dokl.* **22** (1980), 762–765.

[25] A. Khovanskii, *Fewnomials*, Transl. Math. Monogr. **88**, Amer. Math. Soc., Providence, RI, 1991.

[26] Lê Dũng Tráng, Some remarks on relative monodromy, in: *Real and Complex Singularities, Oslo 1976* (P. Holm, ed.), Sijthoff & Noordhoff, 1977, 397–403.

[27] J.-M. Lion, Inégalité de Lojasiewicz en géométrie pfaffienne, *Illinois J. Math.* **44** (2000), 889–900.

[28] J.-M. Lion, C. Miller, and P. Speissegger, Differential equations over polynomially bounded o-minimal structures, to appear in: *Proc. Amer. Math. Soc.*

[29] J.-M. Lion an J.-P. Rolin, Volumes, feuilles de Rolle de feuilletages analytiques et théorème de Wilkie, *Ann. Fac. Sci. Toulouse Math. (6)* **7** (1998), 93–112.

[30] S. Łojasiewicz, Triangulation of semi-analytic sets, *Ann. Scuola Norm. Pisa (3)* **18** (1964), 449–474.

[31] S. Łojasiewicz, Ensembles semi-analytiques. Lecture Notes, IHES, Bures-sur-Yvette, 1965.

[32] C. Miller and P. Speissegger, Pfaffian differential equations over exponential o-minimal structures, *J. Symbolic Logic* **67** (2002), 438–448.

[33] J. Milnor, On the Betti numbers of real varieties, *Proc. Amer. Math. Soc.* **15** (1964), 275–280.

[34] O. A. Oleinik, Estimates of the Betti numbers of real algebraic hypersurfaces (Russian), *Mat. Sb.* **28** (1951), 635–640.

[35] W. F. Osgood, On functions of several complex variables, *Trans. Amer. Math. Soc.* **17** (1916), 1–8.

[36] S. Pericleous and N. Vorobjov, New complexity bounds for cylindrical decompositions of sub-Pfaffian sets, to appear in: *Discrete and Computational Geometry (Goodman-Pollack Festschrift)* (B. Aronov et al., eds.), Springer, 2003. Available at `http://www.bath.ac.uk/~masnnv`

[37] I. G. Petrovskii and O. A. Oleinik, On the topology of real algebraic hypersurfaces (Russian), *Izv. Acad. Nauk SSSR* **13** (1949), 389–402.

[38] J. Renegar, On the computetional complexity and geometry of the first order theory of reals, I–III, *J. Symbolic Comput.* **13** (1992), 255–352.

[39] D. Richardson, Elimination of infinitesimal quantifiers, *J. Pure Appl. Algebra* **139** (1999), 235–253.

[40] M. Rosenlicht, The rank of a Hardy field, *Trans. Amer. Math. Soc.* **280** (1983), 659–671.

[41] P. Speissegger, The Pfaffian closure of an o-minimal structure, *J. Reine Angew. Math.* **508** (1999), 189–211.

[42] R. Thom, Sur l'homologie des variétés algébriques réelles, in: *Differential and Combinatorial Topology*, Princeton University Press, Princeton, 1965, 255–265.

[43] J.-C. Tougeron, Algèbres analytiques topologiquement nœthériennes, Théorie de Hovanskii, *Ann. Inst. Fourier (Grenoble)* **41** (1991), 823–840.

[44] O. Viro, Some integral calculus based on Euler characteristic, in: *Topology and Geometry—Rohlin Seminar*, Lecture Notes in Math. **1346**, Springer, Berlin-New York, 1988, 127–138.

[45] N. Vorobjov, The complexity of deciding consistency of systems of polynomials in exponent inequalities, *J. Symbolic Comput.* **13** (1992), 139–173.

[46] H. Whitney, Elementary structure of real algebraic varieties, *Ann. of Math.* **66** (1957), 545–556.

[47] A. Wilkie, A general theorem of the complement and some new o-minimal structures, *Selecta Math. (N.S.)* **5** (1999), 397–421.

[48] H. R. Wüthrich, Ein Enscheidungsverfahren für die Theorie der reell-abgeschlossenen Körper, Lecture Notes in Comput. Sci. **43**, Springer, 1976, 138–162.

[49] T. Zell, Betti numbers of semi-Pfaffian sets, *J. Pure Appl. Algebra* **139** (1999), 323–338.

[50] T. Zell, Betti numbers of Pfaffian limit sets, preprint, 2003. Available at `http://www.math.purdue.edu/tzell`

# Hamiltonian bifurcations and local analytic classification

V. GELFREICH

*Mathematics Institute*
*University of Warwick*
*Coventry CV4 7AL*
*United Kingdom*

**Abstract**

We discuss the topics related to generic bifurcations near strongly resonant periodic orbits in an analytic Hamiltonian system with two degrees of freedom. The problem is reduced to an one-parameter family of area-preserving maps. Up to any order, the dynamics near the bifurcation can be described by an integrable resonant normal form. In general the corresponding series diverge. In a natural way the study of the divergence leads to problems of local analytic classification and the splitting of invariant manifolds.

There is a remarkable similarity between bifurcations of equilibria for planar vector fields and bifurcations of periodic orbits in dimension 3. This similarity is due to the fact that the Poincaré map near a bifurcating periodic orbit can be formally written as a time-1 map of a planar vector field. Although all Taylor coefficients of the interpolating vector field can be determined, there is no reason to expect convergence of the corresponding series since the dynamics in the 3-dimensional space can be much richer than the dynamics on the plane.

The interpolating planar vector field can be used to construct normal forms for bifurcations of the periodic orbits. In this way the difference between these two different types of bifurcations is moved beyond all algebraic orders. In real-analytic theory this difference is exponentially small [BR01].

Detection of exponentially small quantities requires studying an analytical continuation of the system, which leads to establishing a connection with problems of local analytic classification similar to [É85]. The present note is mainly based on the papers [Gel00b, Gel00a, GL01, GS01, Gel02].

## 1 Resonant periodic orbits

We consider a Hamiltonian system described by an analytic Hamilton function $\mathcal{H}(p_1, p_2, q_1, q_2)$. The dynamics of the system is determined by the system of Hamiltonian equations:

$$\begin{aligned} \dot{p}_i &= \frac{\partial \mathcal{H}}{\partial q_i} \\ \dot{q}_i &= -\frac{\partial \mathcal{H}}{\partial q_i} \end{aligned} \qquad i = 1, 2.$$



*Y. Ilyashenko and C. Rousseau (eds.),*
*Normal Forms, Bifurcations and Finiteness Problems in Differential Equations,* 251–265.

The energy $\mathcal{H}$ is an integral of motion and a constant-energy surface $\{\mathcal{H} = 0\}$ is a 3-dimensional invariant subset of the 4-dimensional phase space. Let the system have a periodic orbit inside the energy level. In order to study the dynamics near the periodic orbit, we take a two-dimensional surface $\Sigma_0$ transversal to the periodic orbit and consider the first-return map $F_0$.

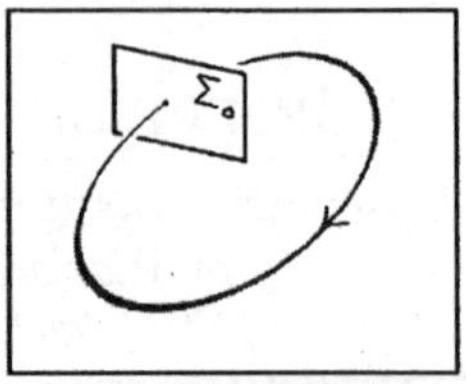


Figure 1

According to the famous Liouville theorem, $F_0$ is area-preserving. The periodic orbit corresponds to a fixed point of $F_0$. We introduce coordinates on $\Sigma_0$, setting the origin to the fixed point. Thus we have

$$F_0(0) = 0.$$

In these coordinates, the derivative $F_0'(0)$ is a $2 \times 2$ matrix with $\det F_0'(0) = 1$, due to the area preservation. The spectrum of the matrix consists of two eigenvalues $\lambda$ and $\lambda^{-1}$, which are often called *multipliers* of the fixed point. The following cases are possible:

(1) $\overline{\lambda} = \lambda \iff \lambda \in \mathbb{R}$; the origin is a hyperbolic fixed point (exclude $\lambda \neq \pm 1$);

(2) $\overline{\lambda} = \lambda^{-1} \iff |\lambda| = 1$; the origin is an elliptic fixed point (exclude $\lambda \neq \pm 1$);

(3) $\lambda = \pm 1$; the origin is a parabolic fixed point.

In the last case, $\lambda$ is a double eigenvalue. Note that sometimes only the case $\lambda = 1$ is called parabolic.

Dynamics near a fixed point can be studied using Birkhoff normal forms: all "non-resonant" terms of a Taylor expansion can be eliminated by a formal coordinate change. Alternatively, non-resonant terms up to any prefixed order can be eliminated by a polynomial change of coordinates [SM71].

Let us see what this theorem implies for our map $F_0$.

**Hyperbolic fixed point** Let $(\xi, \eta)$ denote the normal form coordinates on $\Sigma_0$. In these variables, the map $F_0$ takes a specially simple form,

$$\begin{pmatrix} \xi \\ \eta \end{pmatrix} \mapsto \begin{pmatrix} \xi \cdot E(\xi\eta) \\ \eta / E(\xi\eta) \end{pmatrix} \quad \text{with} \quad E(0) = \lambda.$$

The function $E$ is an analytic function of a single variable – the product of two normal form coordinates. This hyperbolic case is quite special: the coordinate functions and the normal form are analytic. Convergence of the normal form series was proved by Moser (1956), a simpler proof was later provided by Siegel [SM71].

In the normal form coordinates, the hyperbolic fixed point is at the origin, its stable $W^s_{\text{loc}}$ and unstable $W^u_{\text{loc}}$ manifolds, also called separatrices, coincide with the coordinate axes. We immediately conclude that these invariant manifolds are one-dimensional analytic curves.

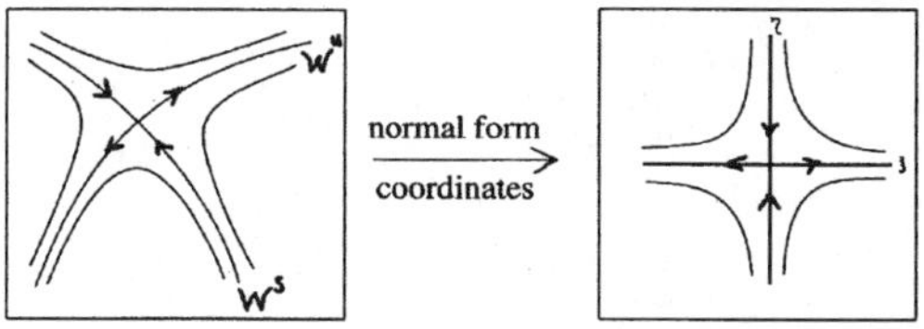


Figure 2

Let us assume that $\lambda$ denotes the largest multiplier of the fixed point: $|\lambda| > 1$. Then $W^s$ coincides with the $\eta$-axis and $W^u$ is the $\xi$-axis. It is convenient to introduce a logarithmic scale on the $\xi$-axis introducing a new parameter, $t = \log \xi$. This parametrization can be transformed to the original coordinates on $\Sigma_0$. Then $W^u$ is parametrized by a solution of the finite-difference equation

$$\mathbf{x}^u(t + \log \lambda) = F_0(\mathbf{x}^u(t)).$$

In the case of $W^s$, it is convenient to let $t = -\log \eta$. Then the parametrizing function $\mathbf{x}^s(t)$ satisfies the same equation.

The normal form coordinates are not defined uniquely, which leaves some freedom in the definition of the parametrizations: the parametrizing functions $\mathbf{x}^s(t)$ and $\mathbf{x}^u(t)$ are defined uniquely up to a substitution $t \mapsto t + \text{const}$.

The local stable and local unstable manifolds can be extended by iterating $F_0$ and its inverse to global invariant manifolds. The stable manifold (as well as the unstable one) does not have any self-intersection. But the stable manifold may intersect the unstable one. In particular, the fixed point belongs to this intersection. All other intersection points are called *homoclinic points.*

**Homoclinic invariant** Let $\mathbf{x}_{\text{homocl}}$ be a homoclinic point: $\mathbf{x}_{\text{homocl}} \in W^u \cap W^s$. Homoclinic points are bi-asymptotic: their forward and backward iterates converge to the fixed point. If the angle between $W^u$ and $W^s$ at a homoclinic point does not vanish, the map has a Smale horseshoe and there is a Cantor set bearing chaotic dynamics.

Unfortunately, an angle is not a natural quantity in Hamiltonian theory: there is no natural way to introduce a Riemann metric in a Hamiltonian system, which is necessary to define an angle. Moreover, angles are not invariant and depend not only on the choice of a homoclinic point but also on the choice of the section $\Sigma_0$.

V. Lazutkin proposed a more natural quantity to measure the transversality of the separatrix splitting, which is usually called the *Lazutkin homoclinic invariant.* Let $t_u$ and $t_s$ denote parameter values corresponding to the homoclinic point: $\mathbf{x}_{\text{homocl}} = \mathbf{x}^u(t_u) = \mathbf{x}^s(t_s)$, then the Lazutkin homoclinic invariant is defined by

$$\omega = \det(\dot{\mathbf{x}}^u(t_u); \dot{\mathbf{x}}^s(t_s)).$$

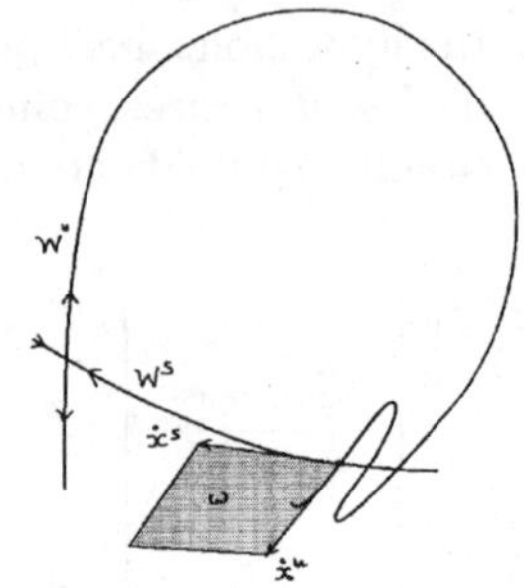

Figure 3

The Lazutkin homoclinic invariant is equal to the area of a parallelogram defined by two vectors tangent to $W^u$ and $W^s$, respectively. It has the following advantages:

(1) $\omega$ is independent of the choice of the canonical coordinates on $\Sigma_0$;

(2) $\omega$ does not depend on the choice of the section $\Sigma_0$;

(3) $\omega$ is constant along a homoclinic trajectory;

(4) $\omega \neq 0$ if and only if the homoclinic trajectory is transversal.

**Elliptic fixed point** In order to describe the normal form, it is convenient to identify $\mathbb{R}^2$ with $\mathbb{C}^1$ and define a complex variable, $z = x + iy$. Then the map $F_0$ takes the form

$$F_0 : z \mapsto \lambda z + \sum_{k+l \geqslant 2} a_{k,l} z^k \overline{z}^l.$$

A resonant term is defined by an equality $\lambda = \lambda^k \overline{\lambda}^l = \lambda^{k-l}$. The terms with $k = l + 1$ are always resonant. We say that the elliptic fixed point is *non-resonant* if $\lambda^n \neq 1$ for all integers $n \neq 0$. Otherwise the periodic orbit is called *resonant.* The least $n$ such that $\lambda^n = 1$ is called the *order of resonance.*

The Birkhoff normal form of a non-resonant elliptic fixed point has the form of a rotation on an angle, which depends on the amplitude,

$$z \mapsto z(\lambda + a_1|z|^2 + a_2|z|^4 + \cdots).$$

The Birkhoff normal form commutes with the rotation $z \mapsto \lambda z$. In the resonant case, $\lambda^n = 1$ and the Birkhoff normal form is equivariant, i.e., it is symmetric with respect to the rotation by the angle $2\pi/n$.

In the resonant case, the resonant normal form contains terms with $k = l + 1 \pmod n$. If $n \in \{1, 2, 3, 4\}$, we say that the periodic orbit is *strongly* resonant. Rigorously speaking, the cases $n = 1$ and 2 are parabolic not elliptic. In each of these four cases, there is an additional resonant term of order 2 ($n = 1$ and 3) or 3 ($n = 2$ and 4), which can influence the stability of the fixed point. In order to study this phenomenon, we note the linear part of $F_0^n$ at 0 is an identity (or Jordan block with double eigenvalue 1 for $n = 1, 2$). Consequently, there is a formal autonomous Hamiltonian $H_n$ such that

$$e^{X_{H_n}} = F_0^n,$$

where $X_H$ stands for the vector field corresponding to a Hamiltonian $H$. We say that $F_0^n$ is *formally interpolated by a flow.* In the case of strong resonances, the highest order terms of $H_n$ are given in Fig. 4.

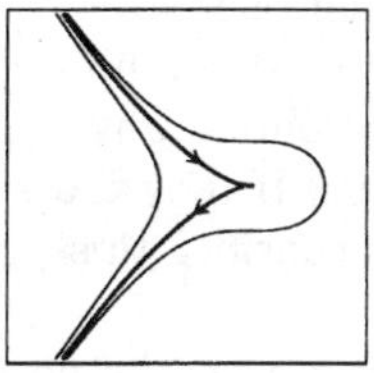

(a) $H_1 = \frac{y^2}{2} + A\frac{x^3}{3} + \cdots$

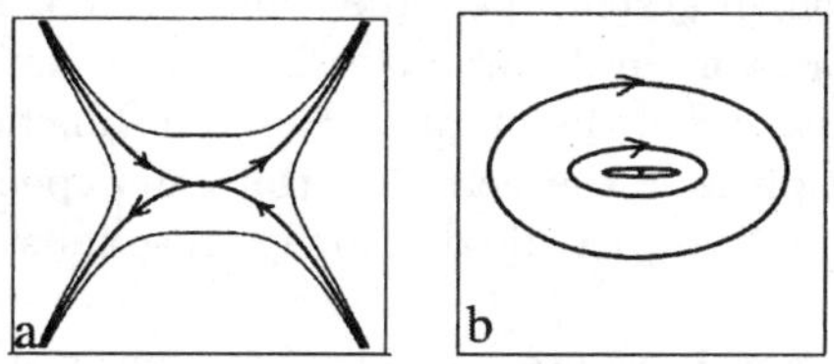


(b) $H_2 = \frac{y^2}{2} \mp A\frac{x^4}{4} + \cdots$

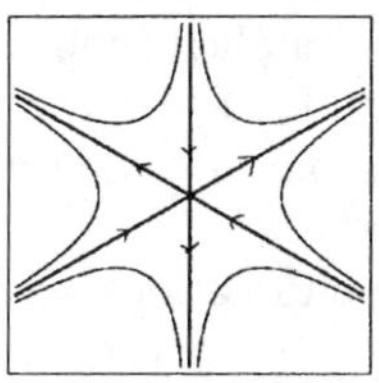

(c) $H_3 = A(x^3 + 3xy^2) + \cdots$

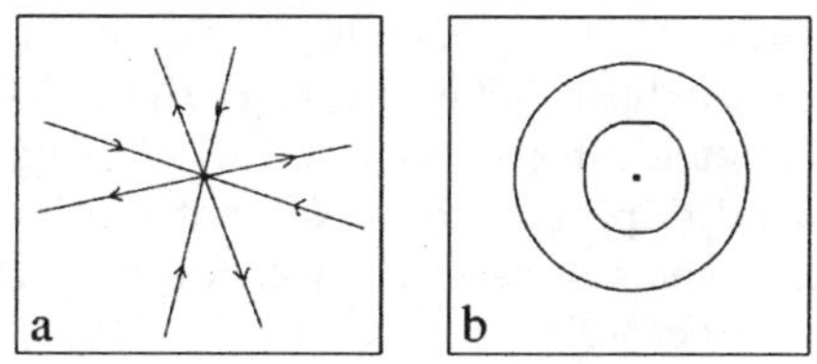


(d) $H_4$

Figure 4

The fixed point of $F_0$ is not hyperbolic but it still has stable and unstable invariant manifolds. In the stable cases with $n = 2$ and 4, these manifolds are complex and do not intersect a real neighborhood of the fixed point. In the unstable cases, these manifolds separate the real neighborhood of the fixed point into parabolic sectors. Prof. Dumortier pointed out that this geometry implies existence of a smooth ($C^\infty$) interpolating vector field [DRR81]. Consequently, in the case of an unstable strongly resonant fixed point, the formal series $H_n$ defines a Taylor series for a smooth Hamiltonian, which exactly interpolates the map $F_0$. Are these series convergent?

It is not very surprising that the answer is negative in general. But the analysis required to solve this problem is not trivial, since any finite jet of the series $H_n$ generates an analytic map which looks similar to $F_0$, but the corresponding formal series are finite and consequently trivially converge.

In the next section we will introduce a complex-valued functional $\Theta(F_0) \in \mathbb{C}$ which has the following properties:

(1) $|\Theta(F_0)|$ is a modulus of local analytic classification;
(2) $\Theta$(time-1 map of an autonomous analytic Hamiltonian flow) $= 0$;
(3) $\Theta(F_0) \neq 0$ generically.

It follows that the series $H_n$ diverge for a "generic" $F_0$. $\Theta(F_0)$ can be considered as a Stokes constant.

## 2 Bifurcations near strong resonances

In a Hamiltonian system, periodic orbits are not isolated but form one-parameter families. The energy is a natural parameter in such a family. Thus we can study bifurcations of a periodic orbit even if the original Hamilton function $\mathcal{H}$ does not contain any explicit parameter. Multipliers of the periodic orbit can depend on the energy. If they cross a strongly resonant value, the periodic orbit may loose its stability. Let us consider this phenomenon more closely.

As in Section 1, we use a Poincaré section to study the dynamics near the periodic orbit. This time we get a family of area-preserving maps $F_\varepsilon$, where we can think about $\varepsilon$ as an energy value.

We assume that $\varepsilon = 0$ corresponds to a strong resonance of order $n = 1, 2$ or 3: $F_0(0) = 0$ and the derivative $F_0'(0)$ has a multiplier $\lambda_0$ such that $\lambda_0^n = 1$. The implicit function theorem can be used to establish the existence of the family of fixed points for $F_\varepsilon$. Their multipliers $\lambda_\varepsilon$ depend on $\varepsilon$, and we can speak about crossing of a strong resonance. The case $n = 4$ is special in several respects, and we will not consider it any further.

We impose some non-degeneracy conditions, which insure the bifurcations we consider in this section to be generic:

(1) $F_0'(0) = \begin{pmatrix} 1 & 1 \\ 0 & 1 \end{pmatrix}$, $n = 1$, or $F_0'(0) = \begin{pmatrix} -1 & 1 \\ 0 & -1 \end{pmatrix}$, $n = 2$;
(2) the main non-linear term is non-degenerate (in the resonant normal form $A \neq 0$);
(3) non-degenerate dependence on $\varepsilon$.

**Proposition 1 ([AKN97])** *There is a formal Hamiltonian $H_n(x, y, \varepsilon)$ such that $F_\varepsilon^n = e^{X_{H_n}}$.*

In each of these cases, a hyperbolic $n$-periodic orbit collides with the fixed point at the exact strong resonance. Near the resonance, separatrices of the hyperbolic periodic orbit form a small loop around the elliptic fixed point. This loop shrinks to a point at the exact resonance. If we consider the time-1 map of the corresponding (truncated at any order) Hamiltonian system defined by $H_n$, we see a closed loop, which belongs to a constant energy level of $H_n$. In the case of the original map, the following theorem implies that exponentially small transversality of separatrices is to be expected generically.

**Theorem 1 ([Gel02])** *In the cases $n = 1, 2(a)$ and 3, $F_\varepsilon$ has at least two primary homoclinic orbits, the corresponding Lazutkin homoclinic invariant is given asymptotically:*

$$\omega = \frac{4\pi}{h^2} e^{-2\pi^2/h} (|\Theta(F_0)| + O(h)),$$

*where $h = \log \mu$, and $\mu$ is the multiplier of the bifurcating $n$-periodic orbit.*

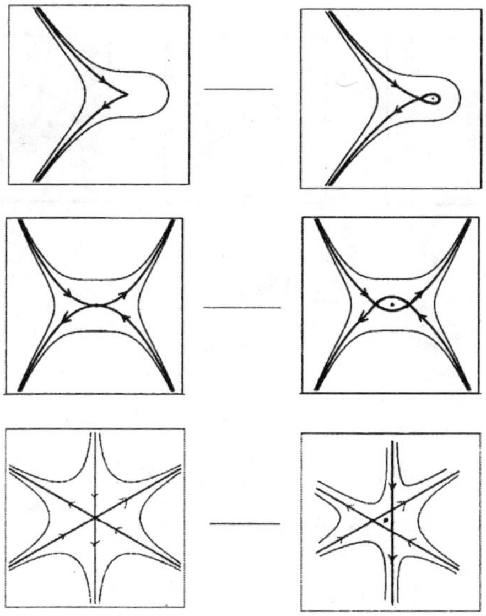

Figure 5: Bifurcations of strongly resonant periodic trajectories, non-degenerate cases of orders $n = 1, 2, 3$.

This theorem shows that the angle between stable and unstable separatrices in a primary homoclinic point is exponentially small compared to $h$. The Floquet exponent $h$ is proportional to a power of $\varepsilon$, thus the Lazutkin homoclinic invariant is exponentially small in $\varepsilon$ too. The proof of this theorem is quite long and difficult, we will discuss its main concepts in Section 4.

**Splitting constant $\Theta(F_0)$** It is an important ingredient of the asymptotic formula. Let us discuss its definition. We consider the finite-difference equation

$$\mathbf{x}(z+1) = F_0^n(\mathbf{x}(z)). \tag{2.1}$$

Let $\mathbf{x}_0 \in \mathbb{R}^2[[z^{-1}]]$ be a formal solution of this equation. The formal solution can also be obtained as a formal separatrix solution of the differential equation $\mathbf{x}' = X_{H_n}|_{\varepsilon=0}$. The formal solutions are defined up to a translation $z \mapsto z + \text{const}$, and there is no further freedom.

We note that in the stable case with $n = 2$, a similar consideration can be performed if we allow coefficients of the formal series to be non-real.

Let us consider analytic solutions $\mathbf{x}_\pm(z)$ which are selected by using the formal solution as an asymptotic boundary condition at infinity in the sectors $D_+$ and $D_- \subset \mathbb{C}$, respectively.

The intersection $D_+ \cap D_-$ consists of two connected components $\Pi_+$ and $\Pi_-$. We compare $\mathbf{x}_+$ and $\mathbf{x}_-$ at infinity inside the sector $\Pi_-$. Since the asymptotic expansions of these two functions coincide, the difference $\mathbf{x}_+(z) - \mathbf{x}_-(z)$ decreases faster than any power of $z$. Actually, it decreases exponentially in $\operatorname{Im} z$, and it is convenient to describe this difference using the

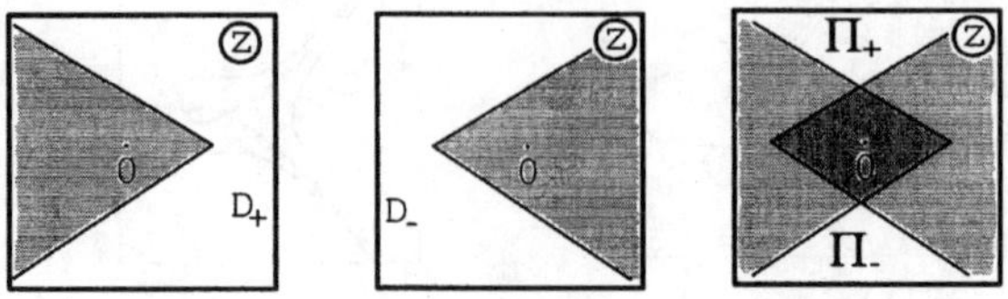


Figure 6

*splitting function*

$$\theta(z) := \det(\mathbf{x}'_-; \mathbf{x}_+ - \mathbf{x}_-)(z). \tag{2.2}$$

Then the splitting constant $\Theta$ is defined by

$$\Theta := \lim_{\operatorname{Im} z \to -\infty} e^{2\pi i z} \theta(z). \tag{2.3}$$

The limit exists for any analytic map $F_0$, which satisfies the non-degeneracy condition discussed at the beginning of this section.

**Elementary properties of $\Theta(F_0)$**

(1) $\Theta$(time-1 map of an autonomous system) $= 0$. (In this case both $\mathbf{x}_+$ and $x_-$ represent a single separatrix solution of the differential equation, and their difference vanishes identically).

(2) $|\Theta|$ is independent of the freedom in the choice of the formal solution, it is invariant with respect to canonical coordinate changes. (Hint: The formal solution is defined up to translating its argument; canonical coordinate changes preserve areas).

(3) $\Theta$ is not defined by any finite jet of $F_0$. (Hint: The normal form of $F_0$ is a time-1 map of an autonomous flow.)

The Hénon map is a natural example for our bifurcation theory. We write it here in the form

$$F_0 : \begin{pmatrix} x \\ y \end{pmatrix} \longmapsto \begin{pmatrix} x + y - x^2 \\ y - x^2 \end{pmatrix},$$

which makes it easy to check that the map satisfies the non-degeneracy conditions. At the beginning of the 90's, I computed the corresponding splitting constant and obtained numerically: $|\Theta| = 2.474\ldots \cdot 10^6$.

**Theorem 2 ([GS01])** *In the case of the Hénon map,* $\Theta \neq 0$.

Taking into account that $\Theta$ depends analytically on $F_0$, we conclude that $\Theta$ does not vanish generically.

## 3 Example: the Hénon map

In this section we study a model equation, which exhibits all main futures of a first order resonance ($n = 1$): the Hénon map with a parabolic fixed point. In this case, the finite-difference equation for separatrices, $\mathbf{x}(z+1) = F_0(\mathbf{x}(z))$, is equivalent to the second order scalar equation:

$$u(z+1) - 2u(z) + u(z-1) = -u^2(z). \tag{3.1}$$

First, we look for a solution in the class of formal power series which decays at infinity:

$$u_0(z) = \sum_{k=1}^{\infty} \frac{a_k}{z^{2k}}$$

and find the coefficients of the formal solution by a substitution into the equation:

$$u_0(z) = -6z^{-2} + \frac{15}{2}z^{-4} - \cdots .$$

The coefficients of the formal solution form a sign-alternating sequence: $a_k = (-1)^k |a_k|$ for all $k$. In this way, we have obtained a unique formal solution in the class of even formal power series $\mathbb{C}[[z^{-2}]]$. The stable and unstable separatrices will be obtained as Borel sums of the formal series. First, we apply the formal Borel transform $\mathcal{B} : z^{n+1} \mapsto \xi^n/n!$, which transforms the divergent series $u_0(z)$ into a convergent series

$$\hat{u}_0(\xi) = \sum_{k=1}^{\infty} \frac{a_k}{(2k-1)!} \xi^{2k-1}.$$

Let us recall some elementary properties of the Borel transform. Assuming $\hat{u} = \mathcal{B}(u)$ and $\hat{v} = \mathcal{B}(v)$, we have

(1) $\mathcal{B}$ is linear;

(2) $\mathcal{B} : \partial_z u \mapsto -\xi\hat{u}(\xi)$;

(3) $\mathcal{B} : u(z+c) \mapsto e^{-c\xi}\hat{u}(\xi)$;

(4) $\mathcal{B} : uv \mapsto \hat{u} * \hat{v}$, where $*$ stands for the convolution:

$$\hat{u} * \hat{v}(\xi) = \int_0^{\xi} \hat{u}(\xi - \xi') * \hat{v}(\xi')\, d\xi'.$$

Applying the Borel transform to (3.1), we obtain a convolutional equation:

$$(e^{-\xi} - 2 + e^{\xi})\hat{u}_0(\xi) = -\hat{u}_0 * \hat{u}_0(\xi),$$

which we rewrite in the form:

$$\alpha(\xi)\hat{u}_0(\xi) = -\hat{u}_0 * \hat{u}_0(\xi), \tag{3.2}$$

where $\alpha(\xi) = 4\sinh^2(\xi/2)$. The equation (3.2) has a unique odd solution analytic in $|\xi| < 2\pi$, which is to coincide with the formal Borel transform $\hat{u}_0(\xi)$ due to uniqueness. A more sophisticated analysis shows that only the zeros of $\alpha$ provide obstacles for the analyticity of $\hat{u}_0$. These zeros are at $\xi_k^* = 2\pi i k$, $k \in \mathbb{Z}$. The function $\hat{u}_0$ can be analytically continued along any path from $\xi = 0$ which avoids the points $\xi_k^*$.

**Theorem 3 ([GS01])** *The function $\hat{u}_0(\xi)$ has an analytic continuation onto the universal cover of $\mathbb{C}\backslash 2\pi i\mathbb{Z}$; it decreases faster than any exponent in $|\operatorname{Re}\xi| > R > 0$. The Laplace transforms*

$$u_\pm(z) = \int_0^{\pm\infty} e^{-z\xi}\hat{u}_0(\xi)\,d\xi \tag{3.3}$$

*define two entire solutions of the finite-difference equation (3.1), each one is asymptotic to the formal solution $\hat{u}_0(z)$ in the sectors $D_-$ or $D_+$, respectively (see Fig. 6).*

The function $u_+$ provides the parametrization of the stable separatrix, and $u_-$ of the unstable one. The difference between these two functions is given by the integral

$$u_+(z) - u_-(z) = \int_{-\infty}^{+\infty} e^{-z\xi}\hat{u}_0(\xi)\,d\xi.$$

The intersection $D_+ \cap D_-$ consists of two connected components $\Pi_+$ and $\Pi_-$. For definiteness, we consider the lower one, $\Pi_-$. In order to estimate the integral we deform the integration path as shown in Fig. 7.

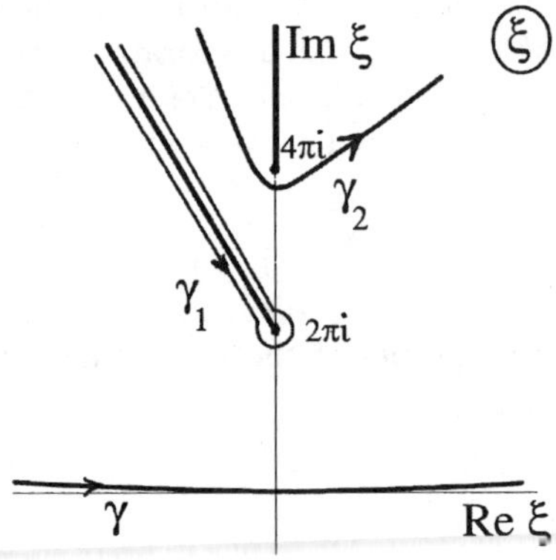


Figure 7

To estimate the integral at infinity in $\Pi_-$, we need information on the "first" singularity of $\hat{u}_0$. In [GS01] we obtained its dtailed description: at $\xi_1^* = 2\pi i$ on the first sheet of the Riemann surface,

$$\hat{u}_0(2\pi i + \xi) = \frac{A_5}{\xi^5} + \frac{A_3}{\xi^3} + \frac{A_1}{\xi} + \hat{h}(\xi)\log\xi + \text{regular function}(\xi), \tag{3.4}$$

where $\hat{h}(\xi)$ is analytic at zero. Let $h_k = \frac{d^k\hat{h}}{d\xi^k}(0)$, then $\hat{h}(\xi) = \sum_{k=1}^{\infty}\frac{h_k}{k!}\xi^k$. Let us consider the formal Laplace transform of this series, also adding a "transform" of the polar part:

$$\varphi_s(z) = \frac{1}{2\pi i}\left(A_5\frac{z^4}{4!} + A_3\frac{z^2}{2!} + A_1 + \sum_{k=1}^{\infty}\frac{h_k}{z^{k+1}}\right).$$

Then an estimate which is rather standard in the theory of Laplace integrals shows that

$$u_+(z) - u_-(z) \stackrel{\text{as}}{=} e^{-2\pi i z}\varphi_s(z). \tag{3.5}$$

The linear operator, $\Delta_{2\pi i} : u_0(z) \mapsto \varphi_s(z)$, called *Écalle derivative* (or *alien derivative*), acts in the space of formal power series in the following way: take a formal power series, make the Borel transform to obtain an analytic function at the origin, continue this function analytically into a neighborhood of $\xi_1^\star = 2\pi i$ (the singularity closest to the origin), take a power series which describes this singularity (polar part + variation around the singularity), and make the formal Laplace transform. The result is a formal power series from $\mathbb{C}[z][[z^{-1}]]$. Obviously this operation is linear. More interesting is the fact that this operation acts by the Leibniz rule on products of power series, and for this reason it is called a derivative. The formal series, which allows to perform these operations in a consistent way, form an algebra.

The Écalle derivative $\Delta_{2\pi i}$ commutes with the translation of the argument $z \mapsto z + m$ with $m \in \mathbb{Z}$, consequently, $\varphi_s$ satisfies the first order variational equation

$$\varphi(z+1) - 2\varphi(z) + \varphi(z-1) = -2u_0(z)\varphi(z). \tag{3.6}$$

The last statement can be checked without application of the resurgent function theory.

In the class of formal power series the variational equation (3.6) has two basic solutions: $\varphi_1(z) := \partial u_0(z)$ and

$$\varphi_2(z) = \sum_{k=-2}^{\infty} \frac{d_k}{z^{2k}} = \frac{z^4}{84} + \frac{17}{840}z^2 + \frac{17}{2240} + \frac{3}{35}z^{-2} + \cdots .$$

**Proposition 2** *A general formal solution of the variational equation* (3.6) *in* $\mathbb{C}[z][[z^{-1}]]$ *is a linear combination of the basic solutions:* $\varphi(z) = \mu\varphi_1(z) + \theta\varphi_2(z)$ *with some* $\mu, \theta \in \mathbb{C}$.

In particular, there are complex constants $\mu$ and $\theta$ such that $\varphi_s = \mu\varphi_1 + \theta\varphi_2$. It is not too difficult to check that this $\theta$ coincides with the splitting constant $\Theta$ from the previous section.

Thus $\Theta$ can be considered as a Stokes constant, which is responsible for the non-uniqueness of the Borel sums of the formal series which describe the separatrix of the non-hyperbolic fixed point.

After establishing the detailed structure of the singularity, it is not too difficult to prove by contradiction that $\Theta \neq 0$. Indeed, the series $\varphi_1$ contains negative powers only, consequently, the polynomial part of $\varphi_s$ is proportional to $\Theta$. If $\Theta$ vanished, the polynomial part of $\varphi_s$ would vanish too. The latter was obtained from the polar part of $\hat{u}_0$ at $\xi_1^\star = 2\pi i$, thus the polar part would vanish too.

We conclude: If $\Theta = 0$, then $\hat{u}_0$ was bounded at $\xi = 2\pi i$.

The series $a_k$ is alternating; consequently, the function $\hat{u}_0$ is purely imaginary on the imaginary axis and $\operatorname{Im}\hat{u}_0(\xi)$ cannot vanish provided $\xi$ belongs to the imaginary axis inside the disk of convergence, $|\xi| < 2\pi$. It follows that $\hat{u}_0 * \hat{u}_0 \neq 0$ on the segment $(0, 2\pi i)$, and its absolute value is bounded from zero near $2\pi i$. On the other hand, $\alpha(2\pi i) = 0$ and the equation (3.2) written in the form $\hat{u}_0 = -\hat{u}_0 * \hat{u}_0/\alpha$ implies that $\hat{u}_0$ is not bounded near $2\pi i$. By contradiction we obtain

$$\Theta(\text{Hénon map}) \neq 0.$$

## 4 Lazutkin theory of exponentially small phenomena

Let us consider an one-parameter family of area-preserving maps $F_\varepsilon(\mathbf{x}) = F_0(\mathbf{x}) + \varepsilon F_1(\mathbf{x}, \varepsilon)$, and assume that $F_0$ has a parabolic fixed point. Under the non-degeneracy conditions, the map $F_\varepsilon$ has two fixed points for $\varepsilon > 0$: an elliptic and a hyperbolic one. These two points collide as $\varepsilon \to +0$. A good idea about this bifurcation can be obtained if we recall that there is a formal Hamiltonian

$$H_\varepsilon = \frac{y^2}{2} - \delta(\varepsilon)\frac{x^2}{2} + A\frac{x^3}{3} + \cdots$$

such that $F_\varepsilon = e^{X_{H_\varepsilon}}$, i.e. formally $F_\varepsilon$ is a time-one map of an autonomous Hamiltonian planar vector field. For definiteness we assume $A > 0$. Then for $\delta > 0$ the Hamiltonian system has a closed homoclinic loop around the elliptic equilibrium.

Let $\mu$ denote the multiplier of the hyperbolic fixed point of the map $F_\varepsilon$. It is convenient to consider the Floquet exponent $h = \log \mu$ instead of $\varepsilon$ as a small parameter of the problem. It is not too difficult to check that $h \sim \varepsilon^{1/4}$. The hyperbolic fixed point of the map has a stable separatrix and an unstable one. These are one-dimensional invariant curves, the initial segments of which are close to the separatrix of the formal interpolating Hamiltonian. Unlike a one-degree-of-freedom Hamiltonian, the separatrices of the map can intersect transversally. The "first" intersection of the separatrices is called a *primary homoclinic point.* Of course it is not defined uniquely. Using the Neishtadt averaging method, it is possible to check that an intersection angle at a primary homoclinic point is exponentially small, namely it is of order $O(e^{-c/\varepsilon^{1/4}})$.

The existence of a transversal homoclinic point implies important dynamical consequences: a map with a transversal homoclinic has a horseshoe and a (small scale) chaotic dynamics. In particular, the presence of a transversal homoclinic point implies divergence of the interpolating Hamiltonian $H_\varepsilon$.

**Lazutkin method** Let $\mathbf{p}_{\mathrm{hyp}} = F_\varepsilon(\mathbf{p}_{\mathrm{hyp}})$ denote the hyperbolic fixed point, and let $\mathbf{p}_{\mathrm{homocl}}$ denote a primary homoclinic point. The existence of a primary homoclinic point can be deduced, following the original Poincaré argument, from the closeness of separatrices to the closed homoclinic loop of the formal interpolating flow and the area-preserving property of the map $F_\varepsilon$. We assume $\mathbf{p}_{\mathrm{homocl}}$ is closed to $\mathbf{x}_0(0)$. The separatrices of $\mathbf{p}_{\mathrm{hyp}}$ can be parametrized by solutions of the finite-difference equation:

$$\mathbf{x}^{u,s}(t+h) = F_\varepsilon(\mathbf{x}^{u,s}(t)),$$

supplied with the following boundary conditions:

$$\mathbf{x}^u(-\infty) = \mathbf{p}_{\mathrm{hyp}}, \qquad \mathbf{x}^s(+\infty) = \mathbf{p}_{\mathrm{hyp}}, \qquad \mathbf{x}^u(0) = \mathbf{x}^s(0) = \mathbf{p}_{\mathrm{homocl}}.$$

Let $\tilde{\mathbf{x}}_0(t)$ denote a separatrix solution of the Hamiltonian system with the Hamilton function:

$$\frac{y^2}{2} - \frac{x^2}{2} + \frac{x^3}{3}.$$

After a properly chosen scaling, we obtain a function $\mathbf{x}_0(t)$, which provides a first order approximation for the separatrices of the map:

$$\mathbf{x}^u(t) \approx \mathbf{x}_0(t) \quad \text{and} \quad \mathbf{x}^s(t) \approx \mathbf{x}_0(t),$$

provided the primary homoclinic point was chosen to be near $\mathbf{x}_0(0)$. In the formula above, we skipped the error terms, which depend on $t$. If $t$ is close to $\pi i$, the first complex singularity of $\mathbf{x}_0(t)$, the approximation error is not sufficiently small, and we use solutions of the finite-difference equation,

$$\mathbf{x}_\pm(\tau + 1) = F_0(\mathbf{x}_\pm(\tau)),$$

to obtain a better approximation. Two solutions, $\mathbf{x}_\pm(\tau)$, are selected by a complex matching principle: the first term of the Laurent expansion of $\mathbf{x}_0(t)$ at $t = \pi i$ is an asymptotic boundary condition at infinity for $\mathbf{x}_\pm(\tau)$, where the new time is centered at the singularity and rescaled:

$$\tau = \frac{t - i\pi}{h}.$$

We have:

$$\mathbf{x}^u(t) \approx \mathbf{x}_-(\tau), \qquad \mathbf{x}^s(t) \approx \mathbf{x}_+(\tau).$$

Thus, the stable and unstable separatrices of the map are approximated by different (provided $\Theta(F_0) \neq 0$) functions. To measure their difference, it is convenient to use the splitting function defined in Section 2:

$$\theta(\tau) = \det(\mathbf{x}'_-; \mathbf{x}_+ - \mathbf{x}_-)(\tau) \approx \Theta e^{-2\pi i \tau}.$$

In a complex domain the error of the approximation is smaller than the splitting of complex separatrices. In this way, we have detected the exponentially small splitting in a complex domain, and we need a tool to come back to the real axis. Let us consider another splitting function:

$$\tilde{\theta}(t) = \det(\dot{\mathbf{x}}^u; \mathbf{x}^s - \mathbf{x}^u)(t).$$

It is analytic and has the following elementary properties:

(1) $\tilde{\theta}(0) = 0$;

(2) $\tilde{\theta}(kh) = 0$, $k \in \mathbb{Z}$;

(3) $\dot{\tilde{\theta}}(0) = \det(\dot{\mathbf{x}}^u; \dot{\mathbf{x}}^s)(0)$ coincides with the Lazutkin homoclinic invariant $\omega$.

We extend this function to complex $t$ in a domain near $\pi i$, where $\mathbf{x}^u(t)$ is approximated by $x_-(\tau)$, and $\mathbf{x}^s(t)$ by $\mathbf{x}_+(\tau)$. Then comparing the definitions of $\tilde{\theta}$ and $\theta$, we have

$$\dot{\tilde{\theta}}(t) \approx h^{-1}\theta'(\tau) \approx \frac{\Theta}{h} e^{-2\pi i \tau} = \frac{\Theta}{h} e^{-2\pi^2/h} e^{-2\pi i t}.$$

The factor $h^{-1}$ in the first equality is due to the change of the independent variable, the second equality uses the definition of $\Theta$, and the last one is again due to the change from the variable $\tau$ back to $t$. If $\tilde{\theta}$ was periodic with the period $h$ and zero mean value, we would get an approximation for its first order Fourier coefficient.

The Fourier coefficients of a function $f(t) = f(t + h)$, periodic in a strip $|\operatorname{Im} t| < \rho$, can be estimated:

$$f(t) = \sum_{k \in \mathbb{Z}} f_k e^{2\pi i k t/h}, \quad |f_k| \leqslant e^{-2\pi |k| \rho/h} \sup_{|\operatorname{Im} t| \leqslant s} |f(t)|.$$

In our case the period $h \sim \varepsilon^{1/4}$ is small compared to the width of the domain $\rho \sim \pi$. Thus, only the first order Fourier terms are of any importance. Then on the real axis we have:

$$\tilde{\theta}(t) \approx \frac{|\Theta| e^{-2\pi^2/h}}{h} \sin\left(\frac{2\pi t}{h} + \arg\Theta\right).$$

The argument of $\Theta$ is not defined uniquely, and the property $\tilde{\theta}(0) = 0$ implies that our choice of the normalization, which came from the complex matching principle, leads to $\arg(\Theta) = \pi k$. We differentiate the splitting function at $t = 0$ and obtain an approximation for the Lazutkin homoclinic invariant:

$$\omega = \dot{\tilde{\theta}}(0) \approx \frac{4\pi|\Theta|}{h^2} e^{-2\pi^2/h}.$$

This estimate is the main statement of Theorem 1. If $\Theta \neq 0$, the primary homoclinic point is transversal for all sufficiently small $\varepsilon$, the map $F_\varepsilon$ has a Smale horseshoe, its topological entropy is positive, and the series for the formal interpolating Hamiltonian $H_\varepsilon$ diverges.

Unfortunately, the splitting function $\tilde{\theta}$ is not exactly periodic, so the Fourier expansion arguments cannot be applied directly. This difficulty may be overcome, since this function is extremely close to a periodic function, which can be constructed using a discrete flow box theorem. Let us discuss the ideas related to this theorem.

**Discrete flow-box theorem** The classical flow-box theorem tells us that a vector field can be transformed into a constant vector field in a neighborhood of a non-singular point. Consider a differential equation $\dot{\mathbf{x}} = g(\mathbf{x})$, $\mathbf{x} \in \mathbb{R}^m$, and let $\mathbf{x}_\star$ be non-singular: $g(\mathbf{x}_\star) \neq 0$. Then in a neighborhood of $\mathbf{x}_\star$ there is a coordinate change which transforms the differential equation into $\dot{\mathbf{y}} = \mathbf{e}_1$, where $\mathbf{e}_1 = (1, 0, \ldots, 0) \in \mathbb{R}^m$.

In the case of a close-to-identity map,

$$\mathbf{x} \longmapsto \mathbf{x} + \varepsilon g(\mathbf{x}) + \varepsilon^2 r(\mathbf{x}, \varepsilon),$$

a similar theorem is valid. Discrete flow box coordinates can be constructed analytically and the map takes the form

$$\mathbf{y} \mapsto \mathbf{y} + \varepsilon \mathbf{e}_1.$$

The coordinate system $\mathbf{y}(\mathbf{x})$ is analytic in an $\varepsilon$-independent neighborhood of $\mathbf{x}_\star$.

In the discrete flow-box coordinates, invariant objects are represented by periodic patterns. In particular, the stable and unstable separatrices are graphs of periodic functions, the splitting function $\tilde{\theta}(t)$ provides an approximation for their difference.

**Conclusion** This was a short outline of the derivation of the asymptotic formula for the exponentially small separatrices splitting near a strongly resonant fixed point of first order. The derivation itself is not too complicated, but the complete proof involves a lot of technical estimates, which provide rigorous meaning to the statements we have made. A complete study of dynamics near a strong resonant remains an open problem. We studied the splitting constant $\Theta$, which is a first non-trivial modulus of the local classification. A complete local analytic classification of maps with a strongly resonant fixed point is not available at the moment. Another open problem is related to extending the theory to a resonance of order 4.

## References

[AKN97] V. I. Arnold, V. V. Kozlov, and A. I. Neishtadt, *Mathematical Aspects of Classical and Celestial Mechanics*, Springer, New York, 1997.

[BR01] H. W. Broer and R. Roussarie, Exponential confinement of chaos in the bifurcation set of real analytic diffeomorphisms, in *Global Analysis of Dynamical Systems* (H. W. Broer et al., eds), IOP, Bristol, 2001.

[DRR81] Fr. Dumortier, P. R. Rodrigues, and R. Roussarie, *Germs of Diffeomorphisms in the Plane*, Lecture Notes in Math. **902**, Springer, New York, 1981.

[É85] J. Écalle, *Les fonctions résurgentes, 1–3*, Publ. Math. Orsay, 1981, 1985.

[Gel00a] V. Gelfreich, Splitting of a small separatrix loop near the saddle-center bifurcation in area-preserving maps, *Phys. D* **136** (3–4) (2000), 266–279.

[Gel00b] V. Gelfreich, Splitting of separatrices near resonant periodic orbits, *Math. Phys. Preprint Arch.* (00–402), 2000.

[Gel02] V. G. Gelfreich, Near strongly resonant periodic orbits in a Hamiltonian system, *Proc. Nat. Acad. Sci. USA* **99** (22) (2002), 13975–13979.

[GL01] V. Gelfreich and V. Lazutkin, Splitting of separatrices: perturbation theory and exponential smallness, *Russian Math. Surveys* **56** (3) (2001), 499–558.

[GS01] V. Gelfreich and D. Sauzin, Borel summation and splitting of separatrices for the Hénon map, *Ann. Inst. Fourier (Grenoble)* **51** (2) (2001), 513–567.

[SM71] C. L. Siegel and J. K. Moser, *Lectures on Celestial Mechanics*, Springer, New York, 1971 (reprinted 1995).

# Confluence of singular points and Stokes phenomena

Alexey GLUTSYUK

*CNRS, Unité de Mathématiques Pures et Appliquées, M.R.*
*École Normale Supérieure de Lyon*
*46 allée d'Italie*
*69364 Lyon Cedex 07*
*France*

Notes taken by Alexey FISHKIN and Victor KLEPTSYN

**Abstract**

We consider a linear analytic ordinary differential equation with complex time having a nonresonant irregular singular point. We study it as a limit of a generic family of equations with confluenting Fuchsian singularities.

In 1984 V. I. Arnold asked the following question: is it true that some operators from the monodromy group of the perturbed (Fuchsian) equation tend to Stokes operators of the nonperturbed irregular equation? Another version of this question was also independently proposed by J.-P. Ramis in 1988. It appears that generically, no monodromy operator tends to a Stokes operator. On the other hand, in the case of Poincaré rank 1, commutators of appropriate noninteger powers of the monodromy operators around singular points tend to Stokes operators [9], and so do the transition operators between appropriately normalized monodromy eigenbases [7].

The present lecture notes contain the results mentioned above and some of their nonlinear analogues.

## 1 Introduction

### 1.1 Brief statements of the presented results, plan of the notes and historical remarks

Consider a linear analytic ordinary differential equation

$$\dot{z} = \frac{A(t)}{t^{k+1}} z, \quad z \in \mathbb{C}^n, \ |t| \leq 1, \ k \in \mathbb{N} \tag{1.1}$$

with a nonresonant irregular singularity of order (the Poincaré rank) $k$ at 0 (or briefly, an irregular equation). This means that $A(t)$ is a holomorphic matrix function such that the matrix $A(0)$ has distinct eigenvalues (denote them by $\lambda_i$). Then the matrix $A(0)$ is diagonalizable, and without loss of generality we suppose that it is diagonal.



*Y. Ilyashenko and C. Rousseau (eds.),*
*Normal Forms, Bifurcations and Finiteness Problems in Differential Equations,* 267–294.

**1.1 Definition** Two equations of type (1.1) are analytically (formally) equivalent, if there exists a change $z = H(t)w$ of the variable $z$, where $H(t)$ is a holomorphic invertible matrix function (respectively, a formal invertible matrix power series), that transforms one equation into the other.

The analytic classification of irregular equations (1.1) is well known [2, 3, 12, 13, 20]: the complete system of invariants for analytic classification consists of a formal normal form (1.4) and Stokes operators (1.6) defined in Section 1.2; the latter are linear operators acting in the solution space of (1.1) comparing appropriate "sectorial canonical solution bases".

On the other hand, an irregular equation (1.1) can be regarded as a result of *confluence of Fuchsian singular points* (recall that a Fuchsian singular point of a linear equation is a first order pole of its right-hand side). Namely, consider a deformation

$$\dot{z} = \frac{A(t,\varepsilon)}{f(t,\varepsilon)} z, \qquad f(t,\varepsilon) = \prod_{i=0}^{k} (t - \alpha_i(\varepsilon)), \tag{1.2}$$

of equation (1.1) that splits the irregular singular point 0 of the nonperturbed equation into $k+1$ Fuchsian singularities $\alpha_i(\varepsilon)$ of the perturbed equation, i.e., $\alpha_i(\varepsilon) \neq \alpha_j(\varepsilon)$ for $i \neq j$. The family (1.2) depends on a parameter $\varepsilon \in \mathbb{R}_+ \cup 0$, $f(t,0) \equiv t^{k+1}$, $A(t,0) \equiv A(t)$.

The *monodromy group* of a Fuchsian equation acts linearly in its solution space by analytic extensions of solutions along closed loops. The analytic equivalence class of a Fuchsian equation is completely determined by the local types of its singularities and the action of its monodromy group.

Everywhere in what follows we denote by $M_i$ the monodromy operator of the perturbed equation (1.2) along a loop going around the singular point $\alpha_i$ (the choice of the corresponding loops will be specified later). The monodromy group of the perturbed equation is generated by appropriately chosen operators $M_i$.

In 1984, V. I. Arnold proposed the following question. Consider a generic deformation (1.2). Is there an operator

$$M_{i_1}^{d_1} \dots M_{i_l}^{d_l} \tag{1.3}$$

from the monodromy group of the perturbed equation that converges to a Stokes operator of the nonperturbed equation?

A version of this question was proposed independently by J.-P. Ramis in 1988.

It appears that already in the simplest case of dimension 2 and Poincaré rank $k = 1$ generically *each operator from the monodromy group* (*except for that along a circuit around both singularities* (*and its powers*)) *tends to infinity*, and none tends to a Stokes operator. In other terms, no word (1.3) with $d_i \in \mathbb{Z}$ tends to a Stokes operator. But if $k = 1$, then appropriate words (1.3) with *noninteger* powers $d_i$ tend to Stokes operators (Theorem 2.12 in Section 2.2). The last two statements are unpublished results by the author [9].

The previous question and its nonlinear analogue for parabolic mappings were studied by J.-P. Ramis, B. Khesin, A. Duval, C. Zhang and J. Martinet (see the historical overview in Section 1.3 and that of recent results below). It was proved by the author [7] in the general case that appropriate branches of the eigenfunctions of the monodromy operators $M_i$ of the perturbed equation tend to appropriate canonical solutions of the nonperturbed equation (Theorem 2.5). In the case of Poincaré rank $k = 1$ this implies (Corollary 2.6 stated in the two-dimensional case) that Stokes operators of the nonperturbed equation are limits of transition operators between appropriate eigenbases of the monodromy operators $M_i$. This corollary has a generalization for higher Poincaré rank and dimension [7]. These results are also extended to a generic resonant case [10].

The conjecture saying that Stokes operators are limit transition operators between monodromy eigenbases of the perturbed equation was first proposed by A. A. Bolibrukh in 1996.

Nonlinear analogues of the previous statements for parabolic mappings (i.e., one-dimensional conformal mappings tangent to identity) and their Écalle-Voronin moduli, saddle-node singularities of two-dimensional holomorphic vector fields and their Martinet-Ramis invariants (sectorial central manifolds in higher dimensions) were obtained by the author in [8] (see Theorem 4.17 in Section 4 for two-dimensional saddle-nodes). Generalizations and other versions of the statement on parabolic mappings were later obtained in the paper [15] by P. Mardesic, R. Roussarie, C. Rousseau, and in two unpublished joint papers by the following authors: (1) X. Buff and Tan Lei; (2) A. Douady, Francisco Estrada, P. Sentenac.

In Section 1.2 we recall the analytic classification of irregular equations (1.1) and the definitions of sectorial canonical solution bases and Stokes operators. Section 1.3 contains a survey of previous results.

In Section 2.1 we state the results on the representation of Stokes operators as limit transition operators between monodromy eigenbases (Theorem 2.5 and Corollary 2.6). In Section 2.2 we state Theorem 2.12 on convergence of appropriate word (1.3) to a Stokes operator. Its proof is given in Section 3.

In Section 4 we state the results from [8] concerning two-dimensional saddle-nodes. One of them (Corollary 4.22) is used in the proof of Theorem 2.5 given in Section 4.3. Corollary 4.22 is proved in Section 4.4.

## 1.2 Analytic classification of irregular equations. Canonical solutions and Stokes operators

Let (1.1) be an irregular equation.

One can ask the following question: is it true that the variables $z = (z_1, \dots, z_n)$ in the equation can be separated, more precisely, that (1.1) is analytically equivalent to a direct sum of one-dimensional linear equations, i.e., a linear equation with a diagonal matrix function on the right-hand side? Generically, the answer is "no". At the same time any irregular equation (1.1) is formally equivalent to a unique direct sum of the type

$$\dot{w}_i = \frac{b_i(t)}{t^{k+1}}\, w_i, \quad i = 1, \dots, n, \tag{1.4}$$

where $b_i(t)$ are polynomials of degree at most $k$, $b_i(0) = \lambda_i$. The normalizing series bringing (1.1) to (1.4) is unique up to left multiplication by a constant diagonal matrix. The system (1.4) is called the *formal normal form* of (1.1) [2, 3, 12, 13, 20].

Generically the normalizing series diverges. At the same time there exists a finite covering $\bigcup_{j=0}^{N} S_j$ of a punctured neighborhood of zero in the $t$-line by radial sectors $S_j$ (i.e., those with the vertex at 0) that have the following property. There exists a unique change of variables $z = H_j(t)w$ over each $S_j$ that transforms (1.1) to (1.4), where $H_j(t)$ is an analytic invertible matrix function on $S_j$ that can be $C^\infty$-smoothly extended to the closure $\overline{S}_j$ of the sector so that its asymptotic Taylor series at 0 coincides with the normalizing series. The preceding statement on existence and uniqueness of sectorial normalization holds in any good sector (see the two following Definitions); the covering consists of good sectors [2, 3, 12, 13, 20].

**Case $k = 1$, $n = 2$, $\lambda_1 - \lambda_2 \in \mathbb{R}$.**

**1.2 Definition** A sector in $\mathbb{C}$ with the vertex at 0 is said to be *good*, if it contains only one imaginary semiaxis $i\mathbb{R}_{\pm}$, and its closure does not contain the other one (see Fig. 1).

**General case.**

**1.3 Definition** Let $k \in N$, $\Lambda = \{\lambda_1, \dots, \lambda_n\} \subset \mathbb{C}$ be an $n$-tuple of distinct numbers, $t$ be the coordinate on $\mathbb{C}$. For a given pair $\lambda_i \neq \lambda_j$ the rays in $\mathbb{C}$ starting at 0 and forming the set $\mathrm{Re}((\lambda_j - \lambda_i)/t^k) = 0$ are called the $(k, \Lambda)$-*imaginary dividing rays* corresponding to the pair $(\lambda_i, \lambda_j)$. A radial sector is said to be $(k, \Lambda)$-*good*, if for any pair $(\lambda_i, \lambda_j)$, $j \neq i$, it contains exactly one corresponding imaginary dividing ray and so does its closure.

**1.4 Remark** In the case, when $k = 1$, $n = 2$, $\lambda_1 - \lambda_2 \in \mathbb{R}$, the imaginary dividing rays are the imaginary semiaxes, and the notions of "good" sector and $(k, \Lambda)$-good sector coincide.

**1.5 Remark** The ratio $\frac{w_i}{w_j}(t)$ of solutions of equations from (1.4) tends either to zero or to infinity, as $t$ tends to zero along a ray distinct from the imaginary dividing rays corresponding to the pair $(\lambda_i, \lambda_j)$. Its limit changes exactly when the ray under consideration jumps over one of the latter imaginary dividing rays.

Consider a covering $\bigcup_{j=0}^{N} S_j$ of a punctured neighborhood of zero by good (or $(k, \Lambda)$-good) sectors numbered counterclockwise, and put $S_{N+1} = S_0$. The standard splitting of the normal form (1.4) into the direct sum of one-dimensional equations defines a canonical base in its solutions space (uniquely up to multiplication of the base functions by constants) with a diagonal fundamental matrix. Denote the latter fundamental matrix by

$$W(t) = \mathrm{diag}(w_1, \dots, w_n).$$

Together with the normalizing changes $H_j$ in $S_j$, it defines the canonical bases $(f_{j1}, \dots, f_{jn})$ in the solution space of (1.1) in the sectors $S_j$ with the fundamental matrices

$$Z^j(t) = H_j(t)W(t), \quad j = 0, \dots, N+1, \tag{1.5}$$

where for any $j = 0, \dots, N$ the branch ("with index $j+1$") of the fundamental matrix $W(t)$ in $S_{j+1}$ is obtained from that in $S_j$ by the counterclockwise analytic extension for any $j = 0, \dots, N$. (We put $S_{N+1} = S_0$. The corresponding branch of $W$ "with index $N+1$" is obtained from that "with index 0" by right multiplication with the monodromy matrix of the formal normal form (1.4).) In the connected component of the intersection $S_j \cap S_{j+1}$ there are two canonical solution bases coming from $S_j$ and $S_{j+1}$. Generically, they do not coincide. The transition between them is defined by a constant matrix $C_j$:

$$Z^{j+1}(t) = Z^j(t)C_j. \tag{1.6}$$

The transition operators (matrices $C_j$) are called *Stokes operators* (*matrices*) (see [2, 3, 12, 13, 20]). The nontriviality of Stokes operators yields the obstruction to analytic equivalence of (1.1) and its formal normal form (1.4).

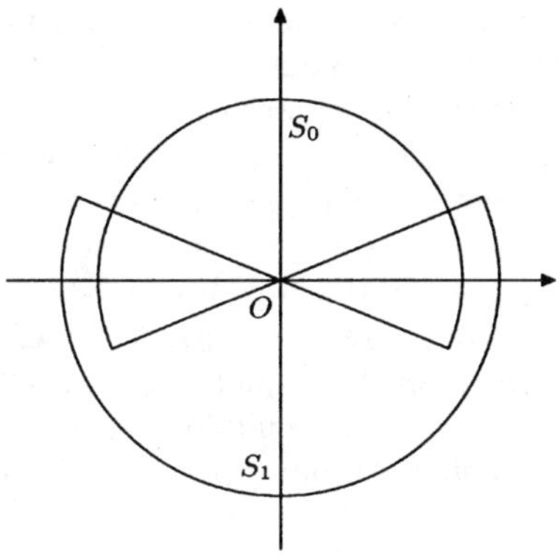


Figure 1

**1.6 Remark** The Stokes matrices (1.6) are well defined up to simultaneous conjugation by one and the same diagonal matrix.

**1.7 Example** Let $k = 1$, $n = 2$. In this case we may assume without loss of generality that $\lambda_1 - \lambda_2 \in \mathbb{R}_+$ (one can achieve this by linear change of the time variable). Then the above covering consists of two sectors $S_0$ and $S_1$ (Fig. 1). The former contains the positive imaginary semiaxis and its closure does not contain the negative one; the latter has the same properties with respect to the negative (respectively, positive) imaginary semiaxis. There are two components of the intersection $S_0 \cap S_1$. So, in this case we have a pair of Stokes operators. *The Stokes matrices* (1.6) *are unipotent:* the one corresponding to the left intersection component is lower-triangular; the other one is upper-triangular [2, 3, 12, 13, 20].

**1.8 Remark** Stokes operators of an irregular equation (1.1) with a diagonal matrix in the right-hand side are identity operators. In this case, (1.1) is analytically equivalent to its formal normal form. In general, *two irregular equations are analytically equivalent, if and only if they have the same formal normal form and the corresponding Stokes matrix tuples are obtained from each other by simultaneous conjugation by one and the same diagonal matrix*, cf. the previous Remark. Thus, the formal normal form and the Stokes matrix tuple taken up to the previous conjugation present the complete system of invariants for analytic classification of irregular equations (see [2, 3, 12, 13, 20]).

## 1.3 Previous results

Earlier, in 1919, R. Garnier [6] had studied some particular deformations of some class of linear equations with nonresonant irregular singularity. He obtained some analytic classification invariants for these equations by studying their deformations. The complete system of analytic classification invariants (Stokes operators and formal normal form) for general irregular differential equations was obtained later in the 70's in the papers by Jurkat, Lutz,

Peyerimhoff [13], Sibuya [20] and Balser, Jurkat, Lutz [3]. Later Jurkat, Lutz and Peyerimhoff extended their results to some resonant cases [14]. In 1985, J.-P. Ramis proved that the Stokes operators and the monodromy operators of a linear ordinary differential equation belong to its Galois group ([18], see also [12]). In 1989 he considered the classical confluenting family of hypergeometric equations and proved convergence of appropriate branches of monodromy eigenfunctions of the perturbed equation to canonical solutions of the nonperturbed one by direct calculation [19]. In the late 80's, B. Khesin also proved a version of this statement, but his result was not published. In 1991, A. Duval [4] proved this statement for the biconfluenting family of hypergeometric equations (where the nonperturbed equation is equivalent to Bessel's equation) by direct calculation. In 1994, C. Zhang [23] had obtained the expression of Garnier's invariants via Stokes operators (for the class of irregular equations considered by Garnier).

The analytic classification of germs of parabolic mappings was obtained separately by J. Écalle [5] and S. M. Voronin [21]. The orbital analytic classification of germs of two-dimensional saddle-node holomorphic vector fields was obtained by J. Martinet and J.-P. Ramis in their joint paper [17]. The analytic classification of two-dimensional saddle-nodes of multiplicity two was recently obtained in the joint paper [22] by S. M. Voronin and Yu. I. Meshcheryakova.

A particular case of the result from [8] concerning parabolic mappings (analogous to the previously mentioned statements on linear equations) was obtained by J. Martinet [16]. For other related results concerning parabolic mappings see the paper by C. Rousseau in the present volume, also [15] and the references therein.

# 2 Main results. Stokes operators and limit monodromy

In the present section we formulate the statements expressing the Stokes operators as limit transition operators between monodromy eigenbases of the confluenting Fuchsian equation (Theorem 2.5 and Corollary 2.6) and as limits of some words (1.3) of noninteger powers of monodromy operators (Theorem 2.12).

## 2.1 Stokes operators as limit transition operators between monodromy eigenbases

We formulate the result from the title of this subsection only in the case when $k = 1$, $n = 2$ (see [7] in the general case). Let $\lambda_i$, $i = 1, 2$, be the eigenvalues of the matrix $A(0)$. Without loss of generality we assume that $\lambda_1 - \lambda_2 \in \mathbb{R}_+$: one can achieve this by linear change of the time variable.

We consider a deformation of (1.1),

$$\dot{z} = \frac{A(t,\varepsilon)}{f(t,\varepsilon)} z, \quad f(t,\varepsilon) = (t - \alpha_0(\varepsilon))(t - \alpha_1(\varepsilon)), \quad f(t,0) \equiv t^2, \quad A(t,0) = A(t), \tag{2.1}$$

where $A(t,\varepsilon)$ and $f(t,\varepsilon)$ depend continuously on a parameter $\varepsilon \geq 0$ so that $\alpha_0(\varepsilon) \neq \alpha_1(\varepsilon)$ for $\varepsilon > 0$. Without loss of generality we assume that $\alpha_0 + \alpha_1 \equiv 0$. We formulate the statement from the title of the subsection for a generic deformation (2.1), see the following Definition.

**2.1 Definition** A family of quadratic polynomials $f(t,\varepsilon)$ depending continuously on a non-negative parameter $\varepsilon$, $f(t,0) \equiv t^2$, with roots $\alpha_i(\varepsilon)$, $i = 0, 1$, $\alpha_0 + \alpha_1 \equiv 0$, is said to be *generic*, if $\alpha_0(\varepsilon) \neq \alpha_1(\varepsilon)$ for $\varepsilon \neq 0$, and the line passing through $\alpha_0(\varepsilon)$ and $\alpha_1(\varepsilon)$ intersects the real axis at an angle bounded away from 0 uniformly in $\varepsilon$. A family (2.1) of *linear equations* is said to be *generic*, if the corresponding family of polynomials $f(t,\varepsilon)$ is generic.

Recall the following:

**2.2 Definition** A singular point $t_0$ of a linear analytic ordinary differential equation $\dot{z} = \frac{B(t)}{t-t_0}z$ is said to be *Fuchsian*, if it is a first order pole of the right-hand side (i.e., the corresponding matrix function $B(t)$ is holomorphic at $t_0$). The *characteristic numbers* of a Fuchsian singularity are the eigenvalues of the corresponding residue matrix $B(t_0)$ (which are equal to the logarithms divided by $2\pi i$ of the eigenvalues of the corresponding monodromy operator).

**2.3 Remark** A family (2.1) of linear equations is generic if and only if the difference of the characteristic numbers at $\alpha_0(\varepsilon)$ (or equivalently, at $\alpha_1(\varepsilon)$) of the perturbed equation is not real for small $\varepsilon$ and, moreover, has argument bounded away from $\pi\mathbb{Z}$ uniformly in $\varepsilon$ small enough. The latter condition implies that the monodromy operator of the perturbed equation around each singular point $\alpha_i$ has distinct eigenvalues (moreover, their modules are distinct), and hence, a well-defined eigenbase in the solution space (for small $\varepsilon$).

The singularities of the perturbed equation from a generic family have imaginary parts of constant (and opposite) signs (by definition). Without loss of generality we assume in what follows that

$$\operatorname{Im}\alpha_0 > 0, \quad \operatorname{Im}\alpha_1 < 0 \quad \text{(see Fig. 2).}$$

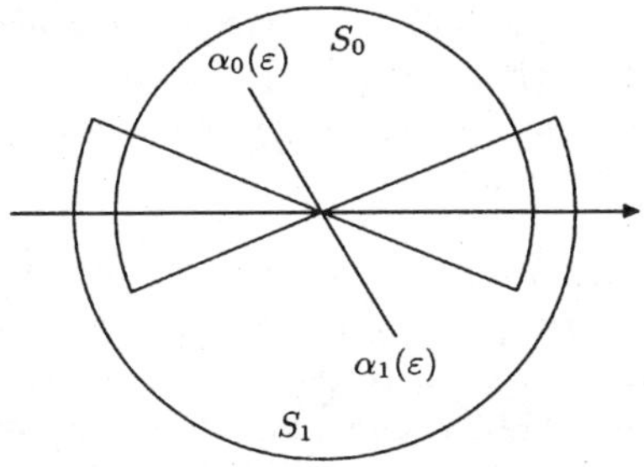


Figure 2

**2.4 Definition** Let (2.1) be a generic family of linear equations (see the previous definition) whose singularity families satisfy the previous inequalities. Let $S_j$, $j = 0, 1$, be a pair of good sectors in the $t$-line such that $\alpha_j(\varepsilon) \in S_j$, $j = 0, 1$, $i\mathbb{R}_+ \subset S_0$, $i\mathbb{R}_- \subset S_1$ (see Fig. 1). The sector $S_j$ is said to be *the sector associated to the singularity family* $\alpha_j$, $j = 0, 1$.

We show that appropriate branches of the eigenfunctions of the monodromy operator $M_i$ around $\alpha_i$ of the perturbed equation converge to canonical solutions of the nonperturbed equation in the corresponding sector $S_i$. This will imply the statement from the title of this subsection.

To formulate the latter statement precisely, consider the auxiliary domain

$$S_i' = S_i \setminus [\alpha_0(\varepsilon), \alpha_1(\varepsilon)], \tag{2.2}$$

which is simply-connected, and the canonical branches of the monodromy eigenfunctions on the domain $S_i'$. In more detail, consider a small circle going around $\alpha_i$, and take a base point on it outside the segment $[\alpha_0(\varepsilon), \alpha_1(\varepsilon)]$. In the space of local solutions of the perturbed equation at the base point consider the monodromy operator $M_i$ acting by the analytic extension of a solution along the circle from the base point to itself in the counterclockwise direction. The eigenfunctions of $M_i$ have well-defined branches (up to multiplication by constants) in the corresponding disc with the segment $[\alpha_0(\varepsilon), \alpha_1(\varepsilon)]$ deleted. Their immediate analytic extension yields their canonical branches on $S_i'$. In other terms, we identify the space of local solutions with the space of solutions on $S_i'$ by immediate analytic extension, consider $M_i$ as an operator acting in the latter space and take its eigenfunctions.

The canonical basic solutions of the nonperturbed equation are numbered by the indices 1 and 2, which correspond to the eigenvalues $\lambda_1, \lambda_2$ of $A(0)$. To state the results previously mentioned, let us define an analogous numbering of the monodromy eigenfunctions at $\alpha_i(\varepsilon)$. The monodromy eigenfunctions are numbered by the eigenvalues of the corresponding residue matrix. The latter eigenvalues are proportional to those of the matrix $A(\alpha_i(\varepsilon), \varepsilon)$, which tend to $\lambda_1$ and $\lambda_2$, as $\varepsilon \to 0$. This induces the numbering of the monodromy eigenfunctions with the indices 1 and 2 corresponding to the limit eigenvalues $\lambda_1$ and $\lambda_2$.

**2.5 Theorem** *Let* (2.1) *be a generic family of linear ordinary differential equations (see Definition* 2.1*), $\alpha_i(\varepsilon)$ its singularity family, let $S_i$ be the corresponding sector (see the previous definition), and $S_i'$ the domain* (2.2)*. Consider the eigenbase on $S_i'$ of the monodromy operator of the perturbed equation around $\alpha_i(\varepsilon)$. The appropriately normalized eigenbase (by multiplication of the basic functions by constants) converges to the canonical solution base* (1.5) *on $S_i$ of the nonperturbed equation.*

**2.6 Corollary** *Let* (2.1) *be a generic linear equation family (see Definition* 2.1*), $\alpha_i$ its singularity families, let $S_i$ be the corresponding sectors (see the previous definition) chosen to cover a punctured neighborhood of zero, and $S_i'$ the corresponding domains* (2.2)*. Let $C_0$, $C_1$ be the corresponding Stokes matrices* (1.6) *of the nonperturbed equation in the left (respectively, right) component of the intersection $S_0 \cap S_1$. Consider the eigenbase on $S_i'$ of the monodromy operator of the perturbed equation around $\alpha_i(\varepsilon)$. Denote by $Z_\varepsilon^i(t)$ the fundamental matrix of this eigenbase. Let $C_0(\varepsilon)$ ($C_1(\varepsilon)$) be the transition matrix between the monodromy eigenbases $Z_\varepsilon^i(t)$, $i = 0, 1$, in the left (respectively, right) component of the intersection $S_0' \cap S_1'$:*

$$\begin{aligned} Z_\varepsilon^1(t) &= Z_\varepsilon^0(t) C_0(\varepsilon) \quad \textit{for} \quad \mathrm{Re}\, t < 0; \\ Z_\varepsilon^0(t) &= Z_\varepsilon^1(t) C_1(\varepsilon) \quad \textit{for} \quad \mathrm{Re}\, t > 0. \end{aligned} \tag{2.3}$$

*For any $i = 0, 1$ and appropriately normalized monodromy eigenbases $Z_\varepsilon^j$, $j = 1, 2$ (the normalization of $Z_\varepsilon^0$ (only) depends on the choice of $i$), $C_i(\varepsilon) \to C_i$ as $\varepsilon \to 0$.*

**2.7 Remark** Theorem 2.5 and Corollary 2.6 extend to the general case of arbitrary Poincaré rank $k$ and dimension $n$ [7], as do the notions of a generic family of linear equations and a sector associated to a singularity family. The statement of Corollary 2.6 in the case of $k = 1$ and arbitrary $n$ remains the same. But for higher $k$ (when the number $k + 1$ of transition matrices is less than that of Stokes matrices) it says that appropriate products of subsequent Stokes matrices (not all the Stokes matrices themselves) are limit transition matrices between appropriate branches of monodromy eigenbases. These limit products of Stokes matrices cover all the Stokes matrices. On the other hand, each element of a Stokes matrix in a limit product can be expressed as a polynomial in the product elements; so, all the Stokes matrices can be recovered from the limit transition matrices.

## 2.2 Stokes operators as limits of commutators of appropriate powers of the monodromy operators

The Stokes and monodromy operators act in different linear spaces: in the solution spaces of the nonperturbed (respectively, perturbed) equations. To formulate the statement from the title of the subsection, let us first identify these solution spaces and specify the loops defining the monodromy operators.

Let (2.1) be a generic family of linear equations. Take the "base point"

$$t_0 = -\frac{1}{2}.$$

**2.8 Remark** The space of local solutions of a linear equation at a nonsingular point $t_0 \in \mathbb{C}$ is identified with the space of initial conditions at $t_0$ (which is common for the nonperturbed and the perturbed equations). This identifies the solution spaces of the latter. The space thus obtained will be denoted by $H_{t_0}$.

**2.9 Remark** Let (1.1) be an irregular equation with $k = 1$, $n = 2$, $\lambda_1 - \lambda_2 \in \mathbb{R}$, and let $S_0$, $S_1$ be good sectors covering a punctured neighborhood of zero in the $t$-line, both containing $\mathbb{R}_-$ and $\mathbb{R}_+$ (see Fig. 1). Let $C_0$, $C_1$ be the Stokes operators (1.6) corresponding to the left (respectively, right) intersection component of the sectors. The operator $C_0$ ($C_1$) is well defined in the space $H_{t_0}$ of local solutions of (1.1) at any point $t_0 \in \mathbb{R}_-$ (respectively, $t_0 \in \mathbb{R}_+$).

Now let us define the monodromy operators acting in the previous space $H_{t_0}$.

**2.10 Definition** Let (2.1) be a generic family of linear equations, $\alpha_i(\varepsilon)$, $i = 0, 1$, be its singularity families. Fix a point $t_0 \in \mathbb{R}$ (independent of $\varepsilon$). Let $l_i$ be a small circle centered at $\alpha_i(\varepsilon)$ whose closed disc is disjoint from $-\alpha_i(\varepsilon)$, $a_i = [t_0, \alpha_i] \cap l_i$, with the segment $[t_0, a_i]$ oriented from $t_0$ to $a_i$. Consider the closed path $\psi_i = [t_0, a_i] \circ l_i \circ [t_0, a_i]^{-1}$, $i = 0, 1$, which starts and ends at $t_0$ (see Fig. 3). Define $M_i : H_{t_0} \to H_{t_0}$ to be the corresponding monodromy operator of the perturbed equation.

We show that commutators of appropriate noninteger powers of the operators $M_i$ (see the following definition) tend to the Stokes operators.

**2.11 Definition** Let $d \in \mathbb{R}$, and let $M : H \to H$ be a linear operator in a finite-dimensional linear space having distinct eigenvalues. The *d-th power of $M$* is the operator having the

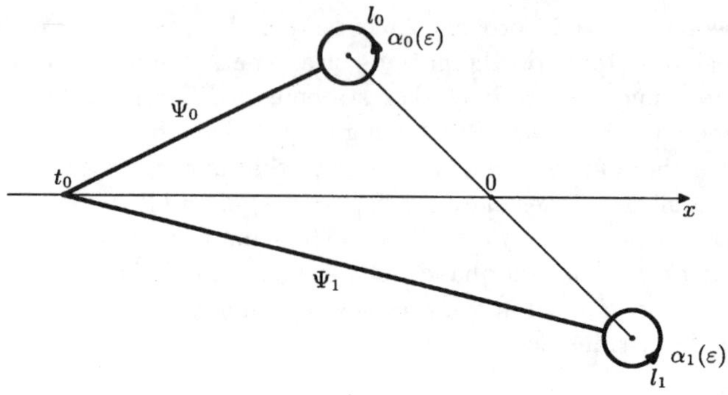


Figure 3

same eigenlines as $M$, whose corresponding eigenvalues are some values of $d$-th powers of those of $M$.

**2.12 Theorem** *Let* (2.1) *be a generic family of linear equations* (*see Definition* 2.1) *and* $\alpha_i(\varepsilon)$, $i = 0, 1$, *its singularity families. Let* $t_0 = \pm 1/2$, $H_{t_0}$ *the corresponding local solution space* (*see Remark* 2.8). *Let* $M_i : H_{t_0} \to H_{t_0}$ *be the corresponding monodromy operators from Definition* 2.10. *Let* $S_i$, $i = 0, 1$, *be the corresponding associated sectors* (*see Definition* 2.4) *forming a covering of a punctured neighborhood of zero, and let* $C_0$, $C_1$ *be the Stokes operators* (1.6) *of the nonperturbed equation corresponding to the left* (*respectively, right*) *component of the intersection* $S_0 \cap S_1$ (*acting in the spaces* $H_{-1/2}$ *and* $H_{1/2}$ *respectively, see Remark* 2.9). *Then for any pair of numbers* $d_0, d_1 > 0$ *such that* $d_0 + d_1 < 1$

$$\begin{aligned} &M_1^{-d_1} M_0^{d_0} M_1^{d_1} M_0^{-d_0} \to C_0 \quad \textit{in the space} \quad H_{-1/2}, \\ &M_0^{-d_0} M_1^{d_1} M_0^{d_0} M_1^{-d_1} \to C_1 \quad \textit{in the space} \quad H_{1/2} \quad \textit{as} \quad \varepsilon \to 0. \end{aligned} \tag{2.4}$$

Theorem 2.12 is proved in the next section.

**2.13 Remark** The statements of Theorem 2.12 imply the same statements in any space $H_{t_0}$, $\operatorname{Re} t_0 < 0$ (respectively, $\operatorname{Re} t_0 > 0$). Theorem 2.12 extends to the case of $k = 1$ and arbitrary dimension [9].

# 3 Convergence of the commutators to Stokes operators. Proof of Theorem 2.12

## 3.1 Projectivization. The plan of the proof of Theorem 2.12

Let us prove convergence of the commutator (2.4); the proof of the convergence of the other commutator from Theorem 2.12 is analogous.

Thus, from now on, we put $t_0 = -1/2$.

For the proof of Theorem 2.12 we consider the projectivization of the space $H_{t_0} = \mathbb{C}^2$. The projectivizations of the monodromy and Stokes operators are Möbius transformations $\overline{\mathbb{C}} \to \overline{\mathbb{C}}$ (denote by $m_i : \overline{\mathbb{C}} \to \overline{\mathbb{C}}$ the projectivizations of the monodromy operators $M_i$, and by $\sigma$ the projectivization of the Stokes operator $C_0$).

Let $d_0, d_1 > 0$, $d_0 + d_1 < 1$. Denote

$$m_i' = m_i^{d_i}.$$

For the proof of (2.4) we show (below and in subsections 3.2, 3.3) that

$$(m_1')^{-1} m_0' m_1' (m_0')^{-1} \to \sigma \quad \text{as} \quad \varepsilon \to 0. \tag{3.1}$$

This means that the commutator (2.4) multiplied by an appropriate constant (depending on the parameter) converges to $C_0$. The commutator (2.4) has unit determinant, as does any commutator and the operator $C_0$ (which is unipotent, see Example 1.7). This together with (3.1) implies that its limit exists and is equal to either $C_0$ or $-C_0$. The fact that it is really equal to $C_0$ will be proved in subsection 3.4.

To sketch the proof of (3.1), let us first recall the following:

**3.1 Definition ([1])** A Möbius transformation is said to be *hyperbolic*, if it has two fixed points one of which is attracting (then the other is repelling). It is said to be *parabolic*, if it has only one fixed point. (Otherwise, it is said to be *elliptic*.)

In what follows, we represent hyperbolic and parabolic transformations by figures as follows. The Riemann sphere $\overline{\mathbb{C}}$ will be drawn in the form of a circle. A hyperbolic transformation with fixed points $a$ and $b$, $a$ being repelling, will be represented by marking $a$ and $b$ at the circle (representing $\overline{\mathbb{C}}$) and an oriented segment going from $a$ to $b$ (see Fig. 4(a)). A parabolic transformation with fixed point $a$, sending $b$ to $c$, will be represented by marking the points $a, b, c$ and the arrow from $b$ to $c$ on the circular arc joining them and disjoint from the fixed point $a$ (see Fig. 4(b)).

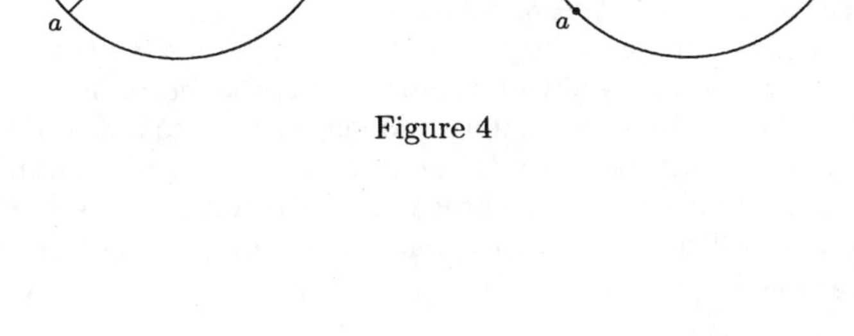


Figure 4

**3.2 Remark** The projectivization of a Stokes operator of an irregular equation is parabolic, since a Stokes operator is unipotent (see Example 1.7). The projectivization of a two-dimensional linear operator having eigenvalues with distinct modulus is hyperbolic: its repelling fixed point corresponds to the eigenfunction with the eigenvalue of the smallest modulus; its multiplier at the repelling fixed point is equal to the ratio of the eigenvalues. Each monodromy operator $M_i$ from Theorem 2.12 has eigenvalues of distinct modulus (see Remark 2.3), so, *its projectivization* $m_i$ *is hyperbolic.*

For the proof of (3.1) we state and prove its analogue (Lemma 3.11 below) for commutators of families of hyperbolic transformations generalizing $m_i' = m_i'(\varepsilon)$. To do this and to motivate the proof, let us first describe the arrangement of the fixed points of $m_0$, $m_1$ and $\sigma$.

**3.3 Proposition** *Let* (1.1) *be a two-dimensional irregular equation,* $\lambda_1$, $\lambda_2$ *be the eigenvalues of the corresponding matrix* $A(0)$, *and* $\lambda_1 - \lambda_2 > 0$. *Let* $S_0$, $S_1$ *be the sectors from Example* 1.7 *(see Fig.* 1), $C_0$ *be the Stokes operator* (1.6) *corresponding to the left component of their intersection, and let* $\sigma$ *be the projectivization of* $C_0$. *Let* $f_{i1}$, $f_{i2}$ *be the canonical solutions of* (1.1) *on the sectors* $S_i$, $i = 0, 1$, $p_{i1}$, $p_{i2}$ *be their projectivizations. Then* $\sigma$ *is a parabolic transformation with the fixed point* $p_{02}$,

$$p_{02} = p_{12}, \quad \sigma(p_{02}) = p_{02}, \quad \sigma(p_{01}) = p_{11}, \quad (\textit{see Fig. } 5(\text{b})).$$

Proposition 3.3 follows from the definition, the unipotence and the lower triangularity of the Stokes matrix $C_0$ (see Example 1.7).

**3.4 Proposition** *Let* (2.1) *be a generic family of linear equations,* $t_0 \in \mathbb{R}$, $M_i$ *be the monodromy operators of the perturbed equation from Definition* 2.10, $f_{i1,\varepsilon}$, $f_{i2,\varepsilon}$ *be their basic eigenfunctions, and* $\lambda_{i1}$, $\lambda_{i2}$ *the corresponding eigenvalues. Then*

$$\mu_0 = \frac{\lambda_{01}}{\lambda_{02}} \to \infty, \quad \mu_1 = \frac{\lambda_{12}}{\lambda_{11}} \to \infty, \quad \textit{as} \quad \varepsilon \to 0. \tag{3.2}$$

**3.5 Corollary** *Under the conditions of Proposition* 3.4, *let* $m_i : \overline{\mathbb{C}} \to \overline{\mathbb{C}}$ *be the projectivizations of* $M_i$, $p_{ij,\varepsilon}$ *be those of* $f_{ij,\varepsilon}$. *Then* $m_i$ *are hyperbolic transformations with fixed points* $p_{i1,\varepsilon}$, $p_{i2,\varepsilon}$. *More precisely,* $p_{02,\varepsilon}$ *is the repelling point of* $m_0$, $p_{11,\varepsilon}$ *is that of* $m_1$ *(see Fig* 5(a)), *the corresponding multipliers are equal to* $\mu_0$, $\mu_1$, *see* (3.2)*: they tend to infinity. Let* $S_i$ *be the sectors associated to the singularities* $\alpha_i$ *of the perturbed equation (see Definition* 2.4), $p_{ij}$ *be the projectivizations of the canonical sectorial solutions on* $S_i$ *of the nonperturbed equation. Then*

$$p_{ij,\varepsilon} \to p_{ij} \quad \textit{as} \quad \varepsilon \to 0 \quad (\textit{see Fig. } 5(\text{b})). \tag{3.3}$$

Statement (3.3) follows from Theorem 2.5.

To motivate the proofs of the convergence of the commutators in (2.4) and (3.1), consider the simplest case, where in the family of equations (2.1) the matrix function family $A(t, \varepsilon)$ is lower-triangular. Then the line $z_1 = 0$ is invariant for each equation of the family. This implies that the monodromy operators $M_0$ and $M_1$ have a common eigenfunction (whose graph lies in the invariant line $z_1 = 0$) and their projectivizations $m_i$ have the common fixed point $p_{02,\varepsilon} = p_{12,\varepsilon}$, repelling for $m_0$ and attracting for $m_1$ (see Fig. 6(a) below). In this case not only does the commutator in (3.1) converge: it does so with arbitrary powers $m_i^{d_i}$,

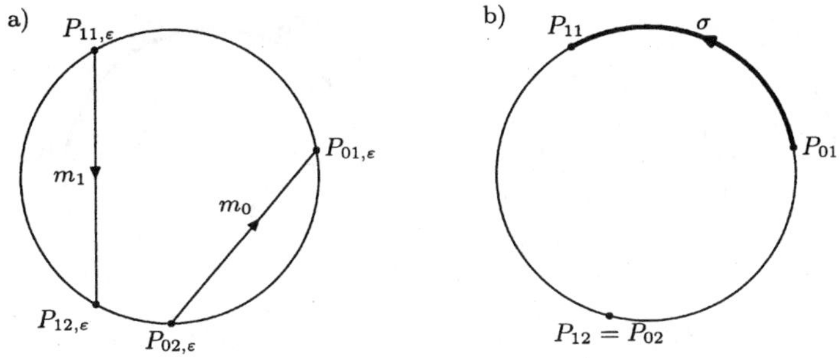


Figure 5

$d_i > 0$, in particular, $m_1^{-1}m_0m_1m_0^{-1} \to \sigma$. This is implied by (3.2), (3.3) and a more general Proposition 3.6 stated below. To formulate it, let us introduce the following notation:

$$h_{a,b,\nu} : \overline{\mathbb{C}} \to \overline{\mathbb{C}} \tag{3.4}$$

is the hyperbolic transformation of the Riemann sphere fixing points $a, b \in \overline{\mathbb{C}}$; $a$ is repelling with the multiplier $\nu$.

**3.6 Proposition** *Let $p$, $p_{01}$, $p_{11}$ be three distinct points of the Riemann sphere, and let $\sigma : \overline{\mathbb{C}} \to \overline{\mathbb{C}}$ be the parabolic transformation fixing $p$ and sending $p_{01}$ to $p_{11}$. Consider three arbitrary families of points $a, b_0, b_1 \in \overline{\mathbb{C}}$ converging to $p_{ij}$ (see Fig. 6):*

$$a \to p, \qquad b_0 \to p_{01}, \qquad b_1 \to p_{11}.$$

*Then in the notation* (3.4)

$$h_{b_1,a,\nu_1}^{-1} h_{a,b_0,\nu_0} h_{b_1,a,\nu_1} h_{a,b_0,\nu_0}^{-1} \to \sigma \quad \textit{as} \quad (a, b_0, b_1) \to (p, p_{01}, p_{11}),\ \nu_0, \nu_1 \to \infty.$$

The proof of Proposition 3.6 is straightforward and can be done by hand (e.g., multiplying the (triangular) matrices of the $h'$ s explicitly). It is omitted to save space.

In the previous case of the lower-triangular matrix $A(t, \varepsilon)$ the families $m_i$ of hyperbolic transformations (and also $m_i^{d_i}$ with arbitrary $d_i > 0$) satisfy the conditions of Proposition 3.6 by (3.2), (3.3). This together with the Proposition implies (3.1).

In the general case, the transformations $m_i$ have distinct fixed points: $p_{02,\varepsilon} \neq p_{12,\varepsilon}$. On the other hand, the latter fixed points are confluent to the fixed point $p = p_{02}$ of $\sigma$. For the proof of (3.1) in the general case we show first that the distance $\mathrm{dist}(p_{02,\varepsilon}, p_{12,\varepsilon})$ is not too large: it decreases as $O(\mu_1^{-1})$ (Corollary 3.8). Then we state and prove a generalization (Lemma 3.11) of Proposition 3.6 for families of hyperbolic transformations $h_{a_0,b_0,\nu_0}$, $h_{b_1,a_1,\nu_1}$ that have no common fixed point, but confluenting families of fixed points $a_0, a_1 \to p$ such that the distance $\mathrm{dist}(a_0, a_1)$ between them decreases fast enough, more precisely, as $o(|\nu_0\nu_1|^{-1})$. We apply Lemma 3.11 to the hyperbolic transformations $m_i' = m_i^{d_i}$ and $\nu_i = \mu_i^{d_i}$. To show the possibility of applying Lemma 3.11 to $m_i'$, it suffices to prove that $\mathrm{dist}(p_{02,\varepsilon}, p_{12,\varepsilon}) = o(|\nu_0\nu_1|^{-1})$. This is the place where we use the inequalities on the exponents $d_i$ from Theorem 2.12.

To estimate the distance $\mathrm{dist}(p_{02,\varepsilon}, p_{12,\varepsilon})$, we use the following

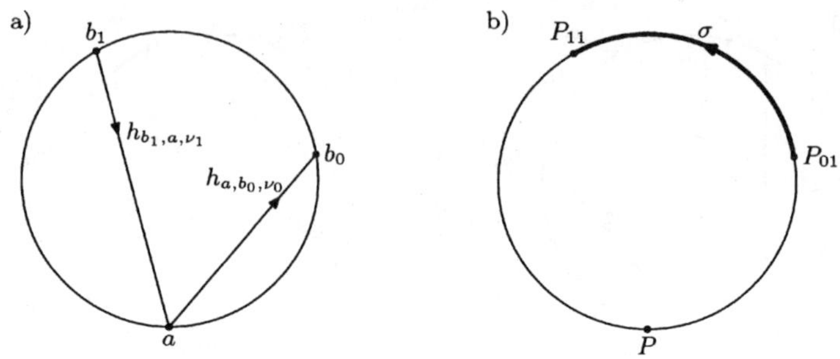


Figure 6

**3.7 Lemma** *Let (2.1) be a generic family of linear equations (see Definition 2.1), $\alpha_i$ be its singularity families, $S_i$ be the corresponding sectors (see Definition 2.4) chosen to cover a punctured neighborhood of zero, $S_i'$ be the corresponding domains from (2.2). Let $C_0$, $C_1$ be the Stokes matrices (1.6) of the nonperturbed equation (corresponding to the left (respectively, right) component of the intersection $S_0 \cap S_1$),*

$$C_0 = \begin{pmatrix} 1 & 0 \\ c_0 & 1 \end{pmatrix}, \quad C_1 = \begin{pmatrix} 1 & c_1 \\ 0 & 1 \end{pmatrix} \quad \text{(see Example 1.7).} \tag{3.5}$$

*Let $M_i$ be the monodromy operator of the perturbed equation around $\alpha_i(\varepsilon)$ acting in the space of solutions on $S_i'$. Let $Z_\varepsilon^i$ be (the fundamental matrix of) its eigenbase. Let $C_0(\varepsilon)$ be the transition matrix (2.3) between the bases $Z_\varepsilon^i$ that converges to $C_0$, as $\varepsilon \to 0$, see Corollary 2.6 (we consider the transition in the left component of the intersection $S_0' \cap S_1'$):*

$$C_0(\varepsilon) = \begin{pmatrix} 1+o(1) & u(\varepsilon) \\ c_0 + o(1) & 1+o(1) \end{pmatrix}, \quad u(\varepsilon) \to 0.$$

*Let $\lambda_{11}, \lambda_{12}$ be the eigenvalues of $M_1$ at $\alpha_1(\varepsilon)$, $\mu_1 = \lambda_{12}/\lambda_{11}$ be the corresponding multiplier of its projectivization. Then the upper triangular element $u(\varepsilon)$ of the matrix $C_0(\varepsilon)$ has the asymptotics*

$$u(\varepsilon) = (-c_1 + o(1))\mu_1^{-1} \quad \text{as} \quad \varepsilon \to 0, \tag{3.6}$$

*where $c_1$ is the upper triangular element of the Stokes matrix $C_1$ in (3.5).*

Lemma 3.7 is proved in subsection 3.2.

**3.8 Corollary** *Let (2.1) be a generic family of linear equations, $t_0 = -1/2$, $M_i$ be the monodromy operators from Definition 2.10, $m_i$ be their projectivizations, $p_{02,\varepsilon}$ be the repelling fixed point of $m_0$, $p_{12,\varepsilon}$ be the attracting fixed point of $m_1$, and let $\mu_1^{-1}$ be the multiplier of the latter attracting fixed point. Then*

$$\text{dist}(p_{02,\varepsilon}, p_{12,\varepsilon}) = O(\mu_1^{-1}) \quad \text{as} \quad \varepsilon \to 0.$$

**3.9 Remark** The multipliers of a hyperbolic transformation at its fixed points are inverse. In particular, in the preceding corollary, $\mu_1$ is the multiplier of $m_1$ at its repelling fixed point $p_{11,\varepsilon}$.

**3.10 Proposition** *Let $M_i$ be the monodromy operators from Definition* 2.10, *$m_i$ their projectivizations, and $\mu_i$ the multipliers at their repelling fixed points. Then*

$$|\mu_0| = |\mu_1|^{1+o(1)} \quad as \quad \varepsilon \to 0.$$

**Proof** Recall that appropriate logarithms of the eigenvalues of the monodromy operators around singularities are equal to $2\pi$ times the corresponding eigenvalues of the residue matrices (i.e., the characteristic numbers). The characteristic numbers at $\alpha_0(\varepsilon)$ are equal to $-(1+o(1))$ times those at $\alpha_1(\varepsilon)$. This together with (3.2) implies that $\ln|\mu_0| = (1+o(1))\ln|\mu_1|$, which proves Proposition 3.10. □

As is shown below, (3.1) is implied by Corollary 3.8, Proposition 3.10, the inequalities on $d_i$ from Theorem 2.12, and the following lemma.

**3.11 Lemma** *Let $p$, $p_{01}$, $p_{11}$ be three distinct points of the Riemann sphere, and $\sigma : \overline{\mathbb{C}} \to \overline{\mathbb{C}}$ the parabolic transformation fixing $p$ and sending $p_{01}$ to $p_{11}$. Consider four arbitrary families of points $a_0, a_1, b_0, b_1 \in \overline{\mathbb{C}}$ converging to $p_{ij}$ (see Fig.* 7*):*

$$a_0, a_1 \to p, \qquad b_0 \to p_{01}, \qquad b_1 \to p_{11}.$$

*Then in the notations* (3.4)

$$h^{-1}_{b_1,a_1,\nu_1} h_{a_0,b_0,\nu_0} h_{b_1,a_1,\nu_1} h^{-1}_{a_0,b_0,\nu_0} \to \sigma, \tag{3.7}$$

*as $a_0, a_1 \to p$, $(b_0, b_1) \to (p_{01}, p_{11})$, $\nu_0, \nu_1 \to \infty$ so that* $\operatorname{dist}(a_0, a_1) = o(|\nu_0\nu_1|^{-1})$.

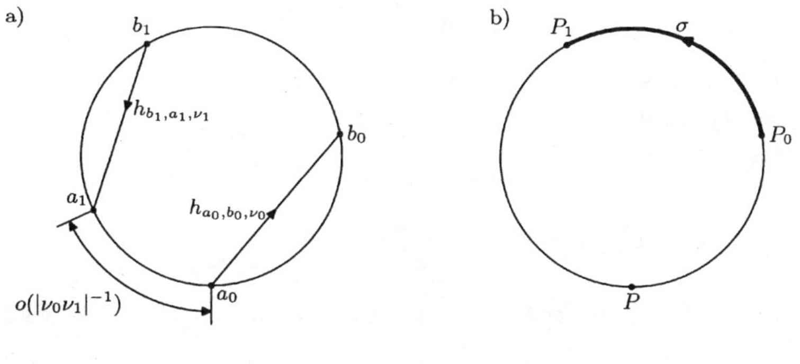


Figure 7

Lemma 3.11 is proved in subsection 3.3.

**Proof of (3.1)** Let us show that the families of hyperbolic transformations $m_i' = m_i^{d_i}$ satisfy the conditions of Lemma 3.11. Their fixed points converge to $p_{ij}$ by (3.3). Their multipliers at the repelling fixed points are equal to $\nu_i = \mu_i^{d_i}$. Now it suffices to prove the last asymptotic formula in (3.7) saying in our case that $\mathrm{dist}(p_{02,\varepsilon}, p_{12,\varepsilon}) = o(|\mu_0^{d_0}\mu_1^{d_1}|^{-1})$. The latter formula follows from Corollary 3.8, Proposition 3.10, positivity of the powers $d_i$ and the inequality $d_0 + d_1 < 1$ from the conditions of Theorem 2.12. This together with Lemma (3.11) proves (3.1). □

## 3.2 The upper triangular element of the transition matrix. Proof of Lemma 3.7

The transition matrix $C_0(\varepsilon)$, which converges to the Stokes matrix $C_0$, $Z_\varepsilon^1 = Z_\varepsilon^0 C_0(\varepsilon)$, compares the monodromy eigenbases $Z_\varepsilon^0$ and $Z_\varepsilon^1$ in the left component of the intersection $S_0' \cap S_1'$, in particular, on a real interval in $\mathbb{R}_-$. It is not changed when we extend the basic functions analytically from $\mathbb{R}_-$ to $\mathbb{R}_+$ along the real line. Denote by $Z_{\varepsilon,+}^i$ the corresponding branch on $\mathbb{R}_+$ of the extended fundamental matrix $Z_\varepsilon^i$, $i = 0, 1$. By construction, $Z_{\varepsilon,+}^0$ is obtained from $Z_\varepsilon^0|_{\mathbb{R}_+}$ by applying the monodromy operator $M_0$; $Z_{\varepsilon,+}^1$ is obtained from $Z_\varepsilon^1|_{\mathbb{R}_+}$ by applying the inverse monodromy operator $M_1^{-1}$:

$$Z_{\varepsilon,+}^1 = Z_\varepsilon^1|_{S_1'} M_1^{-1}; \quad \text{the matrix } M_1 \text{ is diagonal} \tag{3.8}$$

On the other hand, we can choose a renormalization of the eigenbase $Z_{\varepsilon,+}^0$ by multiplication of the basic functions by constants (i.e., changing it to $Z_{\varepsilon,+}^0\Lambda(\varepsilon)$, $\Lambda(\varepsilon) = \mathrm{diag}(l_1(\varepsilon), l_2(\varepsilon))$ so that in the right component of the intersection $S_0' \cap S_1'$ the transition matrix $C_1(\varepsilon)$ between $Z_{\varepsilon,+}^0\Lambda(\varepsilon)$ and $Z_\varepsilon^1$ tends to the Stokes matrix $C_1$:

$$Z_{\varepsilon,+}^0\Lambda(\varepsilon) = Z_\varepsilon^1|_{S_1'} C_1(\varepsilon), \quad C_1(\varepsilon) \to C_1.$$

Substituting (3.8) and (2.3) in the latter formula yields

$$C_0(\varepsilon) = \Lambda(\varepsilon) C_1^{-1}(\varepsilon) M_1^{-1}. \tag{3.9}$$

The matrices $C_i(\varepsilon)$ tend to the Stokes matrices $C_i$, which are unipotent. The matrices $\Lambda(\varepsilon)$, $M_1$ are diagonal and depend on $\varepsilon$. This implies that

$$\Lambda(\varepsilon) = M_1(1 + o(1)) \quad \text{as} \quad \varepsilon \to 0.$$

This together with (3.9) implies (3.6).

## 3.3 Commutators of hyperbolic transformations with close fixed points. Proof of Lemma 3.11

Lemma 3.11 can be proved "by hand" by multiplying explicitly the matrices of the hyperbolic transformations in the commutator (3.7).

Denote the latter commutator by $\Delta$. For the proof of Lemma 3.11 it suffices to show that

$$\Delta(a_0) \to p, \tag{3.10}$$
$$\Delta'(a_0) \to 1, \qquad \Delta(b_0) \to p_{11}:$$

these statements imply that $\Delta$ does not tend to infinity and each of its limit points is a Möbius transformation having fixed point $p$ and sending $p_{01}$ to $p_{11}$ (thus, coinciding with $\sigma$), hence $\Delta \to \sigma$.

Let us prove (3.10) (the proof of the other two statements is analogous). Recall the last asymptotic condition from Lemma 3.11:

$$\operatorname{dist}(a_0, a_1) = o(|\nu_0\nu_1|^{-1}). \tag{3.11}$$

Consider the orbit of the point $a_0$ under consecutive hyperbolic transformations forming the commutator (3.7). Applying $h^{-1}_{a_0,b_0,\nu_0}$ does not move $a_0$. Applying $h_{b_1,a_1,\nu_1}$ moves $a_0$ to a point (denoted by $a_0'$) close to $a_1$; more precisely,

$$\operatorname{dist}(a_0', a_1) = \nu_1^{-1} \operatorname{dist}(a_0, a_1)(1 + o(1)) = o(\nu_0^{-1}\nu_1^{-2}) \tag{3.12}$$

(by (3.11)). Put

$$a_0'' = h_{a_0,b_0,\nu_0}a_0', \qquad a_0''' = h^{-1}_{b_1,a_1,\nu_1}a_0''.$$

For the proof of (3.10) it suffices to show that

$$a_0''' \to p, \quad \text{or equivalently,} \quad \operatorname{dist}(a_0''', a_1) \to 0. \tag{3.13}$$

By (3.11), (3.12),

$$\operatorname{dist}(a_0', a_0) = o(|\nu_0\nu_1|^{-1}).$$

Applying $h_{a_0,b_0,\nu_0}$ to $a_0'$ yields: $\operatorname{dist}(a_0'', a_0) = o(\nu_1^{-1}) \to 0$, hence by (3.11),

$$\operatorname{dist}(a_0'', a_1) = o(\nu_1^{-1}).$$

Applying $h^{-1}_{b_1,a_1,\nu_1}$ to $a_0''$ and using the previous formula yields $\operatorname{dist}(b_0''', a_1) \to 0$. This proves (3.13) and (3.10).

## 3.4 Convergence of projectivizations versus convergence of linear operators. The end of the proof of Theorem 2.12

We have already proved that the projectivization of the commutator (2.4) converges to that of the Stokes matrix $C_0$. Let us show that the commutator itself converges to $C_0$. This is implied by Lemma 3.11 and the following:

**3.12 Proposition** *Under the conditions of Lemma* 3.11 *consider two-dimensional linear operators whose projectivizations are the hyperbolic transformations $h_{x,y,\nu}$ from the commutator* (3.7). *Then the corresponding commutator of linear operators converges to a unipotent operator.*

**Proof** The transformation $\sigma$ is parabolic; thus, it is the projectivization of a (unique) unipotent operator (denote that operator by $C$). The convergence of projectivizations means that the commutator of the linear operators under consideration multiplied by appropriate constant converges to $C$. The commutator has unit determinant, as a commutator, and so does $C$. Therefore, the commutator converges either to $C$, or to $-C$. Let us show that it converges to $C$.

Let $a_0$, $a_1$ be the confluenting fixed points of the hyperbolic transformations. In the case where $a_0 \equiv a_1$, this statement holds by definition: the operators in the commutator have a common eigenline, hence, the corresponding eigenvalue of the commutator is equal to 1, not $-1$, so, the limit is $C$.

In the general case we can consider without loss of generality that the families of points $a_0, a_1$ meet infinitely many times while confluenting. The commutators of linear operators corresponding to the meeting places tend to $C$ by the previous statement. This proves the proposition. □

Thus, by Lemma 3.11 and the above proposition, the commutator (2.4) converges to a unipotent operator whose projectivization is the same as that of the Stokes operator $C_0$, which is also unipotent. Hence, the limit operator coincides with $C_0$. This finishes the proof of Theorem 2.12.

# 4 Nonlinear analogues and proof of Theorem 2.5

In the present section we state the nonlinear analogues of Theorem 2.5 and Corollary 2.6 for two-dimensional saddle-node holomorphic vector fields and their Martinet-Ramis moduli (subsection 4.2). We consider a two-dimensional holomorphic vector field with an elementary degenerate singular point (saddle-node). We study its generic deformation under which the degenerate singularity of the nonperturbed field splits into nondegenerate linearizable singularities of the perturbed field. The Martinet–Ramis invariant (of the orbital analytic classification) of the nonperturbed field is expressed in terms of the limit transition functions between the linearizing charts of the singularities of the perturbed field in [8]. Here we state this result only in the case of multiplicity two (see [8] for its statement for higher multiplicities). The linearizing charts determine the canonical first integrals of the perturbed field. Theorem 4.17 says that appropriate branches of the canonical first integrals of the perturbed field converge to appropriate sectorial canonical integrals of the nonperturbed field. This implies that the components of the Martinet-Ramis invariant are the limit transition functions between the canonical integrals of the perturbed field (Corollary 4.18).

The main result on saddle-nodes (Theorem 4.17) implies Corollary 4.22 saying that the "horizontal" separatrices of the perturbed field converge to the sectorial central manifolds (zeros of canonical integrals) of the nonperturbed field.

The main result on linear equations (Theorem 2.5) is related to its nonlinear analogue for saddle-nodes. Namely, the projectivization transforms the nonperturbed linear equation (1.1) to a holomorphic vector field on $\overline{\mathbb{C}} \times \{|t| < 1\}$ having two saddle-node singularities. A generic deformation of (1.1) is transformed to a generic deformation of the pair of saddle-nodes. It appears that Theorem 2.5 reformulated in terms of the projectivization follows from the previously mentioned Corollary 4.22 on the convergence of the horizontal separatrices of generically perturbed saddle-nodes.

The previously mentioned results concerning saddle-nodes are stated in subsection 4.2. Theorem 2.5 and Corollary 4.22 are proved in subsections 4.3 and 4.4, respectively.

The basic definitions (canonical first integrals and Martinet-Ramis moduli of saddle-nodes), which may be found in [11, 17], are recalled in subsection 4.1.

## 4.1 Two-dimensional saddle-node singularities and their Martinet-Ramis invariants

**4.1 Definition** We say that an isolated singular point of a holomorphic vector field is of *complex saddle-node* type, if the corresponding linearization operator has exactly one zero eigenvalue.

**4.2 Definition** Two holomorphic vector fields are said to be *orbitally analytically equivalent*, if there exists a biholomorphic diffeomorphism of the corresponding phase spaces that maps the complex phase curves of the first vector field into the phase curves of the second one. Orbital analytic equivalence of germs of holomorphic vector fields is defined similarly. The formal orbital equivalence of germs is defined analogously with a formal diffeomorphism, i.e., a two-dimensional formal power series invertible under composition. More precisely, two germs are said to be *formally orbitally equivalent*, if there exists a formal diffeomorphism transforming the first germ to the second one multiplied by a formal nonzero function, i.e., a formal power series with nonzero free term.

**4.3 Remark** Any germ of a holomorphic vector field in $(\mathbb{C}^2, 0)$ with a saddle-node singularity at the origin is orbitally analytically equivalent to the germ at the origin of a vector field of the form

$$\begin{cases} \dot{p} = p + O(|p|^2 + |t|^{k+1}), \\ \dot{t} = t^{k+1}. \end{cases} \tag{4.1}$$

**4.4 Definition** Let $S$ be a radial sector on a complex line with coordinate $t$. For any $r > 0$, we set $S^r = S \cap \{|t| < r\}$.

One can ask the following question: Is it possible to separate variables in the differential equation corresponding to the vector field (4.1) or, more precisely, is it true that the germ of (4.1) is locally orbitally analytically equivalent to the germ of a field corresponding to a differential equation with separated variables? Generally, this question has a negative answer. At the same time, the answer is positive for the formal equivalence. Namely, any saddle-node field (4.1) is formally orbitally equivalent to a unique vector field of the form

$$\begin{cases} \dot{\tilde{p}} = \tilde{p}(1 + \lambda t^k), \\ \dot{t} = t^{k+1}, \end{cases} \qquad \lambda \in \mathbb{C}. \tag{4.2}$$

The corresponding vector field (4.2) is called the *formal normal form* of (4.1) (see [11, 17]).

Generically, the normalizing power series is divergent. On the other hand, there are neighborhoods $U_p$ and $U_t$ of the origin on the axes $p$ and $t$, respectively, and a covering of the punctured neighborhood $U_t$ by $2k$ radial sectors $S_j$ (i.e., sectors with vertex at the origin), $j = 0, \dots, 2k-1$, possessing the following property: for appropriate $r > 0$ in each of the domains $\widetilde{S}_j = U_p \times S_j^r$, there is a holomorphic coordinate transformation

$$(p, t) \mapsto (\tilde{p} = H_j(p, t), t) \tag{4.3}$$

that transforms (4.1) to its normal form (4.2); furthermore, at the origin $H_j(p, t)$ possesses an asymptotic power series in $z$ and $t$ coinciding with the normalizing series (see [11, 17]).

The "nontriviality" of the transition from one normalizing chart (4.3) to another (over the intersection of the sectors of the covering) gives rise to an obstruction for orbital analytic equivalence between the vector field (4.1) and its formal normal form (4.2), and is called the *nonlinear Stokes phenomenon.* This obstruction is the nontriviality of the Martinet-Ramis invariant. We now give its definition. To this end, consider the canonical first integral

$$I(\tilde{p},t)=\tilde{p}t^{-\lambda}\exp\left\{\frac{1}{kt^k}\right\}$$

of the formal normal form (4.2). The integral $I$, together with the sectorial normalizing coordinate transformations $H_j$, induces the first integrals

$$I_j=I\circ H_j \tag{4.4}$$

of (4.1) over the sectors $S_j$ (more precisely, in the domains $\widetilde{S}_j^r$). These integrals are called the *sectorial canonical integrals.* We set $S_{2k}=S_0$, $H_{2k}=H_0$. In the definitions of all the integrals $I_j=I\circ H_j$, $j=0,\dots,2k$, we choose the branches of the (multivalued) function $I$ so that for each $j\leq 2k-1$ its branch over $S_{j+1}$ (corresponding to the index $j+1$) be the analytic extension of its branch over $S_j$ when moving counterclockwise in the $t$-plane. We introduce $2k$ transition functions $\phi_j(\tau)$, $j=0,\dots,2k-1$, comparing the canonical integrals $I_j$ and $I_{j+1}$ over components of the intersections of the corresponding sectors $S_j$ and $S_{j+1}$:

$$I_{j+1}=\phi_j\circ I_j. \tag{4.5}$$

**4.5 Remark** The system of functions $\phi_j$ in (4.5) is determined uniquely up to conjugation by multiplication by a constant, i.e., up to transformations of the form

$$\phi_j(\tau)\mapsto c\phi_j(c^{-1}\tau),\quad \text{where}\quad c\in\mathbb{C}\setminus 0 \text{ does not depend on } j. \tag{4.6}$$

The vector field (4.1) is orbitally analytically equivalent to its formal normal form (4.2) if and only if $\phi_j(\tau)\equiv\tau$ for all $j$. More generally, two germs of vector fields of the form (4.1) are orbitally analytically equivalent if and only if they have the same formal normal form and the corresponding systems of functions $\phi_j$ from (4.5) are obtained one from the other by applying successively a transformation of the form (4.6) and a cyclic shift of order $k$ of the $2k$ indices $j$ (see [11, 17]).

**4.6 Example** Consider the case of multiplicity two, i.e., when $k=1$ in (4.1), (4.2). Then the previous covering consists of the same two good sectors $S_0$ and $S_1$, as in subsection 1.2, in the case of linear equations (see Example 1.7 and Fig. 1). The previous collection $\{\phi_j\}$ consists of two functions $\phi_0$ and $\phi_1$. The function $\phi_1(\tau)$ is holomorphic on $\mathbb{C}$ and has the form $\phi_1(\tau)=\tau+c_1$, $c_1\in\mathbb{C}$. The function $\phi_0(\tau)$ is holomorphic in a neighborhood of the origin and has unit derivative at 0: $\phi_0(\tau)=\tau+o(\tau)$, as $\tau\to 0$.

**4.7 Definition** The equivalence class of a collection of functions $\phi_j$ in (4.5) under transformations (4.6) (and cyclic shifts of order $k$ of the indices, if $k>1$) is called *the Martinet-Ramis orbital analytic classification invariant* of the vector field (4.1).

## 4.2 Confluence of singular points and Martinet-Ramis invariant

We state the result on expressing the Martinet-Ramis invariant via limit transitions between linearizing charts only in the case of multiplicity two, i.e., $k = 1$ (its statement in the general case may be found in [8]). To do this, we introduce some notations and recall the theorem on linearizability of a generic nondegenerate singular point of a two-dimensional holomorphic vector field.

**4.8 Definition** A singular point of a holomorphic vector field is said to be *linearizable* if the corresponding germ of the field is orbitally analytically equivalent to its linear part.

**4.9 Definition** The *characteristic number* of a two-dimensional holomorphic vector field at its singular point is the ratio of the eigenvalues of the corresponding linearization operator.

**4.10 Theorem ([2])** *A singular point of a two-dimensional holomorphic vector field with a finite nonreal characteristic number is linearizable.*

**4.11 Definition** A singular point of a two-dimensional holomorphic vector field is said to be *typical* if in suitable coordinates the corresponding linear part has the form

$$\begin{cases} \dot{p} = \lambda p, \\ \dot{t} = \mu t, \end{cases} \qquad |\lambda| > |\mu|, \qquad \frac{\mu}{\lambda} \notin \mathbb{R}.$$

The *canonical integral* of this linear vector field is its first integral $pt^{-\lambda/\mu}$. The *canonical integral* of a two-dimensional vector field at its typical singular point is obtained from the canonical integral of the corresponding linear part by applying the linearizing coordinate transformation.

**4.12 Remark** If a singular point of a two-dimensional holomorphic vector field is typical in the sense of the preceding definition, then it is linearizable (Theorem 4.10). The corresponding canonical integral is determined uniquely up to a constant multiple.

We consider the following continuous one-parameter deformation (depending on the parameter $\varepsilon \geq 0$) of the saddle-node (4.1) (which corresponds to $\varepsilon = 0$) in the class of holomorphic vector fields:

$$\begin{cases} \dot{p} = p(1 + R(p,t,\varepsilon)) + g(t,\varepsilon)f(t,\varepsilon), \\ \dot{t} = f(t,\varepsilon), \end{cases} \qquad f(t,\varepsilon) = (t - \alpha_0(\varepsilon))(t - \alpha_1(\varepsilon)), \tag{4.7}$$

$$f(t,0) = t^2, \qquad R(0,0,0) = 0, \qquad \alpha_0 + \alpha_1 \equiv 0,$$

where $g$ and $R$ are continuous families of holomorphic functions. Assume that the degenerate singular point 0 of the nonperturbed field splits into two typical singularities $(0, \alpha_i(\varepsilon))$ of the perturbed field, $\alpha_i(\varepsilon) \neq \alpha_l(\varepsilon)$ for $i \neq l$, $\varepsilon \neq 0$. For a generic deformation (4.7) (see the next definition) we shall express the Martinet-Ramis invariant of the nonperturbed field in terms of the limit transition functions comparing the canonical integrals of the perturbed field.

**4.13 Remark** When we restrict ourselves to deformations of the type (4.7) only, we do not loose generality (see [8]).

**4.14 Definition** A vector field family (4.7) is said to be a *generic saddle-node family*, if the corresponding family of polynomials $f(t,\varepsilon)$ is generic (see Definition 2.1).

**4.15 Remark** Suppose that (4.7) is a generic saddle-node family. Then the arguments of the characteristic numbers of the singular points of the perturbed vector field are uniformly bounded away from $\pi\mathbb{Z}$ for all sufficiently small values of the parameter. In particular, for small $\varepsilon$ the singular points of the perturbed field are typical: one eigenvalue of the corresponding linearized operator tends to zero and the other eigenvalue tends to one. Thus the corresponding canonical integrals (see Definition 4.11) are well defined for small $\varepsilon \neq 0$. Conversely, if the characteristic numbers of the perturbed field in a continuous family of vector fields (4.7) satisfy the above estimate, then the families $f(t,\varepsilon)$ and (4.7) are generic.

Recall that the roots $\alpha_i(\varepsilon)$ of a generic family $f(t,\varepsilon)$ of polynomials have imaginary parts of constant sign. Without loss of generality we assume that $\operatorname{Im}\alpha_0 > 0$, then $\operatorname{Im}\alpha_1 < 0$.

A sector in the $t$-line associated to a root family $\alpha_i(\varepsilon)$ of $f(t,\varepsilon)$, $i=0,1$, is defined in the same way as in Definition 2.4.

For a typical family (4.7), we shall show that a branch of the appropriately normalized canonical integral of the perturbed field at the singular point $(0,\alpha_i(\varepsilon))$ converges to the sectorial integral of the nonperturbed field over the corresponding sector $S_i$.

**4.16 Definition** Suppose that $V$ is a domain on the Riemann sphere, $V_\varepsilon$ is a one-parameter family (depending on the parameter $\varepsilon \geq 0$) of domains on the sphere. We say that the family $V_\varepsilon$ *converges* to $V$ as $\varepsilon \to 0$, if it converges to $V$ in the Hausdorff sense, i.e., if the maximal distance from a point of the boundary $\partial V_\varepsilon$ to the boundary $\partial V$ tends to zero, and the same is true for the boundaries $\partial V$ and $\partial V_\varepsilon$ interchanged. By convergence of a family of functions holomorphic in $V_\varepsilon$ depending on the same parameter $\varepsilon$ we mean uniform convergence of these functions on compact subsets of $V$.

**4.17 Theorem** *Suppose that* (4.7) *is a generic saddle-node family of vector fields* (*see Definition* 4.14), $\alpha = \alpha_i(\varepsilon)$ *is a continuous family of $t$-coordinates of their singularities,* $S = S_i$ *is a sector associated to it* (*see Definition* 2.4). *There exist an* $r > 0$, *a neighborhood* $U_p$ *of the origin on the p-axis, and a family* $\Omega_\varepsilon$ *of simply connected domains on the t-axis that contain* $\alpha(\varepsilon)$ *and do not contain* $-\alpha(\varepsilon)$ (*this family depends on the same parameter* $\varepsilon$ *and is defined for all small values* $\varepsilon \neq 0$) *such that the following statements hold:*

(1) *The connected component containing* $\alpha(\varepsilon)$ *of* $\Omega_\varepsilon \cap (S^r \setminus [0,-\alpha(\varepsilon)])$ *converges to* $S^r$ *as* $\varepsilon \to 0$ (*see Definition* 4.16).

(2) *Let* $\Omega'_\varepsilon = \Omega_\varepsilon \setminus [\alpha_0(\varepsilon),\alpha_1(\varepsilon)]$. *The canonical integral* $I_\varepsilon$ *of the perturbed field* (4.7) *at the singular point* $(0,\alpha(\varepsilon))$ (*see Definition* 4.11) *is a multivalued holomorphic function on* $\widetilde{\Omega_\varepsilon} = U_p \times \Omega_\varepsilon$ *branched along the line* $t = \alpha(\varepsilon)$. *This function has a single-valued branch on* $\widetilde{\Omega_\varepsilon}' = U_p \times \Omega'_\varepsilon$. *This branch, when appropriately normalized* (*see Remark* 4.12), *converges to the sectorial canonical integral* (4.4) *of the nonperturbed field on* $\widetilde{S}^r = U_p \times S^r$.

This theorem is proved in [8] for saddle-nodes of arbitrary multiplicity (but for a less general class of deformations (4.7) in the case of multiplicity two). In fact, its version from [8] in the latter case is equivalent to Theorem 4.17.

**4.18 Corollary** *Let (4.7) be a generic saddle-node family of vector fields, $(0, \alpha_i(\varepsilon))$ its singularities, $i = 0, 1$, and $S_i$ the corresponding sectors (see Definition 2.4). Accordingly, suppose that $r > 0$, $U_p$, and $\Omega_\varepsilon(i)$ are the constant, the neighborhood, and the domains $\Omega_\varepsilon$ corresponding to $\alpha = \alpha_i$ from the preceding theorem, $\Omega'_\varepsilon(i) = \Omega_\varepsilon(i) \setminus [\alpha_0(\varepsilon), \alpha_1(\varepsilon)]$, $I_{i,\varepsilon}(t)$ is the canonical integral of the perturbed field at the singular point $(0, \alpha_i(\varepsilon))$ (see Definition 4.11). More precisely, we take its single-valued branch in the domain $\widetilde{\Omega_\varepsilon}'(i) = U_p \times \Omega'_\varepsilon(i)$; we set $I_{2,\varepsilon} = I_{0,\varepsilon}$. Let $C^j$ be the connected component of $S_0^r \cap S_1^r$, $j = 0, 1$ (we assume that $C^0 \supset \mathbb{R}_-$, $C^1 \supset \mathbb{R}_+$), and let $\phi_j$ be the corresponding component (4.5) of the Martinet-Ramis invariant of the non-perturbed field. There exists a family $C(\varepsilon)$ of connected components of $S_0^r \cap S_1^r \cap \Omega'_\varepsilon(0) \cap \Omega'_\varepsilon(1)$ that converges to $C^j$ as $\varepsilon \to 0$ and possesses the following property: the transition function $\Phi_\varepsilon$ between appropriately normalized integrals $I_{l,\varepsilon}$ in $\widetilde{C}(\varepsilon) = U_p \times C(\varepsilon)$, $I_{j+1,\varepsilon} = \Phi_\varepsilon \circ I_{j,\varepsilon}$, is holomorphic in a domain (depending on $\varepsilon$) that converges to the domain of $\phi_j$, and $\Phi_\varepsilon \to \phi_j$ as $\varepsilon \to 0$.*

Corollary 4.18 and its extension to higher multiplicities are contained in [8].

Now we formulate another corollary of Theorem 4.17, on convergence of appropriate separatrices of the perturbed field to the sectorial central manifolds of the saddle-node. We use this corollary further in the proof of Theorem 2.5.

**4.19 Definition** The *sectorial separatrix* of a saddle-node (4.1) over a good sector $S$ is the zero curve of the corresponding canonical sectorial integral (or equivalently, the image of the central manifold of the formal normal form under the inverse of the normalizing change of variables). The *horizontal separatrix* of a typical singular point of a two-dimensional holomorphic vector field is the zero curve of the corresponding canonical integral (see Definition 4.11).

**4.20 Remark** Let (4.1) be a saddle-node vector field, $S$ a good sector (see Definition 1.2), and $\Gamma$ the corresponding sectorial separatrix (see the preceding definition). There exists an $r > 0$ such that $\Gamma$ contains the graph

$$p = q(t) \tag{4.8}$$

of a function $q(t)$ with the following properties:

(i) $q$ is holomorphic in $S^r$ and continuous in its closure;

(ii) this is the unique function satisfying i whose graph is tangent to the field.

**4.21 Remark** Consider a two-dimensional holomorphic vector field in coordinates $(p, t)$ with a typical singularity. Let the eigenline of its linearization operator with the largest eigenvalue be parallel to the $z$-axis. Then the corresponding horizontal separatrix (see Definition 4.19) contains the graph

$$p = q(t) \tag{4.9}$$

of a holomorphic function, the graph contains the singularity. This is the unique graph of a holomorphic function tangent to the field and passing through the singularity.

**4.22 Corollary** *Let (4.1) be a generic saddle-node family. Then the horizontal separatrices at the singularities of the perturbed field converge to the sectorial separatrices of the saddle-node over the corresponding sectors (see the previous Definition). More precisely, let $b(\varepsilon) = (0, \alpha(\varepsilon))$ be a singularity family, $S$ be the sector associated to $\alpha$ (see Definition 2.4). Let $q(t)$*

*be the function whose graph* (4.8) *is contained in the sectorial separatrix over* $S$, *and* $q_\varepsilon(t)$ *the function with graph* (4.9) *contained in the horizontal separatrix of the perturbed field at* $b(\varepsilon)$. *There exist an* $r > 0$ *and a family* $\Omega_\varepsilon$ *of domains in the t-line,* $\alpha(\varepsilon) \in \Omega_\varepsilon$, $-\alpha(\varepsilon) \notin \Omega_\varepsilon$, *satisfying the following statements:*

(1) *the connected component containing* $\alpha(\varepsilon)$ *of the intersection* $(S^r \setminus [0, -\alpha(\varepsilon)]) \cap \Omega_\varepsilon$ *converges to* $S^r$, *as* $\varepsilon \to 0$ (*see Definition* 4.16);

(2) *the function* $q(t)$ *is holomorphic in* $S^r$, $q_\varepsilon$ *is holomorphic in* $\Omega_\varepsilon$, *and* $q_\varepsilon \to q$.

The generalization of the corollary to arbitrary dimension and multiplicity is stated and proved in [8].

## 4.3 Projectivization. Proof of Theorem 2.5

For the proof of Theorem 2.5 we projectivize all the linear equations involved. The projectivization of a linear equation is a tangent line field on the product $\mathbb{P}^1 \times \{|t| < 1\}$ that is the pushforward of the linear equation under the tautological projection $\mathbb{C}^2 \setminus 0 \to \mathbb{P}^1$ (or a holomorphic vector field on the latter product contained in the tangent line field).

The projectivization of a two-dimensional irregular equation (1.1) is a holomorphic vector field on $\mathbb{P}^1 \times \{|t| < 1\}$ having a pair of singularities on the fiber $\mathbb{P}^1 \times 0$ (which correspond to the eigenlines of the matrix $A(0)$, the coordinate lines in our case). These singularities are saddle-nodes of the same order $k$ as the Poincaré rank of the equation under consideration ($k = 1$). The projectivization transforms the graphs of the canonical sectorial solutions of (1.1) to the sectorial separatrices of the corresponding saddle-node singularities of the projectivization (and hence, the solutions themselves to the corresponding functions (4.8)). Indeed, the images of the canonical solutions under the tautological projection are functions holomorphic in the corresponding sectors and continuous in their closures (by construction), and their graphs are tangent to the projectivization. By uniqueness (see Remark (4.20)), they coincide with those defining the corresponding sectorial separatrices.

The projectivization of a perturbed equation from a generic family (2.1) is a holomorphic vector field on the same space $\mathbb{P}^1 \times \{|t| < 1\}$ with four typical singularities: a pair of singularities in each fiber $\mathbb{P}^1 \times \alpha_i(\varepsilon)$, $i = 0, 1$. Analogously, the projectivization transforms the graphs of the monodromy eigenfunctions at the singularities of the perturbed linear equation to the horizontal separatrices of the projectivization.

The projectivization of a generic family of linear equations becomes a generic saddle-node family (locally near each saddle-node singularity of the projectivization of the nonperturbed equation) after applying an appropriate family of changes of the space variable. Now the preceding corollary applied to the family of projectivizations says that the horizontal separatrices converge to the sectorial separatrices of the projectivized nonperturbed equation. This means that the branches in $S_i^r$ of the monodromy eigenfunctions (taken up to multiplication by constants) converge to the canonical basic solutions of the nonperturbed equation (also taken up to multiplication by constants). Therefore, appropriately normalized monodromy eigenfunctions converge to appropriately normalized canonical basic solutions. This proves Theorem 2.5 modulo Corollary 4.22.

## 4.4 Convergence of the horizontal separatrices. A brief proof of Corollary 4.22

We give a brief proof of Corollary 4.22 independent on Theorem 4.17 (the complete text of the proof may be found in [8]).

Let us prove the statements of Corollary 4.22, say, for

$$\alpha = \alpha_0, \ S = S_0 : \quad \text{let us show that} \quad q_\varepsilon \to q.$$

To do this, we show that the functions $q_\varepsilon$ are holomorphic in domains $\Omega_\varepsilon$ large enough (satisfying statement (1) of Corollary 4.22) and form a normal family (i.e., precompact in the topology of uniform convergence on compact sets in $S^r$): more precisely,

$$|q'_\varepsilon(t)| < 1, \ |q_\varepsilon(t)| \leq |t - \alpha_0(\varepsilon)| \quad \text{for any} \quad t \in \Omega_\varepsilon. \tag{4.10}$$

(Recall that by definition, $q_\varepsilon(\alpha_0(\varepsilon)) = 0$.) Then the limit of any convergent sequence $q_{\varepsilon_n}$, $\varepsilon_n \to 0$, is a function holomorphic in the sector $S^r$ and continuous in its closure that vanishes at 0. Therefore, by the uniqueness statement of Remark 4.20, the limit coincides with $q$. This together with normality proves the convergence $q_\varepsilon \to q$.

For the proof of the bounds (4.10) we consider the following family $K$ of tangent cones at the points of the phase plane and the corresponding cone $K'$:

$$K = \{|\dot{p}| < |\dot{t}|\}, \qquad K' = \{|p| < |t - \alpha_0(\varepsilon)|\}.$$

The inequalities (4.10) are equivalent to the inclusions

$$T\Gamma q_\varepsilon \subset K, \qquad \Gamma q_\varepsilon \subset K', \tag{4.11}$$

where $\Gamma q_\varepsilon$ is the graph of the function $q_\varepsilon$.

For the proof of the inclusions (4.11), we consider an appropriate constant multiple

$$v_\theta(\varepsilon) = e^{i\theta}(4.7), \quad \theta \in \mathbb{R} \text{ is independent on } \varepsilon,$$

of the vector field family (4.7). We choose the number $\theta$ so that the singular point $b_0(\varepsilon) = (0, \alpha_0(\varepsilon))$ of the perturbed field $v_\theta(\varepsilon)$ from the new family is hyperbolic with the stable manifold $W^s = \{t = \alpha_0(\varepsilon)\}$ and the unstable manifold being locally the horizontal separatrix $W^u = \Gamma q_\varepsilon$. More precisely,

(1) the eigenvalue of the linearization operator of $v_\theta(\varepsilon)$ at $b_0(\varepsilon)$ at the eigenline tangent to the line $\{t = \alpha_0(\varepsilon)\}$ has a negative real part;

(2) the other eigenvalue has a positive real part;

(3) the previous conditions hold "uniformly": the real part of the former eigenvalue is bounded away from zero; the argument of the latter eigenvalue is bounded away from $\pi/2 + \pi\mathbb{Z}$.

The above conditions will be satisfied if, e.g., $\theta < -\pi/2$ and $\theta$ is close enough to $-\pi/2$.

In the proof of (4.11) we use the fact that for any $\theta$ satisfying (1)–(3)) there exists a bidisc $U$ in the phase space (independent of $\varepsilon$) such that for any $\varepsilon$ small enough the tangent cone field $K$ is invariant under the real flow of the perturbed field $v_\theta(\varepsilon)$: each cone of $K$ is mapped under a positive time flow map strictly inside the cone of $K$ at the image of the point under consideration. This implies that the cone $K'$ is also $v_\theta(\varepsilon)$-invariant.

The inclusions (4.11) hold a priori at the singular point $b_0(\varepsilon)$, and hence in its neighborhood (whose size depends on the parameter). By invariance of $K$, they remain valid in all the trajectories of the field $v_\theta(\varepsilon)$ in the unstable manifold $\Gamma q_\varepsilon$ that go out from the singular point. These trajectories saturate a domain in $\Gamma q_\varepsilon$ bijectively projected onto some domain in the $t$-line (denoted by $\Omega_\varepsilon$). If the bidisc $U$ is chosen in an appropriate way (say, centered at 0 and so that its height in the coordinate $p$ is at least two times greater than its width in the coordinate $t$; denote by $V$ its projection to the $t$-line), then the previous domain $\Omega_\varepsilon$ is saturated by the real trajectories of the quadratic vector field

$$\dot{t} = e^{i\theta}(t - \alpha_0(\varepsilon))(t - \alpha_1(\varepsilon))$$

in the disc $V$ that go out from its repelling singular point $\alpha_0(\varepsilon)$ (see Fig. 8(a)). The family of the domains $\Omega_\varepsilon$ thus constructed satisfies statement (1) of Corollary 4.22, at least for some sector $S$ associated to $\alpha_0$. (In fact one can achieve this for an arbitrary given sector $S$ associated to $\alpha_0$ by appropriate choice of $\theta$.) This proves Corollary 4.22.

In fact, the domain $\Omega_\varepsilon$ converges to a domain (denoted $\Omega$, see Fig. 8(b)) bounded by a cardioid-like curve having a "cusp" at 0 with tangency to the ray $\arg t = \pi - \theta$. Recall that the closure of the sector $S_0$ is disjoint from $i\mathbb{R}_-$. One can achieve that the latter cusp ray be arbitrarily close to $i\mathbb{R}_-$ (so that the limit domain $\Omega$ contains the sector $S_0^r$ for appropriate $r > 0$) by choosing a $\theta < -\pi/2$ close enough to $-\pi/2$.

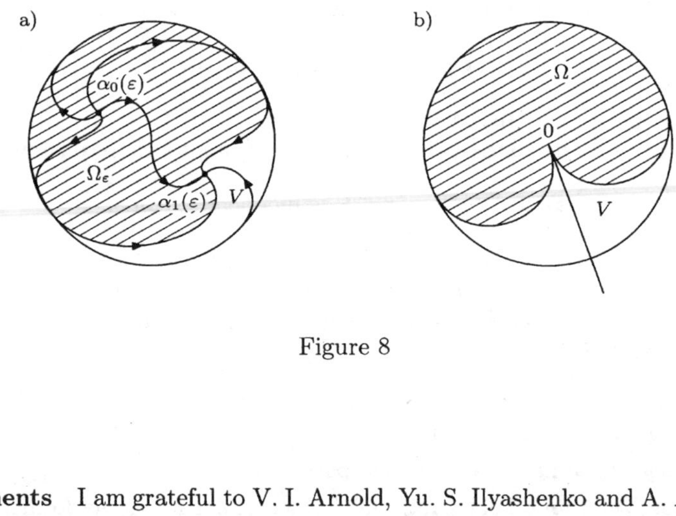


Figure 8

**Acknowlegments** I am grateful to V. I. Arnold, Yu. S. Ilyashenko and A. A. Bolibruch for stating the problems. I am grateful to them and to J.-P. Ramis for helpful discussions. The main result presented in these notes (Theorem 2.5) was obtained during my stay at Labòratoire Émile Picard, Université Paul Sabatier (Toulouse, France). The corresponding papers were written in parts during my visits to the Institute for Advanced Study (IAS, Princeton, USA), the Max Planck Institut für Mathematik (MPIM, Bonn, Germany), Independent University of Moscow, and the Institut des Hautes Études Scientifiques (IHÉS, Bures-sur-Yvette, France). My stay at the last three institutes was supported by the European Post-Doctoral

Institute joint Fellowship of the Max Planck Institut für Mathematik and the Institut des Hautes Études Scientifiques. I wish to thank all these institutions for hospitality and support. During the preparation of the present text for publication I used the notes taken by A. Fishkin and V. Kleptsyn. I wish to thank them for writing the notes and helpful remarks on the text.

The research represented in the notes was supported in part by a grant of the Pro-Mathematica Foundation of the French Mathematical Society, by INTAS grant 93-0570-ext, by CRDF grants RM1-229, RM1-2086, RM1-2358-MO-02, by a fellowship "Séjour Scientifique" of the French Ministry of Foreign Affairs, by a State Fellowship for Young Scientists of the Russian Academy of Sciences, and by Russian Foundation for Basic Research (RFFI) grants 98-01-00455, 02-02-00482.

## References

[1] B. N. Apanasov, N. A. Gusevskii, and S. L. Krushkal, *Kleinian Groups and Uniformization in Examples and Problems*, Transl. Math. Monographs **62**, Amer. Math. Soc., Providence, RI, 1986.

[2] V. I. Arnold and Yu. S. Ilyashenko, Ordinary differential equations, in: *Dynamical Systems–1*, Itogi Nauki i Tekhniki, VINITI, Moscow, 1985; English transl. in: *Dynamical Systems I*, Encyclopaedia Math. Sci., Springer, Berlin, 1988, 1–148.

[3] W. Balser, W. B. Jurkat, and D. A. Lutz, Birkhoff invariants and Stokes' multipliers for meromorphic linear differential equations, *J. Math. Anal. Appl.* **71** (1) (1979), 48–94.

[4] A. Duval, Biconfluence et groupe de Galois, *J. Fac. Sci. Univ. Tokyo Sect. IA Math.* **38** (2) (1991), 211–223.

[5] J. Écalle, Invariants holomorphes simples des transformations de multiplicateur 1, *C. R. Acad. Sci. Paris* Sér. *A–B*, **276** (1973), A375–A378.

[6] R. Garnier, Sur les singularités irregulières des équations différentielles linéaires, *J. Math. Pures Appl. (8)* **2** (1919), 99–198.

[7] A. Glutsyuk, Stokes operators via limit monodromy of generic perturbation, *J. Dynam. Control Systems* **5** (1) (1999), 101–135.

[8] A. Glutsyuk, Confluence of singular points and the nonlinear Stokes phenomena, *Trudy Moskov. Mat. Obshch.* **62** (2001), 54–104 (Russian); English transl. *Trans. Moscow Math. Soc.* 2001, 49–95.

[9] A. Glutsyuk, On the monodromy group of confluenting linear equations, to appear.

[10] A. Glutsyuk, Resonant confluence of singular points and Stokes phenomena, to appear.

[11] Yu. S. Ilyashenko, ed., *Nonlinear Stokes Phenomena*, Adv. Soviet Math. **14**, Amer. Math. Soc., Providence, RI, 1993.

[12] Yu. S. Ilyashenko, and A. G. Khovanskii, Galois groups, Stokes operators and a theorem of Ramis, *Funktsional. Anal. i Prilozhen.* **24** (4) (1990), 31–42, 96 (Russian); English transl. *Funct. Anal. Appl.* **24** (4) (1990), 286–296 (1991).

[13] W. B. Jurkat, D. A. Lutz, and A. Peyerimhoff, Birkhoff invariants and effective calculations for meromorphic linear differential equations, *J. Math. Anal. Appl.* **53** (2) (1976), 438–470.

[14] W. B. Jurkat, D. A. Lutz, and A. Peyerimhoff, Birkhoff invariants and effective calculations for meromorphic linear differential equations, *Houston J. Math.* **2** (2) (1976), 207–238.

[15] P. Mardesic, R. Roussarie, and C. Rousseau, Modulus of analytic classification for unfoldings of generic parabolic diffeomorphisms, Prépublication No 292 (2002), Laboratoire de Topologie et Laboratoire Gevrey de Mathématique Physique, Université de Bourgogne.

[16] J. Martinet, Remarques sur la bifurcation nœud-col dans le domaine complexe, *Astérisque* **150–151** (1987), 131–149.

[17] J. Martinet and J.-P. Ramis, Problèmes de modules pour des équations différentielles non linéaires du premier ordre, *Inst. Hautes Études Sci. Publ. Math.* **55** (1982), 63–164.

[18] J.-P. Ramis, Phénomène de Stokes et filtration Gevrey sur le groupe de Picard-Vessiot, *C. R. Acad. Sci. Paris Sér. I Math.* **301** (5) (1985), 165–167.

[19] J.-P. Ramis, Confluence et resurgence, *J. Fac. Sci. Univ. Tokyo Sect. IA Math.* **36** (3) (1989), 703–716.

[20] Y. Sibuya, Stokes phenomena, *Bull. Amer. Math. Soc.* **83** (1977), 1075–1077.

[21] S. M. Voronin, Analytic classification of germs of conformal maps $(\mathbb{C}, 0) \to (\mathbb{C}, 0)$ with identity linear part, *Funktsional. Anal. i Prilozhen.* **15** (1) (1981), 1–17; English transl. *Funct. Anal. Appl.* **15** (1) (1981), 1–13.

[22] S. M. Voronin and Yu. I. Meshcheryakova, Analytic classification of generic degenerate elementary singular points of germs of holomorphic vector fields on the complex plane (in Russian), *Izv. Vyssh. Uchebn. Zaved. Mat.* (1) (2002), 13–16.

[23] C. Zhang, Quelques études en théorie des équations fonctionnelles et en analyse combinatoire, Thèse, Institut de Recherche Mathématique Avancée, Univ. Louis Pasteur et CNRS (URA 01), 1994.

# Bifurcations of relaxation oscillations

John GUCKENHEIMER*

*Mathematics Department*
*Cornell University*
*Ithaca, NY 14853*
*USA*

**Abstract**

We are still far from a comprehensive theory of bifurcation in dynamical systems with multiple time scales. However, systematic application of geometric methods and study of examples have produced descriptions of varied phenomena. These lectures present a selective summary of some of what has been discovered, concentrating on periodic orbits called relaxation oscillations. We focus upon key aspects of the phenomena, avoiding mathematical details and making the analysis as simple as possible. The dynamics of reduced systems plays a central role in our discussion.

## 1 Introduction

Relaxation oscillations are periodic orbits that occur in dynamical systems with multiple time scales. They are characterized as periodic orbits in which there are fast and slow segments along the periodic orbit. We formulate these concepts in the context of *slow-fast* vector fields that take the form

$$\begin{aligned} \varepsilon\dot{x} &= f(x,y) \\ \dot{y} &= g(x,y) \end{aligned} \tag{1.1}$$

or

$$\begin{aligned} x' &= f(x,y) \\ y' &= \varepsilon g(x,y) \end{aligned} \tag{1.2}$$

with $x \in \mathbb{R}^m$, $y \in \mathbb{R}^n$, $f : \mathbb{R}^m \times \mathbb{R}^n \to \mathbb{R}^m$, and $g : \mathbb{R}^m \times \mathbb{R}^n \to \mathbb{R}^n$. For simplicity, we shall assume throughout these lectures that $f$ and $g$ are $C^\infty$. Here the parameter $\varepsilon \geq 0$ is the ratio of time scales and is assumed to be small. Throughout the paper, we consider families and their solutions that vary continuously as $\varepsilon \to 0$. The limit $\varepsilon = 0$ is singular and looks different for the two systems (1.1) and (1.2). For system (1.1), the limit $\varepsilon = 0$ is a *differential algebraic equation (DAE)* with trajectories evolving on the slow time scale. The system (1.2) emphasizes the fast time scale. The limit $\varepsilon = 0$ is a family of differential equations (called the *layer equations* [29]). The layer equations express the motion of the fast variable $x$ with the variable $y$ acting as a parameter. For fixed $y$, we say that the equation $x' = f(x,y)$ is a *fast subsystem.*

*This research was partially supported by the National Science Foundation and the Department of Energy. Conversations with Warren Weckesser and Katherine Bold contributed to the results about maximal canards.



*Y. Ilyashenko and C. Rousseau (eds.),*
*Normal Forms, Bifurcations and Finiteness Problems in Differential Equations,* 295–316.

For (1.1) or (1.2), solutions of the equation $f = 0$ comprise the *critical manifold* of the system. By definition, solutions of the limit DAE of (1.1) lie on the critical manifold. A family of solutions of (1.1) that approach a solution of the DAE as $\varepsilon \to 0$ are called *slow* trajectories and the flow of the DAE is called the *slow flow.* Trajectories of (1.1) that do not lie in a region of phase space where $f$ is $O(1)$ are *fast.* Continuous curves that are a concatenation of slow trajectories and trajectories of the layer equations are called *degenerate phase curves.* The points where a slow segment ends and a fast segment begins is called a *break* point, and the points where a fast segment ends and a slow segment begins is called an *entry* point. *Relaxation oscillations* are periodic orbits whose limits as $\varepsilon \to 0$ are degenerate phase curves with both fast and slow segments.

The term relaxation oscillation was introduced by van der Pol [31] who studied their properties in the vector field that now bears his name:

$$\begin{aligned} \varepsilon\dot{x} &= y - x + \frac{x^3}{3} \\ \dot{y} &= -x. \end{aligned} \tag{1.3}$$

Fig. 1 shows the limit cycle of (1.3) with $\varepsilon = 10^{-4}$ together with its critical manifold and a pair of solutions of the layer equation that begin at the fold points of the critical manifold.

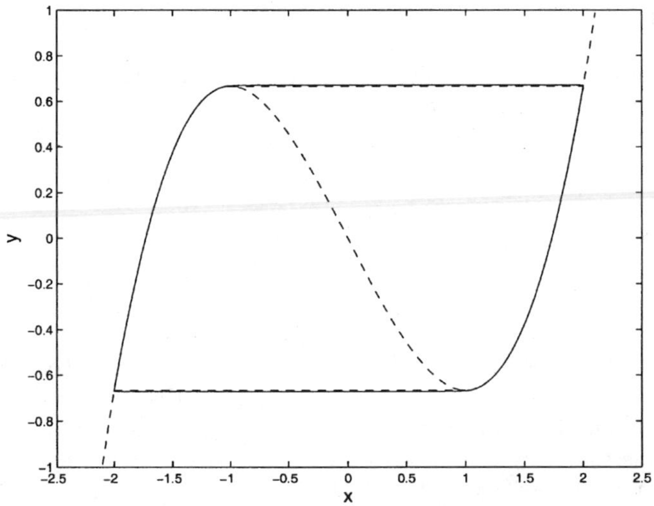


Figure 1: The solid curve is the periodic orbit of the van der Pol equation (1.3) with $\varepsilon = 0.0001$. The dashed lines are the critical manifold and segments of fast jumps from the folds of the critical manifold. The upper jump segment is difficult to resolve from the periodic orbit at this scale.

We have used the forced van der Pol system

$$\begin{aligned} \varepsilon\dot{x} &= y + x - \frac{x^3}{3} \\ \dot{y} &= -x + a\sin(2\pi\theta) \\ \dot{\theta} &= \omega \end{aligned} \tag{1.4}$$

as a case study to investigate the bifurcations of relaxation oscillations [15, 16, 17]. Although this example does not contain all of the different types of generic bifurcations relaxation oscillations can undergo, it has led to a better understanding of how local degeneracies in the slow-fast decomposition of these periodic orbits play a role in the global bifurcation structure. Since the phenomenon of chaos for dissipative dynamical systems was first observed in the forced van der Pol system, we have paid particular attention to the creation and destruction of chaotic invariant sets. Some of this work is summarized in Section 4 to motivate the discussion of maximal canards that follows in Section 5. Finally, in Section 6 we give a simpler presentation of the "classical" results about canards in another extension of the van der Pol equation.

Our goal is to understand bifurcations of relaxation oscillations as additional parameters (i.e., parameters different from $\varepsilon$) are varied in a slow-fast system. Bifurcations of periodic orbits in generic one-parameter systems with a single time scale have been classified into a small number of types. Specifically, Hopf bifurcations occur as a periodic orbit collapses onto an equilibrium point. Saddle-node, period-doubling (also known as flip) and secondary Hopf (also known as Neimark-Sacker) bifurcations occur where the periodic orbit has neutrally stable eigenvalues and changes stability [18]. Finally, there are homoclinic and saddle-node in cycle bifurcations at which the period of the orbits become infinite. The bifurcations of relaxation oscillations are more varied and harder to analyze numerically than those of periodic orbits in systems with a single time scale. Indeed, our understanding is incomplete, even at the conceptual level of drawing plausible scenarios of how bifurcation occurs for relaxation oscillations. This is especially true for bifurcations associated with some types of qualitative change in the decomposition of orbits into slow and fast segments. These lectures survey what we know and what we conjecture. We emphasize examples and discuss recently discovered phenomena. The material about maximal canards is new, appearing here for the first time. Our presentation does not attempt to be fully rigorous or as general as possible. Instead we try to illuminate the basic principles underlying the bifurcations of relaxation oscillations in their simplest manifestations.

## 2 Normal hyperbolicity and slow manifolds

We begin our discussion with a review of basic properties of slow-fast dynamical systems (1.1). Formally, we say that $\gamma_\varepsilon : \mathbb{R} \times \mathbb{R}^+ \to \mathbb{R}^m \times \mathbb{R}^n$ is a trajectory or solution of (1.1) if

- $\gamma$ is continuous in $\varepsilon$ for $\varepsilon \geq 0$ and smooth in $t$ when $\varepsilon > 0$.
- When $\varepsilon > 0$, $\gamma_\varepsilon(t)$ solves (1.1).
- When $\varepsilon = 0$, $\gamma_0(t)$ is the concatenation of solutions of the DAE (1.1) and the layer equations (1.2).

Arnold et al. [2] say that a curve that satisfies the last statement is a "phase curve of the degenerate system." We emphasize the slow time scale in our treatment of relaxation oscillations, defining below a reduced system that collapses the fast segments to discrete time jumps.

Trajectories of slow-fast systems are often observed to flow along sheets of a critical manifold that are attractors of the layer equations. Jumps along fast segments occur where the trajectory encounters a *singularity* of the projection of the critical manifold onto the space of slow variables $y$. This prompts a closer examination of the critical manifolds of (1.1) and the singularities of this projection. Sard's theorem [14] states that the regular values of the map $f : \mathbb{R}^m \times \mathbb{R}^n \to \mathbb{R}^m$ are a set of full measure if $f$ is $C^n$. If 0 is a regular value, then the implicit function theorem implies that the critical manifold $C$, defined by $f = 0$, is indeed an $n$-dimensional manifold. We will assume throughout this paper that this is the case, namely that 0 is a regular value of $f$ and the solutions of $f(x, y) = 0$ constitute a smooth manifold.

The layer equations (1.2) have equilibrium points along the critical manifold. The critical manifold is seldom invariant for the flow of (1.1) when $\varepsilon > 0$, but there are general circumstances in which there is a nearby invariant manifold. The following example illustrates this principle.

### 2.1 Example

$$
\begin{aligned}
\varepsilon \dot{x} &= y - x \\
\dot{y} &= 1.
\end{aligned}
\tag{2.1}
$$

The solutions of this equation are given by

$$
\begin{pmatrix} x(t) \\ y(t) \end{pmatrix} = \begin{pmatrix} y(0) + t - \varepsilon + (x(0) - y(0) + \varepsilon) \exp(-t/\varepsilon) \\ y(0) + t \end{pmatrix}
$$

as can be verified by differentiating these expressions with respect to $t$ and observing that $y(t) - x(t) = \varepsilon - (x(0) - y(0) + \varepsilon) \exp(-t/\varepsilon)$. Solving the equation $y(t) = y(0) + t$ for $t$ and substituting $t = y(t) - y(0)$ into the expression for $x(t)$, we find that

$$
x(t) = y(t) - \varepsilon + (x(0) - y(0) + \varepsilon) \exp(-(y(t) - y(0)/\varepsilon)
$$

on the trajectory with initial point $(x(0), y(0))$. We see that if $x(0) - y(0) + \varepsilon = 0$, then $x(t) = y(t) - \varepsilon$ on the entire solution. Thus the line $y = x + \varepsilon$ is an *invariant manifold* for the vector field (2.1). Moreover, the other solutions of the equation converge to this invariant manifold at the rate $(x(0) - y(0) + \varepsilon) \exp(-t/\varepsilon)$. Note that the invariant manifold is not the solution of the algebraic equation $y - x = 0$ obtained by setting $\varepsilon = 0$ in (2.1), but the distance from the invariant manifold to the line $y = x$ is $O(\varepsilon)$.

The theory of *normal hyperbolicity* [20] has been used by Fenichel [13] to establish the existence of invariant slow manifolds. The fundamental assumption is that the points of the critical manifold are hyperbolic equilibria of their fast subsystems. A technical complication in this theory is that the invariant manifolds have only a finite degree of smoothness, as illustrated by the following example.

### 2.2 Example

$$
\begin{aligned}
\dot{x} &= -r(x + y^r) \\
\dot{y} &= -y.
\end{aligned}
\tag{2.2}
$$

The vector field (2.2) is not written explicitly as a slow-fast system, but observe that if the integer $r$ is large, then $1/r$ plays the role of $\varepsilon$ in a slow-fast system. The solution of the second equation is $y(t) = y(0)\exp(-t)$. Substituting this into the first equation gives $\dot{x} = -r(x + y^r(0)\exp(-rt))$ which has the solution $x = x(0)\exp(-rt) - ry^r(0)t\exp(-rt)$. Eliminating $t$ from the solutions yields

$$x = x(0)(y/y(0))^r + ry^r \ln(y/y(0))$$

when $y(0) \neq 0$, and $y = 0$ when $y(0) = 0$. The $x$-axis is a smooth, invariant manifold at the origin along which the motion $x = x(0)\exp(-rt)$ is fast. However, slow invariant manifolds at the origin are constructed by piecing together a pair of curves $x = x(0)(y/y(0))^r + ry^r \ln(y/y(0))$ for two choices of $y(0)$ of opposite sign. The $r$th derivative of the curves $x = x(0)(y/y(0))^r + ry^r \ln(y/y(0))$ are unbounded as they approach the origin. Thus, the slow manifolds of the system are not infinitely differentiable. Their degree of differentiability increases with $r$. This illustrates an aspect of normal hyperbolicity: the existence and smoothness of slow manifolds depends upon exponential rates of contraction or expansion normal to the manifold and exponential rates of contraction or expansion along the manifold. Note also that there are many slow manifolds in this system. Indeed, any pair of trajectories on opposite sides of the origin together with the origin give a slow manifold.

These two examples highlight properties of slow manifolds in slow-fast systems that make them elusive. Specifically:

(1) Slow manifolds are displaced from the critical manifold by a distance $O(\varepsilon)$.

(2) Slow manifolds are not unique.

(3) Slow manifolds have only a finite degree of smoothness.

The existence of invariant slow manifolds was first proved for attracting manifolds in classical work of Tikhonov [26] and for hyperbolic manifolds by Fenichel [13]:

**2.3 Theorem** *Let $\bar{C}_0$ be a normally hyperbolic, compact manifold with boundary of dimension $n$ contained in the critical manifold of* (1.1)*. There is a continuous family $\bar{C}_\varepsilon$ of $C_r$ invariant manifolds with boundary for* (1.1)*, defined for $\varepsilon \geq 0$ sufficiently small.*

Here invariance of $\bar{C}_\varepsilon$ means that trajectories intersecting $\bar{C}_\varepsilon$ enter or leave $\bar{C}_\varepsilon$ only through its boundary. We remark further that the flow on $\bar{C}_\varepsilon$ is slow and is approximated by the DAE of (1.1).

Associated with the normal hyperbolicity are normalizing "Fenichel" coordinates that make the flow in a neighborhood of the slow manifold preserve a bundle structure. Jones and Kopell introduced one set of normalizing coordinates in their work on the *Exchange Lemma* [22]; a stronger version of these normalizing coordinates was given by Ilyashenko [21]:

**2.4 Theorem** *Let $w$ be a normally hyperbolic point of the critical manifold of* (1.1) *at which the slow vector field defined by $g$ is non-zero. For any $r > 0$ and $\varepsilon > 0$ sufficiently small, there is a neighborhood of $w$ and a $C^r$ change of coordinates that transforms* (1.1) *into a system of the form*

$$\begin{aligned} \varepsilon\dot{x} &= a(y,\varepsilon)x \\ \dot{y} &= 1. \end{aligned} \tag{2.3}$$

The DAE of (1.1) can be reduced to a system of ordinary differential equations on normally hyperbolic critical manifolds. The Fenichel Theorem implies that this system of ODE's is a good approximation for the slow flow along invariant manifolds of (1.1). Thus we take the ODE's on the critical manifold as the backbone for defining a *reduced system* of (1.1) that describes its limit properties as $\varepsilon \to 0$. However, trajectories may depart from the critical manifold at points where the projection of the slow flow onto the space of slow variables $y$ is singular. How this happens is discussed in the next section.

## 3 Singularities, folds and jumps

Denote by $\pi : \mathbb{R}^m \times \mathbb{R}^n \to \mathbb{R}^n$ the projection $\pi(x, y) = y$. Normal hyperbolicity of the critical manifold $C$ fails at singular points of $\pi|C$, the restriction of the projection to $C$. We shall denote the set of these singular points by $C_s$. The equation $\det(D_x f) = 0$ is a *defining equation* for $C_s$. Observe that the DAE of (1.1) may fail to have solutions at points of $C_s$ as illustrated by the following example.

### 3.1 Example

$$\begin{aligned} \varepsilon\dot{x} &= y - x^2 \\ \dot{y} &= -1. \end{aligned} \tag{3.1}$$

When $\varepsilon = 0$, the origin is a minimum of $y$ along the curve $y - x^2 = 0$, but the differential equation says that $\dot{y} = -1$. These two conditions are clearly inconsistent since $\dot{y}$ must be zero to remain on the curve $y - x^2 = 0$. Thus the DAE does not always give an immediate representation of limiting solutions to (1.1) as $\varepsilon \to 0$. We can ask under which circumstances there is a unique way to continue solutions of the DAE at singularities of its critical manifold so as to give the limit behavior of solutions of the slow-fast system (1.1). To deal systematically with this issue, we turn to singularity theory and bifurcation theory.

Singularity theory [14] gives a set of tools for classifying the singularities of smooth maps up to smooth coordinate changes. Arnold et al. [2] discuss application of singularity theory to the critical manifolds of generic slow-fast systems having two slow variables. Generic maps $h : \mathbb{R}^2 \to \mathbb{R}^2$ have only two types of singularities: *folds* and *cusps*. A similar result holds for the critical manifolds of generic slow-fast systems with two slow variables. Folds are characterized by the condition that $D_x f$ has corank one and an inequality on second derivatives of $f$: if $w$ and $v$ are the left and right eigenvectors of $D_x f$, then $w D_{xx} f(v, v) \neq 0$. At folds, there is a coordinate change of the form $(u, v) = (h(x, y), k(y))$ that maps the critical manifold to the set of solutions of $u_1 - v_1^2 = 0$; $u_i = 0$, $2 \leq i \leq m$, but these coordinate changes do not preserve the flow of (1.1). Even in dynamical systems with a single time scale, questions concerning equivalence of vector fields by smooth change of coordinates at equilibrium points are intricate, involving resonances [18] and small divisors [1]. In the case of slow-fast systems, every point of the critical manifold is an equilibrium point of the layer equations, and these questions become still more complicated. There has been little progress thus far in classifying slow-fast systems at points of $C_s$ up to smooth equivalence.

Bifurcation theory for single time scale systems can be applied to the fast dynamics of (1.2). At fold points which are saddle-node equilibria of their fast subsystem with no eigenvalues having a positive real part, there is a single unstable separatrix leaving the saddle-node point [18]. If this separatrix approaches another equilibrium of the fast subsystem as its $\omega$-limit, then the fast time motion of the trajectory starting at the fold point tends to the unstable separatrix as $\varepsilon \to 0$. In the limit, we define a discrete (slow) time *jump* for the reduced system that maps the fold point to the $\omega$-limit of its unstable separatrix.

More formally, we define the reduced system of a slow-fast system (1.1) in the context in which the limit sets of the fast subsystems are all equilibria. If this assumption holds, we say that the slow-fast system has *no rapid oscillations.* We define the *reduced system* of a slow-fast system with no rapid oscillations on its critical manifold by allowing a point $w$ to evolve in one of two ways: it may follow the slow flow of the DAE, or if there is a fast trajectory with $\alpha$-limit $w$ and $\omega$-limit $z$, then it may jump in discrete (slow) time from $w$ to $z$. This definition allows the system to be multi-valued, so the uniqueness property for the evolution of a dynamical system breaks down. For example, at regular points of an unstable sheet of the critical manifold, there is the option of evolving along the slow flow or jumping along a trajectory of the unstable manifold in the fast subsystem. Trajectories of this reduced system have been called *candidates* by the "Strasbourg School" [5]. A comprehensive treatment of reduced systems requires reconsideration of basic properties of dynamical systems, something we do not pursue here. Nonetheless, this definition of the reduced system is a very useful tool for the study of bifurcations of relaxation oscillations.

There is an important situation in which the reduced system is single valued, namely when slow trajectories flow along stable sheets of the critical manifold, reach its boundary transversally at saddle-node points of the fast subsystem and then jump to a new stable sheet of the critical manifold. Nearby trajectories also have these properties. The smoothness of the jump maps has been studied by Levinson [23], Pontryagin [27] and Szmolyan and Wechselberger [30]. The fundamental result is that, in the circumstances described above, phase curves of the degenerate system are the limits of trajectories of the slow-fast system (1.1) as $\varepsilon \to 0$. If these hypotheses are met, then there are asymptotic expansions that describe the $\varepsilon$ dependence of how the slow-fast trajectories approach the degenerate phase curve. As will be evident below, the trajectories do not depend smoothly on $\varepsilon$, complicating the analysis of the jump phenomenon. We incorporate these concepts into the following definition, adding additional requirements on the entry points of trajectories at a jump.

A relaxation oscillation is *simple* if the slow-fast decomposition of its degenerate phase curve $\gamma_0$ satisfies the following properties:

- The slow segments of $\gamma_0$ lie in the closure of the normally hyperbolic, stable sheets of the critical manifold $C$.
- The terminal points of the slow segments lie on the boundary of the normally hyperbolic, stable sheets of the critical manifold $C$ and approach the boundary $C_s$ of $C$ transversally.
- The terminal points of the slow segments are saddle-node points of their fast subsystems.
- The initial points of the slow segments are in the interior of the normally hyperbolic, stable sheets of the critical manifold $C$.
- If the image $J$ of a jump map defined on $C_s$ contains an initial point $w$, then the slow vector field at $w$ is transverse to $J$ in $C_s$.

Simple relaxation oscillations that persist remain simple when the slow-fast systems (1.1) is perturbed. They may encounter bifurcations similar to those present in systems with a single time scale, but there are other possibilities associated with the breakdown of the properties stated above.

A key part of the strategy that we advocate to explore bifurcations of relaxation oscillations is to examine how generic one parameter families of relaxation oscillations cross the boundary of the region of simple relaxation oscillations. This can occur through several different mechanisms, so a goal becomes to investigate these.

In generic slow-fast systems (1.1) with two slow variables ($n = 2$) and one fast variable ($m = 1$), $C_s$ is a curve. The flow of a slow-fast system near generic break points can be transformed by an $\varepsilon$ dependent change of coordinates to a *system of first approximation* [2] of the form

$$\begin{aligned} \dot{x} &= y + x^2 + O(\mu) \\ \dot{y} &= 1 + O(\mu) \\ \dot{z} &= x + O(\mu) \end{aligned} \tag{3.2}$$

with

$$x = O(\mu), \qquad y = O(\mu^2), \qquad z = O(\mu^3), \qquad t = O(\mu^2), \qquad \varepsilon = \mu^3.$$

The limit as $\varepsilon \to 0$ of (3.2) clearly converges. The limit system is invariant under translation along the $z$-axis and the motion in the $(x, y)$ plane is independent of $z$. In the $(x, y)$ plane there is a unique trajectory that remains a finite distance from the half parabola $y + x^2$, $x < 0$. This trajectory divides a region of trajectories whose $x$ coordinates tend to $-\infty$ in finite decreasing time from a region which tends to the half parabola $y + x^2$, $x > 0$ with decreasing time. The dividing trajectories are the rescaled limit of the trajectories that flow along the stable slow manifold. Their $x$-coordinates tend to $\infty$ in finite time, with a limit value of $y > 0$. Since $y$ is scaled by $\varepsilon^{2/3}$ in the system of first approximation, the asymptotic expansions of the flow past a generic fold are not smooth in $\varepsilon$. We shall discuss the system (3.2) further in Section 3 when we consider the properties of maximal canards.

At fold points, coordinates for the critical manifold must mix the slow and fast coordinates since the projection $\pi$ onto the space of slow coordinates is singular. Since the gradient of $f$ is non-zero, $\partial f/\partial y_1$ or $\partial f/\partial y_2$ is non-zero, say $\partial f/\partial y_1 \neq 0$. This implies that there is a function $h(x, y_2)$ so that the critical manifold near the fold point is locally given by $y_1 = h(x, y_2)$. The slow flow can be expressed in terms of $(x, y_2)$ coordinates so that, after rescaling, it extends to the fold curve. The equation $f(x, y) = 0$ is differentiated along trajectories to give the relationship

$$\frac{\partial f}{\partial x}\dot{x} + \frac{\partial f}{\partial y_1}\dot{y}_1 + \frac{\partial f}{\partial y_2}\dot{y}_2 = 0$$

on the critical manifold. Since $\partial f/\partial y_1 \neq 0$, this equation can be solved for $\dot{y}_1$:

$$\dot{y}_1 = -\left(\frac{\partial f}{\partial y_1}\right)^{-1}\left(\frac{\partial f}{\partial x}\dot{x} + \frac{\partial f}{\partial y_2}\dot{y}_2\right).$$

Substituting this into $\dot{y} = g$ and gives

$$-\left(\frac{\partial f}{\partial y_1}\right)^{-1}\frac{\partial f}{\partial x}\dot{x} = g_1 + \left(\frac{\partial f}{\partial y_1}\right)^{-1}\frac{\partial f}{\partial y_2}g_2.$$

Rescaling time, we obtain the system

$$\begin{aligned} x' &= g_1 + \left(\frac{\partial f}{\partial y_1}\right)^{-1} \frac{\partial f}{\partial y_2} g_2 \\ y_2' &= -\left(\frac{\partial f}{\partial y_1}\right)^{-1} \frac{\partial f}{\partial x} g_2, \end{aligned} \tag{3.3}$$

where the functions $(f, g_1, g_2)$ are evaluated at $(x, h(x, y_2), y_2)$. We continue to call (3.3) the slow flow. This system has equilibrium points on the fold curve when $x' = 0$ since $\partial f/\partial x = 0$ defines the fold curve. Generically, $g_1$ vanishes at isolated points of the fold curve. These points are called *folded equilibria*. They are not limits of equilibrium points of the full system. Instead they occur at locations where the flow is tangent to the fold curve, trajectories flowing toward the fold curve on one side of the folded equilibrium and away from the fold curve on the other side of the folded equilibrium.

The dynamics of slow-fast systems (1.1) near generic folded equilibria was first studied by Benoît [4]. He proved the existence of *canards*, trajectories that continue along the unstable sheet of the critical manifold after approaching the folded equilibrium, in the cases that the equilibrium is a saddle or node. The geometry of the flow near a folded node has not been fully characterized and remains as an open problem in the theory of slow-fast systems. The analysis of folded equilibria proceeds in several steps. First, preliminary coordinate changes and rescalings are performed to produce a system of first approximation that is closely related to the *normal form*

$$\begin{aligned} \varepsilon\dot{x} &= y - x^2 \\ \dot{y} &= az + bx \\ \dot{z} &= 1. \end{aligned} \tag{3.4}$$

With the scalings

$$x = O(\mu), \qquad y = O(\mu^2), \qquad z = O(\mu), \qquad t = O(\mu), \qquad \varepsilon = \mu^2,$$

used to obtain the system of first approximation, the normal form retains its form exactly with $\varepsilon = 1$. There is a folded equilibrium at the origin.

The slow flow of the normal form (3.4) after its time rescaling is

$$\begin{aligned} x' &= az + bx \\ z' &= 2x. \end{aligned} \tag{3.5}$$

The system (3.5) is linear. If $a > 0$, then the origin is a saddle; if $a < 0$, then the origin is a node if $b^2 + 8a > 0$ and a focus if $b^2 + 8a < 0$. A node or focus is stable if $b < 0$ and unstable if $b > 0$. In the case of a saddle or a node, there are a pair of invariant lines for the slow flow along the eigenvectors of the system. These lines are limits of a pair of polynomial solutions of the full system (3.4) as $\varepsilon \to 0$. The polynomial solutions are

$$\begin{pmatrix} x \\ y \\ z \end{pmatrix}(t) = \begin{pmatrix} \alpha t \\ \varepsilon\alpha + \alpha^2 t^2 \\ t \end{pmatrix} \tag{3.6}$$

where $\alpha = (b \pm \sqrt{b^2 + 8a})/4$ is a root of the equation $2\alpha^2 - b\alpha - a = 0$.

We now analyze the case of a folded saddle further. When the origin is a folded saddle, the two values of alpha have opposite sign. The polynomial solution of the full system with $\alpha = (b - \sqrt{b^2 + 8a})/4 < 0$ is a *maximal* canard that remains on the slow manifold for all time, flowing from its stable sheet to its unstable sheet. The polynomial solution with $\alpha = (b + \sqrt{b^2 + 8a})/4 > 0$ also lies on the slow manifold, but flows from its unstable sheet to its stable sheet. It has been called a "faux-canard" by Benoît [4].

Note that the distance between the two polynomial solutions is $2\varepsilon\alpha = O(\varepsilon)$. Solutions attracted to the stable sheet of the slow manifold approach it at a rate $O(\exp(-c/\varepsilon))$, except in the vicinity of the fold curve. Therefore, we expect that trajectories with canards will be close enough to the maximal canard that a variational analysis along this canard will describe their general properties. We are especially interested in computing a transition map for trajectories that approach the maximal canard on the stable sheet of the slow manifold and then jump away from the unstable sheet of the slow manifold after following a canard segment for some distance.

The equations (3.4) have a time reversal symmetry $R(x, y, z, t) = (-x, y, -z, -t)$, so the process of jumping from the unstable sheet of the slow manifold is a mirror image of the process by which trajectories approach the stable sheet of the slow manifold. The stable and unstable slow manifolds are not quite two dimensional surfaces, but rather three dimensional sets of thickness $O(\exp(-c/\varepsilon))$ in directions parallel to the fast flow. We want to demonstrate that stable and unstable slow manifolds containing the maximal canard intersect transversally as they reach the plane $z = 0$. Our argument is based upon an analysis of the variational equations of the flow along the maximal canard.

The variational equations along a trajectory $w(t) : \mathbb{R} \to \mathbb{R}^3$ of (3.4) are given by $\dot{\xi}(t) = A(w(t))\xi(t)$, where $A$ is the matrix

$$A = \begin{pmatrix} -2x/\varepsilon & 1\varepsilon & 0 \\ b & 0 & a \\ 0 & 0 & 0 \end{pmatrix}. \tag{3.7}$$

Along the maximal canard trajectory

$$\begin{pmatrix} x \\ y \\ z \end{pmatrix}(t) = \begin{pmatrix} \alpha t \\ \varepsilon\alpha + \alpha^2 t^2 \\ t \end{pmatrix} \tag{3.8}$$

$x = \alpha t$ in the equation (3.7) with $\alpha = (b - \sqrt{b^2 + 8a})/4 < 0$. The equation becomes simpler if we make a change of coordinates along the trajectory. Set

$$B = \begin{pmatrix} 1 & 0 & 1 \\ 0 & 1 & 2\alpha t \\ 0 & 0 & \alpha^{-1} \end{pmatrix} \tag{3.9}$$

and $\xi = B\eta$. The equation (3.7) transforms to $\dot{\eta} = B^{-1}(AB - \dot{B})\eta$. Now

$$B^{-1} = \begin{pmatrix} 1 & 0 & -\alpha \\ 0 & 1 & -2\alpha^2 t \\ 0 & 0 & \alpha \end{pmatrix}, \tag{3.10}$$

giving

$$\dot{\eta} = \begin{pmatrix} -2\alpha t/\varepsilon & 1/\varepsilon & 0 \\ b & 0 & 0 \\ 0 & 0 & 0 \end{pmatrix} \eta. \tag{3.11}$$

Thus the vector $\eta(t) = (0,0,1)^t$ is one solution of the equation (3.11), and the subspace spanned by the first two unit vectors is invariant. We want to follow the evolution of vectors in this subspace, paying particular attention to their direction at $t = 0$. Assuming that $b < 0$, the rescaling

$$\zeta_1 = \sqrt{-b\varepsilon}\eta_1, \qquad \zeta_2 = \eta_2, \qquad \tau = \sqrt{-b/\varepsilon}$$

produces the system

$$\zeta' = \begin{pmatrix} 2\gamma\tau & 1 \\ -1 & 0 \end{pmatrix} \zeta \tag{3.12}$$

which has the single parameter $\gamma = \alpha/b$. The time derivative in (3.12) is with respect to $\tau$. Note that in the case of the folded saddle, $\gamma = \alpha/b > 1/2$ for the negative root $\alpha = (b - \sqrt{b^2 + 8a})/4$ and $b < 0$. If we set $\theta = \arg(\zeta)$, then

$$\begin{aligned} \theta' &= \frac{\zeta_2'\zeta_1 - \zeta_1'\zeta_2}{\zeta_1^2 + \zeta_2^2} = -\cos^2(\theta) - 2\gamma\tau\cos(\theta)\sin(\theta) - \sin^2(\theta) \\ &= -1 - 2\gamma\tau\cos(\theta)\sin(\theta) = -1 - \gamma\tau\sin(2\theta). \end{aligned} \tag{3.13}$$

We want to follow $\theta$ as $\tau$ approaches 0. For most initial conditions, with $\tau \ll 0$, $\theta(0)$ will give the direction of the intersection of the stable slow manifold with the plane $z = 0$ in the $\zeta$ coordinates. Using the time reversal symmetry $R$, the comparable direction for the intersection of the unstable slow manifold with $z = 0$ will be $-\theta(0)$. Therefore, if $\theta(0)$ is not a multiple of $\pi/2$, the stable and unstable slow manifolds will intersect transversally at $z = 0$.

Let us turn to the analysis of the equation $\theta' = -1 - \gamma\tau\sin(2\theta)$. The equation is $\pi$-periodic in $\theta$, so we focus attention on $\theta \in [0, \pi]$. The zeros of the right hand side occur when $\tau = -1/(\gamma\sin(2\theta))$. In the interval $(-1/\gamma, 0]$, $\theta' < 0$. When $\tau < -1/\gamma$, $\theta' = 0$ on the curve $\tau = -1/(\gamma\sin(2\theta))$ and $\theta' < 0$ when $\theta = \pi/2$. Consequently, trajectories that enter the $\theta$ interval between the upper branch of $\tau = -1/(\gamma\sin(2\theta))$ and $\theta = \pi/2$, remain there until they reach $\tau = -1/\gamma$. Most trajectories with initial conditions $\tau \ll 0$ (fixed) and $\pi/4 < \theta < \pi$ (varying) quickly converge to one another with $\theta$ close to $\pi/2$. These trajectories arrive at $\tau = -1/\gamma$ with $\pi/4 < \theta < \pi/2$. Now $\partial\theta'/\partial\gamma = -\tau\sin(2\theta) > 0$ when $\tau < 0$ and $0 < \theta < \pi$. Therefore, if the trajectory with initial condition $\tau \ll 0$ and $\theta = \pi/2$ reaches $\tau = 0$ with $\theta > 0$, then this will be true for all trajectories with the same initial conditions and larger values of $\gamma$.

We use numerical integration to estimate for which value of $\gamma$, trajectories with initial conditions $\tau \ll 0$ and $0 < \theta < \pi$ pass through the origin. These calculations suggest that the critical value of $\gamma$ is $1/2$. In the case of the folded saddle, $\gamma > 1/2$ so we find that $\theta(0) > 0$ for these trajectories, implying transversality of the slow stable and unstable manifolds.

The presence of canards near folded saddles complicates the definition of the reduced system for (1.1). Canard trajectories can jump from the unstable sheet of the critical manifold anywhere along the saddle separatrix of the slow flow. Thus the reduced system becomes multivalued: the trajectory arriving along the stable manifold of the saddle in the slow flow may continue through the saddle. At any point along this stable manifold on the unstable sheet of the critical manifold, the trajectory may do one of three things: continue on the unstable sheet or jump to one of the two sides of the critical manifold along the fast direction until it hits a new sheet of the critical manifold. To understand the effects associated with this behavior better, we turn to the forced van der Pol equation.

# 4 The forced van der Pol equation

The phase space of the forced van der Pol system (1.4) is $\mathbb{R}^2 \times S^1$ and its critical manifold is the surface defined by $y = x^3/3 - x$. The singularities of the critical manifold are the curves defined by $x = \pm 1$, $y = \mp 2/3$. The slow flow is given by

$$\begin{aligned} \theta' &= \omega(x^2 - 1) \\ x' &= -x + a\sin(2\pi\theta). \end{aligned} \tag{4.1}$$

The reduced system is given by this slow flow (4.1) together with jumps from the fold curves $x = \pm 1$ to $x = \mp 2$. Folded singularities of the reduced system occur when $x = a\sin(2\pi\theta) = \pm 1$. Thus, folded singularities occur in the parameter regime $a \geq 1$. There is a narrow strip of the $(a, \omega)$ parameter space bounded by $a = 1$ where there are folded nodes, but the folded singularities are saddles and foci in most of this parameter plane. At the folded saddles, we augment the reduced system with trajectories representing the canards and their jumps. The folded foci do not play a substantial role in the dynamics of the system because the unstable manifolds of the saddles of (4.1) "shield" the foci from trajectories that have undergone a jump: trajectories of the slow flow with initial conditions on $x = \pm 2$ do not reach the folded foci in the parameter regimes we consider.

Our approach to studying the dynamics of the forced van der Pol system has been based upon analysis of return maps for its reduced system [17]. We utilize the symmetry $s(x, y, \theta) = (-x, -y, \theta + 0.5)$ in defining the return map. Slow flow trajectories with initial conditions on $x = 2$ flow until they hit the fold at $x = 1$. Note that if $a > 2$, $x$ first increases on some of these trajectories and there are points where the final requirement for a relaxation oscillation to be simple is violated. Trajectories jump to $x = -2$ when they reach $x = 1$ unless they are on the stable manifold of the folded saddle or node and continue along canards for some distance before they jump. To incorporate the canards most directly into the return map, we define cross-sections to the jumps as suggested by Weckesser [6]. The cross sections are the half-cylinders $\Sigma_1$ and $\Sigma_2$ defined by $x = 1$, $y > -2/3$ and $x = -1$, $y < 2/3$. A *half-return map* $\sigma$ is obtained by flowing from $\Sigma_1$ to $\Sigma_2$ and then applying the symmetry $s$ to get back from $\Sigma_2$ to $\Sigma_1$. Except for the canards, the image of $\sigma$ lies along the circle $y = -2/3$. The canards adjoin to the image of $\sigma$ two additional segments: one from jumps of canards from the unstable sheet of the critical manifold directly to $\Sigma_2$ as $x$ decreases and one from jumps of canards back to the stable sheet of the critical manifold with $x > 0$, where they return to the fold and jump a second time to $\Sigma_2$. The image of the second segment also lies on the circle $y = -2/3$, while the image of the first segment does not. The map $\sigma$ has rank 1: when points jump from $\Sigma_1$ to the critical manifold with $x > 1$, those that land on the same trajectory of the slow flow have the same $\sigma$ image. Thus we idealize $\sigma$ as a one dimensional map with the images of the points in the stable manifold of the folded saddle being entire intervals. Chaotic invariant sets formed from trajectories with canards are present in the full system (1.4) when the images of the canard segments of the reduced system contain a point of the stable manifold of the folded saddle. The analysis of folded saddles in Section 3 gives information about the scale of these chaotic invariant sets. This analysis breaks down when the canard trajectories that return to the stable manifold are *maximal*, meaning that the break point of the canard is at the place that the saddle separatrix returns to the fold curve on the critical manifold. Fig. 2 depicts the stable manifold of the folded saddle and the maximal canard of the slow flow at parameter values for which the maximal canard flows into the stable manifold.

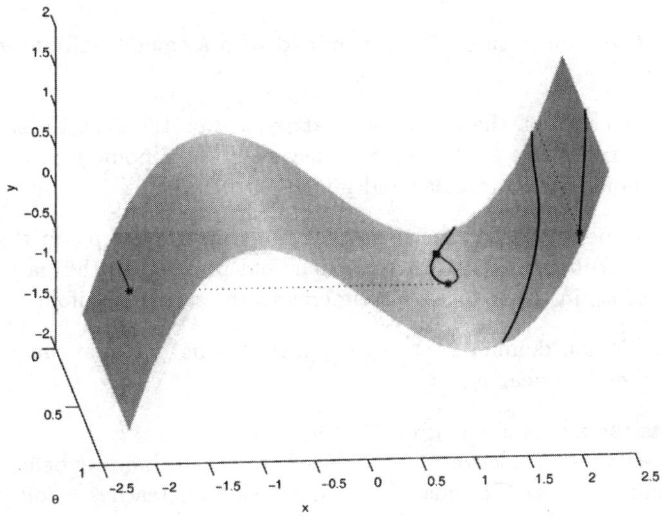


Figure 2: The critical manifold of the forced van der Pol equation (1.4) is drawn as a shaded surface together with the stable manifold and canards of the folded saddle as thick solid curves. The $\theta$ direction is periodic with fundamental domain $[0, 1]$ shown. The images of the canards after jumping to the stable sheets of the critical manifold are drawn as thin solid curves. The jump from the maximal canard and the line $(x, y) = (2, 2/3)$ are drawn as dotted curves. The folded saddle is marked by a square; the maximal canard, its image after jumping to the critical manifold at $x = -2$, and the image of this point under the symmetry $s$ are marked by asterisks. The parameters are $(a, \omega) = (1.1, 1.48)$.

# 5 Maximal canards

The creation of horseshoes in families of two dimensional maps has been extensively studied, with the Hénon map serving as a prototype [19]. The principal results [3, 32] extend Jakobson's Theorem for one dimensional maps [8] into the setting of two dimensional diffeomorphisms with small Jacobian. We have examined canards in vector fields with two slow and one fast variable. Here we focus upon "maximal" canards that reach a fold, connecting two families of canards that jump in opposite directions from an unstable slow manifold. In the vicinity of the maximal canards we study, the vector field has a return map with a fold and small Jacobian. The singular limit of the return map has some aspects that differ from the Hénon family. In particular, the critical point of the limit map is not $C^3$: it has different order from the two sides and different limit directions. The scenario that we describe below for the creation of chaotic invariant sets takes place in the forced van der Pol system.

Consider a one parameter family of slow-fast vector field with two slow variables and one fast variable that has the following properties:

- There is a two dimensional critical manifold with a smooth fold curve $S$ and a folded saddle $p_s$.
- In the reduced system, the canard separatrix of $p_s$ on the unstable sheet of the critical manifold intersects the fold curve $S$ transversally at a point $p_m$, the *maximal canard point*. The point $p_m$ is a regular fold point.
- There is a value of the parameter so that the trajectory of $p_m$ in the reduced system flows to $p_s$, all jumps occurring at regular fold points. As the parameter varies, the trajectory of $p_m$ in the reduced system crosses the stable manifold of $p_s$ transversally.
- Strong contraction dominates strong expansion along maximal canard trajectories beginning and ending near $p_s$.

Fig. 3 depicts the maximal canard trajectories.

Our first objective is to describe in the reduced system the limiting behavior of trajectories near the maximal canards. The maximal canard solution reaches a fold and then jumps, but it does so by following the unstable sheet of the slow manifold. The system of first approximation (3.2) for generic trajectories reaching a fold gives a model for the maximal canard. We derive the reduced system for (3.2) by setting $\varepsilon = 0$, differentiating $y + x^2 = 0$ with respect to time, substituting $-2x\dot{x}$ for $\dot{y}$ and then rescaling time to obtain

$$\begin{aligned} \dot{x} &= 1 \\ \dot{z} &= -2x^2. \end{aligned} \tag{5.1}$$

The equation (5.1) is easily integrated: we normalize the trajectories $(x(t), z(t)) = (t, c - 2t^3/3)$ so that they reach the fold curve at zero time $t$. Thus the trajectory beginning at $(x_0, z_0)$ has $t = x_0$ and $c = z_0 + 2x_0^3/3$. If $x_0 > 0$, we flow trajectories with decreasing time since the time rescaling used to obtain the reduced system is negative on the unstable sheet of the critical manifold. On the critical manifold, $y(t) = -x^2(t)$ so each solution projects onto a semicubical parabola $(-t^2, c - 2t^3/3))$ in the $(x, z)$ plane.

All of the solutions of (3.2) are unbounded, tending to infinity in finite time along the direction parallel to the $x$-axis. We want to see how trajectories in the reduced system follow the canard for varying distances and then jump intersecting a cross-section $x = b > 0$. Without loss of generality, assume that the canard trajectory has $c = 0$ on the unstable sheet $x > 0$ of the critical manifold. Using the formula we computed above, this trajectory is given by the curve $(t, -t^2, -2t^3/3)$ on the critical manifold with $t > 0$. Note that $t$ decreases along the solution curve. If the trajectory jumps in the direction of increasing $x$ at time $t = u$, then it hits the cross-section $x = b > 0$ at the point with coordinates $(-u^2, -2u^3/3)$. If the

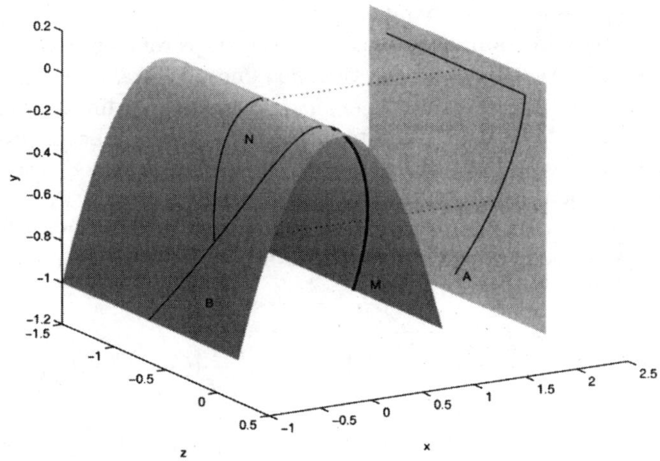


Figure 3: The reduced flow of the system (3.2) showing the behavior of maximal canards at a fold. The canard trajectory is the heavy trajectory labeled $M$. The critical manifold and a cross-section to the jumps are drawn shaded. The images of the canard after jumps are solid curves $A$ and $B$. Dotted lines are shown along jumps from one point along the canard. From the points that jump back to the critical manifold, the reduced trajectories flow along the stable sheet of the critical manifold to its fold (solid curve $N$) where it jumps to the cross-section (dotted line). The image on the cross-section consists of a segment parallel to the fold of the critical manifold and the image of the jumps that proceed directly from the canard to the cross-section.

trajectory jumps in the direction of decreasing $x$ at time $t = u$, then it lands on the stable sheet of the critical manifold at the point $(-u, -2u^3/3)$. The reduced trajectory through the point $(-u, -2u^3/3)$ is $(t, -4u^3/3 - 2t^3/3)$ with $t = -u < 0$ initially. The trajectory then flows along the stable sheet of the critical manifold until it reaches the fold at $x = y = 0$ at time $t = 0$ and with $z = -4u^3/3$. From the fold, it jumps parallel to the $x$-axis and hits the cross-section $x = b$ at the point $(0, -4u^3/3)$. Thus the intersection of the trajectories jumping from canards on the cross-section $x = b$ traces out a curve $C$ which is not smooth at the point coming from the maximal canard. It has one segment along the $z$-axis and one segment lies along a semicubical parabola tangent to the $y$-axis.

We return to the study of the flow map $\sigma$ from a small cross-section $y = a$ of the canard trajectories to the plane $x = b$ for the full system (3.2). We expect the image to lie close to the curve $C$ described above. Both the reduced and full system are translation invariant in the $z$ direction, so we examine the flow of an initial segment $S$ parallel to the $x$ axis that intersects the canard trajectory. Since $\dot{y} = 1$, the evolution of the $x$ coordinate is given by the equation $\varepsilon\dot{x} = x^2 + a + t$. The motion of the initial points that concern us is to flow along the slow manifold $x^2 + y = 0$ to a height $y = u$ where the trajectory jumps from the slow manifold in the direction of increasing or decreasing $x$. In the direction of increasing $x$, the trajectory proceeds nearly parallel to the $x$ axis until it hits the plane $x = b$ near $(-u^2, -2u^3/3)$. In the direction of decreasing $x$, the trajectory approaches the stable sheet of the slow manifold, flows to the fold and then flows nearly parallel to the $x$ axis, hitting the plane $x = b$ near $(0, -4u^3/3)$. We need to estimate $u$, the $y$-coordinate of the break point, in terms of the distance of the initial point from the slow manifold. Let $x_s$ be the $x$ coordinate of a point on the intersection of the canard orbit with the section $y = a$. If $x - x_s = v$ is small, then the deviation of the trajectories on the section $y = a$ from the slow manifold can be estimated by solving the variational equation $\varepsilon\dot{\xi} = 2x(t)\xi$ with initial condition $v$ and $x(t) = \sqrt{-(a+t)}$ the $x$ coordinate of the canard trajectory of the reduced system. The solution of this equation is

$$\xi(t) = v \exp\left(\frac{2\int_0^t \sqrt{-(a+\tau)}\,d\tau}{\varepsilon}\right) = v \exp\left(\frac{4}{3\varepsilon}((-a)^{3/2} - (-(a+t))^{3/2}\right)$$

The jump height can be approximated by solving $|\xi(t)| = 1$ and then using $u = a + t$. This gives an estimate $u_e$ for the value of $u$ at the break point as

$$u_e = -\left(\frac{3\varepsilon}{4}\log|v| + (-a)^{3/2}\right)^{2/3}.$$

Note that the derivative

$$\frac{du_e}{dv} = \frac{-\varepsilon}{2v}\left(\frac{3\varepsilon}{4}\log|v| + (-a)^{3/2}\right)^{-1/3}$$

is unbounded as $|v| \to 0$. This calculation implies that $\sigma$ stretches $S$.

In the vicinity of the maximal canard, we require finer analysis of $\sigma$. We apply the inverse of the scale transformation $(x, y, z, t) = (\varepsilon^{1/3}X, \varepsilon^{2/3}Y, \varepsilon Z, \varepsilon^{1/3}T)$ to determine the scale of the domain near the maximal canard where more analysis of the system of first approximation (3.2) is needed. We want to see that $\sigma(S)$ is a strictly convex curve near the maximal canard, so that the methods used for the study of Hénon-like maps [32] can be applied in this setting. Explicit or asymptotic solutions to (3.2) near the maximal canard are not apparent, so we rely upon numerical integration to study its solutions. Fig. 4 shows a plot of $\sigma(S)$ with $a = -4$, $b = 10$ and the $x$ coordinates of $S$ lying in the interval $[1.93111, 1.93143]$. The circles are the images of 32 points regularly spaced on $S$ at increments of $10^{-5}$. The crosses are the values of the least squares fit to this data by a sixth degree polynomial $y(x) \approx 6.91 + 0.693x - 0.265x^2 - 0.113x^3 - 0.0172x^4 + 0.080x^5 - 0.0395x^6$. The second derivative of this approximating polynomial is negative: the maximum of its second derivative is approximately $-0.398$. These data are strong evidence for the convexity of $\sigma(S)$.

Based on this analysis of the flow near canards of (3.2), we give a conjectural description of the generic properties of flow past a maximal canard that ends in a generic manner near a fold. Let $R_0$ be a rectangle transverse to a canard trajectory, with one side parallel to the unstable sheet of the slow manifold and one side parallel to the strong expanding direction. Let $R_1$ be a section transverse to the fast trajectory which begins at the end $p$ of a maximal canard. We want to examine the flow $\sigma$ from $R_0$ to $R_1$. In the direction parallel to the slow manifold on $R_0$, the dynamics remain slow and there is no fast expansion or contraction

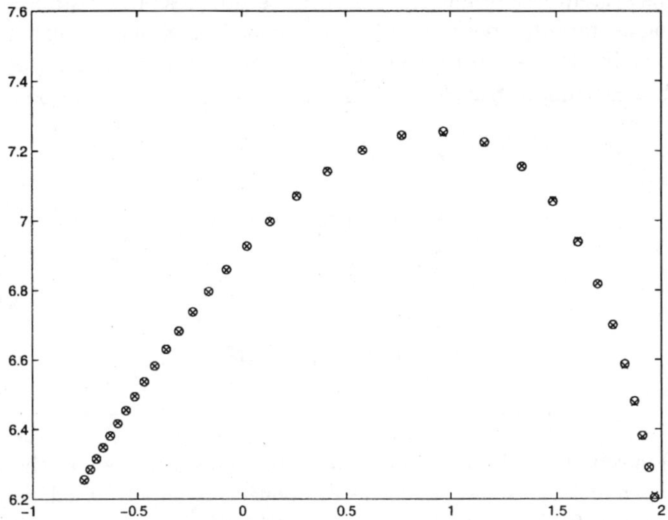


Figure 4: The graph of $\sigma(S)$ on the cross-section. Data points obtained from numerical integration are marked by an $\circ$ while the least squares fit by a polynomial of degree six are marked by $\times$.

prior to a jump. Segments $S$ parallel to the strong expanding direction on $R_0$ are mapped to $R_1$ by $\sigma$ in the following manner. Exponentially short pieces of $S$ follow the unstable slow manifold long enough to contain canards. These short segments are exponentially stretched by an amount that can be estimated from integration of the variational equation for the fast expansion. Denoting by $p_{s,\varepsilon}$ points of $S$ whose trajectories reach the fold of the slow manifold, the limiting image of $\sigma(S)$ as $\varepsilon \to 0$ has a corner at $\sigma(p_{s,0})$ and different asymptotics for jumps in opposite directions. In terms of distance to $\sigma(p_{s,0})$, $\sigma(S)$ is quadratic for jumps directly to $R_1$ and cubic for jumps back to the stable sheet of the slow manifold and then to $R_1$. This behavior breaks down in a neighborhood of $\sigma(p_{s,\varepsilon})$ of size comparable to $\varepsilon^{2/3}$. In this region $\sigma(S)$ is a smooth, strictly convex curve. With a scaling that magnifies this region by amounts $(\varepsilon^{-2/3}, \varepsilon^{-1})$ transverse and parallel to the image of the fold, $\sigma(S)$ has a smooth limit as $\varepsilon \to 0$.

# 6 "Classical" canards

A group in Strasbourg [5, 9] discovered canards in the two-dimensional vector fields

$$\begin{aligned} \varepsilon\dot{x} &= y - \left(\frac{x^3}{3} - x\right) \\ \dot{y} &= a - x. \end{aligned} \tag{6.1}$$

There are Hopf bifurcations that occur at $a = \pm 1$ in this extension of the van der Pol equation. The analysis of this system was originally done with techniques of non-standard analysis and then repeated using asymptotic methods [12] and geometric singular perturbation theory [11]. Here we shall apply simpler geometric methods to recover basic properties of this example.

We begin with a coordinate transformation that fixes the equilibrium point of the system at the origin. Set

$$\begin{aligned} u &= x - a \\ v &= y - \left(\frac{a^3}{3} - a\right) \\ a &= b + 1 \end{aligned} \tag{6.2}$$

to obtain

$$\begin{aligned} \varepsilon \dot{u} &= v - \left(\frac{u^3}{3} + (b+1)u^2 + (b^2 + 2b)u\right) \\ \dot{v} &= -u. \end{aligned} \tag{6.3}$$

The system (6.3) has a Hopf bifurcation at $b = 0$. The formula for the relevant cubic coefficient of its normal form shows that the bifurcation is supercritical, with stable limit cycles appearing for $b < 0$. To study how these limit cycles grow, we consider the *rotational* properties of the vector fields as $b$ varies.

**6.1 Theorem (Duff [10])** *Let $X_b$ be a one parameter family of planar vector fields such that*

- *$X_b$ has a stable, hyperbolic limit cycle $\gamma_0$ that is oriented clockwise when $b = b_0$;*
- *$X \times \frac{\partial X}{\partial b}$ is negative on $\gamma$.*

*Then there is a neighborhood $B$ of $b_0$, such that $b \in B$ implies that $X_b$ has a stable, hyperbolic limit cycle $\gamma_b$. The curves $\gamma_b$ are disjoint and shrink with increasing $b$.*

Taking the right-hand side of (6.3) by $X$, we have

$$X \times \frac{\partial X}{\partial b} = \frac{-1}{\varepsilon}(u^3 + (2b+2)u^2)$$

This is negative in the half plane $u > -(2b+2)$. Applying Duff's theorem, we conclude that the limit cycles grow monotonically with decreasing $b$, at least until they come close to the line $u = -2$.

Further insight into the flow of (6.3) can be obtained by studying its slow manifolds. We are especially interested in two aspects of these manifolds: (1) the amount of expansion and contraction that occurs, and (2) where the stable and unstable parts of the manifold intersect the $v$ coordinate axis. To address (1), we use the variational equation for fast contraction and expansion along the critical manifold. The fast contraction and expansion of (6.3) takes place along the $u$ direction, with a neighborhood of the origin to be avoided. The variational equation in this direction is $\varepsilon\dot{\xi} = -(u^2 + 2(b+1)u + b^2 + 2b)\xi$. We want to compute solutions to this equation along relaxation oscillations approximated by stable and unstable segments of the critical manifold, together with jumps parallel to the $u$-axis. The solution of the variational equation is

$$\xi(t) = \xi(0) \exp\left(\frac{-1}{\varepsilon} \int_0^t (u^2 + 2(b+1)u + b^2 + 2b)\, dt\right).$$

We change the variable of integration to $u$ via $dt = \frac{dt}{dv}\frac{dv}{du}du$ and use the estimate that $\frac{dv}{du} \approx u^2 + 2(b+1)u + b^2 + 2b$ along the regular portions of the critical manifold. When $b = 0$, the integral becomes $\int -u(u+2)^2\, du = -(u^4/4 + 4u^3/3 + 2u^2)$. We compute the integral on the portion of the critical manifold lying over $u$ intervals $(u_0, u_1)$ oriented with decreasing $u$ where $-2 < u_1 < 0 < u_0 < 1$ and $u_0^3/3 - u_0 = u_1^3/3 - u_1$. These intervals correspond to the slow segments of a "duck without head", a relaxation oscillation containing a canard in which the jump from the canard is with increasing $u$. We find that the integrals are all positive. Fig. 5 plots the integrals as a function of $u_0$. We conclude from this computation that the ducks without heads are all strongly attracting. The ducks with heads (relaxation oscillations with decreasing u along jumps from the canards) are even more strongly attracting.

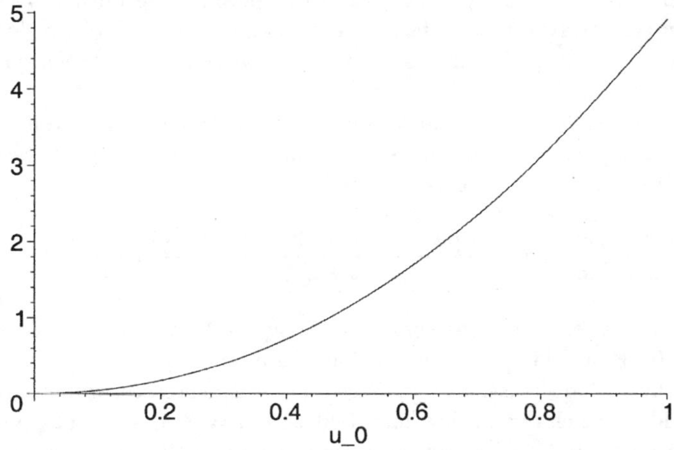


Figure 5: Values of the variational integral that give the magnitude of the exponential contraction for canards without heads as a function of their right hand end point $u_0$ in system (6.3).

Fenichel theory breaks down in the vicinity of the origin because normal hyperbolicity of the critical manifold fails there. Nonetheless, trajectories with initial conditions near the stable part of the critical manifold with $u > 0$ cross $u = 0$ in a tiny, exponentially small interval $I_s$. The same is true of backward trajectories with initial conditions near the unstable part of the critical manifold with $-1 < u < 0$, giving an interval $I_u$. Relaxation oscillations with canards will only occur when $I_u$ and $I_s$ intersect. Thus, we want to estimate the location of these intervals and determine how they vary with $b$. This is done by locating pieces of the slow manifold of the system within a strip of width $O(\varepsilon^2)$.

We observe first that the curve

$$v = h_0(u) = \frac{u^3}{3} + (b+1)u^2 + (b^2+2b)u - \varepsilon\frac{u}{u^2+2(b+1)u+b^2+2b}$$

has the property that

$$\frac{dh_0}{du} - \frac{\dot{v}}{\dot{u}} = \varepsilon\frac{u^2-b^2-2b}{(u+b)^2(u+b+2)^2}.$$

For $(1+b) < u$, this difference is positive, implying that $h$ is transverse to the vector field and, since $\dot{u} < 0$, the trajectories of the vector field cross the graph of $h_0$ from right to left. However, on the curve $v = h_1(u) = h_0(u) + \varepsilon^2$ with $b = 0$, we have

$$\frac{dh_1}{du} - \frac{\dot{v}}{\dot{u}} = -\varepsilon(\frac{u^5+8u^4+24u^3+32u^2+16u-1}{(u+2)^2}) + O(\varepsilon^2).$$

This quantity is negative when $0.056 < u < 1$, so trajectories cross the graph of $h_1$ from left to right in this interval of values of $u$. When $b$ and $\varepsilon$ are sufficiently small, trajectories entering the strip between the graphs of $h_0$ and $h_1$ remain there until $u$ decreases to 0.06. Now $h_0$ has a non-zero derivative with respect to $b$, so the slow manifold varies at a non-zero rate with respect to $b$. In particular, its intersection with $u = 0$ has a negative derivative with respect to $b$.

A similar calculation to the one above demonstrates that the unstable segment of the slow manifold for $u$ in a compact subinterval of $(-2, 0)$ lies in the strip between the graph of $h_0$ and $h_2 = h_0(u) - \varepsilon^2$. When $b = 0$ we compute

$$\frac{dh_2}{du} - \frac{\dot{v}}{\dot{u}} = \varepsilon(\frac{u^5+8u^4+24u^3+32u^2+16u-1}{(u+2)^2}) + O(\varepsilon^2).$$

Next we observe that when $\varepsilon = 0$, the sign of $\frac{dh_0}{db} = u(u+2) + O(b)$ is positive for $u > 0$ and negative for $u < 0$. From this we conclude that the intersection of the right branch of the slow manifold with $u = 0$ decreases at a non-zero rate as $b$ decreases, while the intersection of the unstable, middle branch of the slow manifold with $u = 0$ increases at a non-zero rate as $b$ decreases. Consequently, there is only an exponentially small range of $b$ in which a stable slow manifold from the right can meet an unstable slow manifold. At the initial points of the family of canard orbits, as $v$ decreases on a vertical line the jump points first move from the origin to the maximal canard point near $u = -2$ with jumps toward increasing $u$; then the jump points return toward the origin with jumps toward decreasing $u$. The jump points on the periodic orbits with canards vary in the same way as $b$ decreases. This establishes the basic properties of the family of periodic orbits with canards in system (6.1).

## References

[1] V. I. Arnold, *Geometrical Methods in the Theory of Ordinary Differential Equations*, Springer-Verlag, New York, 1988.

[2] V. I. Arnold, V. S. Afrajmovich, Yu. S. Il'yashenko, and L. P. Shil'nikov, *Dynamical Systems V*, Encyclopaedia Math. Sci., Springer-Verlag, New York, 1994.

[3] M. Benedicks and L. Carleson, The dynamics of the Hénon map, *Ann. of Math.* **133** (1991), 73–169.

[4] É. Benoît, Canards et enlacements, *Inst. Hautes Études Sci. Publ. Math.* **72** (1990), 63–91.

[5] É. Benoît, J. L.Callot, F. Diener, and M. Diener, Chasse au canards, *Collect. Math.* **31** (1981) 37–119.

[6] K. Bold, C. Edwards, J. Guckenheimer, S. Guharay, K. Hoffman, Judith Hubbard, R. Oliva, and W. Weckesser, The forced van der Pol equation II: Canards in the reduced system, preprint, Cornell University, 2002.

[7] M. Cartwright and J. E. Littlewood, On nonlinear differential equations of the second order: II the equation $\ddot{y} - kf(y,\dot{y})\dot{y} + g(y,k) = p(t) = p_1(t) + kp_2(t)$, $k > 0, f(y) \geq 1$, *Ann. of Math.* **48** (1947), 472–494 (Addendum **50** (1949), 504–505).

[8] W. de Melo and S. van Strien, *One-Dimensional Dynamics*, Ergeb. Math. Grenzgeb. (3) **25**, Springer-Verlag, Berlin, 1993.

[9] M. Diener, The canard unchained or how fast/slow dynamical systems bifurcate, *Math. Intelligencer* **6** (1984), 38–48.

[10] G. F. D. Duff, Limit cycles and rotated vector fields, *Ann. of Math. (2)* **57** (1953), 15–31.

[11] F. Dumortier and R. Roussarie, Canard cycles and center manifolds, *Mem. Amer. Math. Soc.* **121** (577) (1996).

[12] W. Eckhaus, Relaxation oscillations, including a standard chase on French ducks, in: *Asymptotic Analysis III* (F. Verhulst, ed.), Lecture Notes in Math. **985**, Springer-Verlag, New York, 1983, 449–494.

[13] N. Fenichel, Persistence and smoothness of invariant manifolds for flows, *Indiana Univ. Math. J.* **21** (1971), 193–225.

[14] M. Golubitsky and V. Guillemin, *Stable Mappings and Their Singularities*, Springer-Verlag, New York, 1973.

[15] J. Guckenheimer, Bifurcation and degenerate decomposition in multiple time scale dynamical systems, in: *Nonlinear Dynamics and Chaos: Where Do We Go From Here?* (J. Hogan et al., eds.), Institut of Physics Publishing, Bristol, 2002, 1–21.

[16] J. Guckenheimer, K. Hoffman, and W. Weckesser, Global analysis of periodic orbits in the forced Van der Pol equation, in: *Global Analysis of Dynamical Systems* (H. Broer, B. Krauskopf, and G. Vegter, eds.), Institut of Physics Publishing, Bristol, 2001, 261–276.

[17] J. Guckenheimer, K. Hoffman, and W. Weckesser, The forced van der Pol equation I: The slow flow and its bifurcations, *SIAM J. Appl. Dyn. Syst.* **2** (2002), 1–35.

[18] J. Guckenheimer and P. J. Holmes, *Nonlinear Oscillations, Dynamical Systems, and Bifurcations of Vector Fields*, Springer-Verlag, New York, 1983.

[19] M. Hénon, A two-dimensional mapping with a strange attractor, *Comm. Math. Phys.* **50** (1976), 69–77.

[20] M. Hirsch, C. Pugh, and M. Shub, *Invariant Manifolds*, Lecture Notes in Math. **583**, Springer-Verlag, Berlin-New York, 1977.

[21] Y. Ilyashenko, Embedding theorems for local maps, slow-fast systems and bifurcation from Morse-Smale to Morse-Williams, *Amer. Math. Soc. Transl. (2)* **180** (1997), 127–139.

[22] C. K. R. T. Jones and N. Kopell, Tracking invariant manifolds with differential forms in singularly perturbed systems, *J. Differential Equations* **108** (1994), 64–88.

[23] N. Levinson, Perturbations of discontinuous solutions of non-linear systems of differential equations, *Acta Math.* **82** (1950), 71–106.

[24] J. E. Littlewood, On nonlinear differential equations of the second order: III. the equation $\ddot{y} - k(1-y^2)\dot{y} + y = bk\mu\cos(\mu t + a)$ for large $k$ and its generalizations, *Acta Math.* **97** (1957), 267–308 (Errata at **98**, p. 110).

[25] J. E. Littlewood, On nonlinear differential equations of the second order: IV. the equation $\ddot{y} - kf(y)\dot{y} + g(y) = bkp(\phi), \phi = t + a$, *Acta Math.* **98** (1957), 1–110.

[26] E. Mischenko and N. Rozov, *Differential Equations with Small Parameters and Relaxation Oscillations*, Plenum, New York, 1980.

[27] L. S. Pontryagin, Asymptotic behavior of the solutions of systems of differential equations with a small parameter in the higher derivatives, *Izv. Akad. Nauk SSSR Ser. Mat.* **21** (1957) 605–626; English translation *Amer. Math. Soc. Transl. Ser. 2* **18** (1961), 295–319.

[28] S. Smale, Differentiable dynamical systems, *Bull. Amer. Math. Soc.* **73** (1967), 747–817.

[29] P. Szmolyan and M. Wechselberger, Canards in $R^3$, *J. Differential Equations* **177** (2001), 419–453.

[30] P. Szmolyan and M. Wechselberger, in preparation.

[31] B. van der Pol, On relaxation oscillations, *Philos. Mag.* **7** (1926), 978–92.

[32] Q. Wang and Lai-Sang Young, From invariant curves to strange attractors, *Comm. Math. Phys.* **225** (2002), 275–304.

# Selected topics in differential equations with real and complex time

Yulij ILYASHENKO*

*Department of Mathematics*
*Cornell University*
*Ithaca, NY 14853*
*USA*

and

*Department of Mathematics*
*Independent and Moscow State Universities and Steklov Institute*
*Moscow*
*Russia*

**Abstract**

These lecture notes contain some results and problems from the theory of dynamical systems in both the real and complex domain. They do not pretend to cover the main streams of the theory. They rather reflect the work of the research group joined around the seminar in Dynamical Systems organized thirty years ago by Yu. Ilyashenko and directed now by him and Dr. Anton Gorodetski.

## 1 Introduction

This survey is split into three parts: real dynamical systems, polynomial dynamical systems, and foliations by analytic curves. The last two parts are strongly overlapping, and some material might be transported from one to another.

There are more than 50 problems here. The list looks uniform though in fact it is not at all homogeneous. Some of the problems are old and important from the point of view of the theory of dynamical systems itself. Some problems may influence the development of certain branches of dynamical systems. Some of them naturally accompany current investigations. We will indicate some principal problems here.

The most important problems mentioned below are the following.

(1) Genericity of the existence of time averages for continuous functions, and coincidence of different types of attractors (Problems 6.8, 6.10). These problems are amidst those playing a central role in the general theory of dynamical systems and go back to Palis, Ruelle, Takens, and others.

*Supported in part by the grants NSF 0100404, RFBR 2-02-00482, and CRDF RM1-2358.



*Y. Ilyashenko and C. Rousseau (eds.),*
*Normal Forms, Bifurcations and Finiteness Problems in Differential Equations,* 317–354.

(2) Hilbert's 16th problem is undoubtedly the central one in the planar theory of differential equations.

Another group of problems is:

(3) Problem 7.1 proposes a general program of investigation of slow-fast systems. It goes back to Guckenheimer.

(4) The Hilbert-Arnold problem 8.1 plays a central role in the multi-parameter theory of planar bifurcations.

(5) Problems 10.7, 11.5, 11.15 about persistence of complex cycles and existence of minimal sets are important for the theory of foliations by analytic curves. They go back to Petrovski-Landis, Anosov and Camacho.

The majority of the results were obtained by the participants of the seminar mentioned above. A reasonable part was obtained by (present or former) undergraduate students. It is my pleasure to mention some names: O. Anosova, P. Bachurin, V. Kleptsyn, V. Moldavskis, D. Pavlov. I omit the names of the older participants of the school whose results, also obtained at the time when they where undergrads, are included below. This is a modest reflection of a characteristic attribute of the Moscow mathematical school: the talented young people are very early involved in the creative work.

These lectures are addressed to beginners rather than experts. Therefore they contain a lot of introductory material and well-known definitions.

# A. Equations with real time

After the discovery of Smale that the structurally stable systems are not dense, the question: *What are the properties of typical dynamical systems?* became and remains one of the dominating questions in the field. Investigations in this direction are enormous. Below we suggest some new approaches to the problem.

## 2 Random dynamical systems as subsystems of smooth ones

### 2.1 Heuristic principle and its application

Consider a closed Riemannian manifold $M$ and a finite set of diffeomorphisms of $M$ into itself. Take a random bi-infinite sequence of the symbols $0, 1, \dots, N-1$:

$$\omega = \dots \omega_{-n} \dots \omega_{-1}\omega_0\omega_1 \dots \omega_n \dots$$

A random dynamical system of diffeomorphisms of a circle generated by this data is the sequence $\{F_n\}$ of products of the given diffeomorphisms; the choice of the factor for the $k$-th place is determined by the $k$-th element of the random sequence. For $n > 0$ this gives:

$$F_n = f_{\omega_n} \circ F_{n-1}, \qquad F_0 = f_{\omega_0}. \tag{2.1}$$

This formula makes sense for $n \leq 0$ as well, and allows us to define $F_{n-1}$ as soon as $F_n$ is known.

At the first glance, properties of random dynamical systems seem to be much richer than those of classical dynamical systems. Yet the following heuristic principle seems to be true:

*Any effect observed for random dynamical systems may be found as well for classical dynamical systems in a phase space of higher dimension.*

As an illustration of this principle we formulate the following

**2.1 Theorem ([24])** *Given a finite interval $I \subset \mathbb{R}$, $0 \in I$, and a closed manifold $M$, $\dim M \geq 3$, there exists an open set $U \subset \mathrm{Diff}^2(M)$ such that any $f \in U$ has a locally maximal invariant set $\Delta \subset M$ with the following properties:*

(i) *There exist two numbers $l_1$ and $l_2 = l_1 + 1$ such that the hyperbolic periodic orbits with stable manifolds of dimension $l_i$ are dense in $\Delta$.*

(ii) *For any $\lambda \in I$ there exists an orbit, dense in $\Delta$, with one of the intermediate Lyapunov exponents equal to $\lambda$.*

*The set $\Delta$ with the property* (i) *for* $\dim M \geq 4$ *may be taken to be a partially hyperbolic attractor.*

The latter statement is proved with quite different tools in [11].

**2.2 Problem** *Prove that the set $\Delta$ with the properties* (i) *and* (ii) *for* $\dim M \geq 4$ *may be taken to be a partially hyperbolic attractor.*

## 2.2 Smooth realizations of random dynamical systems

The way of proving Theorem 2.1 is the following. We first observe the properties like (i) and (ii) in a random dynamical system. Then we consider a *smooth realization* of this system. It gives a particularly degenerate dynamical system, having yet the properties (i) and (ii). After that we *perturb* the system thus constructed and prove that for all small perturbations of the above system the properties (i), (ii) still hold.

The random system may be included in a skew product as follows. Let $\Sigma^N$ be the space of all bi-infinite sequences of symbols $0, 1, \dots, N-1$, and let $\sigma$ be the Bernoulli shift $\Sigma^N \to \Sigma^N$. The map

$$F : \Sigma^N \times M \to \Sigma^N \times M, \ (\omega, \varphi) \mapsto (\sigma\omega, f_{\omega_0}(\varphi)) \tag{2.2}$$

is called a (universal) random dynamical system. In fact, the map $F$ is well determined. But any orbit of a random dynamical system defined at the beginning of the section is an orbit of some point under $F$.

The first factor of the product $\Sigma^N \times M$ is called the *base* and the second one the *fiber*. The metric $d_M$ on the fiber is induced by the Riemannian structure. The metric on $\Sigma^N$ is defined in the following way:

$$d_{\Sigma^N}(\omega, \omega') = 2^{-l}, \quad l = \min\{j \in \mathbb{Z}^+ \mid \omega_j \neq \omega'_j \text{ or } \omega_{-j} \neq \omega'_{-j}\}. \tag{2.3}$$

Note that the restriction of the map $F$, see (2.2), to a fiber over a point $\omega$ does not depend on the whole sequence $\omega$ but only on its element $\omega_0$. We refer to such systems as *step* random dynamical systems. In the theory of random dynamical systems this corresponds to the case of independent and identically distributed random mappings. We will also consider the map

$$G : \Sigma^N \times M \to \Sigma^N \times M, \ (\omega, \varphi) \mapsto (\sigma\omega, f_\omega(\varphi)), \tag{2.4}$$

where $f_\omega : M \to M$ depends on the whole sequence $\omega$. We refer to these maps as *mild* random dynamical systems.

We now find a smooth dynamical system such that the restriction of this system to its locally maximal invariant subset is conjugate to the step random dynamical system $F$ (2.2).

Consider a standard Smale horseshoe realized as a map of pairwise disjoint rectangles. Namely, let $D = D_0 \cup \cdots \cup D_{N-1}$, $D' = D'_0 \cup \cdots \cup D'_{N-1}$, $D_i \cap D_j = \varnothing$, $D'_i \cap D'_j = \varnothing$ for $i \neq j$, $D$ and $D'$ are subject to well-known geometric restrictions [40]. Let $S : D \to D'$ be a map such that $S|_{D_i} : D_i \to D'_i$ is a restriction to $D_i$ of a linear hyperbolic map. It is well known that the map $S$ has an invariant set $\Lambda$ homeomorphic to $\Sigma^N$, and $S|_\Lambda$ is conjugate to the Bernoulli shift $\sigma : \Sigma^N \to \Sigma^N$.

Consider a locally constant function $i : D \to \{0, \ldots, N-1\}$, defined by $i(x) = j \iff x \in D_j$. A skew product

$$\mathfrak{F} : D \times M \to D' \times M, \ \mathfrak{F}(x, \varphi) = (S(x), f_{i(x)}(\varphi)) \tag{2.5}$$

is called a *smooth realization* of the system (2.2).

It is easy to see that $\mathfrak{F}|_{\Lambda\times M}$ is conjugate to the map $F$. The set $\Lambda\times M$ is partially hyperbolic for $\mathfrak{F}$, and its central leaves are fibers of the projection on the first factor: $\Lambda\times M \to \Lambda$, $\Lambda \simeq \Sigma^N$.

The natural question is: what happens after a small perturbation of the map $\mathfrak{F}$?

### 2.3 Perturbations of skew products

Under some mild restrictions added to the construction of the map $\mathfrak{F}$ above, the theory of Hirsh, Pugh and Shub implies that any $C^r$-diffeomorphism $\mathfrak{G}$, $C^r$ close to $\mathfrak{F}$, has an invariant subset $\Delta$ homeomorphic to $\Sigma^N \times M$, the projection $\Phi : (\Delta, \mathfrak{G}) \to (\Sigma^N, \sigma)$ is a semiconjugacy, and the leaves $\Phi^{-1}(\omega)$ are $C^r$-smooth (this is proved under some extra restrictions [23] that we omit here.)

The restriction $\mathfrak{G}|_\Delta$ is conjugate to a *mild* dynamical system (2.4). This system is close to (2.2), but at first glance it is impossible to speak about the similarity of their asymptotic properties, however close they are. Indeed, the difference between the two maps accumulates under iterations, and in the limit there may be no similarity at all.

The situation is saved if the map $f_\omega$ in the definition of the mild skew product depends in a sense regularly on the point $\omega$ of the base. The precise statement is that *the fibers of the set $\Delta$ are Hölder continuous in $\omega$ as elements of the space $C^2$*. The same holds true for the maps $f_\omega$.

Now we can distinguish three steps of the investigation motivated by the heuristic principle above.

(1) Study new properties of step skew products.

(2) Prove that they are preserved for the nearby mild skew products with Hölder dependence of the leaves on the point of the base.

(3) Prove that the fibers of a small perturbation of a suitable smooth realization of the step skew product depend Hölder continuously on the point of the base.

The third step has been done once and for all by A. Gorodetski [23]. The first two steps are to be done individually for a property under investigation. For the properties mentioned in Theorem 2.1 this is done in [24, 25].

The idea of the proof of the Hölder continuity of the fibers is very elegant and transparent. Consider two hyperbolic sets of two diffeomorphisms and a homeomorphism that conjugates them. Then this homeomorphism necessarily is Hölder continuous. This is a classical statement.

Consider now two maps: one acting on the base $\Lambda$ (we denote it by $S$) and the other acting in a functional space where the fibers of the set $\Delta$ are points; this second map is induced by the map $\mathfrak{G}$ that permutes the fibers. These maps are obviously conjugate: the fiber over $\omega$ is sent by $\mathfrak{G}$ to a fiber over $S(\omega)$. The set $\Lambda$ is hyperbolic for $S$. The problem is that the set of fibers of $\Delta$ is by no means hyperbolic because there is no ambient manifold for it to define tangent bundles and contracting and expanding subbundles. The solution lies in the notion of *intrinsic hyperbolicity* introduced by Alekseev and Yakobson in [1]. It appears that in the conjugacy result quoted above, hyperbolicity may be replaced by intrinsic hyperbolicity. On the other hand, the induced map on the set of fibers appears to be intrinsically hyperbolic. This proves the Hölder dependence of the fibers on the point of the base, and consequently, the likely dependence of the maps $f_\omega$ on $\omega$.

### 2.4 Zero Lyapunov exponents

It was recently proved by Bochi [8] that, *among volume preserving diffeomorphisms*, those that have zero Lyapunov exponents are in a sense typical. For *arbitrary* diffeomorphisms it is believed that zero Lyapunov exponents are *not* typical. Yet we hope that the following problem may be positively solved.

**2.3 Problem** *Construct a stably non-hyperbolic measure for diffeomorphisms from some domain in the space of dynamical systems. In more detail, for any map from Theorem* 2.1 *find an ergodic invariant measure with respect to which the map has an intermediate zero Lyapunov exponent.*

The study presented here was motivated by the theory of nonlocal bifurcations, and motivates in turn new problems in the domain. On the other hand, it motivates future study of random dynamical systems, or synonymously, skew products over the Bernoulli shift. We begin with the second topic, then pass to the first one.

## 3 Skew products over the Bernoulli shift

### 3.1 Synchronization of orbits

In the search for new properties of random dynamical systems, numerical experiments with the skew products having fiber a circle were carried out by S. Trifonov, then by M. Nalski. Both discovered an effect of "synchronization of orbits": orbits with different initial conditions approached each other, formed kind of a small cloud, and then became indistinguishable. This effect, under the name "cloud of points", was investigated theoretically by V. Kleptsyn, who obtained the following theorem:

**3.1 Theorem ([52])** *Consider a finite set of orientation preserving diffeomorphisms of a circle (called generators). Suppose that they generate a semigroup with the following properties:*

(1) *minimality: all the orbits of the semigroup are dense in the circle;*

(2) *Morse-Smale: at least one map of the semigroup is Morse-Smale;*

(3) *non-factorization: there is no map of a circle topologically equivalent to a rational rotation that commutes with all the maps of the semigroup.*

*Moreover, the semigroup generated by the maps inverse to the diffeomorphisms under consideration has the minimality property as well.*

*Then for any two points of the circle there exists a set of the full Bernoulli measure in the space of one-sided sequences of $N$ symbols (where $N$ is the number of generators) such that for every sequence $\omega$ from this set, the images of these two points under the maps* (2.1) *approach each other, that is, the distance between the images tends to zero.*

The author is informed that H. Furstenberg proved a similar result (unpublished). The paper [52] is unpublished as well.

The tuples of generators that satisfy the hypotheses of the Kleptsyn theorem are in a sense typical. Namely, the tuples that satisfy all the conditions except for the third one, form an open set in the space of all tuples. The violation of the third condition gives rise to a subset of codimension infinity.

**3.2 Problem** *Are the Lyapunov exponents generically negative under the hypotheses of the Kleptsyn theorem? That is, is it true that for a.e. $x \in S^1$ and $\omega \in \Sigma^N$*

$$\lim_{n\to\infty} \frac{1}{n} \sum_{1}^{n} \log F'_m(x) < 0? \tag{3.1}$$

*Here $F_m$ is the same as in* (2.1).

## 3.2 Harmonic measures

The preceding problem is closely related to the properties of the harmonic measure related to the tuple of the generators.

**3.3 Definition** A measure on a manifold is called *harmonic* with respect to a tuple of diffeomorphisms of this manifold onto itself provided that the measure equals the average of its images under the diffeomorphisms above:

$$\mu = \frac{1}{N} \sum_{1}^{N} f_n^* \mu, \tag{3.2}$$

where $f_n^*\mu$ is defined by

$$f_n^*\mu(U) = \mu f_n^{-1} U$$

for any open set $U \subset B$.

**3.4 Remark** The Cartesian product of the Bernoulli measure on the base and the harmonic measure on the fiber of the skew product (2.2) is an invariant measure of the map (2.2).

Kleptsyn proved that *under the assumptions of Theorem* 3.1, *the harmonic measure on a circle with respect to the generators is unique and homeomorphic to the Lebesgue mesaure.*

**3.5 Problem** *Is harmonic measure absolutely continuous with respect to Lebesgue measure at least for some open set in the space of tuples of generators that satisfy the assumptions of Theorem* 3.1*?*

A positive answer to this question may help to get a positive answer to Problem 3.2.

### 3.3 Topology of skew products

A closely related subject is a topological classification of step skew products (2.2). A sufficient condition of such equivalence is *topological conjugacy of the corresponding generators.* This means that there exists a one-to-one correspondence between the generators, and a homeomorphism of a circle such that the corresponding generators are conjugated by the homeomorphism above. Note that this homeomorphism is one and the same for all generators.

**3.6 Theorem (A. Gorodetski, V. Kleptsyn)** *Under the hypotheses of Theorem* 3.1, *two systems* (2.2) *are topologically equivalent with the base untouched, that is, conjugated by a homeomorphism that sends every fiber to itself, iff the corresponding generators are simultaneously topologically conjugated, that is, conjugated by one and the same homeomorphism of the circle.*

## 4 Nonlocal bifurcations

### 4.1 Bifurcations of several homoclinic surfaces

Dynamical systems with partially hyperbolic invariant sets like those described above may appear in the nonlocal bifurcation of several homoclinic surfaces of a saddle-node cycle [40]. Recall that saddle-node cycle of a multi-dimensional vector field is a periodic orbit with one multiplier equal to 1, and the others not lying on the unit circle. The following theorem was claimed to hold in [24]:

**4.1 Theorem** *There exists an open set of one-parameter families of vector fields in* $\mathbb{R}^n$, $n \geq 4$, *such that for each family from this set:*

(1) *The zero value of a parameter corresponds to a vector field on the boundary of Morse-Smale set with a saddle-node cycle;*

(2) *There is an open subset* $\mathcal{B} \subset \mathbb{R}$ *with zero on its boundary such that the Poincaré map of vector fields which correspond to parameter values from this subset have partially hyperbolic set described in Theorem* 2.1*;*

(3) *Subset* $\mathcal{B}$ *has a positive density at* 0.

The program proclaimed in [24] is now mainly carried out in [25, 23]. Yet the proof of the above theorem is not finished. The main problem is to study the invariant set generated via the bifurcation of several homoclinic surfaces of a saddle-node cycle. The topology of this set is described in [40] Section 5.7; the set is homeomorphic to the product $\Sigma^N \times S^1$, thus being a union of fibers homeomorphic to the circle.

**4.2 Problem** *Prove that the fibers of the invariant set mentioned above are smooth and Hölder continuous in the $C^2$ norm depending on the point in the base.*

### 4.2 Twisted homoclinic surfaces

A saddle-node periodic orbit may have a very complicated set formed by the union of its homoclinic orbits. The diversity of these sets was discovered by Turaev and Shilnikov, and is described in detail in [40]. For instance, homoclinic orbits of the cycle may fill up a compact surface that is adjacent to the cycle on one side as an $m$-leaf surface and on the other side as an $n$-leaf surface. It is called a twisted homoclinic surface of type $(m, n)$.

Up to now, only the bifurcation of a surface of type $(2, 1)$ has been studied in detail. It may be called "from Morse-Smale to Smale-Williams", and generates a strange attractor just after the crossing of the boundary of the domain of the Morse-Smale systems [40]. The next result in this study was obtained by an undergraduate student, D. Pavlov (unpublished):

**4.3 Theorem** *The bifurcation of a twisted homoclinic surface of type* $(3, 2)$ *gives rise to an invariant hyperbolic set of the global Poincaré map . The topological entropy of the restriction of the Poincaré map to this set equals* $\log 3$.

**4.4 Problem** *Carry out a complete study of the bifurcations of twisted homoclinic surfaces of a saddle-node cycle.*

**4.5 Problem** *Study the bifurcations of several twisted homoclinic surfaces of a saddle-node cycle.*

## 5 Rotation numbers and moduli of elliptic curves

About 25 years ago, Arnold [4] found a connection between the rotation numbers of circle diffeomorphisms and moduli of elliptic curves. Consider a lift $f : \mathbb{R} \to \mathbb{R}$ of a diffeomorphism of a circle $S^1 = \mathbb{R}/\mathbb{Z}$. Consider a strip $\Pi_\alpha : \operatorname{Im} z \in [0, \alpha]$. Let us glue the lower and the upper boundary of $\Pi_\alpha$ by the map $f_\alpha : x \mapsto f(x) + i\alpha$. Then identify the points of the cylinder thus obtained that differ by an integer. These identifications transform the strip to an elliptic curve. This curve may be represented as a factor $T_\alpha = \mathbb{C}/\mathbb{Z} + \mu_\alpha \mathbb{Z}$ with some $\mu_\alpha = \mu_\alpha(f)$ in the upper half-plane, well-defined modulo $\mathbb{Z}$. This $\mu_\alpha(f)$ is called a modulus of the elliptic curve $T_\alpha$. When $f(x) = x + \mu$, $\mu_\alpha(f) = \mu + i\alpha$.

The problem is:

*Is it true that $\mu_\alpha$ tends to the rotation number of $f$ as $\alpha \to 0$?*

The known results in this respect are the following:

**5.1 Theorem ([61])** *If $f$ is $C^1$ equivalent to a rigid rotation, then the answer to the above question is affirmative.*

**5.2 Theorem ([42])** *If $f$ is a Morse-Smale diffeomorphism of a circle then the answer to the above question is negative.*

**5.3 Definition** Let $G$ be a group. A map $h : G \to \mathbb{R}$ is a *quasihomomorphism* if their exists a constant $C$ such that

$$|h(g_1 g_2) - h(g_1) - h(g_2)| \leq C. \tag{5.1}$$

**5.4 Example** Let $G$ be the group of the lifted diffeomorphisms of a circle: $(g : \mathbb{R} \to \mathbb{R})$ $g \in G \iff \{g(x+1) = g(x) + 1,\ g \text{ is bijective}\}$. Denote by $\rho(g)$ the rotation number of $g \in G$. Note that it is well defined not only modulo $\mathbb{Z}$, but as a real number. Then the map

$$h : G \to \mathbb{R},\ g \mapsto \rho(g) \tag{5.2}$$

is a quasihomomorphism (with $C = 2$).

**5.5 Theorem (Ghys [16])** *Let $G$ be the same group as in the example above. Then any quasihomomorphism of $G$ sends maps to their rotation numbers, i.e., coincides with the one described in the previous example.*

The following related problems remain unsolved.

**5.6 Problem** *Is the answer to the Arnold's question affirmative for diffeomorphisms with the Liouvillian rotation number?*

By definition this is an irrational rotation number for which the corresponding map $f$ may be *not Lipschitz equivalent* to a rigid rotation.

Consider the limit (if exists):

$$\mu(f) = \lim_{\alpha \to 0} \mu_\alpha(f). \tag{5.3}$$

By the last two theorems, the map

$$f \mapsto \mu(f) \tag{5.4}$$

is not a quasihomomorphism.

**5.7 Problem** *What is the reason that the map* (5.4) *is not a quasihomomorphism? Is the limit* (5.3) *not well defined? Or is it well defined but the relation* (5.1) *is violated?*

# 6 Attractors

## 6.1 Definition and genesis of minimal attractors

In 1985, during the work on the survey [5], Arnold suggested to the author to give a new definition of attractors. "I visited Kolmogorov in the hospital", he said, "and we discussed the possible definition of attractors. It should be a set from which the orbits might escape from time to time but near which they should spend the greatest amount of time." Motivated by this suggestion, I gave a definition of a *statistical limit set*, [5] Section 3.8, see the definition below.

Later on, I tried to compare this set with the closure of the union of supports of invariant measures obtained by the Krylov-Bogolyubov procedure applied to the Lebesgue measure. It appeared that these two sets would coincide provided that the definition of the first one would be slightly changed. The "edited" definition gave rise to minimal attractors. We pass now to a detailed presentation.

Consider a homeomorphism $f$ of a compact metric measure space $(X, \mu)$ into itself; the total measure is 1. We will discuss two main cases.

*Case 1.* $X$ is a compact Riemannian manifold with or without boundary. The distance and the measure are induced by the Riemannian metric, $f$ is smooth.

*Case 2.* $X$ is a Cartesian product of a torus (the fiber) with the metric and measure induced from the covering Euclidian space, and a space of two-sided sequences of, say, $N$ symbols (the base) with a Bernoulli metric and measure. The distance and measure on $X$ are Cartesian products of distances and measures on the base and the fiber. The map $f$ is a step skew product, smooth on the fibers.

For any function $\varphi$ let

$$\varphi_n = \frac{1}{n}\sum_{0}^{n-1} \varphi \circ f^k$$

be its $n$th time average under $f$. For any set $U \subset X$ let $\chi_U$ be the indicator of $U$.

**6.1 Definition** An open set $U$ is *non-essential* for a map $f$ if the sequence of averages of the indicator function of $U$ under the map $f$ converges to zero *in measure.*

**6.2 Definition** An open set $U$ is *strictly non-essential* for a map $f$ if the sequence of averages of the indicator function of $U$ under the map $f$ converges to zero *almost everywhere.*

**6.3 Remark** Any strictly non-essential set is non-essential but not vice versa in general.

**6.4 Definition** A *statistical limit set* is the complement of the union of all the strictly non-essential open sets.

**6.5 Definition** A *minimal attractor* is the complement of the union of all the non-essential open sets.

**6.6 Remark** In general, the minimal attractor is smaller than the statistical limit set, see Remark 6.3. It is, however, nontrivial to exhibit examples when the two sets are different, see Kleptsyn's example below.

The following definition of attractor was given at the same time and is widely used [60].

**6.7 Definition** A *Milnor attractor* of $f$ is a minimal closed set of full measure that contains $\omega$-limit sets for almost all points of $X$. (Milnor himself called it *probabilistic limit set.*)

All the definitions of attractors above were given for Riemannian manifolds, Case 1. For the Case 2 they are presented here for new.

It is easy to give an example of a system for which the minimal attractor is different from Milnor attractor. One of the main problems is:

**6.8 Problem** *Do the Milnor attractor, the statistical limit set and the minimal attractor generically coincide?*

A positive answer for the Case 1 follows from the famous conjecture of Palis (as yet unproved) about the "finitude of attractors" [63]. For Case 2 we abstain from formulating any conjecture. Suffice it to say that it may be easier than Case 1.

**6.9 Problem** *Suppose that a dynamical system (Case 1) has an invariant hyperbolic set. Is it true that this set belongs to a minimal attractor?*

The following dynamical system demonstrates a queer behavior. Consider a vector field in the plane with two saddles and two saddle connections between them. For sufficiently large ratios of eigenvalues in each saddle, this system has a strange property: time averages of continuous functions diverge almost everywhere, provided the function takes different values at different singular points. This is the renowned Bowen example.

**6.10 Problem** *Is it true that there exists an open set in the space of dynamical systems such that for the diffeomorphisms from this set, time averages of continuous functions do not converge almost everywhere?*

The example of non-coincidence of statistical limit sets and minimal attractors due to Kleptsyn is a modification of the Bowen's example. Let us replace one saddle by a saddle-node (one eigenvalue is negative, another one is zero). Let $X$ be the domain bounded by the union $M$ of two connections between the singular points, and let the orbits wind from inside of $X$ onto $M$. Then the statistical limit set will consist of two points, the minimal attractor will be a saddle-node only, and the Milnor attractor will be the boundary of $X$.

The study of the different types of attractors for the dynamical systems mentioned above in Case 2 is quite a new field. It is interesting in itself and may give approaches to the classical smooth case.

## 6.2 Good measures

The following definition shows that the minimal attractor is really nonempty. It uses the famous Krylov-Bogolyubov construction, with the only difference that the initial measure is the Lesbegue measure. Let $f$ be the same as in Case 1, Subsection 6.1.

**6.11 Definition** Let $\mu_n$ be an average of a shifted measure $\mu$:

$$\mu_n = \frac{1}{n}\sum_{1}^{n} f^k \mu.$$

A *good measure* $\mu_\infty$ is an arbitrary limit point of a set $\{\mu_n\}$ in the weak topology.

That is, $\mu_\infty$ is a good measure provided that there exists a sequence $k_n$ such that $\mu_\infty = \lim \mu_{k_n}$ in the weak topology. Any good measure is $f$-invariant. In general, good measures are not uniquely defined.

**6.12 Theorem** *A* minimal attractor *of the map $f$ above is a closure of the union of supports of all the good measures of $f$.*

**6.13 Problem** *Suppose a diffeomorphism f as above has a finite minimal attractor, as well as all the nearby ones. Is it true that in this case the minimal attractor is Lyapunov stable?*

**6.14 Theorem (Kleptsyn, Golenishcheva-Kutuzova)** *Good measures for the Bowen example are not uniquely determined.*

## 6.3 Time averages of continuous functions and minimal attractors

It is not obvious (and not even true, see Bowen's example) that any continuous function has a well-defined time average for Lebesgue almost all initial conditions. An object that always exists is a *weak time average* defined as follows.

Consider any continuous function on a Lebesgue measure space as a distribution. Any sequence of bounded continuous functions contains a subsequence that converges in the sense of distribution theory (i.e. converges weakly).

**6.15 Definition** Any distribution which is a weak limit of some subsequence of a sequence of time averages of a continuous function is called a *weak time average* of this function.

**6.16 Theorem ([6])** *Any weak time average of a continuous function with the support inside a nonessential open set is zero. For any open set which has a nonempty intersection with the minimal attractor there exists a continuous function with the support inside this set such that at least one time average of this function is nonzero.*

This theorem shows that time averages of continuous functions do not depend on the behavior of the function outside the minimal attractor, and depends on its values at any point of the minimal attractor. Hence, minimal attractors are really physically meaningful.

## 6.4 More about non-coincidence of attractors

There are examples of non-coincidence of the Milnor and minimal attractors provided by degenerate systems. A codimension 2 example is elementary [5]. Consider a planar vector field with a hyperbolic saddle having two separatrix loops. The union of these loops forms an eight-shaped figure. Suppose that the orbits in a neighborhood of this figure approach it as time tends to plus infinity. Then the minimal attractor of the system consists of the saddle point only, and the Milnor attractor consists of the entire eight-shaped figure.

An example of codimension 1 (or even $1^{1-\varepsilon}$ in a sense explained below) is provided by a Cherry flow on a two-torus. This is a vector field on a two-torus with the following properties. The field has exactly two singular points: a hyperbolic saddle and an unstable node with a heteroclinic orbit that connects them. It also has an invariant set $M$ such that the unstable separatrices of the saddle are dense in it.

**6.17 Theorem (Kleptsyn)** *The minimal attractor of the Cherry flow is the saddle. The Milnor attractor is the whole set $M$.*

Consider a vector field $v$ that determines a Cherry flow. It appears that for a generic one-parameter perturbation of $v$, Cherry flows correspond to a *measure zero set of cardinality continuum in the parameter space.* Hence, non-coincidence of the Milnor and minimal attractors is a degenerate effect that occurs in locally generic one-parameter families not only for isolated but also for sets of parameter values that have the cardinality of the continuum.

**6.18 Problem** *Is the same true for higher dimensional analogs of the Cherry flow?*

# 7 Slow-fast systems and smooth classification

Recall some classical definitions. A system

$$\dot{x} = f(x, y, \varepsilon), \qquad \dot{y} = \varepsilon g(x, y, \varepsilon) \tag{7.1}$$

is called a *slow-fast system.* It is a one-parameter family with a parameter $\varepsilon$; $x$ is a *fast variable*, and $y$ is a *slow* one. The critical value $\varepsilon = 0$ corresponds to a *fast system*:

$$\dot{x} = f(x, y, 0), \qquad \dot{y} = 0. \tag{7.2}$$

A closely related family with the fast variable being a phase variable, and the slow one being a parameter,

$$\dot{x} = v(x, y), \qquad v(x, y) = f(x, y, 0), \tag{7.3}$$

is also called a fast system.

## 7.1 Decoding fingerprints

Any family of vector fields admits a *slow-fast extension.* Indeed, consider a family (7.3). Any family (7.1) is called a *slow-fast extension* of (7.3). If a bifurcation occurs in the family (7.3), then the corresponding change of the behavior of orbits in the family (7.1) is called a corresponding *dynamical bifurcation.*

When we deal in reality with one-parameter bifurcation problems, the parameter in fact is slowly changing with time. Therefore, dynamical bifurcations seem to be physically even more realistic than classical ones.

**7.1 Problem** *Investigate slow-fast extensions of local and nonlocal bifurcations described in* [5, 40]. *In particular, find out how to decode the "fingerprints": using the observation of the slow-fast system* (7.1), *find the family* (7.3) *for which the given system* (7.1) *is a slow-fast equation.*

This meta-problem goes back to J Guckenheimer [26].

## 7.2 Slow-fast extensions of nonlocal bifurcations in the plane

Consider planar nonlocal bifurcations related to the vanishing of a limit cycle at the critical parameter value. The following cases occur. The cycle vanishes:

(a) on a separatrix loop;

(b) on a heteroclinic orbit of a saddle-node singular point;

(c) on a semistable limit cycle.

**7.2 Problem** *Consider a slow-fast extension over one of the bifurcations* (a)–(c). *Study the rate of change of the period of the slow oscillations close to the cycles of the fast system, as the slow variable passes the critical value of the parameter of the fast system. Find a way to distinguish the bifurcations* (a)–(c) *using these rates of change.*

### 7.3 Sectorial embedding problem

A slow-fast extension of a bifurcation of a semistable limit cycle is similar to that of a bifurcation of a semistable singular point. The latter slow-fast system is a classical one, studied in detail beginning at the time of Andronov and Pontryagin in the middle of the past century. A formal reduction of the first system to the second one may be based on the conjecture that the following problem has an affirmative answer.

**7.3 Problem** *Consider a saddle-node local family on the line:*

$$x \mapsto x + (x^2 + y)\Omega(x, y), \ \Omega(0,0) \neq 0, \tag{7.4}$$

*and a corresponding slow-fast extension*

$$x' = x + (x^2 + y) f(x, y, \varepsilon), \qquad y' = \varepsilon g(x, y, \varepsilon), \qquad f(x, y, 0) = \Omega(x, y). \tag{7.5}$$

*Is it true that there exists a finite cover of a punctured neighborhood of* $0$ *in the phase-parameter space* $(x, y, \varepsilon)$ *such that in any domain of this cover the system* (7.5) *is a time-one shift along the orbits of a system*

$$\dot{x} = (x^2 + y) F(x, y, \varepsilon), \qquad y' = \varepsilon G(x, y, \varepsilon), \tag{7.6}$$

*where* $F$ *and* $G$ *depend on the domain of the cover?*

### 7.4 Generalized true slow surface

Suppose that the fast system (7.3) has a closed attracting invariant manifold $M_y$ for any value of a slow variable $y$ from some open domain $Y$. Suppose that the rate of approach of the orbits of the fast system to this manifold is much larger than the rate of approaching each other for the orbits of the fast system on the invariant manifold.

**7.4 Theorem (O. Anosova [2, 3])** *The slow-fast extension of the family* (7.3) *described above has a family of smooth invariant manifolds* $M_\varepsilon$ *such that*

$$M_\varepsilon \to \bigcup_{y \in K} M_y, \tag{7.7}$$

*where* $K$ *is an arbitrary compact subset of* $Y$. *The family* $M_\varepsilon$ *is smooth in* $\varepsilon$.

The precise statement of this theorem may be found in [2, 3]. The invariant manifolds $M_\varepsilon$ are called *generalized true slow surfaces.* This reduces a slow-fast system to a family of systems on generalized true slow surfaces, and provides a mighty tool for the investigation of some dynamical bifurcations; examples are mentioned in the following problem.

**7.5 Problem** *Study slow-fast extensions of the following bifurcations:*

(a) *a saddle-node singular point with exactly one homoclinic orbit;*

(b) *a saddle-node cycle with a smooth homoclinic torus;*

(c) *a saddle-node cycle with a smooth homoclinic Klein bottle.*

Problem 7.5(c) is especially ripe for study. The fact is that recently A. Borisyuk gave a complete description [12, 13] of the bifurcations that occur on a homoclinic Klein bottle of a saddle-node cycle.

**7.6 Problem** *Study slow-fast extensions of nonlocal bifurcations for which a nontrivial hyperbolic invariant set is generated. The following cases are particularly interesting to begin with:*

- *bifurcations of several homoclinic orbits of a saddle-node singular point;*
- *bifurcation of an attracting twisted homoclinic surface of a saddle-node cycle.*

## 7.5 Canards on the two-torus

Generic slow-fast systems in the plane

$$\dot{x} = f(x, y, \varepsilon), \qquad \dot{y} = \varepsilon g(x, y, \varepsilon), \qquad (x, y) \in \mathbb{R}^2$$

exhibit a particularly simple description of the shape of orbits as $\varepsilon \to 0$. Namely, the orbits tend to the trajectories of a degenerate system combined with alternating segments of phase curves $y = \text{const}$ of the fast system and stable portions of the slow curve $M : f(x, y, 0) = 0$.

In the presence of an auxiliary parameter a new kind of limit behavior of solutions was discovered. Namely, for the two-parameter family,

$$\begin{aligned} \dot{x} &= y - x^2 - x^3 \\ \dot{y} &= \varepsilon(a - y) \end{aligned}$$

there is an exponentially narrow horn $a \in (a_1(\varepsilon), a_2(\varepsilon))$, $\varepsilon_0 > \varepsilon > 0$, $|a_2(\varepsilon) - a_1(\varepsilon)| < e^{-c/\varepsilon}$ such that for $(a, \varepsilon)$ inside this horn solutions are close to an attracting cycle combined with the arcs passing near not only phase curves of the fast system and stable portions of the slow curve, but *also near unstable portions of the slow curve.*

Solutions of this type are called *ducks* or *canards* (ducks in French).

The existence of canard solutions in two-dimensional slow-fast systems was discovered in different situations, see [5] and references therein. But all known examples contained an auxiliary parameter like $a$ in the previous system.

The system on a torus

$$\dot{x} = a - \cos x - \cos y, \qquad \dot{y} = \varepsilon, \qquad (x, y) \in T^2 \tag{7.8}$$

has no auxiliary parameter: $a$ is fixed, and one may replace it by, say, 1.1.

**7.7 Theorem ([27])** *There is a sequence of intervals $C_n \subset \{\varepsilon > 0\}$, $C_n \to 0$ as $n \to \infty$, such that for any $\varepsilon \in C_n$, the system* (7.8) *has an attracting canard cycle.*

The great majority of solutions do attract to this cycle. We demonstrate this new effect using a particular example (7.8). The conjecture is that this effect is locally generic.

**7.8 Problem** *There is an open set of slow-fast systems in $T^2$ with the same property as in Theorem* 7.7.

## 7.6 Semi-linear slow-fast systems

This is the name for the systems linear with respect to the fast variable:

$$\dot{x} = A(y, \varepsilon)x, \qquad \dot{y} = \varepsilon f(y, \varepsilon). \tag{7.9}$$

**7.9 Problem** *Describe loss of stability in system* (7.9) *as the greater eigenvalue of $A$ becomes positive via the slow change of $y$. How to predict this event before it happens, using the (numerical) observation of the system in the stability domain?*

## 7.7 Bogdanov-Takens homoclinic funnel

Local bifurcations of planar vector fields with nonzero nilpotent linear part are described by Bogdanov and Takens, [10, 9, 77]. They found all the phase portraits that occur in generic perturbations of a germ of a vector field with a nonzero nilpotent linearization at a singular point corresponding to zero parameter value in the family

$$\dot{x} = v(x, \varepsilon), \quad x = (x_1, x_2) \in \mathbb{R}^2, \quad \varepsilon \in \mathbb{R}^2, \quad v(x, 0) = (x_2, 0) + \cdots . \tag{7.10}$$

In particular, they found a curve like a branch of a parabola in the parameter plane such that the points on this curve correspond to a vector field with a separatrix loop of a hyperbolic saddle.

Consider now a discrete analog of the Bogdanov-Takens family

$$x \mapsto x + V(x, \varepsilon), \quad V(x, 0) = (x_2, 0) + \cdots . \tag{7.11}$$

An example of such a family is a time-one shift along the phase flow of the orbits of the family (7.10). The maps of this family corresponding to the points of the same parabola-like curve as above, have a saddle fixed point with a homoclinic loop formed by its stable and unstable manifolds. This is a degeneracy of codimension infinity. In general, stable and unstable manifolds intersect transversally. This gives rise to the following problem.

**7.10 Problem** *Is it true that for a generic discrete Bogdanov-Takens family* (7.11) *there exists a funnel in the parameter plane with the vertex at the origin such that the points of the funnel correspond to maps with a homoclinic intersection of stable and unstable manifolds of a hyperbolic saddle. The width of the funnel in the analytic case tends exponentially to zero at the vertex.*

For analytic maps a complementary result is proved [14]: outside some exponentially narrow funnel there are no homoclinic intersections. The question is to prove an affirmative statement: the parameter values that correspond to homoclinic intersections range over some exponentially narrow funnel.

The problem is very close to the topics discussed in the lectures of V. Gelfreich.

## 7.8 Smooth classification of local families and global vector fields

Smooth classification of local families of vector fields and diffeomorphisms is a key tool in the nonlocal bifurcation theory, see for example [40]. For simple local families, normal forms are simple themselves; in particular, they depend on a finite number of parameters and are integrable. More complicated normal forms are characterized by functional moduli. This is true for instance for a saddle-node family on the line [45],

$$x \mapsto x + (x^2 + \varepsilon)f(x,y), \quad f(0,0) \neq 0. \tag{7.12}$$

The following problem deals with slightly more complicated families and is unsolved.

**7.11 Problem** *Find all the (functional?) invariants of the smooth classification for the following 2-parameter families:*

(a) *Unfolding of a saddle-node germ of a map of the line of multiplicity 3:*

$$x \mapsto x + ax^3 + \cdots, \quad a \neq 0;$$

(b) *Unfolding of a slow focus of a planar vector field with the first Lyapunov focus value zero;*

(c) *Bogdanov-Takens family.*

A nonlocal analog of this problem looks as follows. In the paper [78], Takens gave a smooth classification of vector fields near a separatrix polygon as in the Bowen example, see above. The theory of normal forms and desingularizations may be useful in the solution of the following problem.

**7.12 Problem** *Find all the invariants of smooth classification of a vector field in a neighborhood of its polycycle (a separatrix polygon).*

It makes sense to begin with hyperbolic polycycles, that is, polycycles whose vertices are hyperbolic saddles.

# B. Hilbert-type numbers and polynomial dynamical systems

Hilbert's 16th problem asks:

*What may be said about the number and location of limit cycles of planar polynomial vector fields?*

This problem appears to be among the most persistent ones in Hilbert's famous list, second only to Riemann's $\zeta$-function conjecture. Different simplified versions of the problem were suggested in the last three decades. We describe here some results and problems in the field. A more detailed presentation may be found in [38].

## 8 Two simplifications of Hilbert's 16th problem

### 8.1 Hilbert-Arnold problem

In our time, generic finite parameter families are as much respected as polynomial ones. This motivates the following counterpart of Hilbert's 16th problem for smooth vector fields:

**8.1 Problem** *Consider a generic one-parameter family of smooth vector fields on the two-sphere with a compact parameter space. Is it true that the number of limit cycles of the vector fields of this family is uniformly bounded with respect to the parameter?*

Recall that the *cyclicity of a polycycle in a family of equations* is the maximal number of limit cycles that may occur from the polycycle in this family.

**8.2 Problem** *Is it true that a polycycle occurring in a generic k-parameter family of smooth planar vector fields may generate only a finite number of limit cycles, with an upper bound $B(k)$ depending on k only?*

This problem is equivalent to Problem 8.1, [70]. It may be considered as a main problem in the nonlocal planar bifurcation theory. Both problems remain unsolved. Problem 8.2 is partly solved, see Theorems 8.3 and 8.5 below.

A singular point of a planar vector field is called *elementary* if the linearization of the field at this point has at least one nonzero eigenvalue. A polycycle is called *elementary* if it contains elementary singular points only.

Denote by $E(k)$ the maximal number of limit cycles that may bifurcate from an elementary polycycle in a typical $k$-parameter family of smooth planar vector fields.

The existence of the number $E(k)$ was proved in 1995:

**8.3 Theorem ([46])** *For any $k$, the number $E(k)$ exists.*

**8.4 Corollary** *The Hilbert-Arnold problem has a positive solution for families of vector fields having elementary singular points only.*

The proof makes use of the theory of normal forms and the fewnomial theory of Khovanski [50]. Recently V. Kaloshin found an upper estimate on $E(k)$.

**8.5 Theorem ([49])** *For any $k$, $E(k) \leq 2^{25k^2}$.*

The proof is based on the new development of the stratification theory due to Kaloshin. Theorem 8.5 is announced in [39] and proved in [49]. The estimate given by this theorem may be too high. Yet it is one of the first *Hilbert type numbers* (that is, bounds to the number of limit cycles) obtained up to now. Some other Hilbert type numbers are presented below.

## 8.2 Infinitesimal Hilbert's 16th problem

Consider a real polynomial $H$ of degree $n+1$ in the plane. A closed connected component of a level curve $H = t$ is denoted by $\gamma(t)$ and called an *oval* of $H$. These ovals form continuous families.

Let $\omega = Adx + Bdy$ be a real 1-form with polynomial coefficients of degree at most $m$. Let

$$I(t) = \int_{\gamma(t)} \omega. \tag{8.1}$$

**8.6 Problem (Infinitesimal Hilbert's 16th problem)** *Find an upper bound $V(m,n)$ of the number of real zeros of the integral (8.1). The bound should be uniform with respect to the choice of the polynomial $H$, the family of ovals $\{\gamma(t)\}$ and the form $\omega$. It should depend only on the degrees $m, n$.*

Zeros of Abelian integrals (8.1) are related to limit cycles in the following way. Consider a perturbation of an integrable system

$$dH + \varepsilon\omega = 0, \tag{8.2}$$

with $H$ and $\omega$ the same as above. We say that an oval $\gamma(t)$ *generates a limit cycle of* (8.2) if there exists a continuous family of closed curves $l(\varepsilon)$, defined for small $\varepsilon$, such that $l(\varepsilon)$ is a limit cycle of (8.2) for $\varepsilon \neq 0$, and $l(0) = \gamma(t)$.

**Pontryagin Criterion** *If an oval $\gamma(t)$ of the polynomial $H$ generates a limit cycle of* (8.2), *then $I(t) = 0$. On the other hand, if $I(t) = 0$ and $I'(t) \neq 0$, then the oval $\gamma(t)$ generates a limit cycle of* (8.2).

There are numerous results giving upper and lower estimates of the number of zeros of the integral above. There are due to L. Gavrilov, A. Glutsyuk, Yu. Ilyashenko, A. Khovanski, P. Mardesiç, J. Musiño, D. Novikov, G. Petrov, I. Pushkar', A. Varchenko, S. Yakovenko and others. The main references, together with a survey of results, may be found in [38]. We mention here only a few results and problems.

## 8.3 Exactness theorems and spaces of Abelian integrals

**8.7 Definition** A complex polynomial $H$ of degree $n+1$ is called *ultra-Morse* if it has $n^2$ nondegenerate critical points with different critical values and its highest homogeneous form $h$ having no multiple linear factors.

The assumption on critical points and values defines a *Morse polynomial*; the latter requirement defines a narrower class; this motivates the term. Ultra-Morse polynomials form a Zariski open set in the space of all polynomials of given degree.

**8.8 Theorem ([30, 31, 67])** *In* (8.1), *let $\{\gamma(t)\}$ be a family of ovals of a real ultra-Morse polynomial $H$, of degree $n+1 \geq 3$, and let $\omega$ be a polynomial 1-form of degree no greater than $n$. If $I(t) \equiv 0$, then the form $\omega$ is exact: $\omega = df$ for some polynomial $f$.*

Theorem 8.8 implies that the linear space of integrals (7.1) corresponding to the same family of ovals and all of the forms $\omega$ from the theorem, is isomorphic to the factor space of these forms modulo the exact forms. This latter space is isomorphic to the space of polynomials in two variables of degree no greater than $n-1$. The dimension of the latter space equals $N = (n^2+n)/2$. This implies that near arbitrary $N-1$ ovals of an ultra-Morse polynomial one may find the same number of limit cycles of the perturbed equation (8.2).

The following multidimensional analog of Theorem 8.8 has also been obtained. Denote by $H$ a polynomial of $k$ complex variables that satisfies some genericity assumptions that we omit here. Let $\gamma(0)$ be a cycle of real dimension $k-1$ on a surface $H=0$ that is supposed to be nonsingular. Let $\gamma(c)$ be a small perturbation of $\gamma(0)$ on the level surface $H=c$. Denote by $I_m$ the space of integrals of the form

$$I_\alpha(c) = \int_{\gamma(c)} \alpha,$$

where $\alpha$ ranges over the space of all polynomial $(k-1)$-forms with the degree of the coefficients no greater than $m+1$. We are interested in the dimension of $I_m$. Consider a cube $K$ given in the integer lattice $\mathbb{Z}^k$ by the inequalities $0 \leq x_j \leq n-2$. Denote by $O$ the orbit of the cube $K$ under the action of the semigroup generated by a shift $(x_1, x_2, \dots, x_k) \to (x_1+n, x_2, \dots, x_k)$. Let $M(m,k,n)$ be the set of all those points of $O$ that satisfy the inequality $x_1+\dots+x_k \leq m$.

**8.9 Theorem (Pushkar' [68])** *Let the cycle $\gamma(0)$ be non-homological to zero in the projective compactification of $H = 0$. Then*

$$\dim I_m = M(m,k,n).$$

**8.10 Problem (Arnold)** *Is it true that the number of real zeros of the functions from the space $I_m$ equals its dimension minus one (the space $I_m$ is Tchebyshev)?*

## 8.4 Strong infinitesimal Hilbert's 16th problem

An estimate of the number of zeros of Abelian integrals, even if it would have been obtained, would not answer the following question.

**8.11 Problem (Strong infinitesimal Hilbert problem)** *Give an upper bound for the number of limit cycles of a polynomial vector field which is close to a hamiltonian one of degree $n$.*

This is the same as to estimate the number of limit cycles of the equation

$$dH + \omega = 0$$

for $\omega = Adx + Bdy$, where $A, B$ are polynomials of degree at most $n$, *with small coefficients.* In addition to limit cycles generated by the ovals of $H$, the problem requires one to estimate the number of limit cycles generated by polycycles of $H$. For ultra-Morse $H$, any of these polycycles may contain only one vertex, because all the critical points of $H$ belong to different level curves. Therefore, it may be a separatrix loop or an eight-shaped figure. *The restricted version of the problem requires investigation of these cases only.* The general problem deals with more complicated critical level curves of $H$. On the one hand, this problem is closely related to bifurcation theory. On the other hand, it is linked to the following two results which are of independent interest.

**8.12 Theorem ([71, 59])** *A separatrix loop of a hyperbolic saddle that occurs in an analytic family of planar vector fields has but a finite cyclicity. Moreover, it is not greater than $n^4$.*

**8.13 Theorem ([59])** *The multiplicity of any zero of an integral* (8.1) *at any interior point of its domain is no greater than $n^4$.*

**8.14 Problem** *Find an upper bound for the cyclicity of an eight-shaped polycycle of an ultra-Morse hamiltonian in the family of planar polynomial vector fields of degree $n$.*

# 9 Hilbert type numbers for special classes of equations

For different simplified classes of equations an analog of Hilbert's 16th problem has been considered. The results obtained are based on the following theorem from one-dimensional complex analysis.

## 9.1 The Growth-and-Zeros Theorem for holomorphic functions

Consider two sets, $U$ and $K$, located in $\mathbb{C}$ or in a Riemann surface. Suppose that $U$ is simply connected and $K \subset U$ is compact and path-connected. Let $f$ be a holomorphic function in $\bar{U}$. The *Bernstein index* of $f$ with respect to $U$ and $K$ is

$$B_{U,K}(f) = \log \frac{\max_{\bar{U}} |f|}{\max_K |f|}.$$

**9.1 Theorem ([48])** *There exists a geometric constant* $\gamma(U,K)$ *such that*

$$\#\{z \in K | f(z) = 0\} \leq \gamma(U,K) B_{U,K}(f).$$

*The value* $\gamma(U,K)$ *may be taken as* $e^{\rho}$*, where* $\rho$ *is the diameter of* $K$ *in the Poincaré metric of* $U$.

The diameter $\rho$ may be estimated in terms of Euclidean, rather than Lobachevski, geometry. Indeed, let the *intrinsic distance* between two points in $K$ be the infimum of the lengths of paths in $K$ that connect these points. Let the *intrinsic diameter* of $K, d_{\text{int}}(K)$ be the diameter of $K$ in the sense of the intrinsic distance. The definition works for $K$ located both in $\mathbb{C}$ and in a Riemann surface over $\mathbb{C}$; in the latter case the metric is lifted from $\mathbb{C}$ to the Riemann surface. Let $\varepsilon$ be the gap between $K$ and $\partial U$ in the sense of the Euclidean distance lifted from $\mathbb{C}$:

$$\varepsilon = d(K, \partial U).$$

**9.2 Theorem ([37, 22])** *Theorem* 9.1 *holds with*

$$\gamma(U,K) = e^{2d_{\text{int}}(K)/\varepsilon}.$$

## 9.2 Applications to Abel and Liénard equations

Abel equation is an equation on the cylinder polynomial in $y$ and periodic in $x$:

$$\frac{dy}{dx} = y^n + \sum_0^{n-1} a_j(x) y^j, \quad y \in \mathbb{R}^1,\ x \in S^1, \tag{9.1}$$

the coefficients $a_j$ are only assumed to be continuous, unless explicitly stated otherwise.

The general problem is to get an upper bound on the number of limit cycles of (9.1). For small values of $n (n \leq 3)$ this number is no greater than $n$. The explosion comes at $n = 4$: for degree $n \geq 4$, the Abel equation may have an arbitrary number of limit cycles (A. Lins Neto [53], A. Panov [64]).

Therefore, in order to get upper bounds for the number of limit cycles, one should make some hypothesis on the coefficients.

**9.3 Problem** *Find an upper bound on the number of limit cycles of the Abel equation* (9.1)*, in which the coefficients* $a_j$ *are trigonometric polynomials of degree no greater than* $m$*. The bound should be expressed in terms of* $n$ *and* $m$ *only.*

This problem is unsolved even for $m = 1$.

In his list of problems for the 21st century, Smale [76] included Hilbert's 16th problem. As a simplified version, he mentioned the Liénard equation

$$\dot{x} = y - F_n(x), \quad \dot{y} = -x, \quad F_n(x) = x^n + \sum_1^{n-1} a_j x^j, \quad n \text{ odd}. \tag{9.2}$$

The free term of $F_n$ may be taken to be 0 by means of translation along the $y$ axis. The only singular point is zero, hence, there is only one nest of limit cycles. Moreover, there are no limit cycles near infinity. Indeed, an elementary study going back to [56] shows that the point $(1:0)$ at infinity is of the type "repelling node". Hence, there are parabolic sectors with their vertices at this point at infinity that repel all of the orbits.

**9.4 Problem** *Find an upper bound on the number of limit cycles for the Liénard equation* (9.2) *in terms of the degree $n$.*

A first step in the study of simplified versions of Hilbert's 16th problem may be a solution to the *restricted versions* of these simplified problems. For the Abel and Liénard equations, the restriction is the magnitude of the coefficients, and this magnitude enters into the upper bound. For the infinitesimal Hilbert problem, this is a restriction on the polynomial $H$. For the Hilbert-Arnold problem, it is the requirement that all of the singular points of the equation are elementary.

These restricted versions are now solved. The solution of the Hilbert-Arnold problem was mentioned in Section 8.1. The solution of the restricted infinitesimal Hilbert problem is discussed in Section 8.2. For the Abel and Liénard equations, the solutions look like the following.

**9.5 Theorem ([37])** *Consider the Abel equation* (9.1) *with the restriction*

$$|a_j(x)| < C. \tag{9.3}$$

*The number $L$ of limit cycles of equation* (9.1) *satisfying* (9.3) *has an upper bound:*

$$L \leq e^{e^{C^{3n}}}.$$

**9.6 Theorem ([43])** *Consider the Liénard equation* (9.2) *with the restriction*

$$|a_j| < C, \quad C \geq 4. \tag{9.4}$$

*The number $L$ of limit cycles of equation* (9.2) *satisfying* (9.4) *has an upper bound:*

$$L \leq e^{e^{C^{14n}}}.$$

The estimates in these theorems do not pretend to be realistic. Yet they are the only known estimates of their kind.

One of the main goals of the future investigation is to get rid of the restrictions on the coefficients for the Liénard equation, and to replace it by another requirement (say, that the coefficients are trigonometric polynomials of given degree) for the Abel equations. For this we need general results about the analytic continuation of the Poincaré map for these equations. These results may be closely related to the persistence problems discussed below.

**9.7 Problem** *Is it true that the Abel equation*

$$\dot{z} = z^n + \sum_0^{n-1} a_j(t) z^j, \quad z \in \mathbb{C}, \ t \in S^1,$$

*with bounded coefficients* $|a^j(t)| < C$ *has no more than* $A_{\mathbb{C}}(n, C)$ *limit cycles corresponding to the fixed points of the Poincaré map related to one circuit over* $S^1$? *If yes, find* $A_{\mathbb{C}}(n, C)$.

**9.8 Problem** *Is it true that the Poincaré map mentioned in the previous problem has a countable number of periodic orbits (with unbounded periods)?*

## 9.3 Quadratic vector fields

Consider a quadratic vector field with a non-degenerate singular point of center or focus type. A linear coordinate change brings this field to the form

$$\dot{x} = y + \alpha x + p, \qquad \dot{y} = -x + \alpha y + q, \tag{9.5}$$

where $p$ and $q$ are nonzero homogeneous polynomials of degree 2. The 6-dimensional tuple $\lambda$ of the coefficients of the polynomials $p$ and $q$ may be transformed, by a rescaling of the phase plane, to a point of the unit sphere. Dulac found polynomials $V_3, V_5, V_7$ of $\lambda$ such that the equation (9.5) has a center at zero if and only if

$$\alpha = 0, \quad V_3 = V_5 = V_7 = 0. \tag{9.6}$$

The explicit form of the polynomials $V_k$ is of no importance now.

Denote by $\Delta(\delta)$ the set of vector fields (9.5) that are $\delta$-distant from the set of quadratic centers in the following sense:

$$|\alpha| + \Sigma_1^3 |V_{2k+1}(\lambda)| \geq \delta,$$

where the $V_k$ are the polynomials from (9.6). Denote by $H(2, \delta)$ the upper bound of the number of limit cycles of equations of the class $\Delta(\delta)$, that are located in the disk $x^2 + y^2 \leq \delta^{-2}$ and whose distance to all singular points of the equation is no less than $\delta$, after the rescaling mentioned above.

**9.9 Theorem** ([41])

$$H(2, \delta) \leq e^{e^{C\delta^{-1}}}. \tag{9.7}$$

This theorem is proved modulo the problem stated below. The theorem shows that the main difficulties in Hilbert's 16th problem for quadratic vector fields is related to the proximity of limit cycles to the singular points of the equation, both finite and infinite.

Consider the Poincaré map $P(z, \alpha, \lambda)$ of equation (9.5) as a Taylor series in the polar radius $r$ with coefficients polynomial in $\alpha$, $\nu = \exp(2\pi i \alpha)$ and $\lambda$:

$$P(x) - x = \sum_1^7 a_j(\alpha, \nu, \lambda) x^j + O(x^8).$$

It is known that $a_1 = e^{2\pi\alpha} - 1$, and $a_{2l} = 0$, provided $a_{<2l} = 0$. Moreover, the coefficients $a_j$ belong to the ideal generated by $V_j$.

**9.10 Problem** *Find a function $f(\delta)$ that gives a lower bound for the Taylor coefficients of the displacement of the Poincaré map in the following sense. For any operation* (9.5) *representing a vector field from $\Delta(\delta)$,*

$$\sum_{1}^{3} |a_{2j+1}| \geq f(\delta).$$

The solution of this problem may specify the factor $C$ in (9.7).

### 9.4 Bautin index and generation of limit cycles from polynomial centers

Consider a polynomial vector field of degree $n$ with a singular point of center type.

**9.11 Problem** *Give an upper bound to the number of "small amplitude" limit cycles that may occur from a singular point of center type under a perturbation in the class of polynomial vector fields of degree $n$.*

For polynomial vector fields with a singular point having linear part a center, the real parts of the coefficients of the resonant normal form are polynomials in the coefficients of the given vector field. Denote these linear parts by $a_1, \dots, a_m, \dots$. They generate a polynomial ideal denoted by $I$. It is called the *Bautin ideal.* By the Hilbert basis theorem, it has a finite number of generators:

$$I = (a_1, \dots, a_m)$$

for some $m > 1$.

The minimal such $m$ is called *the Bautin index* of the ideal $I$; it depends on $n$ only. It is known [72] that the number of cycles in Problem 9.11 does not exceed the Bautin index. This gives rise to the following problem.

**9.12 Problem** *For any $n$ find the corresponding Bautin index.*

It is only known [7, 80], that $m(2) = 3$. The result about the Bautin index was recently improved in [28] where a *reduced Bautin index* is introduced. It is in general smaller than the Bautin index, and gives an upper bound for the number of the small amplitude limit cycles as well.

**9.13 Problem** *How many limit cycles may be generated from a (large amplitude) oval of a quadratic center?*

## 10 Kupka-Smale property and persistence problems for polynomial dynamical systems

Polynomial dynamical systems consist of polynomial maps and polynomial vector fields. In this setting, polynomials may be substituted by rational functions. For simplicity we will speak about polynomials. Most of the theory is developed in the complex plane, but problems and conjectures make sense for higher dimensional systems as well.

When the dimension of the phase space and the degree of the polynomial maps is fixed, we get a finite-dimensional family of dynamical systems. Local perturbations are impossible in such families: any (small) change of the coefficients that determine the system produces a global change in the whole of the phase space.

## 10.1 Kupka-Smale property for polynomial automorphisms of the plane and Petrovski-Landis strategy

A polynomial automorphism of the plane is a polynomial biholomorphic map of the complex (or real) plane to itself. A map is *Kupka-Smale* if all periodic points are hyperbolic, and the stable and unstable manifolds of any two saddle points are transverse. G. Buzzard, S. Hruska and the author (in preparation) proved that Kupka-Smale maps form a residual set of full Lebesgue measure in the space $\mathcal{P}_d$ of polynomial automorphisms of $\mathbb{C}^2$ of fixed (dynamical) degree $d \geq 2$. Moreover, a heteroclinic point of two saddle periodic orbits may be in a sense globally continued over the whole parameter space. This is the first *persistence theorem* proved in holomorphic dynamics.

The genericity of the Kupka-Smale property for *smooth* real maps is proved with by means of the local perturbation technique. As explained above, this technique is not applicable to polynomial dynamical systems, and we have to use the so-called Petrovski-Landis strategy instead. This strategy is applied to prove that certain geometric features of a system (say, transversality of stable and unstable manifolds of periodic orbits) are inherent in a dense (or even more massive) set of polynomial dynamical systems. The strategy is carried out as follows. Suppose that the property under investigation is not inherent in systems from some open subset of the parameter space. The first step is to prove that, under this assumption, the desired property is violated globally with respect to the parameter. This is formalized in so-called *persistence theorems.* The second step is to prove that there exists a domain, presumably distant from the first one, for which every system has the property under consideration. This contradicts the persistence theorem above and proves the density of the systems with the desired property. The Kupka-Smale theorem below is proved in this way.

Recall that the *dynamical degree* $d$ of a polynomial map $F$ is defined as

$$d = \lim \frac{1}{n} \log \deg F^n.$$

**10.1 Theorem** *There exists a set $KS$ in the space $\mathcal{P}_d$ of all polynomial automorphisms of the real (and complex) plane of dynamical degree $d > 1$ such that $KS$ is residual and of full Lebesgue measure and such that for any $F$ in $KS$, every periodic point of $F$ is hyperbolic and the stable and unstable manifolds of any two saddle periodic points are transverse.*

Consider now some open problems.

**10.2 Problem** *Is the same result true for (multidimensional) polynomial automorphisms of $\mathbb{C}^n$?*

**10.3 Problem** *Is the Kupka-Smale property generic for planar polynomial vector fields?*

The positive answer in the real case is trivial: saddle connections may be destroyed by a rotation of a vector field. In the complex plane the absence of common separatrices (complex saddle connections) of two singular points for a generic field is not yet proved. Note that the leafs of generic complex foliations are dense. It is natural to suggest that the following question has an affirmative answer.

**10.4 Problem** *Is it true that the set of complex planar polynomial vector fields of fixed degree $n > 1$ with a complex saddle connection is dense in the parameter space?*

Let us return to the Kupka-Smale property.

**10.5 Problem** *Is the Kupka-Smale property typical for volume preserving polynomial automorphisms of the plane?*

The proof of Theorem 10.1 is based on a kind of a persistence theorem stated below. The simplest persistence theorem proved up to now looks as follows:

**10.6 Theorem** *Let $p$ and $q$ be saddle periodic points of $F \in \mathcal{P}_d$, and $T$ their heteroclinic point. Let $\gamma : [0,1] \to \mathcal{P}_d \times \mathbb{C}^4$ be any curve such that $\gamma(s) = (F(s), p(s), q(s))$ and $p(s) \in \mathbb{C}^2$, $q(s) \in \mathbb{C}^2$ are saddle periodic points for $F(s)$. Then for any $\varepsilon > 0$ there exists a curve $\gamma' : [0,1] \to \mathcal{P}_d \times \mathbb{C}^4$ such that: $\gamma(0) = \gamma'(0)$ for any $s \in [0,1]$, $|\gamma(s) - \gamma'(s)| < \varepsilon$; the heteroclinic intersection $T$ may be continued over $\gamma'$ in the following sense: there exists a continuous map*

$$\hat{\gamma} : [0,1] \to \mathcal{P}_d \times \mathbb{C}^6, \quad s \mapsto (\gamma'(s), T(s)), \quad \gamma'(s) = (F'(s), p'(s), q'(s))$$

*such that $p'(s)$ and $q'(s)$ are saddle periodic points of the map $F'(s)$, and $T(s)$ is a heteroclinic point for these periodic points, $T(0) = T$.*

Roughly speaking this means that the heteroclinic point may be extended over the parameter plane whenever this extension may be expected.

A stronger persistence theorem is needed to deduce Theorem 10.1 but its statement requires some technical details and is omitted.

## 10.2 Persistence problems

### Persistence of complex limit cycles

Consider the set of polynomial vector fields of degree at most $n$ in the complex plane. Denote by $\mathcal{A}_n^{\mathbb{R}}$ the subset of real vector fields. Fix one of those, say $\alpha_0$, and consider a robust limit cycle $\gamma_0$ of this field in $\mathbb{R}^2$. Take a line $L$ that intersects $\gamma_0$ transversally. Let $z$ be a chart on $L$. For any real equation $\alpha \in \mathcal{A}_n^{\mathbb{R}}$ close to $\alpha_0$ there exists a limit cycle $\gamma$ close to $\gamma_0$ that intersects $L$ at a point $z(\alpha)$. The germ of the function $z(\alpha)$, "the initial condition of the limit cycle on the cross-section" is analytic.

**10.7 Problem** *What may be said about the domain of analytic continuation of the function "initial condition of a limit cycle" of a planar vector field of fixed degree?*

The complexification of the germ $z(\alpha)$ is in fact the initial condition of a complex limit cycle for equation $\alpha$, see Section 10.1 below for the definition of complex limit cycles.

**Conjecture** *The initial condition of a limit cycle of a polynomial vector field as a function of the coefficients may be in a sense globally extended.*

This conjecture goes back to the seminal work of Petrovski-Landis [66]. It is not yet proved, but it gave rise to the persistence problems stated below.

**Persistence of the hypersurface of saddle connections**

Consider a real polynomial vector field $\alpha \in \mathcal{A}_n^{\mathbb{R}}$. Let $R_\theta\alpha$ be the field $\alpha$ rotated through the angle $\theta$; it is still in $\mathcal{A}_n^{\mathbb{R}}$. Fix $\alpha_0 \in \mathcal{A}_n^{\mathbb{R}}$ with a saddle connection. Then for any $\alpha$ close to $\alpha_0$, under some mild restriction on $\alpha_0$, there exists a germ of an analytic function $\theta(\alpha)$ at $\alpha_0$ such that $R_{\theta(\alpha)}\alpha$ has a saddle connection analytically depending on $\alpha$.

**10.8 Problem** *What may be said about the analytic continuation of the germ $\theta(\alpha)$ related to polynomial vector fields with a saddle connection?*

**Domain of the Poincaré map**

Let us consider a simpler construction with no parameter involved. Consider a real polynomial vector field, a limit cycle of this field, and the corresponding Poincaré map.

**10.9 Problem** *What may be said about the domain of the analytic continuation of the Poincaré map of a limit cycle of a fixed polynomial vector field?*

In both problems above some results about global analytic continuation are expected.

**Persistence of the identical cycle**

**10.10 Problem** *Consider a real planar polynomial vector field with a singular point of center type. Is it true that for an open and dense set of initial conditions the leaves with these initial conditions are not simply connected?*

The name of the problem will be explained in Section 11.1.

# C. Foliations by analytic curves

The theory of differential equations in the complex plane is drastically different from its real prototype, though the main notions are similar.

## 11 Polynomial vector fields in the complex plane

### 11.1 Complex cycles and problem of their persistence

A polynomial vector field in $\mathbb{C}^2$ determines complex phase curves in the complement of the set of singular points of the field. We suggest that the components of the field are coprime; hence the set of the singular points is finite. Outside this set, complex phase curves are Riemann surfaces that form a *foliation*: near any nonsingular point the partition of an appropriate neighborhood into the phase curves is biholomorphically equivalent to a partition of a polydisk by disks parallel to one of the coordinate axes. The phase curves are the *leaves* of the foliation. We say that the vector field determines a foliation with singularities by analytic curves on the whole of $\mathbb{C}^2$.

**11.1 Definition** A *complex cycle* is a nontrivial free homotopy class of loops on a leaf of a foliation.

Consider a loop $\gamma$ on a leaf of a foliation defined above. Its tubular neighborhood on the leaf is an annulus $A$. A tubular neighborhood $U$ of $A$ in $\mathbb{C}^2$ is topologically equivalent to a Cartesian product $A \times D$, where $D \subset \mathbb{C}$ is a disk centered at zero. This disk may be considered as a holomorphic cross-section, transversal to the leaves of the foliation, with $0 = D \cap \gamma$. Let $\pi$ be the projection of $U$ to $A$ along $D$. For any $z \in D$ close to zero, the loop $\gamma$ may be lifted to a curve $\gamma_z$ that lies on the leaf passing through $z$ and covers $\gamma$ under the projection $\pi$. Let $\Delta_\gamma(z) \in D$ be the endpoint of $\gamma_z$. The map

$$z \mapsto \Delta_\gamma(z)$$

is called the *holonomy map* of $\gamma$.

For a real closed integral curve $\gamma$ of a real polynomial vector field , its holonomy map is the complexification of the corresponding Poincaré map. Holonomy maps of free homotopic loops are analytically conjugated.

**11.2 Definition** A *complex limit cycle* is a complex cycle for which a corresponding holonomy map has an *isolated* fixed point 0.

**11.3 Definition** A complex limit cycle is called *identical* provided that the corresponding holonomy map is the identity.

Note that, by the uniqueness theorem for analytic functions, any complex cycle in the plane is either a limit cycle or identical. By the Weierstrass preparation theorem, a complex limit cycle cannot vanish under a small perturbation of the vector field.

The following problems are counterparts of Problems 10.7, 10.10:

**11.4 Problem** *Is it true that the initial condition of a complex limit cycle of a polynomial vector field may be analytically globally continued as a function of the coefficients of the field?*

**11.5 Problem** *Suppose that a polynomial vector field has an identical complex cycle. Is it true that almost all leaves of the corresponding foliation are not simply connected?*

The class of all complex planar polynomial vector fields of degree $n$ is denoted by $\mathcal{A}_n$. There are other natural classes of foliations of algebraic origin, for instance, those determined by polynomial vector fields in higher dimensional spaces. We will mostly deal with the class $\mathcal{A}_n$.

## 11.2 Generic properties of the foliations of class $\mathcal{A}_n$

Everywhere in this subsection *genericity* means *almost everywhere in the sense of the Lebesgue measure.*

**11.6 Theorem** *A generic foliation of class $\mathcal{A}_n$ has no algebraic leaves. Each leaf of a generic foliation of class $\mathcal{A}_n$ is dense in $\mathbb{C}^2$.*

The density property was discovered by M. Khudai-Verenov in 1962, [51]. Several improvements may be found in [33, 73, 62].

**11.7 Theorem ([33])** *A generic foliation of class $\mathcal{A}_n$ has at least a countable number of complex limit cycles.*

**11.8 Theorem ([66])** *Any foliation of class $\mathcal{A}_n$ has no more than a countable number of complex limit cycles.*

**11.9 Definition** Two foliations with singularities on a complex manifold $X$ are *topologically conjugate* if there exists a homeomorphism of $X$ onto itself that maps the leaves and the set of singular points of the first foliation to that of the second one. Such a homeomorphism is called *conjugating.*

**11.10 Definition** A complex equation $\alpha$ of class $\mathcal{A}_n$ is *absolutely rigid* if the following holds. There exists a neighborhood of $\alpha$ in $\mathcal{A}_n$ and a neighborhood of the identity in $\mathrm{Hom}(\mathbb{C}P^2, \mathbb{C}P^2)$, the set of homeomorphisms of $\mathbb{C}P^2$ onto itself, such that:

If $\alpha$ is topologically conjugated to a foliation $\beta$ from the first neighborhood, and the conjugating homeomorphism belongs to the second neighborhood, then $\alpha$ is *affinely equivalent* to $\beta$.

This means that there exists an affine coordinate change that maps $\alpha$ to $\beta$.

In the real case, topological equivalence is very robust and does not distinguish the generic real equation (8.1) from the nearby ones. In the complex case it is as delicate as the classification with respect to affine coordinate changes, as the following theorem shows.

**11.11 Theorem ([33])** *A generic foliation of class $\mathcal{A}_n$ is absolutely rigid.*

The following improvements of these results are known. The genericity conditions in the above theorems are that the system is"outside some set of measure zero in the parameter space". This may be improved to "outside some real algebraic subset in the parameter space", as discussed below.

**11.12 Theorem** *There exists a real algebraic subset $\Sigma$ in the space $\mathcal{A}_n$ such that any equation $\alpha$ from the real Zariski open set $\mathcal{A}_n \setminus \Sigma$ has the following properties:*

(i) *it is absolutely rigid;*

(ii) *it has an infinite number of complex limit cycles (for $n \geq 3$);*

(iii) *any leaf of $\alpha$, except for the infinite one, is dense in $\mathbb{C}P^2$.*

Statement (i) is proved in [74]. In [57] it is proved that the set of rigid equations in $\mathcal{A}_n \setminus \Sigma$ is open and dense.

Statement (ii) is proved in [75]. The restriction $n \geq 3$ in (ii) seems to be purely technical; the statement is believed to hold for $n = 2$ as well.

Statement (iii) was never published, though I believe that A. Shcherbakov has a proof.

There is a multidimensional analog of the preceding result for higher dimension. Yet the theorem below gives an analog of the results above for *locally generic* foliations in $\mathbb{C}P^n$.

For any $n$ and $m$ denote by $\mathcal{F}^n(\mathbb{C}P^m)$ the space of all foliations with singularities of $\mathbb{C}P^n$ by analytic curves such that in any affine chart the foliation is given by a polynomial vector field of degree no greater than $n$, and at least in one chart the equality holds.

**11.13 Theorem ([58])** *For any $n \geq 2$, $m \geq 2$ there exists a nonempty open subset $U \subset \mathcal{F}^n(\mathbb{C}P^n)$ such that any element $\alpha \in U$ has a finite number of singularities and is "chaotic", i.e. satisfies:*

- *minimality: each leaf is dense in $\mathbb{C}P^n$;*
- *ergodicity: any measurable union of leaves has zero or total Lebesgue measure;*
- *entropy: the geometric entropy of the foliation induced by $\alpha$ on $\mathbb{C}P^n$, with small balls centered at singular points deleted, is strictly positive; and*
- *rigidity: there exists a neighborhood $\mathcal{V}$ of the identity in the space of all homeomorphisms of $\mathbb{C}P^n$ onto itself such that: if $\alpha$ is topologically conjugated to $\beta \in U$ by some $H \in \mathcal{V}$, then $\alpha$ and $\beta$ are also conjugate by a projective change of coordinates.*

The entropy here is considered in the sense of [17], however, we do not reproduce the definition because of lack of space. This result gives a partial proof of the minimality and rigidity conjectures stated in [34].

**11.14 Problem** *Prove an analog of the preceding theorem for a Zariski open set in $\mathcal{F}^n(\mathbb{C}P^m)$.*

A global problem on *minimal sets*, below, is due to Camacho [15].

*A minimal set of a foliation, with singularities, in* $\mathbb{C}P^n$ is a closed invariant nonempty subset of $\mathbb{C}P^n$ that contains no proper subset with these three properties. Invariance means that the set is either one singular point (in this case it is called *trivial*), or alternatively contains no singular points, and contains the entire leaf passing through each point of the set.

**11.15 Problem ([15])** *Are there foliations, with singularities, on $\mathbb{C}P^2$ that have nontrivial minimal sets?*

Note that the theorems proved for foliations of class $\mathcal{A}_n$ suggest the results that may be expected for other classes of foliations, at least in lower dimensions.

## 11.3 Topology of leaves

**11.16 Problem** *Is it true that for a generic foliation of class $\mathcal{A}_n$ the leaves are topologically either cylinders (a countable number of these) or disks?*

The answer seems to be affirmative. Indeed, if two complex limit cycles are located on the same leaf, then it seems possible to split them by a small perturbation in the class $\mathcal{A}_n$ in such a way that the cycles will be moved to different leaves. On the other hand, there is at most a countable number of complex limit cycles for foliations of class $\mathcal{A}_n$. The perturbation arguments work poorly for polynomial dynamical systems. They work much better in the analytic domain. This motivates the problem:

**11.17 Problem** *Is it true that for a generic foliation determined by a planar vector field whose components are entire functions, the leaves are topologically either cylinders (a countable number of these) or disks?*

The main step in a solution of Problems 11.16, 11.17 should be *proving that typical equations of the class $\mathcal{A}_n$ have no complex cycles with trivial holonomy.* A priori, it may happen that *every* equation from some domain in $\mathcal{A}_n$ has leaves that may be called *hidden multiply connected surfaces* that form continuous families and contain complex cycles whose holonomy map is the identity. Experts do not believe that such families exist. The only result proved in this direction shows that generically an infinite leaf does not belong to any continuous family of nonsimply connected leaves.

**11.18 Theorem ([44])** *A generic equation of class $\mathcal{A}_n$, $n \geq 5$, has no cycle on the infinite leaf whose holonomy map is the identity.*

One may try to use the Petrovski-Landis strategy to prove the absence of "hidden multiply connected surfaces" in the following way. Suppose that these surfaces exist for some domain $U$ in $\mathcal{A}_n$. Suppose that Problem 11.5 has a positive solution. Then any equation $\alpha \in U$ has a family of identical complex cycles that approach the infinite line. Another step of the proof should be to *deduce from that that the equations from $U$ actually have a cycle on the infinite leaf with holonomy map the identity.* This will contradict Theorem 11.18.

In the next section an approach to Problem 11.5 is discussed.

### 11.4 Topological invariance of vanishing holonomy

We conclude this section by a problem on complex foliations that has a global content and local origin. Consider a germ of a holomorphic vector field at a singular point zero. Suppose that its $(n-1)$-jet is zero, and that the Taylor decomposition begins with the polynomial vector field $f, g$ satisfying the following genericity assumption: the polynomial $yf - xg$ has simple linear factors only. By one step of blowing up, such a vector field is transformed into a complex line field having $n+1$ elementary singular points on the pasted-in divisor. This divisor, with singular points deleted, is a leaf of the foliation obtained by the blow-up. The fundamental group of this leaf is free with $n$ generators. The corresponding monodromy group is called the *vanishing holonomy group.*

**11.19 Problem** *Let two germs of the class described above be orbitally topologically equivalent. Is it true that their groups of vanishing holonomy are topologically equivalent as well? This means that there exists a germ of a homeomorphism $(\mathbb{C}, 0) \to (\mathbb{C}, 0)$ that conjugates all the corresponding generators of the two groups simultaneously.*

## 12 Simultaneous uniformization

### 12.1 Statement of the general problem

The following constructions are motivated by the persistence problems above.

**12.1 Definition** A *skew cyclinder* is a triple $(M, B, \pi)$, where $M$ is a complex manifold, $\pi : M \to B$ is an analytic retraction with rank equal to $\dim M - 1$, and with simply connected fibers. The manifold $M$ is called *a total space*, the hypersurface $B$ is the *base* of the *skew cylinder*, and the map $\pi$ is a *projection*.

**12.2 Definition** Two skew cylinders $(M, B, \pi)$, $(\tilde{M}, \tilde{B}, \tilde{\pi})$, are called *equivalent* if there exist biholomorphic maps $H, \rho$ of the total spaces and bases, respectively, that commute with the projections: $H : M \to \tilde{M}$, $\rho : B \to \tilde{B}$, $\rho \circ \pi = \tilde{\pi} \circ H$.

**12.3 Definition** A skew cylinder is called *standard* if $M \supset B \times \hat{\mathbb{C}}$, and $\pi$ is a retraction onto $B \times \{0\}$ along the second factor.

These definitions are related to foliations in the following way.

**12.4 Theorem ([32])** *The union of the universal covers of a foliation with singularities of* $\mathbb{C}^n$ *by analytic curves admits the structure of a skew cylinder provided that the fibers saturate a holomorphic cross-section.*

**12.5 Theorem ([32, 36])** *If the cross-section in the preceding theorem is a Stein manifold, then the union of the universal covers over the leaves is also a Stein manifold.*

**12.6 Problem** *Is it true that a skew cylinder from Theorem* 12.4 *for an equation of class* $\mathcal{A}_n$ *is equivalent to a standard one, that is,* admits a simultaneous uniformization*?*

A more general conjecture appears to be wrong:

**12.7 Theorem (Glutsyuk [20])** *There exists a non-unifomizable skew cylinder whose total space is a* 2*-dimensional Stein manifold.*

A stronger result is obtained in [21]. Yet it may happen that the answer to Problem 12.6 is affirmative, as well as for the next one:

**12.8 Problem** *Is it true that a skew cylinder from Problem* 12.6 *is equivalent to a standard one, and that the latter cylinder is included in a bidisk and* has a continuous boundary, *that is, any fiber is a limit of the nearby ones in the sense of the Hausdorff distance?*

## 12.2 Degeneracy of hidden cylinders

Here we give a sketch of the proof of the affirmative answer to Problem 11.5 assuming that the answer to the last two problems is positive.

Suppose that the answer to Problem 11.5 is negative. Then there exists a compact cross-section $B$ (an embedded disk) transversal to the leaves of the considered foliation and two disks $D_1$, $D_2$ in $B$ such that for any $b \in D_2$ the corresponding leaf $\varphi_b$ has a complex cycle $\gamma(b)$ that depends continuously on $b \in D_2$. Moreover, there exists an arc $\sigma \subset \partial D_1$ such that any leaf passing through the points of this arc is simply connected.

Consider a simultaneous uniformization of the universal covers over the leaves crossing $B$. Let $(M_0, B, \pi_0)$ be the corresponding standard cylinder, with coordinates $z, w$, where $z$ is the coordinate on the base, and $w$ that on the fibers. Let $B = \{w = 0\}$. Let $\bar{\varphi}_b$ be the fiber of this skew cylinder passing through $b$. Denote by $U \subset B$ the set of those $b \in B$ for which the cycle $\gamma(b)$ may be continuously extended. Let $M_1 = \pi_0^{-1}(U)$. There exists a biholomorphic map $H : M_1 \to M_1$ (the *cover shift transformation* corresponding to the cycle $\gamma(b)$) that maps any fiber of $M_1$ to itself conformally. Let

$$H = (z, h(z, w)).$$

Consider the function

$$f = \frac{\partial h}{\partial w}(z,0).$$

Without loss of generality we may assume that $M_1$ contains a bidisk $B \times D$, where $D$ is a disk in the $w$-plane of a small radius $r$. For any $b \in U$, the corresponding fiber is invariant under $H$:

$$\varphi_b^0 = \{h(z,w) \mid w \in \varphi_b^0,\ z = b\}.$$

By the Köbe 1/4-theorem, this fiber contains a disk $\frac{1}{4}|f(z)|r$. All the fibers of the standard cylinder are uniformly bounded. Hence the function $f$ is bounded in $U$.

For simplicity (this is a sketch, not a proof!) suppose that $U$ is itself a disk. It is easy to reduce the general case to this one. By the Fatou theorem, $f$ has radial limits along almost all radii in $U$. By the Privalov-Riesz theorem, these limits are nonzero almost everywhere. Hence, there exists at least one point $b \in \partial U$, an $a \in \mathbb{C}^*$, and a sequence $b_n \to b$ such that

$$a_n = f(b_n) \to a \neq 0, \infty.$$

The conformal maps $h(b_n, \cdot)/a_n$ have a unit derivative at zero. Therefore, by the Köbe distortion (compactness) theorems, this sequence has a subsequence that tends to some map $h(w)$ on every compact subset $K \subset \varphi_b$. Note that any such subset belongs to all the fibers $\varphi_{b_n}$, for $n$ large. This follows from the conjecture that the standard cylinder under consideration has a continuous boundary. We replace the sequence by one of the subsequences mentioned above, and still denote it by $b_n$. Hence the limit

$$h_b(w) = \lim h_{b_n}(w)$$

exists, where $h_{b_n}(w) = h(b_n, w)$ and $h_b(w) = h(b,w)$, and is a conformal mapping of the cover over the fiber $\varphi_b$ onto itself.

There exists a sequence of curves $\bar{\gamma}(b_n) \subset \bar{\varphi}_{b_n}$ that connect $(b_n, 0)$ and $h(b_n, 0)$, and a curve $\bar{\gamma}(b) \subset \varphi_b$ such that

$$\bar{\gamma}(b_n) \to \bar{\gamma}(b)$$

in the sense of the Hausdorff distance. Let $\bar{\pi}$ be the natural projection that sends any universal cover $\bar{\varphi}_b$ to the corresponding leaf $\varphi_b$. Then

$$\bar{\pi}(\bar{\gamma}(b_n)) = \gamma(b_n),$$

an identical cycle on the leaf $\varphi_{b_n}$.

Consider now the curve $\gamma(b) = \bar{\pi}(\gamma(b_n))$. This is a loop on the leaf $\varphi_b$. It is a limit of identical cycles of the foliation, hence an identical cycle itself. This is a general fact from the elementary theory of foliations, see [66]. A continuous family of identical cycles may be analytically continued in a neighborhood of $\gamma(b)$. This contradicts the assumption that the family of cycles under consideration cannot be continued across the boundary of $U$, and proves the persistence of the identical cycle.

To conclude, note that the result of Glutsyuk quoted above, makes an affirmative answer in Problem 12.8 highly debatable. Yet there is hope that the proposed strategy would work if the biholomorphic maps are replaced by spatial quasiconformal ones, and the methods of [18, 19, 54, 55] are applied. But this is quite another story.

## References


[1] V. Alekseev and M. Yakobson, Symbolic dynamics and hyperbolic dynamical systems, *Phys. Rep.* **75** (5) (1981), 287–325.

[2] O. Anosova, On invariant manifolds in singularly perturbed systems, *J. Dynam. Control Systems* **5** (1999), 501–507.

[3] O. Anosova, Invariant manifolds in singularly perturbed systems, *Proc. Steklov Inst. Math.* **236** (2002), 27–32.

[4] V. I. Arnold, *Geometrical Methods in the Theory of Ordinary Differential Equations*, Springer-Verlag, New York, 1983.

[5] V. I. Arnold, ed., *Dynamical Systems V: Bifurcation Theory and Catastrophe Theory*, Encyclopaedia Math. Sci. **5**, Springer, Berlin-New York, 1994 (transl. from the Russian edition 1986).

[6] P. Bachurin, On the relation between time averages and minimal attractors, *Uspekhi Mat. Nauk* **54** (6) (1999), 151–152 (Russian).

[7] N. Bautin, Du nombre de cycles limites naissant en cas de variation des coefficients d'un état d'équilibre du type foyer ou centre, *Dokl. Acad. Sci. URSS (N.S.)* **24** (1939), 669–672; Amer. Math. Soc. Transl. **100** (1954).

[8] Bochi, Genericity of zero Lyapunov exponents, *Ergodic Theory Dynam. Systems* **22** (6) (2002), 1667–1696.

[9] R. Bogdanov, Bifurcations of a limit cycle of a certain family of vector fields on the plane, *Trudy Sem. Petrovsk.* **2** (1976), 23–35.

[10] R. Bogdanov, The versal deformation of a singular point of a vector field on the plane in the case of zero eigenvalues, *Trudy Sem. Petrovsk.* **2** (1976), 37–65.

[11] C. Bonatti and L. Diaz, Persistent nonhyperbolic transitive diffeomorphisms, *Ann. of Math. (2)* **143** (2) (1996), 357–396.

[12] A. Borisyuk, Global bifurcations on the Klein bottle: unimodal case, *Mat. Zametki* **71** (3) (2002), 348–363 (Russian).

[13] A. Borisyuk, Global bifurcations on the Klein bottle: general case, in preparation.

[14] H. Broer, R. Roussarie, and C. Simo, Invariant circles in the Bogdanov-Takens bifurcation for diffeomorphisms, *Ergodic Theory Dynam. Systems* **16** (6) (1996) 1147–1172.

[15] C. Camacho, Problems on limit sets of foliations on complex projective spaces, in: *Proc. Internat. Congress of Mathematicians (Kyoto, 1990)*, Math. Soc. Japan, Tokyo, 1991, 1235–1239.

[16] E. Ghys, Groups acting on the circle, *Enseign. Math. (2)* **47** (2001), 329–407.

[17] É. Ghys, R. Langevin, and P. Walczak, Entropie géométrique des feuilletages, *Acta Math.* **160** (1–2) (1988), 105–142.

[18] A. Glutsyuk, Hyperbolicity of phase curves of a general polynomial vector field in $C^n$, *Funktsional. Anal. i Prilozhen.* **28** (2) (1994), 1–11 (Russian).

[19] A. Glutsyuk, Hyperbolicity of the leaves of a generic one-dimensional holomorphic foliation on a nonsingular projective algebraic variety, *Trudy Mat. Inst. Steklov.* **213** (1997), 90–111 (Russian).

[20] A. Glutsyuk, Nonuniformizable skew cylinders. A counterexample to the simultaneous uniformization problem, *C. R. Acad. Sci. Paris Sér. I Math.* **332** (3) (2001), 209–214.

[21] A. Glutsyuk, On simultaneous uniformization and local nonuniformizability, *C. R. Acad. Sci. Paris Sér. I Math.* **334** (2002), 489–494.

[22] A. Glutsyuk and Yu. Ilyashenko, Restricted version of the Infinitesimal Hilbert 16th problem, in preparation.

[23] A. Gorodetski, Regularity of central leaves of partially hyperbolic sets and applications, to appear.

[24] A. Gorodetski and Yu. Ilyashenko, Some new robust properties of invariant sets and attractors of dynamical systems, *Funct. Anal. Appl.* **33** (2) (1999), 16–32.

[25] A. Gorodetski and Yu. Ilyashenko, Some properties of skew products over a horseshoe and solenoid, *Proc. Steklov Inst. Math.* **231** (2000), 96–118.

[26] J. Guckenheimer, Towards a global theory of singularly perturbed systems, in: *Nonlinear Differential Equations and Chaos (Groningen 1995)* (H. W. Broer et al., eds.), Birkhäuser, Basel, 1996, 213–225.

[27] J. Guckenheimer and Yu. Ilyashenko, The duck and the devil: canards on the staircase, *Moscow Math. J.* **1** (1) (2001), 27–47.

[28] H. Hauser, J.-J. Risler, and B. Teissier, The reduced Bautin index of planar vector fields, *Duke Math. J.* **100** (3) (1999), 425–445.

[29] D. Hilbert, Mathematical problems, *Bull. Amer. Math. Soc. (N.S.)* **37** (2000), 407–436 (reprinted from *Bull. Amer. Math. Soc.* **8** (1902), 437–479).

[30] Yu. Ilyashenko, The appearance of limit cycles under a perturbation of the equation $dw/dz = -R_z/R_w$, where $R(z,w)$ is a polynomial, *Mat. Sb. (N.S.)* **78 (120)** (3) (1969), 360–373 (Russian).

[31] Yu. Ilyashenko, An example of equations $dw/dz = P_n(z,w)/Q_n\,(z,w)$ having a countable number of limit cycles and arbitrarily high Petrovskiĭ-Landis genus, *Mat. Sb. (N.S.)* **80 (122)** (3) (1969), 388–404 (Russian).

[32] Yu. Ilyashenko, Foliations by analytic curves, *Math. Sb.* **88** (4) (1972), 558–577.

[33] Yu. Ilyashenko, Topology of phase portraits of analytic differential equations on a complex projective plane, *Trudy Sem. Petrovsk.* **4** (1978), 83–136 (Russian).

[34] Yu. Ilyashenko, Global and local aspects of the theory of complex differential equations, in: *Proc. Internat. Congress of Mathematicians (Helsinki 1978)*, vol. II, Springer-Verlag, Berlin, 1979, 821–826.

[35] Yu. Ilyashenko (ed.), *Nonlinear Stokes Phenomena*, Adv. in Soviet Math. **14**, Amer. Math. Soc., Providence, RI, 1993.

[36] Yu. Ilyashenko, Covering manifolds for analytic families of leaves of foliations by analytic curves, *Topol. Methods Nonlinear Anal.* **11** (1998), 361–373.

[37] Yu. Ilyashenko, Hilbert-type numbers for Abel equations, growth and zeros of holomorphic functions, *Nonlinearity* **13** (4) (2000), 1337–1342.

[38] Yu. Ilyashenko, Centennial history of Hilbert's 16th problem, *Bull Amer. Math. Soc. (N.S)* **39** (3) (2002), 301–354.

[39] Yu. Ilyashenko and V. Kaloshin, Bifurcation of planar and spatial polycycles: Arnold's program and its development, in: *The Arnoldfest* (E. Bierstone et al., eds.) Fields Inst. Commun. **24**, Amer. Math. Soc., Providence, RI, 1999, 241–271.

[40] Yu. Ilyashenko and Weigu Li, *Nonlocal Bifurcations*, Math. Surveys Monogr. **66**, Amer. Math. Soc., Providence, RI, 1999.

[41] Yu. Ilyashenko and J. Llibre, A restricted version of Hilbert's 16th problem for quadratic vector fields, in preparation.

[42] Yu. Ilyashenko and V. Moldavskis, Morse-Smale circle diffeomorphisms and moduli of elliptic curves, to appear.

[43] Yu. Ilyashenko and A. Panov, Some upper estimates of the number of limit cycles of planar vector fields with applications to the Liénard equation, *Moscow Math. J.* **1** (4) (2001), 583–599.

[44] Yu. Ilyashenko and A. Pyartli, Monodromy group at infinity of generic polynomial vector field in the complex projective plane, *Russian J. Math. Phys.* **2** (3), 275–315.

[45] Yu. Ilyashenko and S. Yakovenko, Nonlinear stokes phenomena in smooth classification problems, in: *Nonlinear Stokes Phenomena* (Yu. Ilyashenko, ed.), Adv. in Soviet Math. **14**, Amer. Math. Soc., Providence, RI, 1993, 235–287.

[46] Yu. Ilyashenko and S. Yakovenko, Finite cyclicity of elementary polycycles in generic families, in: *Concerning the Hilbert 16th Problem*, Amer. Math. Soc. Transl. Ser. 2 **165**, Amer. Math. Soc., Providence, RI, 1995, 21–95.

[47] Yu. Ilyashenko and S. Yakovenko, eds., *Concerning the Hilbert 16th Problem*, Amer. Math. Soc. Transl. Ser. 2 **165**, Amer. Math. Soc., Providence, RI, 1995.

[48] Yu. Ilyashenko and S. Yakovenko, Counting real zeros of analytic functions satisfying linear ordinary differential equations, *J. Differential Equations* **126** (1) (1996), 87–105.

[49] V. Kaloshin, The Existential Hilbert 16th Problem and an estimate for cyclicity of elementary polycycles, *Invent. Math.*, **152**, to appear.

[50] A. Khovanskii, Fewnomials, Transl. Math. Monogr. **88**, Amer. Math. Soc., Providence, RI, 1991.

[51] M. G. Khudai-Verenov, A property of the solutions of a differential equation, *Mat. Sb.* **56** (1962), 301–308 (Russian).

[52] V. Kleptsyn and M. Nalski, Synchronization of orbits for random dynamical systems, in preparation.

[53] A. Lins Neto, On the number of solutions of the equation $dx/dt = \sum_{j=0}^{n} a_j(t)x^j$, $0 \leq t \leq 1$, for which $x(0) = x(1)$, *Invent. Math.* **59** (1) (1980), 67–76.

[54] A. Lins Neto, Simultaneous uniformization for the leaves of projective foliations by curves, *Bol. Soc. Brasil. Mat. (N.S.)* **25** (2) (1994), 181–206.

[55] A. Lins Neto, Uniformization and the Poincaré metric on the leaves of a foliation by curves, *Bol. Soc. Brasil. Mat. (N.S.)* **31** (3) (2000), 351–366.

[56] A. Lins Neto, W. de Melo, and C. Pugh, On Liénard's equation, in: *Geometry and Topology (Rio de Jeneiro 1976)* (J. Palis and M. do Carmo, eds.), Lecture Notes in Math. **597**, Springer, Berlin-New York, 1977, 335–357 .

[57] A. Lins Neto, P. Sad, and B. Scardua, On topological rigidity of projective foliations, *Bull. Soc. Math. France* **126** (3) (1998), 381–406.

[58] F. Loray and J. C. Rebello, Stably chaotic rational vector fields on $\mathbb{CP}^n$, Stony Brook IMS preprint 2000/5.

[59] P. Mardesić, An explicit bound for the multiplicity of zeros of generic Abelian integrals, *Nonlinearity* **4** (3) (1991), 845–852.

[60] J. Milnor, On the concept of attractor, *Comm. Math. Phys.* **99** (1985), 177-195.

[61] V. Moldavskii, Moduli of elliptic curves and rotation numbers of circle diffeomorphisms, *Funct. Anal. Appl.* **35** (2001), 234–236.

[62] I. Nakai, Separatrices for nonsolvable dynamics on $\mathbb{C}, 0$, *Ann. Inst. Fourier (Grenoble)* **44** (2) (1994), 569–599.

[63] J. Palis, A global view of dynamics and a conjecture on the denseness of finitude of attractors, *Astérisque* **261** (2000), 335–347.

[64] A. Panov, Variety of Poincaré mappings for cubic equations with variable coefficients, *Funktsional. Anal. i Prilozhen* **33** (4) (1999), 84–88 (Russian).

[65] G. Petrov, Elliptic integrals and their nonoscillation, *Funktsional. Anal. i Prilozhen.* **20** (1) (1986), 46–49 (Russian).

[66] I. Petrovskii and E. Landis, On the number of limit cycles of the equation $dy/dx = P(x,y)/\,Q(x,y)$, where $P$ and $Q$ are polynomials of 2nd degree, *Mat. Sb. (N.S.)* **37** (79) (1955), 209–250 (Russian).

[67] I. Pushkar', A multidimensional generalization of Ilyashenko's theorem on abelian integrals, *Funktsional. Anal. i Prilozhen.* **31** (2) (1997), 34–44, 95 (Russian).

[68] I. Pushkar', On limit cycles generated via perturbations of Hamiltonian systems, *Russian Math. Surveys* **57** (5) (2002), 161–162.

[69] A. Pyartli, Rational differential equations with a commutative monodromy group at infinity, *Trans. Moscow Math. Soc.* **61** (2000), 67–95.

[70] R. Roussarie, A note on finite cyclicity property and Hilbert's 16th problem, in: *Dynamical Systems Valparaiso 1986* (R. Bamón et al., eds.), Lecture Notes in Math. **1331**, Springer, Berlin-New York, 1986, 161–168.

[71] R. Roussarie, Cyclicité finie des lacets et des points cuspidaux, *Nonlinearity* **2** (1) (1989), 73–117.

[72] R. Roussarie, *Bifurcation of Planar Vector Fields and Hilbert's Sixteenth Problem*, Birkhäuser, Basel, 1998.

[73] A. Shcherbakov, Density of the orbit of a pseudogroup of conformal mappings and generalization of the Khudai-Verenov theorem, *Vestnik Moskov. Univ. Ser. I Mat. Mekh.* **1982** (4), 10–15 (Russian).

[74] A. Shcherbakov, Topological and analytic conjugation of noncommutative groups of germs of conformal mappings, *Trudy Sem. Petrovsk.* **10** (1984), 170–196, 238–239 (Russian).

[75] A. Shcherbakov, E. Rosales-Gonzalez, and L. Ortiz-Bobadilla, Countable set of limit cycles for the equation $dw/dz = P_n(z,w)/Q_n(z,w)$, *J. Dynam. Control Systems* **4** (4) (1998), 539–581.

[76] S. Smale, Mathematical problems for the next century, *Math. Intelligencer* **20** (2) (1998), 7–15.

[77] F. Takens, Forced oscillations and bifurcations, in: *Applications of Global Analysis I, Comm. Math. Inst. Rijksuniv. Utrecht* No. 3–1974, 1–59.

[78] F. Takens, Heteroclinic attractors: time averages and moduli of topological conjugacy, *Bol. Soc. Brasil. Mat. (N.S.)* **25** (1) (1994), 107–120.

[79] A. Varchenko, Estimation of the number of zeros of an abelian integral depending on a parameter, and limit cycles, *Funktsional. Anal. i Prilozhen.* **18** (2) (1984), 14–25 (Russian).

[80] S. Yakovenko, A geometric proof of the Bautin theorem, in: *Concerning the Hilbert 16th Problem* (Yu. Ilyashenko and S. Yakovenko, eds.), Amer. Math. Soc. Transl. Ser. 2 **165**, Amer. Math. Soc., Providence, RI, 1995, 203–219.

# Growth rate of the number of periodic points

Vadim Yu. KALOSHIN
*School of Mathematics*
*Institute for Advanced Study*
*Princeton, NJ*
*and*
*Department of Mathematics*
*California Institute of Technology*
*Pasadena, CA 91125*
*USA*

**Abstract**

Let $M$ be a compact manifold and let $\mathrm{Diff}^r(M)$ be the space of $C^r$ invertible selfmaps of $M$ with the uniform $C^r$ topology. Denote by $P_n(f) = \#\{x \in M : f^n(x) = x\}$ the number of periodic points of period $n$ for each positive integer $n$. In these lectures we investigate how quickly $P_n(f)$ grows with $n$ for a generic $f$. It turns out that for a (Baire) generic $f$ there is no bound on the rate of growth. This, in particular, answers an old question by Smale. Chapter 1 describes this phenomenon with more details. In Chapters 2 and 3 we present three different mechanisms for arbitrarily fast growth of the number of periodic points with $n$. The last chapter gives a precise statement of the deep fact that "with probability one" for any $\delta > 0$ there is $C = C(\delta) > 0$ such that $P_n(f) < \exp(Cn^{1+\delta})$. This fact is in contrast with the superexponential growth of the number of periodic points for a (Baire) generic $f$ which is reminiscent of KAM theory.

# 1 A problem of the growth of the number of periodic points and decay of hyperbolicity for generic diffeomorphisms

## 1.1 Introduction

Let $\mathrm{Diff}^r(M)$ be the space of $C^r$ diffeomorphisms of a finite-dimensional smooth compact manifold $M$ with the uniform $C^r$-topology, where $\dim M \geq 2$, and let $f \in \mathrm{Diff}^r(M)$. Consider the number of periodic points of period $n$

$$P_n(f) = \#\{\text{isolated } x : x = f^n(x)\}. \tag{1.1}$$

The main question of this paper is:

**1.1 Question** How fast can $P_n(f)$ grow with $n$ for a $C^r$ "generic" diffeomorphism $f$?

We put the word "generic" in quotation marks because, as the reader will see, the answer depends on the notion of genericity. We call a diffeomorphism $f \in \mathrm{Diff}^r(M)$ an *Artin-Mazur*



*Y. Ilyashenko and C. Rousseau (eds.),*
*Normal Forms, Bifurcations and Finiteness Problems in Differential Equations,* 355–385.

*diffeomorphism* (or simply *A-M diffeomorphism*) if the number of isolated periodic orbits of $f$ grows at most exponentially fast, i.e. for some number $C > 0$,

$$P_n(f) \le \exp(Cn) \quad \text{for all} \quad n \in \mathbb{Z}_+. \tag{1.2}$$

Artin and Mazur [AM] proved that *for any* $0 \le r \le \infty$, *A-M diffeomorphisms are dense in* $\mathrm{Diff}^r(M)$ *with the uniform* $C^r$*-topology.*

We say that a point $x \in M$ of period $n$ for $f$ is hyperbolic if $Df^n(x)$, the derivative of $f^n$ at $x$, has no eigenvalues with modulus 1. (Notice that a hyperbolic solution to $f^n(x) = x$ must also be isolated.) We call $f \in \mathrm{Diff}^r(M)$ a strongly Artin-Mazur diffeomorphism if for some number $C > 0$,

$$P_n(f) \le \exp(Cn) \quad \text{for all} \quad n \in \mathbb{Z}_+, \tag{1.3}$$

and all periodic points of $f$ are hyperbolic. In [K1] an elementary proof of the following extension of the Artin-Mazur result is given.

**1.2 Theorem** *For any* $0 \le r < \infty$, *strongly A-M diffeomorphisms are dense in* $\mathrm{Diff}^r(M)$ *with the uniform* $C^r$*-topology.*

According to the standard terminology a set in $\mathrm{Diff}^r(M)$ is called *residual* if it contains a countable intersection of open dense sets and a property is called (*Baire*) *generic* if diffeomorphisms with that property form a residual set. In what follows we say:

**1.3 Definition** An open set $\mathcal{N} \subset \mathrm{Diff}^r(M)$ has a *supergrowth* or a *supergrowth is generic in* $\mathcal{N}$ if for any given sequence $a = \{a_n\}_{n \in \mathbb{Z}_+}$ there is a generic set $\mathbb{R}_a$ in $\mathcal{N}$ depending on the sequence $a_n$ with the property if $f \in \mathbb{R}_a$, then for infinitely many $n_k \in \mathbb{Z}_+$ we have $P_{n_k}(f) > a_{n_k}$.

It turns out that the A-M property is *not generic*, as is shown in [K2]. Moreover, the following is proved there:

**1.4 Theorem** *For any* $2 \le r < \infty$ *there is an open set* $\mathcal{N} \subset \mathrm{Diff}^r(M)$ *where supergrowth is generic.*

Notice that Theorem 1.4 implies the first statement, since A-M diffeomorphisms have empty intersection with $\mathbb{R}_a$ for any superexponentially fast growing sequence $a = \{a_n\}_{n \in \mathbb{Z}}$ and any two generic sets must have non-empty intersection. The proof of this result is based on a result of Gonchenko-Shilnikov-Turaev [GST1]. Two slightly different detailed proofs of their result are given in [K2] and [GST2]. The proof in [K2] relies on a strategy outlined in [GST1]. An example of a $C^\infty$ smooth diffeomorphism on a torus with arbitrarily fast growth of the number of periodic points has been constructed earlier by Rozales-Gonsales [RG].

In relation to the problem of growth of the number of periodic points, Smale [Sm] posed the following question (Problem 4.5, p. 765): *Is the dynamical* $\zeta_f$*-function generically rational* (*i.e. is* $\zeta_f$ *rational for a residual set of* $f \in \mathrm{Diff}^r(M)$)*?*[1]

Bowen asked the following question in his book [Bo]: *Let* $h(f)$ *denote the topological entropy of* $f$. *Is the property that* $h(f) = \limsup_{n\to\infty} \log P_n(f)/n$ *generic with respect to the* $C^r$*-topology?*

[1]In [Si] it is shown that for the 3-dimensional torus there is no generic set on which the dynamical $\zeta_f$-function is rational.

Theorem 1.4 implies a negative answer to both questions, since in order to have a positive answer, both require A-M to be generic. The following problems naturally arise from the above results:

**1.5 Problem** Prove that in the space of $C^r$ smooth vector fields on a compact manifold $M$ A-M vector fields are $C^r$-dense for any $1 \le r$. Are they $C^0$- or $C^1$-dense?

**1.6 Problem** For any $1 \le r$ prove that in the space of $C^r$ smooth volume-preserving or symplectic diffeomorphisms, A-M diffeomorphisms are $C^r$-dense.

**1.7 Problem** Construct an analytic diffeomorphism of a compact manifold with a superexponential growth of the number of isolated periodic points[2].

## 1.2 Surface diffeomorphisms, Newhouse phenomenon, and Palis' conjecture

Actually, Theorem 1.4 for diffeomorphisms of a 2-dimensional compact manifold $M^2$ can be extended to the following theorem.

**1.8 Theorem** *For any* $2 \le r < \infty$ *in* $\mathrm{Diff}^r(M^2)$ *outside of an open set of hyperbolic diffeomorphisms there is a* $C^1$*-dense* $C^r$*-open set satisfying Theorem* 1.4. *In other words, outside of hyperbolic diffeomorphisms, supergrowth is* $C^1$*-dense* $C^r$*-generic*[3].

**1.9 Definition** We say that a $C^r$ smooth diffeomorphism $f : M \to M$ has a *homoclinic tangency* (HT for short) if, for some saddle periodic point $p$, its stable and unstable manifolds $W^s(p)$ and $W^u(p)$ respectively have a point of tangency (see Fig. 1).

Notice that diffeomorphisms with HT can occur in any dimension and this definition is dimension independent.

**1.10 Theorem ([Ne])** *For any* $2 \le r$ *the* $C^r$*-closure of surface diffeomorphisms with HT has a nonempty interior in* $\mathrm{Diff}^r(M^2)$.

**1.11 Definition** We say that an open set $\mathcal{N} \subset \mathrm{Diff}^r(M^2)$ is a *Newhouse domain* if diffeomorphisms with HT are dense in $\mathcal{N}$.

**1.12 Theorem ([K2])** *For any open set* $\mathcal{N} \subset \mathrm{Diff}^r(M^2)$ *which is a Newhouse domain Theorem* 1.4 *holds.*

**1.13 Definition** We say that a diffeomorphism $f \in \mathrm{Diff}^r(M)$ is *hyperbolic* if almost every point is attracted to a uniformly hyperbolic attractor and there is only a finite number of those.

**1.14 Conjecture (Palis)** *For any* $1 \le r$ *the* $C^r$*-closure of hyperbolic surface diffeomorphisms and diffeomorphisms with HT coincides with* $\mathrm{Diff}^r(M^2)$.

---

[2]Yoccoz informed the author that without restriction on isolatedness of periodic points he can produce examples of analytic germs with an arbitrarily fast growth of the number of periodic points.

[3]$C^1$-dense $C^r$-generic contains a countable intersection of $C^1$-dense $C^r$-open nested sets.

**1.15 Theorem ([PS])** *For any $1 \leq r$ the $C^1$-closure of hyperbolic surface diffeomorphisms and diffeomorphisms with HT coincides with* $\mathrm{Diff}^r(M^2)$.

Theorems 1.10 and 1.15 along with Remark 1.12 imply Theorem 1.8.

## 1.3 3-dimensional and higher dimensional diffeomorphisms

For 3-dimensional diffeomorphisms, hyperbolic and diffeomorphisms with HT do not form a dense set. There is at least another open set which we shall define.

**1.16 Definition** We say that a $C^r$ smooth diffeomorphism $f : M^3 \to M^3$ has a *heteroclinic cycle* (HC for short) if for two saddle periodic points $p$ and $q$ with different indices $W^u(p)$ and $W^s(q)$ (resp. $W^s(p)$ and $W^u(q)$) are 2-dimensional and have 1-dimensional transversal intersection. Moreover, $W^s(p)$ and $W^u(q)$ (resp. $W^u(p)$ and $W^s(q)$) also intersect.

**1.17 Theorem ([D, BD])** *For any $1 \leq r$ the $C^r$-closure of diffeomorphisms with HC has a nonempty interior in* $\mathrm{Diff}^r(M^3)$.

**1.18 Definition** We say that an open set $\mathcal{B} \subset \mathrm{Diff}^r(M^2)$ has a *blender* if the diffeomorphisms with HC are dense in $\mathcal{B}$.

**1.19 Theorem ([KP])** *For any open set $\mathcal{B} \subset \mathrm{Diff}^r(M^3)$ which has a blender Theorem* 1.4 *holds.*

This theorem also exhibits a different scenario for superexponential growth different from the one in Theorem 1.4 produced by the HT-phenomenon. It is believed that the following conjecture holds true:

**1.20 Conjecture ([P])** *For any $1 \leq r$ the $C^r$-closure of hyperbolic diffeomorphisms and diffeomorphisms with HT and HC coincides with* $\mathrm{Diff}^r(M^3)$ .

Similarly as Theorems 1.10 and 1.15 along with Remark 1.12 imply Theorem 1.4 we have that Theorem 1.19, an analog of Theorem 1.4 for 3-dimensional domains (to be proved), and Conjecture 1.20 imply the following:

**1.21 Theorem** *For any $2 \leq r < \infty$ in $\mathrm{Diff}^r(M^3)$ outside of an open set of hyperbolic diffeomorphisms there is an open $C^1$-dense set satisfying Theorem 1.4. In other words, outside of hyperbolic diffeomorphisms, supergrowth is a $C^1$-dense $C^r$-generic set.*

It is natural to pose the following

**1.22 Conjecture** *For any $\dim M \geq 2$ and any $1 \leq r$ in $\mathrm{Diff}^r(M)$ outside of hyperbolic diffeomorphisms, supergrowth is $C^r$-generic.*

Theorems 1.8 and 1.21 are a justification that Conjecture 1.22 holds true. It would be great if one could find a proof of this conjecture without proving Palis' conjecture or some Palis-type conjecture. So far the proposed way is through these conjectures.

## 1.4 Superexponential growth in dimension 1

In this section we state another surprising result which says that even for 1-dimensional maps with one critical point, superexponentiality of the number of periodic points is possible.

**1.23 Theorem ([KK])** *For any sequence $a = \{a_n\}_{n\in\mathbb{Z}_+}$ there is a $C^r$ smooth unimodal map $f : I \to I$ of an interval such that for some $k_0 \in \mathbb{Z}_+$ and any $k > k_0$ we have $P_{3^k}(f) > a_{3^k}$. Moreover, the map is $C^\infty$ everywhere except at the critical point.*

**1.24 Definition** A $C^r$ smooth $f : I \to I$ has a *non-flat* critical point at $x_0$ if, after a $C^r$ coordinate change $\phi : U_{x_0} \to U_{x_0}$ in a neighborhood of $x_0$, we have that $(\phi \circ f)^{(s)}(x_0) \neq 0$ for some $1 < s \leq r$.

**1.25 Theorem ([MMS])** *Let $f : I \to I$ be a $C^r$ smooth map with no flat critical points and $r \geq 2$. Then for some $n_0 \in \mathbb{Z}_+$ and $\gamma > 0$ we have that*

$$|(f^n)'(x)| > 1 + \gamma$$

*for each periodic point $x$ of period $n > n_0$.*

**1.26 Corollary** *With $f$ as above we have $P_n(f) \leq |I|(2\gamma)^{-1}\|f\|^n_{C^2}$ for each $n > n_0$.*

This Corollary follows from Proposition 1.28. Theorem 1.23 also says that the requirement of non-flatness of critical points is not only sufficient, but also necessary.

## 1.5 Growth of the number of periodic points for prevalent diffeomorphisms and Arnold's problem

### 1.5.1 Arnold's problem

However, it seems unnatural that if you pick a diffeomorphism at random then it may have an arbitrarily fast growth of the number of periodic points. Moreover, Baire generic sets in Euclidean spaces can have zero Lebesgue measure. Phenomena that are Baire generic, but have a small probability are well known in dynamical systems, KAM theory, number theory, etc. (see [O, HSY, K3] for various examples). This partially motivates the problem posed by Arnold [A]:

**1.27 Problem** Prove that "with probability one" $f \in \mathrm{Diff}^r(M)$ is an A-M diffeomorphism.

Arnold suggested the following interpretation of "with probability one": *for a (Baire) generic finite parameter family of diffeomorphisms $\{f_\varepsilon\}$, for Lebesgue almost every $\varepsilon$ we have that $f_\varepsilon$ is A-M.* As Theorem 1.4 shows, a result on the genericity of the set of A-M diffeomorphisms based on (Baire) topology is likely to be extremely subtle, if possible at all. We use instead a notion of "probability one" based on prevalence [HSY, K3], which is independent of Baire genericity. We are also able to state the result in the form Arnold suggested for generic families using this measure-theoretic notion of genericity. For a more detailed discussion of prevalence we refer to [HSY, KH]. Our first main prevalent result is a partial solution to Arnold's problem. It says that *for a prevalent diffeomorphism $f \in \mathrm{Diff}^r(M)$, with $1 < r < \infty$, and all $\delta > 0$ there exists $C = C(\delta) > 0$ such that for all $n \in \mathbb{Z}_+$,*

$$P_n(f) \leq \exp(Cn^{1+\delta}). \tag{1.4}$$

### 1.5.2 Quantitative hyperbolicity and growth of the number of periodic points

The Kupka-Smale theorem (see e.g. [PM]) states that for a generic diffeomorphism all periodic points are hyperbolic and all associated stable and unstable manifolds intersect one another transversally. [K3] shows that the Kupka-Smale theorem also holds on a prevalent set. We shall say *how hyperbolic are the periodic points, as a function of their period, for a Baire generic (resp. prevalent) diffeomorphism f?*

Recall that a linear operator $L : \mathbb{R}^N \to \mathbb{R}^N$ is *hyperbolic* if it has no eigenvalues on the unit circle $\{|z| = 1\} \subset \mathbb{C}$. Denote by $|\cdot|$ the Euclidean norm in $\mathbb{C}^N$. Then we define the *hyperbolicity* of a linear operator $L$ by

$$\gamma(L) = \inf_{\phi\in[0,1)} \inf_{|v|=1} |Lv - \exp(2\pi i\phi)v|. \tag{1.5}$$

We also say that $L$ is $\gamma$-*hyperbolic* if $\gamma(L) \geq \gamma$. In particular, if $L$ is $\gamma$-hyperbolic, then its eigenvalues $\{\lambda_j\}_{j=1}^N \subset \mathbb{C}$ are at least $\gamma$-distant from the unit circle, i.e. $\min_j ||\lambda_j| - 1| \geq \gamma$. The *hyperbolicity* of a periodic point $x = f^n(x)$ of period $n$, denoted by $\gamma_n(x, f)$, equals the hyperbolicity of the derivative $Df^n(x)$ of $f^n$ at the point $x$, i.e. $\gamma_n(x, f) = \gamma(Df^n(x))$. Similarly to the number of periodic points $P_n(f)$ of period $n$, define

$$\gamma_n(f) = \min_{\{x:\ x=f^n(x)\}} \gamma_n(x, f). \tag{1.6}$$

The idea of Gromov and Yomdin [Y] of measuring hyperbolicity is that a $\gamma$-hyperbolic point of period $n$ of a $C^2$ diffeomorphism $f$ has an $M_2^{-2n}\gamma$-neighborhood (where $M_2 = \|f\|_{C^2}$) free from periodic points of the same period[4]. In [KH] we prove the following result.

**1.28 Proposition** *Let $M$ be a compact manifold of dimension $N$, let $f : M \to M$ be a $C^2$ diffeomorphism that has only hyperbolic periodic points, and let $M_2 = \max(\|f\|_{C^2}, 2)$. Then there is a constant $C = C(M) > 0$ such that for each $n \in \mathbb{Z}_+$ we have*

$$P_n(f) \leq C(M_2)^{2nN}\gamma_n(f)^{-N}. \tag{1.7}$$

A similar statement can be proved for $C^{1+\rho}$ diffeomorphisms with $0 < \rho \leq 1$. Proposition 1.28 implies that a lower estimate on the decay of hyperbolicity $\gamma_n(f)$ gives an upper estimate on the growth of the number of periodic points $P_n(f)$. Therefore, a natural question is:

**1.29 Question** How fast can $\gamma_n(f)$ decay with $n$ for a $C^r$ "generic" diffeomorphism $f$?

For Baire generic $f \in \mathrm{Diff}^r(M)$, the existence of a lower bound on the rate of decay of $\gamma_n(f)$ would imply the existence of an upper bound on the rate of growth of the number of periodic points $P_n(f)$, whereas no such bound exists by Theorem 1.4. Thus again we consider genericity in the measure-theoretic sense of prevalence. Our second main prevalent result, which in view of Proposition 1.28 implies the first main prevalent result, is that *for*

[4]In [Y] hyperbolicity is introduced as the minimal distance of eigenvalues to the unit circle. This way of defining hyperbolicity does not guarantee the existence of a $M_2^{-2n}\gamma$-neighborhood free from periodic points of the same period; see [KH].

*a prevalent diffeomorphism* $f \in \mathrm{Diff}^r(M)$, *with* $1 < r < \infty$, *and for any* $\delta > 0$ *there exists* $C = C(\delta) > 0$ *such that*

$$\gamma_n(f) \geq \exp(-Cn^{1+\delta}). \tag{1.8}$$

Now we shall discuss in more detail our definition of prevalence ("probability one") in the space of diffeomorphisms $Diffr(M)$.

### 1.5.3 Prevalence in the space of diffeomorphisms $\mathrm{Diff}^r(M)$

The space of $C^r$ diffeomorphisms $\mathrm{Diff}^r(M)$ of a compact manifold $M$ is a Banach manifold. Locally we can identify it with a Banach space, which gives it a local linear structure in the sense that we can perturb a diffeomorphism by "adding" small elements of the Banach space. As we described in the preceding section, the notion of prevalence requires us to make additive perturbations with respect to a probability measure that is independent of the place at which we make the perturbation. Thus, although there is no unique way to put a linear structure on $\mathrm{Diff}^r(M)$, it is important to make a choice that is consistent throughout the Banach manifold.

The way we make perturbations on $\mathrm{Diff}^r(M)$ by small elements of a Banach space is as in [K1]. First we embed $M$ into the interior of the closed unit ball $B^N \subset \mathbb{R}^N$, which we can do for $N$ sufficiently large by the Whitney Embedding Theorem [W]. Choose any embedding of $M$ into $\mathbb{R}^N$. Then we consider a closed neighborhood $U \subset B^N$ of $M$ and the Banach space $C^r(U)$ of $C^r$ functions from $U$ to $\mathbb{R}^N$. Next we extend every element $f \in \mathrm{Diff}^r(M)$ to an element $\tilde{f} \in C^r(U)$ that is strongly contracting in the directions transverse to $M$. This is done in two steps:

First we consider a manifold $\tilde{M} = M \times [0,1]/\sim$, where the equivalence relation is defined by $(x,0) \sim (f(x),1)$ for all $x \in M$. $\tilde{M}$ is as smooth as $f$ is and carries a naturally defined vector field $X_f$ whose time-one map, restricted to $M \times \{0\}$, coincides with $f$. Such a construction is usually called a *suspension.*

Now embed $\tilde{M}$ into the interior of the closed unit ball $B^N \subset \mathbb{R}^N$. One can extend the vector field $X_f$ to a tube neighborhood of $\tilde{M}$ so it contracts strongly in the directions transverse to $\tilde{M}$. A corresponding Poincaré return map along the extended vector field is the $\tilde{f}$ we need[5].

By Sacker [Sac] and Fenichel [F] (persistence of invariant manifolds), since $\tilde{f}$ has $M$ as an invariant manifold, if we add to $\tilde{f}$ a small perturbation $g \in C^r(U)$, the perturbed map $\tilde{f}+g$ has an invariant manifold in $U$ that is close to $M$. Then $\tilde{f}+g$ restricted to its invariant manifold corresponds in a natural way to an element of $\mathrm{Diff}^r(M)$, which we consider to be the perturbation of $f \in \mathrm{Diff}^r(M)$ by $g \in C^r(U)$. In this way we reduce the problem to the study of maps in $\mathrm{Diff}^r(U)$, the open subset of $C^r(U)$ consisting of those elements that are diffeomorphisms from $U$ to some subset of its interior. The details of this construction are described in [KH].

The perturbation measure $\mu$ that we use here is supported within the analytic functions in $C^r(B^N)$. In this sense we foliate $\mathrm{Diff}^r(B^N)$ by analytic leaves that are compact and overlapping. The main prevalent result then says that *for every analytic leaf* $L \subset \mathrm{Diff}^r(B^N)$ *and every* $\delta > 0$, *for almost every diffeomorphism* $f \in L$ *in the leaf* $L$ *both* (1.4) *and* (1.8)

[5]Thank you, C. Carminati, for pointing out that a process involving only two steps clarifies the way the extension $\tilde{f}$ is constructed.

*are satisfied.* Now we define an analytic leaf as a "Hilbert brick" in the space of analytic functions and a natural Lebesgue product probability measure $\mu$ on it.

### 1.5.4 Formulation of the Main Prevalent Results

Fix a coordinate system $x = (x_1, \dots, x_N) \in \mathbb{R}^N \supset B^N$ and the scalar product $\langle x, y\rangle = \sum_i x_i y_i$. Let $\alpha = (\alpha_1, \dots, \alpha_N)$ be a multiindex from $\mathbb{Z}_+^N$, and let $|\alpha| = \sum_i \alpha_i$. For a point $x = (x_1, \dots, x_N) \in \mathbb{R}^N$ we write $x^\alpha = \prod_{i=1}^N x_i^{\alpha_i}$. Associate to a real analytic function $\phi : B^N \to \mathbb{R}^N$ the set of coefficients of its expansion:

$$\phi_{\vec{\varepsilon}}(x) = \sum_{\alpha \in \mathbb{Z}_+^N} \vec{\varepsilon}_\alpha x^\alpha. \tag{1.9}$$

Denote by $W_{k,N}$ the space of $N$-component homogeneous vector-polynomials of degree $k$ in $N$ variables, and by $\nu(k, N) = \dim W_{k,N}$ the dimension of $W_{k,N}$. According to the notation of the expansion (1.9), denote the coordinates in $W_{k,N}$ by $\vec{\varepsilon}_k = (\{\vec{\varepsilon}_\alpha\}_{|\alpha|=k}) \in W_{k,N}$.

In $W_{k,N}$ we use a scalar product that is invariant with respect to orthogonal transformation of $\mathbb{R}^N \supset B^N$ (see [KH]), defined as follows:

$$\langle \vec{\varepsilon}_k, \vec{\nu}_k\rangle_k = \sum_{|\alpha|=k} \binom{k}{\alpha}^{-1} \langle \vec{\varepsilon}_\alpha, \vec{\nu}_\alpha\rangle, \qquad \|\vec{\varepsilon}_k\|_k = (\langle \vec{\varepsilon}_k, \vec{\varepsilon}_k\rangle_k)^{1/2}. \tag{1.10}$$

Denote by $B_k^N(r) = \{\vec{\varepsilon}_k \in W_{k,N} : \|\vec{\varepsilon}_k\|_k \le r\}$ the closed $r$-ball in $W_{k,N}$ centered at the origin. Let $Leb_{k,N}$ be the Lebesgue measure on $W_{k,N}$ induced by the scalar product (1.10) and normalized by a constant so that the volume of the unit ball is one: $Leb_{k,N}(B_k^N(1)) = 1$.

Fix a nonincreasing sequence of positive numbers $\vec{\mathbf{r}} = (\{r_k\}_{k=0}^\infty)$ such that $r_k \to 0$ as $k \to \infty$ and define a *Hilbert brick* of size $\vec{\mathbf{r}}$ to be

$$\begin{aligned} HB^N(\vec{\mathbf{r}}) &= \{\vec{\varepsilon} = \{\varepsilon_\alpha\}_{\alpha \in \mathbb{Z}_+^N} : \text{for all } k \in \mathbb{Z}_+,\ \|\varepsilon_k\|_k \le r_k\} \\ &= B_0^N(r_0) \times B_1^N(r_1) \times \cdots \times B_k^N(r_k) \times \cdots \subset W_{0,N} \times W_{1,N} \times \cdots \times W_{k,N} \times \cdots. \end{aligned} \tag{1.11}$$

Define a *product Lebesgue probability measure* $\mu_{\vec{\mathbf{r}}}^N$ *associated to the Hilbert brick* $HB^N(\vec{\mathbf{r}})$ *of size* $\vec{\mathbf{r}}$ by normalizing for each $k \in \mathbb{Z}_+$ the corresponding Lebesgue measure $Leb_{k,N}$ on $W_{k,N}$ to the Lebesgue probability measure on the $r_k$-ball $B_k^N(r_k)$:

$$\mu_{k,r}^N = r^{-\nu(k,N)}\, Leb_{k,N} \quad \text{and} \quad \mu_{\vec{\mathbf{r}}}^N = \times_{k=0}^\infty \mu_{k,r_k}^N. \tag{1.12}$$

**1.30 Definition** Let $f \in \mathrm{Diff}^r(B^N)$ be a $C^r$ diffeomorphism of $B^N$ into its interior. We call $HB^N(\vec{\mathbf{r}})$ a *Hilbert brick* of an *admissible size* $\vec{\mathbf{r}} = (\{r_k\}_{k=0}^\infty)$ with respect to $f$ if

(1) for each $\vec{\varepsilon} \in HB^N(\vec{\mathbf{r}})$, the corresponding function $\phi_{\vec{\varepsilon}}(x) = \sum_{\alpha \in \mathbb{Z}_+^N} \vec{\varepsilon}_\alpha x^\alpha$ is analytic on $B^N$;

(2) for each $\vec{\varepsilon} \in HB^N(\vec{\mathbf{r}})$, the corresponding map $f_{\vec{\varepsilon}}(x) = f(x) + \phi_{\vec{\varepsilon}}(x)$ is a diffeomorphism from $B^N$ into its interior, i.e. $\{f_{\vec{\varepsilon}}\}_{\vec{\varepsilon} \in HB^N(\vec{\mathbf{r}})} \subset \mathrm{Diff}^r(B^N)$;

(3) for all $\delta > 0$ and all $C > 0$, the sequence $r_k \exp(Ck^{1+\delta}) \to \infty$ as $k \to \infty$.

**1.31 Remark** The first and second conditions ensure that the family $\{f_{\vec{\varepsilon}}\}_{\vec{\varepsilon}\in HB^N(\vec{\mathbf{r}})}$ lie in an analytic leaf within the class of diffeomorphisms $\mathrm{Diff}^r(B^N)$. The third condition provides us with enough freedom to perturb. It is important for our method to have infinitely many parameters to perturb. If the $r_k$'s were decaying too fast to zero it would make our family of perturbations essentially finite-dimensional.

An example of an admissible sequence $\vec{\mathbf{r}} = (\{r_k\}_{k=0}^{\infty})$ is $r_k = \tau/k!$, where $\tau$ depends on $f$ and is chosen sufficiently small to ensure that condition (B) holds. Notice that the diameter of $HB^N(\vec{\mathbf{r}})$ is then proportional to $\tau$, so that $\tau$ can be chosen as some multiple of the distance from $f$ to the boundary of $\mathrm{Diff}^r(B^N)$.

**1.32 Main Prevalent Theorem** *For any $0 < \rho \le \infty$ and any $C^{1+\rho}$ diffeomorphism $f \in \mathrm{Diff}^{1+\rho}(B^N)$, consider a Hilbert brick $HB^N(\vec{\mathbf{r}})$ of an admissible size $\vec{\mathbf{r}}$ with respect to $f$ and the family of analytic perturbations of $f$*

$$\{f_{\vec{\varepsilon}}(x) = f(x) + \phi_{\vec{\varepsilon}}(x)\}_{\vec{\varepsilon}\in HB^N(\vec{\mathbf{r}})} \tag{1.13}$$

*with the Lebesgue product probability measure $\mu_{\vec{\mathbf{r}}}^N$ associated to $HB^N(\vec{\mathbf{r}})$. Then for every $\delta > 0$ and for $\mu_{\vec{\mathbf{r}}}^N$-a.e. $\vec{\varepsilon}$ there is $C = C(\vec{\varepsilon}, \delta) > 0$ such that for all $n \in \mathbb{Z}_+$*

$$\gamma_n(f_{\vec{\varepsilon}}) > \exp(-Cn^{1+\delta}) \qquad P_n(f_{\vec{\varepsilon}}) < \exp(Cn^{1+\delta}). \tag{1.14}$$

Since a countable intersection of prevalent subsets of a Banach space is prevalent [HSY], the Main Prevalent Theorem implies the results stated in terms of prevalence in the introduction. Indeed, we can decompose $\mathrm{Diff}^r(B^N)$ into a nested countable union of sets $\mathcal{S}_j$ that are each a positive distance from the boundary of $\mathrm{Diff}^r(B^N)$, and for each $j \in \mathbb{Z}^+$ choose an admissible sequence $\vec{\mathbf{r}}_j$ that is valid for all $f \in \mathcal{S}_j$.

Similar results can be proved for $C^{1+\rho}$ smooth 1-dimensional maps. Moreover, thw 1-dimensional case is a good starting point to learn the method [KH].

In [KH] we deduce from the Main Prevalent Theorem the following more general result. Let $\gamma \ge 0$ and $f \in \mathrm{Diff}^{1+\rho}(B^N)$ be a $C^{1+\rho}$ diffeomorphism for some $\rho > 0$. A point $x \in B^N$ is called *$(n, \gamma)$-periodic* if $\|f^n(x) - x\| \le \gamma$, and *$(n, \gamma)$-hyperbolic* if $\gamma_n(x, f) = \gamma(Df^n(x)) \ge \gamma$. Notice that a point can be $(n, \gamma)$-hyperbolic regardless of its periodicity, but this property is of interest primarily for $(n, \gamma)$-periodic points. For positive $C$ and $\delta$ let $\gamma_n(C, \delta) = \exp(-Cn^{1+\delta})$.

**1.33 Theorem** *Given the hypotheses of the Main Prevalent Theorem, for every $\delta > 0$ and for $\mu_{\vec{\mathbf{r}}}^N$-a.e. $\vec{\varepsilon}$ there is $C = C(\vec{\varepsilon}, \delta) > 0$ such that for all $n \in \mathbb{Z}_+$, every $(n, \gamma_n^{1/\rho}(C, \delta))$-periodic point $x \in B^N$ is $(n, \gamma_n(C, \delta))$-hyperbolic. (Here we assume $0 < \rho \le 1$; in a space $\mathrm{Diff}^{1+\rho}(B^N)$ with $\rho > 1$, the statement holds with $\rho$ replaced by 1.)*

This result together with Proposition 1.28 implies the Main Theorem, because any periodic point of period $n$ is $(n, \gamma)$-periodic for any $\gamma > 0$. Also in the statements of the Main Prevalent Theorem and Theorem 1.33, the unit ball $B^N$ can be replaced by a bounded open set $U \subset \mathbb{R}^N$. After scaling, $U$ can be considered as a subset of the unit ball $B^N$.

One can define a distance on a compact manifold $M$ and almost periodic points of diffeomorphisms of $M$. Then one can cover $M = \bigcup_i U_i$ by coordinate charts and define hyperbolicity for almost periodic points using these charts $\{U_i\}_i$ (see [Y] for details). This gives a precise meaning to the following result.

**1.34 Theorem** *Let $\{f_\varepsilon\}_{\varepsilon\in B^m} \subset \mathrm{Diff}^{1+\rho}(M)$ be a generic m-parameter family of diffeomorphisms of a compact manifold M for some $\rho > 0$. Then for every $\delta > 0$ and almost every $\varepsilon \in B^m$ there is a constant $C = C(\varepsilon, \delta)$ such that every $(n, \gamma_n^{1/\rho}(C, \delta))$-periodic point x in $B^N$ is $(n, \gamma_n(C, \delta))$-hyperbolic. (Here again we assume $0 < \rho \le 1$, replacing $\rho$ with 1 in the conclusion if $\rho > 1$.)*

In [KH] we also give a precise meaning to the term *generic.*

# 2 Superexponential growth of the number of periodic points in Newhouse domains

In this section we discuss a somewhat improved version of Theorem 1.12, which says that in a Newhouse domain $\mathcal{N} \subset \mathrm{Diff}^r(M^2)$, supergrowth is $C^r$-generic with any $1 \le r < \infty$. We outline the proof of this from [K2], with the main emphasis being placed on pointing out the original ideas from [GST1]. This represents the *first mechanism* of arbitrary fast growth of the number of periodic points we are going to present. The mechanism for arbitrarily fast growth for 1-dimensional maps and 3-dimensional diffeomorphisms with a heteroclinic cycle is presented in the next chapter. Another extremely important result of this chapter, due to Gonchenko-Shilnikov-Turaev [GST1], is that *it is impossible to describe bifurcations of a surface diffeomorphism exhibiting a homoclinic tangency with any finite-parameter family.*

We point out also that Theorem 1.12 also implies Theorem 1.4 (see [K1] section 4.2). The main result of this chapter is the following:

**2.1 Theorem** *Let $2 \le r < \infty$ and M be a compact 2-dimensional manifold. Let $\mathcal{N} \subset Diffr(M)$ be a Newhouse domain. Then for an arbitrary sequence of positive integers $\{a_n\}_{n=1}^{\infty}$ there exists a residual set $\mathcal{R}_a \subset \mathcal{N}$, depending on the sequence $\{a_n\}_{n=1}^{\infty}$, with the property that $f \in \mathcal{R}_a$ implies that*

$$\limsup_{n\to\infty} P_n(f)/a_n = \infty.$$

*Moreover, there is a dense set $\mathcal{D}$ in $\mathcal{N}$ such that any diffeomorphism $f \in \mathcal{D}$ has a curve of periodic points.*

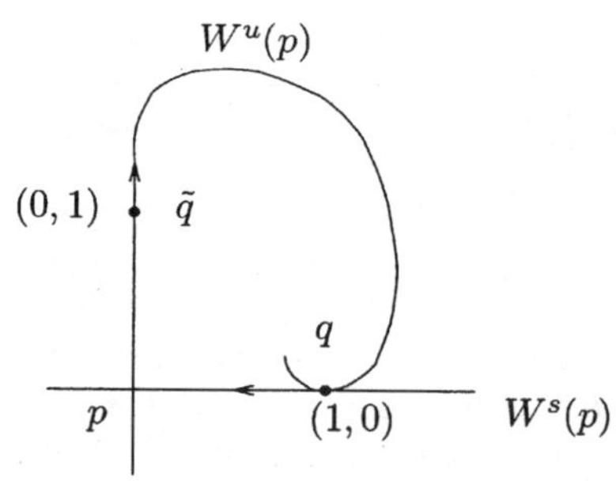


Figure 1: Homoclinic tangency

The key to the proof of this theorem is the following result of Gonchenko-Shilnikov-Turaev [GST1, GST2].

## 2.1 Gonchenko-Shilnikov-Turaev Theorem and degenerate periodic orbits in Newhouse domains

Assume that a $C^r$ diffeomorphism $f$ exhibits a homoclinic tangency. By the Newhouse theorem [Ne] there exists a Newhouse domain in each $C^r$ neighborhood of a diffeomorphism $f$ exhibiting a homoclinic tangency.

Let us define a degenerate periodic point of order $k$ or a $k$-degenerate periodic point. Sometimes, it is also called a *saddlenode periodic orbit of multiplicity* $k+1$.

**2.2 Definition** Let $f$ be a $C^s$ diffeomorphism of a smooth manifold having a periodic orbit $p$ of period $m$. A periodic point $p$ is called *$k$-degenerate*, where $k < s$, if the linear part of $f^m$ at the point $p$ has a multiplier $\nu = 1$ while the other multipliers are different in absolute value from the unit, and a restriction of $f$ to the central manifold in some coordinate system can be written in the form

$$x \mapsto x + l_{k+1}x^{k+1} + o(x^{k+1}). \tag{2.1}$$

Let $s > r$. Then $C^s$ diffeomorphisms are dense in the space $\text{Diff}^r(M)$ and, therefore, in any Newhouse domain $\mathcal{N} \subset \text{Diff}^r(M)$ (see e.g. [PM]).

**2.3 Theorem ([GST1] Theorem 4)** *For any positive integers $s > k \geq r$ the set of $C^s$ diffeomorphisms having a $k$-degenerate periodic orbit is dense in a Newhouse domain $\mathcal{N} \subset$* $\text{Diff}^r(M)$.

This theorem and Theorem 1.10 of Newhouse imply the following extremely important result:

**2.4 Corollary ([GST1])** *Let $f \in$ $\text{Diff}^r(M)$ be a diffeomorphism exhibiting a homoclinic tangency. There is no finite number $s$ such that a generic $s$-parameter family $\{f_\varepsilon\}$ unfolding a diffeomorphism $f_0 = f$ is a versal family of $f_0$, meaning that the family $\{f_\varepsilon\}$ describes all possible bifurcations occurring next to $f$. Indeed, to describe all possible bifurcations of a $k$-degenerate periodic orbit one needs at least $k+1$ parameters and $k$ can be arbitrary large.*

Once Theorem 2.3 is proved, the proof of Theorem 2.1 can be completed by inductive application of the following idea. Let $f$ be a $C^s$ diffeomorphism from a Newhouse domain $\mathcal{N} \subset \text{Diff}^r(M)$ with a $k$-degenerate periodic orbit $p$ of period, say $n$, of $f$ for $s > k \geq r$, then $p$ is a *flat periodic point* along the central manifold with respect to the $C^r$-topology; namely, by a $C^r$-perturbation one can make the restriction to the central manifold be the *identical map*. This allows us either to create a *curve of periodic orbits* or *split* $p$ into *any arbitrary given number of* hyperbolic periodic orbits of the same period (or double the period of $p$) by a small $C^r$-perturbation. Since the created periodic orbits are hyperbolic, they persist under perturbations. Moreover, after a perturbation we are still in a Newhouse domain; one can therefore iterate this procedure of creating a $k$-degenerate periodic orbits and splitting them without destroying what was done in previous stages (see [K2] section 2.7).

In what follows we need a few notions related to a saddle periodic point. These definitions will be needed in the proof of Theorem 2.3.

**2.5 Definition** Let $f$ be a $C^s$ diffeomorphism of a 2-dimensional manifold $M$, and let $p$ be a saddle periodic point of period $m$, namely, $f^m(p) = p$ with eigenvalues $\lambda$ and $\mu$, $\lambda < 1 < \mu$. The *saddle exponent* of $p$ is the number $\rho(p, f) = -(\log \lambda)/\log \mu$. We call $p$ a *$\rho$-shrinking saddle*, where $\rho = \rho(p, f)$. If $\rho$ is greater than some $r$, then $p$ is also called *at least $r$-shrinking.*

A saddle $p$ is called *nonresonant* if for any pair of positive integers $n$ and $m$ the number $\lambda^n \mu^m$ is different from 1.

## 2.2 A scheme of a proof of Theorem 2.3

Theorem 2.3 is stated in [GST1], Thm. 4. A proof of this theorem is outlined there. We present a slightly different outline based on [K2]. [K2] in turn essentially uses ideas given in [GST1]; see also [GST2] for a detailed proof. In what follows, a $C^r$-perturbation means a small $C^r$-perturbation. The proof of Theorem 2.3 consists of four steps.

*The first step.* From the existence of a homoclinic tangency of a dissipative saddle, we deduce the existence (after a $C^r$-perturbation) of a homoclinic tangency of an at least $k$-shrinking saddle, $k > r$.[6]

*The second step.* From the existence of a homoclinic tangency of an at least $k$-shrinking saddle, we create a $k$-floor tower (defined in section 2.3) after a $C^r$-perturbation (see Fig. 2 for $k = 3$).

*The third step.* From the existence of a $k$-floor tower, we show that a $C^r$-perturbation can make a $k$-th order homoclinic tangency.

*The fourth step.* From the existence of a $k$-th order homoclinic tangency we construct by a $C^r$-perturbation a $k$-th order degenerate periodic orbit of an arbitrarily high period.

Notice that the way we construct a $k$-tower is slightly different from the one in [GST1, GST2].

Steps one and four above are done in [K2] by direct calculations (see sections 2.2 and 2.3 there) so we shall skip the description of these steps. Steps two and three are highly original and require the beautiful idea of Gonchenko-Shilnikov-Turaev [GST1] of a *separatrix tower.* In the next section we define a separatrix tower or simply a tower and in the following section explain how towers can produce homoclinic tangencies of arbitrarily large orders.

## 2.3 The second step: construction of a $k$-floor tower

Consider a $C^\infty$ diffeomorphism $f$ with a nonresonant saddle periodic point $p$ exhibiting a homoclinic tangency at a point $q$. First, we give a definition of *a $k$-floor tower.* Recall that $U$ denotes a neighborhood of the homoclinic tangency $q$. Let $\tilde{p}$ be a saddle periodic orbit of $f$, $\tilde{p} \in U$. Then denote by $W^s_{\text{loc}}(\tilde{p})$ (resp. $W^u_{\text{loc}}(\tilde{p})$) the first connected component of the intersection of the stable (resp. unstable) manifold $W^s(\tilde{p})$ (resp. $W^u(\tilde{p})$) with $U$.

**2.6 Definition** A *$k$-floor tower* is a collection of $k$ saddle periodic points $p_1, \dots, p_k$ (of different periods) such that $W^u_{\text{loc}}(p_i)$ is tangent to $W^s_{\text{loc}}(p_{i+1})$ for $i = 1, \dots, k-1$, and $W^u_{\text{loc}}(p_k)$ intersects $W^s_{\text{loc}}(p_1)$ transversally (see Fig. 3 for $k = 3$).

[6] Theorem 2.3 can be done without $k$-shrinking saddles (see [GST2]), but it is easy to find such a saddle ([K2] Corollary 4).

For determines we shall consider towers with saddles posed in turn to the left and to the right, i.e. the saddles $p_1, p_3, \ldots$ are to the right of a point of homoclinic tangency, and the saddles $p_2, p_4, \ldots$ are to the left of homoclinic tangency.

The construction of a $k$-floor tower is an intermediate step in the proof of Theorem 2.3. Notice that Fig. 2 might be misleading, because in order to get the natural shape of a 3-floor tower one should shrink the figure along the horizontal direction by a sufficiently large factor. If one draws a 3-floor tower in natural size such a figure would become unreadable.

In this section we prove the following:

**2.7 Lemma ([K2] Lemma 2)** *For any positive integer $k$ a $C^r$ diffeomorphism $f$ exhibiting a homoclinic tangency for an at least $r$-shrinking saddle periodic orbit $p$ admits a $C^r$-perturbation $\tilde{f}$ such that $\tilde{f}$ has a $k$-floor tower. If $q$ is a point of homoclinic tangency of $f$, then the aforementioned tower of $\tilde{f}$ is located in a neighborhood $U$ of $q$.*

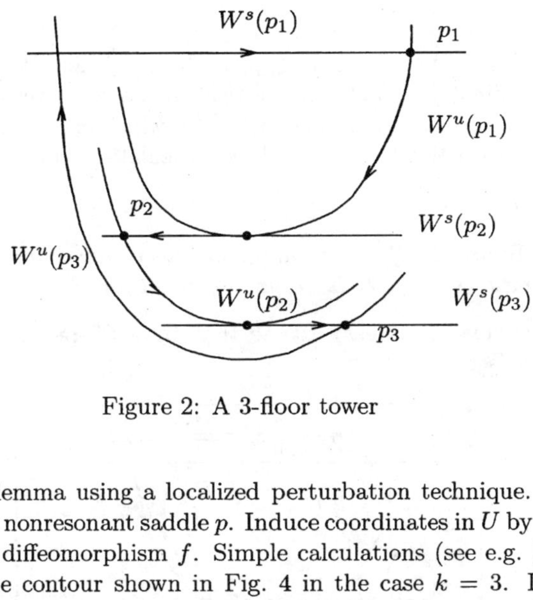


Figure 2: A 3-floor tower

**Proof** We prove this lemma using a localized perturbation technique. As usual consider normal coordinates for a nonresonant saddle $p$. Induce coordinates in $U$ by normal coordinates for the point $p$ and the diffeomorphism $f$. Simple calculations (see e.g. [K2] Proposition 1) give the existence of the contour shown in Fig. 4 in the case $k = 3$. Indeed, consider an increasing sequence of numbers $n_1, \ldots, n_k$ such that for each $i = 1, \ldots, k$ the following two properties hold:

(1) $T_{n_i}$ intersects $f^{n_i+N}(T_{n_i})$ and the latter form a horseshoe (for a precise definition of $T_n$ see [K2] Proposition 1).

(2) $n_{i+1}$ is the largest number such that $T_{n_{i+1}}$ and $f^{n_i+N}(T_{n_i})$ intersect in a horseshoe-like way, i.e., that they bound an open set.

For each $i = 1, \ldots, k$, condition (1) implies existence of a saddle periodic point $p_i \in T_{n_i} \cap f^{n_i+N}(T_{n_i})$ of period $n_i + N$ and condition (2) that $W^s_{\text{loc}}(p_{i+1})$ and $W^u_{\text{loc}}(p_i)$ intersect.

Let $U$ be equipped with normal coordinates. Define the maximal distance in the vertical direction between $W^s_{\text{loc}}(p_i)$ and $W^u_{\text{loc}}(p_i)$, denoted by $s_i$, as the maximum of distance between

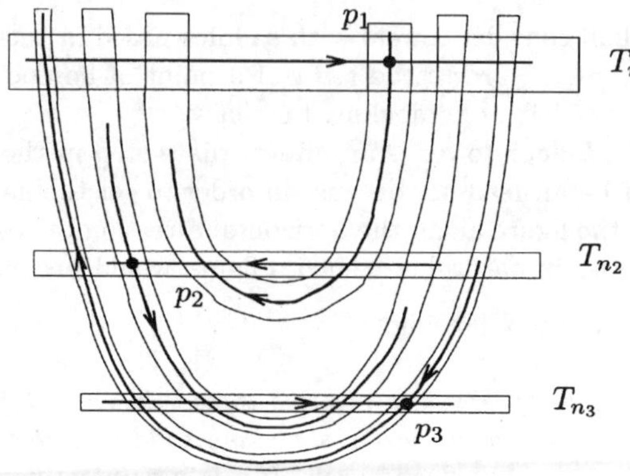


Figure 3: An incomplete 3-floor tower

any two points $x \in W^s_{\text{loc}}(p_i)$ and $y \in W^u_{\text{loc}}(p_i)$, which is below $W^s_{\text{loc}}(p_i)$ such that $x$ and $y$ have the same $\bar{x}$-coordinate. Denote the vertical distance between the centers of $T_{n_i}$ and $T_{n_{i+1}}$ by $t_i$ (see Fig. 4). So, $s_i - t_i$ is the distance by which one should lift $W^u_{\text{loc}}(p_i)$ to create heteroclinic tangency with $W^s_{\text{loc}}(p_{i+1})$. By direct calculation from section 2.2 of [K2] we get for $t_i = \mu^{-n_i} - \mu^{-n_{i+1}}$. □

**2.8 Lemma ([K2] Prop. 2)** *If the saddle $p$ having a homoclinic tangency is at least $r$-shrinking, then the ratio $(s_i - t_i)/t_i^r$ is arbitrarily small for each $i = 1, \dots, k-1$.*

If such a ratio is small it is easy to see that by a $C^r$-perturbation one can create a heteroclinic tangency (see Fig. 4)

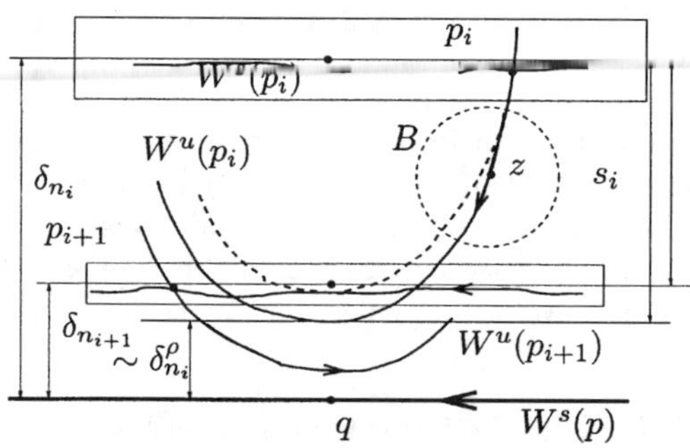


Figure 4: A localized perturbation for a floor of a tower

In order to construct a $k$-floor tower one needs to create a heteroclinic tangency of $W^s_{\text{loc}}(p_{n_{i+1}})$ and $W^u_{\text{loc}}(p_{n_i})$ by a $C^r$-perturbation. We construct it by "bending" $W^u_{\text{loc}}(p_{n_i})$.

Another way to construct a $k$-floor tower, used in [GST1, GST2], is by fixing the eigenvalue $\mu > 1$ and varying the other eigenvalue $\lambda < 1$ of the saddle $p$ exhibiting homoclinic tangency. Simple calculations (see e.g. [K2] Prop. 1) shows that the rectangle $T_n$ is centered at $(1, \mu^{-1})$

and the curvilinear rectangle $f^{n+N}(T_n)$ is at distance $C\delta_n^\rho = C\lambda^n$ from $W^s(p)$; therefore, by changing $\lambda$ one can vary the position of $f^{n+N}(T_n)$ without changing the position of $T_n$. But, in this case one needs some additional geometric argument to construct all heteroclinic tangencies of a $k$-tower simultaneously.

## 2.4 The third step: Construction of a $k$th order tangency

We shall prove that by a perturbation of a $(k+1)$-floor tower one can create a $k$th order homoclinic tangency. Let us start with a 2nd order tangency for a 3-floor tower and then use induction on $k$.

**2.9 Proposition ([GST1, K2])** *A perturbation of a 3-floor tower can create a 2nd order homoclinic tangency.*

**Proof** Even though this lemma is proved in [GST1, GST2], [K2] we present a proof of it, which is simple and elegant. It also gives a pretty good idea of how a $k$-floor tower might produce a $k$-th order homoclinic tangency by a $C^r$-perturbation.

*Step 1.* Let us create a 1st order tangency of $W^u_{\text{loc}}(p_2)$ and $W^s_{\text{loc}}(p_1)$. Start with a 3-tower as in Fig. 2. "Push" $W^u_{\text{loc}}(p_2)$ down to $W^s_{\text{loc}}(p_3)$. Denote by $\gamma$ the tongue (the part) of $W^u_{\text{loc}}(p_2)$ underneath $W^s_{\text{loc}}(p_3)$. The tongue $\gamma$ is in the sector of the saddle hyperbolic point $p_3$; therefore, under iteration of $f$, $\gamma$ will be stretched along $W^u_{\text{loc}}(p_3)$ and for some $s$ $f^s(\gamma)\cap W^s(p_1)\neq\emptyset$. Varying the size of the tongue $\gamma$ we can create a heteroclinic tangency (see Fig. 5.a with $i=0$). Denote a point of tangency by $q^*$. Only two parts of $W^u(p_2)$ are depicted on Fig. 5.a: the starting part of $W^u(p_2)$ at $p_2$, and the image of $\gamma$ after a number of iterations under $f$ (in above notations $f^s(\gamma)$).

Assume that the saddle $p_1$ is nonresonant. Then there exist normal coordinates around $p_1$, linearizing $f$. Induce by $f$ normal coordinates in a neighborhood of $U^*$ of $q^*$. In what follows we shall use these coordinate systems in $U^*$.

*Step 2.* Let us create a 2nd order homoclinic tangency of $W^u_{\text{loc}}(p_1)$ and $W^s_{\text{loc}}(p_1)$. Start with a contour on Fig. 5.a. "Push" $W^u_{\text{loc}}(p_1)$ down to $W^s_{\text{loc}}(p_2)$. Denote by $\gamma^1$ the tongue (the part) of $W^u(p_1)$ underneath $W^s(p_2)$. Some iterate of the tongue $\gamma^1$ of $W^u_{\text{loc}}(p_1)$ comes into $U^*$. $U^*$ has normal coordinates and naturally defined horizontal and vertical directions. Now our goal is, varying the size of $\gamma^1$, to construct a tangency of some iterate of $\gamma^1$ to the horizontal direction in $U^*$.

Fix some coordinates in a neighborhood of $\tilde{U}$ of a tangency $\tilde{q}$ of $W^u_{\text{loc}}(p_1)$ and $W^s_{\text{loc}}(p_2)$. Consider a 1-parameter family of diffeomorphisms $\{f_\varepsilon\}$, where $\varepsilon$ is the maximal distance of $W^u_{\text{loc}}(p_1)\cap\tilde{U}$ and $W^s_{\text{loc}}(p_2)\cap\tilde{U}$ in the vertical direction.

Let $\gamma_\varepsilon = W^u_{\text{loc}}(p_1)\cap\{y\leq 0\}$. Fix $\varepsilon>0$ and $s=s(\varepsilon)$ such that $\gamma^s_\varepsilon = f^s(\gamma^1_\varepsilon)\cap U^*\neq\emptyset$ (see Fig. 5.b). The curve $\gamma^s_\varepsilon$ has two points $q_1$ and $q_2$ of tangency to the horizontal direction. As $\varepsilon$ decreases, $q_1$ and $q_2$ approach one another, and for some critical value $\varepsilon^*$ they collide and $q_1=q_2$ (see Fig. 5.c).

At the point $q_1=q_2$ the local unstable manifold $W^u_{\text{loc}}(p_1)$ has a 2nd order tangency to the horizontal direction. Let this point have coordinates $(\varepsilon_1,\varepsilon_2)$ in $U^*$. Lifting $W^u_{\text{loc}}(p_2)$ by $\varepsilon_2$ we can create a 2nd order tangency at point $(\varepsilon_1,0)$. This completes the proof of the proposition. □

For the construction of higher order tangencies we refer to [GST2, K2] and point out that the idea is the same as for the 3rd order, but requires more delicate calculations (see e.g. [K2]

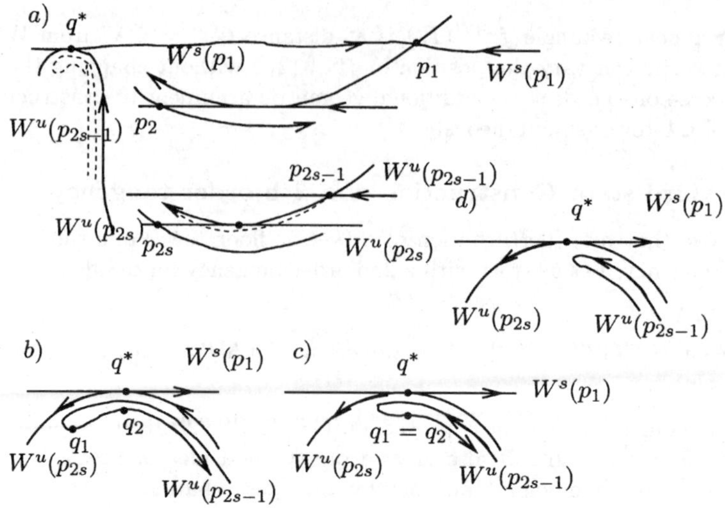


Figure 5: A 2nd (even) order tangency

Prop. 5). This exposition shows how the third step of the proof of Theorem 2.1 is done. As we pointed out above, Steps 1 and 4 can be done by direct calculations, and their description is omitted here (see [GST2, K2] for details).

# 3 Superexponential growth of the number of periodic points for 1-dimensional maps and 3-dimensional heteroclinic cycles

In this chapter we prove Theorem 1.23 in the case $r = 2$ and outline the proof of Theorem 1.19 from [KP]. The main emphasis will be to point out the ideas for the proofs, and we refer to the papers [KK, KP] for more details. These two proofs represent the *second and third mechanisms* of arbitrary fast growth of the number of periodic points we are going to present. These results are indications of the fact that there are many scenarios of how superexponential growth of the number of periodic points can occur outside of hyperbolic diffeomorphisms. Theorems 1.12 and 1.19 support Conjecture 1.22, provided Conjecture 1.20 is true.

## 3.1 Superexponential growth of periodic points for 1-dimensional maps

This section is devoted to 1-dimensional example in the case $r = 2$. In the case $r \geq 2$ the proof is almost the same, just that the calculations are more involved [KK]. Denote by $C^r(I, I)$ the space of such maps with the uniform $C^r$-topology. As we pointed out in the introduction, there is a $C^r$-dense set of Artin-Mazur maps. Moreover, by the Martens-de Melo-van Strien theorem (see Theorem 1.25 and its Corollary), those maps which are not Artin-Mazur form a set of codimension $r$. The main result of the first part of this chapter (Theorem 1.23) is a counterpart of the Martens-de Melo-van Strien theorem. Below we describe the basic idea of the proof of this theorem.

Let $f : I \to I$ be a $C^r$ map. We say that a periodic point $f^n x = x$ is *neutral of order* $k \le r$ if $(f^n)'x = 1$, $(f^n)^{(s)} = 0$ for $s = 2, \dots, k-1$, and $(f^n)^{(k)} \neq 0$.[7]

**3.1 Remark** If $f$ has a neutral periodic point $x = f^n x$ of order $k > r$, then by a $C^r$-perturbation one can create arbitrary many periodic points of period $n$ close to $x$. Thus, if one can create a $C^r$ unimodal map with an infinite number of neutral periodic points $\{p_k\}_{k \in \mathbb{Z}_+}$ of periods $3^k$ whose orbits are isolated, then this proves Theorem 1.23.

### 3.1.1 Fixed point of a renormalization operator with a degenerate critical point

A *unimodal map* is an endomorphism of the interval $I$. It is of the form $f = \phi \circ q_t$, where $\phi \in \mathrm{Diff}^2_+(I)$ is an orientation preserving $C^2$ diffeomorphism of $I$, and $q_t : I \to I$, $t \in [0,1]$, is defined by $q_t(x) = -2t|x|^\alpha + 2t - 1$. The exponent $\alpha > 1$ is called the *critical exponent* of $f$. The map $q_t$ is called the *canonical folding map* with *peak-value* $t \in [0,1]$. The peak-value determines the maximum $q_t(0) = 2t - 1$. The above form for the canonical folding map is not just a choice for convenience, it arises naturally [Ma]. The diffeomorphism $\phi$ is called the *diffeomorphic part* of $f$. Notice that $f(-1) = f(1) = -1$. The collection of unimodal maps with given critical exponent $\alpha > 1$ is denoted by $\mathcal{U}_\alpha$.

Let $\mathcal{U}_\alpha$ be the collection of unimodal maps whose peak-value is high enough such that the unimodal map has a fixed point $p \in (0,1)$. For every $f \in \mathcal{U}_\alpha$ we can consider the first return map to the interval $[-p,p]$. If the peak-value is not too high, the first return map will just be $f^2|_{[-p,p]}$; the unimodal map $f$ is called *renormalizable*. The unimodal map obtained by rescaling this first return-map to $[-p,p]$ is called the *renormalization* of $f$. The operator defined in this way is called the *renormalization operator*. Lanford [L] and, later, Sullivan [S], proved that there is a fixed point for the renormalization operator.

More generally, a unimodal map $f \in \mathcal{U}_\alpha$ is called *renormalizable* if and only if there exists an expanding periodic point $p \in (-1,1)$ such that the first return map to the central interval $C = [-p,p]$ is of the form $f^q : C \to C$ with $f^q(p) = p$ and $q \ge 2$. The first return map to $C$ will be, up to rescaling, a unimodal map. This unimodal map is a *renormalization* of $f$. Notice that a renormalization is completely determined by the periodic point $p$. In particular, when $q = 3$ and the renormalization operator is well-defined. By a theorem of Epstein [E], such an operator has a fixed point $f_\alpha \in \mathcal{U}_\alpha$.[8] Moreover, $f_\alpha$ is real analytic.

Let $\alpha = [2r+2]$, let $f_\alpha \in \mathcal{U}_\alpha$ denote a fixed point of the renormalization operator for $q = 3$, and let 0 be the critical point of $f_\alpha$. Then it has a critical point of order $\ge 2r$ and satisfies the following properties:

- $f_\alpha$ has a fixed point, denoted by $p$;
- $p$ has a neighborhood $[p^-, p^+]$ free from the forward orbit of 0;

[7]For $m$-dimensional maps with $m \ge 2$, a periodic point $x$ is called a *saddle-node of order* $k$ if the absolute values of the eigenvalues of its linearization are all but one different from 1, and along the exceptional eigendirection the first derivative is 1 and the others of order up to $k-1$ are zero.

[8]See Martens [Ma] for a more detailed analysis of such fixed points.

- for each $k \in \mathbb{Z}_+$ there is a periodic point $p_k$ of period $3^k$ and $p_k = \lambda^k p$ for some $\lambda \in (-1,0)$, where $[-\lambda,\lambda]$ is the interval where the renormalization $f_\alpha^3 : [-\lambda,\lambda] \to [-\lambda,\lambda]$ of $f_\alpha$ is defined and equals $f_\alpha$ after rescaling;
- each of the intervals $I_k = [\lambda^k p^-, \lambda^k p^+]$, $k \in \mathbb{Z}_+$, is free from the forward orbit of 0;
- the derivatives along periodic orbits of the $p_k$'s for each integer $s$ satisfy $(f_\alpha^{3^k})^{(s)}(p_k) = \lambda^{k(1-s)} f_\alpha^{(s)}(p)$.

*The Main Idea:* We shall prove that an arbitrarily small $C^r$-perturbation $\tilde{f}$ of $f_\alpha$ for a sufficiently large $k_0$ and all $k > k_0$ has a neutral periodic point of order $r+1$ at $p_k$. Moreover, the orbits $O_k(f_\alpha) = \{f_\alpha^s(p_k)\}_{s=0}^{3^k-1}$ of the $p_k$'s are isolated from each other. Therefore, a $C^r$ small perturbation of $\tilde{f}$ around the $p_k$'s, for each $k > k_0$ can create explosions of the number of periodic points from the $p_k$'s of periods $3^k$.

### 3.1.2 Calculations

In this section we show that, indeed, for a large enough $k$ by a small $C^r$ perturbation inside $I_k$ we can obtain a neutral periodic point $p_k$ of order $r+1$. The crucial point of the proof is that *for any set of constants $D_1, \dots, D_{r+1}$ a small $C^r$ perturbation $\tilde{f}$ of $f_\alpha$ in a small neighborhood of a critical point, say at $p_k$, $k \gg 1$, allows one to change its first $r+1$ derivatives by a constant, i.e. $\tilde{f}^{(s)}(p_k) = D_s f_\alpha^{(s)}(p_k)$, $1 \le s \le r+1$.*

Suppose $f_\alpha$ and some $\tilde{f}$ coincide outside the union of intervals $\bigcup_{k \ge k_0} I_k$. Then, since for each $k \in \mathbb{Z}_+$ the trajectory of $p_k$, excluding $p_k$, does not visit $I_k$, we have $f_\alpha^{(s)}(f^l p_k) = \tilde{f}^{(s)}(f^l p_k)$ for all $s, k \in \mathbb{Z}_+$, and $1 \le l < 3^k$. Then direct calculation proves that there is a change of the first $r+1$ derivatives of $f_\alpha$ to $\{\tilde{f}^{(s)}(p_k) = D_s^k f_\alpha^{(s)}(p_k)\}_{s=1}^{k-1}$ so that $p_k$ becomes neutral periodic of order $r+1$.

To make the calculation in the case of general $r$ easier to comprehend, we discuss the case $r=2$ first. We fix a large $k$ and consider the trajectory $O_k(f_\alpha)$ of $p_k$. All the points of $O_k(f_\alpha)$ except $p_k$ are away from $I_k$. It suffices to $C^2$ perturb $f_\alpha$ to $\tilde{f}$ only inside $I_k$ keeping the trajectory $O_k(f_\alpha)$ of $p_k$ fixed, so that $p_k$ becomes neutral of order 3, and the $C^2$-norm of the perturbation decays to 0 as $k$ tends to infinity.

Put $\tilde{f} = f_\alpha + \delta_k$, where $\delta_k$ is a function supported in $I_k$ such that

- $\delta_k(\lambda^k p) = \delta(\lambda^k p^-) = \delta(\lambda^k p^+) = 0$, $\lambda^k p = p_k$;
- $\delta_k^{(s)}(\lambda^k p^-) = \delta_k^{(s)}(\lambda^k p^+) = 0$ for $s = 0,1,2$;
- $\delta_k^{(s)}(\lambda^k p) \sim \lambda^{(6-s)k}$ for $s = 0,1,2$;
- $\|\delta_k\|_{C^2} \to 0$ as $k \to 0$.

The last item is clearly satisfied because the support of $\delta_k$ is of length $\sim \lambda^k$, and $C^2$-smallness requires $\delta_k$ be of order $o(\lambda^{3k})$.

Straightforward calculation gives

$$
\begin{aligned}
(f_\alpha^{3^k})'(p_k) &= f_\alpha'(f^{3^k-1}p_k)\dots f_\alpha'(p_k) = f_\alpha'(p) \\
(f_\alpha^{3^k})''(p_k) &= (f_\alpha^{3^k})'(p_k)\left(\sum_{l=0}^{3^k-1} \frac{f_\alpha''(f_\alpha^l p_k)}{f_\alpha'(f_\alpha^l p_k)}\right) = \lambda^{-k} f_\alpha''(p).
\end{aligned}
\tag{3.1}
$$

The left-hand side equalities is the 5th remark about properties of renormalization. It is easy to see that a change of $f'_\alpha(p_k)$ and $f''_\alpha(p_k)$ by a constant suffices to satisfy the equations $(\tilde{f}^{3^k})'(p_k) = 1$ and $(\tilde{f}^{3^k})''(p_k) = 0$. To prove a similar statement in the general case $r \geq 2$ more involved calculations are required (see [KK]).

## 3.2 Superexponential growth of periodic points for 3-dimensional heteroclinic cycles

In this section we outline the proof of Theorem 1.19 from [KP]. This theorem represents another mechanism for the creation of superexponential growth of the number of periodic points outside the set of hyperbolic diffeomorphisms. It is also another indication toward Conjecture 1.22.

The initial idea of the proof of Theorem 1.19 is the same as for Theorem 1.12. We shall prove that diffeomorphisms having a degenerate periodic orbits of all orders are dense.

**3.2 Theorem** *Let $1 \leq r < \infty$ and let $M$ be a compact 3-dimensional manifold. Let $\mathcal{B} \subset \mathrm{Diff}^r(M)$ be a domain with a blender. Then for any positive integers $s > k \geq r$ the set of $C^s$ diffeomorphisms having a $k$-degenerate periodic orbit is dense in $\mathcal{B} \subset \mathrm{Diff}^r(M)$.*

Then arguments from Section 2.1 independently of dimension of $M$ show that Theorem 3.2 implies Theorem 1.19. Similarly to Section 2.2 we outline the proof of Theorem 3.2 in the next section. Below we formulate several auxiliary definitions.

**3.3 Definition** In Definition 1.9 of an HC (heteroclinic cycle) we call the periodic points $p, q$ *vertices* of the HC.

We say that such an HC is *$k$-degenerate* if its vertices $p, q$ are $k$-degenerate periodic points for some $k > 0$.

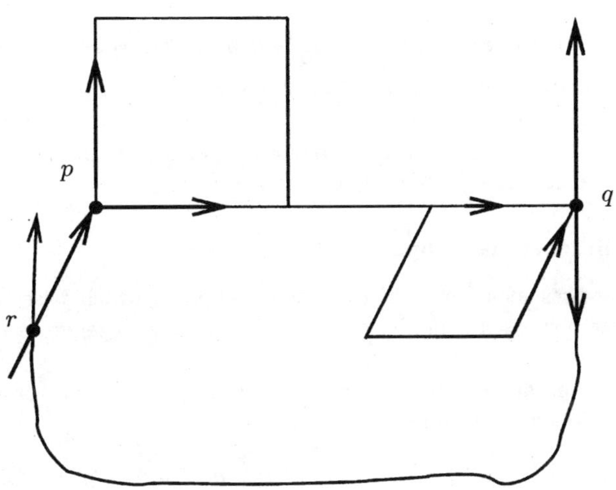


Figure 6: A heteroclinic cycle

### 3.2.1 A scheme of a proof of Theorem 3.2

In what follows we repeatedly use the expression, *to create a dynamic object*, meaning that there is a small $C^r$-perturbation of a starting diffeomorphism (before a creation) so that the perturbed diffeomorphism has a "created" property. For example, Theorem 2.3 implies that from a diffeomorphism in a Newhouse domain one can *create* a homoclinic tangency, i.e. there is a $C^r$ close diffeomorphism with a homoclinic tangency.

The proof of Theorem 3.2 consists of $k$ stages. Each stage, except the last one consists of four steps. Start with a diffeomorphism $f : M^3 \to M^3$ having an HC.

Stage 1: *Construction of a 1-degenerate HC*

*Step 1.* We approximate the dynamics near HC by dynamics of a 1-parameter family of 1-dimensional maps (see Fig. 7).

*Step 2.* Within this family of 1-dimensional maps we can find a 1-degenerate periodic point (or simply a saddlenode). Then using a reduction from Step 1 we deduce the existence of a 1-degenerate periodic point $p_1$ for a $C^r$-perturbation of $f$, denoted by $f_1$.

*Step 3.* Without destroying $p_1$, we create another 1-degenerate periodic point $q_1$ for $f_2$.
*Step 4.* We now create a 1-degenerate HC for $f_3$ with $p_1$ and $q_1$ as vertices.

Stage 2: *Construction of a 2-degenerate HC*

In this stage we basically repeat Stage 1 for higher order degenerate HCs. Namely:

*Step 1.* We approximate the dynamics near HC by dynamics of a 1-paramater family of 1-dimensional maps (see Fig. 7).

*Step 2.* Using such a reduction deduce existence of a 2-degenerate periodic point $p_2$ for a $C^r$-perturbation of $f_4$, denoted by $f_4$.

*Step 3.* Similarly, without destroying $p_2$, we create another 2-degenerate periodic point $q_2$ for $f_5$.

*Step 4.* Now we create for $f_6$ a 2-degenerate HC with $p_2$ and $q_2$ as vertices.

Following this algorithm after the $(k-1)$st stage we have a $(k-1)$-degenerate HC whose vertices are the $(k-1)$-degenerate periodic points $p_{k-1}$ and $q_{k-1}$ as vertices. The first stage of the $k$-th step gives the existence of a $k$-denenerate periodic point $p_k$. This is the required statement which completes the proof of Theorem 3.2.

### 3.2.2 Model 1-dimensional maps

In this subsection we present a family of 1-dimensional maps which arises in the first steps of each stage, and show how such a family can produce a more degenerate periodic orbit than those in the family.

Denote the unit interval by $I$. Fix a small $\varepsilon > 0$. Consider a pair of $C^r$-smooth maps $g$ and $h_\varepsilon$, $r \geq 1$, of the following form:

$$\begin{gathered} g : I \to I, \quad g(0) = 0, \quad g'(0) > 1, \quad g(1) = 1, \quad g'(1) < 1, \\ g(x) > x \text{ for every } x \in I, \quad g \text{ is monotonically increasing}; \\ h_\varepsilon : [1-\varepsilon, 1] \to [0, \varepsilon], \quad h(x) = 1 - x. \end{gathered} \tag{3.2}$$

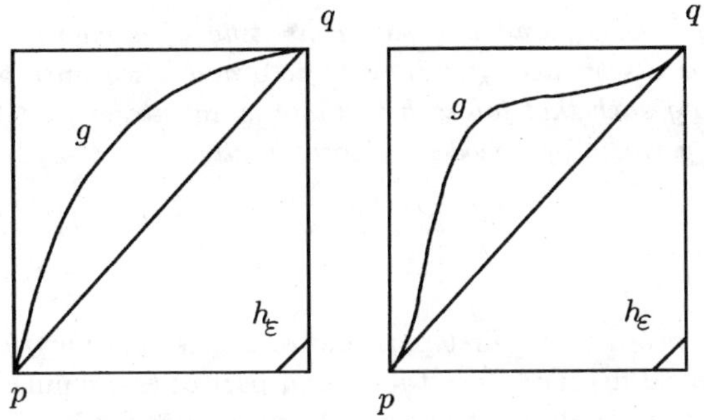


Figure 7: Model 1-dimensional maps

The way the dynamics of this pair is defined is as follows: if $x \in [1-\varepsilon]$ we always apply $g$, and if $x \in [1-\varepsilon, 1]$, then we can apply either $g$ or $h_\varepsilon$.

Heuristically, the connection between an HC in Fig. 6 and this family is the following. Suppose a diffeomorphism $f : M^3 \to M^3$ has an HC. Consider a 1-parameter family $f_\varepsilon$ perturbing $f = f_0$, where $\varepsilon$ measures the oriented minimal distance between the continuation of $W^s(p)$, denoted $W^s(p_\varepsilon)$, and the continuation of $W^u(q)$, denoted $W^u(q_\varepsilon)$ in a small neighborhood of $r$ (see Fig. 6). For every small $\varepsilon$, the continuations $W^s(q_\varepsilon) \cap W^u(p_\varepsilon)$ form a smooth segment, say $J$. Moreover, the dynamics induced on $J$ is smoothly conjugate to a smooth map $\tilde{g}$ of the unit interval $I$ and $\tilde{g}$ satisfies the conditions (3.2) of $g$. Indeed, one end of $J$ is repelling, the other is attracting, with no fixed points in between.

To see heuristically the analog of the map $h_\varepsilon$ in Fig. 6, notice first that for $\varepsilon = 0$ we have a sequence of points on $W^u(q)$, converging to $q$, which under forward iterations approach $p$ or, equivalently, belong to $W^s(p)$. By analogy $h_0(1) = 0$. Now pick $\varepsilon > 0$, then the intersection of the local parts of $W^s(p_\varepsilon)$ and $W^u(q_\varepsilon)$ becomes empty. So trajectories on $W^u(q)$ can start arbitrarily close to $q$, but don't have to approach $p$. In terms of $h_\varepsilon$ we have $h_\varepsilon(1) = \varepsilon \neq 0$. Below we formulate the statement which is necessary for the proof.

In view of the discussion above, one might guess that if we consider a $k$-degenerate HC with vertices $p$ and $q$, then the corresponding family has the same $h_\varepsilon$, but a different $C^r$-smooth $g$, $r \geq k$, where $g$ has the following properties:

$$\begin{gathered} g : I \to I, \quad g(0) = 0, \quad g'(0) = 1, \quad g(1) = 1, \quad g'(1) = 1, \\ g^{(s)}(0) = g^{(s)}(1) = 0 \text{ for } s = 2, \dots, k, \\ g(x) > x, \ x \in I, \quad \text{and } g \text{ is monotonically increasing;} \\ h_\varepsilon : [1-\varepsilon, 1] \to [0, \varepsilon], \quad h(x) = 1 - x. \end{gathered} \tag{3.3}$$

Pairs of maps with conditions (3.2) and (3.3) satisfy the following

**3.4 Lemma ([KP])** *For a pair consisting of a map $g$ and a family $\{h_\varepsilon\}_{\varepsilon>0}$ satisfying* (3.2) *there is a sequence $\varepsilon_n$ decreasing to* 0 *such that for each $\varepsilon_n$ there is an integer $p_n \in \mathbb{Z}_+$ and a point $x \in I$ which is* 1*-degenerate periodic of period $p_n$. More exactly,*

$$h_\varepsilon \circ g^{p_n - 1}(x) = x \quad \textit{and} \quad (h_\varepsilon \circ g^{p_n-1})'(x) = 1.$$

*Similarly, for any $k \in \mathbb{Z}_+$ and any pair consisting of a map $g$ and a family $\{h_\varepsilon\}_{\varepsilon>0}$ satisfying* (3.3) *there is a sequence $\varepsilon_n$ decreasing to* 0 *and a sequence of $C^r$ perturbations $g_n$ which $C^r$ converges to $g$ such that for each $\varepsilon_n$ there is an integer $p_n \in \mathbb{Z}_+$ and a point $x \in I$ which is* 1*-degenerate periodic of period $p_n$. More exactly,*

$$h_\varepsilon \circ g_n^{p_n-1}(x) = x, \qquad (h_\varepsilon \circ g_n^{p_n-1})'(x) = 1, \qquad (h_\varepsilon \circ g_n^{p_n-1})^{(s)}(x) = 0$$

*for every $s = 2, \dots, k+1$.*

We only present a proof of the first part and of the second part for $k = 2$ in order not to overload with technical details. Once the second part of the lemma is stated the proof of it is direct calculations. In [KP] we consider a $2k$-parameter family. The first $k$ parameters unfold the fixed point 0 and the second $k$ unfold 1. One can show that for a sufficiently large period $p$ the following equations are solvable within the family:

- the first derivative along a trajectory of length $p$ is 1;
- the second derivative along the trajectory is zero, and so on;
- the $(k+1)$st derivative is zero.

Moreover, the values of the $2k$ parameters solving these $(k+1)$ equations tend to zero as $k$ tends to infinity.

**Proof of Lemma 3.4** The first part is a trivial consequence of the following pair of remarks. Call a point $x \in I$, *$(g, h_\varepsilon)$-periodic of period $p$* if $h_\varepsilon \circ g_n^{p-1}(x) = x$.

Suppose for some $\varepsilon > 0$ there is a $(g, h_\varepsilon)$-periodic point $x$ of period $p$, then $p$ depends on $\varepsilon$, and as $\varepsilon$ tends to zero $p$ tends to infinity. This follows from the fact that $x$ has to belong to the interval $[0, \varepsilon]$, and as $\varepsilon$ tends to zero the number of iterates by $g$ to get from $[0, \varepsilon]$ to $[1-\varepsilon, 1]$ tends to infinity.

Therefore, for each large enough $p$, there is a segment $[\varepsilon_p^-, \varepsilon_p^+]$ in the $\varepsilon$-parameter space for which a $(g, h_\varepsilon)$-periodic point of period $p$ exists. In order to disappear, any periodic point has to pass through a bifurcation point. At that point (the endpoint of $[\varepsilon_p^-, \varepsilon_p^+]$) the periodic point has to be 1-degenerate. This completes the proof of the first part of the lemma.

To prove the second part of the lemma for $k = 2$ we start with a pair of a $C^r$ map $g$ and a family $\{h_\varepsilon\}_{\varepsilon>0}$ satisfying (3.3) for $k = 2$. Suppose $g''(0) = b_1$ and $g''(1) = b_2$ are nonzero; otherwise, make an arbitrary small perturbation to get it. Let $a = (a_1, a_2) \in \mathbb{R}^2$. Consider the 2-parameter family $g_a(x) = g(x) + a_1 x(x-1)^2 - a_2 x^2(x-1)$. Conditions (3.3) are fulfilled for all small $a_1 > 0$, $a_2 > 0$. Observe the following:

(1) Since $h'_\varepsilon(x) \equiv 1$ where it is defined, $(h_\varepsilon \circ g_n^{p-1})'(x) = 1$ if and only if $(g^{p-1})'(x) = 1$.

(2) $x$ is a $(g, h_\varepsilon)$-periodic point of period $p$ if $x \in [0, \varepsilon]$ and $g_n^{p-1}(x) \in [1-\varepsilon, 1]$.

(3) For any $a = (a_1, a_2)$, $a_1, a_2 > 0$, application of the first part of this lemma shows that for any sufficiently large $p \in \mathbb{Z}_+$ there exists $\varepsilon = \varepsilon(a, p) > 0$ and a 1-degenerate $(g_a, h_\varepsilon)$-periodic point $x$ of period $p$.

(4) Fix $\varepsilon_0 > 0$. Define

$$p_-(x) = \#\{s \geq 0 : g^s(x) \in [0, \varepsilon]\} \quad \text{and} \quad p_+(x) = \#\{0 \leq s \leq p-1 : g^s(x) \in [1-\varepsilon, 1]\}.$$

As $\varepsilon$ tends to zero and $x$ is a 1-degenerate $(g_a, h_\varepsilon)$-periodic point, then $p_\pm(x)$ tends to infinity and $p_+(x)/p_-(x)$ tends to $a_2/a_1$.

(5) Notice that

$$(h_\varepsilon \circ g_a^{p-1})''(x) = (h_\varepsilon \circ g_a^{p_n-1})'(x) \left( \sum_{j=0}^{p-1} \frac{g_a''(g_a^j(x))}{g_a'(g_a^j(x))} \right). \tag{3.4}$$

Notice also that $g_a''(x)\ b_1 + a_1 + a_2$ for $x$ near zero and $g_a''(x)\ b_2 + a_1 + a_2$ for $x$ near one. Therefore, with the notation of part (3) above, as $\varepsilon$ tends to zero and $a_1, a_2$ are small positive, we have

$$\frac{(h_\varepsilon \circ g_a p - 1)''(x)}{p} \to \frac{p_-(x)}{p} \frac{b_2 + a_1 + a_2}{1 + a_1} + \frac{p_+(x)}{p} \frac{b_1 + a_1 + a_2}{1 + a_2} + o(1). \tag{3.5}$$

By part (4), $p_+(x)/p_-(x) \to a_2/a_1$. Thus, monotonically changing the ratio $a_2/a_1$ and keeping it small, we can solve the equation $(h_\varepsilon \circ g_a^{p-1})''(x) = 0$ for a 1-degenerate $(g_a, h_\varepsilon)$-periodic point $x$ of period $p$. This proves that $x$ constructed in this way is actually a 2-degenerate $(g_a, h_\varepsilon)$-periodic point of period $p$. Parts (4) and (5) require some calculations (see [KP] for details). The proof of the lemma for $k = 2$ is complete. □

# 4 Newton interpolation polynomials and their application for perturbation of trajectories

This chapter is in deep contrast with Chapters 2 and 3. There we have constructed arbitrarily fast growth of the number of periodic points. Here the goal is completely different. We estimate the number of periodic points from above. The technique we develop for this is "brand new".

The proof of Main Prevalent Theorem from Section 1.5.4 is extremely long and overloaded with technical details. The main purpose of this chapter is to present an application of Newton Interpolation Polynomials to study the dynamics of recurrent trajectories. This is one of the crucial ideas for the proof of the Main Prevalent Theorem from [KH, K4]. This chapter is based on [HK] part II.

## 4.1 Perturbation of recurrent trajectories by Newton interpolation polynomials

In order to keep the notations and formulas simple as we formalize this approach, we consider the case of 1-dimensional maps, but the reader should always have in mind that our approach is designed for multidimensional diffeomorphisms. Let $f : I \to I$ be a $C^1$ map on the interval $I = [-1, 1]$. Recall that a trajectory $\{x_k\}_{k\in\mathbb{Z}}$ of $f$ is called *recurrent* if it returns arbitrarily close to its initial position — that is, for all $\gamma > 0$ we have $|x_0 - x_n| < \gamma$ for some $n > 0$. A very basic question is how much one should perturb $f$ to make $x_0$ periodic. Here is an elementary Closing Lemma that gives a simple partial answer to this question.

**4.1 Closing Lemma** *Let $\{x_k = f^k(x_0)\}_{k=0}^n$ be a trajectory of length $n+1$ of a map $f : I \to I$. Let $u = (x_0 - x_n)/\prod_{k=0}^{n-2}(x_{n-1} - x_k)$. Then $x_0$ is a periodic point of period $n$ of the map*

$$f_u(x) = f(x) + u \prod_{k=0}^{n-2} (x - x_k). \tag{4.1}$$

Of course, $f_u$ is close to $f$ if and only if $u$ is sufficiently small, meaning that $|x_0 - x_n|$ should be small compared to $\prod_{k=0}^{n-2} |x_{n-1} - x_k|$. However, this product is likely to contain small factors for recurrent trajectories. In general, it is difficult to control the effect of perturbations for recurrent trajectories. The simple reason why is because *one cannot perturb $f$ at two nearby points independently.*

The Closing Lemma above also gives an idea of how much we must change the parameter $u$ to make a point $x_0$ that is $(n, \gamma)$-periodic not be $(n, \gamma)$-periodic for a given $\gamma > 0$. As we described above, this is one way to make a map that is "bad" for the initial condition $x_0$ become "good". To make use of our other alternative we must determine how much we need to perturb a map $f$ to make a given $x_0$ become $(n, \gamma)$-hyperbolic for some $\gamma > 0$.

**4.2 Pertubation of hyperbolicity** *Let $\{x_k = f^k(x_0)\}_{k=0}^{n-1}$ be a trajectory of length $n$ of a $C^1$ map $f : I \to I$. Then for the map*

$$f_v(x) = f(x) + v(x - x_{n-1}) \prod_{k=0}^{n-2} (x - x_k)^2 \tag{4.2}$$

*such that $v \in \mathbb{R}$ and*

$$||(f_v^n)'(x_0)| - 1| = \left| \left| \prod_{k=0}^{n-1} f'(x_k) + v \prod_{k=0}^{n-2} (x_{n-1} - x_k)^2 \prod_{k=0}^{n-2} f'(x_k) \right| - 1 \right| > \gamma \tag{4.3}$$

*we have that $x_0$ is an $(n, \gamma)$-hyperbolic point of $f_v$.*

One more time we can see that the product of distances $\prod_{k=0}^{n-2} |x_{n-1} - x_k|$ along the trajectory is an important quantitative characteristic of how much freedom we have to perturb.

The perturbations (4.1) and (4.2) are reminiscent of Newton interpolation polynomials. Let us put these formulas into a general setting using singularity theory.

Given $n > 0$ and a $C^1$ function $f : I \to \mathbb{R}$ we define an associated function $j^{1,n} f : I^n \to I^n \times \mathbb{R}^{2n}$ by

$$j^{1,n} f(x_0, \dots, x_{n-1}) = (x_0, \dots, x_{n-1}, f(x_0), \dots, f(x_{n-1}), f'(x_0), \dots, f'(x_{n-1})). \tag{4.4}$$

In singularity theory this function is called the *$n$-tuple 1-jet* of $f$. The ordinary 1-jet of $f$, usually denoted by $j^1 f(x) = (x, f(x), f'(x))$, maps $I$ to the *1-jet space* $\mathcal{J}^1(I, \mathbb{R}) \simeq I \times \mathbb{R}^2$. The product of $n$ copies of $\mathcal{J}^1(I, \mathbb{R})$, called the *multijet space*, is denoted by

$$\mathcal{J}^{1,n}(I, \mathbb{R}) = \underbrace{\mathcal{J}^1(I, \mathbb{R}) \times \cdots \times \mathcal{J}^1(I, \mathbb{R})}_{n \text{ times}}, \tag{4.5}$$

and is equivalent to $I^n \times \mathbb{R}^{2n}$ after rearranging coordinates. The $n$-tuple 1-jet of $f$ associates with each $n$-tuple of points in $I^n$ all the information necessary to determine how close the $n$-tuple is to being a periodic orbit, and if so, how close it is to being nonhyperbolic.

The set

$$\Delta_n(I) = \{\{x_0, \dots, x_{n-1}\} \times \mathbb{R}^{2n} \subset \mathcal{J}^{1,n}(I, \mathbb{R}) : \exists\, i \neq j \text{ such that } x_i = x_j\} \tag{4.6}$$

is called the *diagonal* (or sometimes the *generalized diagonal*) in the space of multijets. In singularity theory the space of multijets is defined outside the diagonal $\Delta_n(I)$ and is usually denoted by $\mathcal{J}^1_n(I,\mathbb{R}) = \mathcal{J}^{1,n}(I,\mathbb{R}) \setminus \Delta_n(I)$ (see [GG]). It is easy to see that *a recurrent trajectory* $\{x_k\}_{k\in\mathbb{Z}_+}$ *is located in a neighborhood of the diagonal* $\Delta_n(I)$ *in the space of multijets for a sufficiently large* $n$. If $\{x_k\}_{k=0}^{n-1}$ is a part of a recurrent trajectory of length $n$, then the product of distances along the trajectory,

$$\prod_{k=0}^{n-2} |x_{n-1} - x_k|, \tag{4.7}$$

measures how close $\{x_k\}_{k=0}^{n-1}$ is to the diagonal $\Delta_n(I)$, or how independently one can perturb points of a trajectory. One can also say that (4.7) is a quantitative characteristic of how recurrent a trajectory of length $n$ is. Introduction of this *product of distances along a trajectory* into the analysis of recurrent trajectories is a new point of our paper.

## 4.2 Newton interpolation and blow-up along the diagonal in multijet space

We now present a construction due to Grigoriev and Yakovenko [GY] which puts the "Closing Lemma" and "Perturbation of Hyperbolicity" statements into a general framework. It is an interpretation of Newton interpolation polynomials as an algebraic blow-up along the diagonal in the multijet space. In order to keep the notations and formulas simple we continue in this section to consider only the 1-dimensional case.

Consider the $2n$-parameter family of perturbations of a $C^1$ map $f: I \to I$ by polynomials of degree $2n-1$

$$f_\varepsilon(x) = f(x) + \phi_\varepsilon(x), \qquad \phi_\varepsilon(x) = \sum_{k=0}^{2n-1} \varepsilon_k x^k, \tag{4.8}$$

where $\varepsilon = (\varepsilon_0, \dots, \varepsilon_{2n-1}) \in \mathbb{R}^{2n}$. The perturbation vector $\varepsilon$ consists of coordinates from the Hilbert brick $HB^1(\mathbf{r})$ of analytic perturbations defined in Section 1.5.4. Our goal now is to describe how such perturbations affect the $n$-tuple 1-jet of $f$, and since the operator $j^{1,n}$ is linear in $f$, for the time being we consider only the perturbations $\phi_\varepsilon$ and their $n$-tuple 1-jets. For each $n$-tuple $\{x_k\}_{k=0}^{n-1}$ there is a natural transformation $\mathcal{J}^{1,n} : I^n \times \mathbb{R}^{2n} \to \mathcal{J}^{1,n}(I,\mathbb{R})$ from $\varepsilon$-coordinates to jet-coordinates, given by

$$\mathcal{J}^{1,n}(x_0, \dots, x_{n-1}, \varepsilon) = j^{1,n}\phi_\varepsilon(x_0, \dots, x_{n-1}). \tag{4.9}$$

Instead of working directly with the transformation $\mathcal{J}^{1,n}$, we introduce intermediate $u$-coordinates based on Newton interpolation polynomials. The relation between $\varepsilon$-coordinates and $u$-coordinates is given implicitly by

$$\phi_\varepsilon(x) = \sum_{k=0}^{2n-1} \varepsilon_k x^k = \sum_{k=0}^{2n-1} u_k \prod_{j=0}^{k-1} (x - x_{j(\text{mod}\, n)}). \tag{4.10}$$

Based on this identity, we will define functions $\mathcal{D}^{1,n} : I^n \times \mathbb{R}^{2n} \to I^n \times \mathbb{R}^{2n}$ and $\pi^{1,n} : I^n \times \mathbb{R}^{2n} \to \mathcal{J}^{1,n}(I,\mathbb{R})$ so that $\mathcal{J}^{1,n} = \pi^{1,n} \circ \mathcal{D}^{1,n}$, or in other words the diagram in Fig. 8 commutes. We will show later that $\mathcal{D}^{1,n}$ is invertible, while $\pi^{1,n}$ is invertible away from the diagonal $\Delta_n(I)$ and defines a blow-up along it in the space of multijets $\mathcal{J}^{1,n}(I,\mathbb{R})$.

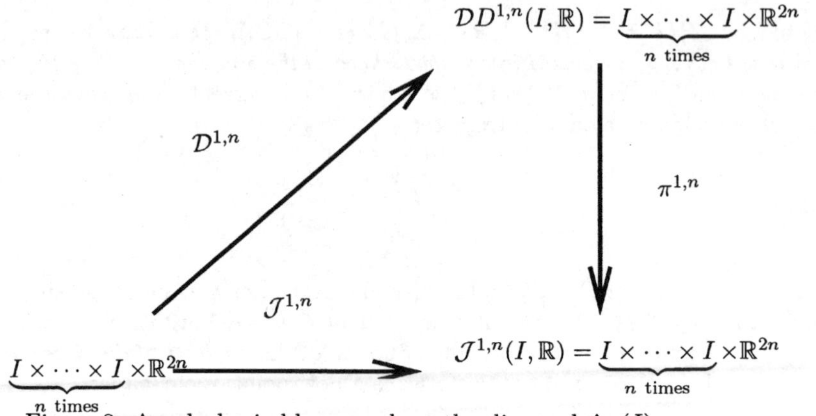

Figure 8: An algebraic blow-up along the diagonal $\Delta_n(I)$

The intermediate space, which we denote by $\mathcal{DD}^{1,n}(I,\mathbb{R})$, is called *the space of divided differences* and consists of $n$-tuples of points $\{x_k\}_{k=0}^{n-1}$ and $2n$ real coefficients $\{u_k\}_{k=0}^{2n-1}$. Here are explicit coordinate-by-coordinate formulas defining $\pi^{1,n} : \mathcal{DD}^{1,n}(I,\mathbb{R}) \to \mathcal{J}^{1,n}(I,\mathbb{R})$.

$$
\begin{aligned}
\phi_\varepsilon(x_0) &= u_0, \\
\phi_\varepsilon(x_1) &= u_0 + u_1(x_1 - x_0), \\
\phi_\varepsilon(x_2) &= u_0 + u_1(x_2 - x_0) + u_2(x_2 - x_0)(x_2 - x_1), \\
&\vdots \\
\phi_\varepsilon(x_{n-1}) &= u_0 + u_1(x_{n-1} - x_0) + \cdots + u_{n-1}(x_{n-1} - x_0)\cdots(x_{n-1} - x_{n-2}), \\
\phi'_\varepsilon(x_0) &= \frac{\partial}{\partial x}\left(\sum_{k=0}^{2n-1} u_k \prod_{j=0}^{k-1}(x - x_{j(\operatorname{mod} n)})\right)\Bigg|_{x=x_0}, \\
&\vdots \\
\phi'_\varepsilon(x_{n-1}) &= \frac{\partial}{\partial x}\left(\sum_{k=0}^{2n-1} u_k \prod_{j=0}^{k-1}(x - x_{j(\operatorname{mod} n)})\right)\Bigg|_{x=x_{n-1}}.
\end{aligned}
\tag{4.11}
$$

These formulas are very useful for dynamics. For a given base map $f$ and initial point $x_0$, the image $f_\varepsilon(x_0) = f(x_0) + \phi_\varepsilon(x_0)$ of $x_0$ depends only on $u_0$. Furthermore, the image can be set to any desired point by choosing $u_0$ appropriately — we say then that it depends nontrivially on $u_0$. If $x_0$, $x_1$, and $u_0$ are fixed, the image $f_\varepsilon(x_1)$ of $x_1$ depends only on $u_1$, and as long as $x_0 \neq x_1$ it depends nontrivially on $u_1$. More generally, for $0 \le k \le n-1$, if pairwise distinct points $\{x_j\}_{j=0}^{k}$ and coefficients $\{u_j\}_{j=0}^{k-1}$ are fixed, then the image $f_\varepsilon(x_k)$ of $x_k$ depends only and nontrivially on $u_k$.

Suppose now that an $n$-tuple of points $\{x_j\}_{j=0}^{n}$, not on the diagonal $\Delta_n(I)$, and Newton coefficients $\{u_j\}_{j=0}^{n-1}$ are fixed. Then the derivative $f'_\varepsilon(x_0)$ at $x_0$ depends only and nontrivially on $u_n$. Likewise for $0 \le k \le n-1$, if distinct points $\{x_j\}_{j=0}^{n}$ and Newton coefficients $\{u_j\}_{j=0}^{n+k-1}$ are fixed, then the derivative $f'_\varepsilon(x_k)$ at $x_k$ depends only and nontrivially on $u_{n+k}$.

As Fig. 9 illustrates, these considerations show that for any map $f$ and any desired trajectory of distinct points with any given derivatives along it, one can choose Newton coefficients $\{u_k\}_{k=0}^{2n-1}$ and explicitly construct a map $f_\varepsilon = f + \phi_\varepsilon$ with such a trajectory. Thus we have shown that $\pi^{1,n}$ is invertible away from the diagonal $\Delta_n(I)$ and defines a blow-up along it in the space of multijets $\mathcal{J}^{1,n}(I, \mathbb{R})$.

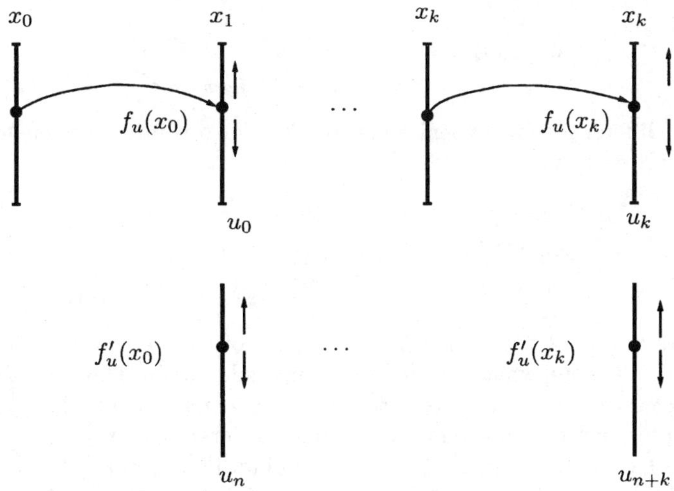


Figure 9: Newton coefficients and their action

Next we define the function $\mathcal{D}^{1,n} : I^n \times \mathbb{R}^{2n} \to \mathcal{DD}^{1,n}(I, \mathbb{R})$ explicitly using so-called divided differences. Let $g : \mathbb{R} \to \mathbb{R}$ be a $C^r$ function of one real variable.

**4.3 Definition** The *first order divided difference* of $g$ is defined as

$$\Delta g(x_0, x_1) = \frac{g(x_1) - g(x_0)}{x_1 - x_0} \tag{4.12}$$

for $x_1 \neq x_0$ and extended by its limit value as $g'(x_0)$ for $x_1 = x_0$. Iterating this construction we define divided differences of the $m$-th order for $2 \leq m \leq r$,

$$\Delta^m g(x_0, \ldots, x_m) = \frac{\Delta^{m-1} g(x_0, \ldots, x_{m-2}, x_m) - \Delta^{m-1} g(x_0, \ldots, x_{m-2}, x_{m-1})}{x_m - x_{m-1}} \tag{4.13}$$

for $x_{m-1} \neq x_m$ and extended by its limit value for $x_{m-1} = x_m$.

A function loses at most one derivative of smoothness with each application of $\Delta$, so $\Delta^m g$ is at least $C^{r-m}$ if $g$ is $C^r$. Notice that $\Delta^m$ is linear as a function of $g$, and one can show that it is a symmetric function of $x_0, \ldots, x_m$; in fact, by induction it follows that

$$\Delta^m g(x_0, \ldots, x_m) = \sum_{i=0}^{m} \frac{g(x_i)}{\prod_{j \neq i} (x_i - x_j)} \tag{4.14}$$

Another identity that is proved by induction will be more important for us, namely

$$\Delta^m\ x^k(x_0,\ldots,x_m) = p_{k,m}(x_0,\ldots,x_m), \tag{4.15}$$

where $p_{k,m}(x_0,\ldots,x_m)$ is 0 for $m > k$, and for $m \leq k$ is the sum of all degree $k-m$ monomials in $x_0,\ldots,x_m$ with unit coefficients,

$$p_{k,m}(x_0,\ldots,x_m) = \sum_{r_0+\cdots+r_m=k-m}\ \prod_{j=0}^{m} x_j^{r_j}. \tag{4.16}$$

The divided differences form coefficients for the Newton interpolation formula. For all $C^\infty$ functions $g:\mathbb{R}\to\mathbb{R}$ we have

$$\begin{aligned} g(x) &= \Delta^0 g(x_0) + \Delta^1 g(x_0,x_1)(x-x_0) + \cdots \\ &\quad + \Delta^{n-1} g(x_0,\ldots,x_{n-1})(x-x_0)\cdots(x-x_{n-2}) \\ &\quad + \Delta^n g(x_0,\ldots,x_{n-1},x)(x-x_0)\cdots(x-x_{n-1}) \end{aligned} \tag{4.17}$$

identically for all values of $x, x_0, \ldots, x_{n-1}$. All terms of this representation are polynomial in $x$ except for the last one which we view as a remainder term. The sum of the polynomial terms is the degree $(n-1)$ *Newton interpolation polynomial* for $g$ at $\{x_k\}_{k=0}^{n-1}$. To obtain a degree $2n-1$ interpolation polynomial for $g$ and its derivative at $\{x_k\}_{k=0}^{n-1}$, we simply use (4.17) with $n$ replaced by $2n$ and the $2n$-tuple of points $\{x_{k(\mathrm{mod}\,n)}\}_{k=0}^{2n-1}$.

Recall that $\mathcal{D}^{1,n}$ was defined implicitly by (4.10). We have described how to use divided differences to construct a degree $2n-1$ interpolating polynomial of the form on the right-hand side of (4.10) for an arbitrary $C^\infty$ function $g$. Our interest then is in the case $g=\phi_\varepsilon$, which as a degree $2n-1$ polynomial itself will have no remainder term and coincide exactly with the interpolating polynomial. Thus $\mathcal{D}^{1,n}$ is given coordinate-by-coordinate by

$$\begin{aligned} u_m &= \Delta^m \left(\sum_{k=0}^{2n-1} \varepsilon_k x^k\right)(x_0,\ldots,x_{m(\mathrm{mod}\,n)}) \\ &= \varepsilon_m + \sum_{k=m+1}^{2n-1} \varepsilon_k p_{k,m}(x_0,\ldots,x_{m(\mathrm{mod}\,n)}) \end{aligned} \tag{4.18}$$

for $m = 0,\ldots,2n-1$. We call the transformation given by (4.18) the *Newton map.* Notice that for fixed $\{x_k\}_{k=0}^{2n-1}$, the Newton map is linear and given by an upper triangular matrix with units on the diagonal. Hence it is Lebesgue volume-preserving and invertible, whether or not $\{x_k\}_{k=0}^{2n-1}$ lies on the diagonal $\Delta_n(I)$.

We call the basis of monomials

$$\prod_{j=0}^{k}(x - x_{j(\mathrm{mod}\,n)}) \quad \text{for} \quad k = 0,\ldots,2n-1 \tag{4.19}$$

in the space of polynomials of degree $2n-1$ the *Newton basis* defined by the $n$-tuple $\{x_k\}_{k=0}^{n-1}$. The Newton map and the Newton basis, and their analogue in dimension $N$, are useful tools for perturbing trajectories and estimating the measure of "bad" parameter values $\vec{\varepsilon} \in HB^N(\mathbf{r})$.

## References


[A] A. Arnold and V. Arnold, *Problems of Arnold*, FAZIS, 2000 (Russian).

[AM] M. Artin and B. Mazur, On periodic orbits, *Ann. of Math.* **81** (1965), 82–99.

[Bo] R. Bowen, On Axiom A diffeomorphisms, CBMS Regional Conf. Ser. in Math. **35**, Amer. Math. Soc., Providence, RI, 1977.

[BD] C. Bonatti and L. Diaz, Persistent nonhyperbolic transitive sets, *Ann. of Math.* **143** (2) (1996), 357–396.

[D] L. Diaz, Robust nonhyperbolic dynamics and heteroclinic cycles, *Ergodic Theory Dynam. Systems* **15** (2) (1995), 291–315.

[E] H. Epstein, Fixed points of composition operators, Procceedings of a NATO Advanced Study Institute on Nonlilenar Evolution, Italy, 1987, 71-100.

[F] N. Fenichel, Persistence and smoothness of invariant manifolds for flows, *Indiana Univ. Math. J.* **21** (1971), 193–226.

[GG] M. Golubitsky and V. Guillemin, *Stable Mappings and Their Singularities*, Springer-Verlag, Berlin-New York, 1973.

[GST1] S. Gonchenko, L. Shil'nikov, and D. Turaev, On models with non-rough Poincaré homoclinic curves, *Phys.* D **62** (1993), 1–14.

[GST2] S. Gonchenko, L. Shil'nikov, and D. Turaev, Homoclinic tangencies of arbitrary order in Newhouse domains, in: *Dynamical systems (Moscow, 1998)*, INT 67, VINITI, Moscow, 1999, 69–128.

[GY] A. Grigoriev and S. Yakovenko, Topology of generic multijet preimages and blow-up via Newton interpolation, *J. Differential Equations* **150** (1998), 349–362.

[G] M. Gromov, On entropy of holomorphic maps, preprint.

[HK] V. Kaloshin and B. Hunt, A streched exponential estimate on growth of the number of periodic points for prevalent diffeomorphisms, I & II, *Electron. Res. Announc. Amer. Math. Soc.* **7** (2001), 17–27 and 28–36.

[HSY] B. Hunt, T. Sauer, and J. Yorke, Prevalence: a translation-invariant almost every for infinite dimensional spaces, *Bull. Amer. Math. Soc. (N.S.)* **27** (1992), 217–238; Prevalence: an addendum, *Bull. Amer. Math. Soc. (N.S.)* **28** (1993), 306–307.

[K1] V. Kaloshin, An extension of the Artin-Mazur Theorem, *Ann. of Math. (2)* **150** (1999), 729–741.

[K2] V. Kaloshin, Generic diffeomorphisms with superexponential growth of number of periodic orbits, *Comm. Math Phys.* **211** (2000), 253–271.

[K3] V. Kaloshin, Some prevalent properties of smooth dynamical systems, *Trudy Mat. Inst. Steklov.* **213** (1997), 123–151.

[K4] V. Kaloshin, Stretched exponential estimate on growth of the number of periodic points for prevalent diffeomorphisms, part 2, in preparation.

[KH] V. Kaloshin and B. Hunt, Stretched exponential estimate on growth of the number of periodic points for prevalent diffeomorphisms, part 1, in preparation.

[KK] V. Kaloshin and O. Kozlovski, An example of a $C^r$-unimodal map with an arbitrary fast growth of the number of periodic points, preprint.

[KP] V. Kaloshin and E. Pujals, Superexponential growth of the number of periodic points is generic for heteroclinic cycles, in preparation.

[L] O. E. Lanford III, A computer-assisted proof of the Feigenbaum conjectures, *Bull. Amer. Math. Soc. (N.S.)* **6** (1984), 427–434.

[Ma] M. Martens, The periodic points of renormalization, *Ann. of Math. (2)* **147** (3) (1998), 435–484.

[MMS] M. Martens, W. de Melo and S. Van Strien, Julia-Fatou-Sullivan theory for real one-dimensional dynamics, *Acta Math.* **168** (1992), 273-318.

[Ne] S. Newhouse, The abundance of wild hyperbolic sets and non-smooth stable sets for diffeomorphisms, *Inst. Hautes Études Sci. Publ. Math.* **50** (1979), 101–151.

[PM] J. Palis and W. de Melo, *Geometric Theory of Dynamical Systems: An Introduction*, Springer-Verlag, New York, 1982.

[O] J. C. Oxtoby, *Measure and Category*, Springer-Verlag, Berlin, 1971.

[P] E. Pujals, private communications.

[PS] E. Pujals and M. Sambarino, Homoclinic tangencies and hyperbolicity for surface diffeomorphisms, *Ann. of Math. (2)* **151** (2000), 961–1023.

[RG] E. Rozales-Gonsales, Growth of periodic orbits of dynamical systems, *Funct. Anal. Appl.* **25** (1991), 254–262.

[Sac] R. J. Sacker, A perturbation theorem for invariant manifolds and Hölder continuity, *J. Math. Mech.* **18** (1969), 705–762.

[Sar] A. Sard, The measure of the critical points of differentiable maps, *Bull. Amer. Math. Soc.* **48** (1942), 883–890.

[Sm] S. Smale, Differentiable Dynamical Systems, *Bull. Amer. Math. Soc.* **73** (1967), 747–817.

[Si] C. Simon, Instability of $\mathrm{Diff}^k(T^3)$ and the nondegenericity of rational zeta functions, *Trans. Amer. Math. Soc.* **174** (1973), 217–242.

[S] D. Sullivan, Bounds, quadratic differentials, and renormalization conjectures, in: *American Mathematical Society Centennial Publications*, vol. 2, Amer. Math. Soc., Providence, RI, 1992, 417–466.

[W] H. Whitney, Differentiable manifolds, *Ann. of Math.* **37** (1936), 645–680.

[Y] Y. Yomdin, A quantitative version of the Kupka-Smale theorem, *Ergodic Theory Dynam. Systems* **5** (1985), 449–472.

# Lectures on meromorphic flat connections

Dmitry NOVIKOV*
*Department of Mathematics*
*Purdue University*
*West Lafayette, IN 47907-1395*
*USA*

Sergei YAKOVENKO†
*Department of Mathematics*
*Weizmann Institute of Science*
*Rehovot 76100*
*Israel*

**Abstract**

These notes constitute an introduction to the analytic local theory of linear integrable Pfaffian differential equations describing meromorphic flat connections. We introduce the mildest type of logarithmic singularities and investigate properties of their residues. In the remaining part we investigate the system of equations describing Abelian integrals and establish a multidimensional Euler-type normal normal form for systems with non-resonant residues on the normal crossing. While most of the results proved in these notes are not new, some of them were of folklore type while other were previously available only in advanced algebraic expositions.

## By way of introduction: the infinitesimal Hilbert problem

One of the challenging problems that recently attracted much attention, is the question about the number of zeros of complete Abelian integrals of algebraic 1-forms over closed ovals of plane algebraic curves. This problem arises as a "linearization" of Hilbert's sixteenth problem on the number of limit cycles of polynomial vector fields, see [Yak01].

There were numerous attempts to solve the infinitesimal Hilbert problem in various, sometimes relaxed, settings. The approach recently suggested by the authors of these notes, suggests to exploit the fact that Abelian integrals satisfy a system of linear ordinary differential equations with rational coefficients, the so-called Picard-Fuchs equations.

The general claim concerning such systems was established in [NY02], see also [Yak01]. Consider a *Fuchsian linear $(n\times n)$-system* on the Riemann sphere $\mathbb{C}P^1$, written in the Pfaffian (coordinate-free) form as $dX = \Omega X$, where $\Omega$ is a meromorphic (rational) matrix 1-form on

*Supported by NSF grant DMS 0200861 and the McDonnell Foundation.

†Gershon Kekst Professorial Chair. The research was supported by the Israeli Science Foundation grant no. 18-00/1.



*Y. Ilyashenko and C. Rousseau (eds.),*
*Normal Forms, Bifurcations and Finiteness Problems in Differential Equations*, 387–430.

$\mathbb{C}P^1$ having only simple poles. After choosing an affine coordinate $t \in \mathbb{C}$ on the Riemann sphere so that infinity is not a pole of $\Omega = A(t)\,dt$, the Pfaffian system can be reduced to a system of *linear ordinary differential equations with rational coefficients* of the form

$$\frac{d}{dt}X = A(t)X, \qquad A(t) = \sum_{j=1}^{r} \frac{A_j}{t - t_j}, \quad \sum_{1}^{r} A_j = 0, \tag{0.1}$$

with the $(n \times n)$-matrix residues $A_1, \dots, A_r \in \operatorname{Mat}_n(\mathbb{C})$ and the singular points $t_1, \dots, t_r \in \mathbb{C}$.

Under certain assumptions on the spectra of monodromy operators (linear transformations of solutions after analytic continuation along closed loops avoiding singular points), solutions of the system (0.1) possess the following property making them remotely similar to algebraic functions. Namely, the number of zeros of any rational combination $f \in \mathbb{C}(X)$ of components of any fundamental matrix solution $X = X(t)$ in any semialgebraic domain $T \subset \mathbb{C} \setminus \{t_1, \dots, t_r\}$ can be explicitly majorized by a computable (though enormously large) function of the dimension $n$ and the degree $r$ of the system (0.1), the degree of $f$ in $\mathbb{C}(X)$ and the *height* $h(\Omega)$ of the system (0.1), defined as $\sum_1^r |A_j|$. It is important to stress that the bound is *uniform* over configurations of the singular points, provided that the condition imposed on the monodromy persists.

The Picard-Fuchs system for Abelian integrals was explicitly derived in [NY01] in a form that can be easily reduced to (0.1). However, the direct application of results from [NY02] is in general impossible (exception occurs in the hyperelliptic case when the methods rather than results of [NY02] can be made to work, see [NY99]). The reason is the *explosion of residues.*

This phenomenon occurs when two or more singular points of (0.1) collide when the additional parameters defining the system, change. In the general system (0.1) the parameters $t_j, A_j$ are completely independent and can be restricted by the condition on the height $h(\Omega) \leqslant$ const. However, in the Picard-Fuchs systems for Abelian integrals, which depend on the coefficients of the equation of the algebraic curve as the parameters, the norms of the matrix residues $|A_i|, |A_j|$ usually tend to infinity if the corresponding singular points $t_i$ and $t_j$ tend to each other. This makes it impossible to apply the principal theorem from [NY02] in order to obtain uniform bounds on the number of zeros of Abelian integrals.

The reason behind this phenomenon is the fact that the system (0.1) for Abelian integrals is an *isomonodromic* family (deformation) if considered as depending on the natural parameters of the problem. This means that the monodromy group of the system remains "the same" even though the parameters vary. While this isomonodromy makes it much easier to verify the spectral condition required in [NY02], it very often forces the residues to explode even when the parametric family of Fuchsian systems (0.1) is not related to Abelian integrals.

The natural way to investigate isomonodromic families of linear systems is to consider them in the multidimensional setting when the the independent variable $t$ and the parameters play the same role. Then the isomonodromy property takes the form of the *flatness condition* (integrability) of the corresponding *meromorphic connection.* The main goal of these notes is to provide the reader with the necessary background in the theory of integrable linear Pfaffian systems. In particular, we discuss briefly the possibility of getting rid of exploding residues by a suitable meromorphic gauge transformation, which (if successfully implemented for the Picard-Fuchs systems) would entail applicability of the technique from [NY02].

# 1 Integrable Pfaffian systems

Let $U$ be a complex $m$-dimensional manifold (in most cases $U$ will be an open subset of $\mathbb{C}^m$ or even more specifically, a small polydisc centered at the origin). Consider a *matrix Pfaffian 1-form* $\Omega = \|\omega_{ij}\|_{i,j=1}^n$, an $n \times n$-matrix whose entries are holomorphic 1-forms $\omega_{ij} \in \Lambda^1(U)$. This matrix defines a systems of linear Pfaffian equations

$$dx_i = \sum_{j=1}^{n} \omega_{ij} x_j, \quad \text{or in the vector form} \quad dx = \Omega x. \tag{1.1}$$

A *solution* of the system (1.1) is a tuple $x = (x_1, \ldots, x_n)$ of holomorphic functions on $U$, $x(\cdot) : U \to \mathbb{C}^n$. Any number of (column) vector solutions of (1.1) can be organized into a matrix. A *fundamental matrix solution* $X = \|x_{ij}\|$ is the holomorphic *invertible* square $n \times n$-matrix solution of the *matrix Pfaffian equation*

$$dX = \Omega X, \qquad X : U \to \mathrm{GL}(n, \mathbb{C}), \tag{1.2}$$

where $dX$ is the matrix 1-form with the entries $dx_{ij}$. Columns of the fundamental matrix solution are linearly independent vector solutions of the system (1.1).

If $m = 1$, after choosing any local coordinate $t = t_1$ on $U$, the system (1.1) becomes a system of linear *ordinary* differential equations with respect to the unknown functions $x_i(t)$. For $m > 1$, (1.1) is a system of partial linear differential equations with respect to any coordinate system $t = (t_1, \ldots, t_m)$, in both cases with holomorphic coefficients.

The "ordinary" case is very well known. In many excellent textbooks, e.g. [Har82, For91, AI88, Bol00a], one can find a detailed treatment of the following issues: local and global existence of solutions, local classification of singularities, local and global theory of systems having only simplest (Fuchsian) singular points.

The main reference for the "partial" case, no less important (in particular, because of its connections with *deformations* of the "ordinary" systems), is the classical book by P. Deligne [Del70]. Yet this book is written in rather algebraic language which naturally determines the choice of the questions addressed therein. The primary goal of these notes was to supply an "analytic translation" of parts of the book [Del70] roughly along the lines characteristic for the "ordinary" point of view. However, the contents of these notes is not limited to re-exposition of [Del70], as we include (with what can be considered as more or less complete proofs) some of the results announced in [YT75, Tak79].

In what follows we will usually assume that the reader is familiar with the "ordinary" case; however, the proofs supplied for the "partial" case will be mostly independent and working also for $m = 1$.

*Notations.* We will systematically use the matrix notation in these notes. The wedge product $\Omega \wedge \Theta$ of two matrix 1-forms $\Omega = \|\omega_{ij}\|$, $\Theta = \|\theta_{ij}\|$ will denote the matrix 2-form with the entries $\sum_{k=1}^n \omega_{ik} \wedge \theta_{kj}$. If $A = \|a_{ij}(t)\|$ is a matrix *function*, then $A\Omega$ is a matrix 1-form with the entries $\sum_j a_{ij}\omega_{jk}$ (sometimes we use the dot notation for this product). The commutator $[A, \Omega]$ means $A\Omega - \Omega A$; similarly, $[\Omega, \Theta] = \Omega \wedge \Theta - \Theta \wedge \Omega$. Note that in the matrix case the bracket is *not* antisymmetric, so that $\Omega \wedge \Omega$ need not necessarily be zero. On the other hand, the associativity $\Omega \wedge (A\Theta) = (\Omega A) \wedge \Theta$ holds.

## 1.1 Local existence of solutions. Integrability

For $m = 1$ the linear system (1.1) (respectively, (1.2)) always admits a solution (resp., fundamental matrix solution) at least locally, near each point $a \in U$. When $m > 1$, an additional *integrability condition* is required.

Indeed, if $X$ is a fundamental matrix solution for (1.2), then $\Omega = dX \cdot X^{-1}$. Using the formula

$$d(X^{-1}) = -X^{-1} \cdot dX \cdot X^{-1}$$

for the differential of the inverse matrix $X^{-1}$, the Leibnitz rule and the associativity, we obtain

$$d\Omega = -dX \wedge d(X^{-1}) = dX \wedge (X^{-1} \cdot dX \cdot X^{-1}) = \Omega \wedge \Omega.$$

This shows that the integrability condition

$$d\Omega = \Omega \wedge \Omega \tag{1.3}$$

is necessary for the existence of fundamental matrix solutions of (1.2).

**1.1 Theorem** *The necessary integrability condition* (1.3) *is sufficient for local existence of fundamental matrix solutions of the system* (1.2) *near each point of holomorphy of the Pfaffian matrix* $\Omega$.

As soon as the local existence theorem is established, it implies in the standard way the structural description of all solutions of (1.1) and (1.2). Any solution $x$ of (1.1) has the form $Xc$, where $c \in \mathbb{C}^n$ is a constant column vector and $X$ is a fundamental matrix solution of the matrix equation (1.2). Any two fundamental matrix solutions $X, X'$ of the same system (1.2) differ by a locally constant invertible right matrix factor, $X' = XC$, $C \in \mathrm{GL}(n, \mathbb{C})$. For any point $a \in U$ and any vector $v$ (resp., any matrix $V$) the *initial value problem* (the Cauchy problem) $x(a) = v$ (resp., $X(a) = V$) has a unique local solution near $a$. These assertions are proved by differentiation of $X^{-1}x$, resp., $X^{-1}X'$.

**Proof** The assertion follows from the Frobenius theorem [War83]. Though it is formulated there in the real smooth category, neither the formulation nor the proof need any change for the complex analytic settings.

Consider the complex analytic manifold $M = U \times \mathrm{Mat}_n(\mathbb{C})$ of dimension $m + n^2$ with the coordinates $(t, X)$, and $n^2$ Pfaffian equations on this manifold, written in the matrix form as

$$\Theta = 0, \quad \text{where} \quad \Theta = \|\theta_{ij}\| = dX - \Omega X. \tag{1.4}$$

By the Frobenius integrability theorem, these Pfaffian equations admit an integral manifold of codimension $n^2$ (i.e., of dimension $m$) through any point $(t_*, X_*) \in M$ if the exterior differentials $d\theta_{ij}$ of the (scalar) forms $\theta_{ij} \in \Lambda^1(M)$ constituting the Pfaffian matrix $\Theta$, belong to the ideal $\langle \theta_{ij} \rangle \subset \Lambda^2(M)$ generated by the forms $\theta_{ij}$ in the exterior algebra. The latter condition can be immediately verified:

$$\begin{aligned} d\Theta &= -d(\Omega X) = -d\Omega \cdot X + \Omega \wedge dX = -d\Omega \cdot X + \Omega \wedge (\Theta + \Omega X) \\ &= (-d\Omega + \Omega \wedge \Omega)X + \Omega \wedge \Theta. \end{aligned}$$

If the integrability condition (1.3) holds, then $d\Theta = \Omega \wedge \Theta$ in $\Lambda^2(M)$, so that each $d\theta_{ij}$ is expanded as $\sum_k \omega_{ik} \wedge \theta_{kj} \in \langle \theta_{1j}, \dots, \theta_{nj} \rangle$ and the assumptions of the Frobenius theorem are satisfied. □

## 1.2 Global solutions. Monodromy and holonomy

For a globally defined system (1.2) on the manifold $U$ represented as the union of local neighborhoods, $U = \bigcup_\alpha U_\alpha$, the corresponding local solutions $X_\alpha$ can sometimes be adjusted to form a global solution. The possibility of doing this depends on the topology of $U$. Indeed, one may look for suitable constant matrices $C_\alpha$ such that the solutions $X'_\alpha = X_\alpha C_\alpha$ will coincide on the pairwise intersections $U_{\alpha\beta} = U_\alpha \cap U_\beta$. For this, the correction terms $C_\alpha$ must satisfy the identities $C_\alpha C_\beta^{-1} = C_{\alpha\beta}$ on any nonempty intersection $U_{\alpha\beta}$, where $C_{\alpha\beta} = X_\alpha^{-1} X_\beta$ are the *constant* matrices arising on these intersections from the local solutions. In topological terms, the problem is reduced to *solvability of a constant matrix cocycle.*

One particular case when this solvability is immediate, appears when $U$ is a small neighborhood of a sufficiently regular (say, compact smooth non-selfintersecting) curve $\gamma$. Then the coordinate neighborhoods $U_\alpha$ can be linearly ordered (indexed by an integer variable $\alpha = 1, \dots, N$, increasing along $\gamma$) and chosen so small that $U_\alpha \cap U_\beta$ is non-void if and only if they have consecutive numbers, $|\alpha - \beta| \leqslant 1$. The solution $\{C_\alpha\}$ of the respective cocycle $\{C_{\alpha\beta}\}$ is obtained by taking appropriate products,

$$C_\alpha = C_{\alpha,\alpha-1} C_{\alpha-1,\alpha-2} \cdots C_{32} C_{21} C_1$$

for any choice of $C_1$ and any $\alpha \leqslant N$.

As a corollary, we conclude that any collection of local solutions $\{X_\alpha\}$ along a simple (smooth non-selfintersecting) parametrized curve $\gamma$ can be modified into a solution $X$ defined and holomorphic in a sufficiently small neighborhood of $\gamma$. This fact is usually stated as the *possibility of unlimited continuation of solutions of any integrable Pfaffian system along any curve.* Moreover, a small variation of the curve $\gamma$ with fixed endpoints results in the same solution, which means that *the result of continuation depends only on the homotopy class of the curve* $\gamma$.

However, if the curve $\gamma$ is a closed non-contractible loop, the result of continuation of any solution $X$ along $\gamma$ in general differs from the initial solution. Let $a \in U$ be a point and $\pi_1(U, a)$ the *fundamental group* of $U$ with the base point $a$. If $X_0$ is a fundamental matrix solution near the point $a$, then for any $\gamma \in \pi_1(U, a)$ the result $X_\gamma$ of analytic continuation of $X_0$ along $\gamma$ differs from $X_0$ by a constant right matrix factor $M_\gamma$, called the *monodromy matrix*: $X_\gamma = X_0 \cdot M_\gamma$. Thus in case $U$ is topologically nontrivial (not simply connected), solutions of an integrable Pfaffian system exist on $U$ globally, but only as multivalued matrix functions. By construction, every branch of any fundamental matrix solution is everywhere nondegenerate.

The correspondence $\gamma \mapsto M_\gamma$ is an (anti)*representation* of the fundamental group: $M_{\gamma\gamma'} = M_{\gamma'} M_\gamma$ for any two loops $\gamma, \gamma' \in \pi_1(U, a)$. The image of this representation is the *monodromy group* of the integrable system (1.2). It is defined modulo a simultaneous conjugacy of all monodromy matrices: if another fundamental solution $X'_0 = X_0 C$ is used for continuation, then $M_\gamma$ will be replaced by $C^{-1} M_\gamma C$. Considered in invariant terms as an automorphism of the space of solutions of the system (1.1), induced by continuation along $\gamma$, $M_\gamma$ is the *monodromy operator.*

The closely related notion of *holonomy* does not depend on the choice of the fundamental solution. For a curve $\gamma$ connecting any two points $a, b \in U$, the holonomy is the linear transformation $F_{ab\gamma} : \{a\} \times \mathbb{C}^n \to \{b\} \times \mathbb{C}^n$, sending $(a, v)$ to $(b, x_\gamma(b))$, where $x_\gamma$ is the unique solution of (1.1) over $\gamma$, defined by the initial condition $x(a) = v$. If $\gamma$ is closed and

$a = b$, then $F_{aa\gamma}$ is an automorphism of $\{a\} \times \mathbb{C}^n$; choosing a different base point $a'$ close to $a$ results in a conjugate automorphism $F_{a'a'\gamma}$ close to $F_{aa\gamma}$.

**1.2 Remark** To reinstate the ultimate rigor, note that the groups $\pi_1(U,a)$ and $\pi_1(U,a')$ are canonically isomorphic for all $a'$ sufficiently close to $a$. This allows us to identify the same loop $\gamma$ in the above two groups and write $F_{a'a'\gamma}$ instead of $F_{a'a'\gamma'}$.

## 1.3 Geometric language: connections and flatness

Starting from the Pfaffian matrix $\Omega$, for any vector field $v$ on $U$ one can define the *covariant derivation* $\nabla_v$, a differential operator on the linear space of vector functions $\{x(\cdot) : U \to \mathbb{C}^n\}$, by the formula

$$\nabla_v x = i_v(dx - \Omega x), \tag{1.5}$$

where the right hand side is a vector function, the value taken by the vector-valued 1-form $dx - \Omega x$ on the vector $v$ tangent to $U$. The derivative $\nabla_v x$ depends linearly on $v$ and satisfies the Leibniz rule with respect to $x$: for any holomorphic function $f$, $\nabla_{fv} x = f\nabla_v x$ and $\nabla_v(fx) = f\nabla_v x + (\nabla_v f)\cdot x$, where $\nabla_v f = i_v\, df$ is the Lie derivative of $f$ along $v$, and $i_v$ the antiderivative (substitution of $v$ as the argument of a differential form).

The vector function $x$ is called $\nabla$-*horizontal* (or simply horizontal when the connection is defined by the context) along the smooth (parametrized) curve $\gamma : [0,1] \to U$, $s \mapsto t(s)$, if $\nabla_{\dot\gamma} x = 0$, where $\dot\gamma = \frac{d\gamma}{ds} \in \mathbb{C}^m$ is the velocity vector of $\gamma$. For any curve $\gamma$ there exists a unique function $x(\cdot)$ horizontal along this curve with any preassigned initial condition $x(\gamma(0)) = c \in \mathbb{C}^n$. This follows from the existence/uniqueness of solutions of *ordinary* differential equations with the real time variable $s \in [0,1]$.

Horizontal functions realize *parallel transport*, a collection of linear maps between different spaces $\{t(s)\} \times \mathbb{C}^n$, $s \in [0,1]$, in particular, between the points $a = t(0)$ and $b = t(1)$. By definition, $T_{ss'}$ is a map sending $c \in \mathbb{C}^n$ into $x(s') \in \mathbb{C}^n$, where $x(\cdot)$ is horizontal along $\gamma$ and satisfies the boundary condition $x(s) = c$. In general (i.e., without the integrability assumption (1.3)), the result of the parallel transport between two points $a$ and $b$ in $U$ depends on the curve $\gamma$ connecting them. In particular, the parallel transport over a small closed loop may well be nontrivial. An example of the parallel transport is the holonomy construction introduced above.

Consider two *commuting* vector fields $v, w$ and the parallel transport $T_\varepsilon : \mathbb{C}^n \to \mathbb{C}^n$ along four sides of an $\varepsilon$-small curvilinear parallelogram formed by flow curves of $v$ and $w$. One can show, using standard "calculus of infinitesimals", that the difference between $T_\varepsilon$ and the identity is $\varepsilon^2$-proportional to the commutator of the two differential operators $[\nabla_v, \nabla_w] = \nabla_v\nabla_w - \nabla_w\nabla_v$. If this commutator is identically zero for any two commuting vector fields, then the parallel transport along any sufficiently small loop is identical and therefore the global transport from $\{a\} \times \mathbb{C}^n$ to $\{b\} \times \mathbb{C}^n$ depends only on the homotopy class of the path $\gamma$ connecting $a$ with $b$.

In differential geometry the rule associating a differential operator $\nabla_v$ with any vector field $v$ on $U$ is called a *connection* (more precisely, affine connection). The *curvature* of the connection is the tensor $\nabla_v\nabla_w - \nabla_w\nabla_v - \nabla_{[v,w]}$, coinciding with the above commutator $[\nabla_v, \nabla_w]$ when $[v,w] = 0$. Connections with zero curvature are called *flat*. As follows from this definition, for flat connections the horizontal vector-functions can be defined without reference to any specific curve.

One can immediately verify by direct computation that the connection (1.5) constructed from a Pfaffian matrix $\Omega$ satisfying (1.3), is flat. Any local solution $x(\cdot)$ of the corresponding system (1.1) is horizontal along any sufficiently short curve $\gamma$ entirely belonging to the domain of this local solution.

Though we will not explore further the geometric aspects, the Pfaffian linear systems satisfying the integrability condition, will be often referred to as *flat connections.* The matrix $\Omega$ is the *connection matrix* or *connection form* in this language. The language of connections becomes especially convenient when discussing *holomorphic vector bundles* that only locally have the cylindric structure $U \times \mathbb{C}^n$ but may be globally nontrivial. However, we will not discuss the global questions in these notes, see [For91, Bol00a].

### 1.4 Gauge transform, gauge equivalence

Two matrix 1-forms $\Omega$ and $\Omega'$, both holomorphic on $U$, are called (holomorphically) *gauge equivalent*, or simply equivalent or conjugate, if there exists a holomorphic and holomorphically invertible matrix function $H : U \to \mathrm{GL}(n, \mathbb{C})$, such that

$$\Omega' = dH \cdot H^{-1} + H\Omega H^{-1}. \tag{1.6}$$

This definition reflects the change $x \mapsto H(t)x$ of the dependent variables in the respective linear system (1.1) or (1.2): if $\Omega$ is integrable and $X$ its (local) fundamental matrix solution of the first system, then $X' = HX$ is a fundamental matrix solution for the second system which is therefore also automatically integrable, $\Omega' = dX' \cdot (X')^{-1}$.

Clearly, this is an equivalence relationship between Pfaffian systems. The fact that $H$ is single-valued on $U$ means that the monodromy of solutions of gauge equivalent systems is the same: after continuation along any loop both $X$ and $X' = HX$ acquire the same right matrix factor $M_\gamma$. The converse statement is also trivially true.

**1.3 Theorem** *Two linear integrable holomorphic Pfaffian systems* (1.2) *on the same manifold are holomorphically gauge equivalent if and only if their monodromy groups coincide.*

**Proof** Without loss of generality we may assume that the base point $a$ and the two fundamental (multivalued) solutions $X, X'$ of the two systems are chosen so that their monodromy matrices $M_\gamma$ are the same for all loops $\gamma$. But then the matrix ratio $H = X' \cdot X^{-1}$ is single-valued (unchanged by continuation along any loop). Being holomorphic and holomorphically invertible, it realizes the holomorphic gauge equivalence between the two systems. □

## 2 Meromorphic flat connections

Integrable Pfaffian systems on multiply connected manifolds naturally appear if the Pfaffian matrix $\Omega$ is meromorphic (e.g., in a polydisc $\{|t_1| < 1, \dots, |t_m| < 1\}$ in $\mathbb{C}^n$). However, the corresponding theory of flat connections with singularities is much richer than that of connections holomorphic on $U \setminus \Sigma$, where $\Sigma$ is the polar locus of $\Omega$.

In what follows we will assume that the reader is familiar with some very basic facts from the local theory of analytic sets (dimension, irreducibility, etc.). The books [GR65] and [Chi89] contain all necessary information.

## 2.1 Polar locus. First examples

Let $f$ be a holomorphic function on $U$ and $\Sigma = \{f = 0\}$ the null set, an analytic hypersurface. We do not assume that $\Sigma$ is smooth (i.e., that $df$ is nonvanishing on $\Sigma$). However, the standing assumption will be that the locus $\Sigma_0 = \{f = 0,\ df = 0\} \subset \Sigma$ of *non-smooth* points is an analytic variety of codimension at least 2 in $U$. In particular, $f$ must be *square-free*, i.e., *not* divisible, even locally, by a square (or any higher power) of a holomorphic function. One may show that this condition is also sufficient to guarantee that the non-smooth part of $\Sigma$ is small in the above sense.

A function $g$ is said to be meromorphic in $U$ with the polar set in $\Sigma$, if $g$ is holomorphic outside $\Sigma$ and for any $a \in \Sigma$ the product $f^r g$ is holomorphic near $a$ for a sufficiently large natural $r \in \mathbb{N}$. The same construction defines also meromorphic differential forms. The matrix 1-form $\Omega$ is meromorphic with the polar locus $\Sigma$, if all its entries are meromorphic 1-forms and $\Sigma$ is the minimal analytic hypersurface with this property. The minimal natural $r$ such that $f^r\Omega$ is holomorphic near $a \in \Sigma$, is called the order of pole of $\Omega$ at $a$. If $\Sigma = \Sigma_1 \cup \Sigma_2 \cup \cdots$ is reducible, the order of poles along different components $\Sigma_j$ may be different.

While analytic hypersurfaces of one-dimensional manifolds are very easy to describe (they are locally finite unions of points), the geometry of analytic hypersurfaces is much richer even locally.

**2.1 Example (Smooth hypersurface, smooth point)** If $df$ does not vanish on $\Sigma = \{f = 0\}$, then the latter is smooth. By the implicit function theorem, $\Sigma$ can be represented locally near any of its points as $\{t_1 = 0\}$. The fundamental group of the complement is the free cyclic group $\mathbb{Z}$ generated by a small loop around zero in the $t_1$-line, intersecting $\Sigma$ transversally.

**2.2 Example (Isolated singularity)** If $df$ vanishes only at an isolated point of $\Sigma$, this case is called an isolated singularity. Its treatment is different for $m = 2$ and $m > 2$. The hierarchy of isolated singularities of functions and their null hypersurfaces is well known (at least, the first several steps of classification), see [AGV85].

**2.3 Example (Cuspidal point)** The cuspidal point (*cusp*) is the simplest isolated singularity of a holomorphic curve, which is locally irreducible. It corresponds to the function $f(t_1, t_2) = t_1^2 - t_2^3$. The fundamental group of the complement $(\mathbb{C}^2, 0) \setminus \Sigma$ is the *trefoil group* generated by two loops $\gamma_1, \gamma_2$ with the identity $\gamma_1^2\gamma_2^3 = \mathrm{id}$, see [Ful95, §22b].

**2.4 Example (Normal crossing)** This example, by far most important for applications, corresponds to few (no more than $m$) smooth hypersurfaces intersecting transversally. Locally near each point, $\Sigma$ can be represented as the locus $\{t_1 \cdots t_k = 0\}$ with $1 < k \leqslant m$ (the case $k = 1$ is nonsingular). The complement $U \setminus \Sigma = \{0 < |t_i| < 1,\ i = 1, \dots, k\}$ is contractible on the $k$-torus $\mathbb{T}^k = \prod_{i=1}^k \{|t_i| = 1\}$. Therefore the fundamental group $\pi_1(U \setminus \Sigma)$ is the cyclic group $\mathbb{Z}^k$ with $k$ commuting generators corresponding to small loops around the local components $\Sigma_i = \{t_i = 0\}$.

This example is especially important since for any analytic hypersurface $\Sigma$ one can construct a *resolution* (blow-up), a holomorphic map $F : U' \to U$ between two holomorphic manifolds, such that $\Sigma' = F^{-1}(\Sigma)$ has only normal crossings and $F$ is bijective between

$U \setminus \Sigma$ and $U' \setminus \Sigma'$. This is the famous Hironaka desingularization theorem, see Section 7, below.

**2.5 Remark** Speaking in more abstract language, one should describe the singular locus $\Sigma$ in terms of sheaves of ideals in the rings of holomorphic germs. However, we prefer to use less invariant but more elementary language and always consider $\Sigma$ together with its square-free local equations. It may cause some technical language problems when dealing with blow-ups (cf. with Section 7.2), but in most cases simplicity of the exposition justifies our choice.

## 2.2 Monodromy. Euler system

The monodromy of a meromorphic system in $U$ with polar locus $\Sigma$ is defined as the monodromy of its restriction on $U \setminus \Sigma$ where it is holomorphic. In particular, if $U$ is a polydisc and $\Sigma = \{t_1 \cdots t_k = 0\}$ is the standard "coordinate cross", the monodromy group is generated by $k$ *commuting* matrices $M_1, \dots, M_k$.

**2.6 Theorem** *Any collection of commuting invertible matrices $M_1, \dots, M_k$ can be realized as the monodromy group of a Pfaffian system of the special form*

$$\Omega_0 = A_1 \frac{dt_1}{t_1} + \cdots + A_k \frac{dt_k}{t_k} \tag{2.1}$$

*with constant pairwise commuting matrices $A_1, \dots, A_k \in \operatorname{Mat}_n(\mathbb{C})$.*

The system (2.1) will be referred to as the (generalized) *Euler system*; the usual Euler system corresponds to $k = m = 1$.

**Proof** The matrix function $t_j^{A_j} = \exp(A_j \ln t_j)$ depending only on the variable $t_j$, has the monodromy matrix $M_j = \exp 2\pi i A_j$. This identity can be resolved with respect to $A_j$ by several methods. One method suggests to reduce $M_j$ to the Jordan block-diagonal form and for each block of the form $\lambda(E+N)$ with $\lambda \in \mathbb{C} \setminus \{0\}$ and nilpotent $N$, define the logarithm by the formula $\ln[\lambda(E+N)] = (\ln\lambda)E + [N - N^2/2 + N^3/3 - \cdots]$. Since $N$ is nilpotent, the series converges. One can choose arbitrarily the branch of logarithm $\ln\lambda$ for all eigenvalues of $M_j$ independently of each other.

An alternative approach is based on the matrix Cauchy integral formula for the logarithm, see [Gan59],

$$A_j = \frac{1}{2\pi i} \oint_{\partial D} (\zeta E - M_j)^{-1} \ln\zeta \, d\zeta, \tag{2.2}$$

where $D \subseteq \mathbb{C}$ is any simply connected domain containing all eigenvalues of all matrices $M_1, \dots, M_k$, but *not* containing the origin $\zeta = 0$. This approach has the advantage that commutativity of the matrices $A_j$ is transparent when the $M_j$ commute.

Regardless of the choice of the matrices $A_j$ the commutative product $Y(t) = t_1^{A_1} \cdots t_k^{A_k}$ has all monodromy matrices equal to $M_j$ as required. It remains to verify by direct computation that $dY \cdot Y^{-1}$ has the required form (2.1). □

## 2.3 Regular singularities

In the "ordinary" (univariate) case a *regular* singular point is defined as a point $a \in \Sigma$ of the polar locus of $\Omega$, such that any fundamental solution $X(t)$ grows *moderately* (i.e., no faster than polynomially in $|t-a|^{-1}$ as $t \to a$ along any non-spiraling curve in the $t$-plane $\mathbb{C}^1$, for instance, a ray). If the monodromy around the point $a$ is trivial (identical), this is tantamount to the requirement that the fundamental solution is meromorphic at $a$. Note that if the function is multivalued, say $f(t) = \ln t$, then its growth along spirals slowly approaching $a = 0$, e.g., $s \mapsto s\exp(i/s)$, $s \in [1, 0)$, may well be exponential in $1/|t|$ despite the moderate growth of $|f(t)|$ along all rays.

In the multivariate case the definition may be given in parallel terms. An integrable Pfaffian system (1.2) is said to have a *regular singularity* (on its polar locus $\Sigma$), if for any simply connected bounded *semianalytic* subset $S \subseteq U \setminus \Sigma$ any fundamental solution $X(t)$ satisfies the estimate

$$|X(t)| \leqslant c\,|f(t)|^{-r}, \quad t \in S.$$

The constants $c > 0$ and $r < +\infty$ may depend on the solution and the domain $S$, while the semianalyticity assumption is aimed at excluding uncontrolled spiraling of $S$ around $\Sigma$.

Alternatively, one can use holomorphic parametrized *probe curves* $z : (\mathbb{C}^1, 0) \to (U, \Sigma)$ not contained entirely in the polar locus. Each such curve defines an "ordinary" system with the matrix 1-form $z^*\Omega$ (the pullback) on $(\mathbb{C}^1, 0)$. For a system regular in the sense of the previous definition, all such probe restrictions will exhibit regular singularities. It turns out that the converse is also true and even in the stronger sense: it is sufficient to consider only probe curves transversal to $\Sigma$ and only at smooth points of the latter.

**2.7 Theorem (see [Del70])** *If the pullback $z^*\Omega$ is regular for any probe curve $z : (\mathbb{C}, 0) \to (U, \Sigma)$ transversal to $\Sigma$ at any smooth point $a = z(0)$ of the latter, then the integrable Pfaffian system* (1.2) *has a regular singularity on $\Sigma$.*

**Proof** We sketch the proof only in the particular case when $\Sigma$ is a normal crossing $\{t_1 \cdots t_k = 0\}$. It was shown (Theorem 2.6) that one can always find an Euler system (2.1) and its solution $Y(t)$ with the same monodromy matrix factors as the fundamental solution $X(t)$ of the system. Clearly, $Y$ is regular on $\Sigma$. The matrix ratio $H(t) = X(t)Y^{-1}(t)$ is hence single-valued. By the assumption of the theorem, for any smooth point $a \in \Sigma$ the function $H$ may have a pole of some finite order $r_a \in \mathbb{N}$. By the Baire category theorem, there exists $r \in \mathbb{N}$ such that $f^r H$ is holomorphic at all smooth points of $\Sigma$. Since the non-smooth points constitute a thin set (of analytic codimension at least 2), $f^r H$ is in fact holomorphic everywhere in $U$. It remains to note that since both $H$ and $Y$ grow moderately on semianalytic sets off $\Sigma$, so does their product $X = HY$. □

**2.8 Remark** The proof in the general case differs only in one instance: one has to construct explicitly just one matrix function $Y$ with moderate growth on semianalytic subsets, having the same monodromy as $X$. This can be done using the resolution of $\Sigma$ to normal crossings, see [Del70] for details.

## 2.4 Holomorphic and meromorphic gauge equivalence

For systems with singularities one can consider two different gauge equivalence relationships. We say that two integrable Pfaffian systems with meromorphic matrices $\Omega$ and $\Omega'$ having *the same* polar locus $\Sigma$, are *holomorphically* gauge equivalent if (1.6) holds for some matrix function $H$ holomorphic and holomorphically invertible everywhere *including* $\Sigma$. The two systems are said to be *meromorphically* gauge equivalent, if $H$ is holomorphic and holomorphically invertible only outside $\Sigma$, while having at worst poles of finite order on $\Sigma$ for both $H$ and $H^{-1}$. Note that the meromorphic equivalence is stronger than holomorphic gauge equivalence on the holomorphy set $U \setminus \Sigma$, occurring in Theorem 1.3. However, for regular systems the difference disappears.

**2.9 Theorem** *Two integrable Pfaffian systems* (1.2) *having only regular singularities, are meromorphically equivalent if and only if their monodromy groups are the same.*

**Proof** The matrix ratio of two fundamental solutions is a single-valued matrix function having moderate growth on the polar locus, hence meromorphic. □

**2.10 Corollary** *Any integrable Pfaffian system* (1.2) *with a regular singularity on the coordinate cross* $\{t_1 \cdots t_k = 0\} \subset \mathbb{C}^m$ *is meromorphically gauge equivalent to a suitable Euler system* (2.1). □

The holomorphic gauge classification is considerably more difficult. It will be addressed in more details in Section 6, below.

## 2.5 Formal and convergent solutions

For future purposes we need to introduce formal gauge equivalence; here we explain it and discuss some basic convergence results.

Assume that $\Sigma$ is the coordinate cross $\Sigma = \{t_1 \cdots t_k = 0\}$ and consider the algebra of formal Laurent series in the variables $t = (t_1, \ldots, t_k)$ involving only non-negative powers of $t_{k+1}, \ldots, t_m$. Over this algebra one can naturally define formal matrix functions, formal meromorphic differential forms, formal vector fields, etc.

Everywhere in these notes we will grade formal Taylor or Laurent series and forms, assigning the same weights $w_j \in \mathbb{R}_+$ to the differentials $dt_j$ of the independent variables and to the variables themselves so that $\frac{dt_j}{t_j}$ always has weight zero. In most cases (though not always) it will be sufficient to choose $w_1 = \cdots = w_m = 1$. In such cases we try to distinguish between the degree and the weighted (quasihomogeneous) degree.

Any meromorphic Pfaffian system with the polar locus on $\Sigma = \{t_1 \cdots t_k = 0\}$, can be expanded as a formal Laurent series, $\Omega = \sum_{s \geqslant -r} \Omega_s$, where each $\Omega_s$ is a *homogeneous* matrix form of degree $s$.

Solutions of (integrable) Pfaffian systems can also be constructed recurrently as infinite sums of homogeneous terms. In general it is not sufficient to consider only Laurent polynomials: already in the "ordinary" case $n = 1$ one has to introduce fractional powers $t^a$, $a \notin \mathbb{Z}$, and logarithms $\ln t$. However, the important fact is that in the regular case any such formal solution always converges. We will use only the simplest version of the corresponding result, dealing only with Laurent formal solutions not involving fractional powers.

**2.11 Theorem** *If $\Omega$ is an integrable Pfaffian form with a regular singularity on the coordinate cross $\{t_1 \cdots t_k = 0\} \subset (\mathbb{C}^m, 0)$, then any formal (Laurent) solution of the system* (1.1) *converges, i.e., is meromorphic.*

**Proof** The assertion of the Theorem is invariant under meromorphic gauge equivalence: two equivalent forms $\Omega, \Omega'$ both satisfy or do not satisfy simultaneously the property that any formal solution necessarily converges.

By Corollary 2.10, any regular system is meromorphically equivalent to an Euler system (2.1), hence it is sufficient to prove the theorem for Euler systems only.

A (vector) formal Laurent series $\sum h_r x_r$ with vector coefficients $x_r \in \mathbb{C}^n$ and scalar (quasi)homogeneous terms $h_r$ of degree $r$ satisfies the system $dx = \Omega_0 x$ with $\Omega_0$ as in (2.1), if and only if each term $h_r x_r$ satisfies this system. This follows from the fact that both $d$ and multiplication by $\Omega_0$ preserve the weight (quasihomogeneous degree) of vector-valued 1-forms. If we choose the weights of the variables $t_j$ independent over $\mathbb{Z}$, the only quasihomogeneous Laurent polynomials will be monomials, $h_r = t^{\mathbf{a}} = t_1^{a_1} \cdots t_m^{a_m}$, $\mathbf{a} = (a_1, \dots, a_m) \in \mathbb{Z}^m$, $a_1 w_1 + \cdots + a_m w_m = r$, modulo a constant multiplier.

The product $x(t) = t^{\mathbf{a}} x_r$ satisfies the linear system $dx = \Omega_0 x$ if and only if

$$t^{\mathbf{a}} x_r \cdot \sum_1^m a_j \frac{dt_j}{t_j} = t^{\mathbf{a}} \cdot \sum_1^m (A_j x_r) \frac{dt_j}{t_j},$$

which is possible only if

$$a_j x_r = A_j x_r, \quad j = 1, \dots, m$$

(we assume that $A_{k+1} = \cdots = A_m = 0$). But these identities imply that the exponents $a_j \in \mathbb{Z}$ are bounded (they range over the spectra of the residue matrices $A_j$). Thus there may be only finitely many nonzero terms in the apriori infinite sum $x = \sum h_r x_r$ which is thus automatically convergent. This proves that the solution $x = x(t)$ must be meromorphic. □

This theorem will always guarantee that there are no differences between formal and convergent (analytic) results as far as only regular singularities are allowed.

## 2.6 Flat connections and isomonodromic deformations

It was already noted in Section 2.3 that a meromorphic flat connection $\Omega$ on $U$ with the polar locus $\Sigma \subset U$ induces a (necessarily flat) connection on any embedded holomorphic curve $Z \subset U$. If $Z$ is parametrized by a chart $z : \mathbb{C} \to Z$, the condition describing horizontal sections of the induced connection takes the form of an ordinary differential equation. This equation exhibits singularities at the points of intersection $Z \cap \Sigma$; they are regular if $\Sigma$ was a regular singularity of the initial connection.

The flatness of the restricted connection follows automatically from the assumption that $Z$ is of (complex) dimension 1. Integrability (flatness) of the initial multivariate connection $\Omega$ results in the fact that the monodromy of the restriction of $\Omega$ on $Z$ is in some sense *independent of* $Z$.

More precisely, assume that the complex $m$-dimensional analytic manifold $U$ on which the flat meromorphic connection $\Omega$ lives, is fibered over another complex $(m-1)$-dimensional manifold $P$ (the "parameter space") by a full rank holomorphic map $p : U \to P$. Assume that for some value $\lambda_0 \in P$ the map $p$ restricted to $U \setminus \Sigma$ is topologically trivial over a small neighborhood $V \ni \lambda_0$: it is sufficient to require that the fiber $p^{-1}(\lambda_0)$ transversally intersects $\Sigma$ only at finitely many smooth points. Under this assumption the nearby fibers $Z_\lambda = p^{-1}(\lambda)$ are all homeomorphic to $Z_{\lambda_0}$ (hence to each other); though this homeomorphism is not canonical, it allows for canonical identification of the fundamental groups $\pi_1(Z_\lambda, a_\lambda)$ with $\pi_1(Z_{\lambda_0}, a_{\lambda_0})$ for any continuous section $a : P \to U \setminus \Sigma$ of the bundle $p$. Needless to say, the topology of the fibers $Z_\lambda$ may be arbitrary; in particular, they can be all conformally equivalent to the Riemann sphere or any other Riemann surface. Everything depends on the global structure of $U$ and $P$.

After all these technical precautions, it is obvious that the monodromy groups of the restricted connections $\Omega_\lambda = \Omega|_{Z_\lambda}$ are conjugate to each other. It is sufficient to choose a (multivalued) fundamental solution $X$ on $p^{-1}(V)$ of the integrable system (1.2) and restrict this solution on each one-dimensional fiber.

Conversely, one may show (cf. [Bol97]), that a reasonably defined *isomonodromic parametric family*, or *isomonodromic deformation* of systems of linear ordinary differential equations can be obtained from a suitable meromorphic flat connection on some multidimensional manifold, e.g., the cylinder $\mathbb{C}^1 \times$ (polydisc).

This link between meromorphic flat connections on higher dimension manifolds and isomonodromic deformations is (for us) a primary source of motivation for studying the former. It is important to stress that speaking about isomonodromic deformation means *introducing the additional bundle structure* on $U$. However, at least locally (near a smooth point $a \in \Sigma$) one can always introduce a local coordinate system $(z, \lambda) \in (\mathbb{C}^1 \times \mathbb{C}^{m-1}, 0)$ on $U$ so that $p(z, \lambda) = \lambda$ and $\Sigma = \{z = 0\}$. When performing calculations, we will always use such *adapted* local coordinates.

# 3 Logarithmic forms, logarithmic poles, residues

In the "ordinary" univariate case, there is a simple sufficient condition guaranteeing regularity of a meromorphic connection matrix $\Omega$ on its polar locus $\Sigma$. The point $a \in \Sigma \subset \mathbb{C}^1$ is called *Fuchsian*, if $\Omega$ has a first order pole at $a$. In the local chart $t = t_1$ the point $t = a$ is Fuchsian, if its principal polar (non-holomorphic) part is an Euler system,

$$\Omega = A\,\frac{dt}{t-a} + \text{holomorphic matrix 1-form.} \tag{3.1}$$

The constant matrix $A \in \operatorname{Mat}(n, \mathbb{C})$ is the *residue* of $\Omega$ at $a \in \Sigma$.

In the multivariate case the order condition on the pole *does not* guarantee that its principal part is an Euler system (2.1) similarly to (3.1). Before proceeding with the definition of an analog of Fuchsian poles for the multivariate case in Section 4, we need some analytic preparatory work.

## 3.1 Preparation: de Rham division lemma

Consider the exterior algebra $\Lambda^\bullet(U)$ of holomorphic forms on $U \subset \mathbb{C}^m$. If $\xi = \sum_1^m c_i(t)\, dt_i \in \Lambda^1(U)$ is a given nonzero 1-form, then a $k$-form $\omega$, divisible by $\xi$, i.e., representable as

$$\omega = \theta \wedge \xi, \quad \theta \in \Lambda^{k-1}(U), \tag{3.2}$$

necessarily satisfies the condition

$$\omega \wedge \xi = 0 \in \Lambda^{k+1}(U). \tag{3.3}$$

It turns out that the necessary divisibility condition (3.3) is very close to being sufficient. Note that the identity (3.2) can be considered as a system of linear algebraic (non-homogeneous) equations on the coefficients of the form $\theta$ in the basis $dt_1, \dots, dt_m$; the condition (3.3) is a necessary condition of solvability of this system. G. de Rham [dR54] described the forms $\xi$ for which (3.3) implies divisibility. We will need only a simple particular case of his result.

**3.1 Lemma** *If one of the coefficients of the form $\xi$, e.g., $c_1(t)$, is non-vanishing (invertible) at some point $a \in U$, then any holomorphic $k$-form $\omega$, $k < m$, satisfying (3.3), is divisible by $\xi$ and the ratio $\theta$ may be chosen holomorphic at this point.*

*More generally, if $c_1$ is allowed to vanish but not identically, $c_1(t) \not\equiv 0$, and $\omega$ is holomorphic or meromorphic with the poles only on $\Sigma_1 = \{c_1 = 0\} \subset U$, then $\theta$ can also be chosen meromorphic with the polar locus on $\Sigma_1$.*

**Proof** If $c_1$ is holomorphically invertible, then without loss of generality we may assume that $\xi = dt_1 + \sum_2^m c_i\, dt_i$ with the other coefficients $c_i$ holomorphic at $a \in U$. Consider the 1-forms $\xi_1 = \xi$, $\xi_i = dt_i$, $i = 2, \dots, m$. These holomorphic forms and their exterior products constitute a basis in the exterior algebra, so that any holomorphic $k$-form can be uniquely written as the sum of wedge monomials

$$\sum_{i_1 < \cdots < i_k} c_{i_1 \cdots i_k}(t)\, \xi_{i_1} \wedge \cdots \wedge \xi_{i_k}$$

with holomorphic coefficients.

For $\xi = \xi_1$ the assertion of the lemma is obvious. Condition (3.3) means that in the expansion of the $k$-form $\omega$ only monomials involving $\xi_1$ may appear as a factor. Each such monomial $c_{1 i_2 \cdots i_k} \xi_1 \wedge \xi_{i_2} \wedge \cdots \wedge \xi_{i_k}$ is divisible by $\xi_1$, the ratio being (among other possibilities) a monomial $\pm c_{1 i_2 \cdots i_k}(t)\, \xi_{i_2} \wedge \cdots \wedge \xi_{i_k}$ with a holomorphic coefficient. This completes the proof when $c_1$ is invertible.

The case when $c_1$ vanishes but admits a meromorphic inverse, is treated in the same way, the only exception being that the form $\xi_1 = dt_1 + \sum_2^m c_i\, dt_i$ will have meromorphic coefficients with the polar locus $\Sigma_1$. □

**3.2 Remark** The natural question appears, whether one can use the non-uniqueness of the division to ensure that the polar locus of the meromorphic coefficients of $\xi$ is smaller than a hypersurface.

**3.3 Theorem (Moussu [Mou75])** *If $\xi = \sum_1^m c_i(t)\, dt_i$ is a holomorphic 1-form on $(\mathbb{C}^m, 0)$ with an isolated singularity at the origin (i.e., all the coefficients $c_1, \dots, c_m$ vanish simultaneously only at the origin $t = 0$), then each holomorphic $k$-form with $k < m$, satisfying the necessary condition (3.3), is divisible by $\xi$ with a holomorphic "ratio" form $\theta$.* □

The assertion of Theorem 3.3 for $m$-forms fails. Any such form trivially satisfies (3.3), but $\omega = c(t) dt_1 \wedge \cdots \wedge dt_m$ is holomorphically divisible by $\xi$ if and only if the holomorphic function $c(t)$ belongs to the ideal $\langle c_1, \dots, c_m \rangle$. In particular, $c(\cdot)$ must vanish at the singular point $t = 0$ of $\xi$.

### 3.2 Logarithmic poles. Residues

A meromorphic $k$-form $\omega \in \Lambda^k(U \setminus \Sigma)$ is said to have a *logarithmic pole* on the analytic hypersurface $\Sigma = \{f = 0\}$, if both $\omega$ and $d\omega$ have a first order pole on $\Sigma$:

$$f\omega,\ f\,d\omega \text{ both extend holomorphically on } \Sigma. \tag{3.4}$$

By our standing assumption on $f$, we may assume that one of the partial derivatives, say, $\partial f/\partial t_1$, is vanishing on an analytic hypersurface

$$\Sigma_1 = \{\partial f/\partial t_1 = 0\} \subset U$$

such that $\Sigma \cap \Sigma_1$ has (complex) codimension at least 2 in $U$.

**3.4 Lemma** *If a k-form $\omega$ has a logarithmic pole on $\Sigma$, then it can be represented as*

$$\omega = \theta \wedge \frac{df}{f} + \eta, \tag{3.5}$$

*where $\theta \in \Lambda^{k-1}(U \setminus \Sigma_1)$ and $\eta \in \Lambda^k(U \setminus \Sigma_1)$ are both holomorphic on $\Sigma \setminus \Sigma_1$.*

The (meromorphic) $(k-1)$-form $\theta$ is called the *residue* of $\omega$ on $\Sigma$ and denoted by $\operatorname{res}_\Sigma \omega$. While the representation (3.5) is not unique, the restriction of $\theta$ on $\Sigma$ is unique.

**Proof** The form $df \wedge \omega = d(f\omega) - f\,d\omega$ is holomorphic on $\Sigma$ by definition of the logarithmic pole, and vanishes after exterior multiplication by $df$ (obviously outside $\Sigma$, hence everywhere). By the Division Lemma 3.1,

$$df \wedge \omega = df \wedge \eta, \tag{3.6}$$

where $\eta$ is a $k$-form holomorphic everywhere outside $\Sigma_1$.

The difference $f\omega - f\eta$ is holomorphic on $\Sigma \setminus \Sigma_1$ since $f\omega$ is, and by (3.6), this difference vanishes after exterior multiplication by $df$. Again by virtue of the Division Lemma, we have

$$f\omega - f\eta = \theta \wedge df \tag{3.7}$$

with $\theta \in \Lambda^{k-1}(U \setminus \Sigma_1)$ as required.

To show the uniqueness of the restriction of $\theta$ on $\Sigma \setminus \Sigma_1$, it is sufficient to show that $\omega \equiv 0$ implies vanishing of this restriction. In this case $\theta \wedge df = -f\eta$, and multiplying this identity by $df$ we see that $df \wedge \eta \equiv 0$. Again by the Division Lemma, $\eta$ must be divisible by $df$, $\eta = \zeta \wedge df$. We conclude then that $(\theta - f\zeta) \wedge df = 0$. But $df \neq 0$ on $\Sigma \setminus \Sigma_1$, while the restriction of $f\zeta$ to $\Sigma \setminus \Sigma_1$ vanishes since $\zeta$ was holomorphic there. This implies that the restriction of $\theta$ must vanish on $\Sigma \setminus \Sigma_1$, as asserted. □

Note that $\operatorname{res}\omega$ does not depend on the choice of $f$ used to describe the logarithmic pole $\Sigma$: multiplying $f$ by an invertible factor contributes a holomorphic additive term to the logarithmic derivative $f^{-1}df$ and nothing to the residue.

Recall that for a (scalar) meromorphic 1-form $\omega$ on a Riemann surface, locally represented as $\omega = g\,\frac{dt}{t-a} + \cdots$, the residue $g(a) = \operatorname{res}_a \omega$ at the pole $t = a$ can be defined via the Cauchy integral $\frac{1}{2\pi i}\oint_\gamma \omega$, where $\gamma$ is a small loop encircling the pole, and does not depend on the choice of the chart $t \in \mathbb{C}$. In the multidimensional case if $\omega$ has a logarithmic pole on $\Sigma$ and $\omega = g\,\frac{df}{f} + \cdots$, then for any probe curve $z : (\mathbb{C}^1, 0) \to (U, a)$ through a smooth point $a \in \Sigma$, the residue of the restricted form $z^*\omega$ at the origin is $g(a)$, independently of the choice of the probe curve. In other words, the residue can be detected by studying probe curves, if and only if the pole on $\Sigma$ is logarithmic.

### 3.3 Logarithmic complex

Denote by $\Lambda^k(\log\Sigma) \subseteq \Lambda^k(U\setminus\Sigma)$ the collection of meromorphic $k$-forms having a logarithmic pole on the given polar locus $\Sigma \subset U$, and let

$$\begin{aligned}\Lambda^\bullet(\log\Sigma) &= \bigcup_{k\geqslant 0}\Lambda^k(\log\Sigma)\\ &= \{\omega : f\omega \text{ and } f\,d\omega \text{ in } \Lambda^\bullet(U)\}\\ &= \{\omega : f\omega \text{ and } df\wedge\omega \text{ in } \Lambda^\bullet(U)\}.\end{aligned} \tag{3.8}$$

We will refer to $\Lambda^\bullet(\log\Sigma)$ as *logarithmic forms* if the locus $\Sigma$ is clear from the context.

Clearly, logarithmic forms constitute a (graded) module over the ring of holomorphic functions on $U$. It is obvious that the exterior derivative $d : \Lambda^k \to \Lambda^{k+1}$ preserves logarithmic forms: indeed, if both $f\omega$ and $f\,d\omega$ are holomorphic (on $\Sigma$), then $f\,d\omega$ and $f\,d^2\omega = 0$ also are. It is less obvious that logarithmic forms are closed under taking exterior (wedge) products.

**3.5 Lemma** *If $\omega, \omega' \in \Lambda^\bullet(\log\Sigma)$ are two logarithmic forms (of different degrees, in general), then their product $\omega\wedge\omega'$ is also a logarithmic form.*

**Proof** By Lemma 3.4, each of these forms can be represented as $\omega = \theta\wedge df/f + \eta$, $\omega' = \theta'\wedge df/f + \eta'$, where the forms $\theta, \theta', \eta, \eta'$ of appropriate degrees are holomorphic everywhere outside the auxiliary locus $\Sigma_1$ intersecting the polar $\Sigma$ by a thin set (of codimension 2 at least). Computing the wedge product, we conclude that $\omega\wedge\omega'$ has at most a first order pole on $\Sigma\setminus\Sigma_1$ (i.e., that $f\,\omega\wedge\omega'$ extends holomorphically on $\Sigma\setminus\Sigma_1$). On the other hand, $f\,\omega\wedge\omega'$ is holomorphic outside $\Sigma$, since both $\omega$ and $\omega'$ are holomorphic there. Thus $f\,\omega\wedge\omega'$ is holomorphic outside $\Sigma\cap\Sigma_1$. Since the intersection $\Sigma\cap\Sigma_1$ is thin, $f\,\omega\wedge\omega'$ is holomorphic everywhere in $U$.

The wedge product $df\wedge(\omega\wedge\omega') = df\wedge\eta\wedge\eta'$ is also holomorphic outside $\Sigma_1\cap\Sigma$, hence everywhere for exactly the same reason. The second line of (3.8) allows to conclude that $\omega\wedge\omega' \in \Lambda^\bullet(\log\Sigma)$. □

Together with the previously remarked $d$-closeness, Lemma 3.5 implies that *logarithmic forms constitute a graded exterior algebra* which is made into a cochain complex by restriction of the exterior derivative $d$.

Logarithmic forms exhibit the mildest singularity along $\Sigma$. In particular, a logarithmic 0-form (function) $g \in \Lambda^0(\log\Sigma)$ is necessarily holomorphic. The term "logarithmic pole" is motivated by the fact that for any meromorphic function $g$, holomorphic and invertible outside $\Sigma$, its logarithmic derivative $g^{-1}\,dg$ belongs to $\Lambda^1(\log\Sigma)$. Moreover, one can show [Del70] that when $\Sigma$ is a normal crossing, the module $\Lambda^\bullet(\log\Sigma)$ is in fact generated by logarithmic derivatives of meromorphic functions as an exterior algebra over the ring of holomorphic functions in $U$.

### 3.4 Holomorphy of residues. Saito criterion

The local representation established in Lemma 3.4 means that the residue of a logarithmic 1-form $\omega$ is a function $g = \operatorname{res}\omega$ which is a priori only meromorphic on $\Sigma$, eventually exhibiting poles on an auxiliary hypersurface $\Sigma_1\cap\Sigma \subsetneq \Sigma$. The polar set $\Sigma_1\cap S$ depends only on $\Sigma$ and not on the form(s).

By the stronger form (Theorem 3.3) of the Division Lemma, the residue of a logarithmic 1-form $\omega$ is *holomorphic* on the polar locus $\Sigma$ of this form, if $m > 2$, and $\Sigma$ has an isolated singularity.

This is, in particular, the case of smooth divisors $\Sigma = \{f = 0\}$ when $df|_\Sigma$ is non-vanishing (in this case $\Sigma_1 = \varnothing$). However, for other analytic hypersurfaces the residue may indeed exhibit singularities (in particular, be locally unbounded) near non-smooth points of $\Sigma$.

**3.6 Example** The 1-form

$$\omega = \frac{1}{y-x}\left(\frac{dx}{x} - \frac{dy}{y}\right), \quad (x,y) \in \mathbb{C}^2 \tag{3.9}$$

on the complex 2-plane $\mathbb{C}^2$, has a logarithmic singularity on the union of three lines through the origin $\Sigma = \{xy(x-y) = 0\}$. The residues on each of the lines $\{x = 0\}$, $\{y = 0\}$ and $\{y = x\}$ are equal to $1/y$, $1/x$ and $-2/x$ correspondingly: all three exhibit a "polar" singularity at the origin.

However, a similar example with only two lines in the plane $\mathbb{C}^2$ is impossible. We will show that if $\Sigma$ is a transversal intersection of smooth analytic hypersurfaces (normal crossing), the residue of any logarithmic form is necessarily holomorphic. The exposition below is based on the seminal paper by K. Saito [Sai80].

**3.7 Lemma** *Let $\Sigma \subset U$ be an analytic hypersurface in an $m$-dimensional analytic manifold, and $\omega_1, \dots, \omega_m \in \Lambda^1(\log \Sigma)$ logarithmic 1-forms.*

*If the (holomorphic by Lemma 3.5) $m$-form $f\,\omega_1 \wedge \cdots \wedge \omega_m$ is non-vanishing on $\Sigma$, then the forms $\omega_i$ form a basis of $\Lambda^1(\log \Sigma)$ over the ring of holomorphic functions: any logarithmic 1-form $\omega$ can be expanded as*

$$\omega = h_1\,\omega_1 + \cdots + h_m\omega_m, \quad h_i \in \Lambda^0(U), \tag{3.10}$$

*with the coefficients $h_1(t), \dots, h_m(t)$ holomorphic on $\Sigma$.*

**Proof** The forms $\omega_i$ are necessarily linearly independent outside $\Sigma$, so the decomposition (3.10) exists with the coefficients $h_i$ that are at worst meromorphic on $\Sigma$. The same argument shows that the $h_i$ are uniquely defined.

To prove that each $h_i$ admits holomorphic extensions on $\Sigma$, we multiply (3.10) by the wedge product $\Xi_i \in \Lambda^{m-1}(\log \Sigma) = \bigwedge_{j\neq i} \omega_j$ of all the forms except $\omega_i$. The coefficient $h_i$ can be obtained by Cramer's rule as a ratio of two $m$-forms, logarithmic by Lemma 3.5,

$$h_i = \frac{\omega \wedge \Xi_i}{\omega_j \wedge \Xi_i} = \pm\frac{f\,\omega \wedge \Xi_i}{f\,\omega_1 \wedge \cdots \wedge \omega_m}.$$

Both forms on the right hand side are holomorphic by definition of logarithmic poles, and the denominator is nonvanishing on $\Sigma$ near $a$ by assumption. □

As a corollary, we can immediately prove that for a divisor with normal crossings, the residues must be holomorphic functions along each component, though *not necessarily coinciding on the intersections.*

**3.8 Theorem** *If $\Sigma = \{t_1 \cdots t_k = 0\}$, $k \leqslant m$ is a normal crossing divisor, than any 1-form $\omega$ with logarithmic pole on $\Sigma$ can be expressed as*

$$\omega = \sum_{i=1}^{k} g_i(t)\,\frac{dt_i}{t_i} + \eta, \tag{3.11}$$

*where the 1-form $\eta$ and the functions $g_i$ are holomorphic on $U$ including the polar locus $\Sigma$.*

**Proof** The 1-forms $\frac{dt_1}{t_1}, \dots, \frac{dt_k}{t_k}, dt_{k+1}, \dots, dt_m$ satisfy the assumptions of Lemma 3.7: their wedge product is $(1/f)$ times the standard "volume form" $dt_1 \wedge \cdots \wedge dt_m$. □

### 3.5 Closed logarithmic forms

One particular case is worth being singled out. Let $\Sigma = \bigcup_1^k \Sigma_j$ be a polar locus represented as the union of *irreducible components*. The irreducibility implies, among other things, that the smooth part of each hypersurface $\Sigma_j \setminus \bigcup_{i\neq j} \Sigma_i$ is connected.

**3.9 Lemma** *If $\omega \in \Lambda^1(\log \Sigma)$ is a closed logarithmic form, $d\omega = 0$, then the residue of $\omega$ is constant on any component of the polar locus $\Sigma$. Any such form with at least one nonzero residue admits representation*

$$\omega = \sum_{j=1}^{k} a_j\,\frac{df_j}{f_j}, \qquad a_1, \dots, a_k \in \mathbb{C}, \tag{3.12}$$

*where $f_j$ are suitably chosen equations of the components $\Sigma_j$.*

**Proof** After differentiation and multiplication by $f$, the representation $\omega = g\,df/f + \eta$ (3.5) yields the identity between the holomorphic forms

$$0 = dg \wedge df + f\,d\eta, \qquad g \in \Lambda^0(U),\ \eta \in \Lambda^1(U).$$

This implies immediately that $dg \wedge df$ vanishes on $\Sigma$ near every smooth point, so that $dg$ is everywhere proportional to $df$ and vanishes on vectors tangent to $\Sigma$. Therefore $g = \text{const}$ along the smooth part of $\Sigma$. Subtracting from $\omega$ the principal part $\sum_j a_j\,df_j/f_j$ which is already a closed logarithmic form, we obtain a nonsingular closed form $\eta$ which can always be expressed as $df_0/f_0$ for a suitable holomorphically invertible function $f_0$. This term $f_0$ can be incorporated into any of the equations corresponding to a nonzero residue, $a_j\,df_j/f_j + df_0/f_0 = a_j\,df_j'/f_j'$, if we put $f_j' = f_j \exp(f_0/a_j)$. □

## 4 Flat connections with logarithmic poles

Rather naturally, a meromorphic flat connection with the connection matrix $\Omega$ is said to have a *logarithmic pole* on $\Sigma = \{f = 0\}$, if both $f\Omega$ and $f\,d\Omega$ extend holomorphically on $\Sigma$. Near any smooth point $a \in \Sigma$ it can be represented as

$$\Omega = A(t)\frac{df}{f} + \text{holomorphic matrix 1-form}, \qquad A = \operatorname{res}_\Sigma \Omega,$$

where the *residue matrix function* $A(\cdot)$ is uniquely defined and holomorphic on the smooth part $\Sigma'$ of $\Sigma$. Abusing the notation, we will write $\Omega \in \Lambda^1(\log \Sigma)$, meaning that $\Omega$ has a logarithmic pole on $\Sigma$.

If the pole of $\Omega$ is known to be logarithmic, then for any probe curve $z : (\mathbb{C}^1, 0) \to (U, a)$, passing through a smooth point $a \in \Sigma$ and not entirely embedded in $\Sigma$, the induced connection $z^*\Omega$ has a Fuchsian singularity and its residue can be immediately computed,

$$\operatorname{res}_0 z^*\Omega = (\operatorname{res}_\Sigma \Omega)(a), \qquad a = z(0).$$

Indeed, if (locally) $f = t_1$ and $a = 0$, then a form having a first order pole on the hyperplane $\{t_1 = 0\}$ can be represented as

$$\omega = A_1(t_2, \dots, t_m)\frac{dt_1}{t_1} + \sum_{i=2}^{m} A_i(t_2, \dots, t_m)\frac{dt_i}{t_1} + \cdots$$

(the dots denote holomorphic terms). The pole is logarithmic only if $A_2 = \cdots = A_m \equiv 0$, and $A_1(0)$ is the residue in this case.

In this section we will study the behavior of the residue(s) of *flat* meromorphic connections satisfying the integrability condition (1.3).

## 4.1 Residues and holonomy

Consider a flat logarithmic connection $\Omega \in \Lambda^1(\log \Sigma)$. Let $a \in \Sigma'$ be a point on the smooth part $\Sigma'$ of the polar locus $\Sigma$. For any point $t \notin \Sigma$ sufficiently close to $a$, one can unambiguously define a small positive simple loop $\gamma_t$ around $\Sigma'$ beginning and ending at $t$ so that the holonomy operator $F_t$ corresponding to the loop $\gamma_t$ analytically depends on $t \notin \Sigma$.

**4.1 Lemma** *The operator $F_t$ has a limit $F_a = \lim_{t\to a} F_t$ related to the residue $A(a) = \operatorname{res}_a \Omega$ of the flat logarithmic connection $\Omega$ by the identity $F_a = \exp 2\pi i A(a)$.*

**Proof** The assertion is almost obvious in the "ordinary" case $m = 1$, where it is sufficient to compute the principal asymptotic term of the expansion of solutions of a Fuchsian system. The multivariate assertion follows now from comparing the restrictions of $\Omega$ on various probe curves $z : (\mathbb{C}^1, 0) \to (U, a)$, and the independence of the residue of the choice of the probe curve. □

It should be remembered, however, that while all operators $F_t$, $t \notin \Sigma$, are conjugate to each other, it may well happen (even in the "ordinary" case) that the limit $F_a$ does *not* belong to the same conjugacy class. Examples can be found in [Del70, Gan59]. What can be asserted is that the characteristic polynomials for $F_t$ and $F_a$ coincide.

## 4.2 Conjugacy of the residues

The following result can be considered as a matrix generalization of Lemma 3.9.

**4.2 Theorem** *If $\Omega \in \Lambda^1(\log \Sigma)$ is a flat connection matrix with a logarithmic singularity on $\Sigma$, then the residue matrix $A(a) = \operatorname{res}_a \Omega$ varies in the same conjugacy class of* $\mathrm{GL}(n, \mathbb{C})$ *as long as the point $a$ varies continuously along the smooth part of $\Sigma$.*

**Proof** We give two different proofs of this theorem, one based on topological considerations, the other achieved by direct explicit computation.

*First proof.* For any two sufficiently close smooth points $a, a' \in \Sigma'$ on $\Sigma$, one can construct two small loops $\gamma, \gamma'$ as in Section 4.1. Moreover, we choose the base points $t, t'$ for these loops at the same "distance" from $\Sigma'$ so that $f(t) = f(t')$. Then these loops are conjugate in the fundamental group by a path $\sigma$ connecting $t$ with $t'$; without loss of generality we may assume that the whole path $\sigma$ belongs to the level curve $f = \text{const}$.

The corresponding holonomy operators are conjugated by the holonomy operator $F_{tt'\sigma}$. The singular term $A\,df/f$ vanishes on $\sigma$, thus the operator $F_{tt'\sigma}$ has a uniform invertible limit $C = C(a, a') \in \mathrm{GL}(n, \mathbb{C})$ as $t \to a$, $t \to a'$, respecting the above assumption $f(t) = f(t')$. By Lemma 4.1, both $F_t$ and $F_{t'}$ have uniform limits $F_a$ and $F_{a'}$, and the operator $C$ conjugates them. It remains to notice that the matrix exponential map $\exp : A \mapsto \exp A$ is a covering of $\mathrm{GL}(n, \mathbb{C})$, so the two close conjugated exponentials correspond to two conjugated logarithms $A(a)$ and $A(a')$.

*Second proof.* The same assertion may be derived directly from the integrability condition. Consider the $(d, \wedge)$-closed subalgebra

$$\mathcal{V} = f\Lambda^\bullet(U) + df \wedge \Lambda^\bullet(U)$$

of holomorphic matrix $k$-forms vanishing after restriction on $\Sigma$.

Then the logarithmic connection form $\Omega$ can be written as

$$\Omega = A\frac{df}{f} + \Theta + f(\cdots) + df \wedge (\cdots) = A\frac{df}{f} + \Theta \mod \mathcal{V},$$

where $\Theta$ is a holomorphic matrix 1-form on the smooth part $\Sigma' \subseteq \Sigma$, and $A$ a holomorphic matrix function on $\Sigma'$.

The integrability condition $d\Omega = \Omega \wedge \Omega$ yields the identity

$$dA \wedge \frac{df}{f} + d\Theta = [\Theta, A] \wedge \frac{df}{f} + \Theta \wedge \Theta \mod \mathcal{V}$$

From this identity we derive, equating forms of different type (tangent to $\Sigma'$ and having a normal component $df$), the following two identities,

$$dA = [\Theta, A] \mod \mathcal{V}, \tag{4.1}$$

$$d\Theta = \Theta \wedge \Theta \mod \mathcal{V}. \tag{4.2}$$

Condition (4.2) means that the connection induced on $\Sigma'$ by restriction of the holomorphic matrix form $\Theta$, is integrable. Therefore there exists a holomorphic matrix solution $H : \Sigma' \to \mathrm{GL}(n, \mathbb{C})$ of the equation $dH = \Theta H$ restricted on $\Sigma'$, normalized by the condition $H(a) = E$. We claim that if $\Sigma'$ is connected, then the only solution of the equation (4.1) with the initial condition $A(a) = C$ is $A = HCH^{-1}$, where $C = A(a)$ is the *constant matrix*, the residue at the point $a \in \Sigma'$. This would imply that $A(\cdot)$ remains in its conjugacy class along connected components of $\Sigma'$, as asserted.

Direct verification shows that $A = HCH^{-1}$ is indeed a solution of (4.1):

$$\begin{aligned} dA &= dH \cdot CH^{-1} + HC \cdot d(H^{-1}) \\ &= dH \cdot H^{-1}HCH^{-1} - HC(H^{-1}\,dH \cdot H^{-1}) = \Theta A - A\Theta = [\Theta, A]. \end{aligned}$$

The initial condition $A(a) = C$ is satisfied since $H(a) = E$. □

Theorem 4.2 guarantees that for a logarithmic connection, all eigenvalues of the residue matrix (and hence all coefficients of its characteristic polynomial) are locally constant along the smooth part of the singular locus. This will allow us to make in a consistent way assumptions about resonances later in Section 6.

## 4.3 Residues on normal crossings

If $\Sigma = \{f_1 \cdots f_k = 0\}$ is a normal crossing of $k \leqslant m$ smooth hypersurfaces $\Sigma_i = \{f_i = 0\}$ and $\Omega \in \Lambda^1(\log \Sigma)$, then

$$\Omega = \sum_{i=1}^{k} A_i \frac{df_i}{f_i} + \text{holomorphic terms}, \tag{4.3}$$

where the residue $A_i$ associated with the $i$th component $\Sigma_i$ is a matrix function on $\Sigma_i$, holomorphic also on the intersections $\Sigma_i \cap \Sigma_j$, though not necessarily coinciding with the respective $A_j$ there (thus, strictly speaking, the residue is *not* a matrix function defined on the whole of $\Sigma$ but rather on its *normalization*).

**4.3 Theorem** *If $\Sigma = \bigcup \Sigma_i$ is a hypersurface with normal crossings, $\Omega \in \Lambda^1(\log \Sigma)$ and $A_i = \operatorname{res}_{\Sigma_i} \Omega$, then on $\Sigma_i \cap \Sigma_j$ the residues $A_i$ commute,*

$$[A_i, A_j] = 0 \quad \text{on} \quad \Sigma_i \cap \Sigma_j. \tag{4.4}$$

**Proof** Substituting the representation for $\Omega$ into the integrability condition (1.3), we obtain

$$\sum_s dA_s \wedge \frac{df_s}{f_s} + \cdots = \sum_{i,j} A_i A_j \frac{df_i \wedge df_j}{f_i f_j} + \text{poles of first order.}$$

Collecting the principal polar (non-holomorphic) terms of order 2, i.e., multiplying both parts of $f_i f_j$ and restricting to $(\Sigma_i \cap \Sigma_j) \setminus \bigcup_{k \neq i,j} \Sigma_k$, we see that

$$0 = A_i A_j \, df_i \wedge df_j + A_j A_i \, df_j \wedge df_i = [A_i, A_j] \, df_j \wedge df_i.$$

Since $df_i$ and $df_j$ are linearly independent, $[A_i, A_j] = 0$ on $\Sigma_i \cap \Sigma_j$ outside $\Sigma_0 = \bigcup_{k \neq i,j} \Sigma_k$. Since the "bad locus" $\Sigma_i \cap \Sigma_j \cap \Sigma_0$ is thin and $A_i, A_j$ are holomorphic on $\Sigma_i$ and $\Sigma_j$ respectively, $A_i$ and $A_j$ commute everywhere on $\Sigma_i \cap \Sigma_j$. □

One can attempt to derive this result from the commutativity of the holonomy operators corresponding to small loops around $\Sigma_i$ and $\Sigma_j$ as in the first proof of Theorem 4.2. However, this commutativity proves only that the exponentials of the residues $\exp 2\pi i A_i$ and $\exp 2\pi i A_j$ commute on the intersection $\Sigma_i \cap \Sigma_j$. However, this implies commutativity of the residues themselves only when the latter are nonresonant (no two eigenvalues of either $A_i$ or $A_j$ differ by an integer), see e.g., [Sch99]. The direct computation presented above, works in all cases.

The arguments proving Theorem 4.2 can almost literally be applied to the form (4.3), proving that each residue matrix $A_j$ holomorphic on the respective component $\Sigma_j$, remains within the same conjugacy class. Moreover, the conjugacy matrix function $H : \Sigma \to \mathrm{GL}(n, \mathbb{C})$ defined as the solution of the system $dH = \Theta H$, integrable on each $\Sigma_j$, after extension from $\Sigma$ to $U$ conjugates the form (4.3) with the form with constant matrices $A_j$, normalizing the principal part of $\Omega$ as follows.

**4.4 Theorem** *A flat connection with logarithmic poles on a normal crossing $\Sigma = \Sigma_1 \cup \cdots \cup \Sigma_k$, $k \leqslant m$, is holomorphically gauge equivalent to a system* (4.3) *with the constant residues $A_j|_{\Sigma_j} = \mathrm{const}_j \in \mathrm{Mat}_n(\mathbb{C})$ pairwise commuting with each other.* □

If $\Sigma$ is not a normal crossing, then the commutativity of the residues is no longer valid.

**4.5 Example** Let $l_1, \dots, l_k$, $k \geqslant 3$, be any number of pairwise different *linear forms* on the plane $\mathbb{C}^2 = \{(t_1, t_2)\}$, and $A_1, \dots, A_k$ any collection of constant matrices. Consider the matrix 1-form $\Omega = \sum_1^k A_j \frac{dl_j}{l_j}$. Assume that none of the factors $l_j$ is proportional to $t_2$, so that without loss of generality we may assume

$$l_j = t_1 - \lambda_j t_2, \quad \lambda_j \in \mathbb{C}, \ j = 1, \dots, k.$$

Using the function $z(t) = t_2/t_1$, and the identity $l_j = t_1(z - \lambda_j)$, so that $dl_j/l_j = dt_1/t_1 + d(z - \lambda_j)/(z - \lambda_j)$, the matrix $\Omega$ can be written

$$\Omega = B\frac{dt_1}{t_1} + \sum_{j=1}^{k} A_j \frac{dz}{z - \lambda_j}, \qquad B = \sum_{j=1}^{k} A_j.$$

This transformation is in fact a blow-up considered in more detail in Section 7.1.

If the matrix $B = \sum A_j$ commutes with each of the matrices $A_j$, the form $\Omega$ is flat. Indeed, $d\Omega = 0$ since all $A_j$ are constant, whereas $\Omega \wedge \Omega$ reduces to a sum of commutators involving $B$ only,

$$\Omega \wedge \Omega = \sum_j [B, A_j] \frac{dt_1 \wedge dz}{t_1(z - \lambda_j)}$$

(pairwise commutators $[A_i, A_j]$ disappear since $dz \wedge dz = 0$). Therefore both sides of (1.3) are zeros.

Clearly, when $k > 2$, one can construct an example of a flat connection with non-commuting residues: for example, it is sufficient to take an arbitrary collection of matrices $A_1, \dots, A_{k-1}$ and put $A_k = -(A_1 + \cdots + A_{k-1})$ to ensure that $B = 0$. For $k = 2$ this is impossible: $0 = [A_1, B] = [A_1, A_1 + A_2] = [A_1, A_2]$.

**4.6 Remark** Note that for $k > 2$ the fundamental group of $U \setminus \Sigma$ is indeed noncommutative. Indeed, the projectivization map $\mathbb{C}^2 \setminus \{0\} \to \mathbb{C}P^1$ restricted on $U \setminus \Sigma$ is a topological bundle over $K = \mathbb{C}P^1 \setminus \{\lambda_1, \dots, \lambda_k\}$ with the fiber $\mathbb{C} \setminus \{0\}$ which is homotopically equivalent to the circle $\mathbb{S}^1$. The exact homotopy sequence

$$0 = \pi_2(K) \to \pi_1(\mathbb{S}^1) \to \pi_1(U \setminus \Sigma) \to \pi_1(K) \to \pi_0(\mathbb{S}^1) = 0$$

implies that the factor of $\pi_1(U \setminus \Sigma)$ by $\mathbb{Z} = \pi_1(\mathbb{S}^1)$ is $\pi_1(K)$ which is isomorphic to the free group with $k - 1$ generators.

## 4.4 Schlesinger equations

Assume that $U = \mathbb{C}^1 \times \mathbb{C}^k = \{(z, \lambda_1, \dots, \lambda_k)\}$ is the affine space equipped with the natural projection $p : (z, \lambda) \mapsto \lambda$, and let $\Sigma = \bigcup_{j=1}^k \{z - \lambda_j = 0\}$ be the union of bisector hyperplanes, all of them *transversal* to the fibers $\{\lambda = \mathrm{const}\}$ of this projection. A flat connection $\Omega$ with poles on $\Sigma$ can be identified, according to Section 2.6, with an isomonodromic deformation of a linear "ordinary" system with singularities at the points $\lambda_1, \dots, \lambda_k \in \mathbb{C}$, parametrized by the location of these points.

We want to construct a flat meromorphic connection with poles on $\Sigma$: any such connection will automatically correspond to an isomonodromic deformation, as explained in Section 2.6. The extra requirement is that after restriction on each fiber $\lambda = \text{const}$, the connection would have only simple (Fuchsian) poles at all finite points $\lambda_1, \dots, \lambda_k$ *and* at the point $z = \infty$ after the compactification.

We start with the formal expression

$$\Omega = \sum_1^k A_j \frac{df_j}{f_j}, \qquad f_j(z, \lambda) = z - \lambda_j. \tag{4.5}$$

with holomorphic matrix functions $A_j(z, \lambda)$. However, if the restriction of $\Omega$ on each fiber $\lambda = \text{const}$ is to have a Fuchsian singularity after compactification at $z = \infty$, then necessarily $A_j$ must be *constant* along this fiber. Actually, to *avoid* appearance of singularities at infinity, we will assume that the residue $\operatorname{res}_{z=\infty} \Omega|_{\lambda=\text{const}} = -\sum A_j(\lambda)$ vanishes identically in $\lambda$,

$$A_j = A_j(\lambda), \quad j = 1, \dots, k, \qquad \sum_1^k A_j(\lambda) \equiv 0. \tag{4.6}$$

We claim that there exists at least one flat connection of the form specified by (4.5)–(4.6). By construction, this connection has logarithmic singularities on the hyperplane $\Sigma_j$. However, it will in general have additional singularities on a union of hyperplanes *parallel* to the $z$-direction.

**4.7 Theorem** *The connection* (4.5)–(4.6) *is flat if and only if the matrix functions* $A_1(\lambda), \dots, A_k(\lambda)$ *satisfy the system of quadratic partial differential equations*

$$dA_s = -\sum_{j \neq s} [A_s, A_j] \frac{d\lambda_s - d\lambda_j}{\lambda_s - \lambda_j}, \quad \forall s = 1, \dots, k. \tag{4.7}$$

**Proof** Indeed, the integrability condition (1.3) for the system (4.5) implies the identity between matrix 2-forms

$$\sum_s dA_s \wedge \frac{df_s}{f_s} = \sum_{i,j} A_i A_j \frac{df_i \wedge df_j}{f_i f_j} = \sum_{i<j} [A_i, A_j] \frac{df_i \wedge df_j}{f_i f_j}. \tag{4.8}$$

Let $v = \partial/\partial z$ be the unit vector field on $U$ tangent to the vertical fibers $\lambda = \text{const}$, and $i_v$ the corresponding antiderivation (substitution of $v$ as the first argument in a differential form, scalar or matrix). Our assumptions on $A_j$ and $f_j$ imply that

$$i_v dA_j = 0, \qquad i_v df_j = 1, \qquad j = 1, \dots, k,$$

so that

$$i_v(dA_s \wedge df_s) = (i_v\, dA_s) df_s - dA_s (i_v\, df_s) = -dA_s, \qquad i_v(df_i \wedge df_j) = df_i - df_j.$$

Applying $i_v$ to both parts of (4.8) and using these identities, we arrive at the following identity between (matrix) 1-forms:

$$-\sum_s \frac{dA_s}{f_s} = \sum_{i,j} A_i A_j \frac{df_i - df_j}{f_i f_j}.$$

Equating the principal polar parts of both sides on each of the hyperplanes $\Sigma_s$ (more precisely, on $\Sigma_s \setminus \bigcup_{j\neq s} \Sigma_j$), we conclude that after restriction to each hyperplane $\Sigma_s$ the following identities are satisfied,

$$-dA_s = \sum_j A_s A_j \frac{df_s - df_j}{f_j} + \sum_i A_i A_s \frac{df_i - df_s}{f_i} = \sum_j [A_s, A_j] \frac{df_s - df_j}{f_j}.$$

It remains to note that $df_s - df_j = d\lambda_s - d\lambda_j$ and that the restriction of $f_j = z - \lambda_j$ to $\Sigma_s = \{z - \lambda_s = 0\}$ is equal to $\lambda_s - \lambda_j$. This shows the necessity of the condition (4.7).

To prove sufficiency of condition (4.7), we rewrite it in the form

$$dA_s = -\sum_{j\neq s} [A_s, A_j] \frac{df_j - df_s}{f_j - f_s},$$

wedge multiply each equation by $df_s/f_s$ from the right and add the results together. Then the left hand side of the sum coincides with the left hand side of the equality (4.8) which is the flatness condition for $\Omega$. It remains to show that the right hand sides also coincide, i.e., that after obvious transformations and renaming the summation variable in (4.8),

$$\sum_{s\neq j} [A_s, A_j] \frac{df_s \wedge df_j}{(f_j - f_s) f_s} = \sum_{s<j} [A_s, A_j] \frac{df_s \wedge df_j}{f_i f_j}. \tag{4.9}$$

Note that the sum on the left involves unordered pairs $(s, j)$, while the sum on the right is extended only on the ordered pairs. After grouping the terms $(s, j)$ and $(j, s)$ together, the identity (4.9) follows from the identity

$$\frac{1}{(f_j - f_s) f_s} + \frac{1}{(f_s - f_j) f_j} = \frac{1}{f_s f_j}.$$

Thus (4.7) implies (4.8) as asserted. □

A computation similar to that proving Theorem 1.1, shows that the system (4.7), known as the *Schlesinger equations*, is also *integrable* so that *locally* (off the *collision locus* $C = \bigcup_{i\neq j} \{\lambda_i - \lambda_j = 0\} \subset \mathbb{C}^k$) it admits solutions $(\lambda_1, \dots, \lambda_k) \mapsto (A_1(\lambda), \dots, A_k(\lambda))$. However, since this system is *nonlinear* (quadratic), its solutions in general explode. The *movable poles* of solutions depend on the initial condition. In [Bol00b] it is proved that if the initial configuration of the singular points and residue matrices correspond to a Fuchsian equation with an irreducible monodromy, then the singularities of solutions of the Schlesinger equation off the collision locus occur on an analytic hypersurface $M \subset \mathbb{C}^m$ and have second order poles on $M$.

Thus the attempt to construct a flat connection on $\mathbb{C}P^1 \times \mathbb{C}^m$ with logarithmic singularities on the hyperplanes $\Sigma_j$, implicitly encoded in the representation (4.5), leads to the creation of an additional singular locus $C \cup M$ on which it is in general non-logarithmic.

## 4.5 First order non-logarithmic poles and isomonodromic deformations of Fuchsian systems

The gap between flat *logarithmic* connections and flat meromorphic connections exhibiting a *first order pole* on a smooth analytic hypersurface $\Sigma$, is not very wide and disappears in the *non-resonant* case (see below).

Indeed, assuming that the hypersurface $\Sigma$ in suitable local coordinates $z, \lambda$ has the form $z = 0$, any form with a first order pole on it, can be written as

$$\Omega = \frac{A(\lambda)\,dz + \Theta}{z} + \cdots, \qquad \Theta = \sum_{1}^{m-1} B_i(\lambda)\,d\lambda_i, \tag{4.10}$$

the dots as usual denoting holomorphic terms. For every point $\lambda$ the operator $A(\lambda)$ can be described as the residue of the restriction of $\Omega$ on the "normal" curve $Z = \{\lambda = \text{const}\}$ to $\Sigma$. In absence of the additional structure of "isomonodromic deformation" this normalcy makes no particular sense, yet for any other curve $Z'$ transversal to $\Sigma$ at the same point, the residue of the restriction will be conjugate to $A(\lambda)$ (cf. with the first proof of Theorem 4.2; recall that $\Omega$ is assumed to be flat). Thus the spectrum (and more generally the characteristic polynomial) of the matrix $A(\lambda)$ make an invariant sense and are locally constant along the smooth connected components of $\Sigma$.

**4.8 Theorem** *If no two eigenvalues of the matrix* $A|_\Sigma$ *of a flat connection* (4.10) *differ by* 1, *then necessarily* $\Theta = 0$ *and the connection is in fact logarithmic.*

**Proof** Keeping polar terms of second order from the integrability condition for (4.10) yields

$$-\frac{dz}{z^2} \wedge \Theta = -\frac{1}{z^2}(A\,dz \wedge \Theta + \Theta \wedge A\,dz) + \cdots,$$

where the dots denote the terms that have a first order pole (or are holomorphic) on $\Sigma$, so that

$$\Theta = A\Theta - \Theta A = [A, \Theta].$$

For a diagonal $A = \operatorname{diag}\{\alpha_i\}$ this means the identity $\theta_{ij} = \theta_{ij}(\alpha_i - \alpha_j)$ for the matrix elements $\theta_{ij}$ of $\Theta$. If none of the differences $\alpha_i - \alpha_j$ is equal to 1, then this is possible only when all $\theta_{ij}$ vanish so that $\Theta = 0$.

In the general case when $A$ may have coinciding eigenvalues and be non-diagonalizable, we use the well-known fact [Lan69]: *the matrix equation of the form* $AX - XA' = B$ *has a unique solution for any square matrix* $B$ *if and only if* $A$ *and* $A'$ *have no common eigenvalues.* The equation $\Theta = [A, \Theta]$ reduces to the equation $(A - E)\Theta - \Theta A = 0$ which has only the trivial solution, since adding the identity matrix $E$ to $A$ shifts all eigenvalues by 1. □

The same argument can be applied to the meromorphic matrix form

$$\Omega = A(\lambda)\frac{dz}{z} + \frac{1}{z^r}(\Theta + \cdots) \tag{4.11}$$

having a pole of order $r \geqslant 2$. In order for this assumption to make sense, we have to assume the structure of isomonodromic deformation (i.e., the bundle over $\lambda$-space making the $z$-direction exceptional, see Section 2.6). Then (4.11) means that the singular point at $z = 0$

remains Fuchsian for all values $\lambda$ of the parameters of the deformation. The residue $A(\lambda)$ in this case becomes unambiguously defined.

Exactly the same computation as above leads to the equation $r\Theta = [A, \Theta]$, $2 \leqslant r \in \mathbb{Z}$ and proves the following claim.

**4.9 Theorem** *If no two eigenvalues of the residue matrix $A(\lambda)$ differ by a nonzero integer, then any Fuchsian isomonodromic deformation (4.11) is in fact logarithmic.* □

The condition that no two eigenvalues of the residue differ by a nonzero integer, is very important. It will be referred to as the *non-resonance condition*.

It is the assertion of Theorem 4.9 that explains the a priori choice (4.5) of the connection form when discussing isomonodromic deformations of Fuchsian systems in Section 4.4. The connection form (4.5) has logarithmic singularities on each hyperplane $\{z - \lambda_j = 0\}$, at least off the diagonal (on the open set $\{\lambda_i \neq \lambda_j,\ i \neq j\} \subseteq \mathbb{C}P^1 \times \mathbb{C}^k$).

Indeed, if all residues of a Fuchsian system on $\mathbb{C}P^1$ are non-resonant, then the connection (4.5)–(4.6) satisfying the Schlesinger equations (4.7), is essentially the only possible isomonodromic deformation of this system. In the resonant case one may well have poles of higher order. This order is nevertheless bounded by the maximal integer difference $\max_{i,j} |\alpha_i - \alpha_j|$ between the eigenvalues. The complete proof of this result and the description of isomonodromic deformations can be found in [Bol97].

# 5 Abelian integrals and Picard-Fuchs systems

One of the most important examples of flat meromorphic connections exhibiting only logarithmic poles, is the *Gauss-Manin connection* described by the *Picard-Fuchs system* of differential equations involving *Abelian integrals* $\int_\Gamma \sigma$ of polynomial 1-forms along 1-cycles on nonsingular algebraic curves $\Gamma = \{H = 0\} \subset \mathbb{C}^2$, considered as functions of the parameters (coefficients of the form $\sigma$ and the polynomial equation $H = 0$ of the curve).

## 5.1 Definitions

Let $P = P(w_1, w_2) \in \mathbb{C}[w] = \mathbb{C}[w_1, w_2]$ be a homogeneous complex polynomial of degree $r = \deg P$ in two complex variables which has an *isolated critical point* at the origin $w = 0$. For this, it is necessary and sufficient to require that $P$ be *square-free*, i.e., has no repeating linear factors in the complete factorization.

Consider a *general bivariate polynomial* with the principal part $P$,

$$H(w, t) = P(w) + \sum_{\deg w^{\mathbf{a}} < r} t_{\mathbf{a}} w^{\mathbf{a}}, \tag{5.1}$$

where by $t_{\mathbf{a}}$, $\mathbf{a} \in \mathbb{Z}^2_+$, are denoted the complex coefficients before the lower degree monomials $w^{\mathbf{a}} = w_1^{a_1} w_2^{a_2}$.

The list of these parameters can be flattened to simplify our notation: if $m$ is the dimension of the parameter space (which can be easily computed knowing the degree $r$), then we write $t = (t_1, \dots, t_m)$ and use the notation $H_t = H(\cdot, t)$ for $t \in \mathbb{C}^m$ to denote the bivariate polynomial with the corresponding non-principal part. The coefficients will always be ordered so that $t_1$ is the *free term* of $H$.

For most of the values of $t \in \mathbb{C}^m$ the zero level curve

$$\Gamma_t = \{H_t = 0\} \subset \mathbb{C}^2, \quad t \in \mathbb{C}^m, \tag{5.2}$$

is a non-singular (smooth) algebraic curve. The exceptional values belong to the *discriminant*, the algebraic hypersurface

$$\begin{aligned} \Sigma &= \{t \in \mathbb{C}^m : 0 \text{ is a critical value for } H_t\} \\ &= \{t \in \mathbb{C}^m : \exists w_* \in \mathbb{C}^2 \text{ such that } dH_t(w_*) = 0,\ H_t(w_*) = 0\}. \end{aligned} \tag{5.3}$$

It can easily be verified, see e.g. [Yak01], that the assumption on the principal part $P$ ensures the local *topological triviality* of the family of the curves $\{\Gamma_t\}$ for $t \notin \Sigma$. This means that the curves $\Gamma_t$ vary continuously with $t$ outside $\Sigma$ so that any 1-cycle $\delta$ (represented by a closed curve) on a nonsingular affine curve $\Gamma_t$, $t \notin \Sigma$, can be locally continued as a uniquely defined element $\delta(t)$ in the first homology group of $\Gamma_t$. In other words, a *flat connection* on the *homology bundle* is defined.

**5.1 Remark** If the principal part $P$ is not square-free (has one or several lines entirely consisting of critical points), the affine curves $\Gamma_t$ may change their topological type for some values $t \notin \Sigma$. The reason is that after projective compactification, $\Gamma_t$ may be non-transversal to the infinite line and thus undergo "bifurcations at infinity". These exceptional *atypical values* form an algebraic subvariety in $\mathbb{C}^m$.

The constructed connection, albeit flat, still has a nontrivial *monodromy.* If $\gamma : [0,1] \to \mathbb{C}^m$ is a closed path in the $t$-space, avoiding the discriminant locus $\Sigma$, then the result of the continuation of $\delta(t)$ along $\gamma$ may well be nontrivial, $\delta(\gamma(0)) \neq \delta(\gamma(1))$. It will be described in more detail later.

Any polynomial 1-form $\sigma \in \Lambda^1(\mathbb{C}^2)$ on $\mathbb{C}^2$ can be restricted on the curve $\Gamma_t$, $t \notin \Sigma$, and integrated along any cycle on this curve. A (complete) *Abelian integral* is by definition the multivalued analytic function on $\mathbb{C}^m \setminus \Sigma$ obtained by integration of $\sigma$ along a continuous (horizontal) section $\delta(t)$ of the homological bundle.

Denote by $n$ the rank of the first homology group of a generic level curve $\Gamma_t$: this means that there can be found $n$ cycles $\delta_1, \dots, \delta_1$ generating all other cycles over $\mathbb{Z}$. Again by topological triviality, the result of continuation $(\delta_1(t), \dots, \delta_n(t))$ will be a *framing* of the (first homology groups of) level curves $\{\Gamma_t\}$. A collection of polynomial 1-forms $\sigma_1, \dots, \sigma_n$ is called a *coframe*, if the *period matrix*

$$X(t) = \begin{pmatrix} \oint\limits_{\delta_1(t)} \sigma_1 & \cdots & \oint\limits_{\delta_n(t)} \sigma_1 \\ \vdots & \ddots & \vdots \\ \oint\limits_{\delta_1(t)} \sigma_n & \cdots & \oint\limits_{\delta_n(t)} \sigma_n \end{pmatrix} \tag{5.4}$$

is not identically degenerate, $\det X(t) \not\equiv 0$. This definition in fact does not depend on the choice of the frame $\{\delta_i(t)\}_1^n$.

It is worth mentioning that in many sources the Abelian integrals are considered as analytic functions of the variable $t_1 \in \mathbb{C}^1$ only (the free term), being defined as integrals $\int_{H(w)=t_1} \sigma$ for a given *fixed* bivariate polynomial $H(w_1, w_2)$. By this definition, the integral is ramified over the set of complex critical values of $H$ (and only over this set under the assumption on the principal homogeneous part of $H$). An abundant wealth of information has been accumulated about the behavior of Abelian integrals defined in this way; some of it will be used to derive multivariate properties.

## 5.2 Ramification of Abelian integrals and Picard-Lefschetz formulas

The part of the discriminant locus $\Sigma$ corresponding to polynomials $H(w,t) = P(w)+\sum t_{\mathbf{a}}w^{\mathbf{a}}$ which have only one nondegenerate (Morse) critical point on the zero level curve $\Gamma_t = \{H_t = 0\}$, is smooth. If $t_* \in \Sigma$ is such a point, then for all sufficiently close $t \in \mathbb{C}^m$ the polynomial $H_t$ has a unique critical value $f(t)$ which is close to zero and holomorphically depends on $t$. The equation $\{f = 0\} \subset \mathbb{C}^m$ locally defines $\Sigma$ near the smooth point $t_*$.

All these assertions are easy to verify. Denote by $w_* \in \mathbb{C}^2$ the critical point of $H_* = H(\cdot, t_*)$. For nearby values of $t$ the critical point $w_t \in \mathbb{C}^2$ of $H_t(\cdot,\cdot)$ is determined by the two equations $\frac{\partial H}{\partial w_i}(w_t) = 0$, $i = 1,2$, by virtue of the nondegeneracy condition

$$\det \frac{\partial^2 H}{\partial w_i \partial w_j}(w_*, t_*) \neq 0.$$

By the implicit function theorem, the point $w_t$ analytically depends on $t$, and therefore so also does the value $f(t) = H(w_t, t)$. To see that $df(t_*) \neq 0$, notice that

$$\frac{\partial f}{\partial t_1} = \sum_{i=1}^{2} \frac{\partial H}{\partial w_i} \cdot \frac{\partial w_i}{\partial t_1}(w_t, t) + \frac{\partial H}{\partial t_1}(w_t, t) = \frac{\partial H}{\partial t_1}(w_t, t) = 1 \neq 0.$$

This computation shows that projection of $\Sigma$ along $t_1$ is smooth at smooth points of the former and in part explains the exceptional role played by the coefficient $t_1$.

To describe branching of the period matrix near smooth points of $\Sigma$, a special frame of cycles has to be chosen. If $t_* \in \Sigma$ is a smooth point on the discriminant and $w_* = (w_{*1}, w_{*2}) \in \mathbb{C}^2$ the corresponding critical point of $H_* = H(\cdot, t_*)$, then by the Morse lemma one can introduce local coordinates on an open set $U \subset \mathbb{C}^2$ near $w_*$ so that

$$H(w,t) = (w_1 - w_{*1})^2 + (w_2 - w_{*2})^2 + f(t), \quad w \in U,\ t \text{ near } t_*.$$

There is a unique (modulo orientation and homotopy equivalence) simple loop on the "bottle-neck", the local level curve $\{H_t = 0\}$ (topologically a cylinder if $f(t) \neq 0$) that is not contractible and shrinks as $f(t) \to 0$. This loop is called the *vanishing cycle* $\delta_1(t)$, and by its construction the integral $\oint_{\delta_1(t)} \sigma$ of *any* polynomial 1-form $\sigma \in \Lambda^1(\mathbb{C}^2)$ is single-valued and bounded near $t_*$. Being thus holomorphic on $\Sigma$, this integral in fact vanishes on $\Sigma$, whence comes the term "vanishing".

Other cycles on $\Gamma_t$, not reducible to a multiple of $\delta_1$, can be ramified over $\Sigma$. To describe ramification, recall that on all nonsingular level curves $\Gamma_t$, $t \notin \Sigma$, the *intersection index* between cycles is well defined. This intersection index, an integer number, is well-defined and locally constant as $t$ varies outside $\Sigma$. It turns out, see [AGV88], that after $t$ goes around $\Sigma$ near a smooth point of the latter, any other cycle $\delta(t)$ is transformed by adding an integral multiple of the vanishing cycle $\delta_1$, the multiplier being the intersection index $(\delta, \delta_1) \in \mathbb{Z}$ between $\delta$ and $\delta_1$:

$$\delta \mapsto \delta + (\delta, \delta_1)\delta_1. \tag{5.5}$$

This formula is known by many names, from *Dehn twist* to (most frequently) *Picard-Lefschetz formula*. Its appearance is rather natural: indeed, on the level of homology the impact of the parallel transport must be an additive contribution supported by a small neighborhood of the critical point only (since far away from $w_*$ the curves $\Gamma_t$ must form a topologically trivial family). On the other hand, it was already remarked that the only nontrivial cycle(s) living in a small neighborhood of a Morse critical point, are (multiples) of the vanishing cycle. The detailed proof can be found in [AGV88].

## 5.3 Nondegeneracy of the period matrix

The Picard-Lefschetz formula has numerous corollaries. First, the determinant of the period matrix $\det X(t)$ is a holomorphic function on $\mathbb{C}^m$ vanishing on the hypersurface $\Sigma$. Indeed, after going around any smooth component of $\Sigma$, independently of the choice of the coframe forms $\sigma_i$, the first column of $X$ remains intact whereas any other column is transformed by adding an integer multiple of the first column. It remains to notice that all integrals in the first column vanish on $\Sigma$, while all other entries are bounded as $t$ tends to $\Sigma$ along non-spiraling domains.

This description means that if the frame $\delta(t)$ of the homology bundle is chosen so that the first cycle $\delta_1(t)$ is vanishing, then locally near each smooth point $a \in \Sigma$, the period matrix can be represented as

$$X(t) = V(t)Y(t), \qquad Y(t) = \begin{pmatrix} f & c_2 f \ln f & \cdots & c_n f \ln f \\ & 1 & & \\ & & \ddots & \\ & & & 1 \end{pmatrix}, \tag{5.6}$$

where $f(t) = 0$ is the local equation for $\Sigma$, the complex constants $c_2, \dots, c_n \in \mathbb{C}$ are proportional to the respective intersection indices, $c_j = (\delta_j, \delta_1)/2\pi i$, $j = 2, \dots, n$, and $V(t)$ is a holomorphic matrix function near $a \in \Sigma$ (depending, in general, on the point $a$).

For an arbitrary choice of the coframe $\{\sigma_i\}_{i=1}^n$, the matrix function $V(t)$ may be degenerate on some analytic hypersurfaces possibly intersecting $\Sigma$. However, in the particular case when the forms are chosen of minimal possible degrees compatible with the nondegeneracy of $X$, the matrix function $V(t)$ is holomorphically invertible. More precisely, the "proper" choice of the coframe can be described as follows.

Recall that $r = \deg H = \deg P$ denotes the degree of the homogeneous polynomial $P$, the principal homogeneous part of $H(\cdot, t)$. In the case when the weights of both $w_1$ and $w_2$ are equal to 1, one can compute the rank $n$ of the (co)homology of the affine curves $\Gamma_t$:

$$n = (r-1)^2.$$

Not accidentally, this number coincides with the number of critical points of $H_t$ counted with multiplicity, which in turn is equal to the (algebraic) dimension of the quotient algebras

$$n = \dim_{\mathbb{C}} \Lambda^2(\mathbb{C}^2)/dP \wedge \Lambda^1(\mathbb{C}^2) = \dim_{\mathbb{C}} \Lambda^2(\mathbb{C}^2)/dH_t \wedge \Lambda^1(\mathbb{C}^2). \tag{5.7}$$

The following result concerning Abelian integrals as functions of one complex variable, belongs to the folklore. With various degrees of generality, completeness and precision it appeared in many sources, among them the recent paper [Gav98]. Other sources, together with complete proofs, can be found in [Nov02].

**5.2 Lemma ([Nov02] and references therein)** *Assume that the monomial* 1*-forms* $\sigma_1, \dots, \sigma_n \in \Lambda^1(\mathbb{C}^2)$ *satisfy the condition*

$$\sum_1^n \deg \sigma_i = n \deg H. \tag{5.8}$$

*Then the determinant of the corresponding period matrix* $X$ (5.4) *as a function of* $t_1$ *is a polynomial of degree no greater than* $n$.

*This polynomial is not identically zero if the differentials* $d\sigma_1, \dots, d\sigma_n$ *are linearly independent modulo the ideal* $dP \wedge \Lambda^1(\mathbb{C}^2)$, *i.e., if no nontrivial linear combination of* $d\sigma_i$ *is divisible by the* 1*-form* $dP$. □

Any collection of monomial 1-forms meeting the condition (5.8) and whose differentials $d\sigma_i$ are linearly independent modulo $dP \wedge \Lambda^1(\mathbb{C}^2)$, will be referred to as the *basic coframe*.

**5.3 Remark** The 1-forms constituting a basic coframe can always be chosen *monomial.* Moreover, the (monomial) primitives of $(r-1)^2$ monomial 2-forms $w_1^{a_1} w_2^{a_2}\, dw_1 \wedge dw_2$ with $0 \leqslant a_{1,2} \leqslant r-2$, constitute a basic coframe for *almost all* principal homogeneous polynomials of the same degree $r$, though certainly not for all.

Application of Lemma 5.2 immediately shows that if the forms $\sigma_i$ constitute a basic coframe, then in the local representation (5.6) of the period matrix (5.4), the matrix term $V(t) = X(t)Y^{-1}(t)$, $Y = Y_f(t)$ as in (5.6), must be holomorphically invertible on $\Sigma$ for any choice of the local equation $f$. Indeed, almost any complex 1-dimensional line in $\mathbb{C}^m$ parallel to the $t_1$-direction, intersects $\Sigma$ in exactly $n = (r-1)^2$ smooth points corresponding to $n$ (in general, distinct) critical values of the polynomial $H_t$. The determinant $\det X$ restricted on this line, necessarily has roots at the points of this intersection. These roots are simple if and only if $\det V \neq 0$ at these points, since $\det Y(t) = f(t)$ already has a simple root there. On the other hand, the number of roots cannot exceed the degree of $\det X$ in $t_1$, since the latter is not identically zero. Thus the number of roots of $\det X(t_1, \text{const})$ is exactly $n$ and all of them are simple. Therefore $\det V(t)$ is nonvanishing at almost all (except for a thin set of complex codimension 2 or more) smooth points of $\Sigma$.

## 5.4 Global equation of $\Sigma$

If the monomial forms $\sigma_i$ constitute a basic coframe, then any polynomial 2-form on $\mathbb{C}^2$ can be divided by $dH_t$ with a remainder that is a linear combination of $d\sigma_i$. In particular, this applies to the forms $H_t\, d\sigma_i$:

$$H_t\, d\sigma_i = \eta_i \wedge dH_t + \sum_{j=1}^n R_{ij}\, d\sigma_j, \quad \eta_i \in \Lambda^1(\mathbb{C}^2), \ R_{ij} \in \mathbb{C}, \tag{5.9}$$

where $t \in \mathbb{C}^m$ is considered as a parameter. The process of division is analyzed in detail in [Yak02] where it is shown, in particular, that the result (both the incomplete ratios $\eta_i$ and the coefficients of the remainders) depends *polynomially* on $t$, giving rise to a matrix polynomial $R(t) = \|R_{ij}(t)\|$. This is a consequence of the fact that $H_t$ depends on $t$ polynomially while its principal homogeneous part does not depend on $t$ at all.

**5.4 Lemma** *The singular locus* $\Sigma \subset \mathbb{C}^m$ *is given by one polynomial equation* $\det R(t) = 0$.

**Proof** The identity (5.9) evaluated at a critical point $w_t \in \mathbb{C}^2$ of $H_t$ with the critical value $0 = H_t(w_t) = H(w_t, t)$, means that the nonzero vector of coefficients of the forms $d\sigma_i$ at this point belongs to the null space of the matrix $R(t)$, since both $dH_t$ and $H_t$ vanish at $w_t$. □

In fact, if $\{\sigma_i\}_{i=1}^n$ is a basic coframe, then the determinant of the corresponding period matrix $X(t)$ is proportional to $\det R(t)$:

$$\det X(t) = \text{const} \cdot \det R(t),$$

the constant depending on the principal part $P$ of the polynomial $H$ and on the coframe. This constant was computed in [Glu00] for a special choice of the basic coframe.

## 5.5 Finite singularities of the Abelian integrals are logarithmic

**5.5 Theorem** *Assume that* $H(w,t) = P(w) + \sum t_{\mathbf{a}} w^{\mathbf{a}}$, $t \in \mathbb{C}^m$, *is a general polynomial with the square-free principal homogeneous part* $P$ *of degree* $r$, *and that* $\sigma_1, \dots, \sigma_n$, $n = (r-1)^2$, *are monomial forms constituting a basic coframe as described in Lemma* 5.2.

*Then the period matrix* $X(t)$ *satisfies an integrable Pfaffian system* (1.2) *with the matrix* $\Omega = dX \cdot X^{-1}$ *having logarithmic poles on the algebraic hypersurface* $\Sigma \subset \mathbb{C}^m$. *The residue matrix* $A = \operatorname{res}_\Sigma \Omega$ *is conjugate to* $\operatorname{diag}\{1, 0, \dots, 0\}$.

**Proof** It is sufficient to show that the "logarithmic derivative" $dX \cdot X^{-1}(t)$ has a logarithmic pole near any smooth point of $\Sigma = \{f = 0\}$, $f = \det R$, where $R$ is the matrix polynomial described in Lemma 5.4.

By (5.6), locally near $\Sigma$ there exists a representation $X = VY$, where $V$ is holomorphically invertible and $Y$ explicitly given by

$$Y(t) = \operatorname{diag}\{f, 1, \dots, 1\} \cdot \begin{pmatrix} 1 & c_2 \ln f & \cdots & c_n \ln f \\ & 1 & & \\ & & \ddots & \\ & & & 1 \end{pmatrix}.$$

Since $V$ is holomorphically invertible, $\Omega$ is holomorphically gauge equivalent to

$$\Omega' = dY \cdot Y^{-1} = \begin{pmatrix} df/f & c_2\, df & \cdots & c_n\, df \\ & 0 & & \\ & & \ddots & \\ & & & 0 \end{pmatrix}$$

which obviously has a logarithmic pole on $\{f = 0\}$: $f\Omega'$ is clearly holomorphic on $\Sigma$, while $d\Omega' = \Omega' \wedge \Omega' = 0$. As $\Omega$ and $\Omega'$ are holomorphically gauge equivalent near $\Sigma$, the form $\Omega$ is also logarithmic on smooth parts of $\{f = 0\}$. Since both $\Omega$ and $f$ are globally defined, the standard arguments involving thin sets of codimension $\geqslant 2$ in $\mathbb{C}^m$ then imply that $\Omega$ has a logarithmic singularity on the whole of $\Sigma \subset \mathbb{C}^m$. □

It is worth a remark that for a different choice of the coframe the period matrix $X'(t)$ differs from the period matrix for a basic coframe $X(t)$ by a *polynomial* matrix factor,

$X' = C(t)X(t)$, $C \in \mathrm{GL}(n, \mathbb{C}[t])$. This factor is in general non-invertible along an algebraic hypersurface $\Sigma'$ different from $\Sigma$ (and depending on the choice of the coframe), and the corresponding "logarithmic derivative" $\Omega' = dX' \cdot (X')^{-1}$ has no reasons to be logarithmic on $\Sigma' \setminus \Sigma$.

The elementary arguments proving Theorem 5.5 can be considerably generalized. A. Bolibrukh established in [Bol77] necessary and sufficient conditions for a matrix function of the form $X(t) = V(t)f(t)^A$ to have a logarithmic singularity on $\{f = 0\}$. As a particular case of his result, using elementary arguments one can show that the "logarithmic derivative" $dX \cdot X^{-1}$ of a *holomorphic* matrix function $X(t)$ has a logarithmic pole on $\Sigma = \{\det X = 0\}$ at those smooth points of the latter hypersurface, where the determinant has a simple (first order) zero.

Concluding this section, note that a large part of the results described here can be modified for the multivariate quasihomogeneous case where the polynomial $H = H(w_1, \dots, w_p)$ in $p \geqslant 2$ complex variables is a sum of the principal quasihomogeneous part $P(w_1, \dots, w_p)$ (subject to the nondegeneracy condition that $dP$ has an isolated singularity) and lower degree monomials with indeterminate coefficients $t_{\mathbf{a}} w^{\mathbf{a}}$, $\deg w_{\mathbf{a}} < \deg P$. The weights assigned to the independent variables $w_j$ need not necessarily be equal to each other. Some of the details can be found in [Yak02], where it is shown that the system (1.2) for the period matrix of a basic coframe has the form

$$R(t)\,dX = R'(t)X,$$

where $R(t)$ is as above and $R'(t)$ is another matrix polynomial.

## 6 Poincaré-Dulac theorem for flat logarithmic connections

In this section we make the first step towards holomorphic gauge classification of integrable connections with logarithmic poles. While the result itself is not new (see [YT75]; it can also be found in [Bol77]), the proof that we give below, close to [YT75] is more in the vein of the traditional "ordinary" approach based on the Poincaré-Dulac technique. It admits immediate generalizations for some types of polar loci different from normal crossings.

### 6.1 "Ordinary" normalization of univariate systems

It is well known that a system of linear ordinary differential equations

$$\dot{x} = A(t)x, \qquad A(t) = t^{-1}(A_0 + tA_1 + \cdots) \tag{6.1}$$

near the singular point $t = 0 \in \mathbb{C}^1$, can be simplified by an appropriate *holomorphic* gauge transformation $x \mapsto H(t)x$, $H = E + tH_1 + \cdots$, so as to keep only *resonant* terms. The latter are defined by the residue matrix $A_0$ of the system depending on the arithmetic properties of this residue. In particular, if $A_0$ is *nonresonant*, i.e., *no two eigenvalues of the residue matrix differ by a nonzero integer*, then there are no resonant terms and the system (6.1) can be made into the Euler system $\dot{x} = t^{-1}A_0x$ by a suitable holomorphic gauge transformation; the corresponding Pfaffian matrix is $A_0\,dt/t$, see (2.1).

The proof consists of an application of the Poincaré-Dulac method of construction of a *formal gauge transformation* conjugating (6.1) with its normal form. Then it can be proved (relatively easily) that the formal gauge transformation in fact always converges.

We implement the same construction in the general Pfaffian case and prove a multi-dimensional analog of the Poincaré-Dulac theorem. Consider a flat connection with the holomorphic pole on the normal crossing $\Sigma = \{t_1 \cdots t_k = 0\} \subset U = (\mathbb{C}^m, 0)$,

$$\Omega = \sum_{s=1}^{k} A_s(t)\,\frac{dt_s}{t_s} + \text{holomorphic terms.} \tag{6.2}$$

By Theorem 4.4, the holomorphic residues $A_s$ corresponding to the smooth components $\Sigma_s = \{t_s = 0\}$, can without loss of generality be assumed constant, $A_s = \text{const} \in \operatorname{Mat}_n(\mathbb{C})$. By Theorem 4.3, the residue matrices $A_s$ commute with each other.

Following the persistent pedagogical tradition, we assume for simplicity that all the matrices $A_1, \ldots, A_k$ are diagonal: this will automatically be the case if one of them has only simple (pairwise different) eigenvalues. In this case the matrix 1-form

$$\Omega_0 = \sum_{s=1}^{k} A_s \frac{dt_s}{t_s} = \operatorname{diag}\{\alpha_i\}_{i=1}^{n} \tag{6.3}$$

will be diagonal with the 1-forms $\alpha_i = \sum_{s=1}^{k} a_i^s \frac{dt_s}{t_s} \in \Lambda^1(\log \Sigma)$ on the diagonal. Note that *any* diagonal flat connection is necessarily *closed*, $d\Omega = 0$, since $d\alpha_i = \alpha_i \wedge \alpha_i = 0$. The numbers $a_i^s$ are (constant by Lemma 3.9) *residues* of closed logarithmic forms $\alpha_i$ on the irreducible components $\Sigma_s \subset \Sigma$ of the polar locus.

The closed logarithmic diagonal matrix form $\Omega_0 = \operatorname{diag}\{\alpha_1, \ldots, \alpha_n\}$ will be called *resonant*, if some difference $\alpha_i - \alpha_j \in \Lambda^1(\log \Sigma)$ has *all* integer residues. This happens if and only if this difference is the logarithmic derivative $dh/h$ of a *meromorphic* function $h$ holomorphically invertible outside $\Sigma$. If we denote by $\mathbf{a}_i = (a_i^1, \ldots, a_i^k, 0, \ldots, 0) \in \mathbb{C}^n$ the vector of residues of the form $\alpha_i$, then the diagonal matrix $\Omega$ is resonant if and only if $\mathbf{a}_i - \mathbf{a}_j \in \mathbb{Z}^n$ for some $i \neq j$. Otherwise $\Omega$ is *non-resonant.*

Non-resonance of a collection of several residues is a rather weak arithmetic condition compared to non-resonance of only one residue: for example, if only one of the matrices $A_s$ is nonresonant (no two eigenvalues differ by an integer), the entire collection will be nonresonant in the sense of this definition.

**6.1 Theorem** *Assume that the closed diagonal matrix form $\Omega_0$ with the logarithmic pole on the normal crossing $\Sigma = \Sigma_1 \cup \cdots \cup \Sigma_k$ is non-resonant.*

*Then any flat logarithmic form* $\Omega_0 +$ (holomorphic terms) *is holomorphically gauge equivalent to the principal part $\Omega_0$.*

The proof of this theorem occupies Sections 6.2 through 6.4

## 6.2 Poincaré-Dulac method: derivation and simplification of the homological equation

Recall that we grade meromorphic functions and differential forms on $(\mathbb{C}^m, 0)$, assigning the same weights to differentials $dt_j$ and the corresponding variables $t_j$. Then the exterior derivative $d$ will be degree-compatible: for any monomial form $\omega$ we have $\deg d\omega = \deg \omega$ unless $d\omega = 0$. In principle the weights assigned to different $t_j$, can (and will) be chosen unequal, so instead of homogeneous forms it is more accurate to speak about *quasi*homogeneity. Regardless of the choice of the weights, the logarithmic residual terms $A_j \frac{dt_j}{t_j}$ always have degree 0.

The connection form $\Omega$ can be explicitly expanded as a sum of (quasi)homogeneous terms,

$$\Omega = \Omega_0 + \Omega_r + \cdots, \qquad \Omega_0 = \sum_{j=1}^{k} A_j \frac{dt_j}{t_j}, \qquad \deg \Omega_r = r,$$

where the "additive dots" denote (as customary) terms of degree higher than that of the terms explicitly specified, the first nontrivial term being $\Omega_r$. Consider a formal matrix function $H = E + H_r + \cdots$, $\deg H_r = r$, whose inverse and differential expand as

$$H = E + H_r + \cdots, \qquad H^{-1} = E - H_r + \cdots, \qquad dH = dH_r + \cdots,$$

Applying to $\Omega$ the formal gauge transform with the matrix $H$, we obtain the new matrix form

$$\Omega' = \Omega_0 + \Omega_r + dH_r + H_r\Omega_0 - \Omega_0 H_r + \cdots.$$

If the homogeneous term $H_r$ can be found such that

$$dH_r - \Omega_0 H_r + H_r \Omega_0 = -\Omega_r, \tag{6.4}$$

then the homogeneous term $\Omega_r$ will be eliminated from the expansion of $\Omega$. If this step can be iterated for all higher degrees, then the proof of Theorem 6.1 will be achieved on the formal level. In analogy with the "ordinary" case, the equation (6.4) is called the *homological equation*.

The term $\Omega_r$ on the right hand side of the homological equation cannot be arbitrary: the condition of flatness of $\Omega$ imposes a restriction on $\Omega_r$. Indeed, from (1.3) and the identity $d\Omega_0 = 0$ it follows that

$$d\Omega_r + \cdots = \Omega_0 \wedge \Omega_r + \Omega_r \wedge \Omega_0 + \cdots.$$

This implies that on the level of $r$-homogeneous terms

$$d\Omega_r = \Omega_0 \wedge \Omega_r + \Omega_r \wedge \Omega_0. \tag{6.5}$$

One can immediately see that the equation (6.4) can be solved only for $\Omega_r$ meeting the condition (6.5). Indeed, any connection obtained by a holomorphic gauge transform $H = E + H_r + \cdots$ from the flat Euler connection $\Omega_0$, must necessarily be flat. We will show that in fact the homological equation (6.4) is always solvable with respect to $H_r$ for $\Omega_r$ satisfying the necessary condition (6.5).

Note the similarity between equation (6.4) and the necessary condition (6.5) for its solvability: both can be written using the operator $[\Omega_0, \cdot]$ of commutation with the principal part $\Omega_0$, if we agree that the "commutator" of two matrix 1-forms has the properly adjusted sign $[\Omega, \Theta] = \Omega \wedge \Theta + \Theta \wedge \Omega$. Using this "extended" commutator, (6.4) and (6.5) take the respective forms

$$dH - [\Omega_0, H] = -\Omega_r, \qquad d\Omega_r - [\Omega_0, \Omega_r] = 0. \tag{6.6}$$

While this similarity may still be regarded as artificial, the fact that $\Omega_0$ is diagonal allows to express the identities (6.6) using the same *scalar* differential operator. Indeed, since

$\Omega_0 = \operatorname{diag}\{\alpha_1, \dots, \alpha_n\}$, both the homological equation and its solvability condition (6.5) split into scalar equations on the matrix elements $\omega = \omega_{ij} \in \Lambda^1(U)$ of $\Omega_r$ and $h = h_{ij} \in \Lambda^0(U)$ of $H_r$. If we denote by

$$\alpha_{ij} = \alpha_i - \alpha_j \in \Lambda^1(\log \Sigma)$$

the corresponding closed logarithmic forms, then (6.4) and (6.5) will take the scalar form which requires to find a solution $h = h_{ij}$ of the equation

$$dh_{ij} - \alpha_{ij} h_{ij} = -\omega_{ij}, \tag{6.7}$$

for any homogeneous 1-form $\omega_{ij}$ meeting the condition

$$d\omega_{ij} - \alpha_{ij} \wedge \omega_{ij} = 0. \tag{6.8}$$

## 6.3 Solvability of the scalar homological equation

Denote provisionally by $\hat{\Lambda}^p$ the collection of $p$-forms with coefficients being Laurent polynomials in $t_1, \dots, t_k$ and the (usual) Taylor polynomials in $t_{k+1}, \dots, t_m$. Let $\alpha \in \Lambda^1(\log \Sigma)$ be a closed logarithmic form with the residues $a_s = \operatorname{res}_{\Sigma_s} \alpha$, $s = 1, \dots, k$.

Consider the formal differential operator $\nabla$ acting on forms of all ranks,

$$\nabla \colon \hat{\Lambda}^p \to \hat{\Lambda}^{p+1}, \qquad \nabla = d - \alpha \wedge \cdot, \qquad \alpha = \sum_1^k a_s \frac{dt_s}{t_s}. \tag{6.9}$$

Note that $\nabla$ is degree-preserving for any quasihomogeneous grading of $\Lambda^\bullet$.

It can be immediately verified that

$$\nabla^2 = d^2 - d(\alpha \wedge \cdot) - \alpha \wedge d + (\alpha \wedge \alpha) \wedge \cdot = -(d\alpha) \wedge \cdot = 0,$$

that is, the diagram

$$0 \longrightarrow \hat{\Lambda}^0 \xrightarrow{\nabla} \hat{\Lambda}^1 \xrightarrow{\nabla} \hat{\Lambda}^2 \xrightarrow{\nabla} \cdots \xrightarrow{\nabla} \hat{\Lambda}^m \xrightarrow{\nabla} 0 \tag{6.10}$$

is a cochain complex. In algebraic terms, the question of solvability of the equation (6.7) under the assumption (6.8) reduces essentially to proving that this complex is exact in the term $\Lambda^1$, i.e., $\ker \nabla = \nabla(\hat{\Lambda}^0)$ in $\hat{\Lambda}^1$. Somewhat more accurately, one also has to verify that the solution $h$ will be a Taylor polynomial whenever $\omega \in \ker \nabla$ is. The need to consider Laurent polynomials is caused by the appearance of denominators in $\alpha$ so that in general $\nabla$ takes Taylor polynomials to Laurent polynomials.

The answer depends on the 1-form $\alpha \in \hat{\Lambda}^1$, more precisely, on arithmetic properties of its residues $a_s \in \mathbb{C}$.

**6.2 Lemma** *Assume that the vector of residues* $\mathbf{a} = (a_1, \dots, a_k, 0, \dots, 0) \in \mathbb{C}^m$ *of* $\alpha$ *is not integer,* $\mathbf{a} \notin \mathbb{Z}^m$.

*Then the equation* $\nabla h = \omega$ *has a unique solution* $h \in \hat{\Lambda}^0$ *if and only if the form* $\omega \in \hat{\Lambda}^1$ *satisfies the condition* $\nabla \omega = 0$. *This solution is a quasihomogeneous Laurent polynomial of the same degree as* $\omega$, *if the latter is a quasihomogeneous Laurent polynomial* 1*-form.*

Conditions for holomorphic solvability are slightly less stringent.

**6.3 Lemma** *If the vector of residues* $\mathbf{a} = (a_1, \dots, a_m)$ *is not nonnegative integer,*

$$\mathbf{a} \notin \mathbb{Z}_+^m, \tag{6.11}$$

*then any holomorphic* 1*-form* $\omega \in \Lambda^1$ *satisfying the condition* $\nabla\omega = 0$*, is uniquely representable as* $\omega = \nabla h$*, with* $h \in \Lambda^0$ *being a holomorphic function.*

In other words, under the assumption that $\mathbf{a} \notin \mathbb{Z}^m$, the cohomology of the complex (6.10) is trivial in the terms $\hat{\Lambda}^0$ and $\hat{\Lambda}^1$. Both results are proved by the same simple computation.

**Proof of both Lemmas** Because of the choice of the form $\alpha$, the operator $\nabla$ preserves the degrees of all forms *independently of the choice of the weights* $w_j$ assigned to the individual variables $t_j$. Thus it is sufficient to prove the lemma only for quasihomogeneous forms.

Choose the weights linearly independent over $\mathbb{Z}$ (as in the proof of Theorem 2.11). Then the only quasihomogeneous holomorphic forms are those that have the structure

$$\omega = t^{\mathbf{b}}\gamma, \qquad \gamma = \sum_1^m c_j \frac{dt_j}{t_j}, \qquad c_j \in \mathbb{C}, \qquad \mathbf{b} = (b_1, \dots, b_m) \in \mathbb{Z}^m$$

(in the holomorphic case we necessarily have $\mathbf{b} \in \mathbb{Z}_+^m$). Since $d(t^{\mathbf{b}}) = t^{\mathbf{b}}\beta$ (see below), the condition $\nabla\omega = 0$ means that

$$t^{\mathbf{b}}\beta \wedge \gamma - t^{\mathbf{b}}\alpha \wedge \gamma = t^{\mathbf{b}}(\beta - \alpha) \wedge \gamma = 0, \qquad \beta = \sum_1^m b_j \frac{dt_j}{t_j}.$$

The equality $(\beta - \alpha) \wedge \gamma = 0$ is possible if and only if $(\mathbf{b} - \mathbf{a}) \wedge \mathbf{c} = 0$, where $\mathbf{c} = (c_1, \dots, c_n) \in \mathbb{C}^n$. If

$$\mathbf{b} - \mathbf{a} \neq 0 \in \mathbb{C}^n, \tag{6.12}$$

the last equality is possible if and only if $\mathbf{c}$ is proportional to $\mathbf{a} - \mathbf{b}$, that is, only if $\gamma = \alpha - \beta$ up to a scalar multiplier. Then

$$\omega = t^{\mathbf{b}}(\alpha - \beta) = \alpha t^{\mathbf{b}} - d(t^{\mathbf{b}}) = \nabla h, \qquad h = -t^{\mathbf{b}} \in \Lambda^0,$$

as asserted. The solvability condition (6.12) for any $\mathbf{b} \in \mathbb{Z}^m$ (resp., for any $\mathbf{b} \in \mathbb{Z}_+^m$) follows from the non-resonance conditions on $\mathbf{a}$ from the assumptions of each lemma.

To see that the equation $\nabla h = 0$ has a *unique* solution $h = 0$, if $\alpha$ is non-resonant, it is sufficient again to consider only the monomial case $h = t^{\mathbf{b}}$, $\mathbf{b} \in \mathbb{Z}_+^m$. Then $0 = d(t^{\mathbf{b}}) - t^{\mathbf{b}}\alpha = t^{\mathbf{b}}(\beta - \alpha)$ is possible only if $\mathbf{b} - \mathbf{a} = 0$, that is, never (unless $h = 0$). □

In the resonant case the above computation allows one to describe in simple terms the (co)kernel of the operator $\nabla$.

### 6.4 Convergence of the formal gauge transform

Application of Lemma 6.3 proves that in the nonresonant case at each step $r$ the homological equation (6.4) can be solved as a formal gauge transform conjugating the system (6.2) to its principal (Euler) part $\Omega_0$, see (6.3).

In order to complete the proof of Theorem 6.1, we have to prove that the formal series converges.

Note that the gauge transform $H$ conjugating $\Omega$ with $\Omega_0$, if it exists, satisfies the Pfaffian system

$$dH = \Omega_0 H - H\Omega. \tag{6.13}$$

Though this is *not* a matrix Pfaffian equation (1.2), it is still a system of $n^2$ Pfaffian equations on the components of the matrix function $H = H(t)$ arranged in any order. The "Pfaffian matrix" $\boldsymbol{\Omega}$ of this system of size $n^2 \times n^2$ is built from the entries of the matrix forms $\Omega$ and $\Omega_0$. Therefore, $\boldsymbol{\Omega}$ has a logarithmic pole on $\Sigma$. Flatness (integrability) of $\boldsymbol{\Omega}$ outside $\Sigma$ follows either from a straightforward computation or from the fact that the corresponding Pfaffian system has locally $n^2$ linearly independent solutions near any nonsingular point.

If the holomorphic matrix form $\Omega$ is formally gauge equivalent to the Euler matrix $\Omega_0$, then the holomorphic Pfaffian system (6.13) admits a formal solution. By Theorem 2.11, this formal solution is automatically convergent. This proves that under the assumptions of Theorem 6.1, $\Omega$ and $\Omega_0$ are holomorphically gauge equivalent. □

## 6.5 Resonant normal form

In the resonant case, when some of the differences $\alpha_i - \alpha_j$ have all residues nonnegative integers, Theorem 6.1 is no longer valid, since not all homological equations are solvable.

In the same way as with the "ordinary" Poincaré-Dulac theorem, one can remove from the expansion of $\Omega$ all *non-resonant* terms, leaving only those that correspond to the resonances: if

$$\alpha_i - \alpha_j = \sum_{s=1}^{m} a_s \frac{dt_s}{t_s}, \qquad a_s \in \mathbb{Z}_+, \tag{6.14}$$

then the $(i, j)$th matrix element of $\Omega$ is allowed to contain a linear combination

$$t^{\mathbf{a}}\gamma, \qquad \gamma = \sum_{s=1}^{m} c_s \frac{dt_s}{t_s}, \qquad c_s \in \mathbb{C}, \tag{6.15}$$

with any complex coefficients $c_s$, different from zero if $a_s \geqslant 1$.

The corresponding *resonant normal form*, a full analog of the Poincaré-Dulac normal form known in the "ordinary" case, is obtained by exactly the same arguments as those proving Theorem 6.1. Convergence of the formal gauge transform follows again from Theorem 2.11. Modulo minor technical details (assuming $\Omega_0$ diagonal rather than upper-triangular, which does not change the assertion), this is the main result announced in [YT75].

## 6.6 Uniqueness of the normalizing gauge transform

**6.4 Theorem** *In the assumptions of Theorem* 6.1, *the only holomorphic gauge transform conjugating the normal form $\Omega_0$ with itself, is a constant diagonal matrix.*

*Consequently, the gauge transformation putting $\Omega$ into the normal form $\Omega_0$, is uniquely defined by the normalizing condition $H(0) = E$.*

**Proof** If in (6.4) $\Omega_r = 0$, then the matrix elements of the homogeneous matrix monomial $H_r = \|h_{ij}\|$ satisfy the equations

$$dh_{ij} - h_{ij}(\alpha_i - \alpha_j) = 0, \qquad i \neq j,$$
$$dh_{ii} = 0, \qquad i = 1, \dots, n, \qquad \deg h_{ii} = r.$$

The off-diagonal terms $h_{ij}$, $i \neq j$, must be zero by Lemma 6.3. The diagonal terms must be constant, which is possible only if $r = 0$. □

## 7 Polar loci that are not normal crossings

Unlike in the "ordinary" case, in the multivariate case there exists a possibility of locally nontrivial changes of the independent variable, in particular, *blow-ups*. Iteration of such transformations allows to simplify the local structure of the polar locus $\Sigma$.

### 7.1 Blow-up

By a *blow-up* of an analytic set $S \subset U$ of codimension $\geqslant 2$ in an analytic manifold $U$ we broadly mean a holomorphic map $F : U' \to U$ between two analytic manifolds, which is biholomorphically invertible outside $S$ and such that the preimage $S' = F^{-1}(S)$ of $S$ is an analytic (possibly singular) hypersurface in $U'$. The most important example is "compactification" of the map

$$F : \mathbb{C}^2 \mapsto \mathbb{C}^2, \qquad F(t_1, s) = (t_1, t_2), \qquad t_2 = st_1. \tag{7.1}$$

Consider the domains ("complex strips")

$$U_1' = \{(t_1, s) : |t_1| < \varepsilon,\ s \in \mathbb{C}^1\}, \qquad U_2' = \{(s', t_2) : |t_2| < \varepsilon,\ s' \in \mathbb{C}^1\},$$

together with the maps $F_{1,2}$,

$$U_1' \xrightarrow{F_1} (\mathbb{C}^2, 0) \xleftarrow{F_2} U_2', \qquad F_1(t_1, s) = (t_1, st_1), \quad F_2(s', t_2) = (s't_2, t_2).$$

If we identify points $(t_1, s)$ and $(s', t_2)$ of the disjoint union $U_1' \sqcup U_2'$ satisfying the identities $s = t_2/t_1 = 1/s'$, the result will be an (abstract) analytic manifold $U'$ which can be described as the "complex Möbius band". The maps $F_1, F_2$ induce a well-defined map $F : U' \to U = (\mathbb{C}^2, 0)$. This map will be a blow-up of the origin $(0, 0) \in U$: the preimage $S' = F^{-1}(0)$ in each chart will be a (smooth) analytic hypersurface $\{t_1 = 0\}$, resp., $\{t_2 = 0\}$, called the *exceptional divisor*.

Note that the complexified Möbius band $U'$ obtained as patching of two "complex strips" $\{|t_i| < \varepsilon\} \times \mathbb{C}$, is not a cylinder (even topologically), and the exceptional divisor *cannot* be globally described as the zero locus of a function holomorphic on $U'$. Indeed, such a function when restricted on $U' \setminus S'$ and pushed forward on $U = (\mathbb{C}^2, 0)$, would be holomorphic outside the origin and bounded, hence holomorphic at the origin also. But the zero locus of a holomorphic germ cannot consist of an isolated point. Keeping in mind these global complications, we will nevertheless proceed by doing computations in only one of the charts $U_i'$, assuming that the map $F$ has the form (7.1).

**7.1 Remark** Consider the *tautological line bundle* over $\mathbb{C}P^1$ whose fiber over a point $(t_1 : t_2) \in \mathbb{C}P^1$ is identified with the line $\mathbb{C} \cdot (t_1, t_2) \subset \mathbb{C}^2$ on the plane. Then the blow-up $U'$ can be described in these terms as the neighborhood of the zero section of this bundle. After deleting the zero section itself the total space of this bundle corresponds to the punctured neighborhood $(\mathbb{C}^2, 0) \setminus \{0\}$ equal to the union of all punctured lines $\mathbb{C}^* \cdot (t_1, t_2)$, $\mathbb{C}^* = \mathbb{C} \setminus \{0\}$.

The above construction admits a multidimensional analog, whose explicit description we will not use. The importance of these bow-ups is immense. Iterating them, one may, among other things, resolve singularities of any analytic hypersurface to normal crossings.

**7.2 Theorem (Special case of Hironaka's theorem)** *For any germ of an analytic hypersurface $\Sigma \subset (\mathbb{C}^m, 0)$ there exists a holomorphic blow-up map $F : U' \to (\mathbb{C}^m, 0)$, holomorphically invertible outside an analytic set of codimension $\geqslant 2$ in $\mathbb{C}^m$, such that $F^{-1}(\Sigma)$ is an analytic hypersurface having only normal crossings in $U'$.* □

Unfortunately, there is no way to predict what the result of the desingularization will be globally. Implementation of the known algorithms is very labor-consuming in general. Therefore we will restrict further discussion entirely to the two-dimensional case.

**7.3 Example (continuation of Example 4.5)** Blowing up the logarithmic matrix form $\Omega = \sum_{j=1}^{k} A_j \frac{dl_j}{l_j}$ on the plane $(\mathbb{C}^2, 0)$ with constant residues $A_j$ and the linear functions $l_j$ satisfying the assumption $dl_i \wedge dl_j(0) \neq 0$, we obtain a connection form on the Möbius band which is flat provided that the sum $B = \sum_j A_j$ commutes with each $A_j$. The polar locus of this form consists of the exceptional divisor $\Sigma_0$ and $k$ "parallel" lines $\Sigma_1, \dots, \Sigma_k$ normally crossing it. The computation carried out in Example 4.5, shows that the blow-up of $\Omega$ is logarithmic, $B$ being the residue along $\Sigma_0$. If this blow-up is flat, Theorem 4.3 ensures that $[B, A_j] = 0$. This explains in geometric terms why the sufficient condition of flatness, obtained in Example 4.5, is also necessary.

However, blow-ups of a logarithmic form need not necessarily be logarithmic.

## 7.2 Blow-up of logarithmic poles may be non-logarithmic

The reason why the definition of logarithmic poles is *not invariant* by transformations that are not locally biholomorphic, is rather simple. If $F : U' \to U$ is a holomorphic map and $\Sigma \subset U$ a (singular) analytic hypersurface defined by an equation $\{f = 0\}$, then the preimage $\Sigma' = F^{-1}(\Sigma)$ is *not* defined by the equation $\{f' = 0\}$, where $f' = F^* f$. More precisely, the function $f'$ may violate our standing assumption that all equations should be locally square-free, as shows the following example.

**7.4 Example** Let $f(t_1, t_2) = t_1 t_2 (t_2 - t_1)$, the equation of three lines through the origin. Then $F^* f(t_1, s) = t_1^3 s(s-1)$ defines the union of three lines and the exceptional divisor $\{t_1 = 0\}$, the latter with multiplicity 3 (the third line $\{s = \infty\}$ is visible only in the second chart $U_2'$, see Section 7.1). Thus the holomorphy of $f'\omega' = F^*(f\,\omega)$ that holds if $\omega \in \Lambda^\bullet(\log \Sigma)$, does not imply that $F^*\omega$ has a first order pole on $\Sigma'$.

Indeed, the pullback of the logarithmic form (3.9) on $\mathbb{C}^2$ is (modulo the obvious change of notation)

$$\frac{1}{t_1} \frac{ds}{s(s-1)},$$

which has logarithmic poles on $\{s = 0\}$ and $\{s = 1\}$ but a non-logarithmic pole along $\{t_1 = 0\}$.

### 7.3 Blow-up of *closed* logarithmic forms is logarithmic

However, if the logarithmic form $\omega \in \Lambda^1(\log \Sigma)$ is closed, then $F^*\omega$ is logarithmic, $F^*\omega \in \Lambda^1(\log \Sigma')$, $\Sigma' = F^{-1}(\Sigma)$.

For a *logarithmic derivative* $df/f$ of a *holomorphic* function this follows from the fact that $F^*f$ is again holomorphic and the discussion in Section 3.3 explaining that $df/f$ has always a logarithmic residue regardless of the order of vanishing of $f$ on $\Sigma$. Note that the residues of logarithmic derivatives of meromorphic functions are *always integers*: $\operatorname{res}_{\Sigma_i}(df/f) \in \mathbb{Z}$ for any smooth component $\Sigma'$ of the polar locus $\Sigma$.

A general closed logarithmic form can be represented by virtue of Lemma 3.9 as $\omega = \sum_j a_j\, df_j/f_j +$ (holomorphic terms), with constant residues $a_j \in \mathbb{C}$. Its pullback is

$$F^*\omega = \sum_j a_j \frac{df_j'}{f_j'} + \text{(holomorphic terms)}, \qquad f_j' = F^* f_j,$$

which immediately means that $\omega' = F^*\omega$ has a logarithmic pole. This representation implies also that the residue of $\omega'$ on the exceptional divisor $\Sigma_0$ is an integer combination of the residues,

$$\operatorname{res}_{\Sigma_0} \omega' = \sum_j \nu_j\, a_j, \qquad \nu_j = \operatorname{res}_{\Sigma_0} \frac{df_j'}{f_j'} \in \mathbb{Z}_+. \tag{7.2}$$

In fact, the above arguments admit a generalization for any logarithmic form with *analytic residues*: if $\Sigma = \{f_1 \cdots f_k = 0\}$, $f_j$ being irreducible equations for components of $\Sigma$, and $\omega = \sum a_j(t)\, \frac{df_j}{f_j} +$ (holomorphic terms) with analytic functions $a_j(\cdot)$ defined everywhere in $U$, then $F^*\omega$ will also have logarithmic poles.

### 7.4 Solvability of the general $\nabla$-equation

As an example of application of the blow-up technique, we can prove the following theorem generalizing Lemma 6.3.

Assume that $U = (\mathbb{C}^m, 0)$ and that $\Sigma = \bigcup_1^k \Sigma_j$ is the representation of an analytic hypersurface as a union of irreducible components (among other things, this assumption means that the smooth part of $\Sigma$ lying inside each $\Sigma_j$, is locally connected). Consider a closed logarithmic 1-form $\alpha = \sum_1^k a_j \frac{df_j}{f_j}$ and the corresponding $\nabla$-equation

$$\nabla h = \omega, \qquad \nabla = d - \alpha \wedge \cdot, \qquad \omega \in \Lambda^1(U). \tag{7.3}$$

**7.5 Theorem** *Assume that $\alpha \in \Lambda^1(\log \Sigma)$ is a closed logarithmic form whose residues $a_j = \operatorname{res}_{\Sigma_j} \alpha \in \mathbb{C}$ on each irreducible component are independent over nonnegative integers: $\sum \nu_j a_j \notin \mathbb{Z}_+$ for any $\nu_1, \dots, \nu_k \in \mathbb{Z}_+$.*

*Then the necessary condition $\nabla\omega = 0$ for solvability of the equation* (7.3) *is also sufficient, and the holomorphic solution $h$ is unique.*

**Proof** Consider the blow-up $F : U' \to U$ of $\Sigma$ such that $\Sigma' = F^{-1}(\Sigma)$ is a hypersurface with normal crossings in $U'$ which is a neighborhood of $\Sigma'$.

The pullback $\alpha' = F^*\alpha$ will be again a closed logarithmic form with the residues $a_j$ on the strict transforms $\hat{\Sigma}_j$ of $\Sigma_j$. The residues of $\alpha'$ on components of the exceptional divisor are nonnegative integral combinations $\sum \nu_j a_j$, $\nu_j \in \mathbb{Z}_+$, of the residues $a_j$ by (7.2).

In such a situation Lemma 6.3 allows to assert that near any point $a \in \Sigma'$ the $\nabla$-equation $dh' - h'\alpha' = \omega'$, $\omega' = F^*\omega$, meeting the necessary condition of solvability $d\omega' - \alpha' \wedge \omega' = 0$, admits a unique holomorphic solution $h'_a \in \Lambda^0(U', a)$. Being unique, all these solutions are patches of a globally defined solution $h' \in \Lambda^0(U')$ in a neighborhood of $\Sigma'$.

The holomorphic function $h = (F^{-1})^* h'$ can be pushed forward on $(\mathbb{C}^m, 0)$ everywhere outside the critical locus $S$ of the blow-up $F$. Since $S$ is thin, $h$ extends as a holomorphic solution of the initial $\nabla$-equation (7.3). □

## 7.5 Example: normalization on the union of pairwise transversal curves

Application of one or several bow-ups allows to reduce many questions about flat logarithmic connections with arbitrary polar loci to those with normal crossings. As an instructive example, consider the following situation.

Assume that the polar locus of a flat meromorphic connection $\Omega$ is a finite union of smooth pairwise transversal curves $\Sigma_j = \{f_j = 0\}$, $j = 1, \dots, k$, passing through the origin in $(\mathbb{C}^2, 0)$:

$$df_i \wedge df_j(0) \neq 0.$$

Assume that $\Omega$ has a logarithmic pole on $\Sigma = \bigcup_{j=1}^k \Sigma_j$.

**7.6 Theorem** *If the residues $A_j = \operatorname{res}_{\Sigma_j} \Omega$ are bounded on $\Sigma_j$ near the origin, and no two eigenvalues of the sum $A_0 = \sum A_j(0)$ differ by an integer number, then the form $\Omega$ is holomorphically conjugate to a form "with constant coefficients"*

$$\Omega_0 = \sum_{j=1}^k A_j(0)\, \frac{df_j}{f_j}.$$

**7.7 Remark** The existence of the limits $A_j(0) = \lim_{t \to 0,\ t \in \Sigma_j} A_j(t)$, and hence of their sum $A_0 = \sum A_j(0)$, follows from the removable singularity theorem.

**Proof** First we notice that the residue matrix functions, holomorphic on $\Sigma_j \setminus \{0\}$ and bounded, are forced to remain holomorphic on $\Sigma_j$ and can be extended as holomorphic functions on $(\mathbb{C}^2, 0)$. This means that

$$\Omega = \sum_1^k A_j(t)\, \frac{df_j}{f_j} + (\text{holomorphic terms}).$$

Consider the standard blow-up $F : U' \to (\mathbb{C}^2, 0)$ described in Section 7.1. Since $df_j(0) \neq 0$, $f'_j = F^* f_j$ is a holomorphic function that vanishes on the union $\Sigma'_j = \Sigma_0 \cup \hat{\Sigma}_j$, where $\Sigma_0$ is the exceptional divisor and $\hat{\Sigma}_j$ the *strict preimage* of $\Sigma_j$, a smooth curve transversal to $\Sigma_0$, obtained as the closure of the preimage $F^{-1}(\Sigma_j \setminus \{0\})$. The logarithmic derivative $df_j/f_j$ has residues equal to 1 on both $\Sigma_0$ and $\hat{\Sigma}_j$.

The pullback $\Omega' = F^*\Omega$ is another matrix form, also flat and having logarithmic poles on $\Sigma'$ which now has only normal crossings. The residues of $\Omega' = F^*\Omega$ on $\hat{\Sigma}_j$ coincide with the pullback $F^* A_j$, while the residue on $\Sigma_0$ is equal to the constant matrix $A_0 = \sum A_j(0)$. By our assumption, $A_0$ is diagonal(izable), and by Theorem 4.4, after a suitable holomorphic gauge transform the residues $A_j$ can also be assumed constant (equal to their limit values $A_j(0)$). Commuting with the diagonal matrix $A_0$ with distinct eigenvalues, all $A_j(0)$ are also diagonal.

The "complex Möbius band" $U'$ can be covered by the union of local charts $U'_s$ such that in each chart the preimage $F^{-1}(\Sigma)$ is the union of one or more coordinate hyperplanes. By Theorem 6.1, the connection form $\Omega'$ is holomorphically gauge equivalent to a *diagonal* flat logarithmic connection $\Omega'_s$. The corresponding local gauge transforms $H_s$, being uniquely defined by the normalizing condition $H'_s|_{\Sigma'} = E$ (Theorem 6.4), coincide on the intersections of the charts and hence define a globally defined on $U'$ holomorphic gauge transformation $H' : U' \to \mathrm{GL}(n, \mathbb{C})$. By $F$, this transformation descends as a holomorphic holomorphically invertible matrix function $H : (\mathbb{C}^2, 0)\setminus\{0\} \to \mathrm{GL}(n, \mathbb{C})$ and, since the origin has codimension 2 in the plane, $H$ extends as a holomorphic gauge equivalence on the entire neighborhood of the origin.

By construction, the action of $H$ on $\Omega$ is a diagonal flat logarithmic connection, $\Omega = \mathrm{diag}\{\alpha_1, \dots, \alpha_n\}$, which means that each $\alpha_j \in \Lambda^1(\log \Sigma)$ is a closed logarithmic form. By Lemma 3.9, this completes the proof. □

A very similar theorem is proved by different arguments by K. Takano [Tak79]. He deals with logarithmic connections on $(\mathbb{C}^m, 0)$ with any $m \geqslant 2$ rather than with the two-dimensional case, but imposes more stringent conditions on the smooth components $\Sigma_j$ that have to be linear hyperplanes. Actually, the arguments given above can be modified to prove also the Takano theorem in its original formulation: the knowledge that $\Sigma_j$ are hyperplanes ensures that the complete desingularization of $\Sigma$ to normal crossings, has an especially simple structure.

## 7.6 Concluding remarks

Theorem 7.6 apparently admits generalizations for other types of polar loci, different from normal crossings, also in more dimensions $m > 2$. However, the assumption that the residues are holomorphic, is crucial and in general does not hold for an arbitrary flat connection with logarithmic poles only. An interesting problem is to develop a holomorphic and meromorphic gauge classification of singularities with non-holomorphic residues (either locally unbounded residues or bounded residues that are not extendible holomorphically from singular components $\Sigma_j$ onto $(\mathbb{C}^m, 0)$). One can list here two problems that, if solved, would advance considerably several problems, in particular related to the infinitesimal Hilbert problem (see the Introduction).

**7.8 Problem** For a given germ of an analytic hypersurface $\Sigma \subset (\mathbb{C}^m, 0)$, find necessary and sufficient conditions on the linear representation $M. : \pi_1((\mathbb{C}^m, 0) \setminus \Sigma) \to \mathrm{GL}(n, \mathbb{C})$ to be realizable as the monodromy group of a flat connection having only logarithmic poles on $\Sigma$.

This local problem is a counterpart of Hilbert's 21st problem. Quite surprisingly, its global analog is nontrivial already when $\Sigma$ is a union of two smooth hypersurfaces with normal crossings in $U = \mathbb{C}P^2$ and $n = 1$ (A. Bolibrukh, [Bol80]).

**7.9 Problem** When is the germ $\Omega$ of a flat connection with logarithmic poles on $\Sigma$ meromorphically gauge equivalent to a flat logarithmic connection $\Omega'$ having locally bounded residues, possibly on a larger hypersurface $\Sigma'$?

## References

[AGV85] V. I. Arnol'd, S. M. Guseĭn-Zade, and A. N. Varchenko, *Singularities of Differentiable Maps, vol. I (The Classification of Critical Points, Caustics and Wave Fronts)*, Birkhäuser, Boston, MA, 1985. MR **86f:58018**

[AGV88] V. I. Arnol'd, S. M. Guseĭn-Zade, and A. N. Varchenko, *Singularities of Differentiable Maps, vol. II, Monodromy and Asymptotics of Integrals*, Birkhäuser, Boston, MA, 1988. MR **89g:58024**

[AI88] V. I. Arnol'd and Yu. S. Ilyashenko, Ordinary differential equations, in: *Dynamical Systems, I*, Springer, Berlin, 1988, translated from *Current problems in Mathematics. Fundamental Directions, Vol. 1*, Akad. Nauk SSSR, Vsesoyuz. Inst. Nauchn. i Tekhn. Inform., Moscow, 1985, 7–149. MR **87e:34049**

[Bol77] A. A. Bolibrukh, The fundamental matrix of a Pfaffian system of Fuchs type, *Izv. Akad. Nauk SSSR Ser. Mat.* **41** (5) (1977), 1084–1109, 1200. MR **58 #18517**

[Bol80] A. A. Bolibrukh, Example of an unsolvable Riemann-Hilbert problem on $\mathbf{C}P^2$, in: *Geometric Methods in Problems of Algebra and Analysis*, **2** (Russian), Yaroslav. Gos. Univ., Yaroslavl', 1980, 60–64, 161. MR **82k:32066**

[Bol97] A. A. Bolibrukh, On isomonodromic deformations of Fuchsian systems, *J. Dynam. Control Systems* **3** (4) (1997), 589–604. MR **99c:34003**

[Bol00a] A. A. Bolibrukh, *Fuchsian Differential Equations and Holomorphic Vector Bundles*, Modern Lecture Courses, Moscow Center for Continuous Mathematical Education, Moscow, 2000, (Russian).

[Bol00b] A. A. Bolibrukh, On orders of movable poles of the Schlesinger equation, *J. Dynam. Control Systems* **6** (1) (2000), 57–73. MR **2000m:34203**

[Chi89] E. M. Chirka, *Complex Analytic Sets*, Math. Appl. (Soviet Ser.) **46**, Kluwer, Dordrecht, 1989. MR **92b:32016**

[Del70] P. Deligne, *Équations différentielles à points singuliers réguliers*, Lecture Notes in Math. **163**, Springer-Verlag, Berlin , 1970. MR **54 #5232**

[dR54] G. de Rham, Sur la division de formes et de courants par une forme linéaire, *Comment. Math. Helv.* **28** (1954), 346–352. MR **16,402d**

[For91] O. Forster, *Lectures on Riemann Surfaces*, Grad. Texts in Math. **81**, Springer-Verlag, New York, 1991. MR **93h:30061**

[Ful95] W. Fulton, *Algebraic Topology*, Grad. Texts in Math. **153**, Springer-Verlag, New York, 1995. MR **97b:55001**

[Gan59] F. R. Gantmacher, *The Theory of Matrices*, Vols. 1, 2, Chelsea, New York, 1959. MR **21 #6372c**

[Gav98] L. Gavrilov, Petrov modules and zeros of Abelian integrals, *Bull. Sci. Math.* **122** (8) (1998), 571–584. MR **99m:32043**

[Glu00] A. Glutsuk, An explicit formula for the determinant of the Abelian integral matrix, Preprint `math.DS/0004040`, April 2000.

[GR65] R. Gunning and H. Rossi, *Analytic Functions of Several Complex Variables*, Prentice-Hall, Englewood Cliffs, NJ, 1965. MR **31 #4927**

[Har82] P. Hartman, *Ordinary Differential Equations*, 2nd ed., Birkhäuser, Boston, MA, 1982.

[Lan69] P. Lancaster, *Theory of Matrices*, Academic Press, New York, 1969. MR **39 #6885**

[Mou75] R. Moussu, Le théorème de de Rham sur la division des formes, *C. R. Acad. Sci. Paris Sér. A–B* **280** (6) (1975), Aii, A329–A332. MR **53 #9282**

[Nov02] D. Novikov, Modules of Abelian integrals and Picard-Fuchs systems, *Nonlinearity* **15** (5) (2002), 1435–1444.

[NY99] D. Novikov and S. Yakovenko, angential Hilbert problem for perturbations of hyperelliptic Hamiltonian systems, *Electron. Res. Announc. Amer. Math. Soc.* **5** (1999), 55–65 (electronic). MR **2000a:34065**

[NY01] D. Novikov and S. Yakovenko, Redundant Picard-Fuchs system for abelian integrals, *J. Differential Equations* **177** (2) (2001), 267–306. MR **1 876 646**

[NY02] D. Novikov and S. Yakovenko, Quasialgebraicity of Picard–Vessiot fields, ArXiv preprint `math.DS/0203210` (2002), 1–38, to appear in *Mosc. Math. J.*

[Sai80] K. Saito, Theory of logarithmic differential forms and logarithmic vector fields, *J. Fac. Sci. Univ. Tokyo Sect. IA Math.* **27** (2) (1980), 265–291. MR **83h:32023**

[Sch99] C. Schmoeger, Remarks on commuting exponentials in Banach algebras, *Proc. Amer. Math. Soc.* **127** (5) (1999), 1337–1338. MR **99h:46090**,

[Tak79] K. Takano, A reduction theorem for a linear Pfaffian system with regular singular points, *Arch. Math. (Basel)* **31** (3) (1978/79), 310–316. MR **80i:58005**

[War83] F. Warner, *Foundations of Differentiable Manifolds and Lie Groups*, Springer-Verlag, New York, 1983 (corrected reprint of the 1971 edition). MR **84k:58001**

[Yak01] S. Yakovenko, Quantitative theory of ordinary differential equations and tangential Hilbert 16th problem, ArXiv preprint `math.DS/0104140` (2001), to appear in: *Asymptotic Series, Differential Algebra and Finiteness Theorems (Montreal, June–July, 2000)*, CRM Proc. Lecture Notes, 78 pp.

[Yak02] S. Yakovenko, Bounded decomposition in the Brieskorn lattice and Pfaffian Picard-Fuchs systems for Abelian integrals, *Bull. Sci. Math.* **126** (7) (2002), 535–554.

[YT75] M. Yoshida and K. Takano, Local theory of Fuchsian systems I, *Proc. Japan Acad.* **51** (4) (1975), 219–223. MR **52 #3648**

# Normal forms, bifurcations and finiteness properties of vector fields

Christiane ROUSSEAU*
*Département de mathématiques et de statistique*
*and*
*Centre de recherches mathématiques*
*Université de Montréal*
*C.P. 6128, Succursale Centre-ville*
*Montréal, Qué., H3C 3J7*
*Canada*

**Abstract**

These lectures make the bridge between the three themes of the present book. Analytic objects have local finiteness properties. Compactness of the phase space and parameter space allows one to derive global finiteness properties. One of the important tools for studying local properties is normal forms, which are also a tool to study bifurcations. In many generic situations the changes of coordinates to normal forms diverge. We explain this phenomenon in the case of a generic parabolic fixed point of a diffeomorphism by constructing a complete modulus of analytic classification for a generic unfolding of the diffeomorphism: the limit dynamics is too complicated to be encoded in the simple normal form. So the point of view of bifurcations sheds a new light on the divergence of the normalizing series. Finally we recall the program to prove the finiteness part of Hilbert's 16th problem for quadratic systems started in 1991, we discuss its progress and what remains to be done.

## 1 Introduction and perspectives

The text of these lectures makes the bridge between the three themes of the present book. Although not exclusively, the main context will be analytic vector fields and analytic families of vector fields. The following ideas will be leading our mathematical developments:

- *Analytic objects have local finiteness properties.* In our context we will be mainly interested to the finite cyclicity of graphics of planar vector fields.
- *Compactness of the space phase yields to global finiteness properties.*
- *Compactness of the parameter space yields uniform finiteness properties.*

The main tools to study a vector field locally are:

*Supported by the Natural Sciences and Engineering Research Council of Canada and the Fonds FCAR of the Ministère de l'Éducation du Québec.



*Y. Ilyashenko and C. Rousseau (eds.),*
*Normal Forms, Bifurcations and Finiteness Problems in Differential Equations,* 431–470

(1) The flow-box theorem in the neighborhood of a singular point;

(2) Blow-ups and normal forms near singular points. We can consider blow-ups and normal forms of a single vector field as well as blow-ups and normal forms of families of vector fields.

Let us consider for a moment the problem of the finite cyclicity of graphics of planar vector fields. The standard method is to "calculate a return map" and to show that it has a finite number of fixed points. This return map can be written as a composition of regular transitions and Dulac maps in the neighborhood of the singular points. In the very generic situations, rough normal forms are sufficient to conclude finite cyclicity, while in the more degenerate cases more refined normal forms are needed. For instance, to prove the finite cyclicity of a homoclinic point through a saddle with nonzero trace (i.e. of codimension 1) a $C^1$ normal form is sufficient to conclude that the homoclinic loop has cyclicity one. When, however, we consider homoclinic loops of higher codimension more refined normal forms are needed. $C^k$ normal forms are sufficient to treat the generic cases while analycity arguments are needed to treat the homoclinic loops with identity return map inside finite parameter analytic families of vector fields.

Let us consider a family $(v_\lambda)_{\lambda\in\Lambda}$ of vector fields with $v_{\lambda_0}(x_0) = 0$. A general principle which we want to push further is the following: *The more generic $v_{\lambda_0}$ is, the more we know about $v_\lambda$ in the neighborhood of $x_0$ for $\lambda$ near $\lambda_0$.* As an example we can consider a vector field $v_{\lambda_0}$ with two pure imaginary eigenvalues at $x_0$ and such that the first Lyapunov quantity is negative. Then we know that $v_\lambda$ has no (resp. exactly one) limit cycle in the neighborhood of $x_0$ for values of $\lambda$ for which the singular point $x_\lambda$ (such that $x_{\lambda_0} = x_0$) is attracting (resp. repelling). If however the first $k-1$ Lypaunov quantities vanish and the $k$-th Lyapunov quantity is nonzero many more phase portraits are allowed. Finally, when $v_{\lambda_0}$ has a center at the origin the only thing we can say about $x_\lambda$ is that it is a monodromic singular point.

The subject of normal forms was started by Poincaré, and normal forms have been studied extensively ever since. If the singular points are not too degenerate, reasonably good results exist for:

- formal normal forms;
- $C^\infty$ normal forms for a single vector field;
- $C^k$ normal forms for families of vector fields.

**What about analytic normal forms?**

(i) They are the tools for problems of analytic classification under analytic changes of coordinates.

(ii) Understanding the analytic type of a vector field in the neighborhood of an elementary singular point is an essential ingredient of the proof by Écalle and Ilyashenko of the Dulac conjecture, namely that a polynomial vector field has a finite number of limit cycles.

(iii) In many situations formal normalizing changes of coordinates diverge.

**Some examples of orbital normal forms for which the formal normalizing change of coordinate can be divergent:**

(1) The resonant saddle:

$$\begin{aligned}\dot{x} &= x(1 + au^k)\\ \dot{y} &= y\left(-\frac{p}{q} + (1-a)u^k\right)\end{aligned} \tag{1.1}$$

where $u = x^p y^q$ is the resonant monomial.

(2) The saddle-node of codimension $n$:

$$\begin{aligned}\dot{x} &= x^{n+1}\\ \dot{y} &= \pm y(1 + ax^n).\end{aligned} \tag{1.2}$$

(3) The non-resonant saddle:

$$\begin{aligned}\dot{x} &= x\\ \dot{y} &= \quad -\lambda y\end{aligned} \tag{1.3}$$

where $\lambda$ is a Liouvillian irrational positive number.

(4) The diffeomorphism $f : (\mathbb{C}, 0) \to (\mathbb{C}, 0)$ with a parabolic fixed point:

$$f(z) = z + z^{n+1} + o(z^n + 1). \tag{1.4}$$

The common point to all these situations is that *divergence is the rule and convergence is the exception.* Such systems have been studied thoroughly for fixed values of the eigenvalues or of the multiplier: they form natural "families" of systems. In a series of papers by Ilyashenko [18, 19] we find theorems of the form: *Either the change of coordinates to normal form converges for all systems in the family or it diverges for the majority of systems (for almost all systems in the case of finite parameter families of systems)*. Later, in the resonant cases, Écalle [12] has given a more precise description of the subspace of parameter space where we can expect convergence.

As mentioned above, in all these situations divergence is the rule and convergence is the exception. The question we want to consider is: "WHY?"

The answer proposed is in the following spirit: our normal form is a kind of "model".

- Divergence is the rule and convergence is the exception *because the "model" is too poor to encode the rich dynamics of the system.*
- So the rich dynamics is encoded in the divergent normalizing transformation. *How to read it?*
- If we can read this rich dynamics it will become obvious that the poor dynamics of the model is an exceptional case.
- If divergence is the generic situation then, by the principle described above, we should expect to be able to describe some properties of local perturbations of the system.

Let us go back to the past and look at the following citation of Abel [37] (Œuvres complètes de Niels Henrik Abel, tome 2, in the letters to Holmboe, pages 256–257):

> Les séries divergentes sont en général quelque chose de bien fatal et c'est une honte qu'on ose y fonder aucune démonstration. On peut démontrer tout ce qu'on veut en les employant, et ce sont elles qui ont fait tant de malheurs et qui ont enfanté tant de paradoxes. (...) Enfin mes yeux se sont dessillés d'une manière frappante, car à l'exception des cas les plus simples, par exemple les séries géométriques, il ne se trouve

dans les mathématiques presque aucune série infinie dont la somme soit déterminée d'une manière rigoureuse, c'est-à-dire que *la partie la plus essentielle des mathématiques est sans fondement.* Pour la plus grande partie les résultats sont justes il est vrai, mais c'est là une chose bien étrange. Je m'occupe à en chercher la raison, problème très intéressant.

The author has been fascinated by this complete change in the dynamics of the subject: Nowadays divergent series occupy a marginal place within mathematics, and we remember Abel as a mathematician who banished the use of divergent series.

To study why generically so many series diverge, the point of view which is taken here is to *unfold*: the bifurcation point of view will help us understand the normal forms. Indeed many situations where we encounter divergence are limit situations, for instance:

(i) Irregular singular points of linear differential equations occur when regular points coallesce.

(ii) A parabolic point of a diffeomorphism of $(\mathbb{C}, 0)$ is a fixed point of multiplicity $n+1 > 1$.

(iii) A saddle-node is a multiple singular point.

(iv) The holonomy of a separatrix of a non integrable resonant saddle with eigenvalues $1 : -1$ has a parabolic fixed point (hence a multiple fixed point) at the origin.

In these lectures we will address the following questions.

**Questions:**

(1) If the change of coordinates to normal form near $x_0$ diverges for $v_{\lambda_0}$ (i.e. we are in the generic situation):

- What can we say of $v_\lambda$ in the neighborhood of $x_0$ for $\lambda$ close to $\lambda_0$? (We look for the properties which depend only on the germ of $v_{\lambda_0}$ at $x_0$.)
- What analytic families $v_\lambda$ can unfold $v_{\lambda_0}$? More precisely, can we give the moduli space of the space of such analytic families?

(2) Can we understand the dynamics of $v_{\lambda_0}$ from the fact that it is the limit of the $v_\lambda$ for $\lambda \to \lambda_0$?

Let us illustrate this on an example:

**The saddle-node** By an analytic change of coordinates we can bring an analytic system with a saddle-node of multiplicity 2 to the normal form

$$\begin{aligned} \dot{x} &= x^2 \\ \dot{y} &= y(1 + ax) + x^2 R(x, y). \end{aligned} \tag{1.5}$$

The formal normal form is our "model" (Fig. 1(a)):

$$\begin{aligned} \dot{x} &= x^2 \\ \dot{y} &= y(1 + ax). \end{aligned} \tag{1.6}$$

One obstruction to bring (1.5) to (1.6) is the non-existence of an analytic center manifold. We now unfold: the "model family" for the unfolding is (Fig. 1(b))

$$\begin{aligned} \dot{x} &= x^2 - \varepsilon \\ \dot{y} &= y(1 + a(\varepsilon)x). \end{aligned} \tag{1.7}$$

To understand the dynamics we need to consider $x, y \in \mathbb{C}$. Fig. 2 represents the model and model family (only the $x$-variable is drawn as a 2-dimensional variable).

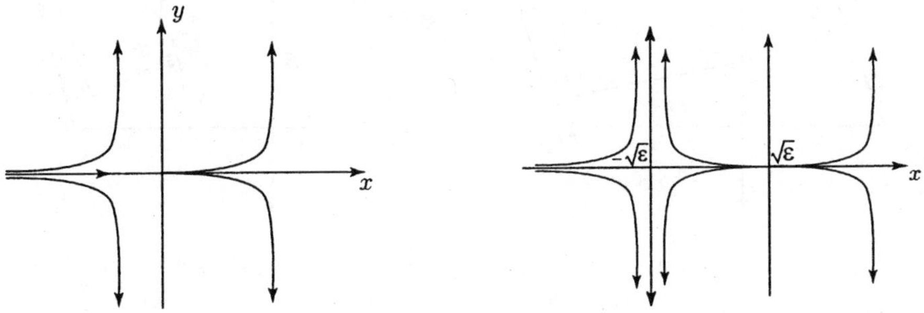


(a) $\varepsilon = 0$ (b) $\varepsilon \neq 0$

Figure 1: The "model"

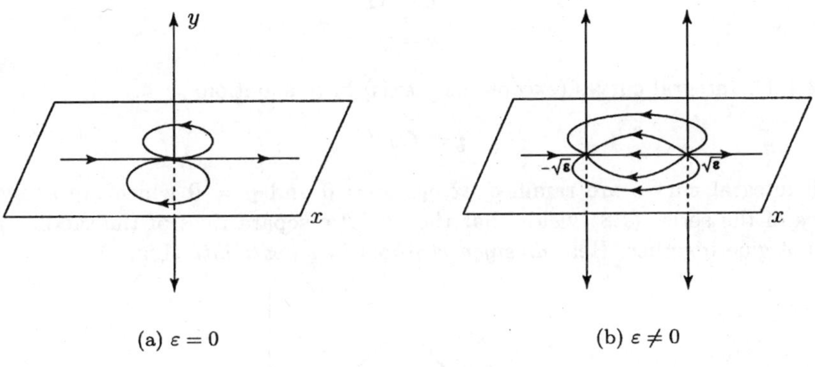


(a) $\varepsilon = 0$ (b) $\varepsilon \neq 0$

Figure 2: The "model"

Let us now suppose that (1.5) has no analytic center manifold. The formal equation of the center manifold,

$$y = \sum_{n=2}^{\infty} b_n x^n, \tag{1.8}$$

is 1-summable in all directions except $\mathbb{R}^+$ [27]. This yields a "ramified" center manifold as in Fig. 3(a). To simplify we can suppose $a(\varepsilon) \in \mathbb{R}$, although this is not essential. Let now $\varepsilon \in \mathbb{R}^+$. The saddle point at $(-\sqrt{\varepsilon}, 0)$ always has an invariant stable manifold which is necessarily ramified at the node as in Fig. 3(b).

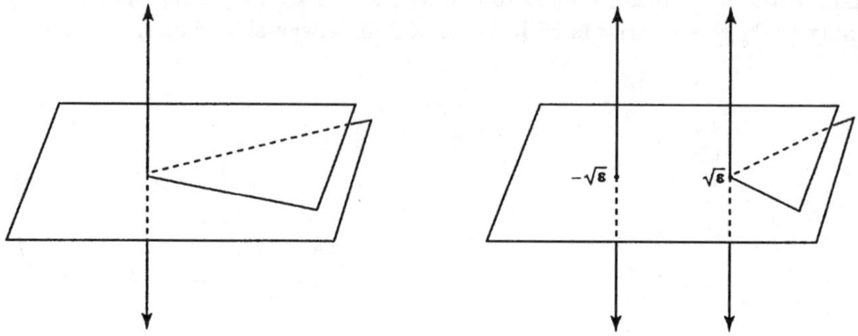


(a) $\varepsilon = 0$ (b) $\varepsilon \neq 0$

Figure 3: The invariant manifold

Let us now look at the local model at the node: it is known since Poincaré that the change of coordinates to the local model is always convergent. We have two cases:

- **Non-resonant node** The normal form is linear

$$\begin{aligned} \dot{x} &= \lambda x \\ \dot{y} &= y, \end{aligned} \tag{1.9}$$

when $\lambda \notin 1/\mathbb{N}$. Integral curves (except the $y$-axis) have equation

$$y = Cx^{1/\lambda}. \tag{1.10}$$

Hence all integral curves are ramified except $x = 0$ and $y = 0$ which are analytic. The divergence of the series (1.8) yields that the analytic separatrices of the saddle and of the node do not glue together. Here *divergence means incompatibility* (Fig. 4).

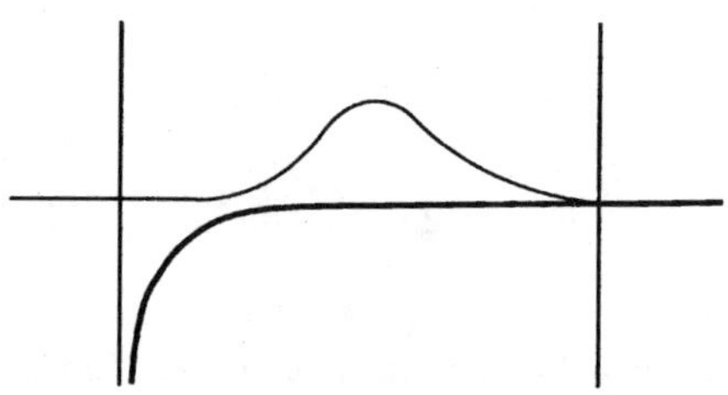

Figure 4: Incompatibility of the invariant analytic manifolds

- **Resonant node** Here we have two cases. The exceptional case is when the node is linearizable

$$\begin{aligned} \dot{x} &= \frac{1}{n}x \\ \dot{y} &= y. \end{aligned} \tag{1.11}$$

In this case all integral curves (except $x = 0$) are of the form $y = Cx^n$. They are analytic, thus non-ramified. Hence this case is necessarily excluded and we need be in the generic situation where the resonant monomial has a nonzero coefficient:

$$\begin{aligned}\dot{x} &= \frac{1}{n}x \\ \dot{y} &= y + bx^n,\end{aligned} \tag{1.12}$$

with $b \neq 0$. In this case the only non-ramified integral curve is $x = 0$.

The conclusion is the following: *For some particular values of $\varepsilon$ where one of the singular points is resonant, the incompatibility implies some generic property of the unfolded point itself.* This observed phenomenon is very general: it is some form of *parametric resurgence.*

Similar resurgence phenomena occur at the saddle for parameter values for which it is resonant. We will learn to read them from the Martinet-Ramis modulus of analytic classification of the saddle-node.

We can of course extend to complex values of the parameter $\varepsilon$. When the ratio of the eigenvalues at $(\pm\sqrt{\varepsilon}, 0)$ is not real the singular points are linearizable and have exactly two analytic trajectories, all other integral curves being ramified. If the saddle-node has no analytic center manifold, then we observe in the unfolding that *the two (weak) analytic separatrices of $(\pm\sqrt{\varepsilon}, 0)$ do not coincide.* This was already observed by Glutsyuk [14]. *But we cannot make a full turn in the parameter space as $\sqrt{\varepsilon}$ and $-\sqrt{\varepsilon}$ are exchanged. We cannot expect a description of the phenomenon analytic in $\varepsilon$.*

The study of all these phenomena is done through the description of a complete modulus of analytic classification for generic analytic families unfolding a diffeomorphism with a generic parabolic point

$$f_\varepsilon(z) = z + (z^2 - \varepsilon)(1 + h(\varepsilon, z)), \tag{1.13}$$

presented in detail in [25]. This allows to give an interpretation of all ingredients of the Écalle-Voronin modulus of analytic classification of $f_0$.

## 1.1 The Écalle-Voronin modulus of a generic diffeomorphism with a parabolic point

We describe the modulus of a germ of diffeomorphism

$$f_0(z) = z + z^2 + (1 - a)z^3 + o(z^3). \tag{1.14}$$

The modulus is composed of three parts:

- *The discrete part* $k = 1$: $k + 1$ is the multiplicity of the parabolic point. (Here we limit ourselves to the generic codimension 1 case.)
- *The formal part* $a \in \mathbb{C}$: in the unfolding $a$ is unfolded as $a(\varepsilon)$, depending analytically of $\varepsilon$, and is interpreted as a *shift* between the two singular points. Indeed, the two singular points may not be resonant simultaneously.
- *The analytic part*: we describe precisely the diffeomorphism by describing its orbit space. The diffeomorphism is topologically equivalent to the time-one map of the flow of the vector field $z^2\partial/\partial z$ (see Fig. 5).

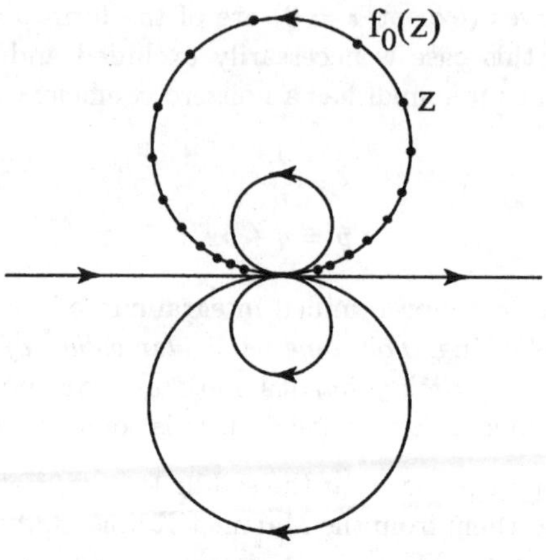


Figure 5: The diffeomorphism $f_0$

To describe the orbit space we give ourselves two "fundamental domains". To construct a *fundamental domain* we take a closed curve $l$ through the origin such that $l \cap f_0(l) = \{0\}$. The fundamental domain is the domain between $l$ and $f_0(l)$, where we identify $l$ and $f_0(l)$ through $z \equiv f_0(z)$. Then each orbit has at most one point in a fundamental domain. There is no way to represent all orbits by a unique fundamental domain, and two fundamental domains are necessary to represent the whole space of orbits. But then some orbits have representatives in the two fundamental domains. To describe the orbit space we must identify points in the two fundamental domains which belong to the same orbit. This is done via a pair of maps defined respectively in the upper end and lower end of the fundamental domains (Fig. 6).

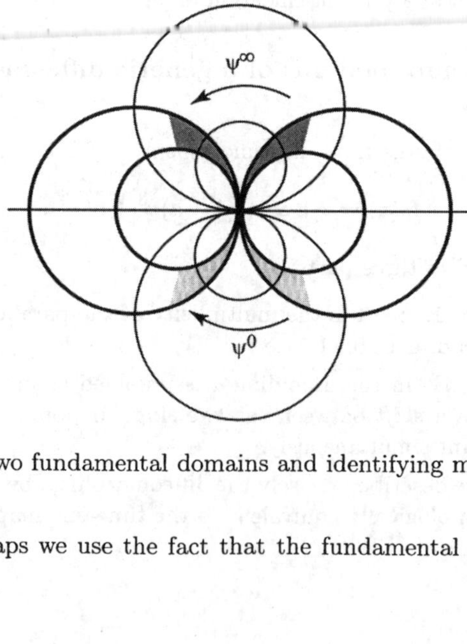


Figure 6: The two fundamental domains and identifying maps $\psi^0$ and $\psi^\infty$

To describe these maps we use the fact that the fundamental domains are conformally equivalent to spheres $S^{\pm}$ with 0 and $\infty$ identified. Then there exist analytic diffeomorphisms $\psi^0, \psi^\infty : S^+ \to S^-$ defined in the neighborhood of 0 and $\infty$ such that the image of $w^+ \in S^+$ by $\psi^0$ or $\psi^\infty$ is a point of $S^-$ belonging to the same orbit as $w^+$. The diffeomorphisms $\psi^0$ and $\psi^\infty$ are defined up to choices of coordinates on the spheres preserving 0 and $\infty$, i.e. up to composition by linear maps in the source and target space. Then

$$(\psi_1^0, \psi_1^\infty) \equiv (\psi_2^0, \psi_2^\infty) \quad \text{iff} \quad \exists C, C' \in \mathbb{C}^* \begin{cases} \psi_1^0(w) = C\psi_2^0(C'w) \\ \psi_1^\infty(w) = C\psi_2^\infty(C'w) \end{cases} . \tag{1.15}$$

The analytic part of the modulus is given by the equivalence class $[\psi^0, \psi^\infty]$.

• The formal part $a$ (mod $1/2\pi i$) can be recovered from the analytic part since $\frac{(\widetilde{\psi}^\infty)'(0)}{(\psi^0)'(0)} = \exp(-4\pi^2 a)$, where $\widetilde{\psi}^\infty$ is $\psi^\infty$ viewed in the coordinate $1/w$. So the Écalle-Voronin modulus is $(k = 1, a, [\psi^0, \psi^\infty])$.

To interpret the analytic part of the modulus we must describe the "normal form" of $f_0$, which we call the *model diffeomorphism.* It is given by the time-one map of the flow of the vector field

$$\frac{z^2}{1 + az}\frac{\partial}{\partial z}. \tag{1.16}$$

There always exists a coordinate change conjugating $f_0$ to the model diffeomorphism. The formal change of coordinate can be summed to an analytic one if and only if $\psi^0$ and $\psi^\infty$ are linear. The reason why the time-one map of the flow of (1.16) is a model is because this diffeomorphism is "fully iterable": we can take its $n$-root for any $n$ and, more generally, define $f^c$, where $f^c$ is the $c$-th iterate of $f$ and $c$ is a complex number.

**An interpretation of the analytic part of the modulus** The nonlinearities of $\psi^0$ and $\psi^\infty$ control *parametric resurgence phenomena* in the unfolding (1.13). Then

• $\psi^0$ (resp. $\psi^\infty$) controls $-\sqrt{\varepsilon}$ (resp. $\sqrt{\varepsilon}$) when its multiplier is on the unit circle.

• If $\widetilde{\psi}^\infty$ is the expression of $\psi^\infty$ in the variable $\widetilde{w} = 1/w$ and

$$\widetilde{\psi}^\infty(\widetilde{w}) = \widetilde{w} + \sum_{n=2}^{\infty} b_n \widetilde{w}^n \tag{1.17}$$

then $\forall p, q, \in \mathbb{N}$ with $1 \le q < p$ there exist polynomials

$$L_{p,q}^\infty(b_2, \dots, b_{p+1}) \tag{1.18}$$

such that if $L_{p,q}^\infty(b_2, \dots, b_{p+1}) \neq 0$ then $f_\varepsilon$ is non-linearizable at $\sqrt{\varepsilon}$ as soon as the multiplier $\lambda = f_\varepsilon'(\sqrt{\varepsilon})$ is of the form

$$\lambda = \exp\left(2\pi i \frac{p}{n}\right) \tag{1.19}$$

with $n$ large and

$$n \equiv q \pmod{p}. \tag{1.20}$$

• If $\psi^\infty$ is nonlinear, then at least one of the $L_{p,q}^\infty(b_2, \dots, b_{p+1}) \neq 0$.

• A similar result exists with $\psi^0$ controlling $-\sqrt{\varepsilon}$. If $2\pi i a \notin \mathbb{Z}$ we cannot simultaneously scale $(\widetilde{\psi}^\infty)'(0) = 1$ and $(\psi^0)'(0) = 1$.

### 1.2 An answer to a question of Martinet-Ramis

In [27] Martinet-Ramis study the modulus of a saddle-node. They show that the analytic type of a saddle-node is characterized by the analytic type of the holonomy of the strong separatrix, which is a diffeomorphism with a parabolic fixed point. However, not all diffeomorphisms with a parabolic fixed point can be realized as the holonomy map of the strong separatrix of a saddle-node, yielding that the moduli space of the holonomy maps of strong separatrices of saddle-nodes is strictly smaller than the moduli space of diffeomorphims with a parabolic fixed point. Martinet and Ramis ask the following question: (the equation (2) to which they refer is our equation (1.5)):

> Un phénomène qui reste un peu surprenant à nos yeux est que les holonomies produites par les équations (2) ne sont pas arbitraires: on obtient seulement une "petite partie du module d'Écalle". (Nous nous proposons de montrer dans un article ultérieur qu'il n'en est plus de même dans le cas des équations résonantes "non dégénérées" ($\lambda = -p/q \neq 0$) : le module des classes d'équivalence analytiques d'équations différentielles s'identifie complètement au "module d'Écalle").

Indeed, the modulus of the holonomy of the strong separatrix of a saddle-node has the form $([\psi^0, \psi^\infty])$, where $\widetilde{\psi}^\infty$ (which is $\psi^\infty$ viewed in the coordinate $1/w$) has the form

$$\widetilde{\psi}^\infty(\widetilde{w}) = \frac{\widetilde{w}}{1 - b\widetilde{w}}. \tag{1.21}$$

The answer comes from the following theorem:

**1.1 Theorem ([25])** *The only analytic function $\widetilde{\psi}^\infty(\widetilde{w})$ of the form* (1.17) *for which*

$$L^\infty_{p,q}(b_2, \dots, b_{p+1}) = 0 \quad \textit{for all } p > 1 \textit{ and all } q, \tag{1.22}$$

*is the Möbius function* (1.21).

*Hence the Möbius function is the only function which can reflect the very simple dynamics of the node in the unfolding.* Any other function would force the node to have a more complex dynamics, in particular to be nonlinearizable for non-resonant cases.

## 2 Modulus of analytic classification of generic families unfolding a diffeomorphism with generic parabolic point

We now present the results of [25]. We consider a generic diffeomorphism $f_0(z)$ with a parabolic fixed point as in (1.14). An unfolding is *generic* if

$$\left.\frac{\partial f_\varepsilon}{\partial \varepsilon}\right|_{\varepsilon=0} \neq 0. \tag{2.1}$$

We classify germs of analytic families of diffeomorphims under *weak equivalence.*

## 2.1 Preliminaries

**2.1 Definition** Two germs $f_{1,\varepsilon_1}(z)$ and $f_{2,\varepsilon_2}(z)$ of analytic families of diffeomorphims are *weakly equivalent* if there exists a germ of bijective map $K = (h, H)$:

$$(\varepsilon_1, z) \mapsto (h(\varepsilon_1), H(\varepsilon_1, z)), \tag{2.2}$$

fibered over the parameter space where:

(i) $h$ is a germ of holomorphic diffeomorphism $(\mathbb{C}, 0) \to (\mathbb{C}, 0)$.

(ii) there exists a representative $H(\varepsilon_1, z)$ for $(z, \varepsilon_1) \in U \times V$ such that

- $H_{\varepsilon_1} = H(\varepsilon_1, \cdot)$ is analytic in $z$ for $z \in U$;
- $H_{\varepsilon_1}$ conjugates $f_{1,\varepsilon_1}$ and $f_{2,h(\varepsilon_1)}$ over $U$, for all $\varepsilon_1 \in V$, namely:

$$f_{2,h(\varepsilon_1)}(H_{\varepsilon_1}(z)) = H_{\varepsilon_1}(f_{1,\varepsilon_1}(z)). \tag{2.3}$$

We will show that if two generic families unfolding a diffeomorphism with a generic parabolic point are weakly equivalent, then they are equivalent in a stronger sense.

**"Preparation" of an unfolding $f_\varepsilon(z)$ of $f_0(z)$.** We can suppose that the two singular points are located at $\pm\sqrt{\varepsilon}$, i.e. $f_\varepsilon(z)$ has the form (1.13). We want to compare our family with the "model family" which is given by the time-one map of the flow of the vector field

$$\frac{z^2 - \varepsilon}{1 + a(\varepsilon)z} \frac{\partial}{\partial z}. \tag{2.4}$$

(From now on we will not write the dependence of $a(\varepsilon)$ and simply denote $a = a(\varepsilon)$.) The singular points of (2.4) are $\pm\sqrt{\varepsilon}$. Their eigenvalues are

$$\mu_\pm = \pm \frac{2\sqrt{\varepsilon}}{1 \pm a\sqrt{\varepsilon}} \tag{2.5}$$

and satisfy

$$\begin{aligned} \frac{1}{\sqrt{\varepsilon}} &= \frac{1}{\mu_+} - \frac{1}{\mu_-} \\ a &= \frac{1}{\mu_+} + \frac{1}{\mu_-}, \end{aligned} \tag{2.6}$$

i.e. in the model family, $a$ and $\varepsilon$ are analytic invariants.

We make a coordinate change $(z, \varepsilon) \mapsto (\overline{z}, \overline{\varepsilon})$ in $f_\varepsilon(z)$:

$$\overline{f}_{\overline{\varepsilon}}(\overline{z}) = \overline{z} + (\overline{z}^2 - \varepsilon)\overline{h}(\overline{z}, \overline{\varepsilon}), \tag{2.7}$$

so that $\overline{\varepsilon}$ becomes an analytic invariant, i.e.

$$\frac{1}{\sqrt{\overline{\varepsilon}}} = \frac{1}{\ln \lambda_+} - \frac{1}{\ln \lambda_-}, \tag{2.8}$$

where $\lambda_\pm$ are the multipliers at the singular points $\pm\sqrt{\overline{\varepsilon}}$ of (2.7). Then for the model family (2.4) (in which we replace $(z, \varepsilon)$ by $(\overline{z}, \overline{\varepsilon})$) we choose

$$a(\varepsilon) = \frac{1}{\ln \lambda_+} + \frac{1}{\ln \lambda_-}. \tag{2.9}$$

The new family $\overline{f}_{\overline{\varepsilon}}(\overline{z})$ is called "prepared". $\overline{\varepsilon}$ is the canonical parameter of the family.

From now on we will limit ourselves to prepared families (1.13). Because the parameter is an analytic invariant, *weak equivalence between prepared families must preserve the parameter.*

Here are some general principles to guide us in the definition of a complete modulus of analytic classification for prepared families:

- We must cover a whole neighborhood of the origin in parameter space.
- We try to have as much continuity and smoothness in $\varepsilon$ as possible.
- But it may be difficult to make a full turn in $\varepsilon$, so it is natural to work with the universal covering of the $\varepsilon$-space.

## 2.2 Description of the modulus of analytic classification for the family

(i) We give ourselves $\delta > 0$ arbitrarily small.

(ii) We consider the universal covering of the $\varepsilon$-space with projection $\pi$, and call $\hat{\varepsilon}$ the corresponding variable so that $\pi(\hat{\varepsilon}) = \varepsilon$.

(iii) We describe the orbit space of a prepared family (1.13) for all neighborhoods $\{z; |z| < r\}$ with $r \leq r_0$, where $r_0$ depends on $\delta$, for all values of $\hat{\varepsilon}$ such that

$$\begin{cases} |\hat{\varepsilon}| < \rho < Kr \\ \arg \hat{\varepsilon} \in (-\pi + \delta, 3\pi - \delta) \end{cases} \tag{2.10}$$

(Fig. 7), where $K > 0$ is a positive constant depending on $r_0$ and $\delta$. This neighborhood covers only once the Siegel direction $\mathbb{R}^-$. We call this sector $V_\delta$.

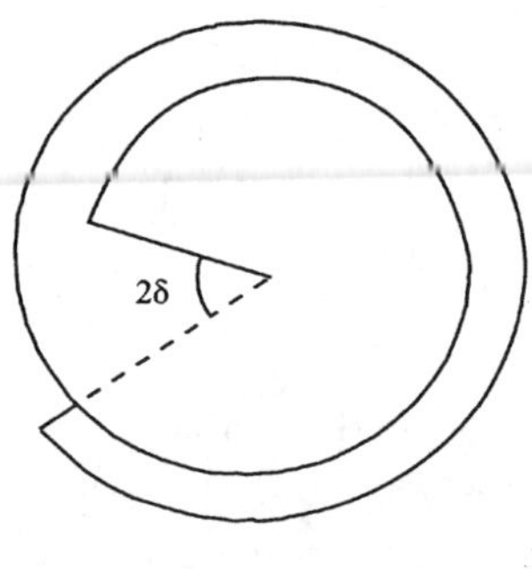


Figure 7: The sector $V_\delta$ in $\hat{\varepsilon}$-space

A similar description can be done in a second sector

$$\begin{cases} |\hat{\varepsilon}| < \rho < Kr \\ \arg \hat{\varepsilon} \in (-3\pi + \delta, \pi - \delta). \end{cases} \tag{2.11}$$

The union of the two neighborhoods covers a full neighborhood of 0 in the $\sqrt{\varepsilon}$-space.

(iv) For all $\hat{\varepsilon} \in V_\delta$ the orbit space is described by two crescents $C_\varepsilon^\pm$ which:

• are fundamental domains for $f_\varepsilon$;

• whose quotient by $f_\varepsilon$ have the conformal structure of spheres $S_\varepsilon^\pm$, the points $\sqrt{\varepsilon}$ (resp. $-\sqrt{\varepsilon}$) corresponding to $\infty$ (resp. 0);

• in the two ends of the crescents, (i.e. near 0 and $\infty$ on the spheres $S_\varepsilon^\pm$) points are identified by germs of analytic diffeomorphisms $\psi_\varepsilon^\infty$ and $\psi_\varepsilon^0$ (Fig. 8).

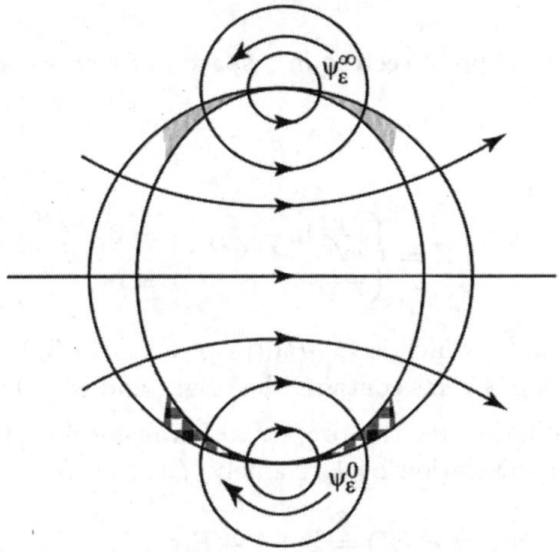


Figure 8: The two crescents and diffeomorphisms $\psi_\varepsilon^\infty$ and $\psi_\varepsilon^0$

(v) It is possible to choose the $C_\varepsilon^\pm$ so that they depend analytically on $\hat{\varepsilon}$ for $\hat{\varepsilon} \neq 0$ and continuously on $\hat{\varepsilon}$ near $\hat{\varepsilon} = 0$.

(vi) The cordinates on the spheres can be chosen so that the functions $\psi_\varepsilon^\infty$ and $\psi_\varepsilon^0$ depend analytically on $\hat{\varepsilon}$ for $\hat{\varepsilon} \neq 0$ and continuously on $\hat{\varepsilon}$ near $\hat{\varepsilon} = 0$.

(vii) We define the usual equivalence relation

$$(\psi_{1,\varepsilon}^0, \psi_{1,\varepsilon}^\infty) \equiv (\psi_{2,\varepsilon}^0, \psi_{2,\varepsilon}^\infty) \quad \text{iff} \quad \exists C, C' \in \mathbb{C}^* \begin{cases} \psi_{1,\varepsilon}^0(w) = C\psi_{2,\varepsilon}^0(C'w) \\ \psi_{1,\varepsilon}^\infty(w) = C\psi_{2,\varepsilon}^\infty(C'w) \end{cases} \tag{2.12}$$

(viii) For all $\hat{\varepsilon} \in V_\delta$, $[(\psi_\varepsilon^0, \psi_\varepsilon^\infty)]$ is a complete modulus of analytic classification for $f_\varepsilon$.

**2.2 Theorem ([25])** *The family* $\{[(\psi_{\hat{\varepsilon}}^0, \psi_{\hat{\varepsilon}}^\infty)]\}_{\hat{\varepsilon}\in V_\delta}$, *together with* $a(0)$, *is a complete modulus of analytic classification of the family* $\{f_\varepsilon\}$.

**Proof** We just give an idea of the proof: it is a generalization of ideas of Lavaurs and Shishikura (their method applies to smaller sectors in $\varepsilon$-space as in Fig. 9).

It uses the construction of Fatou coordinates embedding $f_\varepsilon$ into the time-one map of (2.4) by means of the Ahlfors-Bers theorem [24].

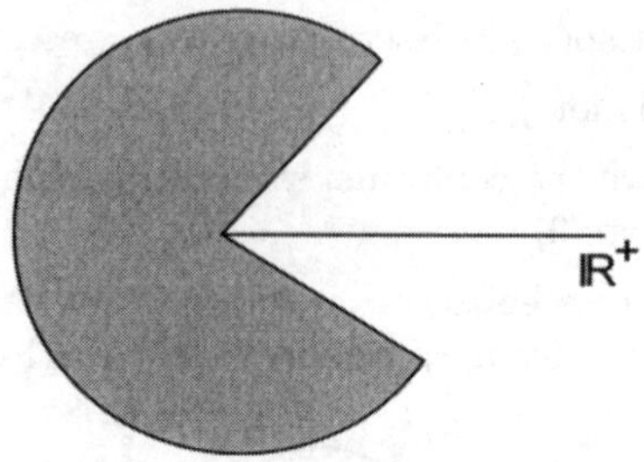


Figure 9: Type of sectors in $\varepsilon$-space in Shishikura's paper

• We lift $f_\varepsilon$ by means of

$$Z = \begin{cases} \frac{1}{2\sqrt{\varepsilon}} \ln \frac{z-\sqrt{\varepsilon}}{z+\sqrt{\varepsilon}}, & \varepsilon \neq 0 \\ -\frac{1}{z}, & \varepsilon = 0. \end{cases} \tag{2.13}$$

The image of $U \setminus \{\pm\sqrt{\varepsilon}\}$ is $\hat{U}$, which is $\mathbb{C} \setminus B_0$ (resp. $\mathbb{C} \setminus \cup_{k\in\mathbb{Z}} B_k$), where the $B_k$ are disjoint simply connected domains of $\mathbb{C}$. $B_0$ contains the origin and is called the *fundamental hole.*

• This change of coordinates has the property of transforming the diffeomophism $f_\varepsilon$ to a small perturbation of the translation by 1. Namely, $f_\varepsilon(z)$ becomes

$$F_\varepsilon(Z) = Z + 1 + R(\varepsilon, Z), \tag{2.14}$$

with $R(\varepsilon, Z)$ and $\partial/\partial Z(R(\varepsilon, Z))$ small.

• We consider *admissible lines* $l$, which are lines such that $l$ and $F_\varepsilon(l)$ generate fundamental domains $\hat{C}_\varepsilon$ for $F_\varepsilon$ (Fig. 10).

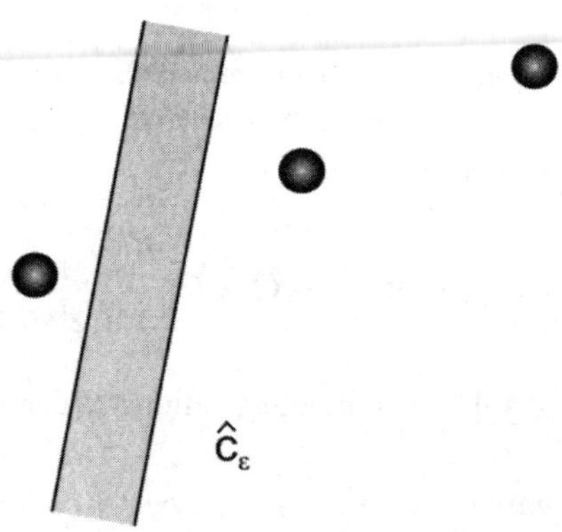


Figure 10: A fundamental domain generated by an admissible line

• Families of admissible lines can be chosen so as to depend continuously on $\hat{\varepsilon} \in V_\delta$ (and analytically of $\hat{\varepsilon} \neq 0$) as in Fig. 11.

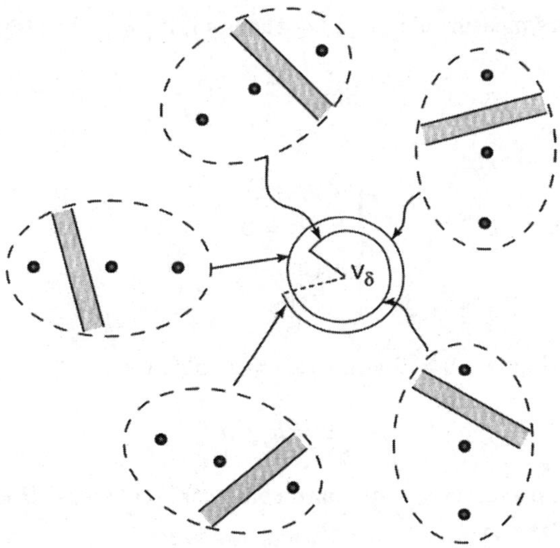


Figure 11: Admissible lines depending continuously on $\hat{\varepsilon}$

• The admissible lines generate *translation domains* $Q(l)$:

$$Q(l) = \{Z|\ \exists n \in \mathbb{Z},\ F_\varepsilon^n(Z) \in \hat{C}_\varepsilon \text{ and } \forall m \in [0,n],\ F_\varepsilon^m \in \hat{U}\}. \tag{2.15}$$

Note that a translation domain is independent of the admissible line which generates it, as long as the admissible line is deformed within $\hat{U}$ while remaining admissible.

• On a translation domain $Q$ we construct a Fatou coordinate $\Phi_\varepsilon$ conjugating $F_\varepsilon$ with $Z \mapsto Z+1$:

$$\Phi_\varepsilon(F_\varepsilon(Z)) = \Phi_\varepsilon(Z) + 1. \tag{2.16}$$

Fatou coordinates are holomorphic diffeomorphisms. Moreover, they are uniquely defined up to a translation.

The construction is done in two steps:

(i) We first construct in a very rough way a homeomorphism $h: Q \to \mathbb{C}$ conjugating $F_\varepsilon$ with $Z \mapsto Z+1$. If $l = \{Z_1 + Ye^{i\theta}|Y \in \mathbb{R}\}$, then

$$h^{-1}(X,Y) = (1-X)(Z_1 + Ye^{i\theta}) + XF_\varepsilon(Z_1 + Ye^{i\theta}). \tag{2.17}$$

(ii) The mapping $h^{-1}$ is quasi-conformal. Indeed let

$$\mu = \frac{\partial h^{-1}/\partial \overline{Z}}{\partial h^{-1}/\partial Z}. \tag{2.18}$$

We extend $\mu$ in a periodic way to all of $\mathbb{C}$ by the translation $Z \mapsto Z+1$. The map $\mu$ is in $L^\infty(\mathbb{C})$ and has $L^\infty$-norm $||\mu||_\infty < 1$. So $\mu$ is a Beltrami field. It induces $\tilde{\mu}$ on $\mathbb{S}^2$ by means of $Z \mapsto \exp(-2\pi i Z)$, and we let $\tilde{\mu}(0) = \tilde{\mu}(\infty) = 0$.

By the Ahlfors-Bers measurable mapping theorem, if $\mu_0$ is the standard conformal structure on $S^2$, then

$$\exists!\tilde{h}_1 : \mathbb{S}^2 \to \mathbb{S}^2 \tag{2.19}$$

such that

$$\begin{cases} \tilde{h}_1^*\mu_0 = \tilde{\mu} \\ \tilde{h}_1(0) = 0 \\ \tilde{h}_1(\infty) = \infty \\ \tilde{h}_1(1) = 1. \end{cases} \tag{2.20}$$

We lift $\tilde{h}_1$ to $h_1$ such that $h_1(0) = 0$ and $h_1(1) = 1$. Then

$$\Phi = h_1 \circ h \tag{2.21}$$

is quasi-conformal and preserves the standard conformal structure. Hence $\Phi$ is conformal and conjugates $F_\varepsilon$ with $Z \mapsto Z + 1$.

• If we choose a base point $Z_0(\varepsilon)$ depending holomorphically on $\varepsilon$, then the unique Fatou coordinate satisfying $\Phi_\varepsilon(Z_0(\varepsilon)) = 0$ depends continuously on $\hat{\varepsilon}$ near $\hat{\varepsilon} = 0$ and holomorphically on $\hat{\varepsilon} \neq 0$. □

There exist two different ways of constructing the modulus:

I. We take two Fatou coordinates $\Phi_\varepsilon^\pm$ associated to two admissible lines passing between holes and located on the two sides of the fundamental hole (Fig. 12).

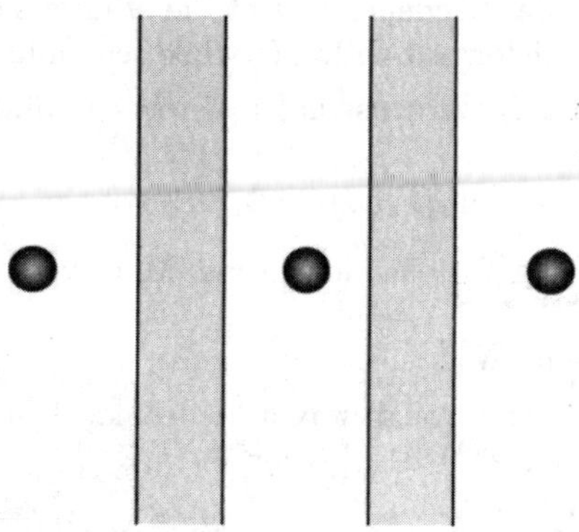

Figure 12: Two fundamental domains on each side of the fundamental hole

The modulus is defined as

$$\Psi_\varepsilon = \Phi_\varepsilon^+ \circ (\Phi_\varepsilon^-)^{-1}. \tag{2.22}$$

It is defined on a domain as in Fig. 13.

Using $Z \mapsto \exp(-2\pi i Z)$, this yields the two diffeomorphisms $\psi_\varepsilon^\infty$ and $\psi_\varepsilon^0$ of Fig. 8, which are the unfolding of the Écalle-Voronin modulus.

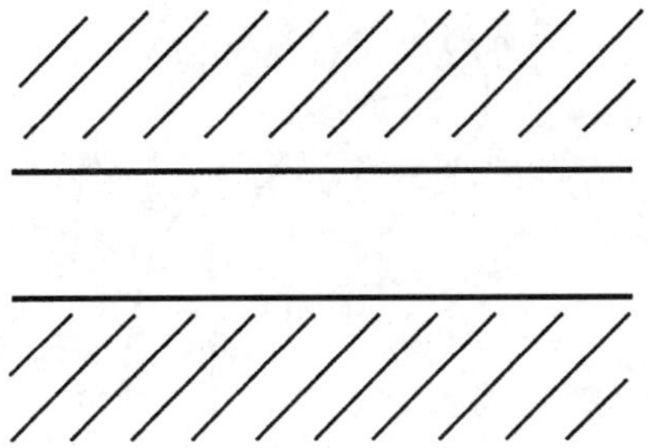

Figure 13: The domain of the modulus $\Psi_\varepsilon$

II. The second method does not pass to the limit when $\varepsilon \to 0$ but it allows one to study the dynamics of $\sqrt{\varepsilon}$. Indeed, we have a global transition diffeomorphism, called the *Lavaurs map*

$$L_\varepsilon : S_\varepsilon^- \to S_\varepsilon^+, \tag{2.23}$$

sending a point $w \in S_\varepsilon^-$ to the first point of $S_\varepsilon^+$ belonging to the orbit of $w$ under $f_\varepsilon$ when iterating $f_\varepsilon$ positively. As the Lavaurs map is global and sends 0 to 0, and $\infty$ to $\infty$ it is linear:

$$w \mapsto L_\varepsilon(w) = \tau(\varepsilon)w \tag{2.24}$$

with $\tau(\varepsilon) \neq 0$. This second point of view allows one to define *renormalized return maps* and to describe the *parametric resurgence.*

**The renormalized return maps** They are defined as $k_\varepsilon^\infty, k_\varepsilon^0 : S_\varepsilon^+ \to S_\varepsilon^+$ (Fig. 14) where

$$\begin{aligned} k_\varepsilon^\infty(w) &= \tau(\varepsilon)\psi_\varepsilon^\infty(w), \\ k_\varepsilon^0(w) &= \tau(\varepsilon)\psi_\varepsilon^0(w). \end{aligned} \tag{2.25}$$

The derivatives of $k_\varepsilon^\infty$ at $\infty$ (resp. $k_\varepsilon^0$ at 0) are analytic invariants:

$$\begin{aligned} (k_\varepsilon^0)'(0) &= \exp(4\pi^2/\mu_-), \\ (\widetilde{k_\varepsilon^\infty})'(0) &= \exp(4\pi^2/\mu_+), \end{aligned} \tag{2.26}$$

where $\widetilde{k}^\infty$ is the expression of $k^\infty$ in the coordinate $1/w$. (Note that $\infty$ (resp. 0) corresponds to $\sqrt{\varepsilon}$ (resp. $-\sqrt{\varepsilon}$)).

**The parametric resurgence** We will limit ourselves to describe the parametric resurgence at $-\sqrt{\varepsilon}$ by means of $\psi_\varepsilon^0$. Let us first remark that it is possible to choose coordinates on $S_\varepsilon^+$ and $S_\varepsilon^-$ so that

$$\psi_\varepsilon^0(w) = w + \sum_{n>1} b_n(\varepsilon)w^n, \tag{2.27}$$

with the $b_n(\varepsilon)$ depending continuously on $\varepsilon$. In particular, for $\varepsilon = 0$,

$$\psi^0(w) = w + \sum_{n>1} b_n w^n, \tag{2.28}$$

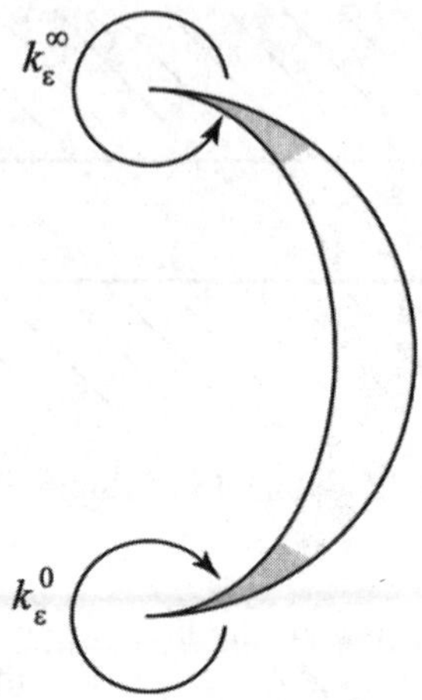


Figure 14: The return maps on $S_\varepsilon^+$

where $b_n = b_n(0)$, is half of the Écalle-Voronin modulus. Hence if $\psi^0$ is nonlinear, then $\psi_\varepsilon^0$ is nonlinear for all small $\varepsilon$.

**The particular case where $\lambda_- = \exp(\mu_-) = \exp(-2\pi i/n)$, with $n$ large** In this case,

$$k_\varepsilon^{0\prime}(0) = 1. \tag{2.29}$$

Hence $k_\varepsilon^0$ nonlinear implies that $k_\varepsilon^0$ non-linearizable.

**The general case** We consider the resonances

$$\lambda_- = \exp(\mu_-) = \exp\left(-\frac{2\pi i p}{n}\right), \qquad n \equiv q \pmod p \tag{2.30}$$

with $n$ large. Then

$$k_\varepsilon^{0\prime}(0) = \exp\left(2\pi i \frac{q}{p}\right). \tag{2.31}$$

**2.3 Theorem ([25])** *If*

$$g(w) = \exp\left(2\pi i \frac{q}{p}\right) \psi^0(w) \tag{2.32}$$

*is non-linearizable, then $k_\varepsilon^0$ is non-linearizable for $\varepsilon$ such that*

$$\begin{cases} \lambda_- = \exp\left(-2\pi i p/n\right) \\ n \text{ large} \\ n \equiv q \pmod p. \end{cases} \tag{2.33}$$

*This occurs as soon as one of the coefficients of the normal form of $g(w)$ does not vanish. The $k$-th coefficient of the normal form of $g(w)$ is a quasi-homogeneous polynomial*

$$L_{p,q}^k(b_2, \ldots, b_{kq+1}). \tag{2.34}$$

## 2.3 The two ways to normalize to the model

The Fatou coordinates allow us to find a normalizing transformation to the model.

*We have embedded the diffeomorphism into the model in the middle (between* $\pm\sqrt{\varepsilon}$*) and read the incompatibility when turning around* $\pm\sqrt{\varepsilon}$*:* we have found a change of coordinates transforming $f_\varepsilon(z)$ to the model over a ramified neighborhood of $\pm\sqrt{\varepsilon}$ for $\hat{\varepsilon} \in V_\delta$ (Fig. 15).

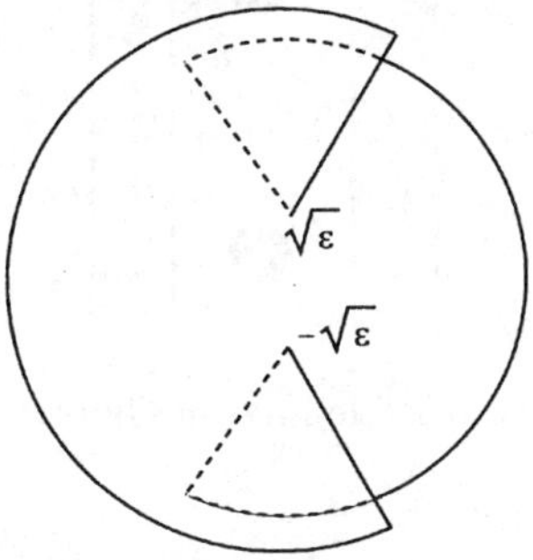


Figure 15: The domain on which the diffeomorphism is embedded in the model

We compare this point of view with that of Glutsyuk [14] which was already introduced by Martinet [26].

Glutsyuk works for values of $\varepsilon$ in a sector $V$

$$\begin{cases} |\varepsilon| < r \\ \arg(\varepsilon) \in (-\pi + \delta, \pi - \delta). \end{cases} \tag{2.35}$$

For any such $\varepsilon \in V$, the diffeomorphism $f_\varepsilon$ is linearizable in the neighborhood of $\pm\sqrt{\varepsilon}$. Glutsyuk finds two changes of coordinates defined in neighborhoods of each singular point transforming $f_\varepsilon(z)$ to the linear model in the neighborhood of $\pm\sqrt{\varepsilon}$. He reads the incompatibility on the intersection of the two neighborhoods (Fig. 16).

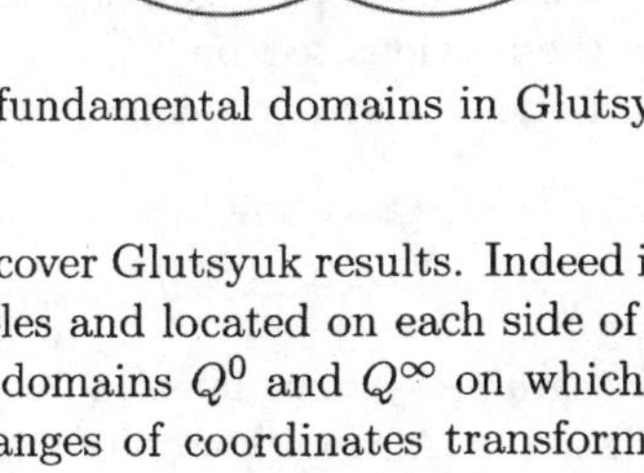


Figure 16: The fundamental domains in Glutsyuk point of view

Our method allows us to recover Glutsyuk results. Indeed it suffices to take two admissible lines parallel to the lines of holes and located on each side of the line of holes (Fig. 17).

They generate translation domains $Q^0$ and $Q^\infty$ on which we have Fatou coordinates $\Phi_\varepsilon^0$ and $\Phi_\varepsilon^\infty$. $\Phi_\varepsilon^0$ and $\Phi_\varepsilon^\infty$ are changes of coordinates transforming the diffeomorphism to the

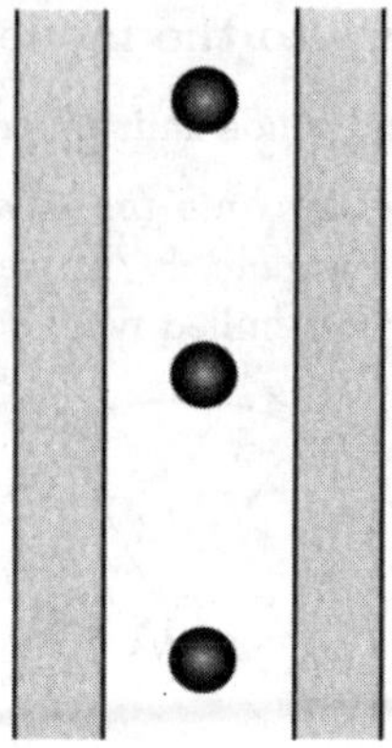

Figure 17: The admissible lines in Glutsyuk's point of view

model diffeormorphism. As the model is linearizable near $\pm\sqrt{\varepsilon}$ this is equivalent to finding linearizing changes of coordinates. The *Glutsyuk modulus* is then given by

$$\Psi_\varepsilon^G = \Phi_\varepsilon^0 \circ (\Phi_\varepsilon^\infty)^{-1}. \tag{2.36}$$

The Glutsyuk modulus is equivalent to the modulus $[(\psi_\varepsilon^0, \psi_\varepsilon^\infty)]$ described before (see [25]).

*In the Glutsyuk point of view we normalize near the singular points $\pm\sqrt{\varepsilon}$ and read the incompatibility in the middle.*

## 3 Applications

One motivation for the previous study comes from work of the author with P. Mardešić and C. Christopher [9, 4]. In these papers we tried to understand how the strata of integrable and linearizable systems at the origin are organized in a family of polynomial systems

$$\begin{aligned} \dot{x} &= x + \sum_{i+j=2}^{n} a_{ij}x^iy^j \\ \dot{y} &= -\lambda y + \sum_{i+j=2}^{n} b_{ij}x^iy^j, \end{aligned} \tag{3.1}$$

where $\lambda \in \mathbb{R}^+$ and $a_{ij}, b_{ij} \in \mathbb{C}$. We observed strange behaviours which we can now explain with the new tools developed in the previous section.

In [4] we study the family of Lotka-Volterra systems:

$$\begin{aligned} \dot{x} &= x(1 + ax + by) \\ \dot{y} &= y(-\lambda + cx + dy), \end{aligned} \tag{3.2}$$

where we want to identify all integrable systems for all $\lambda = p/q \in \mathbb{Q}^+$. A characterization exists for all $\lambda = n/2$ and $2/n$ [13, 15]. Also all Darboux integrable systems are known [2].

As the system is simple we could calculate saddle quantities for several values of $\lambda \in \mathbb{Q}^+$ and arrive at the following conjecture.

**3.1 Conjecture** ([4]) The following two mechanisms are sufficient to describe all systems (3.2) with an integrable point at the origin:

(i) The system has one extra invariant algebraic curve apart from the two axes. (Limit cases are possible when one of the axes or the equator is a multiple invariant algebraic curve) Such systems are classified in [2].

(ii) The system is integrable by the "monodromy argument".

**The monodromy argument** This argument deals with the monodromy group of one of the two axes which are the separatrices of the origin. Let us describe it on the $x$-axis. The system has three singular points on $y = 0$, namely $(0, 0)$, $P_1 = (-1/a, 0)$, and $P_\infty$ which is the point at infinity.

• It is known that the origin is integrable if and only if its monodromy map $M_0$ is linearizable.

• $M_0$ is conjugate to $M_\infty^{-1} \circ M_1^{-1}$ where $M_1$ (resp. $M_\infty$) is the monodromy of $P_1$ (resp. $P_\infty$), see Fig. 18.

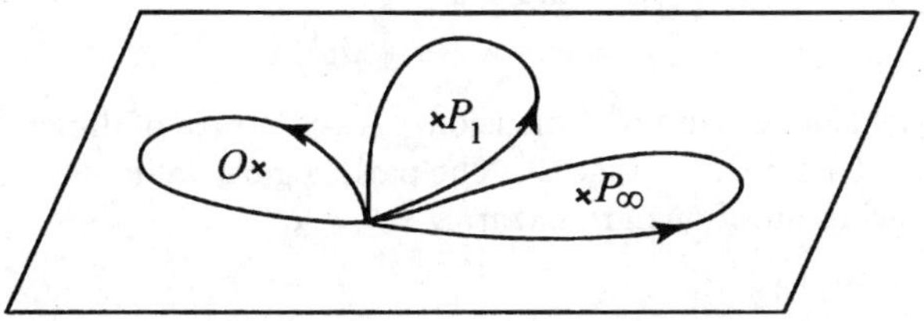


Figure 18: The monodromy group of the $x$-separatrix

• Hence a first sufficient condition for the linearizability of $M_0$ is given by $M_\infty \equiv id$ and $M_1$ linearizable. $M_\infty \equiv id$ as soon as it is a resonant node with eigenvalues $\mu_1, \mu_2$ with $\mu_2/\mu_1 \in \mathbb{N}$ (we have no problem of resonance as the node has two analytic separatrices). Moreover, $M_1$ is linearizable as soon as it is a non-resonant node or a resonant node with a zero resonant monomial.

• Similarly, a second sufficient condition for the linearizability of $M_0$ is given by $M_1 \equiv id$ and $M_\infty$ linearizable. This is realized if the eigenvalues $\nu_1, \nu_2$ of $P_1$ satisfy $\nu_2/\nu_1 \in \mathbb{N}$ and $M_\infty$ linearizable. To realize $M_\infty$ linearizable we have two choices: either $P_\infty$ is a node (we have no problem of resonance as the node has two analytic separatrices) or it is a saddle and we play the monodromy argument on the line at infinity.

**3.2 Theorem** ([4]) *These two mechanisms are sufficient to provide necessary and sufficient conditions for integrability in the Lotka-Volterra system* (3.2) *for the following rational values of* $\lambda$*:*

• $\lambda = n, 1/n$. *This case was already studied in* [13] *but other arguments were used to prove that the conditions were necessary and sufficient.*

• $\lambda = n/2, 2/n$. *This case was already studied in* [15] *but other arguments were used to prove that the conditions were necessary and sufficient.*

• $\lambda = p/q$, $p + q \leq 12$.

The paper [4] also lists general conditions of integrability. One is the following:

**Condition (*)** The system (3.2) is integrable if

$$(*) \begin{cases} a = -c \\ \lambda > 1, \ \lambda \neq 1 + 1/m \text{ with } m \in \mathbb{N}. \end{cases} \tag{3.3}$$

If $\lambda = 1 + 1/m$, an additional condition is necessary for integrability.

Let us explain condition (*). The condition $a = -c$ guarantees that $P_\infty$ is a node with eigenvalues in the ratio 1 : 2, yielding $M_\infty \equiv \text{id}$. The condition $\lambda > 1$ guarantees that $P_1$ is a node. If $\lambda \neq 1 + 1/m$ the node is non-resonant, yielding that $M_1$ is linearizable. If $\lambda = 1 + 1/m$ then the node is resonant with eigenvalues in the ratio $n : 1$, and we have to require that the resonant monomial has a zero coefficient.

In practice, when $b, c \neq 0$, we can scale $b = c = 1$ yielding $a = -1$, and the family depends only on the parameter $d$:

$$\begin{aligned} \dot{x} &= x(1 - x + y) \\ \dot{y} &= x(-\lambda + x + dy). \end{aligned} \tag{3.4}$$

Then condition (*) yields a stratum of dimension 1 (resp. strata of dimension 0) when $\lambda > 1$, $\lambda \neq 1+1/m$ (resp. $\lambda = 1+1/m$), see Fig. 19. The pathological values $\lambda = 1+1/m$ accumulate to $\lambda = 1$. So the following questions are natural.

**Questions:**

(1) What happens for $\lambda = 1$? Indeed the integrable points for $\lambda = 1+1/m$ of Fig. 19 show a nice pattern of sequences with accumulation points on $\lambda = 1$.

(2) What happens for $\lambda < 1$?

**Computational answers:**

(1) Calculations yield that for $\lambda = 1$ the origin is integrable only if $d = -1$. So all limit points except one are not integrable.

(2) Calculations on the side $\lambda < 1$, in particular for $\lambda = 1 - 1/m$, show that points seem generically non-integrable, so the pattern seems not to be symmetric.

**Explanation:**

Except for $d = -1$, the map $M_0$ has a generic parabolic point as studied in the previous section. The limit points at $\lambda = 1$ in Fig. 19 are *semi-normalizable*: If $(\psi^0, \psi^\infty)$ is the Écalle-Voronin modulus of $M_0$, then we have

$$\begin{cases} \psi^\infty \text{ linear} \\ \psi^0 \text{ non-linear.} \end{cases} \tag{3.5}$$

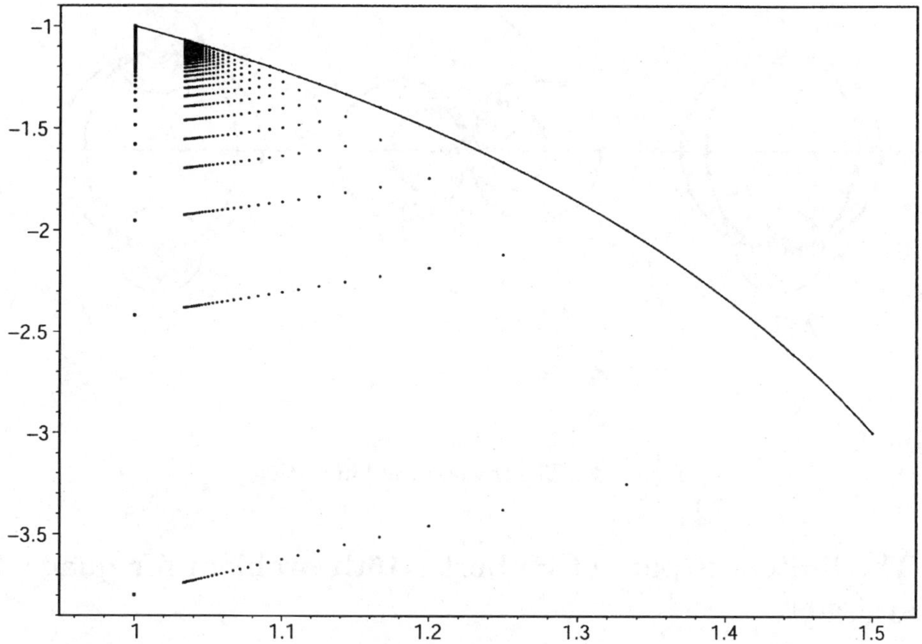


Figure 19: The values of $d$ for which the system (3.4) is integrable for $\lambda = 1 + 1/n$. The system is integrable on the curve $d = \lambda/(\lambda - 2)$.

At $\lambda = 1$ we have a *transcritical bifurcation.* Indeed $M_0$ has two fixed points for $\lambda \neq 1$ and a double fixed point for $\lambda = 1$. The first fixed point is the origin and the second fixed point corresponds to an invariant manifold [21]. These two points "pass through each other" at $\lambda = 1$, as is usual in a transcritical bifurcation (Fig. 20):

- For $\lambda < 1$, $\psi^\infty$ controls the invariant manifold while $\psi^0$ controls the origin. Hence the origin is necessarily nontrivial by Theorem 2.3 for $\lambda = 1 - 1/m$ with $m$ large, while the invariant manifold may or may not be trivial.
- For $\lambda > 1$, $\psi^\infty$ controls the origin which can hence be integrable if $d(\lambda)$ is well chosen, while $\psi^0$ controls the invariant manifold which is necessarily nontrivial for $\lambda = 1 + 1/m$ by Theorem 2.3.

### Another consequence

For $\lambda = 1$ we can say more about the points which are not limit points of sequences of integrable systems for $\lambda = 1 + 1/m$. These points are approached by integrable saddles as soon as $\lambda \neq 1 + 1/m$. If $(\psi^0, \psi^\infty)$ is the Écalle-Voronin modulus of the limit point, then we know that the unfolding $\psi^\infty$ controls the origin for $\lambda > 1$. Hence necessarily $\psi^\infty$ is a Möbius function by Theorem 1.1.

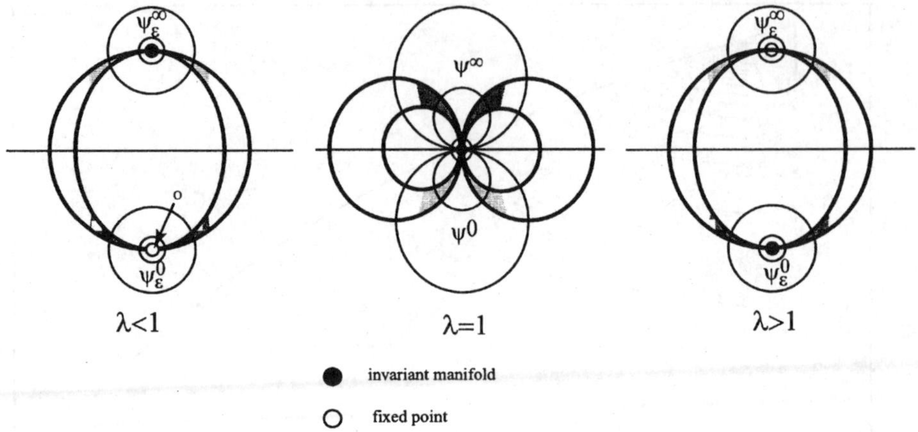


Figure 20: The transcritical bifurcation

# 4 The finiteness part of Hilbert's 16th problem for quadratic systems

Hilbert's 16th problem asks for the maximum number $H(n)$ of limit cycles of a polynomial planar vector field of degree $\leq n$. It is still open for any $n$. Since the simultaneous proofs of Écalle and Ilyashenko in 1987 it is known that a single vector field has a finite number of limit cycles. The finiteness part of Hilbert's 16th problem deals with the existence of a uniform bound for the number of limit cycles of a polynomial system of degree $\leq n$:

**The finiteness part of Hilbert's problem for polynomial vector fields** For any $n \in \mathbb{N}$ there exists an $M(n) \in \mathbb{N}$ such that any polynomial vector field

$$\begin{aligned} \dot{x} &= P(x,y) \\ \dot{y} &= Q(x,y) \end{aligned} \tag{4.1}$$

with $P, Q \in \mathbb{R}[x,y]$ and $\max\{\deg(P), \deg(Q)\} \leq n$, has at most $M(n)$ limit cycles. It is usually given in the condensed form

$$H(n) < \infty. \tag{4.2}$$

In [10] a strategy was developed to attack the finiteness part in the particular case $n = 2$. The idea of the strategy is the following. It is possible to compactify the phase space to the Poincaré sphere. Hence, if there is an infinite number of limit cycles, then they need to accumulate on some graphics. This idea is sufficient to prove that any individual polynomial vector field has a finite number of limit cycles. For quadratic systems the proof is elementary [1]. Moreover, the coefficients of the system are parameters. Rescaling time, it is possible to consider the parameter space of nonzero vector fields of degree $\leq 2$ as a compact space. Limit cycles depend on the parameters and accumulate on "limit periodic sets", which we want to look at as "local" objects. If all these limit periodic sets have *finite cyclicity*, a "local finiteness property", then we can expect a uniform bound for the number of limit cycles.

**Notation for the figures** In this section double arrows on an invariant manifold in the figures indicate hyperbolicity in that direction, i.e. the corresponding eigenvalue is nonzero.

**4.1 Definition** A limit periodic set $\Gamma$ of a vector field $v_{\lambda_0}$ has *finite cyclicity inside a family of vector fields* $\{v_\lambda\}_{\lambda\in\Lambda}$ if there exist $\varepsilon, \delta > 0$, and $N \in \mathbb{N}$ such that any $v_\lambda$ with $|\lambda - \lambda_0| < \delta$ has at most $N$ limit cycles $\gamma_1, \ldots, \gamma_m$ such that $d_H(\gamma_i, \Gamma) < \varepsilon$, where $d_H$ is the Hausdorff distance between compact sets.

**4.2 Theorem ([10])** *$H(2) < \infty$ if and only if all* 121 *limit periodic sets surrounding the origin in the family*

$$\begin{aligned} \dot{x} &= \lambda x - \mu y + a_1x^2 + a_2xy + a_3y^2 \\ \dot{y} &= \mu x + \lambda y + b_1x^2 + b_2xy + b_3y^2, \end{aligned} \tag{4.3}$$

$(\lambda, \mu) \in \mathbb{S}^1$, $(a_1, a_2, a_3, b_1, b_2, b_3) \in \mathbb{S}^5$, *have finite cyclicity inside the family* (4.3).

Although the paper [10] was published in 1994, the program has been under way since 1991. It progresses well. The purpose of the present section is to present its actual status with a discussion on what is done and what remains to be done. Since 1994 no error was found in the list of graphics of [10]. The author however feels that the list should have contained 125 graphics. Indeed, the list of [10] is a list up to topological equivalence. Four graphics with elliptic point appear in two forms (Fig. 21) which in the blow-ups are not topologically equivalent. The graphics appear in Fig. 22.

Figure 21: The two forms of an hh-graphic through an elliptic point

Table 1 lists the different classes of graphics and what is done in each case.

The different classes will be discussed below together with references for proofs. The category "open but probably easy" is obviously subjective and reflects the belief of the author that no new methods but merely elementary calculations are enough to prove finite cyclicity. The calculations may not be short. The author has not checked all proofs. This would be premature at this stage of the program, if we consider the fact that unifying proofs of many cases could appear before the program is completed.

### Strategies to attack the program and avoid making mistakes

(1) Start from the most generic graphics and follow with the more degenerate. As degenerate graphics bifurcate into generic graphics everything has to be coherent.

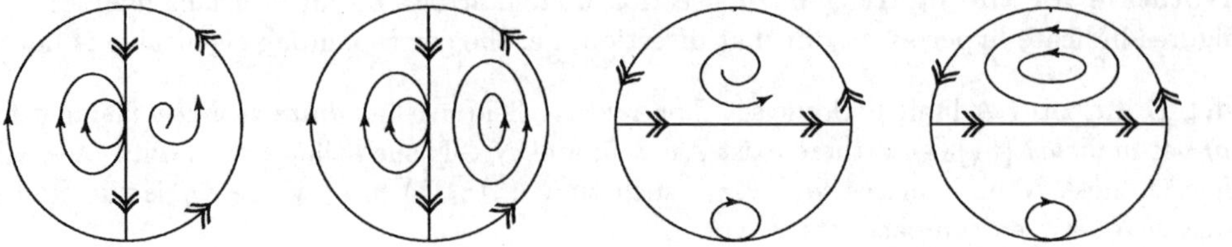

(a) $(H_4^1)$ (b) $(H_5^1)$ (c) $(H_{12}^3)$ (d) $(H_{13}^3)$

Figure 22: The four additional cases

| Class of graphics | Done | Open but probably easy | Open |
|---|---|---|---|
| Hyperbolic graphics | 10 (Mourtada) | | |
| Elementary non hyperbolic graphics | 47 | 1 | |
| Graphics through a nilpotent saddle | 2 | | 2 (center cases) |
| Graphics through a nilpotent elliptic point with pp-transition | 16 | | 4 (center cases) |
| Graphics through a nilpotent elliptic point with hp-transition | 6 | 4 | |
| Graphics through a nilpotent elliptic point with hh-transition (mult. 3) | 2 | 4 | 6 (center cases) |
| The four additional cases | | | 4 |
| Graphics through a nilpotent elliptic point with multiplicity 4 | | | 4 |
| Degenerate graphics | | | 13 |
| Total = 125 | 83 | 9 | 33 |

Table 1

(2) Some graphics appear in families: the strategy is to study simultaneously all graphics in a family including the bordering graphics. Everything has to be coherent. Examples will be discussed below.

## 4.1 Hyperbolic graphics

Nearly all cases have been done, one by one, by elementary methods. More recently, Mourtada announced

**4.3 Theorem ([29, 30, 31, 32])** *Any hyperbolic polycycle has finite cyclicity inside any analytic finite-parameter family of vector fields.*

When working with elementary methods the center cases are usually difficult problems. For instance, the "triangle" (Fig. 23) can lie at the intersection of three strata: indeed, it has three invariant lines, yielding Darboux integrability. It can be time-reversible when there is a symmetry axis, and it can simultaneously be Hamiltonian.

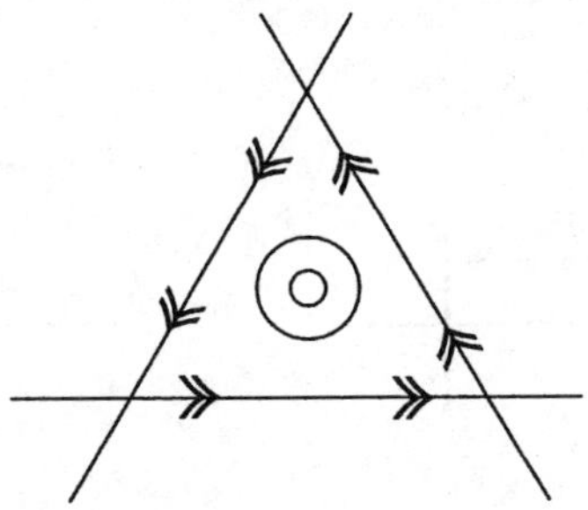

Figure 23: The triangle

## 4.2 Elementary non-hyperbolic graphics

**4.4 Definition** A graphic is *elementary* if all its singular points are *elementary* (i.e. with one or two nonzero eigenvalues), either hyperbolic saddles or *semi-hyperbolic points* (i.e. points with exactly one zero eigenvalue).

The finite cyclicity of these graphics are usually easier problems and all are treated with elementary methods.

- The case $(I_{13}^2)$ is treated in [41].
- 46 cases are treated in different papers to which the author contributed: [11, 5, 6, 34].
- The graphic $(I_{16a}^2)$ is still open.

### Ingredients used

(1) $C^k$ normal forms near the saddle-nodes,

$$\begin{aligned} \dot{x} &= x^2 + \varepsilon \\ \dot{y} &= \pm y(1+ax), \end{aligned} \tag{4.4}$$

and semi-hyperbolic saddles

$$\begin{aligned} \dot{x} &= x^3 + \varepsilon_2 x + \varepsilon_1 \\ \dot{y} &= -y(1+ax^2). \end{aligned} \tag{4.5}$$

It is remarkable that the $C^k$ normal forms have been sufficient even in the center cases!

(2) We have two types of transition maps near a semi-hyperbolic point:

• **Central transition** (Fig. 24(a)). The Dulac map is linear when it exists (it does not exist for $\varepsilon = 0$!):

$$D_\varepsilon(y) = m(\varepsilon)y, \qquad \lim_{\varepsilon \to 0} m(\varepsilon) = 0. \tag{4.6}$$

• **Stable-center transition** (Fig. 24(b)). The Dulac map $D_\varepsilon$ is flat in $(x, \varepsilon)$ and has additional flatness properties [11].

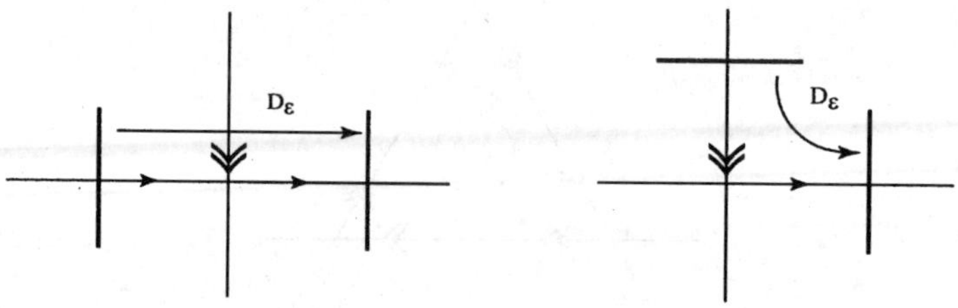


(a) central transition (b) stable-center transition

Figure 24: The Dulac maps for a saddle-node

(3) Khovanskii's method is used as soon as the Poincaré return map contains the composition of a stable-center transition with a center-unstable transition. The principle of *Khovanskii's method* is that the Dulac map is a solution of a Pfaff equation:

• **Case of a central transition** The Pfaff equation is:

$$x\,dy - y\,dx = 0. \tag{4.7}$$

• **Case of a stable-center (center-unstable) transition** The Pfaff equation is:

$$F(x)\,dy \mp y\,dx = 0, \tag{4.8}$$

where the normal form is

$$\begin{aligned} \dot{x} &= F(x) \\ \dot{y} &= \mp y. \end{aligned} \tag{4.9}$$

(4) Except for $(I_{13}^2)$, which is treated in [41], all center cases are not located at the intersection of strata of centers: the *Bautin ideal* is radical allowing an elementary treatment [34].

Hence:

• we first make a proof in the generic case (e.g. $(I_4^2)$, Fig. 25(a));

• we then transform it into a proof for the center case by a Bautin type argument (e.g. $(I_5^2)$, Fig. 25(b)).

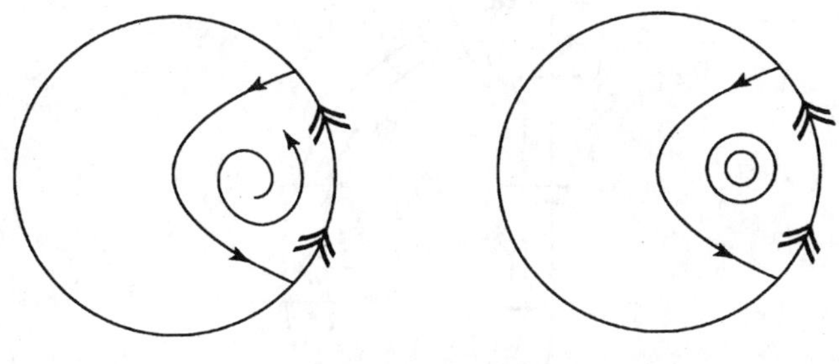

(a) $(I_4^2)$ (b) $(I_5^2)$

Figure 25: The graphics $(I_4^2)$ and $(I_5^2)$

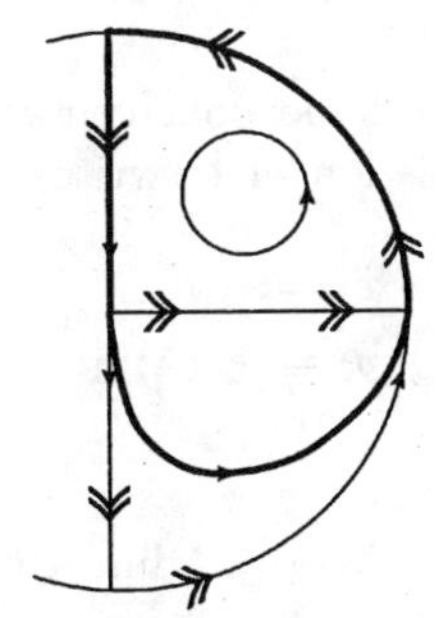

Figure 26: The graphic $(I_{16a})$

**The remaining open case** $(I_{16a}^2)$ There is a continuum of graphics $(I_{16a}^2)$ (Fig. 26): they occur inside a family.

The bordering graphic $(I_{16b}^2)$ is treated in [34] and $(H_6^3)$ is treated in [6]. Moreover, a proof of finite cyclicity is given in [6] for the corresponding generic graphics $(I_{14a}^2)$ and $(I_{15a}^2)$ appearing when breaking the center condition. As the Bautin ideal is radical we "should" be able, with mere calculations, to reduce the center case to the generic case via a Bautin type argument.

We still believe that such calculations can be done, but that they will take more than one or two pages.

**A remark on the genericity condition used to make the proof in the generic cases** Note that any of the graphics $(I_{14a}^2)$, $(I_{15a}^2)$ and $(I_{16a}^2)$ is of the form appearing in Fig. 27. We take sections near the singular points as in Fig. 27 and calculate the different Dulac maps in the neighborhood of the singular points and regular transitions.

The sections are taken in normalizing $C^k$ normal forms in the neighborhood of the singular points, allowing an explicit calculation of the Dulac maps. In all this discussion $\lambda$ is the multi-parameter of the problem, and we suppose that the graphic occurs at $\lambda = 0$. We discuss the

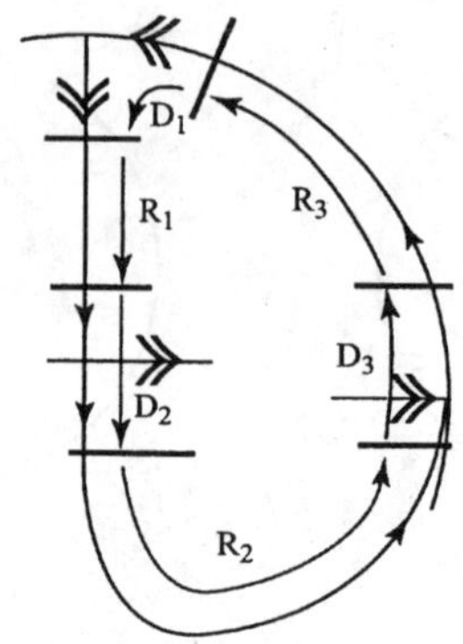


Figure 27: The transition maps for $(I^2_{14a})$, $(I^2_{15a})$ and $(I^2_{16a})$

easier case, namely the one where the hyperbolicity ratio at $P_1$ is $r(0)$, with $r(0)$ irrational (the other cases are similar but a little longer to write). Then the normal form near $P_1$ is

$$\begin{aligned} \dot{x}_1 &= x_1 \\ \dot{y}_1 &= -r(\lambda)y_1. \end{aligned} \tag{4.10}$$

Moreover, we have

$$D_2(x_2) = M(\lambda)x_2, \qquad \lim_{\lambda\to 0} M(\lambda) = +\infty, \tag{4.11}$$

$$D_3(x_3) = m(\lambda)x_3, \qquad \lim_{\lambda\to 0} m(\lambda) = 0. \tag{4.12}$$

It is shown in [16] that the $C^k$-changes of coordinates to normal form are not unique. Using this freedom allows to simplify the regular transitions. Because the connection along the equator is fixed we can suppose that $R_3$ is the identity, and $R_1(y_1) = y_1 + \delta$ is a translation. Hence the Poincaré return map is identically zero for some $\lambda$ if

$$R_{2,\lambda}(x_3) = [m(\lambda)^{-1/r(\lambda)} M(\lambda)^{-1} x_3 - \delta]^{1/r(\lambda)}, \tag{4.13}$$

i.e. $R_{2,\lambda}$ is the $1/r$-th power of an affine map. A generic condition on $R_{2,0}$ for $\lambda = 0$ is that $R_{2,0}$ is not the $1/r$-th power of an affine map at the level of a finite jet.

**How to check this condition in practice?** A simple method, *the analytic extension principle,* exists because we deal with analytic vector fields. Indeed our graphic occurs in a family of graphics. Let us look near the boundary graphic which is the hemicycle (Fig. 28). The transition maps $f$ and $g$ along the invariant line and the equator are nonlinear as soon as the hyperbolicity ratio of the point $P_4$ is different from 1.

Indeed in [6] we showed through an explicit calculation that $f$ and $g$ have a non-vanishing second derivative. In the particular case where the hyperbolicity ratio is equal to 1 it is possible to prove that $P_4$ is non-integrable (the first saddle quantity is nonzero).

- So graphics very close to the hemicycle satisfy the genericity condition.

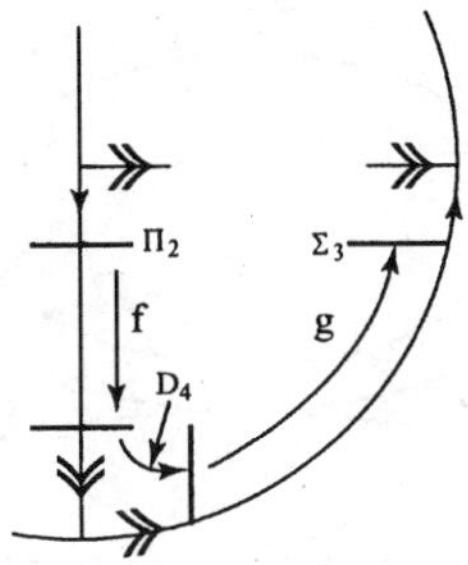


Figure 28: The lower part of the hemicycle

• The analycity extension principle guarantees that we can choose the sections $\Pi_2$ and $\Sigma_3$ analytic and parametrized by analytic coordinates. Hence we can "push" the genericity condition to the whole family of graphics.

The analycity extension principle relies on the following theorems proved in [7].

**4.5 Theorem** *A saddle-node of an analytic vector field can be brought in the neighborhood of the origin to the normal form*

$$\begin{aligned} \dot{x} &= x^2 \\ \dot{y} &= \pm y(1+ax) \end{aligned} \tag{4.14}$$

*by a* $C^\infty$ *change of coordinates which is analytic outside* $x = 0$.

**4.6 Theorem** *For a* $C^\infty$ *unfolding of a germ of analytic vector field with a saddle-node at the origin there exists a finitely smooth orbital equivalence with the polynomial normal form*

$$\begin{aligned} \dot{x} &= x^2 - \varepsilon \\ \dot{y} &= \pm y(1+ax). \end{aligned} \tag{4.15}$$

*For* $\varepsilon = 0$ *this equivalence is analytic outside the strong manifold.*

## 4.3 Graphics through a nilpotent saddle

**4.7 Theorem (Zhu [40])** *A "convex" graphic through a nilpotent saddle of codimension* 3 *(Fig.* 29*) has finite cyclicity if the return map has a derivative different from* 1.

The two cases appearing in Fig. 30 have been shown to have finite cyclicity:

• the transition occurs along an invariant parabola allowing to calculate $P'(0)$.

• A few adjustments to the proof allow to treat the case with an additional saddle-node.

• The point can be of codimension 4, but the transition is fixed along the equator so the proof of the Theorem 4.7 can be easily adjusted to this case.

The two cases which remain to be done are two center cases. Moreover, they can lie at the intersection of strata of centers.

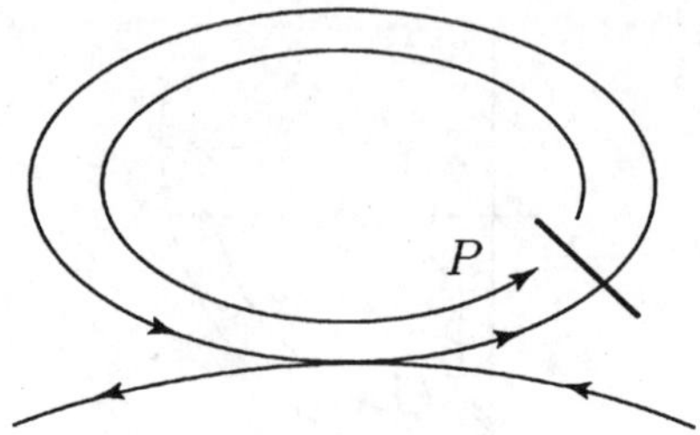


Figure 29: A "convex" graphic through a nilpotent saddle

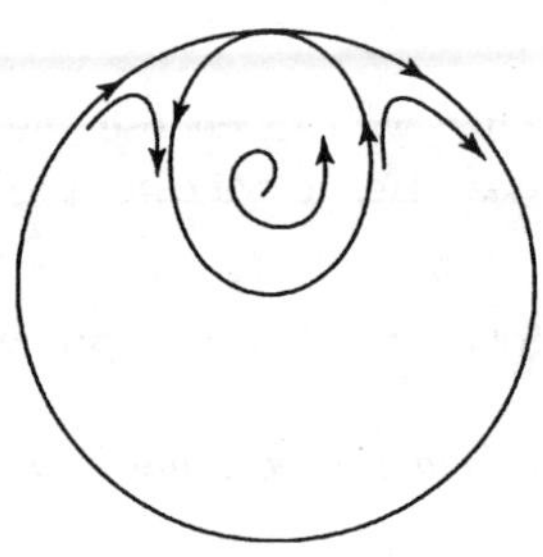

(a) $(I^1_{12})$

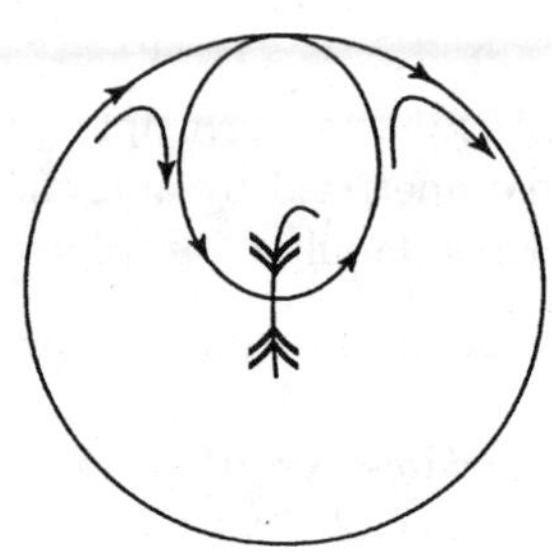

(b) $(I^1_{13})$

Figure 30: The two graphics with nilpotent saddle and finite cyclicity

## 4.4 The three types of transition through an elliptic nilpotent point of multiplicity 3

They appear in Fig. 31: on the left we draw the graphics. On the right the same graphics are drawn when the elliptic point has been blown up. They are called pp-graphic (resp. hp-graphic, hh-graphic), using the names of [22]. In the names, "p" refers to a parabolic sector and "h" to a hyperbolic sector. Hence an hp-graphic starts between two hyperbolic sectors to end in a parabolic sector.

**4.8 Theorem (Zhu [40])** *A graphic through a nilpotent elliptic point of multiplicity* 3 *has finite cyclicity under one of the following conditions:*

- *hp-graphic: no condition;*
- *pp-graphic: the transition map in normal coordinates near the singular points on the blow-up circle (Fig.* 32*) is nonlinear;*
- *hh-graphic: the point is of codimension* 3 *and the return map has a derivative different from* 1.

We will discuss the fact that these hypotheses, which look so different for the three cases, are indeed consistent.

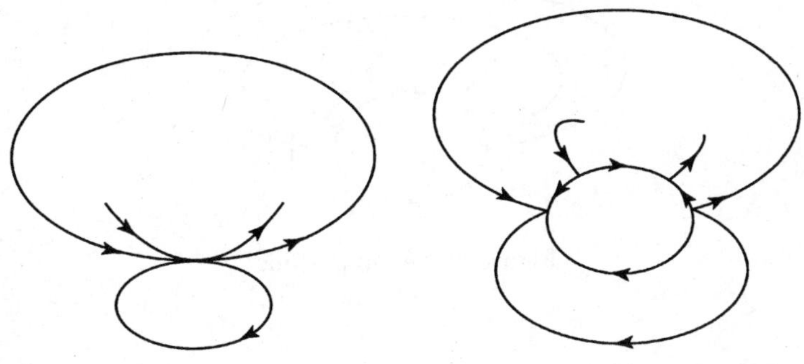

(a) (b)

pp-graphic

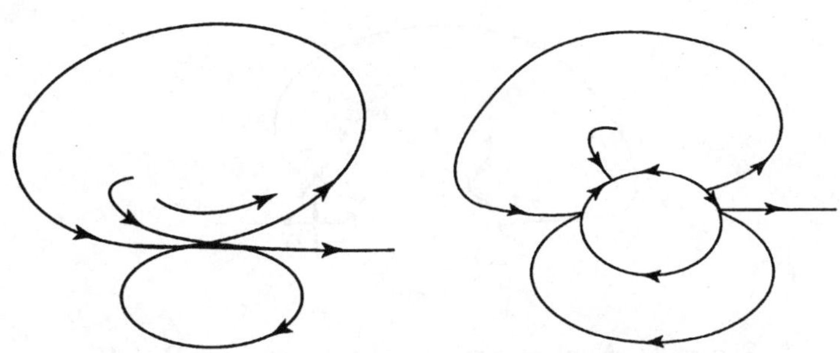

(c) (d)

hp-graphic

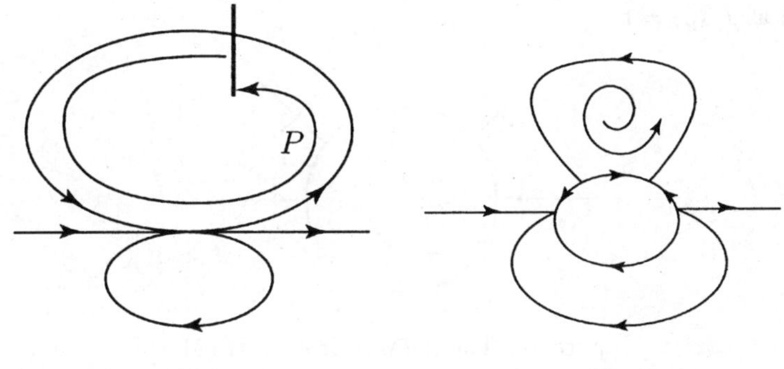


(e) (f)

hh-graphic

Figure 31: The three types of graphics through an elliptic point

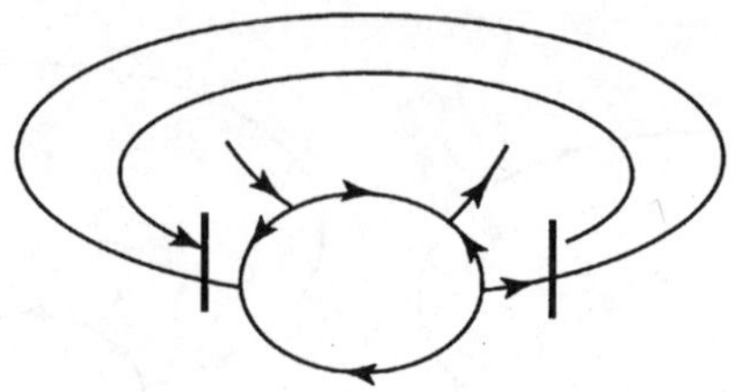

Figure 32: A pp-graphic

**Consistency of the hypotheses**

• An hp-graphic contains pp-graphics in its neighborhood and can bifurcate into a pp-graphic. All these pp-graphics satisfy the genericity condition. This comes from the nonlinearity of the passage from $\Sigma$ to $\Pi$ in Fig. 33.

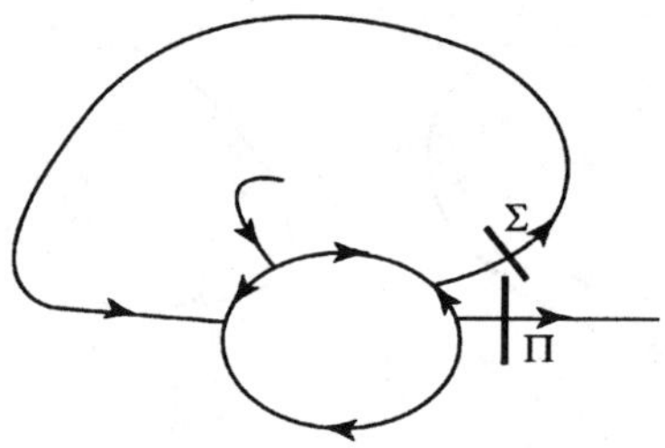


Figure 33: Nonlinearity of the passage from $\Sigma$ to $\Pi$

• Similarly, an hh-graphic can bifurcate into hp- and pp-graphics. The pp-graphics produced are generic because the two passages from $\Sigma_i$ to $\Pi_i$ in Fig. 34 are nonlinear and nearly symmetric to one another. The transition map from $\Pi_2$ to $\Pi_1$ remains nonlinear (there is no compensation) as $P'(0) \neq 1$.

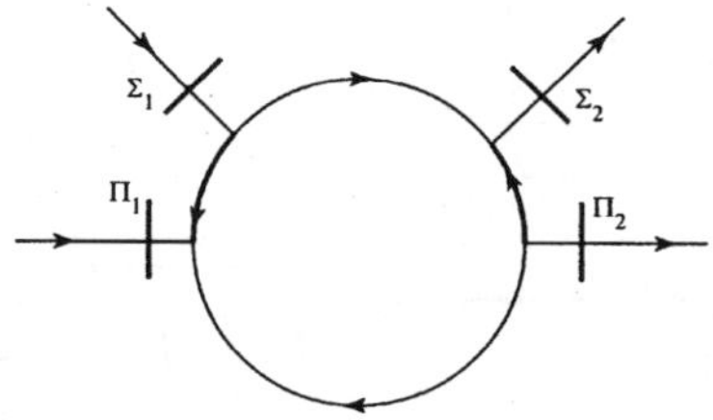


Figure 34: The passage from $\Sigma_i$ to $\Pi_i$

**What remains to be done for quadratic systems?** Essentially to adapt the proof of Theorem 4.8 when either the nilpotent point is of multiplicity 4 or there are additional singular points on the graphics: up to two hyperbolic saddles and/or one saddle-node. Also all center cases need to be done.

### Case of pp-graphics

- All adaptations have been done in [35] for the graphics with extra singular points and not surrounding a center.
- Four center cases remain to be done.

### Case of hp-graphics

- Adaptations to the graphics with an extra saddle-node have been done.
- Adaptations to the graphics with an extra saddle remain to be done. If the saddles have had a hyperbolicity ratio $r < 1$ the adaptation would have been a few lines. In quadratic systems we always have the bad case $r > 1$. We still hope to solve the four remaining cases with mere calculations.

### Case of hh-graphics

- As for graphics through a nilpotent saddle, the transition occurs along an invariant parabola, allowing to calculate $P'(0)$.
- An adaptation is needed for the codimension 4 case. The codimension 4 type I corresponds to $\varepsilon = 0$, $b > 2\sqrt{2}$, in the normal form

$$\begin{aligned}\dot{x} &= y\\ \dot{y} &= -x^3 + bxy + \varepsilon x^2 y + o(|x,y|^5).\end{aligned} \tag{4.16}$$

The condition $\varepsilon \neq 0$ was used to show that the transition maps of Fig. 35 are nonlinear.

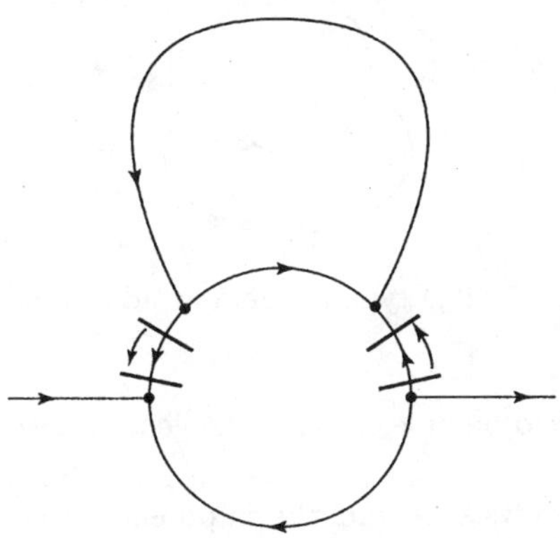

Figure 35: The two transition maps along arcs of the blow-up circle

For quadratic systems with $\varepsilon = 0$ the four singular points on the blow-up circle are non integrable as soon as $P'(0) \neq 1$, hence we believe that the proofs can be adapted.

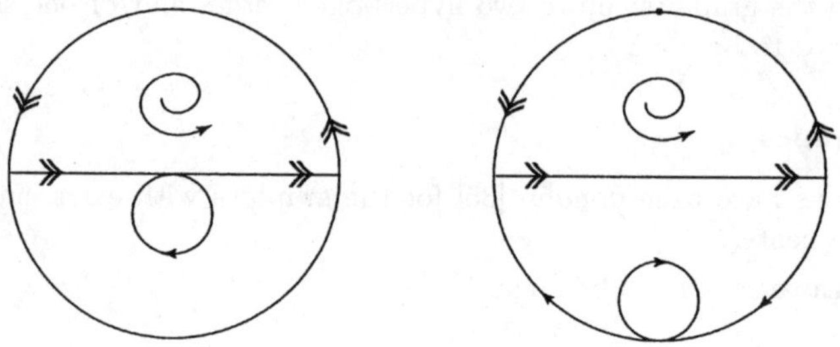

(a) $(H_4^1)$ (b) $(H_8^3)$

Figure 36: The graphics $(H_4^1)$, $(H_8^3)$

- An adaptation is needed for hemicycles (two extra saddles) (Fig. 36) with the nilpotent point on the invariant line or at infinity. We may have an additional saddle-node.
- The center cases have to be done.

## 4.5 The four additional cases

These are four graphics with a nilpotent elliptic point of codimension 4 type II ($b = 2\sqrt{2}$ in (4.16)) (see Fig. 37).

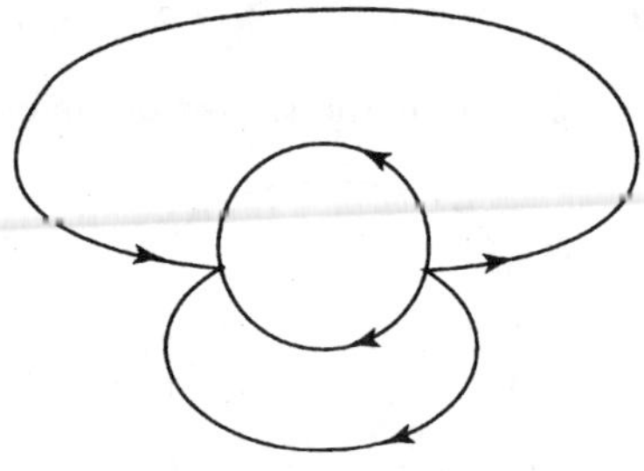

Figure 37: The type of the four additional cases

A priori we could expect to be able to treat the generic cases by the standard technique of the blow-up of the family.

As we work in quadratic systems and the nilpotent focus does not exist in quadratic systems (there is a fold in parameter space) we can take the following family as a normal form near the elliptic point

$$\begin{aligned}\dot{x} &= y + \left(\frac{1}{2} + \mu_4\right)x^2 + \mu_2\\ \dot{y} &= \mu_1 + \mu_3 y + xy.\end{aligned} \tag{4.17}$$

The blow-up of the family is the following

$$\begin{cases} x = r\overline{x} \\ y = r^2\overline{y} \\ \mu_1 = r^3\rho^3\overline{\mu}_1 \\ \mu_2 = r^2\rho^2\overline{\mu}_2 \\ \mu_3 = r\rho\overline{\mu}_3, \end{cases} \tag{4.18}$$

$(\overline{x}, \overline{y}, \rho) \in \mathbb{S}^2$, $(\overline{\mu}_1, \overline{\mu}_2, \overline{\mu}_3) \in \mathbb{S}^2$, $r, \mu_4$ small.

## 4.6 The graphics with a nilpotent point of multiplicity 4 and the degenerate graphics

Nilpotent points of multiplicity 4 have a normal form

$$\begin{aligned} \dot{x} &= y \\ \dot{y} &= x^4 + bxy + \cdots . \end{aligned} \tag{4.19}$$

Degenerate graphics have one or two lines of singular points.

A priori we expect that the method of the blow-up of the family adapted by Dumortier and Roussarie ([8, 9]) for the Canard cycles can work for the generic cases. But the blow-up will generate lines of singular points in the blown-up family. So even for the graphics with a nilpotent point of multiplicity 4 we expect to apply the methods developed to study singular perturbations.

## 4.7 The status of all graphics

For the sake of completeness we have added the status of all graphics of the family. The names refer to the grahics of [10].

### Graphics done

- Hyperbolic graphics: $(F_1^1)$, $(F_2^1)$, $F_2^2)$, $(F^3)$, $(H_1^1)$, $(H_2^1)$, $(I_1^2)$, $(I_2^2)$, $(I_6^2)$, $(I_7^2)$;
- semi-hyperbolic graphics: $(F_2^1)$, $(F_3^1)$, $(F_4^1)$, $(F_3^2)$, $(F_4^2)$, $(H_3^1)$, $(H_8^1)$, $(H_9^1)$, $(H_{10}^1)$, $(H_{11}^1)$, $(H_{12}^1)$, $(I_3^2)$, $(I_4^2)$, $(I_5^2)$, $(I_8^2)$, $(I_9^2)$, $(I_{10}^2)$, $(I_{11}^2)$, $(I_{12}^2)$, $(I_{13}^2)$, $(H_1^3)$, $(H_2^3)$, $(H_3^3)$, $(H_4^3)$, $(I_{14a}^2)$, $(I_{14b}^2)$,$(H_5^3)$, $(I_{15a}^2)$, $(I_{15b}^2)$, $(H_6^3)$, $(I_{16a}^2)$, $(I_{19}^2)$, $(I_{20}^2)$, $(I_{21}^2)$, $(I_{22}^2)$, $(I_{27}^2)$, $(I_{28}^2)$, $(I_{29}^2)$, $(I_{30}^2)$, $(I_{31}^2)$, $(I_{32}^2)$, $(I_{33}^2)$, $(I_{34}^2)$, $(I_{35}^2)$, $(I_{36}^2)$, $(I_{37}^2)$, $(I_{38}^2)$;
- graphics with a nilpotent saddle: $(I_{12}^1)$, $(I_{13}^1)$;
- pp-graphics with a nilpotent elliptic point: $(H_6^1)$, $(F_{6a}^1)$, $(H_7^3)$, $(I_{17a}^2)$, $(H_9^3)$, $(I_{5a}^1)$, $(H_{10}^3)$, $(I_{18a}^2)$, $(I_{23}^2)$, $(I_{7a}^1)$, $(I_{24}^2)$, $(I_{8a}^1)$, $(I_{25}^2)$, $(I_{9a}^2)$, $(I_{39}^2)$, $(I_{10a}^1)$;
- hp-graphics with a nilpotent elliptic point: $(F_{6b}^1)$, $(I_{5b}^1)$, $(I_{7b}^1)$, $(I_{8b}^1)$, $(I_{10b}^1)$, $(I_{11a}^1)$;
- hh-graphics with a nilpotent elliptic point: $(I_{9b})$, $(I_{11b}^1)$.

### Graphics to be done and probably easy

- Semi-hyperbolic graphic: $(I_{16a}^2)$;
- hp-graphics with a nilpotent elliptic point: $(I_{17b}^2)$, $(I_{18b}^2)$, $(I_{26}^2)$, $(I_{40}^2)$;
- hh-graphics with a nilpotent elliptic point: $(H_4^1)$, $(H_8^3)$, $(H_{12}^3)$, $(H_{15}^3)$.

**Open cases**

- Graphics with a nilpotent saddle: $(F_5^1)$, $(I_{14}^1)$;
- pp-graphics with a nilpotent elliptic point: $(H_7^1)$, $(F_{7a}^1)$, $(H_{11}^3)$, $(I_{6a}^1)$;
- hh-graphics with a nilpotent elliptic point: $(H_5^1)$,$(F_{7b}^1)$, $(H_{13}^1)$, $(I_{6b}^1)$, $(H_{13}^3)$, $(H_{14}^3)$;
- The four additional cases: $(H_4^1)$, $(H_5^1)$, $H_{12}^3)$, $(H_{13}^3)$;
- Graphics with a nilpotent point of multiplicity 4: $(I_1^1)$, $(I_2^1)$, $(I_3^1)$, $(I_4^1)$;
- Degenerate graphics: $(DH_1)$, $(DF_{1a})$, $(DF_{1b})$, $(DH_2)$, $(DF_{2a})$, $(DF_{2b})$, $(DH_3)$, $(DF_{3a})$, $(DF_{3b})$, $(DH_4)$, $(DF_{4a})$, $(DF_{4b})$, $(DH_5)$.

## References


[1] R. Bamon, Quadratic vector fields in the plane have a finite number of limit cycles, *Inst. Hautes Études Sci. Publ. Math* **64** (1986), 111–142.

[2] L. Cairo, H. Giacomini, and J. Llibre, Liouvillian first integrals for the planar Lotka-Volterra system, preprint UAB Barcelona (2002).

[3] C. Christopher, P. Mardešić, and C. Rousseau, Normalizable, integrable and linearizable points in complex quadratic systems in $\mathbb{C}^2$, preprint CRM-2687 (2002), to appear in *J. Dynam. Control Systems.*

[4] C. Christopher and C. Rousseau, Normalizable, integrable and linearizable saddle points in the Lotka-Volterra system, preprint CRM-2921 (2003).

[5] F. Dumortier, M. El Morsalani, and C. Rousseau, Hilbert's 16th problem for quadratic systems and cyclicity of elementary graphics, *Nonlinearity* **9** (1996), 1209–1261.

[6] F. Dumortier, A. Guzmán, and C. Rousseau, Finite cyclicity of elementary graphics surrounding a focus or center in quadratic systems, to appear in *Qual. Theory Dyn. Syst.*

[7] F. Dumortier, Y. Ilyashenko, and C. Rousseau, Normal forms near a saddle-node and applications to the finite cyclicity of graphics, *Ergodic Theory Dynam. Systems* **22** (2002), 783–818.

[8] F. Dumortier and R. Roussarie, Canard cycles and center manifolds. With an appendix by Chengzhi Li, *Mem. Amer. Math. Soc.* **121** (1996), no. 577.

[9] F. Dumortier and R. Roussarie, Multiple canard cycles in generalized Liénard equations, *J. Differential Equations* **174** (2001), 1–29.

[10] F. Dumortier, R. Roussarie, and C. Rousseau, Hilbert's 16-th problem for quadratic vector fields, *J. Differential Equations* **110** (1994), 86–133.

[11] F. Dumortier, R. Roussarie, and C. Rousseau, Elementary graphics of cyclicity 1 and 2, *Nonlinearity* **7** (1994), 1001–1043.

[12] J. Écalle, *Les fonctions résurgentes*, Publ. Math. Orsay, 1985.

[13] A. Fronville, A. Sadovski, and H. Żołądek, Solution of the 1 : 2 resonant center problem in the quadratic case, *Fund. Math.* **157** (1998), 191–207.

[14] A. A. Glutsyuk, Confluence of singular points and nonlinear Stokes phenomenon, *Trans. Moscow Math. Soc.* **62** (2001), 49–95.

[15] S. Gravel and P. Thibault, Integrability and linearizability of the Lotka-Volterra system for $\lambda \in \mathbb{Q}$, *J. Differential Equations* **184** (2002), 20–47.

[16] A. Guzmán and C. Rousseau, Genericity conditions for finite cyclicity of elementary graphics, *J. Differential Equations* **155** (1999), 44–72.

[17] H. Hukuhara, T. Kimura, and T. Matuda, *Équations différentielles ordinaires du premier ordre dans le champ complexe*, Math. Soc. of Japan, 1961.

[18] Y. Ilyashenko, Divergence of series reducing an analytic differential equation to linear normal form at a singular point, *Funct. Anal. Appl.* **13** (1979), 227–229.

[19] Y. Ilyashenko, In the theory of normal forms of analytic differential equations, divergence is the rule and convergence the exception when the Bryuno conditions are violated, *Moscow Univ. Math. Bull.* **36** (1981), 11–18.

[20] Y. Ilyashenko, Nonlinear Stokes phenomena, in: *Nonlinear Stokes Phenomena* (Y. Ilyashenko, ed.), Adv. in Soviet Math. **14**, Amer. Math. Soc., Providence, RI, 1993, 1–55.

[21] Y. S. Ilyashenko and A. S. Pyartli, Materialization of Poincaré resonances and divergence of normalizing series, *J. Soviet Math.* **31** (1985), 3053–3092.

[22] A. Kotova and V. Stanzo, On few-parameter generic families of vector fields on the two-dimensional sphere, in: *Concerning the Hilbert 16th Problem*, Amer. Math. Soc. Transl. Ser. 2 **165**, Amer. Math. Soc., Providence, RI, 1995, 155–201.

[23] P. Lavaurs, Systèmes dynamiques holomorphes: explosion de points périodiques paraboliques, Ph.D. Thesis, Université de Paris-Sud, 1989.

[24] O. Lehto, *Quasi-conformal mappings in the Plane*, Springer, New York, 1973.

[25] P. Mardešić, R. Roussarie and C. Rousseau, Modulus of analytic classification for unfoldings of generic parabolic diffeomorphisms, preprint CRM-2860 (2002).

[26] J. Martinet, Remarques sur la bifurcation nœud-col dans le domaine complexe, *Astérisque* **150-151** (1987), 131–149.

[27] J. Martinet and J.-P. Ramis, Problèmes de modules pour des équations différentielles non linéaires du premier ordre, *Inst. Hautes Études Sci. Publ. Math.* **55** (1982), 63–164.

[28] J.-F. Mattei and R. Moussu, Holonomie et intégrales premières, *Ann. Sci. École Norm. Sup. (4)* **13** (1980), 469–523.

[29] A. Mourtada, Projection de sous-ensembles quasi-réguliers de Dulac-Hilbert. Un cas nœthérien, prépublication, Laboratoire de Topologie, Université de Bourgogne, 1997.

[30] A. Mourtada, Projection de sous-ensembles quasi-réguliers d'Hilbert II, prépublication, Laboratoire de Topologie, Université de Bourgogne, 1999.

[31] A. Mourtada, Projection de sous-ensembles quasi-réguliers d'Hilbert III, prépublication, Laboratoire de Topologie, Université de Bourgogne, 2001.

[32] A. Mourtada, Projection de sous-ensembles quasi-réguliers d'Hilbert IV: le cas général, prépublication, Laboratoire de Topologie, Université de Bourgogne, 2002.

[33] R. Pérez-Marco and J.-C. Yoccoz, Germes de feuilletages holomorphes à holonomie prescrite, *Astérisque* **222** (1994), 345–371.

[34] C. Rousseau, G. Świrszcz, and H. Żołądek, Cyclicity of graphics with semi-hyperbolic points inside quadratic systems, *J. Dynam. Control Systems* **4** (1998), 149–189.

[35] C. Rousseau and H. Zhu, PP-graphics with a nilpotent elliptic singularity in quadratic systems and Hilbert's 16th problem, preprint CRM-2869 (2002).

[36] M. Shishikura, Bifurcations of parabolic fixed points, in: *The Mandelbrot Set, Theme and Variations* (Tan Lei, ed.), London Math. Soc. Lecture Note Ser. **274**, Cambridge Univ. Press, 2000, 325–363.

[37] L. Sylow and S. Lie, *Œuvres complètes de Niels Henrik Abel*, Grøndlahl & Søn, Christiania, Johnson Reprint Corp., 1965.

[38] S. M. Voronin, Analytic classification of germs of conformal maps $(\mathbb{C},0) \to (\mathbb{C},0)$ with identical linear part, *Funct. Anal. Appl.* **16** (1982).

[39] J.-C. Yoccoz, Théorème de Siegel, nombres de Bruno et polynômes quadratiques, *Astérisque* **231** (1995), 1–88.

[40] H. Zhu, and C. Rousseau, Finite cyclicity of graphics through a nilpotent singularity of elliptic or saddle type, *J. Differential Equations* **178** (2002), 325–436.

[41] H. Żołądek, The cyclicity of triangles and segments in quadratic systems, *J. Differential Equations* **122** (1995), 137–159.

# Aspects of planar polynomial vector fields: global versus local, real versus complex, analytic versus algebraic and geometric

Dana SCHLOMIUK*

*Département de mathématiques et de statistique*
*and*
*Centre de recherches mathématiques*
*Université de Montréal*
*C.P. 6128, Succursale Centre-ville*
*Montréal, Qué., H3C 3J7*
*Canada*

**Abstract**

The theory of planar polynomial vector fields was initiated by Poincaré and Darboux towards the end of the 19th century. There are several famous open problems on these systems, among them the second part of Hilbert's 16th problem. Apart from occurring often in applications and in theoretical studies, these vector fields are interesting objects in themselves due to their mixed nature: analytic, algebraic and geometric. They form a fertile soil for intertwining diverse methods and for blending the multiple aspects into a unified picture. The goal of this article is to discuss global problems on these systems, and to show this interplay at work on specific methods and results.

## 1 Introduction

In this article we consider planar polynomial vector fields and their associated differential equations

$$\frac{dx}{dt} = p(x,y), \qquad \frac{dy}{dt} = q(x,y), \tag{1.1}$$

where $p(x,y)$ and $q(x,y)$ are polynomials in $x$, $y$ with real coefficients. Let us call *degree* of a system (1.1), the number $n = \max(\deg(P), \deg(Q))$. The study of such systems was initiated by Poincaré in two articles, [60] which appeared in 1881 and [61] in 1885. Darboux published an article in 1878 [14] where he initiated the theory of algebraic differential equations in the complex projective plane. Each real planar polynomial vector field generates a differential equation on the complex projective plane to which Darboux's work can be applied. We may thus say that the founders of the theory of these systems were Poincaré and Darboux. Part of Poincaré's motivation for the study of these equations was his work on celestial mechanics. The equations appearing in celestial mechanics, in particular for the three-body problem, are

*Partially supported by NSERC and by the Education Ministry of Québec.



*Y. Ilyashenko and C. Rousseau (eds.),*
*Normal Forms, Bifurcations and Finiteness Problems in Differential Equations,* 471–509.

very complicated and Poincaré decided to begin by studying simpler nonlinear differential equations: those defined by polynomials. But Poincaré was also interested in their algebraic nature as his later articles [63, 64] indicate.

Nowadays we are interested in the study of planar polynomial vector fields for two reasons. Firstly because they occur very often in applications. Indeed, such equations appear in modelling chemical reactions, population dynamics, shock waves and in many other areas of applied mathematics. Secondly they also appear in theoretical studies. For example, the main result in Dumortier's thesis says that in generic smooth $n$-parameters families of vector fields on a two-dimensional manifold, all singularities with characteristic orbits are topologically determined by their finite jet. Another example appears in the Takens-Bogdanov bifurcation where in the blow-up process for the family, the origin is replaced by a polynomial system whose global behaviour intervenes in the study.

But planar polynomial vector fields and, more generally, algebraic differential equations over the projective space are interesting objects of study for their own sake. Indeed, due to their analytic, algebraic and geometric nature they form a fertile soil for intertwining diverse methods, and success in finding solutions to problems in this area depends very much on the capacity we have to blend the diverse aspects into a unified whole. One of the goals of the present article is to illustrate this on specific problems.

In the 125 years since the work of Darboux appeared, the theory of planar polynomial vector fields has developed around famous problems, some of which have a global character and are still open today. As global problems, they need to be treated from a geometric viewpoint, in the spirit of Poincaré's and Darboux's work. At the same time, the real equations must be considered along with their complexified versions, and there is an interplay between real and complex methods. These multiple aspects and methods are part of the beauty of the subject which is a small part of mathematics where its unity can be glimpsed.

The article is organised as follows: In Section 2 we briefly present the relation between polynomial vector fields on the real affine plane and algebraic differential equations on the complex projective plane. In Section 3 we state the most significant problems which shaped the theory of planar polynomial vector fields so far. There is no doubt that from all the problems on planar vector fields, the most famous one is the second part of Hilbert's 16th problem and we devote the Section 4 to a discussion of this problem. Hilbert placed this problem in a very interesting mathematical context and we think it is worthwhile to remind the reader about this context. This problem failed to be solved even in its simplest case, the quadratic one. Section 5 is devoted to some geometrical concepts needed for the global theory of such systems. Section 6 is devoted to algebraic invariants and comitants. We relate them to concepts introduced in Section 5 and apply them to a specific global problem. In Section 7 we again consider the concepts in Section 5, apply them to an initial stage in the classification problem of quadratic differential systems, and we draw some conclusions for future work.

## 2 Algebraic differential equations in the complex projective plane

The topic of Darboux's memoir [14] is the algebraic integration of differential equations of the first degree and first order on the complex projective plane. Darboux's memoir was described by Poincaré in [62] as "admirable" and a "masterly work". But Poincaré also observed in [62] that this Memoir "did not gain the attention of the geometers as much as it deserved". It was on Poincaré's initiative that to stimulate interest in this topic, the Académie des Sciences proposed it as a subject for the Grand Prix des Sciences Mathématiques. Two memoirs were chosen: Painlevé's (Grand Prix 1890) [57] and L. Autonne's (Mention honorable) [5]. Following the publication of these memoirs Poincaré contributed two articles on this topic, [63, 64].

Starting from ideas of Clebsch, Darboux considered the notion of a *Clebsch connex*, i.e. a connection between the points and lines of the projective plane expressible by an equation of the form $F(u,v,w;X,Y,Z)=0$, where $X$, $Y$, $Z$ are homogeneous coordinates of points, $u$, $v$, $w$ are homogeneous coordinates of lines, and $F$ is a homogeneous polynomial in $u$, $v$, $w$ and respectively in $X$, $Y$, $Z$. Every differential equation establishes a relation between an arbitrary point of one of the curves defined by the equation and its tangent at this point. Darboux defines a differential equation as a Clebsch connex on the complex projective plane satisfying the following conditions: The point of homogeneous coordinates $(X,Y,Z)$ lies on the line of coordinates $(u,v,w)$ and the line of coordinates $(u,v,w)$ is tangent to the curve defined by $(X,Y,Z)$, i.e.:

$$uX+vY+wZ=0, \tag{2.1}$$

$$udX+vdY+wdZ=0. \tag{2.2}$$

From the above two equations we have:

$$(YdZ-ZdY)/u=(ZdX-XdZ)/v=(XdY-YdX)/w. \tag{2.3}$$

Using the homogeneity of F in $u$, $v$, $w$, Darboux obtained the differential equation:

$$F(YdZ-ZdY,ZdX-XdZ,XdY-YdX;X,Y,Z)=0. \tag{2.4}$$

Darboux considers only equations of first degree in $u$, $v$, $w$, i.e.

$$L(X,Y,Z)u+M(X,Y,Z)v+N(X,Y,Z)w=0 \tag{2.5}$$

where $L$, $M$, $N$ are complex polynomials of the same degree, homogeneous in $X$, $Y$, and $Z$. Hence we obtain a differential equation

$$L(YdZ-ZdY)+M(ZdX-XdZ)+N(XdY-YdX)=0. \tag{2.6}$$

This equation can be written in determinant form (the Clebsch Form):

$$\begin{vmatrix} L & M & N \\ X & Y & Z \\ dX & dY & dZ \end{vmatrix}=0. \tag{CF}$$

Darboux observed that there are an infinite number of similar equations leading to the same integral manifolds. Indeed, if $A$ is any homogeneous polynomial of degree $m-1$ where $m$ is the degree of the polynomials $L$, $M$ and $N$, then the differential equation in Clebsch form associated to the polynomials $L'=L+AX$, $M'=M+AY$, $N'=N+AZ$ clearly has the same left hand side as the one associated to $L$, $M$, $N$. If $Z$ divides $N$, then $N=ZA$ and taking $L'=L-AX$, $M'=M-AY$, $N'=N-AZ$ from (CF) we see that $Z=0$ is an integral manifold. The following theorem can be easily established:

**2.1 Theorem** *Let $L$, $M$, $N$ be homogeneous polynomials of degree $m$ in $X$, $Y$, $Z$ over $\mathbb{C}$. Then there exist homogeneous polynomials $P$, $R$, $Q$ over $\mathbb{C}$ of degree $m+1$ such that the integral curves of the equation* (CF) *associated to $L$, $M$, $N$ satisfy the equation*

$$PdX + QdY + RdZ = 0, \tag{2.7}$$

*and $P$, $Q$, and $R$ are subject to the identity*

$$PX + QY + RZ = 0. \tag{2.8}$$

*Conversely, given $P$, $Q$, and $R$ satisfying the above identity in $\mathbb{C}[X,Y,Z]$ and given the above equation, we can construct an associated equation* (CF)*, with the same integral curves.*

Starting with an equation in Clebsch form (CF), we can pass to a differential equation (or to a vector field) on the affine plane. Indeed, for $Z \neq 0$ we can pass to the affine chart by $x = X/Z$, $y = Y/Z$. Since $Z \neq 0$ we may assume $Z = 1$ and hence

$$\begin{vmatrix} L(x,y,1) & M(x,y,1) & N(x,y,1) \\ x & y & 1 \\ dx & dy & 0 \end{vmatrix} = 0,$$

or equivalently

$$dy/dx = (M'(x,y) - N'(x,y)y)/(L'(x,y) - N'(x,y)x) = p(x,y)/q(x,y), \tag{2.9}$$

and to this equation we can associate the differential system

$$dx/dt = L'(x,y) - N'(x,y)x = p(x,y), \qquad dy/dt = M'(x,y) - N'(x,y)y = q(x,y) \tag{2.10}$$

with

$$M'(x,y) = M(x,y,1), \qquad L'(x,y) = L(x,y,1), \qquad N'(x,y) = N(x,y,1),$$

and $\max(\deg(p), \deg(q)) \leq m+1$.

Conversely a differential system (1.1) yields a differential equation on the complex projective plane. Clearly, the system (1.1) defines a foliation with singularities on $\mathbb{C}^2$. The affine plane $\mathbb{C}^2$ is compactified on the complex projective space $\mathbb{CP}^2 = (\mathbb{C}^3 \setminus \{0\})/\sim$, where $(X,Y,Z) \sim (X',Y',Z')$ if and only if $(X,Y,Z) = \lambda(X',Y',Z')$ for some complex number $\lambda \neq 0$. The equivalence class of $(X,Y,Z)$ will be denoted by $[X:Y:Z]$.

The foliation defined by equation (1.1) on $\mathbb{C}^2$ can be extended to a singular foliation on $\mathbb{CP}^2$ and the 1-form $\omega_1 = xdy - ydx$ can be extended to a meromorphic 1-form $\omega$ on $\mathbb{CP}^2$ which yields an equation $\omega = 0$:

$$\omega = A(X,Y,Z)dX + B(X,Y,Z)dY + C(X,Y,Z)dZ = 0, \tag{2.11}$$

whose coefficients $A$, $B$, $C$ are homogeneous polynomials of degree $m+1$ and satisfy the relation:

$$A(X,Y,Z)X + B(X,Y,Z)Y + C(X,Y,Z)Z = 0. \tag{2.12}$$

Indeed, consider the map $i : \mathbb{C}^3 \setminus \{Z=0\} \to \mathbb{C}^2$, given by $i(X,Y,Z) = (X/Z, Y/Z) = (x,y)$ and suppose that $\max(\deg(p), \deg(q)) = m > 0$. Since $x = X/Z$ and $y = Y/Z$ we have:

$$dx = (ZdX - XdZ)/Z^2, \qquad dy = (ZdY - YdZ)/Z^2,$$

the pull-back form $i^*(\omega_1)$ has poles at $Z = 0$ and the resulting equation can be written as

$$i^*(\omega_1) = q(X/Z, Y/Z)(ZdX - XdZ)/Z^2 - p(X/Z, Y/Z)(ZdY - YdZ)/Z^2 = 0.$$

After multiplying the above 1-form by $Z^{m+2}$, the 1-form $\omega = Z^{m+2}i^*(\omega_1)$ in $\mathbb{C}^3 \setminus \{Z \neq 0\}$ has homogeneous polynomial coefficients of degree $m+1$, and for $Z \neq 0$ the equations $\omega = 0$ and $i^*(\omega_1) = 0$ have the same solutions. Therefore the differential equation $\omega = 0$ can be written as (2.11) where

$$\begin{aligned} A(X,Y,Z) &= ZQ(X,Y,Z) = Z^{m+1}q(X/Z, Y/Z), \\ B(X,Y,Z) &= -ZP(X,Y,Z) = -Z^{m+1}p(X/Z, Y/Z), \\ C(X,Y,Z) &= YP(X,Y,Z) - XQ(X,Y,Z). \end{aligned} \tag{2.13}$$

Clearly $A$, $B$ and $C$ are homogeneous polynomials of degree $m+1$ satisfying (2.12).

To study the foliation with singularities in the neighborhood of the line $Z = 0$, defined by the differential equation (2.11) subject to (2.12) with $A$, $B$, $C$ satisfying the above conditions, we consider the two charts of $\mathbb{CP}^2$: $(u, z) = (Y/X, Z/X)$, $X \neq 0$, and $(v, w) = (X/Y, Z/Y)$, $Y \neq 0$, covering this line. We note that in the intersection of the charts $(x, y) = (X/Z, Y/Z)$ and $(u, z)$ (respectively $(v, w)$) we have the change of coordinates $x = 1/z$, $y = u/z$ (respectively $x = v/w$, $y = 1/w$). Except for the point $[0 : 1 : 0]$, the foliation defined by equations (2.11), (2.12) with $A$, $B$, $C$ as in (2.13) yields in the neighborhood of the line $Z = 0$ the foliation defined by the phase curves of the vector field associated with the system

$$\dot{u} = uP(1,u,z) - Q(1,u,z) = C(1,u,z), \qquad \dot{z} = zP(1,u,z), \tag{2.14}$$

and except for the point $[1 : 0 : 0]$, the foliation defined by (2.13) in the neighborhood of the line $Z = 0$, yields the foliation defined in this neighborhood by the integral curves of the vector field associated with the system

$$\dot{v} = vQ(v,1,w) - P(v,1,w) = -C(v,1,w), \qquad \dot{w} = wP(v,1,w). \tag{2.15}$$

In a similar way we can associate a real foliation with singularities on $\mathbb{RP}^2$ to a real planar polynomial vector field. A *singular point* of an equation (2.11) where $A$, $B$, $C$ satisfy the identity (2.11), is a common solution of the equations $A = 0$, $B = 0$, and $C = 0$.

**2.2 Theorem ([14])** *The number of singular points of an equation (2.12) where $A$, $B$, $C$ are homogeneous polynomials of degree $m+1$ over $\mathbb{C}$, subject to the above-mentioned identity, is $m^2 + m + 1$.*

# 3 Open problems on planar polynomial vector fields

Some developments on planar polynomial vector fields focused around a few famous problems stated by Poincaré, Painlevé, and Hilbert which are still open today as well as around several problems which arose from Darboux's work in connection with the existence of algebraic invariant curves of polynomial differential equations. All these problems are interrelated. In its affine version, an invariant algebraic curve can be defined as a curve $f(x, y) = 0$ such that $f$ is a polynomial in $x$, $y$ with real or complex coefficients and such that $f$ is a divisor of $Df$ over $\mathbb{C}$, where $D = p\partial/\partial x + q\partial/\partial y$.

Poincaré introduced a number of notions (center, limit cycle, polycycle, first return map, etc.) and he founded the qualitative theory of differential equations. One of the theorems proved in his article [61] is the following:

**3.1 Theorem** *A polynomial system with a singularity possessing complex eigenvalues which are not real has a center at this singularity if and only if the system has a nonconstant local analytic first integral in a neighborhood of the singularity.*

This theorem was generalized by Lyapunov for analytic differential equations. A problem which arises naturally from Poincaré's work on the center is the following one:

**The problem of the center (Poincaré 1885)** Find necessary and sufficient conditions for a planar polynomial differential system (1.1) of degree $m = \max(\deg(p), \deg(q))$ to possess a center. (For references and a discussion on this problem see [71].)

This problem is still unsolved for systems of degree greater than two. Two other problems which are open are:

**The problem of Poincaré** (cf. [62, 63]) Find necessary and sufficient conditions for a planar polynomial differential equation with real or complex coefficients to have a first integral which is a rational function over $\mathbb{C}$.

**The problem of Painlevé** (cf. [57]) Assume that $g$ is a non-negative integer. This problem asks for necessary and sufficient conditions for a planar polynomial differential equation to have a polynomial general integral $p(x,y) - Kq(x,y) = 0$, where $K$ is an arbitrary constant, yielding curves of genus $g$.

### Problems emerging from the work of Darboux

Darboux proved a theorem saying that a planar polynomial vector field of degree $m$ possessing $m(m+1)/2$ invariant algebraic curves has an integrating factor, and if it has more invariant curves, then it has a Darboux first integral, i.e. a first integral of the form $f_1^{\lambda_1} \cdots f_k^{\lambda_k}$ for some complex numbers $\lambda_i$. This is only a sufficient condition for Darboux integrability. Indeed, for example in the case of quadratic systems we may have an invariant line and an invariant conic and if these are in general position, then the system has a Darboux integral. This is a case not covered by Darboux's theorem. It is covered by a theorem of Christopher and Kooij (cf. [41]) which again gives only sufficient conditions for Darboux integrability. There is much activity in problems related to Darboux's memoir (see for example [12]).

The problems mentioned above were discussed in [71] with particular emphasis on the problem of the center.

All the above problems are of an algebraic nature. The next problem has a strongly transcendental content. There is no doubt that the most famous problem on planar polynomial vector fields or differential equations is:

### The second part of Hilbert's 16th problem

For this reason we devote an entire section to this problem, where apart from stating it we shall take the time to discuss this problem at length. We only mention here that all the above problems are interelated and connections among them are indicated in Diagram 1.

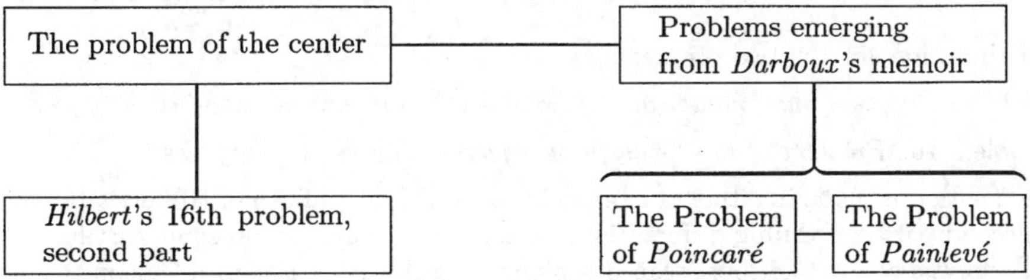


Diagram 1

Each one of the problems mentioned in this section has a global character. This is clear for the problems of Poincaré, Painlevé, the problems resulting from the work of Darboux as well as for Hilbert's 16th problem second part, as we see in the next section. But also the problem of the center has a global character. Indeed, for each fixed $n$ this problem asks to determine within the class of all systems of degree $n$, the whole class of systems which have a center.

## 4 Hilbert's 16th problem, second part

There are 23 problems on Hilbert's list and the second part of his 16th problem is one of the few which remains unsolved. More work was devoted to this problem than to any of the above-mentioned ones but this problem is still unsolved even in its simplest case, i.e. for quadratic differential equations. This motivates our discussion below. Recently Il'yashenko published an article [36] "Centennial history of Hilbert's 16th problem" in which he surveys work done in this direction for a period of over one hundred years. Surprisingly, there is no mention in this article of the original context in which Hilbert placed this problem. Instead, via a number of generalizations the problem is placed into a rather different realm: the generic families of $C^\infty$ vector fields with compact parameter space, and results in this and other directions are given. In particular there are results on the number of limit cycles which can bifurcate from an elementary polycycle (for this notion consult [22]) of a system in the family. This is clearly a bifurcation theory problem and, generally speaking, Il'yashenko's article appears to have a very strong bifurcation theory flavor. Bifurcation theory has indeed an important role to play in a solution to Hilbert's 16th problem in its original form as well as and in its later formulations. However, taking the problem out of its original context alters unequivocally its content. In this process we gain freedom from the limitations of considering polynomial systems by moving the problem into a more general realm: $C^\infty$ families of systems. But we also loose something and it is worthwhile to remind the reader about what Hilbert really had in mind when he formulated this problem and this is what we do below. The context in which Hilbert placed his 16th problem, second part, is very beautiful and ties it in with his 15th problem. The wide vision of Hilbert is inspiring and must be recalled.

Hilbert grouped together his 15th and 16th problems, placing before his statement of Problem 15, a preamble which says:

> From the boundary region between algebra and geometry, I will mention two problems. The one concerns enumerative geometry and the other the topology of algebraic surfaces.

The headings for the two problems are:

Problem 15: *Rigorous foundations of Schubert's enumerative calculus.*

Problem 16: *Problem of the topology of algebraic curves and surfaces.*

No mention is made in either the preamble or in the title above of Hilbert's 16th problem, of its second part concerning differential equations. Although this second part has a preponderant analytic content, Hilbert placed it along with the other two problems in this algebraic and geometric context. He states this second part as follows:

> In connection with this purely algebraic problem, I wish to bring forward a question which, it seems to me, may be attacked by the same method of continuous variation of coefficients, and whose answer is of corresponding value for the topology of families of curves defined by differential equations. This is the question as to the maximum number and position of Poincaré boundary cycles (cycles limites) for a differential equation of the first order and degree of the form $dy/dx = Y/X$ where $X$ and $Y$ are rational integral functions of the $n$th degree in $x$ and $y$.

Hilbert's way of stating the problem ("...as to the maximum...") is informal. But one thing is clear: his problem has a doubly global content: he is interested in the whole class of planar polynomial differential equations and for each such equation in the number of limit cycles of the equation in the whole plane. As a problem which has a doubly global content, this problem has a strong geometric content. Furthermore, in the problem he also asks for the position of these limit cycles, clearly a geometric issue. Recent work proves he had indeed much foresight in placing this problem in a geometric context and next to his 15th problem on Schubert's enumerative calculus which he states as follows:

> To establish rigourously and with an exact determination of the limits of validity those geometrical numbers which Schubert especially has determined on the basis of the so-called principle of special position, or conservation of numbers, by means of the enumerative calculus developed by him.

Schubert's Calculus is an enumeration theory of numbers with geometrical meaning. As an example of a geometrical number we have the number $(1!2!\cdots d!)/(n-d)!\cdots n!$ of $d$-planes in $n$-space meeting $h = (d+1)(n-d)$ general $(n-d-1)$-planes (cf. [48]). For a discussion of this problem we refer to Steven Kleiman's article [48]. In the explanation Hilbert gave for this problem, Hilbert shows that he is not only interested in guaranteeing that it is possible to carry out the process to obtain the (geometric) numbers but much more: he is interested in effective computability. This is an interesting point for us here and for the view we take on Hilbert's 16th problem, second part. Here is what Hilbert said after stating his 15th problem:

> Although the algebra of today guarantees, in principle, the possibility of carrying out the processes of elimination, yet for the proof of the theorems of enumerative geometry decidedly more is requisite, namely, the actual carrying out of the process of elimination in the case of equations of special form in such a way that the degree of the final equations and multiplicities of their solutions may be foreseen.

We see here that Hilbert was interested in effective computability of the geometric numbers and his viewpoint is thus very modern. This point escaped being noticed by some of the people interested in his problems. For example in 1977 in [40], I. Kaplansky says: "Today's interpretation of Hilbert's 15th problem is this:"

> One is to give a suitable definition of intersection-multiplicities for algebraic varieties and prove the expected properties.

Kaplansky observes: "We note the statement in the introduction to Weil's book [89] that the theory he develops constitutes a complete solution of the 15th problem." Indeed, in his introduction to [89] Weil says: *"Our results include all that is required for a rigourous treatment of so-called "enumerative geometry", thus providing a complete solution of Hilbert's fifteenth problem."*

It is interesting to consider Steven Kleiman's view on Problem 15 stated in [48] (page 455), where referring to Schubert's calculus Kleiman says:

> The development of intersection theory is unquestionably a very great accomplishment, which is necessary for a rigorous foundation of Schubert's enumerative calculus, yet to claim that the development of an intersection theory provides a complete solution to Hilbert's fifteenth problem is to give to the problem a narrow interpretation, for the development of an intersection theory is only the first step on the road toward the validation of the geometrical numbers of classical enumerative geometry, and in both the statement and the explanation of the problem Hilbert makes clear his interest in the efficient production of accurate geometrical numbers.

Placed in this light, Hilbert 16th problem may be interpreted as a problem on the effective computation of geometric numbers about planar polynomial vector fields. In particular, one of the most difficult numbers to compute here is the *Hilbert number* $H(n)$ which is the cardinal number, upper uniform bound of all the numbers of limit cycles appearing in a planar polynomial vector field of maximum degree $n$.

Ian Stewart who wrote a short article in Nature (cf. [87]), also says that in the second part of his 16th problem Hilbert was interested in effective numbers. Referring to the maximum number of limit cycles, Stewart says: *"...But Hilbert asks for more. He asked for the precise value..."*

This problem is of course way too difficult in the general case and it is difficult even in the quadratic case. This is why seemingly more accessible problems such as the problem of proving that $H(n)$ is finite, became the goal.

There are good reasons for placing Hilbert 16th problem, second part, next to his problem on enumeration of geometrical numbers. Indeed, there are close connections between Problem 15 and the second part of Hilbert's 16th problem. Consider for example the bifurcation diagram of the class of quadratic systems. The bifurcation set of this class contains an algebraic part formed by bifurcation points of singularities of the systems. This is an algebraic hypersurface and the way the components of this hypersurface intersect (their intersection numbers), is relevant for the bifurcation diagram of this family of differential systems.

Hilbert may have had the intuition that every given planar polynomial vector field $X$ has a finite number of limit cycles and that for fixed $n$, the set of the numbers of limit cycles of systems of degree less than or equal to $n$ has a finite maximum say $N$. Assuming this part were proved the question is: What is this maximum number $N$?

The three steps resulting from this are: (1) Prove that for an individual polynomial vector field the number of limit cycles is finite. (2) Prove that $H(n)$ is finite. (3) Find this function $H(n)$.

So far only the part (1) of this program is done. This is the most important result obtained since Hilbert stated the problem. This result was first stated as a theorem by Dulac 1923 (cf. [17]) but Il'yashenko found an error in the proof (cf. [35]). The theorem was finally proved in the late 1980's independently by Écalle and Il'yashenko (cf. [35, 24]):

**4.1 Theorem (Individual finiteness theorem)** *Every individual polynomial vector field has a finite number of limit cycles.*

Actually, Écalle and Il'yashenko proved a more general finiteness result:

**4.2 Theorem (Analytic individual finiteness theorem)** *Every analytic vector field on a closed two-dimensional surface has a finite number of limit cycles.*

This is a more general theorem, as any planar polynomial vector field yields by central projection a vector field on the upper and also on the lower hemisphere and there exists an analytic vector field on the whole sphere whose phase curves on the upper hemisphere coincide with those obtained from the polynomial vector field (cf. [27]). From the above theorem it thus follows that the induced analytic vector field on the whole sphere has a finite number of limit cycles.

According to the individual finiteness theorem we have $H(n) \leq \aleph_0$. The problem of proving the *finiteness of* $H(n)$ appeared also under the name of the *existential part of Hilbert's 16th problem* [22] simply because if one chooses to define the Hilbert number as the (supposedly finite) uniform bound of all the numbers of limit cycles appearing in polynomial differential systems of maximum degree $n$, then such a number may not exist. So the definition we previously gave of the Hilbert number $H(n)$ is preferable. We call the problem of proving the finiteness of $H(n)$, the *global finiteness problem* in contrast to the individual finiteness problem which was solved.

An analytic version of Hilbert's 16th problem is the *global finiteness problem* (*analytic case*): For any analytic family of line fields on the sphere with compact parameter base, the number of limit cycles is uniformly bounded by a finite number.

As Kaloshin mentions in [38], analytic families of vector fields are difficult to analyze and so in the 1980's Arnol'd proposed to consider generic families of smooth vector fields as a first step towards understanding families of analytic vector fields and a *smooth analog of the global finiteness problem* (*analytic case*) was formulated (see [36, 38, 39]):

**The Hilbert-Arnol'd problem** *Prove that any generic $C^\infty$ family of vector fields with compact parameter space has a finite uniform bound for the number of limit cycles.*

Other more limited versions were also formulated (cf. [36], where the history and developments on these problems are given). We note that one of the motivations for making these numerous formulations was the difficulty in attacking either the analytic, or the polynomial problems. In the smooth case progress was possible and for this case we even have explicit bounds, though large (cf. [37, 39, 38]). [39] makes precise, generic conditions which were rather hidden in [37] and produces the best bounds.

Geometric elements intervene in the smooth version of Hilbert's 16th problem for generic families of vector fields as above stated. They are subservient to the main analytic theme.

The smooth version of the second part of Hilbert's 16th problem retains aspects of the original problem but others are lost. The original problem deals which the whole class of planar polynomial vector fields, and because this is a truly global problem the generic vector fields in this class are as important as the nongeneric ones. In this very concrete global problem geometry, algebra and analysis have each a role to play, adding to the special appeal of this problem.

Among the various formulations of Hilbert's 16th problem, there is another one which has a similar appeal:

**The tangential (or infinitesimal) Hilbert's 16th problem** *which asks for the number of zeroes of a complete Abelian integral (integral of a polynomial* 1*-form) over ovals of a planar algebraic curve* $H = h$*, where* $H$ *is a polynomial.*

This problem involves a number of diverse tools from either analysis, group representations, commutative algebra, algebraic and integral geometry, and it produced some of the most interesting developments in this area of differential equations, in particular fascinating connections between Hilbert's 16th and 21st problems, as Bolibrukh (cf. [8]) and Yakovenko [91] indicate in their articles in [73]. The work done in the direction of the tangential Hilbert's 16th problem by Yakovenko and his collaborators is beautiful. We also mention here the beautiful work of the Bulgarian school (of Horozov, Gavrilov and Iliev [33, 34, 26]) on quadratic systems, and of Roussarie (see his article in this volume).

Except for the research done on the tangential Hilbert's 16th problem, which has many contact points with other areas of mathematics, most articles devoted to Hilbert's 16th problem, second part, have a predominant bifurcation theory flavour. Hilbert's 16th problem stimulated much research in the direction of bifurcation theory, more specifically on finiteness problems, as [36] rightly indicates. But the original problem of Hilbert has a beauty of its own, and for many reasons deserves to be pursued.

The most significant result on Hilbert's 16th problem for polynomial systems, obtained after the proof of Écalle and Il'yashenko, is Mourtada's result in [51].

**4.3 Theorem (Mourtada [51])** *The cyclicities (cf.* [67, 22]*) of the generic polycycles (i.e. polycycles with hyperbolic singularities) in the family of planar polynomial vector fields of maximum degree* $n$ *are uniformly bounded.*

This is a general deep result and due to compactness arguments (cf. [67]) it solves the generic Hilbert's 16th problem for polynomial vector fields. The proof is very original, using differential algebraic techniques and a combination of interesting ideas. However, to comprehend this proof one needs to plough through four manuscripts (cf. [50, 51]), so far in preprint form only.

The year 2000 was declared International Mathematical Year by UNESCO. In preparation for a meeting entitled "Perspectives in Mathematics" V. I. Arnol'd (organiser of the meeting together with M. Gromov), wrote a letter addressed to a number of mathematicians on behalf of the IMU, asking them to describe some great problems for this century. One of the responses came from Smale who gave a list of problems (cf. [83]), and his 13th problem is a new version of Hilbert's 16th problem:

*"Is there a bound $K$ on the number of limit cycles in a polynomial system of degree $n$ which is less than or equal to $n^q$ where $q$ is a universal constant?"*

In this article referring to Écalle and Il'yashenko's proofs of the individual finiteness theorem Smale says:

> Subsequently, two long works have appeared, independently, giving proofs of Dulac's assertion [Écalle, 1992] and [Il'yashenko, 1991]. These papers have yet to be thoroughly digested by the mathematical community.

The very best results obtained so far and the most general, for the original Hilbert 16th problem, have not been assimilated.

At the beginning of his article Smale said that he chose the problems using the following criteria: (1) simple statement, preferably precise, (2) his personal acquaintance with the problem, (3) a belief that the question, its solution, partial results are likely to have great importance for mathematics in the XXIst century. As he placed Hilbert's 16th problem on his list, he considers this problem important. When referring to this problem he also adds a rather stunning remark:

> Except for the Riemann Hypothesis, it seems to be the most elusive one of Hilbert's list.

Smale placed Hilbert's 16th problem side by side with what is perhaps the most celebrated problem in mathematics, a problem which generated enormously interesting developments. It is clear that the connections of the Riemann hypothesis go deep and extend far, and it is hard for any problem to come close to it in this respect. It is also clear that the currents of ideas ignited by this hypothesis moved without any field limitations, and their contributors had a very wide vision which no doubt influenced the turn taken by the developments. The beauty of the mathematics generated, the weft of connections of the results could be an inspiration for work in any direction in mathematics. The Riemann Hypothesis has been an open problem for almost 150 years, ever since Riemann stated it in [66], an eight page article, and the only one he wrote in number theory, an article which changed the subject by its implications.

The *Riemann Hypothesis* says that all nontrivial zeroes of $\zeta(s)$ are located on the line $\mathrm{Re}(s) = 1/2$.

The real function

$$\zeta(s) = \sum_{n\geq 1} \frac{1}{n^s} \tag{4.1}$$

was considered by Euler. Riemann showed that this function can be continued analytically beyond the line $\mathrm{Re}(s) = 1$ to a meromorphic function having a single pole at 1. The meromorphic function $\zeta(s)$ is known as the Riemann $\zeta$-function. He sensed that the distribution of primes is deeply related to the properties of this complex zeta function. Riemann used analytic continuation to prove the functional relation of $\zeta(s)$ which relates the zeta function to the Gamma function $\Gamma$. Using this relation and the properties of $\Gamma$ one can show that for $\mathrm{Re}(s) \leq 0$ the only zeroes of the $\zeta(s)$ are the negative even integers (the so called trivial zeroes of $\zeta(s)$).

The Riemann $\zeta$-function has an Euler product:

$$\zeta(s) = \prod_{p \text{ prime}} \frac{1}{1-p^{-s}}. \tag{4.2}$$

The Riemann zeta-function was later generalized and we now have a large class of more general zeta-functions, called $L$-functions because Dirichlet denoted by the letter $L$ the first one of these, the Dirichlet $L$-function. Like the $\zeta$-function, these $L$-functions are defined by series or by Euler products, admit analytic continuation and are meromorphic functions on the complex plane. They encode important information about the objects in connections to which they are defined, in particular in their zeroes and their residues. One such $L$-function is the $L$-function attached to an elliptic curve over $\mathbb{Q}$ which may be thought as a curve (E)

$$y^2 = x^3 + ax + b, \tag{E}$$

where $a$, $b$ are rational and the discriminant $\Delta$ of the cubic polynomial is nonzero. To such a curve together with its point at infinity we can give the structure of an abelian group which was proved to be finitely generated by Mordell in 1922. Is this group finite? A fascinating conjecture, stated in terms of the $L$-function attached to (E), was formulated by Birch and Swinnerton-Dyer (cf. [10]).

Let $p$ be a prime and let $N_p$ be the number of solutions mod $p$ of the equation and let $a_p = p - N_p$. Then the *local L-function* attached to (E) at $p$, where $p$ is not a divisor of $\Delta$ is:

$$L(E,p,s) = [1 - a_p p^{-s} + p^{1-2s}]^{-1}. \tag{4.3}$$

$L(E,p,s)$ could also be defined for the finite number of primes $p$ which divide $\Delta$ (for reference cf. [15]).

Then the *L-function of* (E) attached to (E) and defined in terms of an Euler product is:

$$L(E,s) = \prod_{p \text{ prime}} L(E,p,s). \tag{4.4}$$

**4.4 Conjecture (Birch and Swinnerton-Dyer [10])** *$L(E,s)$ has a zero at $s = 1$ of order exactly the rank of the group of $E$.*

Fore more information and references consult [15].

Some of the most interesting conjectures stated in the XXth century are formulated in terms of $L$-functions. In the Diagram 2 we indicate four directions of developments where $L$-functions played a crucial role.

The Riemann $\zeta$-function and the Riemann Hypothesis, an analysis problem, generated these amazing developments in algebraic geometry, number theory, diophantine geometry and automorphic representations. This of course is due to its being a centrally situated problem in mathematics, but also it is due to the wide vision of contributors and the free movement of ideas beyond any field restriction.

Planar polynomial vector fields are dynamical systems but to perceive them uniquely from this angle is limiting, missing part of their essence and hampering development of their theory. Indeed, as dynamical systems they are very special systems and the prevalent generic viewpoint pushes them on the side. This may explain in part why Hilbert's 16th problem as well as other problems are still unsolved even in their simplest case, the quadratic one. But Poincaré's work shows that he regarded these systems as interesting objects of study from several viewpoints, and his appreciation of the work of Darboux which he qualifies as "admirable" emphasizes this point. This area is rich with problems, very hard, it is true, but exactly for this reason an open mind and a free flow of ideas is necessary. It is to be hoped that in the future there will be a better understanding of this area which lies at a crossroads of dynamical systems, algebra and geometry and where algebraic and geometric problems go hand in hand with those of dynamical systems.

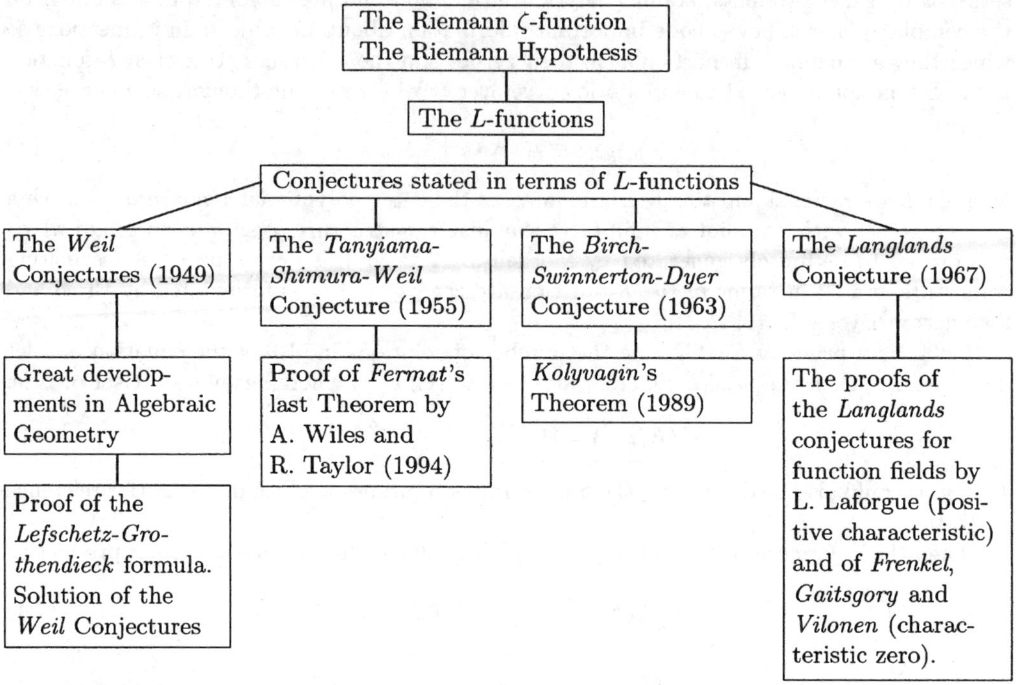


Diagram 2

## 5 Some global concepts useful in the study of polynomial vector fields

We are interested in global problems about planar vector fields and in Section 3 we mentioned several such problems. In this section we give some global concepts encoding globally local information.

First let us briefly recall *the Poincaré compactification on the sphere* of the systems [60]. A planar polynomial vector field, placed on $z = 1$ in $\mathbb{R}^3$, yields by central projection a vector field on the uppper hemisphere and also on the lower hemisphere of $S^2$, and it was shown in [27] that there exists an analytic vector field on the whole sphere whose phase portrait on the upper hemisphere coincides with the one induced by the polynomial vector field. So the vector field is compactified on the union of the upper hemisphere and the equator.

A global concept, introduced by Markus in [49] for planar $C^1$ systems, is the *separatrix configuration* formed by all the separatrices of the system, together with an orbit for each open connected component (called *canonical region*) of the complement in the plane of the union of all separatrices. Markus showed that the separatrix configuration is a topological invariant; more precisely, he showed that two $C^1$ systems are topologically equivalent if and only if their corresponding separatrix configurations are topologically equivalent. This concept clearly applies to planar polynomial vector fields and, more generally, to their compactifications on the sphere. In the separatrix configuration all topological features of a phase portrait are retained. These configurations could have a high degree of complexity and to help in distinguishing them simpler concepts leading to simpler invariants are needed. To define them we shall use the algebro-geometric concepts of *r-cycle* and *divisor* on the affine and projective complex planes. It is essential here to work over $\mathbb{C}$ and not $\mathbb{R}$ as the complex foliation with singularities (see Section 2) has a dominant role in the classification problem of planar polynomial differential systems (see [46]). $r$-Cycles on a projective variety are purely geometric objects. They are defined as follows:

If $V$ is an irreducible algebraic variety over a field $K$, a *cycle* of dimension $r$ or *r-cycle* on $V$ is a formal sum $\sum_W n_W W$, where $W$ is a subvariety of $V$ of dimension $r$ which is not contained in the singular locus of $V$, $n_W \in \mathbb{Z}$, and only a finite number of $n_W$'s are non-zero.

An $(n-1)$-cycle, where $n$ is the dimension of $V$, is called a *divisor*. These notions which occur frequently in algebraic geometry (cf. [28]) were used for the purpose of classifying planar quadratic differential systems in [72, 74, 46].

Since in our case the role of $V$ will be played by the real or complex affine or projective planes or by a plane algebraic curve over $\mathbb{C}$, we shall only need to consider zero-cycles and divisors.

Let $C$ be an $r$-cycle (in our case $r \leq 1$). We define the *support* of $C$ as the set $\mathrm{Supp}(C) = \{W | n_W \neq 0\}$. We denote by $\mathrm{Max}(C)$ the maximum value of the coefficients $n_W$ in $C$. For every $m \leq \mathrm{Max}(C)$ let $s(m)$ be the number of the coefficients $n_W$ in $C$ which are equal to $m$. We call *type* of the cycle $C$ the set of ordered couples $(s(m), m)$, where $1 \leq m \leq \mathrm{Max}(C)$, and we call the *degree* of a cycle of $\mathbb{C}^2$, (respectively $\mathbb{R}^2$, $\mathbb{CP}^2$, $\mathbb{RP}^2$) the sum of its coefficients $n_W$. If $J$ is a cycle we shall denote by $\deg(J)$ its degree.

We consider systems:

$$\frac{dx}{dt} = p(x,y), \qquad \frac{dy}{dt} = q(x,y), \tag{5.1}$$

with $p$ and $q$ relatively prime polynomials in $\mathbb{R}[x,y]$ with $\max(\deg(f), \deg(g)) = m$.

We use divisors and zero-cycles to encode globally, the multiplicities of the singularities, finite and infinite of the complex foliation with singularities on $\mathbb{CP}^2$ attached to the system. The total number (including multiplicities) of these singularities is $n^2+n+1$, where $n$ is the degree of the polynomial system.

To such a system we can associate several zero–cycles and divisors. For this we need to recall here the notion of intersection number of two algebraic curves:

The intersection number of two affine algebraic curves $C : f(x,y) = 0$ and $C' : g(x,y) = 0$ over $\mathbb{C}$ at a point $a$ in $\mathbb{C}^2$ is the number:

$$I_a(f,g) = \dim_{\mathbb{C}} \mathbf{O}_a/(f,g),$$

where $\mathbf{O}_a$ is the local ring of the affine complex plane $\mathbf{A}^2(\mathbb{C}) = \mathbb{C}^2$ at $a$, i.e. $\mathbf{O}_a$ is the ring of rational functions $u(x,y)/v(x,y)$ which are defined at $a$, i.e. $v(a) \neq 0$.

This local notion which is invariant under affine coordinate changes (cf. [25]) can also be applied for projective curves by using local affine charts, and the resulting concept is invariant under projective transformations.

The information about the multiplicities of singularities is encoded in a number of divisors and zero–cycles to be defined below. For this we recall the following notations used in Section 2, i.e.:

$$P(X,Y,Z) = Z^m p(X/Z, Y/Z), \qquad Q(X,Y,Z) = Z^m q(X/Z, Y/Z).$$

and

$$C(X,Y,Z) = YP(X,Y,Z) - XQ(X,Y,Z)$$

The following divisors and zero-cycles encode multiplicities of singularities:

(i) A zero-cycle which encodes the information regarding the intersection numbers of the real projective curves $P = 0$ and $Q = 0$ in $\mathbb{CP}^2$:

$$D(P,Q) = \sum_{W \in \mathbb{CP}^2} I_W(P,Q)W.$$

(ii) A divisor on the line at infinity $Z = 0$ which encodes the multiplicities of the intersection numbers of $P = 0$ with $Q = 0$ in $\mathbb{CP}^2$ which lie on $Z = 0$:

$$D(P,Q;Z) = \sum_{\substack{W \in \mathbb{CP}^2 \\ W \in \{Z=0\}}} I_W(P,Q)W.$$

(iii) A zero-cycle which encodes the information regarding the intersection numbers of the affine algebraic curves $p = 0$ and $q = 0$ in $\mathbb{C}^2$:

$$D(p,q) = D(P,Q) - D(P,Q;Z).$$

(iv) A divisor which encodes the multiplicities of real or complex singular points at infinity whenever $Z$ does not divide $C$, which counts how many real or complex singular points at infinity will bifurcate at infinity in a perturbation of the system:

$$D(C,Z) = \sum_{W \in \mathbb{CP}^2} I_W(C,Z)W.$$

(v) A zero-cycle which encodes the information regarding the intersection numbers of all the real or complex singularities $W$ of the foliation with singularities in $\mathbb{CP}^2$ associated to a system (5.1):

$$D(S) = D(p,q) + D(P,Q;Z) + D(C,Z) = \sum_{W \in \mathbb{CP}^2} I_W(A,B,C)W,$$

whenever $Z$ does not divide $C$, and $A$, $B$, and $C$ are defined as in Section 2.

In this last equality we have used

$$I_a(A,B,C) = \begin{cases} I_a(P,Q) = I_a(p,q) & \text{if } a \text{ is finite,} \\ I_a(P,Q) + I_a(C,Z) & \text{if } a \text{ is infinite.} \end{cases} \tag{5.2}$$

We also want to encode globally the information about the topology of the local configuration of phase curves around real singularities. Each singular point of the resulting analytic vector field on the sphere has a local topological index. The indices of two isolated opposite singular points are identical. So we define the topological index $i(W)$ of a (real) isolated singular point on the equator as the index of one of the two opposite singular points on the sphere. We may then collect these indices on the real projective plane in a zero-cycle. However for the sake of uniformity we may consider these indices as being defined on $\mathbb{C}^2$ or on $\mathbb{CP}^2$ by making the convention that for a complex non-real singular point, its topological index is zero.

(vi) We define a zero-cycle $I$ which encodes the topological indices of all the isolated singular points $W$ of the foliation in $\mathbb{CP}^2$ associated to system (5.1) as follows:

$$I = \sum_{W \in \mathbb{CP}^2} i(W)W,$$

where $i(W)$ is the topological index of the singularity as defined above.

The degrees and the types of all these cycles are clearly affine invariant.

The above cycles must intervene at the start of global classification problems of polynomial differential systems. Other cycles may be needed (see [46] or [74]). As one goes further into classification problems more cycles are needed, in particular the divisor to be defined below, naturally has a role to play. According to the Écalle-Il'yashenko individual finiteness theorem, the number of limit cycles of the vector field is finite. We can therefore construct the following (more general) divisor over $\mathbb{R}^2$, using analytic subsets $W$ in place of algebraic subvarieties:

$$Lc = \sum_{W \in \mathbb{R}^2} m_c(W)W,$$

where $W$ here is a limit cycle of the differential system and $m_c(W)$ is its multiplicity.

# 6 Connections between the theory of algebraic invariants and planar polynomial vector fields

Planar polynomial vector fields are analytic dynamical systems and problems on these systems may have a strong transcendental content such as in the case of Hilbert's 16th problem. Analytic methods, naturally play a major role in solving them. But as elswhere in analytic problems, ideas from geometry (or algebra) may have a surprisingly important role.

Thinking about the explicit expression of analytic equations discussed by Cartwright and Levinson, Smale looked at the analytic problem from his own background which was geometric and ended up by introducing what is now called the Smale horseshoe. In [84], Smale says:

> It was necessary to translate Levinson's analytic arguments into my own geometric way of thinking. (...) The horseshoe is a natural consequence of a geometrical way of looking at the equations of Cartwright-Levinson. It helps understand the mechanism of chaos, and explain the widespread unpredictability in dynamics.

We begin this section with a very brief description and a bit of history of algebraic invariant theory and then connect this with our planar polynomial differential equations.

In very general terms, in the theory of algebraic invariants one considers a collection $M$ of mathematical objects and an equivalence relation "$\sim$" on $M$.

An *invariant* of the relation "$\sim$" is then a map $f : M \to B$, which is constant on equivalence classes.

In classification problems we encounter exactly this setting: we are given a set of objects $M$ which we want to classify by a specific equivalence relation on $M$. We can then ask: could one find a natural structure for the set $M/\sim$ of equivalence classes of $M$?

Thus in the problem of the topological classification of the set **QS** of quadratic differential systems, $M$ is the set **QS** and the equivalence relation is the topological equivalence: two systems $(S)$ and $(S')$ are topologicaly equivalent if there exists a homeomorphism $f : \mathbb{R}^2 \to \mathbb{R}^2$ such that $f$ carries orbits to orbits preserving (or reversing) their orientation.

A particular case of this general setting is the case when we have a group $G$ acting on a set $M$, via a map $f : G \times M \to M$, $f(g, x) = gx$, and the equivalence relation is $a \sim b$ if there exists a group element $g$ such that $b = ga$. The equivalence classes are then the orbits of the group action where an orbit is a set $\mathrm{Orb}(x) = \{gx \mid g \in G\}$ for $x \in M$. Every such action induces a map $r$: $G \to \mathrm{Bij}(M)$, $r_g$: $M \to M$, which is a homomorphism of $G$ into the group $\mathrm{Bij}(M)$ of all bijections on $M$.

**Example 1** Consider the set of all quadratic polynomials $Q(x) = x^2 + 2bx + c$ over $\mathbb{C}$ on which acts the group of transations, $x = x' + d$, i.e. for every complex number $d$ and quadratic polynomial $Q(x)$ we obtain a quadratic polynomial $R(x') = Q(x' + d)$, i.e. $R(x') = (x' + d)^2 + 2b(x' + d) + c = x'^2 + 2(b + d)x' + d^2 + 2bd + c$. To each such $Q(x)$ we can associate its discriminant $D(Q) = b^2 - c$. Then $D(R) = (b + d)^2 - (d^2 + 2bd + c) = b^2 - c = D(Q)$. So the discriminant is an invariant of these quadratic polynomials with respect to translations.

**Example 2** In 1801, Gauss considered (in *Disquisitiones Arithmeticae*) the discriminant function $D(Q) = b^2 - ac$ as an invariant with respect to the action of the special linear group $Sl(2, \mathbb{Z})$ on two-forms $Q(x, y) = ax^2 + 2bxy + cy^2$ over $\mathbb{Z}$.

In the classical theory of algebraic invariants the role of $M$ is played by a finite-dimensional vector space $V$, $\dim_k(V) = n$, over a field $k$, the role of $G$ is played by a group of automorphioms of $V$, and the role of invariants is played by polynomial functions in $n$ variables over $k$. Thus in Example 1, V is the two-dimensional vector space of quadratic polynomials $Q(x) = x^2 + 2bx + c$ which we can identify with $\mathbb{C}^2$, and the group $G$ is played by the transformations $T$ of $V$ induced by the translations $\tau : x = x' + d$, i.e. $T(Q) = R$, where $R(x') = Q(\tau(x'))$. Identifying $Q$ with $(b, c)$, we then have the induced map $T : \mathbb{C}^2 \to \mathbb{C}^2$, i.e. $T(b, c) = (b + d, d^2 + 2bd + c)$. Consider the ring $R = \mathbb{C}[x_1, x_2]$ of polynomials in two variables over $\mathbb{C}$. An invariant of a quadratic polynomial $Q$ by the action of the group $G$ is a polynomial function $\mathbb{C}^2 \to \mathbb{C}$ given by a polynomial $f$ in $R$ such that for any transformation $T$ given as above, we have $f(T(b, c)) = f(b, c)$ for all $(b, c)$ in $\mathbb{C}^2$.

We recall that a *character* of a group $G$ is a homomorphism of $G \to \mathbb{C}^*$ of the group $G$ into the multiplicatif group of $\mathbb{C}^* = \mathbb{C} - \{0\}$.

More generally, a *relative invariant* for a linear action of a group $G$ on a finite-dimensional vector space $V$ over a field $k$, of weight $\chi$, where $\chi$ is a character of $G$, is a map $U : V \to \mathbb{C}$ such that for any $g$ in $G$ and any $x$ in $V$ we have $U(gx) = \chi(g)U(x)$. An *absolute invariant* is a relative invariant such that $\chi = 1$. Now if $\dim_k(V) = n$, such functions depend on $n$ variables which in invariant theory are taken to be polynomial functions, i.e. given by elements $f$ of $R = \mathbb{C}[x_1, \ldots, x_n]$. The set of all such polynomial invariants is denoted by $k[V]^G$.

Boole launched the theory of algebraic invariants in [9]. Initially this theory was developed for $n$-forms in $m$ variables over a field $k$. Work by British (Boole, Cayley, Sylvester, Salmon), German (Hesse, Aronhold, Clebsch, Gordan), and later French (Hermite, Jordan, Laguerre) and Italian (Brioschi, Trudi, Corrado Segre) mathematicians contributed to develop this theory. After an initial period when people searched for methods of constructing invariants the question arose as to the possibility of obtaining all invariants by using polynomials in a finite number of them, the generators.

Paul Gordan gave the first fundamental theorem of the invariant theory of binary forms: all invariants and comitants (notion to be defined below) are explicitly constructible as polynomials in a finite number of them over the underlying field. In the 1890's Hilbert (cf. [30]) proved this theorem in the more general case of $n$-forms in $m$ variables over a field $k$. Unlike Gordan's constructive proof, Hilbert gave an existential proof based on abstract reasoning, which had had a lot of impact for the development of abstract algebra. But his proof provoked much criticism from Gordan who said: "Das ist Theologie und nicht Mathematik". In his second paper [31] Hilbert gave a constructive proof. He proved that all invariants of $n$-forms in $m$ variables could be obtained from a finite number of them and that there is a finite basis for relations (syzygies) among invariants of $n$-ary forms. Apart from theorems in invariant theory, in these papers Hilbert proved many others, for example the Hilbert basis theorem and the Hilbert Nullstellensatz. Hilbert continued to have an interest in invariant theory and his 14th problem given in 1900 asks for a generalization of his results for any subgroup of $GL(n,k)$. This problem was solved in the negative by a counterexample given by Nagata in 1958 (cf. [16]). For more information on invariant theory consult [16, 65, 55].

Referring to Hilbert's results in invariant theory, Herman Weyl said [90]: "Here is only one man to mention — Hilbert. His papers (1890/92) mark a turning point in the history of invariant theory. He solves the problem and thus almost kills the subject." Weyl was careful enough to place here the word "almost". Indeed, as the authors say in [16]: "Invariant theory has already been pronounced dead several times, and like the phoenix it has been again and again rising from its ashes". Not only did invariant theory survive but evolved further and new problems were stated. In its modern day version, formulated by Mumford [52] as geometric invariant theory, it is a topic in mainstream mathematics in the form of moduli theory. The language used is the language of algebraic geometry, a subject with close connections with invariant theory right from its beginning. A goal of this modern theory is to describe the structure of $M/\sim$, the *moduli space* of $M$, under the equivalence relation "$\sim$". Mumford was interested in spaces of orbits $V/G$, where $V$ is an algebraic variety, acted upon by an algebraic group $G$ via an algebraic action.

Classification results on low degree planar polynomial vector fields, which are chart independent, were obtained by Sibirsky and his school [81, 82]. The theory of algebraic invariants for differential equations developed by Sibirsky is based on the classical algebraic invariant theory. In these classification problems of quadratic systems, the differential systems are written in tensorial form:

$$dx^j/dt = a^j + a^j_\alpha x^\alpha + a^j_{\alpha\beta}x^\alpha x^\beta, \qquad (S)$$

where $a^j, a^j_\alpha, a^j_{\alpha\beta}$, are real numbers and $j, \alpha, \beta \in \{1, 2\}$.

**Example 3** Let $M$ be the set **QS** of quadratic systems and $G$ be the group $\mathit{Aff}(2,\mathbb{R})$ of real affine transformations with its induced action on **QS**. We make explicit this action below.

$\forall g \in \mathit{Aff}(2,\mathbb{R})$, $g:\mathbb{R}^2 \to \mathbb{R}^2$ we have:

$$g:\begin{pmatrix}\tilde{x}\\ \tilde{y}\end{pmatrix} = A\begin{pmatrix}x\\ y\end{pmatrix} + B;\qquad g^{-1}:\begin{pmatrix}x\\ y\end{pmatrix} = A^{-1}\begin{pmatrix}\tilde{x}\\ \tilde{y}\end{pmatrix} - A^{-1}B,$$

where $A = ||A_{ij}||$ is a $2\times 2$ nonsingular matrix, $B$ is a $2\times 1$ matrix over $\mathbb{R}$. For any $S \in \mathbf{QS}$ we can form its transformed system $\tilde{S} = gS$:

$$\frac{\partial \tilde{x}}{\partial t} = \tilde{P}(\tilde{x},\tilde{y}),\qquad \frac{\partial \tilde{y}}{\partial t} = \tilde{Q}(\tilde{x},\tilde{y}), \tag{$\tilde{S}$}$$

where

$$\begin{pmatrix}\tilde{P}(\tilde{x},\tilde{y})\\ \tilde{Q}(\tilde{x},\tilde{y})\end{pmatrix} = A\begin{pmatrix}(P\circ g^{-1})(\tilde{x},\tilde{y})\\ (Q\circ g^{-1})(\tilde{x},\tilde{y})\end{pmatrix}.$$

The map

$$\begin{aligned} \mathit{Aff}(2,\mathbb{R})\times \mathbf{QS} &\to \mathbf{QS}\\ (g,S) &\to \tilde{S} = gS \end{aligned}$$

satisfies the axioms for a left group action. For any subgroup $G \subseteq \mathit{Aff}(2,\mathbb{R})$ we have an induced action of $G$ on $\mathbf{QS}$. We can identify the set $\mathbf{QS}$ of systems $(S)$ with $\mathbb{R}^{12}$ via the bijection $\mathbf{QS} \to \mathbb{R}^{12}$ which associates to each system $(S)$ the ordered 12-tuple of its coefficients.

The action of $\mathit{Aff}(2,\mathbb{R})$ on $\mathbf{QS}$ yields an action of this group on $\mathbb{R}^{12}$ via the above identification. For $g \in \mathit{Aff}(2,\mathbb{R})$ let $r_g : \mathbb{R}^{12} \to \mathbb{R}^{12}$ be the map which corresponds to $g$ via this action. It can easily be checked that $r_g$ is linear and the map $r : \mathit{Aff}(2,\mathbb{R}) \to GL(12,\mathbb{R})$ thus obtained is a group homomorphism. For each subgroup $G$ of $\mathit{Aff}(2,\mathbb{R})$, $r$ induces a representation of $G$ onto its image $\mathcal{G}$ in $GL(12,\mathbb{R})$.

**6.1 Definition** A polynomial $U(a,x,y) \in \mathbb{R}[a,x,y]$ is called a *comitant* of systems in $\mathbf{QS}$ with respect to the subgroup $G$ of $\mathit{Aff}(2,\mathbb{R})$, if there exists $\sigma \in \mathbb{Z}$ such that for every $(g,\boldsymbol{a}) \in G\times\mathbb{R}^{12}$ the following identity holds in $\mathbb{R}[x,y]$:

$$U(r_g(\boldsymbol{a}), g(x,y)) \equiv (\det g)^{-\sigma} U(\boldsymbol{a},x,y),$$

where $\det g = \det A$. If the polynomial $U$ does not explicitly depend on $x$ and $y$ then it is called *invariant.* The number $\sigma \in \mathbb{Z}$ is called the *weight*[1] of the comitant $U(a,x,y)$. An invariant is called an *absolute invariant* whenever $\sigma = 0$. (*Covariant* is another name occurring in the literature for the term comitant.) If $G = GL(2,\mathbb{R})$ (or $G = \mathit{Aff}(2,\mathbb{R})$) then the comitant $U(a,x,y)$ of systems in $\mathbf{QS}$ is called *GL-comitant* (respectively, *affine comitant*).

We indicate below how these notions were applied to a specific classification problem. As mentioned in Section 5, to every system in $\mathbf{QS}$ we can associate its compactification on $\widetilde{H}_+$, the upper hemisphere completed by the equator, and its phase portrait in $\widetilde{H}_+$.

**6.2 Problem** Classify topologically all real quadratic differential systems with respect to the behavior of their phase curves in the neighborhood of infinity.

[1] As in this definition the character $\chi$ is always of the form $(\det g)^{-\sigma}$ we may call here weight just the exponent $\sigma$, which agrees with the use of the term "weight" in classical invariant theory.

In this problem our set of objects $M$ is the class **QS** of real quadratic differential systems. The equivalence relation we consider is the following: two systems $(S)$ and $(S')$ in **QS** are *topologically equivalent in the neighborhood of infinity* if and only if there exists a neighborhood of the circle at infinity in $\widetilde{H}_+$ on which their phase portraits extended with the phase curves on the equator are topologically equivalent.

In this problem we need to exhibit all possible phase portraits in the vicinity of infinity of systems in **QS**, and for each such portrait we need to specify the region in the 12-parameter space where this portrait occurs (i.e. a complete set of conditions on the parameters which yield systems with the given phase portrait). Since affinely equivalent systems have phase portraits which are also topologically equivalent, the topological equivalence classes are unions of affine equivalence classes. It is for this reason that invariants and comitants can be used in the above classification problem, and in fact they completely solve the problem.

This global topological classification problem which is in fact entirely algebraic, was solved in [54]. The authors used a method which is computationally very powerful: no matter how the system in **QS** may be presented, i.e. in whatever normal form, one can compute the invariants and comitants and determine its phase portrait in the vicinity of infinity, and these computations can be implemented on a computer. On the other hand, this work is very technical and clearly not easily accessible for someone, not knowledgeable in the use of Sibirsky's theory of invariants, and in the West few people know this theory. Furthermore, the authors use 20 invariants and comitants which are tensorially defined and in terms of these they define others, by using polynomial expressions, resulting in still more complicated invariants. These invariants and comitants were chosen from the arsenal of classical invariant theory and slightly adapted for use here. The final result stated in [54] (on page 478) is therefore hermetically sealed for nonspecialists of classical invariant theory and Sibirsky's theory for differential equations.

Considering the value of the result, it is necessary to give a much more accessible classification in such a way that the global geometry of the systems becomes transparent. For this to occur one needs to construct invariants and comitants which are especially adapted to fit in with the geometry of the systems. This is what the authors did in [74]. The most important global affine invariant of the systems is the number $N_{\mathbb{C}}(S)$ of distinct singularities on the complex projective plane of the complex foliation with singularities associated to the system $(S)$ (see Section 2). In order of importance for this specific classification problem the next few invariants we considered are: the degree $\deg(I)$ of the index divisor $I$, the infimum $\inf(I)$ of all the coefficients appearing in $I(S)$, and the number $N_{\mathbb{R},\infty}(S)$ of distinct real infinite singularities of the system $(S)$. Other invariants are constructed using more divisors, some defined on the oriented equator with respect to a chosen orientation (see [74]).

We give below some ideas from [74][2].

The following notation is of use:

$$\nu_S = \max(D(P,Q;Z)) = \max\{I_p(P,Q) \mid p \in \mathrm{Supp}(D(P,Q;Z))\};$$
$$\text{for every } m \leq \nu_S, \; s(m) = \#\{p \in \{Z=0\} \mid I_p(P,Q) = m\}.$$

The *type* of the divisor $D(P,Q;Z)$ is then the set

$$\{(s(m), m) \mid m \leq \nu_S\}.$$

[2]A new version of this work, to be submitted for publication, is completed.

We also introduce for systems in **QS** the following notations:

$$\Delta_S = \deg D(P,Q;Z), \qquad M_C = \max(D(C,Z)) = \max\{I_p(C,Z) | p \in \mathrm{Supp}(D(C,Z))\}.$$

Let $T(2,\mathbb{R})$ be the subgroup of $A\!f\!f(2,\mathbb{R})$ formed by translations. Consider the representation $r$ of $T(2,\mathbb{R})$ into $GL(2,\mathbb{R})$, i.e. $\forall \tau \in T(2,\mathbb{R})$, $\tau : x = \tilde{x} + \alpha,\ y = \tilde{y} + \beta$ we consider as before the induced map $r_\tau : \mathbb{R}^{12} \to \mathbb{R}^{12}$ and let its image in $GL(2,\mathbb{R})$ be the subgroup $\mathcal{T} \subset GL(12,\mathbb{R})$.

As can easily be verified, the following polynomials are $GL$-comitants of systems in **QS**:

$$\begin{aligned}
K(a,x,y) &= \frac{1}{4}\,\mathrm{Jacob}(p_2,q_2), \quad \mu_0(a) = \mathrm{Res}_\gamma(p_2,q_2), \quad \left(\gamma = \frac{x}{y} \text{ or } \gamma = \frac{y}{x}\right),\\
H(a,x,y) &= \frac{1}{4}\,\mathrm{Discrim}_\gamma(\alpha p_2(x,y) + \beta q_2(x,y))\Big|_{\{\alpha=y,\beta=-x\}},\\
C_i(a,x,y) &= y p_i(x,y) - x q_i(x,y)\ (i=1,2), \quad \eta(a) = \mathrm{Discrim}_\gamma(C_2),\\
M(a,x,y) &= \frac{1}{4}\,\mathrm{Hess}(C_2) = \frac{1}{4}\left[\frac{\partial^2 C_2}{\partial x^2}\frac{\partial^2 C_2}{\partial y^2} - \left(\frac{\partial^2 C_2}{\partial x \partial y}\right)^2\right],\\
L(a,x,y) &= 2K - 4H - M, \quad \kappa = \mathrm{Discrim(2K\text{-}H)},\\
K_1(a,x,y) &= p_1(x,y)q_2(x,y) - p_2(x,y)q_1(x,y),\\
\kappa_1 &= \frac{\partial^2 C_1}{\partial x^2}\frac{\partial^2 M}{\partial y^2} - 2\frac{\partial^2 C_1}{\partial x \partial y}\frac{\partial^2 M}{\partial x \partial y} + \frac{\partial^2 C_1}{\partial y^2}\frac{\partial^2 M}{\partial x^2},
\end{aligned} \tag{6.1}$$

where $p_i(x,y)$, $q_i(x,y)$, $i = 0,1,2$ are the homogeneous parts of degree $i$ of the polynomials $p(x,y)$ and $q(x,y)$. Of the above polynomials only $\eta$, $\mu_0$, $\kappa$ and $\kappa_1$ are invariants.

The topological classification of all phase portraits of quadratic systems in the neighborhood of infinity was given in [54] where the authors gave lists $L_i$ of conditions formulated in terms of algebraic invariants and comitants of **QS**, and for each such set $L_i$ the phase portrait in a neighborhood of the circle at infinity of those systems satisfying the list $L_i$ was indicated. The conditions in $L_i$ when simultaneously taken are affinely invariant but the geometric meaning of the invariants appearing in the conditions in $L_i$ remained unclear.

In [74] the same classification was done in terms of geometric conditions. We also provided a complete dictionary connecting both ways the geometry of the systems and conditions expressed in terms of algebraic invariants and comitants chosen so as to fit in with the geometry of the systems.

We illustrate this by an example stated in the proposition below:

**6.3 Proposition ([74])** *The set of points in the* 12*-parameter real space where the systems have the phase portrait indicated in Fig.* 1 *can be described geometrically by the structure of the (multiple) divisor* $\mathcal{D}$ *on* $Z = 0$, *with values in* $\mathbb{Z}^3$ *which sums up the information contained in the three separate divisors* $I$, $D(P,Q;Z)$ *and* $D(C,Z)$ *for a system* $S \in$ **QS***:*

$$\mathcal{D}_S = \sum_{p \in \{Z=0\}} \begin{pmatrix} i(p) \\ I_p(P,Q) \\ I_p(C,Z) \end{pmatrix} p$$

*(where $Z = 0$ is the complex projective line). More precisely we have:*

$$\mathcal{D}_S = \begin{pmatrix}2\\2\\2\end{pmatrix} p + \begin{pmatrix}0\\1\\1\end{pmatrix} q \Longleftrightarrow \begin{bmatrix}\eta = \kappa = 0,\ M \neq 0,\\ \mu_{0,1,2} = \kappa_1 = 0,\\ L = 0,\ \mu_3 K_1 < 0\end{bmatrix} \Longleftrightarrow$$

$\Longleftrightarrow$ *The phase portrait of $S$ in the neighborhood of infinity is as in Fig. 1 with the same or reversed orientation.*

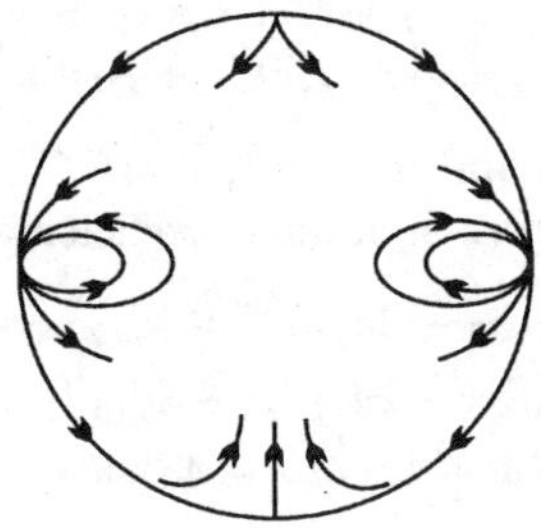

Figure 1

**6.4 Remark** In the left hand side of the first implication of the above sequence of implications we actually just mean the structure of the divisor. Indeed, the actual position of the two points $p$ and $q$ is irrelevant for the above implications. Only the numbers indicated in the column of their coefficients which yield an element in $\mathbb{Z}^3$ count.

In view of the definitions of $\eta$ and $M$ above, the following lemma follows immediately:

**6.5 Lemma** *For $C_2(a, x, y) \not\equiv 0$ the types of the divisors $D_S(C, Z)$ for systems in* **QS** *are indicated in Table 1 and determined by the corresponding conditions indicated in this table.*

| $M_C$ | Type of $D_S(C, Z)$ | Necessary and sufficient conditions on the comitants | Notation for the conditions |
|---|---|---|---|
| 1 | $p + q + r$ | $\eta > 0$ | $(\mathcal{I}_1)$ |
| | $p + q^c + r^c$ | $\eta < 0$ | $(\mathcal{I}_2)$ |
| 2 | $2p + q$ | $\eta = 0, \quad M \neq 0$ | $(\mathcal{I}_3)$ |
| 3 | $3p$ | $M = 0$ | $(\mathcal{I}_4)$ |

Table 1

*Here we write $p + q^c + r^c$ if two of the points, i.e. $q^c, r^c$, are complex but not real and the indicated inequalities determine sets in $\mathbb{R}^{12}$; for example $\eta > 0$ means $\{\mathbf{a} \in \mathbb{R}^{12} \mid \eta(\mathbf{a}) > 0\}$.*

We introduce the polynomials $\mu_i(a, x, y)$, $i = \overline{0, 3}$, using the following resultants:

$$\mathrm{Res}_X(P, Q) = \mu_0 Y^4 + \mu_{10} Y^3 Z + \mu_{20} Y^2 Z^2 + \mu_{30} Y Z^3 + \mu_{40} Z^4; \tag{6.2}$$

$$\mathrm{Res}_Y(P, Q) = \mu_0 X^4 + \mu_{01} X^3 Z + \mu_{02} X^2 Z^2 + \mu_{03} X Z^3 + \mu_{04} Z^4 \tag{6.3}$$

where $\mu_{ij} = \mu_{ij}(a) \in \mathbb{R}[a_{00}, \ldots, b_{02}]$ (think of $a_{ij}, b_{ij}$ as the coefficients of the differential system (1.1)) and

$$\begin{aligned}
\mu_0(a) &= \mu_0(a); \\
\mu_1(a, x, y) &= \mu_{10}x + \mu_{01}y; \\
\mu_2(a, x, y) &= \mu_{20}x^2 + \mu_{11}xy + \mu_{02}y^2; \\
\mu_3(a, x, y) &= \mu_{30}x^3 + \mu_{21}x^2y + \mu_{12}xy^2 + \mu_{03}y^3.
\end{aligned}$$

The polynomial $\mu_0$ is $GL$-invariant and $\mu_i$, $i = \overline{1, 3}$, are $GL$-comitants. Their coefficients $\mu_{i0}$ (respectively $\mu_{0i}$, $i = \overline{1, 3}$) are the corresponding coefficients in (6.2) (respectively (6.3)) and

$$\begin{aligned}
\mu_{11} &= 2d_{25}d_{26} - 4d_{25}d_{35} + 2d_{34}d_{35} - 3d_{23}d_{46} + 4d_{45}d_{16} - 4d_{56}d_{14}, \\
\mu_{21} &= 4d_{13}d_{45} + 4d_{12}d_{46} + 2d_{14}d_{35} - 2d_{14}d_{26} + 4d_{15}d_{34} - 4d_{15}d_{25}, \\
\mu_{12} &= d_{23}d_{26} + 2d_{23}d_{35} + 4d_{13}d_{46} + 4d_{12}d_{56} + 4d_{15}d_{35} - 4d_{15}d_{26} + 2d_{16}d_{34} - 2d_{16}d_{25}.
\end{aligned}$$

Here we denote by $d_{jl}$ the minors of the $2 \times 6$ matrix made up of the coefficients of the right-hand sides of system in **QS**, which are formed by the columns $j$ and $l$.

The following lemma can be proved easily by using a theorem proved in [6].

**6.6 Lemma** *For every $s \in \{0, 1, 2, 3\}$, $\deg(D(P, Q; Z)) = s$ is equivalent to the condition: $\mu_i = 0$ and $\mu_s \neq 0$.*

Clearly this gives the geometrical meaning of the conditions $\mu_i = 0$ for $s \in \{0, 1, 2, 3\}$ and hence of the role of the $\mu_i$ in determining the geometry of the systems.

**6.7 Lemma**

$$D(P, Q; Z) = D(C, Z) = 2p + q \iff \mu_0 = \mu_1 = \mu_2 = \kappa = \kappa_1 = L = 0 \text{ and } \mu_3 \neq 0$$

Clearly this lemma gives the geometrical meaning of the condition on the right side of the double implication sign.

**Idea of the proof of the Proposition** We start by working from the bottom line up of the divisor incorporated in $\mathcal{D}_S$, namely $D(C, Z) = \sum I_p(C, Z)p$. From the first one of the above lemmas, for systems in **QS** with $C_2(x, y) \not\equiv 0$, this divisor is of the form determined by the conditions in the Table 1 given before. We see that our case corresponds to the third line in this table, i.e. $M_C = 2$, $D(C, Z) = 2p + q$, with necessary and sufficient conditions: $\eta = 0$ and $M \neq 0$.

For the second part of the proof we consider the middle divisor, i.e. $D(P, Q; Z) = 2p + q$ and use Lemma 6.7. We first consider a normal form for all systems such that their divisor $D(C, Z)$ is of the form $D(C, Z) = 2p + q$. This normal form is the following one:

$$\begin{cases}
\dfrac{dx}{dt} = k + cx + dy + gx^2 + hxy, \\
\dfrac{dy}{dt} = l + ex + fy + (g - 1)xy + hy^2.
\end{cases} \tag{$S_3$}$$

We now impose on the coefficients of this normal form the conditions given by the invariants and comitants in Lemma 6.7. For the systems $(S_3)$ (we keep here the same notation as used in [74]), we obtain $\kappa = -h^2 = 0$ and then $L = gx^2$. So, condition $L = 0$ yields $g = 0$ and this implies $\kappa_1 = -2d = 0$. Then $\mu_0 = \mu_1 = \mu_2 = 0$, $\mu_3 = -c(cf + k)x^2y \neq 0$ and $K_1 = -cx^2y \neq 0$. After the translation of the point $(e, f)$ to the origin of coordinates, a system $(S_3)$ becomes:

$$\frac{dx}{dt} = k + cx, \qquad \frac{dy}{dt} = l - xy. \tag{6.4}$$

Thus, here $\mu_3 = -ckx^2y$ and $\text{sign}(\mu_3 K_1) = \text{sign}(k)$. To finish the proof we consider the upper divisor, i.e. the index divisor $I$ and impose the condition indicated on the first line of $\mathcal{D}_S$, i.e. $I = 2p + 0q$. For the system (6.4) the singular points at infinity are $p = [0 : 1 : 0]$ and $q = [1 : 0 : 0]$ with total multiplicities equal to 4 and 2, respectively. The point $q$ is a saddle-node while for the point $p$ we have only two possible local phase portraits obtained after the blow-up: one (with index 0) being a saddle-node and another one (with index 2) having two elliptic sectors. These two possibilities depend on the sign of the parameter $k$. More precisely, if $k < 0$ we have the phase portrait in Fig. 1 and in view of $\text{sign}(\mu_3 K_1) = \text{sign}(k)$ we obtain the remaining condition in the middle of the sequence of implications in the statement of our proposition. □

**6.8 Theorem ([54])** *There are exactly* 40 *topologically distinct phase portraits of quadratic systems (S) in the neighborhood of infinity, with $C_2(x, y)$ not identically zero. For each one of these phase portraits, affinely-invariant necessary and sufficient conditions are given and these are expressed in terms of algebraic invariants and comitants.*

**6.9 Theorem ([74])** *For each one of the* 40 *topologically distinct phase portraits in the neighborhood of infinity encountered for quadratic differential systems with $C_2(x, y)$ not identically zero, the geometric meaning of the necessary and sufficient conditions on the parameters (expressed in terms of invariants and comitants) for the phase portrait to occur is given in terms of the structure (types) of divisors on the complex projective line at infinity or on the oriented equator of the Poincaré sphere.*

As indicated in the above theorem, for some of the phase portraits, additional divisors were needed.

Analogous results for the finite singular points of systems in **QS** were recently obtained in work done by Artés, Llibre and Vulpe, and this work is now being written up for publication. The work described in this section is important in the study of the bifurcation diagram of the class **QS**. Indeed, for this diagram one needs to use various charts and normal forms and both for the bifurcation diagram associated to these pieces as well as in the gluing up process the work described in this section is essential.

# 7 On Hilbert's 16th problem for quadratic systems

We interpret Hilbert's 16th problem as the problem of *determining* $H(n)$ as well as the possible relative positions of the limit cycles in systems of maximum degree $n$. The problem of finding the value of $H(n)$ is very difficult even in the quadratic case. But it is natural to start doing this case even though motivation to "prove results for $n$" prevails. The following quotation from Hilbert's address in Paris in 1900 is here relevant:

> In dealing with mathematical problems, specialization plays, as I believe, a still more important part than generalization. Perhaps in most cases where we seek in vain the answer to a question, the cause lies in the fact that problems simpler and easier than the one at hand have been either not at all or incompletely solved.

As it is expected that quadratic differential systems will yield over two thousand phase portraits, this particular case is rich enough to demand organization. Hilbert's 16th problem was not attacked head on, by trying to solve it at least in the quadratic case. Instead, a program for proving the finiteness part, i.e. that $H(2)$ is finite, was stated in [22] where it is shown how to reduce the finiteness part, via compactness arguments, to the problem of proving the finite cyclicity of graphics in the family. The authors followed the initial article [22] with [23] and subsequently numerous rather technical articles dealing with perturbations in the quadratic family of quadratic systems with specific types of polycycles or more generally graphics were published in which these were proven to have finite cyclicity. So far the general result was not obtained. On the positive side we stress that in [22] the authors isolated the 121 graphics for which one needs to show finite cyclicity in order to solve the finiteness problem for the quadratic case. Also on the positive side we mention that out of these 121 cases, only about 37 still remain to be established. On the negative side: the most interesting cases turned out, as expected, to be the most difficult ones, and to this day they remain untouched. For the detailed picture in this direction we refer to Rousseau's article in this volume.

This work on showing that $H(2)$ is finite has helped refine some bifurcation theory tools. But this program has not been devised for gaining global understanding of the class of quadratic systems and naturally has done little in this direction. Anyone working with specific families of quadratic systems has intimate knowledge about such families which not only persuades us of the finiteness of $H(2)$ but also gives insight as to why this is true and even suggests the value for $H(2)$. The most interesting quadratic perturbations are those of systems which are the most degenerate ones among the systems with center and these are exactly the cases so far untouched by the finiteness program. What we want are precise things about the global geometry of this class and answers to precise questions such as for example:

**Question** Is it true that the systems with the maximum number of limit cycles cluster around systems with center, in the sense that the regions of systems with these limit cycles have always in their boundary at least one system with center?

This seems to be true and if this is so, then clearly the most interesting quadratic perturbations are those of systems with center. Among these, the ones that are most degenerate are the most interesting because they are those which are boundary points of the largest number of generic regions. Exactly these cases are so far untouched by the finiteness program.

It may take a long time to solve Hilbert's 16th problem in the quadratic case, but even preliminary steps in this direction bring light in understanding these systems. We mention here the beautiful work done in this direction by the Bulgarian school (work done by Horozov, Gavrilov and Iliev). For some of the results and references, consult [26, 34, 33]. The global geometry of this class will slowly begin to take shape and will get us closer to a solution to the following more extensive problem:

**The classification problem** Classify topologically all real planar quadratic differential systems.

This family of systems depends on twelve parameters but due to the group action of real affine transformations and positive time rescaling, the class ultimately depends on five parameters. In [13] Coppel hoped that the characterization of the phase portraits of this class could be done by means of algebraic inequalities on the coefficients of the specific normal forms of the systems. But as one could see (cf. [68] and [20] pp. 118–119, also cf. [46]) algebraic inequalities do not suffice and analytic as well as non–analytic ones must also be considered. In most of the work on specific families of quadratic systems done in the past, the classifications (see for example [3]) were done by means of inequalities expressed in terms of the coefficients of the normal forms used and the results were stated by giving lists of inequalities on the coefficients of the specific normal forms considered. Ever since Felix Klein stated his Erlangen Program, a property is called geometric when it is invariant under some group action. In particular here we have the action of the real affine group. But most of these classifications are tied to specific normal forms and so they are not geometric (or intrinsic) in the above sense. They are thus not readily applicable to other normal forms given in different coordinate charts.

Work done by the Moldavian school of Sibirsky produced classifications in terms of algebraic invariants and some of the classifications obtained are indeed affinely invariant. Although the work done by Sibirsky and by his former students is very valuable, until recently this work has had little impact in the West because it is very technical. Also because of the belief that the theory of algebraic invariants cannot do much for solving what is essentially a transcendental problem. But this is not quite right as we see below.

In the initial stages of the topological classification problem for quadratic systems, the work of the Moldavian school proved to be an invaluable asset. The class essentially depends on five parameters, a large number for drawing bifurcation diagrams. We are bound to use several charts which must be glued to obtain the final picture. It is in choosing the charts and in the gluing process that the methods of algebraic invariants intervene. This method gives us the tools to start a program for attacking Hilbert's 16th problem and the classification problem for quadratic systems. This program looks very much like an onion: we start from a nucleus formed by systems with center which we know, and then go on to the second layer formed by systems with zero first and second Poincaré-Lyapunov constants, and then continue to the next layers. According to the theorem of Bautin if the first three Poincaré-Lyapunov constants of a quadratic system with a singularity which is a center for its linearization are zero, then this singularity is a center. This program could be started because the conditions for systems with center as well as their bifurcation diagram are known [70, 58].

The conditions for center for quadratic systems were first found by Dulac and appeared in [19]. However, these conditions are spread out through the paper, interspersed with various coordinate changes and case by case discussions. Furthermore, Dulac defines a (nondegenerate) center (cf. [18]) as a singular point of a real or complex differential system whose eigenvalues are not zero with a quotient which is rational and negative and which has a local analytic first integral. In [19] he gives these conditions for the case when the quotient of the eigenvalues is $-1$. Dulac showed much foresight in giving this definition of a (nondegenerate) center but his conditions do not have a compact form and are not readily applicable to real systems as he worked with a normal form in which the linearization at the singular point has a saddle. After many failed attempts to give conditions for center in compact convenient form, we find in [79, 80] such a compact and correct set of conditions. In [88] the phase portraits of all quadratic systems with a center are given and in [70] and [58] we find the bifurcation diagram for this class done in the adequate parameter space, the projective space, and the role of invariant algebraic curves in this bifurcation highlighted.

Thus the first step in this program was to classify topologically (cf. [85]) the systems with zero first and second Poincaré-Lyapunov constants. A bifurcation diagram (not entirely complete) for this class was done in [2], a master's thesis under the direction of Llibre. Also a bifurcation diagram (again not complete) of this class was later done in [1]. Interest shown for the classification eventually led to a publication (cf. [4]) based on [2]. This work again was incomplete because the uniqueness of the limit cycle in the three phase portraits with limit cycles was not proved. Secondly, the bifurcation curve of saddle-to-saddle connections was only conjectured to be analytical. The main theorem in the article said that there are at least 20 topologically distinct phase portraits. This lower bound was later corrected in [46]. This error was found with the help of the invariants introduced in [46]. [2, 1, 4] were done in the spirit of Coppel's paper [13], resulting in nonintrinsic classifications. As the discussion of the systems was not geometric, this also lead to an insufficient amount of summing up the results. Thus the parameter space was considered to be $\mathbb{R}^3$ when it could have been considered as $\mathbb{RP}^2$; the lists of phase portraits, the explanation for notations and the bifurcation diagram (on sections of $\mathbb{R}^3$) appeared on different pages. Naturally, the global geometry of the class was hard to grasp. In [46] the study of this class was done anew. Apart from corrections to the previous works, this paper gives the complete bifurcation diagram for this class, including the proof of the analyticity of the saddle-to-saddle connection curve. Furthermore this work is done in geometric terms, based on invariants defined by using divisors and zero-cycles.

We shall give below some of the ideas from [46].

We are interested in the class of quadratic systems with a singularity which has a center for its linearisation, with zero first and second Poincaré-Lyapunov constants and more specifically in the systems with a third order weak focus (as defined below). Using a translation of axes we may assume the singularity to be placed at the origin. The following proposition can easily be proved:

**7.1 Proposition** *Any quadratic system with a singularity at the origin which is a center for its linearization, can be written in the normal form*

$$\dot{x} = -y + lx^2 + rxy + ny^2, \qquad \dot{y} = x + ax^2 + bxy, \tag{7.1}$$

*by a linear change of variables and a rescaling of the independent variable. Furthermore, the transformation into this normal form depends continuously on the parameters.*

Calculations of Li [42] yield that $V_1 = L_1$, $V_2 = L_2 \pmod{V_1}$, $V_3 = L_3 \pmod{V_1, V_2}$ where $V_1, V_2, V_3$ are the first three Poincaré-Lyapunov constants (of the singularity) and

$$\begin{aligned} L_1 &= r(l+n) - a(b+2l), \\ L_2 &= ra(5a-r)[(l+n)^2(n+b) - a^2(b+2l+n)], \\ L_3 &= ra^2[2a^2 + n(l+2n)][(l+n)^2(n+b) - a^2(b+2l+n)]. \end{aligned}$$

The origin is then a *third order weak focus* if $V_1 = V_2 = 0$ and $V_3 \neq 0$, or equivalently $L_1 = L_2 = 0$ and $L_3 \neq 0$. Using the above proposition and $L_1 = 0, L_2 = 0, L_3 \neq 0$ we obtain:

**7.2 Corollary** *A planar quadratic vector field with a weak focus of third order may be written in the following form where the weak focus is placed at the origin:*

$$\dot{x} = p(x,y) = -y + lx^2 + 5axy + ny^2, \qquad \dot{y} = q(x,y) = x + ax^2 + (3l+5n)xy, \tag{7.2}$$

*with* $L_3 = 5a^3[2a^2 + n(l+2n)][3(l+n)^2(l+2n) - a^2(5l+6n)] \neq 0$.

The above equations depend on three real parameters $l, a, n$ but since for a nonlinear system, a rescaling does not change the phase portrait, actually the parameter space is $\mathbb{RP}^2$. So we have a parameter $\lambda = [a : l : n]$, where we denote by $[a : l : n]$ the equivalence class of $(a, l, n)$ in $\mathbb{RP}^2$. We only need to consider $n \geq 0$. The systems (7.2) have a symmetry. Indeed, a system (7.2) corresponding to the parameter $\lambda = (a, l, n)$ with $a < 0$ is topologically equivalent to a system (7.2) corresponding to the parameter $\lambda' = (-a, l, n)$ through the symmetry $(x, y, t) \to (-x, y, -t)$. Due to this, we only consider the parameters $[a : l : n] \in \mathbb{RP}^2$ with $a \geq 0$ and $n \geq 0$. Since we can picture the projective space on the disk with two diametrically opposed points identified, in view of the symmetry, the bifurcation diagram could be pictured on half a disc ($a \geq 0$), placing $n = 0$ on the circumference. We can compute the three cycles $D(P, Q; Z)$, $D(p, q)$ and $D(C, Z)$ with respect to this normal form. $D(P, Q; Z)$ involves the common roots of $p_2(x, y)$ and $q_2(x, y)$. We define

$$\Omega = \text{Resultant}[p_2(x,1), q_2(x,1), x] = n[l(3l+5n)^2 - 3a^2(5l+8n)] = n\overline{\Omega}.$$

**7.3 Proposition** *For all values of the parameter $\lambda \in \mathbb{RP}^2$, the divisor $D(P, Q; Z)$ is well defined and we have*

$$D(P,Q;Z) = \begin{cases} 0 & \text{if } \Omega(\lambda) \neq 0, \\ [3l+5n : -a : 0] & \text{if } \overline{\Omega}(\lambda) = 0, \\ [0:1:0] & \text{if } n = 0. \end{cases}$$

*Hence the bifurcation curve for this divisor is* $\Omega(\lambda) = n\overline{\Omega}(\lambda) = 0$.

We see this curve pictured in Fig. 2. Analogously we can compute the other two cycles.

For the divisor $D(C; Z)$, we need the common points of $C = 0$ and $Z = 0$. We denote by $c(x) = C(x, 1, 0) = p_2(x, 1) - xq_2(x, 1)$, where $p_2$ and $q_2$ are the degree 2 homogeneous parts of $p$, respectively $q$, or one of them zero. Let $\Delta$ be the discriminant of the cubic polynomial $c(x)$. We have:

$$\Delta = 125a^4 + a^2(25l^2 + 170ln + 262n^2) + n(2l+5n)^3 = \text{Resultant}[c(x), c'(x), x]/(4a).$$

The bifurcation curve of the divisor $D(C, Z)$ is clearly the curve $\Delta = 0$. This curve is pictured in Fig. 3. Calculations show that the bifurcation curve for the zero-cycle $D(p, q)$ is the curve $\delta(l + 2n) = 0$. This curve is shown in Fig. 4. We picture in Fig. 5 the images on the disk of the three components $\mathcal{L}_1$, $\mathcal{L}_2$, $\mathcal{L}_3$ (dotted curves) of the curve $L_3 = 0$ on which we have centers. The bifurcation curve due to the multiplicity of singularities is the algebraic curve given by the equation $\bar{\Omega}n(l + 2n)\delta\Delta = 0$ whose trace on the half disk is pictured in Fig. 6.

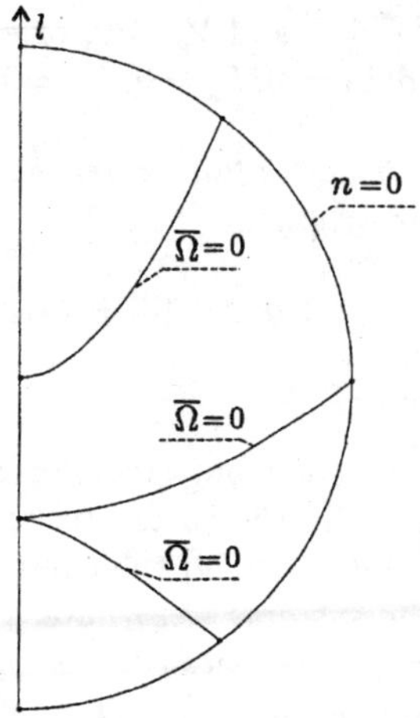


Figure 2

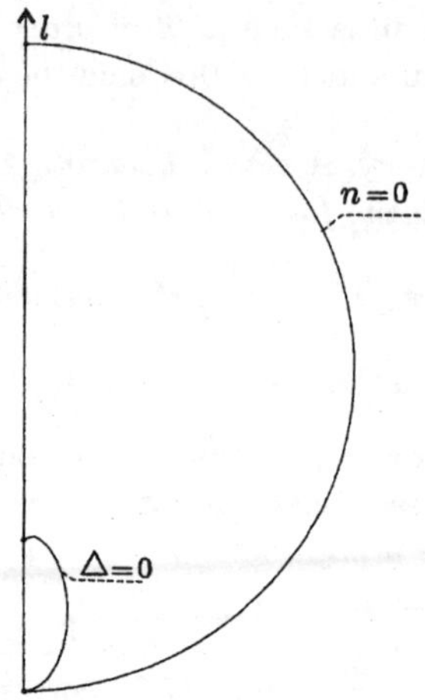


Figure 3

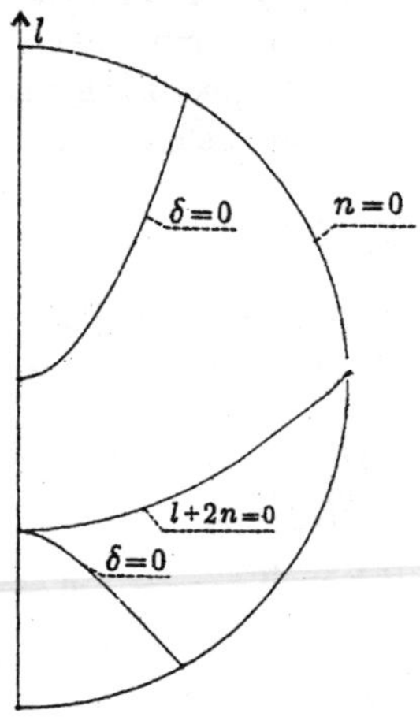


Figure 4

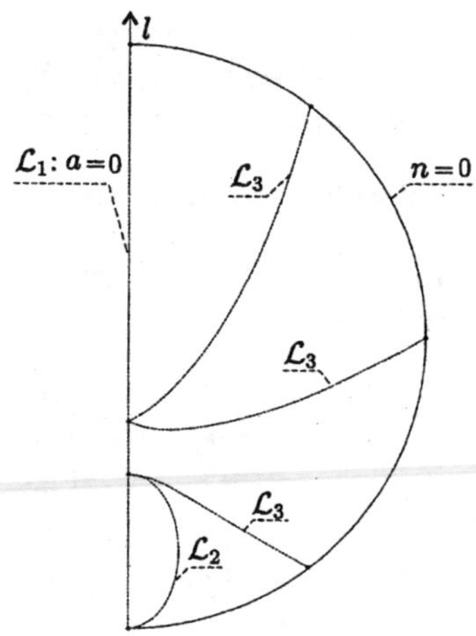


Figure 5

The main step for obtaining the bifurcation diagram of the systems is to first start with placing on the half disc the curves where we have centers (the dotted curves in Fig. 6) and by placing on them their corresponding phase portraits. Secondly to obtain the phase portraits in the regions adjacent to these portraits with center we observe how the portraits change when we perturb them slightly. Two important theorems help in the final stages, the first one of which is a result on general quadratic systems obtained by Zhang Pingguang [92] ([93] is an English version of [92]).

**7.4 Theorem** *If there are limit cycles surrounding two foci of a quadratic system, then around one of the foci there is at most one limit cycle.*

Another result is of Li [43]:

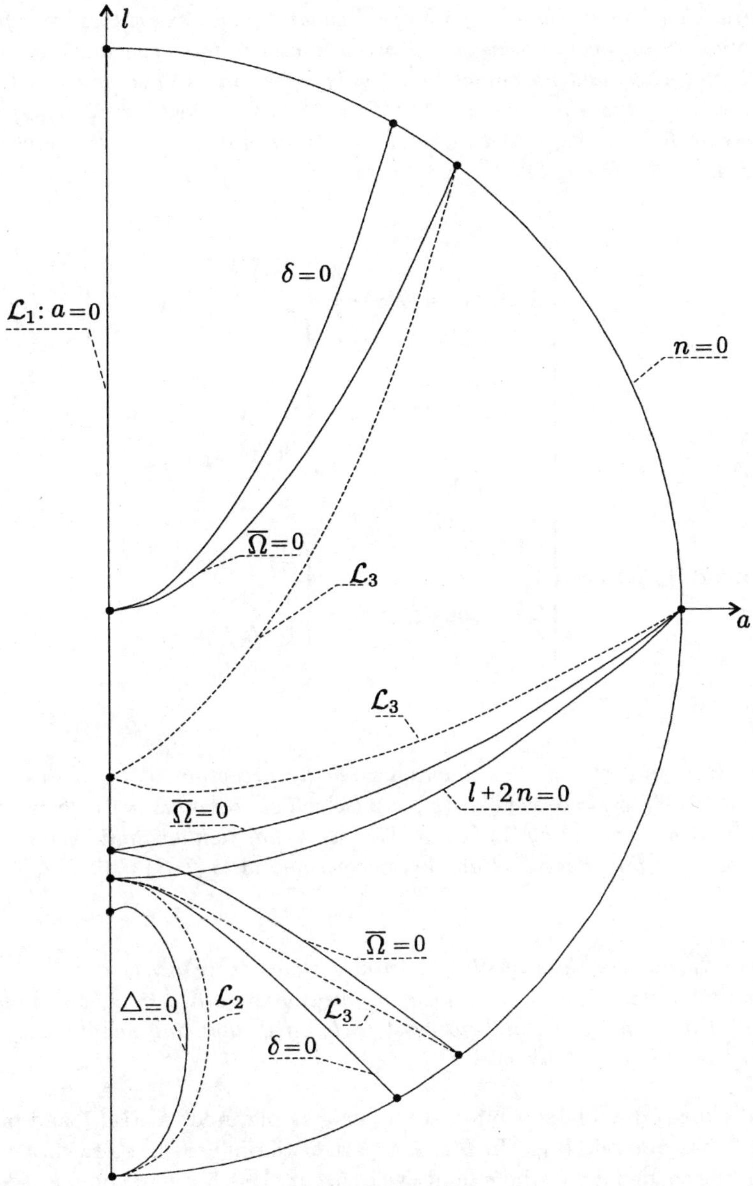


Figure 6

**7.5 Theorem** *There are no limit cycles of systems* (7.2) *surrounding the weak focus of third order, see* [43].

We give below two of the results obtained in [47].

**7.6 Theorem** *Consider the family $QW3$ of all quadratic systems with a weak focus of third order. There are 18 topologically distinct phase portraits in this class and among these, only three have limit cycles (and no graphic) and only three others have graphics (and no limit cycle). The values of the affine invariant $\mathcal{J}(S) = (\mathbb{N}_{\mathbb{C}}, \mathbb{N}_{\mathbb{R},f}, \deg(I_f), \mathbb{N}_{\mathbb{R},\infty})(S)$ (for $\mathbb{N}_{\mathbb{C}}$ and $\mathbb{N}_{\mathbb{R},\infty}$ see Section 6; $f$ in $\mathbb{N}_{\mathbb{R},f}$ and in $I_f$ stands for finite) given in the following diagram yield a partition of the family $QW3$[3] as follows:*

$$\mathbb{N}_{\mathbb{C}}(S) = \begin{cases} 7 \textit{ and } \mathbb{N}_{\mathbb{R},f}(S) = \begin{cases} 2 \textit{ and } \deg(I_f(S)) = \begin{cases} 0 \quad (W_1) \\ 2 \textit{ and } \mathbb{N}_{\mathbb{R},\infty}(S) = \begin{cases} 1 \quad (W_{14}), \\ 3, \end{cases} \end{cases} \\ 4 \textit{ and } \deg(I_f(S)) = \begin{cases} -2 \quad (W_8), \\ 0 \quad (W_5), \\ 2 \quad (W_9), \end{cases} \end{cases} \\ 6 \textit{ and } \mathbb{N}_{\mathbb{R},f}(S) = \begin{cases} 1 \quad (W_2), \\ 2, \\ 3 \textit{ and } \deg(I_f(S)) = \begin{cases} -1 \quad (W_3), \\ 0 \quad (W_7), \\ 1 \quad (W_6), \\ 2 \quad (W_{10}), \end{cases} \end{cases} \\ 5 \quad (W_4). \end{cases}$$

*For each value of $\mathcal{J}(S)$ in this diagram with the exception of two cases:* (i) $\mathcal{J}(S) = (7,2,2,3)$, *and* (ii) $\mathbb{N}_{\mathbb{C}}(S) = 6$ *and* $\mathbb{N}_{\mathbb{R},f}(S) = 2$ *(which occurs on $\Delta = 0$), there corresponds a unique phase portrait denoted by $W_i$ (as in [47]) which has neither limit cycles, nor graphics, with the exception of the phase portrait $W_{14}$ corresponding to $\mathcal{J}(S) = (7,2,2,1)$ which has a unique limit cycle.*

**7.7 Theorem** *Inside the class of all quadratic systems there exists a neighborhood of the class of quadratic systems with a weak focus of third order, which may have graphics (in the sense of [22]) but no polycycle and no limit cycle, such that any quadratic system in this neighborhood has at most 4 limit cycles.*

The saddle-to-saddle analytic bifurcation curve is pictured in Fig. 7 and on this curve we have the phase portrait $W_{18}$. In Fig. 7 a part of the half disk is drawn, containing the entire region, the shaded area, where limit cycles occur. It is bordered on one side by the line $a = 0$ on which we have centers and on the other by the bifurcation curve on which we have the three phase portraits with graphics. For more details and the full bifurcation diagram we refer to [46] where (in the new version of the paper) the proof of the analyticity of the saddle-to-saddle connection curve is also given.

[3]We point out that in the cases other than (i) and (ii), whenever just the first component (or the first two components) of $\mathcal{J}(S)$ suffice to yield a single phase portrait, we do not write the values of the remaining invariants.

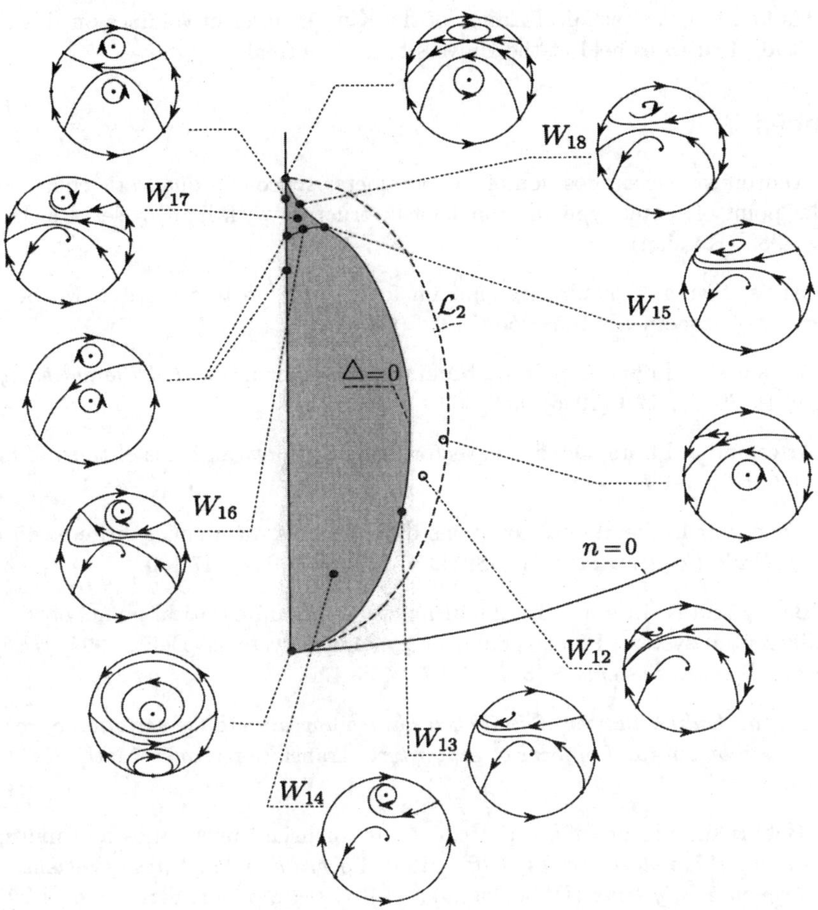


Figure 7

The next stage of the program outlined in this section is being pursued by Artés, Llibre and the author who are using the method of algebraic invariants with special invariants constructed by Vulpe, helpful in the first stages of the classification problem.

For a general setting for the geometry of the whole class of planar quadratic vector fields, we refer the reader to [69] where, roughly speaking, this class is viewed as a principal fiber bundle over pencils of conics. This is an initial stage in the problem of determining the structure of the *moduli space* of this class of differential systems.

**Acknowledgements** The author is very grateful to Nicolae Vulpe from whom she learnt what she knows about the theory of algebraic invariants of differential equations as developed by the school of Sibirsky. She also thanks him for reading this manuscript and making observations and for technical help with the Tex manuscript. The author also thanks Gert Sabidussi for his excellent work as Director of the Advanced Study Institute, his professionalism and his fine work as technical Editor of the Kluwer series of volumes on this and past Advanced Study Institutes held at the Université de Montréal.

## References

[1] E. A. Andronova, Decomposition of the parameter space of a quadratic equation with a singular point of center type and topological structure with limit cycles, Ph. D. Thesis, Gorky, 1988 (Russian).

[2] J. C. Artés, Sistemes quadràtics amb un focus feble de tercer ordre, M. Sc. Thesis, Universitat Autònoma de Barcelona, 1984 (Catalan).

[3] J. C. Artés and J. Llibre, Quadratic Hamiltonian vector fields, *J. Differential Equations* **107** (1994), 80–95; **129** (1996), 559–560.

[4] J. A. Artés and J. Llibre, Quadratic vector fields with a weak focus of third order, *Publ. Mat.* **41** (1997), 7–39.

[5] L. Autonne, Sur la théorie des équations différentielles du premier ordre et du premier degré, *J. École Polytechnique* **61** (1891), 35-122; **62** (1892), 47-180.

[6] V. A. Baltag and N. I. Vulpe, Total multiplicity of all finite critical points of the polynomial differential system, Planar nonlinear dynamical systems (Delft, 1995), *Differential Equations Dynam. Systems* **5** (3–4) (1997), 455–471.

[7] N. N. Bautin, *On the number of limit cycles which appear with the variation of coefficients from an equilibrium point of focus or center type*, Translations Amer. Math. Soc. **1** (1962), 396–413.

[8] A. A. Bolibrukh, Linear differential equations, fuchsian inequalities and multiplicities of zeroes, in: *Asymptotic Series, Differential Algebra and Finiteness Problems in Nonlinear Dynamical Systems* (D. Schlomiuk, ed.), CRM Monogr. Ser., Amer. Math. Soc., Providence, RI, 30 pages (to appear) .

[9] G. Boole, Exposition of a general theory of linear transformations, *Cambridge Math. J.* **3** (1841), 1–20, 106–119.

[10] B. Birch and H. Swinnerton-Dyer, Notes on elliptic curves I, II, *J. Reine Angew. Math.* **212** (1963), 7–25.

[11] L. S. Chen and M. S. Wang, The relative position and number of limit cycles of the quadratic differential systems, *Acta Math. Sinica* **22** (1979), 751–758.

[12] C. Christopher and J. Llibre, Algebraic aspects of integrability for polynomial systems, *Qual. Theory Dyn. Syst.* **1** (1999), 71–95.

[13] W. A. Coppel, A survey of quadratic systems, *J. Differential Equations* **2** (1966), 293–304.

[14] G. Darboux, Mémoire sur les équations différentielles algébrique du premier ordre et du premier degré (Mélanges), *Bull. Sci. Math. Sér. 2* **2** (1878), 60–96; 123–144; 151–200.

[15] H. Darmon, Elliptic curves and $p$-adic uniformisation, 1998 Coxeter-James prize lecture, CMS Notes, Vol. 31, No. 2, Canadian Math. Soc., 1999, 8–16.

[16] J. A. Dieudonné and J. B. Carrell, *Invariant Theory Old and New*, Academic Press, New York, 1971.

[17] H. Dulac, Sur les cycles limites, *Bull. Soc. Math. France* **51** (1923), 45–188.

[18] H. Dulac, Recherches sur les points singuliers des équations différentielles, *J. École Polytechnique Sér. 2* **9** (1904), 1–25.

[19] H. Dulac, Détermination et intégration d'une certaine classe d'équations différentielles ayant pour point singulier un centre (Mélange No. 1), *Bull. Sci. Math.* **51** (1923), 45–188.

[20] F. Dumortier and P. Fiddelaers, Qualitative models for generic local 3-parameter bifurcations on the plane, *Trans. Amer. Math. Soc.* **326** (1991), 101–126.

[21] F. Dumortier, C. Herssens, and L. Perko Local bifurcations and a survey of bounded quadratic systems, *J. Differential Equations* **165** (2000), 430–467.

[22] F. Dumortier, R. Roussarie, and C. Rousseau, Hilbert's 16th problem for quadratic vector fields, *J. Differential Equations* **110** (1994), 66–133.

[23] F. Dumortier, R. Roussarie, and C. Rousseau, Elementary graphics of cyclicity 1 and 2, *Nonlinearity* **7** (1994), 1001–1043.

[24] J. Écalle, *Introduction aux fonctions analysables et preuve constructive de la conjecture de Dulac*, Hermann, Paris, 1992.

[25] W. Fulton, *Algebraic Curves. An Introduction to Algebraic Geometry*, Benjamin, New York, 1969.

[26] L. Gavrilov, The infinitesimal 16th Hilbert problem in the quadratic case, *Invent. Math.* **143** (2001), 449–497.

[27] E. A. González Velasco, Generic properties of polynomial vector fields at infinity, *Trans. Amer. Math. Soc.* **143** (1969), 201–222.

[28] R. Hartshorne, *Algebraic Geometry*, Grad. Texts in Math. **52**, Springer, 1977.

[29] D. Hilbert, Mathematische Probleme, Lecture at the Second International Congress of Mathematicians, Paris 1900; reprinted in: *Mathematical Developments Arising from Hilbert Problems* (F. E. Browder, ed.), Proc. Symp. Pure Math. **28**, Amer. Math. Soc., Providence, RI, 1976, 1–34.

[30] D. Hilbert, Über die Theorie der algebraischen Formen, *Math. Ann.* **36** (1890), 473–534; also *Gesammelte Abhandlungen*, vol. 2, Springer-Verlag, Berlin, 1933, 199–257.

[31] D. Hilbert, Über die vollen Invariantensysteme, *Math. Ann.* **36** (1893), **42** (1893), 313-373; also *Gesammelte Abhandlungen*, vol. 2, Springer-Verlag, Berlin, 1933, pp. 287–344.

[32] D. Hilbert, *Theory of Algebraic Invariants*, Cambridge Univ. Press, New York, 1933.

[33] E. Horozov and I. D. Iliev, On saddle-loop bifurcations of limit cycles in perturbations of quadratic hamiltonian systems, *J. Differential Equations* **113** (1994).

[34] I. D. Iliev, The cyclicity of the period annulus of the quadratic Hamiltonian triangle, *J. Differential Equations* **128** (1996).

[35] Yu. Il'yashenko, *Finiteness Theorems for Limit Cycles*, Transl. Math. Monogr. **94**, Amer. Math. Soc., Providence, RI, 1991.

[36] Yu. Il'yashenko, Centennial history of Hilbert's 16th problem, *Bull. Amer. Math. Soc. (N.S.)* **39** (3) 301–354.

[37] Yu. Il'yashenko and S. Yakovenko, Finite cyclicity of elementary polycycles in generic families, in: *Concerning the Hilbert's 16th Problem*, Amer. Math. Soc. Transl. Ser. 2 **165**, Amer. Math. Soc., Providence, RI, 1995, 21–95.

[38] V. Kaloshin, Around Hilbert-Arnol'd Problem, in: *Asymptotic Expansions, Differential Algebra and Finiteness Problems in Nonlinear Dynamical Systems* (D. Schlomiuk, ed.), CRM Monogr. Ser., Amer. Math. Soc., Providence, RI (to appear).

[39] V. Kaloshin, The Hilbert-Arnol'd Problem and estimates of cyclicity of planar and spacial polycycles, *Funct. Anal. Appl.* **35** (2001), 78–81.

[40] I. Kaplanski, *Hilbert's Problems*, Preliminary version of a book, Chicago, 1977.

[41] R. Kooij and C. Christoper, Algebraic invariant curves and the integrability of polynomial systems, *Appl. Math. Lett.* **6** (4) (1993), 51–53.

[42] Chengzhi Li, Two problems of planar quadratic systems, *Sc. Sinica Ser. A* **26** (1983), 471–481.

[43] Chengzhi Li, Non-existence of limit cycles around a weak focus of order three for any quadratic system, *Chinese Ann. Math. Ser. B* **7** (1986), 174–190.

[44] Chengzhi Li, J. Llibre, and Zhifen Zhang, Weak focus, limit cycles and bifurcations for bounded quadratic systems, *J. Differential Equations* **115** (1995), 193–223.

[45] Weigu Li, J. Llibre, M. Nicolau, and Xiang Zhang, On the differentiability of first integrals of two-dimensional flows, to appear in *Proc. Amer. Math. Soc.*

[46] J. Llibre and D. Schlomiuk, The geometry of quadratic differential systems with a weak focus of third order, CRM Barcelona, preprint 486, November 2001. A new version of this work is due to appear in *Canadian J. Math.*

[47] V. A. Lunkevich and K. S. Sibirskii, Integrals of general quadratic differential systems in cases of a center, *Differential Equations* **8** (1982), 563–568.

[48] S. L. Kleiman, *Problem 15. Rigorous Foundation of Schubert's Enumerative Calculus*, Proc. Sympos. Pure Math. **28**, Amer. Math. Soc., Providence, RI, 1976, 445–482.

[49] L. Markus, Global structure of ordinary differential equations in the plane, *Trans. Amer. Math. Soc.* **76** (1954), 127–148.

[50] A. Mourtada, *Projections de sous-ensembles quasi-réguliers d'Hilbert I, II, III. Le cas général*, Rapports de recherche 124 (1997); 207 (1999); (2001); Laboratoire de Topologie, Univ. de Bourgogne.

[51] A. Mourtada, *Projections de sous-ensembles quasi-réguliers d'Hilbert IV. Le cas général.* Rapport de recherche 276 (2002); Laboratoire de Topologie, Univ. de Bourgogne.

[52] D. Mumford, *Geometric Invariant Theory*, Springer, 1965; 2nd ed. 1981, with J. Fogarty; 3rd ed. 1992, with J. Fogarty and F. Kirwan.

[53] D. A. Newman, Classification of continuous flows on 2-manifolds, *Proc. Amer. Math. Soc.* **48** (1975), 73–81.

[54] I. Nikolaev and N. Vulpe, Topological classification of quadratic systems at infinity, *J. London Math. Soc.* **55** (1997), 473–488.

[55] Peter J. Olver, *Classical Invariant Theory*, London Math. Soc. Stud. Texts **44**, Cambridge Univ. Press, 1999.

[56] P. Painlevé, Sur les intégrales algébriques des équations différentielles du premier ordre, *C. R. Acad. Sci. Paris* **110**, 945–948; *Oeuvres*, vol. 2, Éditions CNRS, Paris, 1974, 236.

[57] P. Painlevé, Mémoire sur les équations différentielles du premier ordre, *Ann. Sci. École Norm. Sup. (3)* **8** (1891), 9–59, 103–140, 201–226, 276–284; **9** (1892), 9–30, 101–144, 283–308.

[58] J. Pal and D. Schlomiuk, Summing up the dynamics of quadratic Hamiltonian systems with a center, *Canad. J. Math.* **49** (1997), 583–599.

[59] J. Pal and D. Schlomiuk, Intersection multiplicity and limit cycles in quadratic differential systems with a weak focus, preprint, Univ. de Montréal, September 1999.

[60] H. Poincaré, Mémoire sur les courbes définies par les équations différentielles, *J. de Mathématiques (3)* **7** (1881), 375–422; **8** (1882), 251–296; *Oeuvres de Henri Poincaré*, vol. 1, Gauthier-Villars, Paris, 1951, 3–84.

[61] H. Poincaré, Mémoire sur les courbes définies par les équations différentielles, *J. Math. Pures Appl.* (4) **1** (1885), 167–244; *Oeuvres de Henri Poincaré*, vol. 1, Gauthier–Villars, Paris, 1951, 95–114.

[62] H. Poincaré, Sur l'intégration algébrique des équations différentielles, *C. R. Acad. Sci. Paris* **112** (1891), 761–764.

[63] H. Poincaré, Sur l'intégration algébrique des équations différentielles du premier ordre et du premier degré, *Rend. Circ. Mat. Palermo* **1** (1891), 169–191.

[64] H. Poincaré, Sur l'intégration algébrique des équations différentielles du premier ordre et du premier degré, *Rend. Circ. Mat. Palermo* **11** (1897), 193–239.

[65] V. L. Popov, *Groups, Generators, Syzygies, and Orbits in Invariant Theory*, Transl. Math. Monogr. **100**, Amer. Math. Soc., Providence, RI, 1992.

[66] B. Riemann, Ueber die Anzahl der Primzahlen unter einer gegebenen Grösse, *Monatsber. Berliner Akademie*, Nov. 1859, 145–153.

[67] R. Roussarie, A note on finite cyclicity and Hilbert's 16th problem, in: *Dynamical Systems Valparaiso 1986* (R. Bamón, R. Labarca, and J. Palis Jr., eds.), Lecture Notes in Math. **1331**, Springer, Berlin-New York, 1988, 161–168.

[68] R. Roussarie, Smoothness property for bifurcation diagrams, *Publ. Mat.* **41** (1997), 243–268.

[69] R. Roussarie and D. Schlomiuk, On the geometric structure of the class of planar quadratic differential systems, to appear in *Qual. Theory Dyn. Syst.* **3** (2002), 93–121.

[70] D. Schlomiuk, Algebraic particular integrals, integrability and the problem of the center, *Trans. Amer. Math. Soc.* **338** (1993), 799–841.

[71] D. Schlomiuk, Algebraic and geometric aspects of the theory of polynomial vector fields, in: *Bifurcations and Periodic Orbits of Vector Fields* (D. Schlomiuk, ed.), NATO Adv. Study Inst. Ser. C **408**, Kluwer, Dordrecht, 1993, 429–467.

[72] D. Schlomiuk and J. Pal, On the geometry in the neighborhood of infinity of quadratic differential systems with a weak focus, *Qual. Theory Dyn. Syst.* **2** (2001), 1–43.

[73] D. Schlomiuk (ed.), *Asymptotic Series, Differential Algebra and Finiteness Problems in Nonlinear Dynamical Systems*, CRM Monogr. Ser., Amer. Math. Soc., Providence, RI (to appear).

[74] D. Schlomiuk and N. Vulpe, Geometry of quadratic differential systems in the neighborhood of the line at infinity, Centre de recherche mathématique, rapport de recherche CRM-2831, Univ. de Montréal, 2001.

[75] H. C. H. Schubert, *Kalkül der abzählenden Geometrie*, Teubner, Leipzig, 1879.

[76] Shi Songling, A concrete example of the existence of four limit cycles for planar quadratic systems, *Sci. Sinica Ser. A* **23** (1980), 153–158.

[77] Shi Songling, A method of constructing cycles without contact around a weak focus, *J. Differential Equations* **41** (1981), 301–312.

[78] Shi Songling, On the structure of Poincaré-Lyapunov constants for the weak focus of polynomial vector fields, *J. Differential Equations* **52** (1984), 52–57.

[79] K. S. Sibirsky, On conditions for the presence of a center and a focus, *Kisinev. Gos. Univ. Uč. Zap.* **11** (1954), 115–117 (Russian).

[80] K. S. Sibirsky, The principle of symmetry and the problem of the center, *Kisinev. Gos. Univ. Uč. Zap.* **11** (1955), 27–34 (Russian).

[81] K. S. Sibirsky, *Algebraic Invariants of Differential Equations and Matrices*, Shtiintsa, 1976 (Russian).

[82] K. S. Sibirsky, *Introduction to the Algebraic Theory of Invariants of Differential Equations* (Transl. from Russian), Nonlinear Science: Theory and Applications, Manchester Univ. Press, Manchester, 1988.

[83] S. Smale, Mathematical problems for the next century, *Math. Intelligencer* **20** (2) (1998). Also in: *Mathematics: Frontiers and Perspectives* (Arnold et al., eds.) I.M.U., Amer. Math. Soc., 2000, 271–294.

[84] S. Smale, Finding a horseshoe on the beaches of Rio, *Math. Intelligencer* **20** (1) (1998), 39–44.

[85] J. Sotomayor, *Liçóes de equaçóes diferenciais ordinárias*, Projecto Euclides, IMPA, Rio de Janeiro, 1979.

[86] J. Sotomayor and R. Paterlini, Quadratic vector fields with finitely many periodic orbits, in: *Geometric Dynamics* (J. Palis, Jr., ed.), Lecture Notes in Math. **1007** (1983), 753–766.

[87] I. Stewart, Hilbert's 16th problem, *Nature* **326** (March 1987), 243.

[88] N. I. Vulpe, Affine-invariant conditions for the topological discrimination of quadratic systems with center, *Differential Equations* **19** (1983), 273–280.

[89] A. Weil, *Foundations of Algebraic Geometry*, Amer. Math. Soc. Colloq. Publ. **29**, 1946.

[90] H. Weyl, *The Classical Groups: Their Invariants and Representations*, Princeton Univ. Press, Princeton, NJ, 1939; 2nd ed. 1946.

[91] S. Yakovenko, Quantitative theory of ordinary differential equations and tangential Hilbert 16th problem, in: *Asymptotic Expansions, Differential Algebra and Finiteness Problems in Nonlinear Dynamical Systems* (D. Schlomiuk, ed.), CRM Monogr. Ser., Amer. Math. Soc., Providence, RI (to appear).

[92] Zhang Pingguang, On the distribution and number of limit cycles for quadratic systems with two foci, *Acta Math. Sinica* **44** (2001), 37–44 (Chinese).

[93] Zhang Pingguang, On the distribution and number of limit cycles for quadratic systems with two foci, *Qual. Theory Dyn. Syst.* **3** (2002), 1–28.

[94] H. Żołądek, Quadratic systems with center and their perturbations, *J. Differential Equations* **109** (1994), 223–273.

# Index